The World of
MATHEMATICS
Volume 1

Edited by
James R. Newman

DOVER PUBLICATIONS, INC.
Mineola, New York

Bibliographical Note

This Dover edition, first published in 2000, is an unabridged republication of the 4-volume work originally published by Simon & Schuster, New York, in 1956.
Page iii constitutes a continuation of this copyright page.

Library of Congress Cataloging-in-Publication Data

The world of mathematics / edited by James R. Newman.
    p. cm.
  Originally published: New York : Simon and Schuster, 1956.
  Includes bibliographical references and indexes.
  ISBN 0-486-41153-2 (pbk. : v. 1)—ISBN 0-486-41150-8 (pbk. : v. 2)—ISBN 0-486-41151-6 (pbk. : v. 3)—ISBN 0-486-41152-4 (pbk. : v. 4)
  1. Mathematics. I. Newman, James Roy, 1907–1966.

QA7 .W67 2000
510—dc21
                                               00-027006

Manufactured in the United States of America
Dover Publications, Inc., 31 East 2nd Street, Mineola, N.Y. 11501

TO
Ruth, Brooke AND Jeff

THE EDITOR wishes to express his gratitude for permission to reprint material from the following sources:

Messrs Allen & Unwin, Ltd., for "Definition of Number," from *Introduction to Mathematical Philosophy*, by Bertrand Russell.

Cambridge University Press and the Royal Society for "Isaac Newton," by E. N. Da C. Andrade, and "Newton, the Man," by John Maynard Keynes, from *Newton Tercentenary Celebrations*; and for "The Sand Reckoner," from *Works of Archimedes* (with the kind permission of Lady Heath).

Harvard University Press for "Greek Mathematics," from *Greek Mathematical Works*, translated by Ivor Thomas and published by the Loeb Classical Library.

Professor Otto Koehler for "The Ability of Birds to 'Count,' " from the *Bulletin of Animal Behavior*, No. 9, translated by Professor W. Thorpe.

Library of Living Philosophers, Evanston, Illinois, for "My Mental Development," by Bertrand Russell, from *The Philosophy of Bertrand Russell*, edited by Paul Schilpp.

The Macmillan Company for "Mathematics as an Element in the History of Thought," from *Science and the Modern World*, by Alfred North Whitehead. © 1925 by The Macmillan Company.

McGraw-Hill Book Company for "The Analyst," by Bishop Berkeley, "On the Binomial Theorem for Fractional and Negative Exponents," by Sir Isaac Newton, and "The Declaration of the Profit of Arithmeticke," by Robert Recorde, from *Source Book in Mathematics*, by D. E. Smith, © 1929 by McGraw-Hill Book Company, Inc.; and for "The Queen of Mathematics," from *Mathematics, Queen and Servant of the Sciences*, © 1951 by Eric Temple Bell.

Methuen & Company, Ltd., for *The Great Mathematicians*, by Herbert Westren Turnbull.

National Council of Teachers of Mathematics for "From Numbers to Numerals and From Numerals to Computation," from *Numbers and Numerals*, by David Eugene Smith and Jekuthiel Ginsburg.

Oxford University Press for "Topology," from *What Is Mathematics?*, by Richard Courant and Herbert Robbins. © 1941 by Richard Courant.

Princeton University Press for "Dürer as a Mathematician," from *Albrecht Dürer*, by Erwin Panofsky, © 1943 by Princeton University Press; and for *Symmetry*, by Hermann Weyl, © 1952 by Princeton University Press.

St. Martin's Press for "Calculating Prodigies," from *Mathematical Recreations and Essays*, by W. W. Rouse Ball. Published by Macmillan and Company, London.

Scientific American for "Projective Geometry," by Morris Kline, © 1955 by Scientific American, Inc.; for "The Rhind Papyrus," by James R. Newman, © 1952 by Scientific American, Inc.; and for "Srinivasa Ramanujan," by James R. Newman, © 1948 by Scientific American, Inc.

Simon and Schuster, Inc., for "Gauss, the Prince of Mathematicians" and "Invariant Twins, Cayley and Sylvester," from *Men of Mathematics*, © 1937 by E. T. Bell.

All four volumes of *The World of Mathematics* are available from Dover Publications. In addition to this volume, they are:

The World of Mathematics, edited by James R. Newman
Volume 2 (720 pp.)   ISBN 0-486-41150-8

> Parts V–VII:
> Mathematics and the Physical World
> Mathematics and Social Science
> The Laws of Chance

The World of Mathematics, edited by James R. Newman
Volume 3 (624 pp.)   ISBN 0-486-41151-6

> Parts VIII–XVII:
> Statistics and the Design of Experiments
> The Supreme Art of Abstraction: Group Theory
> Mathematics of Infinity
> Mathematical Truth and the Structure of Mathematics
> The Mathematical Way of Thinking
> Mathematics and Logic
> The Unreasonableness of Mathematics
> How to Solve It
> The Vocabulary of Mathematics
> Mathematics as an Art

The World of Mathematics, edited by James R. Newman
Volume 4 (464 pp.)   ISBN 0-486-41152-4

> Parts XVIII–XXVI:
> The Mathematician
> Mathematical Machines: Can a Machine Think?
> Mathematics in Warfare
> A Mathematical Theory of Art
> Mathematics of the Good
> Mathematics in Literature
> Mathematics and Music
> Mathematics as a Culture Clue
> Amusements, Puzzles, Fancies

In these days of conflict between ancient and modern studies, there must surely be something to be said for a study which did not begin with Pythagoras and will not end with Einstein, but is the oldest and youngest of all.
—G. H. HARDY (*A Mathematician's Apology*)

Introduction

A PREFACE is both a greeting and a farewell. I have been so long occupied with this book that taking leave of it is difficult. It is more than fifteen years since I began gathering the material for an anthology which I hoped would convey something of the diversity, the utility and the beauty of mathematics. At the beginning it seemed that the task would be neither too arduous nor unduly prolonged, for I was acquainted with the general literature of the subject and I had no intention of compiling a mammoth sourcebook. I soon discovered I was mistaken in my estimate. Popular writings on the nature, uses and history of mathematics did not yield the variety I had expected; thus it was necessary to go through an immense technical and scholarly literature to search out examples of mathematical thought which the general reader could follow and enjoy. Digestible essays were hard to find on the foundations and philosophy of mathematics, the relation of mathematics to art and music, the applications of mathematics to social and economic problems. Moreover, while I had not planned to write introductions to the separate selections, it became clear as I proceeded that many of the pieces, though illuminating when read in context, had little meaning when they stood alone. Settings had to be provided, explanations of how the pieces came to be written and of the place they occupied in the growth of mathematical thought. So the work I expected to finish in two years has taken the greater portion of two decades; what was envisaged as a volume of moderate size has assumed dimensions which even a self-indulgent author must acknowledge to be extended.

I have tried in this book to show the range of mathematics, the richness of its ideas and multiplicity of its aspects. It presents mathematics as a tool, a language and a map; as a work of art and an end in itself; as a fulfillment of the passion for perfection. It is seen as an object of satire, a subject for humor and a source of controversy; as a spur to wit and a leaven to the storyteller's imagination; as an activity which has driven men to frenzy and provided them with delight. It appears in broad view as a body of knowledge made by men, yet standing apart and independent of them. In this collection, I hope, will be found material to suit every taste and capacity.

Many of the selections are long. I dislike snippets and fragments that tease. Understanding mathematical logic, or the theory of relativity, is not *an indispensable attribute of the cultured mind. But if one wishes to learn anything about these subjects, one must learn something. It is necessary to master the rudiments of the language, to practice a technique, to follow step by step a characteristic sequence of reasoning and to see a problem through from beginning to end. The reader who makes this effort will not be disappointed. Some of the selections are difficult. But what is surprising is how many can be understood without unusual aptitude or special training. Of course those who are bold enough to tackle the more formidable subjects will gain a special reward. There are few gratifications comparable to that of keeping up with a demonstration and attaining the proof. It is for each man an act of creation, as if the discovery had never been made before; it inculcates lofty habits of mind.*

An anthology is a work of prejudice. This is no less true when the subject is mathematics than when it is poetry or fiction. Magic squares, for example, bore me, but I never tire of the theory of probability. I prefer geometry to algebra, physics to chemistry, logic to economics, the mathematics of infinity to the theory of numbers. I have shunned topics, skimped some and shown great hospitality to others. I make no apology for these prejudices; since I pretend to no mathematical skill, I have felt at liberty to present the mathematics I like.

MANY persons have helped to make this book. I owe a great deal, more than I can express, to my friend and former colleague Robert Hatch for his editorial counsel. No damp idea got by him, no potted phrase. In substance and style the work bears his mark. My teacher and friend Professor Ernest Nagel has been generous beyond measure not only in giving advice and criticism but in writing specially for this book a brilliant essay on symbolic logic. Sam Rosenberg read what I wrote and improved it with his wit. My wife, as always, has been wise, patient and encouraging. Dr. Ralph Shaw, Professor of Bibliography at Rutgers University and former Librarian of the Department of Agriculture, gave invaluable assistance in the preparation of the manuscript. And I wish to thank my publishers— I mention especially the contributions of Jack Goodman, Tom Torre Bevans, and Peter Schwed—both for their forbearance in waiting until 1956 to publish a book scheduled for 1942, and for their imagination and skill in the difficult task of design and manufacture.

J. R. N.

Table of Contents

VOLUME ONE

PART III: Arithmetic, Numbers and the Art of Counting

PART IV: Mathematics of Space and Motion

PART I

General Survey

1. The Nature of Mathematics *by* PHILIP E. B. JOURDAIN

COMMENTARY ON
PHILIP E. B. JOURDAIN

PHILIP E. B. JOURDAIN (1879–1919), whose little book on the nature of mathematics is here reproduced in its entirety, was a logician, a philosopher and a historian of mathematics. To each of these subjects he brought a fresh outlook and a remarkably penetrating and creative intelligence. He was not yet forty when he died and from adolescence had been afflicted by a terrible paralytic ailment (Friedrick's ataxia) which gradually tightened its grip upon him. Yet he left behind a body of work that influenced the development of both mathematical logic and the history of science.

Jourdain, the son of a Derbyshire vicar, was educated at Cheltenham College and at Cambridge. The few years during which he was able to enjoy the normal pleasures of boyhood—long walks were his special delight—are described in a poignant memoir by his younger sister, Millicent, who suffered from the same hereditary disease. In 1900 the brother and sister went to Heidelberg to seek medical help. While at the hospital he began in earnest his study of the history of mathematics. "We had," wrote Millicent, "what was to be nearly our last bit of walking together here." The treatment was unavailing and when they returned to England, Jourdain could no longer walk or stand or even hold a pencil without difficulty. Nevertheless, he undertook with great energy and enthusiasm the first of a series of mathematical papers which established his reputation. Among his earlier writings were studies of Lagrange's use of differential equations, the work of Cauchy and Gauss in function theory, and conceptual problems of mathematical physics.[1] Between 1906 and 1912 he contributed to the *Archiv der Mathematik und Physik* a masterly group of papers on the mathematical theory of transfinite numbers, a subject in which he was always deeply interested. In the same period the *Quarterly Journal of Mathematics* published a group of essays on the development of the theories of mathematical logic and the principles of mathematics. Jourdain was an editor of *Isis* and the *Monist*, in whose pages appeared his articles on Leibniz, Napier, Hooke, Newton, Galileo, Poincaré and Dedekind. He edited reprints of works by De Morgan, Boole, Georg Cantor, Lagrange, Jacobi, Gauss and Ernst Mach; he wrote a brilliant and witty book, *The Philosophy of Mr. B*rtr*nd R*ss*ll*, dealing with Russell's analysis of the problems of logic and the foundations of mathematics; he took out a patent covering an invention of a "silent engine" (I have been unable to

---

[1] Bibliographies of Jourdain's writings appear in *Isis*, Vol. 5, 1923, pp. 134–136, and in the *Monist*, Vol. 30, 1920, pp. 161–182.

discover what this machine was) and he wrote poems and short stories which never got published. In 1914, at the height of his powers, he was producing enough "to keep two typists busy all day."

The distinctive qualities of Jourdain's thought were its independence and its cutting edge. He was renowned for his broad scholarship in the history and philosophy of science, but he was more than a scholar. Never content with comprehending all that others had said about a problem, he had to work it through in his own way and overcome its difficulties by his own methods. This led him to conclusions peculiarly his own. They are not always satisfactory but they always deserve close attention: Jourdain rarely failed to uncover points overlooked by less subtle and original investigators.

The Nature of Mathematics reflects his excellent grasp of the subject, his at times oblique but always rewarding approach to logic and mathematics, his wit and clear expression. He had sharpened his thinking on some of the hardest and most baffling questions of philosophy and had achieved an orderly understanding of them which he was fully capable of imparting to the attentive reader. The book is not a textbook collection of methods and examples, but an explanation of "how and why these methods grew up." It discusses concepts which are widely used even in elementary arithmetic, geometry and algebra—negative numbers, for example—but far from widely comprehended. It presents also a careful treatment of "the development of analytical methods and certain examinations of principles." There are at least two other excellent popularizations of mathematics, A. N. Whitehead's celebrated *Introduction to Mathematics* [2] and the more recent *Mathematics for the General Reader* by E. C. Titchmarsh.[3] Both books can be recommended strongly, the first as a characteristic, immensely readable work by one of the greatest of twentieth-century philosophers; the second as a first-class mathematician's lucid, unhurried account of the science of numbers from arithmetic through the calculus. Jourdain's book follows a somewhat different path of instruction in that it emphasizes the relation between mathematics and logic. It is the peer of the other two studies and has for the anthologist the additional appeal of being unjustly neglected and out of print. "I hope that I shall succeed," says Jourdain in his introduction, "in showing that the process of mathematical discovery is a living and a growing thing." In this attempt he did not fail.

[2] Oxford University Press, New York, 1948.
[3] Hutchinson's University Library, London, n. d.

Pure mathematics consists entirely of such asseverations as that, if such and such a proposition is true of anything, then such and such another proposition is true of that thing. It is essential not to discuss whether the first proposition is really true, and not to mention what the anything is of which it is supposed to be true. . . . If our hypothesis is about anything and not about some one or more particular things, then our deductions constitute mathematics. Thus mathematics may be defined as the subject in which we never know what we are talking about, nor whether what we are saying is true.
—BERTRAND RUSSELL

1   The Nature of Mathematics

By PHILIP E. B. JOURDAIN

CONTENTS

INTRODUCTION

AN eminent mathematician once remarked that he was never satisfied with his knowledge of a mathematical theory until he could explain it to the next man he met in the street. That is hardly exaggerated; however, we must remember that a satisfactory explanation entails duties on both sides. Any one of us has the right to ask of a mathematician, "What is the use of mathematics?" Any one may, I think and will try to show, rightly suppose that a satisfactory answer, if such an answer is anyhow possible, can be given in quite simple terms. Even men of a most abstract science, such as mathematics or philosophy, are chiefly adapted for the ends of ordinary life; when they think, they think, at the bottom, like other men. They are often more highly trained, and have a technical facility for thinking that comes partly from practice and partly from the use of the contrivances for correct and rapid thought given by the signs and rules for dealing with them that mathematics and modern logic provide. But there is no real reason why, with patience, an ordinary person should not understand, speaking broadly, what mathematicians do, why they do it, and what, so far as we know at present, mathematics is.

Patience, then, is what may rightly be demanded of the inquirer. And this really implies that the question is not merely a rhetorical one—an expression of irritation or scepticism put in the form of a question for the sake of some fancied effect. If Mr. A. dislikes the higher mathematics because he rightly perceives that they will not help him in the grocery business, he asks disgustedly, "What's the use of mathematics?" and does not wait for an answer, but turns his attention to grumbling at the lateness of his dinner. Now, we will admit at once that higher mathematics is of no more use in the grocery trade than the grocery trade is in the navigation of a ship; but that is no reason why we should condemn mathematics as entirely useless. I remember reading a speech made by an eminent surgeon, who wished, laudably enough, to spread the cause of elementary surgical instruction. "The higher mathematics," said he with great satisfaction to himself, "do not help you to bind up a broken leg!" Obviously they do not; but it is equally obvious that surgery does not help us to add up accounts; . . . or even to think logically, or to accomplish the closely allied feat of seeing a joke.

To the question about the use of mathematics we may reply by pointing out two obvious consequences of one of the applications of mathematics: mathematics prevents much loss of life at sea, and increases the commercial prosperity of nations. Only a few men—a few intelligent philosophers and more amateur philosophers who are not highly intelligent—would doubt if these two things were indeed benefits. Still, probably, all of us act as if we thought that they were. Now, I do not mean that mathematicians go about with life-belts or serve behind counters; they do not usually do so. What I mean I will now try to explain.

Natural science is occupied very largely with the prevention of waste of the labour of thought and muscle when we want to call up, for some purpose or other, certain facts of experience. Facts are sometimes quite useful. For instance, it is useful for a sailor to know the positions of the stars and sun on the nights and days when he is out of sight of land. Otherwise, he cannot find his whereabouts. Now, some people connected with a national institution publish periodically a *Nautical Almanac* which contains the positions of stars and other celestial things you see through telescopes, for every day and night years and years ahead. This *Almanac*, then, obviously increases the possibilities of trade beyond coasting-trade, and makes travel by ship, when land cannot be sighted, much safer; and there would be no *Nautical Almanac* if it were not for the science of astronomy; and there would be no practicable science of astronomy if we could not organise the observations we make of sun and moon and stars, and put hundreds of observations in a convenient form and in a little space—in short, if we could not economise our mental or bodily activity by remembering or carrying about two or three little formulæ instead of fat books full of details; and, lastly, we could not economise this activity if it were not for mathematics.

Just as it is with astronomy, so it is with all other sciences—both those of Nature and mathematical science: the very essence of them is the prevention of waste of the energies of muscle and memory. There are plenty of things in the unknown parts of science to work our brains at, and we can only do so efficiently if we organise our thinking properly, and consequently do not waste our energies.

The purpose of this little volume is not to give—like a text-book—a collection of mathematical methods and examples, but to do, firstly, what text-books do not do: to show how and why these methods grew up. All these methods are simply means, contrived with the conscious or unconscious end of economy of thought-labour, for the convenient handling of long and complicated chains of reasoning. This reasoning, when applied to foretell natural events, on the basis of the applications of mathematics, as sketched in the fourth chapter, often gives striking results. But the methods of mathematics, though often suggested by natural events, are purely logical. Here the word "logical" means something more than the traditional doctrine consisting of a series of extracts from the science of reasoning, made by the genius of Aristotle and frozen into a hard body of doctrine by the lack of genius of his school. Modern logic is a science which has grown up with mathematics, and, after a period in which it moulded itself on the model of mathematics, has shown that not only the reasonings but also conceptions of mathematics are logical in their nature.

In this book I shall not pay very much attention to the details of the elementary arithmetic, geometry, and algebra of the many text-books, but shall be concerned with the discussion of those conceptions—such as that of negative number—which are used and not sufficiently discussed in these books. Then, too, I shall give a somewhat full account of the development of analytical methods and certain examinations of principles.

I hope that I shall succeed in showing that the process of mathematical discovery is a living and a growing thing. Some mathematicians have lived long lives full of calm and unwavering faith—for faith in mathematics, as

I will show, has always been needed—some have lived short lives full of burning zeal, and so on; and in the faith of mathematicians there has been much error.

Now we come to the second object of this book. In the historical part we shall see that the actual reasonings made by mathematicians in building up their methods have often not been in accordance with logical rules. How, then, can we say that the reasonings of mathematics are logical in their nature? The answer is that the one word "mathematics" is habitually used in two senses, and so, as explained in the last chapter, I have distinguished between "mathematics," the methods used to discover certain truths, and "Mathematics," the truths discovered. When we have passed through the stage of finding out, by external evidence or conjecture, how mathematics grew up with problems suggested by natural events, like the falling of a stone, and then how something very abstract and intangible but very real separated out of these problems, we can turn our attention to the problem of the nature of Mathematics without troubling ourselves any more as to how, historically, it gradually appeared to us quite clearly that there is such a thing at all as Mathematics—something which exists apart from its application to natural science. History has an immense value in being suggestive to the investigator, but it is, logically speaking, irrelevant. Suppose that you are a mathematician; what you eat will have an important influence on your discoveries, but you would at once see how absurd it would be to make, say, the momentous discovery that 2 added to 3 makes 5 depend on an orgy of mutton cutlets or bread and jam. The methods of work and daily life of mathematicians, the connecting threads of suggestion that run through their work, and the influence on their work of the allied work of others, all interest the investigator because these things give him examples of research and suggest new ideas to him; but these reasons are psychological and not logical.

But it is as true as it is natural that we should find that the best way to become acquainted with new ideas is to study the way in which knowledge about them grew up. This, then, is what we will do in the first place, and it is here that I must bring my own views forward. Briefly stated, they are these. Every great advance in mathematics with which we shall be concerned here has arisen out of the needs shown in natural science or out of the need felt to connect together, in one methodically arranged whole, analogous mathematical processes used to describe different natural phenomena. The application of logic to our system of descriptions, which we may make either from the motive of satisfying an intellectual need (often as strong, in its way, as hunger) or with the practical end in view of satisfying ourselves that there are no hidden sources of error that may ultimately lead us astray in calculating future or past natural events, leads at once to those modern refinements of method that are regarded with disfavour by the old-fashioned mathematicians.

In modern times appeared clearly—what had only been vaguely suspected before—the true nature of Mathematics. Of this I will try to give some account, and show that, since mathematics is logical and not psychological in its nature, all those petty questions—sometimes amusing and often tedious—of history, persons, and nations are irrelevant to Mathematics in itself. Mathematics has required centuries of excavation, and the process of excavation is not, of course, and never will be, complete. But we see enough now of what has been excavated clearly to distinguish between it and the tools which have been or are used for excavation. This confusion, it should be noticed, was never made by the excavators themselves, but only by some of the philosophical onlookers who reflected on what was being done. I hope and expect that our reflections will not lead to this confusion.

CHAPTER I

THE GROWTH OF MATHEMATICAL SCIENCE IN ANCIENT TIMES

IN the history of the human race, inventions like those of the wheel, the lever, and the wedge were made very early—judging from the pictures on ancient Egyptian and Assyrian monuments. These inventions were made on the basis of an instinctive and unreflecting knowledge of the processes of nature, and with the sole end of satisfaction of bodily needs. Primitive men had to build huts in order to protect themselves against the weather, and, for this purpose, had to lift and transport heavy weights, and so on. Later, by reflection on such inventions themselves, possibly for the purposes of instruction of the younger members of a tribe or the newly-joined members of a guild, these isolated inventions were classified according to some analogy. Thus we see the same elements occurring in the relation of a wheel to its axle and the relation of the arm of a lever to its fulcrum— the same weights at the same distance from the axle or fulcrum, as the case may be, exert the same power, and we can thus class both instruments together in virtue of an analogy. Here what we call "scientific" classification begins. We can well imagine that this pursuit of science is attractive in itself; besides helping us to communicate facts in a comprehensive, compact, and reasonably connected way, it arouses a purely intellectual interest. It would be foolish to deny the obvious importance to us of our bodily needs; but we must clearly realise two things:—(1) The intellectual need is very strong, and is as much a fact as hunger or thirst; sometimes it is even stronger than bodily needs—Newton, for instance, often forgot to take food when he was engaged with his discoveries. (2) Practical results of value often follow from the satisfaction of intellectual needs. It was the satisfaction of certain intellectual needs in the cases of Maxwell and Hertz that ultimately led to wireless telegraphy; it was the satisfaction of some of Faraday's intellectual needs that made the dynamo and the electric telegraph possible. But many of the results of strivings after intellectual satisfaction have as yet no obvious bearing on the satisfaction of our bodily needs. However, it is impossible to tell whether or no they will always be barren in this way. This gives us a new point of view from which to consider the question, "What is the use of mathematics?" To condemn branches of mathematics because their results cannot obviously be applied to some practical purpose is short-sighted.

The formation of science is peculiar to human beings among animals. The lower animals sometimes, but rarely, make isolated discoveries, but never seem to reflect on these inventions in themselves with a view to rational classification in the interest either of the intellect or of the indirect furtherance of practical ends. Perhaps the greatest difference between man and the lower animals is that men are capable of taking circuitous paths for the attainment of their ends, while the lower animals have their minds so filled up with their needs that they try to seize the object they want, or remove that which annoys them, in a direct way. Thus, monkeys often vainly snatch at things they want, while even savage men use catapults or snares or the consciously observed properties of flung stones.

The communication of knowledge is the first occasion that compels distinct reflection, as everybody can still observe in himself. Further, that which the old members of a guild mechanically pursue strikes a new member as strange, and thus an impulse is given to fresh reflection and investigation.

When we wish to bring to the knowledge of a person any phenomena or processes of nature, we have the choice of two methods: we may allow the person to observe matters for himself, when instruction comes to an end; or, we may describe to him the phenomena in some way, so as to save him the trouble of personally making anew each experiment. To describe an event—like the falling of a stone to the earth—in the most comprehensive and compact manner requires that we should discover what is constant and what is variable in the processes of nature; that we should discover the same law in the moulding of a tear and in the motions of the planets. This is the very essence of nearly all science, and we will return to this point later on.

We have thus some idea of what is known as "the economical function of science." This sounds as if science were governed by the same laws as the management of a business; and so, in a way, it is. But whereas the aims of a business are not, at least directly, concerned with the satisfaction of intellectual needs, science—including natural science, logic, and mathematics—uses business methods consciously for such ends. The methods are far wider in range, more reasonably thought out, and more intelligently applied than ordinary business methods, but the principle is the same. And this may strike some people as strange, but it is nevertheless true: there appears more and more as time goes on a great and compelling beauty in these business methods of science.

The economical function appears most plainly in very ancient and modern science. In the beginning, all economy had in immediate view the satisfaction simply of bodily wants. With the artisan, and still more so with the investigator, the most concise and simplest possible knowledge of a given province of natural phenomena—a knowledge that is attained with the least intellectual expenditure—naturally becomes in itself an aim; but though knowledge was at first a means to an end, yet, when the mental motives connected therewith are once developed and demand their satisfaction, all thought of its original purpose disappears. It is one great object of science to replace, or save the trouble of making, experiments, by the reproduction and anticipation of facts in thought. Memory is handier than experience, and often answers the same purpose. Science is communicated by instruction, in order that one man may profit by the experience of another and be spared the trouble of accumulating it for himself; and thus, to spare the efforts of posterity, the experiences of whole generations are stored up in libraries. And further, yet another function of this economy is the preparation for fresh investigation.[1]

The economical character of ancient Greek geometry is not so apparent as that of the modern algebraical sciences. We shall be able to appreciate this fact when we have gained some ideas on the historical development of ancient and modern mathematical studies.

The generally accepted account of the origin and early development of geometry is that the ancient Egyptians were obliged to invent it in order to restore the landmarks which had been destroyed by the periodical inundations of the Nile. These inundations swept away the landmarks in the valley of the river, and, by altering the course of the river, increased or decreased the taxable value of the adjoining lands, rendered a tolerably accurate system of surveying indispensable, and thus led to a systematic study of the subject by the priests. Proclus (412–485 A.D.), who wrote a summary of the early history of geometry, tells this story, which is also told by Herodotus, and observes that it is by no means strange that the invention of the sciences should have originated in practical needs, and that, further, the transition from perception with the senses to reflection,

[1] *Cf.* pp. 5, 13, 15, 16, 42, 71.

and from reflection to knowledge, is to be expected. Indeed, the very name "geometry"—which is derived from two Greek words meaning *measurement of the earth*—seems to indicate that geometry was not indigenous to Greece, and that it arose from the necessity of surveying. For the Greek geometricians, as we shall see, seem always to have dealt with geometry as an abstract science—to have considered *lines* and *circles* and *spheres* and so on, and not the rough pictures of these abstract ideas that we see in the world around us—and to have sought for propositions which should be absolutely true, and not mere approximations. The name does not therefore refer to this practice.

However, the history of mathematics cannot with certainty be traced back to any school or period before that of the Ionian Greeks. It seems that the Egyptians' geometrical knowledge was of a wholly practical nature. For example, the Egyptians were very particular about the exact orientation of their temples; and they had therefore to obtain with accuracy a north and south line, as also an east and west line. By observing the points on the horizon where a star rose and set, and taking a plane midway between them, they could obtain a north and south line. To get an east and west line, which had to be drawn at right angles to this, certain people were employed who used a rope ABCD, divided by knots or marks at B and C, so that the lengths AB, BC, CD were in the proportion $3:4:5$. The length BC was placed along the north and south line, and pegs P and Q inserted at the knots B and C. The piece BA (keeping it stretched all the time) was then rotated round the peg P, and similarly the piece CD was rotated round the peg Q, until the ends A and D coincided; the point thus indicated was marked by a peg R. The result was to form a triangle PQR whose angle at P was a right angle, and the line PR would give an east and west line. A similar method is constantly used at the present time by practical engineers, and by gardeners in marking tennis courts, for measuring a right angle. This method seems also to have been known to the Chinese nearly three thousand years ago, but the Chinese made no serious attempt to classify or extend the few rules of arithmetic or geometry with which they were acquainted, or to explain the causes of the phenomena which they observed.

The geometrical theorem of which a particular case is involved in the method just described is well known to readers of the first book of Euclid's *Elements*. The Egyptians must probably have known that this theorem is true for a right-angled triangle when the sides containing the right angle are equal, for this is obvious if a floor be paved with tiles of that shape. But these facts cannot be said to show that geometry was then studied as a science. Our real knowledge of the nature of Egyptian geometry depends mainly on the Rhind papyrus.

The ancient Egyptian papyrus from the collection of Rhind, which was written by an Egyptian priest named Ahmes considerably more than a thousand years before Christ, and which is now in the British Museum, contains a fairly complete applied mathematics, in which the measurement of figures and solids plays the principal part; there are no theorems properly so called; everything is stated in the form of problems, not in general terms, but in distinct numbers. For example: to measure a rectangle the sides of which contain two and ten units of length; to find the surface of a circular area whose diameter is six units. We find also in it indications for the measurement of solids, particularly of pyramids, whole and truncated. The arithmetical problems dealt with in this papyrus—which, by the way, is headed "Directions for knowing all dark things"—contain some very interesting things. In modern language, we should say that the first part deals with the reduction of fractions whose numerators are 2 to a sum of fractions each of whose numerators is 1. Thus $\frac{2}{29}$ is stated to be the sum of $\frac{1}{24}$, $\frac{1}{58}$, $\frac{1}{174}$, and $\frac{1}{232}$. Probably Ahmes had no rule for forming the component fractions, and the answers given represent the accumulated experiences of previous writers. In one solitary case, however, he has indicated his method, for, after having asserted that $\frac{2}{3}$ is the sum of $\frac{1}{2}$ and $\frac{1}{6}$, he added that therefore two-thirds of one-fifth is equal to the sum of a half of a fifth and a sixth of a fifth, that is, to $\frac{1}{10} + \frac{1}{30}$.

That so much attention should have been paid to fractions may be explained by the fact that in early times their treatment presented considerable difficulty. The Egyptians and Greeks simplified the problem by reducing a fraction to the sum of several fractions, in each of which the numerator was unity, so that they had to consider only the various denominators: the sole exception to this rule being the fraction $\frac{2}{3}$. This remained the Greek practice until the sixth century of our era. The Romans, on the other hand, generally kept the denominator equal to twelve, expressing the fraction (approximately) as so many twelfths.

In Ahmes' treatment of multiplication, he seems to have relied on repeated additions. Thus, to multiply a certain number, which we will denote by the letter "a," by 13, he first multiplied by 2 and got $2a$, then he doubled the results and got $4a$, then he again doubled the result and got $8a$, and lastly he added together a, $4a$, and $8a$.

Now, we have used the sign "a" to stand for any number: not a particular number like 3, but *any* one. This is what Ahmes did, and what we learn to do in what we call "algebra." When Ahmes wished to find a number such that it, added to its seventh, makes 19, he symbolised the number by the sign we translate "heap." He had also signs for our "+," "−," and "=".[2] Nowadays we can write Ahmes' problem as: Find the number x

---

[2] In this book I shall take great care in distinguishing signs from what they signify. Thus 2 is to be distinguished from "2": by "2" I mean the *sign*, and the sign written without inverted commas indicates the thing signified. There has been, and is, much confusion, not only with beginners but with eminent mathematicians between a sign

such that $x + \dfrac{x}{7} = 19$. Ahmes gave the answer in the form $16 + \frac{1}{2} + \frac{1}{8}$.

We shall find that algebra was hardly touched by those Greeks who made of geometry such an important science, partly, perhaps, because the almost universal use of the abacus [3] rendered it easy for them to add and subtract without any knowledge of theoretical arithmetic. And here we must remember that the principal reason why Ahmes' arithmetical problems seem so easy to us is because of our use from childhood of the system of notation introduced into Europe by the Arabs, who originally obtained it from either the Greeks or the Hindoos. In this system an integral number is denoted by a succession of digits, each digit representing the product of that digit and a power of ten, and the number being equal to the sum of these products. Thus, by means of the local value attached to nine symbols and a symbol for zero, any number in the decimal scale of notation can be expressed. It is important to realise that the long and strenuous work of the most gifted minds was necessary to provide us with simple and expressive notation which, in nearly all parts of mathematics, enables even the less gifted of us to reproduce theorems which needed the greatest genius to discover. Each improvement in notation seems, to the uninitiated, but a small thing: and yet, in a calculation, the pen sometimes seems to be more intelligent than the user. Our notation is an instance of that great spirit of economy which spares waste of labour on what is already systematised, so that all our strength can be concentrated either upon what is known but unsystematised, or upon what is unknown.

Let us now consider the transformation of Egyptian geometry in Greek hands. Thales of Miletus (about 640–546 B.C.), who, during the early part of his life, was engaged partly in commerce and partly in public affairs, visited Egypt and first brought this knowledge into Greece. He discovered many things himself, and communicated the beginnings of many to his successors. We cannot form any exact idea as to how Thales presented his geometrical teaching. We infer, however, from Proclus that it consisted of a number of isolated propositions which were not arranged in a logical sequence, but that the proofs were deductive, so that the theorems were not a mere statement of an induction from a large number of special instances, as probably was the case with the Egyptian geometri-

---

and what is signified by it. Many have even maintained that *numbers* are the *signs* used to represent them. Often, for the sake of brevity, I shall use the word in inverted commas (say "*a*") as short for "what we call '*a*,'" but the context will make plain what is meant.

[3] The principle of the abacus is that a number is represented by counters in a series of grooves, or beads strung on parallel wires; as many counters being put on the first groove as there are units, as many on the second as there are tens, and so on. The rules to be followed in addition, subtraction, multiplication, and division are given in various old works on arithmetic.

cians. The deductive character which he thus gave to the science is his chief claim to distinction. Pythagoras (born about 580 B.C.) changed geometry into the form of an abstract science, regarding its principles in a purely abstract manner, and investigated its theorems from the immaterial and intellectual point of view. Among the successors of these men, the best known are Archytas of Tarentum (428–347 B.C.), Plato (429–348 B.C.), Hippocrates of Chios (born about 470 B.C.), Menaechmus (about 375–325 B.C.), Euclid (about 330–275 B.C.), Archimedes (287–212 B.C.), and Apollonius (260–200 B.C.).

The only geometry known to the Egyptian priests was that of surfaces, together with a sketch of that of solids, a geometry consisting of the knowledge of the areas contained by some simple plane and solid figures, which they had obtained by actual trial. Thales introduced the ideal of establishing by exact reasoning the *relations* between the different parts of a figure, so that some of them could be found by means of others in a manner strictly rigorous. This was a phenomenon quite new in the world, and due, in fact, to the abstract spirit of the Greeks. In connection with the new impulse given to geometry, there arose with Thales, moreover, scientific astronomy, also an abstract science, and undoubtedly a Greek creation. The astronomy of the Greeks differs from that of the Orientals in this respect: the astronomy of the latter, which is altogether concrete and empirical, consisted merely in determining the duration of some periods or in indicating, by means of a mechanical process, the motions of the sun and planets; whilst the astronomy of the Greeks aimed at the discovery of the geometrical laws of the motions of the heavenly bodies.

Let us consider a simple case. The area of a right-angled field of length 80 yards and breadth 50 yards is 4000 square yards. Other fields which are not rectangular can be approximately measured by mentally dissecting them—a process which often requires great ingenuity and is a familiar problem to land-surveyors. Now, let us suppose that we have a circular field to measure. Imagine from the centre of the circle a large number of radii drawn, and let each radius make equal angles with the next ones on each side of it. By joining the points in succession where the radii meet the circumference of the circle, we get a large number of triangles of equal area, and the sum of the areas of all these triangles gives an approximation to the area of the circle. It is particularly instructive repeatedly to go over this and the following examples mentally, noticing how helpful the abstract ideas we call "straight line," "circle," "radius," "angle," and so on, are. We all of us know them, recognise them, and can easily feel that they are trustworthy and accurate ideas. We feel at home, so to speak, with the idea of a square, say, and can at once give details about it which are *exactly* true for it, and *very nearly* true for a field which we know is *very nearly* a square. This replacement in thought by an abstract geometrical object economises labour of thinking and imagining by leading us to concentrate our thoughts on that alone which is essential for our purpose.

Thales seems to have discovered—and it is a good thing to follow these discoveries on figures made with the help of compasses and ruler—the proof of what may be regarded as the obvious fact that the circle is divided into halves by its diameter, that the angles at the base of a triangle with two equal sides—an "isosceles" triangle—are equal, that all the triangles described in a semi-circle with two of their angular points at the ends of the diameter and the third anywhere on the circumference contain a right angle, and he measured the distance of vessels from the shore, presumably by causing two observers at a known distance apart to measure the two angles formed by themselves and the ship. This last discovery is an application of the fact that a triangle is determined if its base and base angles are given.

When Archytas and Menaechmus employed mechanical instruments for solving certain geometrical problems, "Plato," says Plutarch, "inveighed against them with great indignation and persistence as destroying and perverting all the good there is in geometry; for the method absconds from incorporeal and intellectual to sensible things, and besides employs again such bodies as require much vulgar handicraft: in this way *mechanics* was dissimilated and expelled from geometry, and, being for a long time looked down upon by philosophy, became one of the arts of war." In fact, manual labour was looked down upon by the Greeks, and a sharp distinction was drawn between the slaves, who performed bodily work and really observed nature, and the leisured upper classes, who speculated and often only knew nature by hearsay. This explains much of the naïve, hazy, and dreamy character of ancient natural science. Only seldom did the impulse to make experiments for oneself break through; but when it did, a great progress resulted, as was the case with Archytas and Archimedes. Archimedes, like Plato, held that it was undesirable for a philosopher to seek to apply the results of science to any practical use; but, whatever might have been his view of what ought to be the case, he did actually introduce a large number of new inventions.

We will not consider further here the development of mathematics with other ancient nations, nor the chief problems investigated by the Greeks; such details may be found in some of the books mentioned in the Bibliography at the end. The object of this chapter is to indicate the nature of the science of geometry, and how certain practical needs gave rise to investigations in which appears an abstract science which was worthy of being cultivated for its own sake, and which incidentally gave rise to advantages of a practical nature.

There are two branches of mathematics which began to be cultivated by the Greeks, and which allow a connection to be formed between the spirits of ancient and modern mathematics.

The first is the method of geometrical analysis to which Plato seems to have directed attention. The analytical method of proof begins by assuming that the theorem or problem is solved, and thence deducing some result. If the result be false, the theorem is not true or the problem is incapable of solution: if the result be true, if the steps be reversible, we get (by reversing them) a synthetic proof; but if the steps be not reversible, no conclusion can be drawn. We notice that the leading thought in analysis is that which is fundamental in algebra, and which we have noticed in the case of Ahmes: the calculation or reasoning with an unknown entity, which is denoted by a conventional sign, as if it were known, and the deduction at last, of some relation which determines what the entity must be.

And this brings us to the second branch spoken of: algebra with the later Greeks. Diophantus of Alexandria, who probably lived in the early half of the fourth century after Christ, and probably was the original inventor of an algebra, used letters for unknown quantities in arithmetic and treated arithmetical problems analytically. Juxtaposition of symbols represented what we now write as "+," and "−" and "=" were also represented by symbols. All these symbols are mere abbreviations for words, and perhaps the most important advantage of symbolism—the power it gives of carrying out a complicated chain of reasoning almost mechanically—was not made much of by Diophantus. Here again we come across the economical value of symbolism: it prevents the wearisome expenditure of mental and bodily energy on those processes which can be carried out mechanically. We must remember that this economy both emphasises the unsubjugated—that is to say, unsystematised—problems of science, and has a charm—an æsthetic charm, it would seem—of its own.

Lastly, we must mention "incommensurables," "loci," and the beginnings of "trigonometry."

Pythagoras was, according to Eudemus and Proclus, the discoverer of "incommensurable quantities." Thus, he is said to have found that the diagonal and the side of a square are "incommensurable." Suppose, for example, that the side of the square is one unit in length; the diagonal is longer than this, but it is not two units in length. The excess of the length of the diagonal over one unit is not an integral submultiple of the unit. And, expressing the matter arithmetically, the remainder that is left over after each division of a remainder into the preceding divisor is not an integral submultiple of the remainder used as divisor. That is to say, the rule given in text-books on arithmetic and algebra for finding the greatest common measure does not come to an end. This rule, when applied to integer numbers, always comes to an end; but, when applied to certain lengths, it does not. Pythagoras proved, then, that if we start with a line of any length, there are other lines whose lengths do not bear to the first length the ratio of one integer to another, no matter if we have all the integers to choose from. Of course, any two fractions have the ratio of two integers to one another. In the above case of the diagonal, if the diagonal *were* in length some number x of units, we should have $x^2 = 2$, and it can be proved that no fraction, when "multiplied"—in the sense to be given in the next chapter—by itself gives 2 exactly, though there are fractions which give this result more and more approximately.

On this account, the Greeks drew a sharp distinction between "numbers," and "magnitudes" or "quantities" or measures of lengths. This distinction was gradually blotted out as people saw more and more the advantages of identifying numbers with the measures of lengths. The invention of analytical geometry, described in the third chapter, did most of the blotting out. It is in comparatively modern times that mathematicians have adequately realised the importance of this logically valid distinction made by the Greeks. It is a curious fact that the abandonment of strictly logical thinking should have led to results which transgressed what was then known of logic, but which are now known to be readily interpretable in the terms of what we now know of Logic. This subject will occupy us again in the sixth chapter.

The question of *loci* is connected with geometrical analysis, and is difficult to dissociate from a mental picture of a point in motion. Think of a point under restrictions to move only in some curve. Thus, a point may move so that its distance from a fixed point is constant; the peak of an angle may move so that the arms of the angle pass—slipping—through two fixed points, and the angle is always a right angle. In both cases the moving point keeps on the circumference of a certain circle. This curve is a "locus." It is evident how thinking of the locus a point can describe may help us to solve problems.

We have seen that Thales discovered that a triangle is determined if its base and base angles are given. When we have to make a survey of either an earthly country or part of the heavens, for the purpose of map-making, we have to measure angles—for example, by turning a *sight*, like those used on guns, through an angle measured in a circular arc of metal —to fix the relative directions of the stars or points on the earth. Now, for terrestrial measurements, a piece of country is approximately a flat surface, while the heavens are surveyed as if the stars were, as they seem to be, scattered on the inside of a sphere at whose centre we are. Secondly, it is a network of *triangles*—plane or spherical—of which we measure the angles and sometimes the sides: for, if the angles of a triangle are known, the *proportionality* of the sides is known; and this proportionality cannot be concluded from a knowledge of the angles of a rectangle, say. Hipparchus (born about 160 B.C) seems to have invented this practical science of the complete measurement of triangles from certain data, or, as it is called, "trigonometry," and the principles laid down by him were worked out by Ptolemy of Alexandria (died 168 A.D.) and also by the Hindoos and Arabians. Usually, only angles can be measured with accuracy, and so the question arises: given the magnitude of the angles, what can be concluded as to the kind of proportionality of the sides. Think of a circle described round the centre O, and let AP be the arc of this circle which measures the angle AOP. Notice that the ratio of AP to the radius is the same for the angle AOP whatever value the radius may have. Draw PM perpendicular to OA. Then the figure $OPMAP$ reminds one of a stretched bow, and hence are derived the names "sine of the arc AP" for the line PM, and "cosine" for OM. Tables of sines and cosines of arcs (or of angles, since the arc fixes the angle if the radius is known) were drawn up, and thus the sides PM and OM could be found in terms of the radius, when the arc was known. It is evident that this contains the essentials for the finding of the proportions of the sides of plane triangles. Spherical trigonometry contains more complicated relations which are directly relevant to the position of an astronomer and his measurements.

Mathematics did not progress in the hands of the Romans: perhaps the genius of this people was too practical. Still, it was through Rome that mathematics came into medieval Europe. The Arab mathematical textbooks and the Greek books from Arab translations were introduced into Western Europe by the Moors in the period 1150–1450, and by the end of the thirteenth century the Arabic arithmetic had been fairly introduced into Europe, and was practised by the side of the older arithmetic founded on the work of Boethius (about 475–526). Then came the Renascence. Mathematicians had barely assimilated the knowledge obtained from the Arabs, including their translations of Greek writers, when the refugees who escaped from Constantinople after the fall of the Eastern Empire (1453) brought the original works and the traditions of Greek science into Italy. Thus by the middle of the fifteenth century the chief results of Greek and Arabian mathematics were accessible to European students.

The invention of printing about that time rendered the dissemination of discoveries comparatively easy.

CHAPTER II

THE RISE AND PROGRESS OF MODERN MATHEMATICS—ALGEBRA

MODERN mathematics may be considered to have begun approximately with the seventeenth century. It is well known that the first 1500 years of the Christian era produced, in Western Europe at least, very little knowledge of value in science. The spirit of the Western Europeans showed itself to be different from that of the ancient Greeks, and only slightly less so from that of the more Easterly nations; and, when Western mathematics began to grow, we can trace clearly the historical beginnings of the use, in a not quite accurate form, of those conceptions—*variable* and *function*—which are characteristic of modern mathematics. We may say, in anticipation, that these conceptions, thoroughly analysed by reasoning as they are now, make up the difference of our modern views of Mathematics from, and have caused the likeness of them to, those of the ancient Greeks. The Greeks seem, in short, to have taken up a very similar position towards the mathematics of their day to that which logic forces us to take up towards the far more general mathematics of to-day. The generality of character has been attained by the effort to put mathematics more into touch with natural sciences—in particular the science of motion. The main difficulty was that, to reach this end, the way in which mathematicians expressed themselves was illegitimate. Hence philosophers, who lacked the real sympathy that must inspire all criticism that hopes to be relevant, never could discover any reason for thinking that what the mathematicians said was true, and the world had to wait until the mathematicians began logically to analyse their own conceptions. No body of men ever needed this sympathy more than the mathematicians from the revival of letters down to the middle of the nineteenth century, for no science was less logical than mathematics.

The ancient Greeks never used the conception of *motion* in their systematic works. The idea of a *locus* seems to imply that some curves could be thought of as generated by moving points; the Greeks discovered some things by helping their imaginations with imaginary moving points, but they never introduced the use of motion into their final proofs. This may have been because the Eleatic school, of which one of the principal representatives was Zeno (495–435 B.C.), invented some exceedingly subtle puzzles to emphasize the difficulty there is in the conception of motion. We shall return in some detail to these puzzles, which have not been appreciated in all the ages from the time of the Greeks till quite modern times. Owing to this lack of subtlety, the conception of variability was freely introduced into mathematics. It was the conceptions of *constant*, *variable*, and *function*, of which we shall, from now on, often have occasion to speak, which were generated by ideas of motion, and which, when they were logically purified, have made both modern mathematics and modern logic, to which they were transferred by mathematical logicians— Leibniz, Lambert, Boole, De Morgan, and the numerous successors of Boole and De Morgan from about 1850 onwards—into a science much more general than, but bearing some close analogies with, the ideal of Greek mathematical science. Later on will be found a discussion of what can be meant by a "moving point."

Let us now consider more closely the history of modern mathematics. Modern mathematics, like modern philosophy and like one part—the speculative and not the experimental part—of modern physical science, may be considered to begin with René Descartes (1596–1650). Of course, as we should expect, Descartes had many and worthy predecessors. Perhaps the greatest of them was the French mathematician François Viète (1540–1603), better known by his Latinized name of "Vieta." But it is simpler and shorter to confine our attention to Descartes.

Descartes always plumed himself on the independence of his ideas, the breach he made with the old ideas of the Aristotelians, and the great clearness and simplicity with which he described his ideas. But we must not underestimate the part that "ideas in the air" play; and, further, we know now that Descartes' breach with the old order of things was not as great as he thought.

Descartes, when describing the effect which his youthful studies had upon him when he came to reflect upon them, said:

"I was especially delighted with the mathematics, on account of the certitude and evidence of their reasonings: but I had not as yet a precise knowledge of their true use; and, thinking that they but contributed to the advancement of the mechanical arts, I was astonished that foundations so strong and solid should have had no loftier superstructure reared on them."

And again:

"Among the branches of philosophy, I had, at an earlier period, given some attention to logic, and, among those of the mathematics, to geometrical analysis and algebra—three arts or sciences which ought, as I conceived, to contribute something to my design. But, on examination, I found that, as for logic, its syllogisms and the majority of its other precepts are of avail rather in the communication of what we already know, or even in speaking without judgment of things of which we are ignorant, than in the investigation of the unknown: and although this science contains indeed a number of correct and very excellent precepts, there are, nevertheless, so many others, and these either injurious or superfluous, mingled with the former, that it is almost quite as difficult to effect a severance of the true from the false as it is to extract a Diana or a Minerva from a rough block of marble. Then as to the analysis of the ancients and the algebra of the moderns: besides that they embrace only matters highly abstract, and, to appearance, of no use, the former is so exclusively restricted to the consideration of figures that it can exercise the understanding only on condition of greatly fatiguing the imagination; and, in the latter, there is so complete a subjection to certain rules and formulas, that there results an art full of confusion and obscurity, calculated to embarrass, instead of a science fitted to cultivate the mind. By these considerations I was induced to seek some other method which would comprise the advantages of the three and be exempt from their defects. . . .

"The long chains of simple and easy reasonings by means of which geometers are accustomed to reach the conclusions of their most difficult demonstrations had led me to imagine that all things to the knowledge of which man is competent are mutually connected in the same way, and that there is nothing so far removed from us as to be beyond our reach, or so hidden that we cannot discover it, provided only that we abstain from accepting the false for the true, and always preserve in our thoughts the order necessary for the deduction of one truth from another. And I had little difficulty in determining the objects with which it was necessary to begin, for I was already persuaded that it must be with the simplest and easiest to know, and, considering that, of all those who have hitherto sought truth in the sciences, the mathematicians alone have been able to find any demonstrations, that is, any certain and evident reasons, I did not doubt but that such must have been the rule of their investigations. I resolved to begin, therefore, with the examination of the simplest objects, not anticipating, however, from this any other advantage than that to be found in accustoming my mind to the love and nourishment of truth and to a distaste for all such reasonings as were unsound. But I had no intention on that account of attempting to master all the particular sciences commonly denominated 'mathematics'; but observing that, however different their objects, they all agree in considering only the various relations or proportions subsisting among those objects, I thought it best for my purpose to consider these proportions in the most general form possible, without referring them to any objects in particular, except such as would most facilitate the knowledge of them, and without by any means restricting them to these, that afterwards I might thus be the better able to apply them to every other class of objects to which they are legitimately applicable. Perceiving further that, in order to understand these relations, I should have sometimes to consider them one by one and sometimes only to bear in mind or embrace them in the aggregate, I thought that, in order

the better to consider them individually, I should view them as subsisting
between straight lines, than which I could find no objects more simple or
capable of being more distinctly represented to my imagination and senses;
and, on the other hand, that, in order to retain them in the memory, or
embrace an aggregate of many, I should express them by certain characters the briefest possible. In this way I believed that I could borrow all
that was best both in geometrical analysis and in algebra, and correct all
the defects of the one by help of the other."

Let us, then, consider the characteristics of algebra and geometry.

We have seen, when giving an account, in the first chapter, of the
works of Ahmes and Diophantus, that mathematicians early saw the advantage of representing an unknown number by a letter or some other
sign that may denote various numbers ambiguously, writing down—much
as in geometrical analysis—the relations which they bear, by the conditions of the problem, to other numbers, and then considering these relations. If the problem is determinate—that is to say, if there are one or
more definite solutions which can be proved to involve only numbers
already fixed upon—this consideration leads, by the use of certain rules
of calculation, to the determination—actual or approximate—of this solution or solutions. Under certain circumstances, even if there is a solution,
depending on a variable, we can find it and express it in a quite general
way, by rules, but that need not occupy us here.

Suppose that you know my age, but that I do not know yours, but
wish to. You might say to me: "I was eight years old when you were
born." Then I should think like this. Let x be the (unknown) number of
years in your age at this moment and, say, 33 the number of years in my
age at this moment; then in essentials your statement can be translated by
the equation "$x - 8 = 33$." The meaning of the signs "$-$," "$=$," and "$+$"
are supposed to be known—as indeed they are by most people nowadays
quite sufficiently for our present purpose. Now, one of the rules of algebra
is that any term can be taken from one side of the sign "$=$" to the other
if only the "$+$" or "$-$" belonging to it is changed into "$-$" or "$+$," as
the case may be. Thus, in the present case, we have: "$x = 33 + 8 = 41$."
This absurdly simple case is chosen intentionally. It is essential in mathematics to remember that even apparently insignificant economies of
thought add up to make a long and complicated calculation readily performed. This is the case, for example, with the convention introduced by
Descartes of using the last letters of the alphabet to denote unknown
numbers, and the first letters to denote known ones. This convention is
adopted, with a few exceptional cases, by algebraists to-day, and saves
much trouble in explaining and in looking for unknown and known quantities in an equation. Then, again, the signs "$+$," "$-$," "$=$" have great

merits which those unused to long calculations cannot readily understand. Even the saving of space made by writing "xy" for "$x \times y$" ("x multiplied by y") is important, because we can obtain by it a shorter and more readily surveyed formula. Then, too, Descartes made a general practice of writing "powers" or "exponents" as we do now; thus "x^3" stands for "xxx" and "x^5" for some less suggestive symbol representing the continued multiplication of five x's.

One great advantage of this notation is that it makes the explanation of logarithms, which were the great and laborious discovery of John Napier (1550–1617), quite easy. We start from the equation "$x^m x^n = x^{m+n}$." Now, if $x^p = y$, and we call p the "logarithm of y to the base x"; in signs: "$p = \log_x y$"; the equation from which we started gives, if we denote x^m by "u" and x^n by "v," so that $m = \log_x u$ and $n = \log_x v$, that $\log_x (uv) = \log_x u + \log_x v$. Thus, if the logarithms of numbers to a given base (say $x = 10$) are tabulated, calculations with large numbers are made less arduous, for *addition replaces multiplication*, when logarithms are found. Also subtraction of logarithms gives the logarithm of the quotient of two numbers.

Let us now shortly consider the history of algebra from Diophantus to Descartes.

The word "algebra" is the European corruption of an Arabic phrase which means *restoration and reduction*—the first word referring to the fact that the same magnitude may be added to or subtracted from both sides of an equation, and the last word meaning the process of simplification. The science of algebra was brought among the Arabs by Mohammed ben Musa (Mahomet the son of Moses), better known as Alkarismi, in a work written about 830 A.D., and was certainly derived by him from the Hindoos. The algebra of Alkarismi holds a most important place in the history of mathematics, for we may say that the subsequent Arab and the early medieval works on algebra were founded on it, and also that through it the Arabic or Indian system of decimal numeration was introduced into the West. It seems that the Arabs were quick to appreciate the work of others—notably of the Greek masters and of the Hindoo mathematicians—but, like the ancient Chinese and Egyptians, they did not systematically develop a subject to any considerable extent.

Algebra was introduced into Italy in 1202 by Leonardo of Pisa (about 1175–1230) in a work based on Alkarismi's treatise, and into England by Robert Record (about 1510–1558) in a book called the *Whetstone of Witte* published in 1557. Improvements in the method or notations of algebra were made by Record, Albert Girard (1595–1632), Thomas Harriot (1560–1621), Descartes, and many others.

In arithmetic we use *symbols of number*. A symbol is any sign for a quantity, which is not the quantity itself. If a man counted his sheep by pebbles, the pebbles would be symbols of the sheep. At the present day, when most of us can read and write, we have acquired the convenient habit of using marks on paper, "1, 2, 3, 4," and so on, instead of such things as pebbles. Our "1 + 1" is abbreviated into "2," "2 + 1" is abbreviated into "3," "3 + 1" into "4," and so on. When "1," "2," "3," &c., are used to abbreviate, rather improperly, "1 mile," "2 miles," "3 miles," &c., for instance, they are called signs for *concrete* numbers. But when we shake off all idea of "1," "2," &c., meaning one, two, &c., of anything in particular, as when we say, "six and four make ten," then the numbers are called *abstract* numbers. To the latter the learner is first introduced in treatises on arithmetic, and does not always learn to distinguish rightly between the two. Of the operations of arithmetic only addition and subtraction can be performed with concrete numbers, and without speaking of more than one sort of 1. Miles can be added to miles, or taken from miles. Multiplication involves a new sort of 1, 2, 3, &c., standing for *repetitions* (or *times*, as they are called). Take 6 miles 5 times. Here are two kinds of units, 1 mile and 1 time. In multiplication, one of the units must be a number of repetitions or times, and to talk of multiplying 6 feet by 3 feet would be absurd. What notion can be formed of 6 feet taken "3 feet" times? In solving the following question, "If 1 yard cost 5 shillings, how much will 12 yards cost?" we do not multiply the 12 *yards* by the 5 *shillings*; the process we go through is the following: Since each yard costs 5 shillings, the buyer must put down 5 shillings as often (as many times) as the seller uses a one-yard measure; that is, 5 shillings is taken 12 times. In division we must have the idea either of repetition or of *partition*, that is, of cutting a quantity into a number of equal parts. "Divide 18 miles by 3 *miles*" means, find out how many *times* 3 miles must be repeated to give 18 miles: but "divide 18 miles by 3" means, cut 18 miles into 3 equal parts, and find how many miles are in each part.

The symbols of arithmetic have a *determinate connection*; for instance, 4 is always 2 + 2, whatever the things mentioned may be, miles, feet, acres, &c. In algebra we take symbols for numbers which have no determinate connection. As in arithmetic we draw conclusions about 1, 2, 3, &c., which are equally true of 1 foot, 2 feet, &c., 1 minute, 2 minutes, &c.; so in algebra we reason upon numbers in general, and draw conclusions which are equally true of all numbers. It is true that we also use, in kinds of algebra which have been developed within the last century, letters to represent things other than numbers—for example, *classes* of individuals with a certain property, such as "horned animals," for logical purposes; or certain geometrical or physical things with directions in space, such as "forces"—and signs like "+" and "−" to represent ways of combination of the things, which are analogous to, but not identical with, addition and subtraction. If "a" denotes "the class of horned animals" and "b" denotes "the class of beasts of burden," the sign "ab" has been used to denote "the class of horned beasts of burden." We see that here $ab = ba$, just as in the multiplication of numbers, and the above operation has been called, partly for this reason, "logical multiplication," and denoted in the above way. Here we meet the practice of mathematicians—and of all scientific men—of using words in a wider sense for the sake of some analogy. This habit is all the more puzzling to many people because mathematicians are often not conscious that they do it, or even talk sometimes as if they thought that they were generalising *conceptions* instead of words. But, when we talk of a "family tree," we do not indicate a widening of our conception of trees of the roadside.

We shall not need to consider these modern algebras, but we shall be constantly meeting what are called the "generalisations of number" and transference of methods to analogous cases. Indeed, it is hardly too much to say that in this lies the very spirit of discovery. An example of this is given by the extension of the word "numbers" to include the names of *fractions as well.* The occasion for this extension was given by the use of arithmetic to express such quantities as distances. This had been done by Archimedes and many others, and had become the usual method of procedure in the works of the mathematicians of the sixteenth century, and plays a great part in Descartes' work.

Mathematicians, ever since they began to apply arithmetic to geometry, became alive to the fact that it was convenient to represent points on a straight line by numbers, and numbers by points on a straight line. What is meant by this may be described as follows. If we choose a unit of length, we can mark off points in a straight line corresponding to 0 units —which means that we select a point, called "the origin," to start from,— 1 unit, 2 units, 3 units, and so on, so that "the point m," as we will call it for short, is at a distance of m units from the origin. Then we can divide up the line and mark points corresponding to the fractions $\frac{1}{2}$, $\frac{2}{3}$, $\frac{7}{8}$, $\frac{1}{13}$, $\frac{5}{3}$, or the point between 1 and 2 which is the same distance from 1 as $\frac{2}{3}$ is from 0, and so on. Now, there is nothing here to distinguish fractions from numbers. Both are treated exactly in the same way; the results of addition, subtraction, multiplication, and division [4] are interpretable, in much the same way as new points whether the "a" and "b" in

[4] The operation of what is called, for the sake of analogy, "multiplication" of fractions is defined in the manner indicated in the following example. If $\frac{3}{4}$ of a yard costs 10d., how much does $\frac{7}{8}$ of a yard cost? The answer is $\frac{10 \times 4 \times 7}{3 \times 8}$ pence, and we define $\frac{4 \times 7}{3 \times 8}$ as $\frac{1}{\frac{3}{4}}$ "multiplied by" $\frac{7}{8}$, by analogy with what would happen if $\frac{3}{4}$ were 1 and $\frac{7}{8}$ were, say, 3.

"$a + b$," "$a - b$," "ab," and so on, stand for numbers or fractions, and we have, for example,

$$a + b = b + a, \, ab = ba, \, a \, (b + c) = ab + ac,$$

always. Because of this very strong analogy, mathematicians have called the fractions "numbers" too, and they often speak and write of "generalisations of numbers," of which this is the first example, as if the conception of number were generalised, and not merely the *name* "number," in virtue of a great and close and important analogy.

When once the points of a line were made to represent numbers, there seemed to be no further difficulty in admitting certain "irrational numbers" to correspond to the end-points of the incommensurable lines which had been discovered by the Greeks. This question will come up again at a later stage: there are necessary discussions of principle involved, but mathematicians did not go at all deeply into questions of principle until fairly modern times. Thus it has happened that, until the last sixty years or so, mathematicians were nearly all bad reasoners, as Swift remarked of the mathematicians of Laputa in *Gulliver's Travels*, and were unpardonably hazy about first principles. Often they appealed to a sort of faith. To an intelligent and doubting beginner, an eminent French mathematician of the eighteenth century said: "Go on, and faith will come to you." It is a curious fact that mathematicians have so often arrived at truth by a sort of instinct.

Let us now return to our numerical algebra. Take, say, the number 8, and the fraction, which we will now call a "number" also, ⅛. Add 1 to both; the greater contains the less exactly 8 times. Now this property is possessed by *any* number, and not 8 alone. In fact, if we denote the number we start with by "a," we have, by the rules of algebra, $\dfrac{a + 1}{1/_a + 1} = a$. This is an instance of a general property of numbers proved by algebra.

Algebra contains many rules by which a complicated algebraical expression can be reduced to its simplest terms. Owing to the suggestive and compact notation, we can easily acquire an almost mechanical dexterity in dealing with algebraical symbols. This is what Descartes means when he speaks of algebra as not being a science fitted to cultivate the mind. On the other hand, this art is due to the principle of the economy of thought, and the mechanical aspect becomes, as Descartes foresaw, very valuable if we could use it to solve geometrical problems without the necessity of fatiguing our imaginations by long reasonings on geometrical figures.

I have already mentioned that the valuable notation "x^m" was due to Descartes. This was published, along with all his other improvements in algebra, in the third part of his *Geometry* of 1637. I shall speak in the next chapter of the great discovery contained in the first two parts of this work; here I will resume the improvements in notation and method made by Descartes and his predecessors, which make the algebraical part of the *Geometry* very like a modern book on algebra.

It is still the custom in arithmetic to indicate addition by juxtaposition: thus "2½" means "2 + ½." In algebra, we always, nowadays, indicate addition by the sign "+" and multiplication by juxtaposition or, more rarely, by putting a dot or the sign "×" between the signs of the numbers to be multiplied. Subtraction is indicated by "−".

Here we must digress to point out—what is often, owing to confusion of thought, denied in text-books—that, where "*a*" and "*b*" denote numbers, "*a* − *b*" can only denote a number if *a* is equal to or greater than *b*. If *a* is equal to *b*, the number denoted is zero; there is really no good reason for denying, say, that the numbers of Charles II.'s foolish sayings and wise deeds are equal, if a well-known epitaph be true. Here again we meet the strange way in which mathematics has developed. For centuries mathematicians used "negative" and "positive" numbers, and identified "positive" numbers with signless numbers like 1, 2, and 3, without any scruple, just as they used fractionary and irrational "numbers." And when logically-minded men objected to these wrong statements, mathematicians simply ignored them or said: "Go on; faith will come to you." And the mathematicians were right, and merely could not give correct reasons— or at least always gave wrong ones—for what they did. We have, over again, the fact that criticism of the mathematicians' procedure, if it wishes to be relevant, must be based on thorough sympathy and understanding. It must try to account for the rightness of mathematical views, and bring them into conformity with logic. Mathematicians themselves never found a competent philosophical interpreter, and so nearly all the interesting part of mathematics was left in obscurity until, in the latter half of the nineteenth century, mathematicians themselves began to cultivate philosophy—or rather logic.

Thus we must go out of the historical order to explain what "negative numbers" means. First, we must premise that when an algebraical expression is enclosed in brackets, it signifies that the whole result of that expression stands in the same relation to surrounding symbols as if it were one letter only. Thus, "*a* − (*b* − *c*)" means that from *a* we are to take *b* − *c*, or what is left after taking *c* from *b*. It is not, therefore, the same as *a* − *b* − *c*. In fact we easily find that *a* − (*b* − *c*) is the same as *a* − *b* + *c*. Note also that "(*a* + *b*) (*c* + *d*)" means (*a* + *b*) multiplied by (*c* + *d*).

Now, suppose *a* and *b* are numbers, and *a* is greater than *b*. Let *a* − *b* be *c*. To get *c* from *a*, we carry out the operation of taking away *b*. *This operation, which is the fulfilment of the order: "Subtract b," is a "negative number."* Mathematicians call it a "number" and denote it by "−*b*" simply because of analogy: the same rules for calculation hold for "negative numbers" and "positive numbers" like "+*b*," whose meaning is now clear too, as do for our signless numbers; when "addition," "subtraction," &c., are redefined for these operations. The way in which this redefinition must take place is evident when we represent integers, fractions, and positive and negative numbers by points on a straight line. To the right of 0 are the integers and fractions, to the left of 0 are the negative numbers, and to the right of 0 stretch the series of positive numbers, +*a* coinciding with *a* and being symmetrically placed with −*a* as regards 0. Also we determine that the operations of what we call "addition," &c., of these new "numbers" *must lead to the same results* as the former operations of the same name. Thus the same symbol is used in different senses, and we write

$$a + b - b = a + 0 = (+a) + (+b) + (-b) = +a = a.$$

This is a remarkable sequence of quick changes.

We have used the sign of equality, "=". It means originally, "is the same as." Thus $3 + 1 = 4$. But we write, by the above convention, "$a = +a$," and so we sacrifice exactness, which sometimes looks rather pedantic, for the sake of keeping our analogy in view, and for brevity.

Let us bear this, at first sight, puzzling but, at second sight, justifiable peculiarity of mathematicians in mind. It has always puzzled intelligent beginners and philosophers. The laws of calculation and convenient symbolism are *the* things a mathematician thinks of and aims at. He seems to identify different things if they both satisfy the same laws which are important to him, just as a magistrate may think that there is not much difference between Mr. A., who is red-haired and a tinker and goes to chapel, and Mr. B., who is a brown-haired horse-dealer and goes to church, if both have been found out committing petty larceny. But their respective ministers of religion or wives may still be able to distinguish them.

Any two expressions connected by the sign of equality form an "equation." Here we must notice that the words "Solve the equation $x^2 + ax = b$" mean: find the value or values of x such that, a and b being given numbers, $x^2 + ax$ becomes b. Thus, if $a = 2$ and $b = -1$, the solution is $x = -1$.

As we saw above, Descartes fixed the custom of employing the letters at the beginning of the alphabet to denote known quantities, and those at the end of the alphabet to denote unknown quantities. Thus, in the above example, a and b are some numbers supposed to be given, while x is sought. The question is solved when x is found in terms of a and b and fixed numbers (like 1, 2, 3); and so, when to a and b are attributed any fixed values, x becomes fixed. The signs "a" and "b" denote ambiguously, not uniquely like "2" does; and "x" does not always denote ambiguously when a and b are fixed. Thus, in the above case, when $a = 2$, $b = -1$, "x"

denotes the one negative number -1. What is meant is this: In each member of the class of problems got by giving a and b fixed values independently of one another, there is an unknown x, which may or may not denote different numbers, which only becomes known when the equation is solved. Consider now the equation $ax + by = c$, where a, b, and c are known quantities and x and y are unknown. We can find x in terms of a, b, c, and y, or y in terms of a, b, c, and x; but x is only fixed when y is fixed, or y when x is fixed. Here in each case of fixedness of a, b, and c, x is undetermined and "variable," that is to say, it may take any of a whole class of values. Corresponding to each x, one y belongs; and y also is a "variable" depending on the "independent variable" x. The idea of "variability" will be further illustrated in the next chapter; here we will only point out how the notion of what is called by mathematicians the "functional dependence" of y on x comes in. The variable y is said to be a "function" of the variable x if to every value of x corresponds one or more values of y. This use has, to some extent, been adopted in ordinary language. We should be understood if we were to say that the amount of work performed by a horse is a function of the food that he eats.

Descartes also adopted the custom—if he did not arrive at it independently—advocated by Harriot of transferring all the terms of an equation to the same side of the sign of equality. Thus, instead of "$x = 1$," "$ax + b = c$," and "$3x^2 + g = hx$," we write respectively "$x - 1 = 0$," "$ax + (b - c) = 0$," and "$3x^2 - hx + g = 0$." The point of this is that all equations of the same degree in the unknown—we shall have to consider cases of more unknowns than one in the next chapter—that is to say, equations in which the highest power of x (x or x^2 or x^3 . . .) is the same, are easily recognisable. Further, it is convenient to be able to speak of the expression which is equated to 0 as well as of the equation. The equations in which x^2, and no higher power of x, appears are called "quadratic" equations—the result of equating a "quadratic" function to 0; those in which x^3, and no higher power, appears are called "cubic"; and so on for equations "of the fourth, fifth . . ." degrees. Now the quadratic equations, $3x^2 + g = 0$, $ax^2 + bx + c = 0$, $x^2 - 1 = 0$, for example, are different, but the differences are unimportant in comparison with this common property of being of the same degree: all can be solved by modifications of one general method.

Here it is convenient again to depart from the historical order and briefly consider the meaning of what are called "imaginary" expressions. If we are given the equation $x^2 - 1 = 0$, its solutions are evidently $x = +1$ or $x = -1$, for the square roots of $+1$ are $+1$ and -1. But if we are given the equation $x^2 + 1 = 0$, analogy would lead us to write down the two solutions $x = +\sqrt{-1}$ and $x = -\sqrt{-1}$. But there is no positive or negative "number" which we have yet come across which, when multiplied by itself, gives a negative "number." Thus "imaginary numbers" were rejected by Descartes and his followers. Thus $x^2 - 1 = 0$ had two solutions, but $x^2 + 1 = 0$ none; further, $x^3 + x^2 + x + 1 = 0$ had one solution $(x = -1)$, while $x^3 - x^2 - x + 1 = 0$ had two $(x = 1, x = -1)$, and $x^3 - 2x^2 - x + 2 = 0$ had three $(x = 1, x = -1, x = 2)$. Now, suppose, for a moment, that we can have "imaginary" roots and $(\sqrt{-1})\,(\sqrt{-1}) = -1$, and also that we can speak of *two* roots when the roots are identical in a case like the equation $x^2 + 2x + 1 = 0$, or $(x + 1)^2 = 0$, which has two identical roots $x = -1$. Then, in the above five equations, the first two quadratic ones have two roots each $(+1, -1,$ and $+ \sqrt{-1}, - \sqrt{-1}$ respectively), and the three cubics have three each $(-1, + \sqrt{-1}, - \sqrt{-1};$ $+1, -1, +1$; and $+1, -1, +2$ respectively). In the general case, the theorem has been proved that every equation has as many roots as (and not merely "no more than," as Descartes said) its degree has units. For this and for many other reasons like it in enabling theorems to be stated more generally, "imaginary numbers" came to be used almost universally. This was greatly helped by one puzzling circumstance: true theorems can be discovered by a process of calculation with imaginaries. The case is analogous to that which led mathematicians to introduce and calculate with "negative numbers."

For the case of imaginaries, let a, b, c, and d be any numbers, then

$$\begin{aligned}
(a^2 + b^2)(c^2 + d^2) &= (a + b\sqrt{-1})(a - b\sqrt{-1}) \\
&\quad (c + d\sqrt{-1})(c - d\sqrt{-1}) \\
&= (a + b\sqrt{-1})(c + d\sqrt{-1}) \\
&\quad (a - b\sqrt{-1})(c - d\sqrt{-1}) \\
&= [(ac - bd) + \sqrt{-1}(ad + bc)] \\
&\quad [(ac - bd) - \sqrt{-1}(ad + bc)] \\
&= (ac - bd)^2 + (ad + bc)^2.
\end{aligned}$$

We get, then, an interesting and easily verifiable theorem on numbers by calculation with imaginaries, and imaginaries disappear from the conclusion. Mathematicians thought, then, that imaginaries, though apparently uninterpretable and even self-contradictory, *must* have a logic. So they were used with a faith that was almost firm and was only justified much later. Mathematicians indicated their growing security in the use of $\sqrt{-1}$ by writing "i" instead of "$\sqrt{-1}$" and calling it "the complex unity," thus denying, by implication, that there is anything really imaginary or impossible or absurd about it.

The truth is that "i" is not uninterpretable. It represents an operation, just as the negative numbers do, but is of a different kind. It is geometrically interpretable also, though not in a straight line, but in a plane. For this we must refer to the Bibliography; but here we must point out that, in this "generalisation of number" again, the words "addition," "multiplication," and so on, do not have exactly the same, but an analogous, meaning to those which they had before, and that "complex numbers" form a domain like a plane in which a line representing the integers, fractions, and irrationals is contained. But we must leave the further development of these questions.

It must be realised that the essence of algebra is its generality. In the most general case, every symbol and every statement of a proposition in algebra is interpretable in terms of certain operations to be undertaken with abstract things such as numbers or classes or propositions. These operations merely express the relations of these things to one another. If the results at any stage of an algebraical process can be interpreted—and this interpretation is often suggested by the symbolism—say, not as operations with operations with integers, but as other operations with integers, they express true propositions. Thus $(a + b)^2 = a^2 + 2ab + b^2$ expresses, for example, a relation holding between those operations with integers that we call "fractionary numbers," or an analogous relation between integers. The language of algebra is a wonderful instrument for expressing shortly, perspicuously, and suggestively, the exceedingly complicated relations in which abstract things stand to one another. The motive for studying such relations was originally, and is still in many cases, the close analogy of relations between certain abstract things to relations between certain things we see, hear, and touch in the world of actuality round us, and our minds are helped in discovering such analogies by the beautiful picture of algebraical processes made in space of two or of three dimensions made by the "analytical geometry" of Descartes, described in the next chapter.

CHAPTER III

THE RISE AND PROGRESS OF MODERN MATHEMATICS—ANALYTICAL GEOMETRY AND THE METHOD OF INDIVISIBLES

WE will now return to the consideration of the first two sections of Descartes' book *Geometry* of 1637.

In Descartes' book we have to glean here and there what we now recognise as the essential points in his new method of treating geometrical questions. These points were not expressly stated by him. I shall, however, try to state them in a small compass.

Imagine a curve drawn on a plane surface. This curve may be considered as a *picture* of an algebraical equation involving x and y in the following way. Choose any point on the curve, and call "x" and "y" the numbers that express the perpendicular distances of this point, in terms of a unit of length, from two straight lines (called "axes") drawn at right angles to one another in the plane mentioned. Now, as we move from point to point of the curve, x and y both vary, *but there is an unvarying relation which connects x and y, and this relation can be expressed by an algebraical equation called "the equation of the curve," and which contains, in germ as it were, all the properties of the curve considered.* This constant relation between x and y is a relation like $y^2 = 4ax$. We must distinguish carefully between a constant relation between variables and a relation between constants. We are always coming across the former kind of relation in mathematics; we call such a relation a "function" of x and y —the word was first used about fifty years after Descartes' *Geometry* was published, by Leibniz—and write a function of x and y in general as "$f(x, y)$." In this notation, no hint is given as to any particular relation x and y may bear to each other, and, in such a particular function as $y^2 - 4ax$, we say that "the *form* of the function is constant," and this is only another way of saying that the relation between x and y is fixed. This may be also explained as follows. If x is fixed, there is fixed one or more values of y, and if y is fixed, there is fixed one or more values of x. Thus the equation $ax + by + c = 0$ gives one y for each x and one x for each y; the equation $y^2 - 4ax = 0$ gives two y's for each x and one x for each y.[5]

Consider the equation $ax + by + c = 0$, or, say, the more definite instance $x + 2y - 2 = 0$. Draw axes and mark off points; having fixed on a unit of length, find the point $x = 1$ on the x-axis, on the perpendicular to this axis measure where the corresponding y, got by substituting $x = 1$ in the above equation, brings us. We find $y = \frac{1}{2}$. Take $x = \frac{1}{3}$, then $y = \frac{5}{6}$; and so on. We find that all the points on the parallels to the y-axis lie on one straight line. This straight line is determined by the equation $x + 2y - 2 = 0$; every point off that straight line is such that its x and y are not connected by the relation $x + 2y - 2 = 0$, and every point of it is such that its x and y are connected by the relation $x + 2y - 2 = 0$. Similarly we can satisfy ourselves that every point on the circumference of a circle of radius c units of length, described round the point where the axes cross, is such that $x^2 + y^2 = c^2$, and every point not on this circumference does not have an x and y such that the constant relation $x^2 + y^2 = c^2$ is satisfied for it.

There are two points to be noticed in the above general statement. Firstly, I have said that the curve "*may* be expressed," and so on. By this I mean that it is possible—and not necessarily always true—that the curve

---

[5] We also denote a function of x by "$f(x)$" or "$F(x)$" or "$\phi(x)$," &c. Here "f" is a sign for "function of," not for a number, just as later we shall find "sin" and "Δ" and "d" standing for functions and not numbers. This may be regarded as an extension of the language of early algebra. The equation $y = f(x)$ is in a good form for graphical representation in the manner explained below.

may be so considered. We can imagine curves that cannot be represented by a finite algebraical equation. Secondly, about the fundamental lines of reference—the "axes" as they are called. One of these axes we have called the "*x*-axis," and the distance measured by the number *x* is sometimes called "the abscissa"; while the line of length *y* units which is perpendicular to the end of the abscissa farthest from the origin, and therefore parallel to the other axis ("the *y*-axis") is called "the ordinate." The name "ordinate" was used by the ancient Roman surveyors. The lines measured by the numbers *x* and *y* are called the "co-ordinates" of the point determining and determined by them. Sometimes the numbers *x* and *y* themselves are called "co-ordinates," and we will adopt that practice here.

Sometimes the axes are not chosen at right angles to one another, but it is nearly always far simpler to do so, and in this book we always assume that the axes are rectangular. The whole plane is divided by the axes into four partitions, the co-ordinates are measured from the point—called "the origin"—where the axes cross. Here the interpretation in geometry of the "negative quantities" of algebra—which so often seems so puzzling to intelligent beginners—gives us a means of avoiding the ambiguity arising from the fact that there would be a point with the same co-ordinates in each quadrant into which the plane is divided.

Consider the *x*-axis. Measure lengths on it from the origin, so that to the origin *(O)* corresponds the number 0. Let *OA*, measured from left to right along the axis, be the unit of length; then to the point *A* corresponds the number 1. Then let lengths *AB*, *BC*, and so on, all measured from left to right, be equal to *OA* in length; to the points *B*, *C*, and so on, correspond the numbers 2, 3, and so on. Further to the point that bisects *OA* let the fraction ½ correspond; and so on for the other fractions. In this way half of the *x*-axis is nearly filled up with points. But there are points, such as the point *P*, such that *OP* is the length of the circumference of a circle, say of unit diameter. For picturesqueness, we may imagine this point *P* got by rolling the circle along the *x*-axis from *O* through one revolution. The point *P* will fall a little to the left of the point 3⅐ and a little to the right of the point 3⁷⁄₅₀, and so on; the point *P* is not one of the points to which names of fractions have been assigned by the process sketched above. This can be proved rigidly. If it were not true, it would be very easy to "square the circle."

There are many other points like this. There is no fraction which, multiplied by itself, gives 2; but there is a length—the diagonal of a square of unit side—which is such that, if *we were to assume that a number corresponded to every point on OX*, it would be a number *a* such that $a^2 = 2$. We will return to this important question of the correspondence of points and lines to numbers, and will now briefly recall that "negative numbers" are represented, in Descartes' analytical geometry, on the *x*-axis, by the points to the left of the origin, and, on the y-axis, by the points below the origin. This was explained in the second chapter.

Algebraical geometry gave us a means of classifying curves. All straight lines determine equations of the first degree between x and y, and all such equations determine straight lines; all equations of the second degree between x and y, that is to say, of the form

$$ax^2 + bxy + cy^2 + dx + ey + f = 0,$$

determine curves which the ancient Greeks had studied and which result from cutting a solid circular cone, or two equal cones with the same axis, whose only point of contact is formed by the vertices. It is somewhat of a mystery why the Greek geometricians should have pitched upon these particular curves to study, and we can only say that it seems, from the present standpoint, an exceedingly lucky chance. For these "conic sections"—of which, of course, the circle is a particular case—are all the curves, and those only, which are represented by the above equation of the second degree. The three great types of curves—the "parabola," the "ellipse," and the "hyperbola"—all result from the above equation when the coefficients a, b, c, d, e, f satisfy certain special conditions. Thus, the equation of a circle—which is a particular kind of ellipse—is always of the form got from the above equation by putting $b = 0$ and $c = a$.

It may be mentioned that, long after these curves were introduced as sections of a cone, Pappus discovered that they could all be defined in a plane as loci of a point P which moves so that the proportion that the distance of P from a fixed point (S) bears to the perpendicular distance of P (PN) to a fixed straight line is constant. As this proportion is less than equal to or greater than 1, the curve is an ellipse, parabola, or hyperbola, respectively.

It will not be expected that a detailed account should here be given of the curves which result from the development of equations of the second or higher degrees between x and y. I will merely again emphasize some points which are, in part, usually neglected or not clearly stated in textbooks. The letters "$a, b, \ldots x, y$," here stand for "numbers" in the extended sense. We have seen in what sense we may, with the mathematicians, speak of fractionary, positive, and negative "numbers," and identify, say, the positive number $+2$ and the fraction $\frac{2}{1}$ with the signless integer 2. Well, then, the above letters stand for numbers of that class which includes in this sense the fractionary, irrational, positive and negative numbers, but excludes the imaginary numbers. We call the numbers of this class "real" numbers. The question of irrational numbers will be discussed at greater length in the sixth chapter, but enough has been said to show how they were introduced. In mathematics it has, I think, always happened that conceptions have been used long before they were formally introduced, and used long before this use could be logically justified or its nature clearly explained. The history of mathematics is the history of a faith whose justification has been long delayed, and perhaps is not accomplished even now.

These numbers are the measurements of length, in terms of a definite unit, like the inch, of the abscissæ and ordinates of certain points. We speak of such points simply by naming their co-ordinates, and say, for example, that "the distance of the point (x, y) from the point (a, b) is the positive square root of $(x - a)^2 + (y - b)^2$."

Notice that x^2, for example, is the length of a *line*. It is natural to make, as algebraists before Descartes did, x^2 stand *primarily* for the number of square units in a square whose sides are x units in length, but there is no necessity in this. We shall often use the latter kind of measurement in the fourth and fifth chapters.

The equation of a straight line can be made to satisfy two given conditions. We can write the equation in the form

$$x + \frac{b}{a} y + \frac{c}{a} = 0,$$

and thus have two ratios, $\frac{b}{a}$ and $\frac{c}{a}$, that we can determine according to the conditions. The equation $ax + by + c = 0$ has apparently *three* "arbitrary constants," as they are called, but we see that this greater generality is only apparent. Now we can so fix these constants that two conditions are fulfilled by the straight line in question. Thus, suppose that one of these conditions is that the straight line should pass through the origin—the point $(0, 0)$. This means simply that when $x = 0$, then $y = 0$. Putting them, $x = 0$ and $y = 0$ in the above equation, we get $\frac{c}{a} = 0$, and thus one of the constants is determined. The other is determined by a new condition that, say, the line also passes through the point $(\frac{1}{3}, 2)$. Substituting, then, in the above equation, we have, as $\frac{c}{a} = 0$ as we know already,

$\frac{1}{3} + \frac{2b}{a} = 0$, whence $\frac{b}{a} = -\frac{1}{6}$. Hence the equation of the line passing through $(0, 0)$ and $(\frac{1}{3}, 2)$ is $x - \frac{1}{6}y = 0$, or $y - 6x = 0$. Instead of having to pass through a certain point, a condition may be, for example, that the perpendicular from the origin on the straight line should be of a certain length, or that the line should make a certain angle with the x-axis, and so on.

Similarly, the circle whose equation is written in the form

$$(x - a)^2 + (y - b)^2 = c^2$$

is of radius c and centre (a, b). It can be determined to pass through any *three* points, or, say, to have a determined length of radius and position of centre. Fixation of centre is equivalent to two conditions. Thus, suppose the radius is to be of unit length: the above equation is $(x - a)^2 + (y - b)^2 = 1$. Then, if the centre is to be the origin, both a and b are determined to be 0, and this may also be effected by determining that the circle is to pass through the points $(\frac{1}{2}, 0)$ and $(-\frac{1}{2}, 0)$, for example.

Now, if we are to find the points of intersection of the straight line $2x + 2y = 1$ and the circle $x^2 + y^2 = 1$, we seek those points which are common to both curves, that is to say, all the pairs of values of x and y which satisfy *both* the above equations. Thus we need not trouble about the geometrical picture, but we only have to apply the rules of algebra for finding the values of x and y which satisfy two "simultaneous" equations in x and y. In the above case, if (X, Y) is a point of intersection, we have $Y = \dfrac{1 - 2X}{2}$, and therefore, by substitution in the other equation,

$X^2 + \left(\dfrac{1 - 2X}{2}\right)^2 = 1$. This gives a quadratic equation

$$8X^2 - 4X - 3 = 0$$

for X, and, by rules, we find that X must be either $\frac{1}{4}(1 + \sqrt{7})$ or $\frac{1}{4}(1 - \sqrt{7})$. Hence there are *two* values of the abscissa which are given when we ask what are the co-ordinates of the points of intersection; and the value of y which corresponds to each of these x's is given by substitution in the equation $2x + 2y = 1$.

Thus we find again the fact, obvious from a figure, that a straight line cuts [a circle] at two points at most. We can determine the points of intersection of any two curves whose equations can be expressed algebraically, but of course the process is much more complicated in more general cases. Here we will consider an important case of intersection of a straight line.

Think of a straight line cutting a circle at two points. Imagine one point fixed and the other point moved up towards the first. The intersecting line approaches more and more to the position of the tangent to the circle at the first point, and, by making the movable point approach the other closely enough, the secant will approach the tangent in position as nearly as we wish. Now, a tangent to a curve at a certain point was defined by the Greeks as a straight line through the point such that between it and the curve no other *straight line* could be drawn. Note that other *curves* might be drawn: thus various circles may have the same tangent at a common point on their circumference, but no circle—and no curve met with in elementary mathematics—has more than one tangent at a point. Descartes and many of his followers adopted different forms of definition which really involve the idea of a *limit,* an idea which appears boldly in the infinitesimal calculus. A tangent is the *limit* of a secant as the points of intersection approach infinitely near to one another; it is a produced side of the polygon with infinitesimal sides that the curve is supposed to be; it is the direction of motion at an instant of a point moving in the curve considered. The equation got from that of the curve by substituting for *y* from the equation of the intersecting straight line has, if this straight line is a tangent, two equal roots. In the above case, this equation was quadratic. In the case of a circle, we can easily deduce the well-known property of a tangent of being perpendicular to the radius; and see that this property has no analogue in the case of other curves.

We must remember that, just as *plane* curves determine and are determined by equations with *two* independent variables *x* and *y,* so surfaces— spheres, for instance—in three-dimensional space determine and are determined by equations with *three* independent variables, *x,* *y,* and *z.* Here *x,* *y,* and *z* are the co-ordinates of a point in space; that is to say, the numerical measures of the distances of this point from three fixed planes at right angles to each other. Thus, the equation of a sphere of radius *d* and centre at (a, b, c) is $(x - a)^2 + (y - b)^2 + (z - c)^2 = d^2$.

We may look at analytical geometry from another point of view which we shall find afterwards to be important, and which even now will suggest to us some interesting thoughts. The essence of Descartes' method also appears when we represent *loci* by the method. Consider a circle; it is the locus of a point (P) which moves in a plane so as to preserve a constant distance from a fixed point (O). Here we may think of *P* as varying in position, and make up a very striking picture of what we call a *variable* in mathematics. We must, however, remember that, by what we call a "variable" for the sake of picturesqueness, we do not necessarily mean something which varies. Think of the point of a pen as it moves over a sheet of writing paper; it occupies different positions with respect to the paper at different times, and we understandably say that the pen's point moves. But now think of a point in space. A geometrical point—which is not the bit of space occupied by the end of a pen or even an "atom" of matter—is merely a mark of position. We cannot, then, speak of a point moving; the very essence of point is to *be* position. The motion of a point of *space,* as distinguished from a point of matter, is a fiction, and is the supposition that a given point can now be one point and now another. Motion, in the ordinary sense, is only possible to matter and not to space.

Thus, when we speak of a "variable position," we are speaking absurdly if we wish our words to be taken literally. But we do not really so wish when we come to think about it. What we are doing is this: we are using a picturesque phrase for the purpose of calling up an easily imagined thought which helps us to visualise roughly a mathematical proposition which can only be described accurately by a prolix process. The ancient Greeks allowed prolixity, and it was only objected to by the uninitiated. Modern mathematics up to about sixty years ago successfully warred against prolixity; hence the obscurity of its fundamental notions and processes and its great conquests. The great conquests were made by sacrificing very much to analogy: thus, entities like the integer 2, the ratio $2/1$, and the real number which is denoted by "2" were identified, as we have seen, because of certain close analogies that they have. This seems to have been the chief reason why the procedure of the mathematicians has been so often condemned by logicians and even by philosophers. In fact, when mathematicians began to try to find out the nature of Mathematics, they had to examine their entities and the methods which they used to deal with them with the minutest care, and hence to look out for the points when the analogies referred to break down, and distinguish between what mathematicians had usually failed to distinguish. Then the people who do not mind a bit what Mathematics *is,* and are only interested in what it *does,* called these earnest inquirers "pedants" when they should have said "philosophers," and "logic-choppers"—whatever they may be—when they should have said "logicians." We have tried to show why ratios or fractions, and so on, are called "numbers," and apparently said to be something which they are not; we must now try to get at the meaning of the words "constant" and "variable."

By means of algebraic formulæ, rules for the reconstruction of great numbers—sometimes an infinity—of facts of nature may be expressed very concisely or even embodied in a *single* expression. The essence of the formula is that it is an expression of a *constant* rule among *variable* quantities. These expressions "constant" and "variable" have come down into ordinary language. We say that the number of miles which a certain man walks per day is a "variable quantity"; and we do not mean that, on a particular day, the number was not fixed and definite, but that on different days he walked, generally speaking, different numbers of miles. When, in mathematics, we speak of a "variable," what we mean is that we are considering a class of definite objects—for instance, the class of men alive at the present moment—and want to say something about *any one* of them indefinitely. Suppose that we say: "If it rains, Mr. A. will take his umbrella out with him"; the letter "A" here is what we call the sign of the "variable." We do not mean that the above proposition is about a *variable man.* There is no such thing; we say that a man varies in health and so *in time,* but, whether or not such a phrase is strictly correct, the meaning we would have to give the phrase "a variable" in the above sentence is not one and the same man at different periods of his own existence, but one and the same man who is different men in turn. What we mean is that if "A" denotes any man, and not Smith or Jones or Robinson *alone,* then he takes out his umbrella on certain occasions. The statement is not always true; it depends on A. If "A" stands for a bank manager, the statement may be true; if for a tramp or a savage, it probably is not. Instead of "A," we may put "B" or "C" or "X"; the kind of mark on paper does not really matter in the least. But we attach, by convention, certain meanings to certain signs; and so, if we wrote down a mark of exclamation for the sign of a variable, we might be misunderstood and even suspected of trying to be funny. We shall see, in the seventh chapter, the importance of the variable in logic and mathematics.

"Laws of nature" express the dependence upon one another of two or more variables. This idea of dependence of variables is fundamental in all scientific thought, and reaches its most thorough examination in mathematics and logic under the name of "functionality." On this point we must refer back to the second chapter. The ideas of function and variable were not prominent until the time of Descartes, and names for these ideas were not introduced until much later.

The conventions of analytical geometry as to the signs of co-ordinates in different quadrants of the plane had an important influence in the transformation of trigonometry from being a mere adjunct to a practical science. In the same notation as that used at the end of the first chapter, we may conveniently call the number $\dfrac{AP}{OP}$, which is the same for all lengths of OP, by the name "u," for short, and define $\dfrac{PM}{OP}$ and $\dfrac{OM}{OP}$ as the "sine of u," and the "cosine of u" respectively. Thus "sin u" and "cos u," as we write them for short, stand for numerical functions of u. Considering O as the origin of a system of rectangular co-ordinates of which OA is the x-axis, so that u measures the angle POA and $\dfrac{x}{r}$ and $\dfrac{y}{r}$ are cos u and sin u respectively. Now, even if u becomes so great that POA is successively obtuse, more than two right angles . . . , these definitions can be preserved, if we pay attention to the signs of x and y in the various quadrants. Thus sin u and cos u become separated from geometry, and appear as numerical functions of the variable u, whose values, as we see on reflection, repeat themselves at regular intervals as u becomes larger and larger. Thus, suppose that OP turns about O in a direction opposite to that in which the hands of a clock move. In the first quadrant, sin u

and cos u are $\dfrac{y}{r}$ and $\dfrac{x}{r}$; in the second they are $\dfrac{y}{r}$ and $\dfrac{-x}{r}$; in the third they

are $\dfrac{-y}{r}$ and $\dfrac{-x}{r}$; in the fourth they are $\dfrac{-y}{r}$ and $\dfrac{x}{r}$; in the fifth they are

$\dfrac{y}{r}$ and $\dfrac{x}{r}$ again; and so on. Trigonometry was separated from geometry

mainly by John Bernoulli and Euler, whom we shall mention later.

We will now turn to a different development of mathematics.

The ancient Greeks seem to have had something approaching a general method for finding areas of curvilinear figures. Indeed, infinitesimal methods, which allow indefinitely close approximation, naturally suggest themselves. The determination of the area of any rectilinear figure can be reduced to that of a rectangle, and can thus be completely effected. But this process of finding areas—this "method of quadratures"—failed for areas or volumes bounded by curved lines or surfaces respectively. Then the following considerations were applied. When it is impossible to find the exact solution of a question, it is natural to endeavour to approach to it as nearly as possible by neglecting quantities which embarrass the combinations, if it be foreseen that these quantities which have been neglected cannot, by reason of their small value, produce more than a trifling error in the result of the calculation. For example, as some properties of curves with respect to areas are with difficulty discovered, it is natural to consider the curves as polygons of a great number of sides. If a regular polygon be supposed to be inscribed in a circle, it is evident that these two figures, although always different, are nevertheless more and more alike according as the number of the sides of the polygon increases. Their perimeters, their areas, the solids formed by their revolving round a given axis, the angles formed by these lines, and so on, are, if not respectively equal, at any rate so much the nearer approaching to equality as the number of sides becomes increased. Whence, by supposing the number of these sides very great, it will be possible, without any perceptible error, to assign to the circumscribed circle the properties that have been found belonging to the inscribed polygon. Thus, if it is proposed to find the area of a given circle, let us suppose this curve to be a regular polygon of a great number of sides: the area of any regular polygon whatever is equal to the product of its perimeter into the half of the perpendicular drawn from the centre upon one of its sides; hence, the circle being considered as a polygon of a great number of sides, its area ought to equal the product of the circumference into half the radius. Now, this result is exactly true. However, the Greeks, with their taste for strictly correct reasoning, could not allow themselves to consider curves as polygons of an "infinity" of sides. They were also influenced by the arguments of Zeno, and thus regarded the use of "infinitesimals" with suspicion.

Zeno showed that we meet difficulties if we hold that time and space are infinitely divisible. Of the arguments which he invented to show this, the best known is the puzzle of Achilles and the Tortoise. Zeno argued that, if Achilles ran ten times as fast as a tortoise, yet, if the tortoise has (say) 1000 yards start, it could never be overtaken. For, when Achilles had gone the 1000 yards, the tortoise would still be 100 yards in front of him; by the time he had covered these 100 yards, it would still be 10 yards in front of him; and so on for ever; thus Achilles would get nearer and nearer to the tortoise, but never overtake it. Zeno invented some other subtle puzzles for much the same purpose, and they could only be discussed really satisfactorily by quite modern mathematics.

To avoid the use of infinitesimals, Eudoxus (408–355 B.C.) devised a method, exposed by Euclid in the Twelfth Book of his *Elements* and used by Archimedes to demonstrate many of his great discoveries, of verifying results found by the doubtful infinitesimal considerations. When the Greeks wished to discover the area bounded by a curve, they regarded the curve as the fixed boundary to which the inscribed and circumscribed polygons approach continually, and as much as they pleased, according as they increased the number of their sides. Thus they exhausted in some measure the space comprised between these polygons and the curve, and doubtless this gave to this operation the name of "the method of exhaustion." As these polygons terminated by straight lines were known figures, their continual approach to the curve gave an idea of it more and more precise, and "the law of continuity" serving as a guide, the Greeks could eventually arrive at the exact knowledge of its properties. But it was not sufficient for geometricians to have observed, and, as it were, guessed at these properties; it was necessary to verify them in an unexceptionable way; and this they did by proving that every supposition contrary to the existence of these properties would necessarily lead to some contradiction: thus, after, by infinitesimal considerations, they had found the area (say) of a curvilinear figure to be *a*, they verified it by proving that, if it is not *a*, it would yet be greater than the area of some polygon inscribed in the curvilinear figure whose area is palpably greater than that of the polygon.

In the seventeenth century, we have a complete contrast with the Grecian spirit. The method of discovery seemed much more important than correctness of demonstration. About the same time as the invention of analytical geometry by Descartes came the invention of a method for finding the areas of surfaces, the positions of the centres of gravity of variously shaped surfaces, and so on. In a book published in 1635, and in certain later works, Bonaventura Cavalieri (1598–1647) gave his "method

of indivisibles" in which the cruder ideas of his predecessors, notably of
Kepler (1571–1630), were developed. According to Cavalieri, a line is
made up of an infinite number of points, each without magnitude, a
surface of an infinite number of lines, each without breadth, and a volume
of an infinite number of surfaces, each without thickness. The use of this
idea may be illustrated by a single example. Suppose it is required to find
the area of a right-angled triangle. Let the base be made up of n points
(or indivisibles), and similarly let the side perpendicular to the base be
made of na points, then the ordinates at the successive points of the base
will contain a, $2a$ . . . , na points. Therefore the number of points in the
areas is $a + 2a + $ . . . $ + na$; the sum of which is $\frac{1}{2}(n^2a + na)$. Since n
is very large, we may neglect $\frac{1}{2}na$, for it is inconsiderable compared with
$\frac{1}{2}n^2a$. Hence the area is composed of a number $\frac{1}{2}(na)n$ of points, and
thus the area is measured in square units by multiplying half the linear
measure of the altitude by that of the base. The conclusion, we know
from other facts, is exactly true.

Cavalieri found by this method many areas and volumes and the centres
of gravity of many curvilinear figures. It is to be noticed that both
Cavalieri and his successors held quite clearly that such a supposition that
lines were composed of points was literally absurd, but could be used as a
basis for a direct and concise method of abbreviation which replaced with
advantage the indirect, tedious, and rigorous methods of the ancient
Greeks. The logical difficulties in the principles of this and allied methods
were strongly felt and commented on by philosophers—sometimes with
intelligence; felt and boldly overcome by mathematicians in their strong
and not unreasonable faith; and only satisfactorily solved by mathematicians—not the philosophers—in comparatively modern times.

The method of indivisibles—whose use will be shown in the next
chapter in an important question of mechanics—is the same in principle
as "the integral calculus." The integral calculus grew out of the work of
Cavalieri and his successors, among whom the greatest are Roberval
(1602–1675), Blaise Pascal (1623–1662), and John Wallis (1616–1703),
and mainly consists in the provision of a convenient and suggestive notation for this method. The discovery of the infinitesimal calculus was completed by the discovery that the inverse of the problem of finding the areas
of figures enclosed by curves was the problem of drawing tangents to these
curves, and the provision of a convenient and suggestive notation for this
inverse and simpler method, which was, for certain historical reasons,
called "the differential calculus."

Both analytical geometry and the infinitesimal calculus are enormously
powerful instruments for solving geometrical and physical problems. The
secret of their power is that long and complicated reasonings can be
written down and used to solve problems almost mechanically. It is the

merest superficiality to despise mathematicians for busying themselves, sometimes even consciously, with the problem of economising thought. The powers of even the most god-like intelligences amongst us are extremely limited, and none of us could get very far in discovering any part whatever of the Truth if we could not make trains of reasoning which we have thought through and verified, very ready for and easy in future application by being made as nearly mechanical as possible. In both analytical geometry and the infinitesimal calculus, all the essential properties of very many of the objects dealt with in mathematics, and the essential features of very many of the methods which had previously been devised for dealing with them are, so to speak, packed away in a well-arranged (and therefore readily got at) form, and in an easily usable way.

CHAPTER IV

THE BEGINNINGS OF THE APPLICATION OF MATHEMATICS TO NATURAL SCIENCE—THE SCIENCE OF DYNAMICS

THE end of very much mathematics—and of the work of many eminent men—is *the simple and, as far as may be, accurate description* of things in the world around us, of which we become conscious through our senses.

Among these things, let us consider, say, a particular person's face, and a billiard ball. The appearance to the eye of the ball is obviously much easier to describe than that of the face. We can call up the image—a very accurate one—of a billiard ball in the mind of a person who has never seen it by merely giving the colour and radius. And, unless we are engaged in microscopical investigations, this description is usually enough. The description of a face is a harder matter: unless we are skilful modellers, we cannot do this even approximately; and even a good picture does not attempt literal accuracy, but only conveys a correct impression—often better than a model, say in wax, does.

Our ideal in natural science is to build up a working model of the universe out of the sort of ideas that all people carry about with them everywhere "in their heads," as we say, and to which ideas we appeal when we try to teach mathematics. These ideas are those of *number, order,* the numerical measures of *times* and *distances,* and so on. One reason why we strive after this ideal is a very practical one. If we have a working model of, say, the solar system, we can tell, in a few minutes, what our position with respect to the other planets will be at all sorts of far future times, and can thus *predict certain future events.* Everybody can see how useful this is: perhaps those persons who see it most clearly are those sailors who use the *Nautical Almanac.* We cannot make the earth tarry in its revolution round its axis in order to give us a longer day for finishing some important piece of work; but, by finding out the unchanging laws concealed in the phenomena of the motions of earth, sun, and stars, the mathematician can construct the model just spoken of. And the mathematician is completely master of his model; he can repeat the occurrences in his universe as often as he likes; something like Joshua, he can make his "sun" stand still, or hasten, in order that he may publish the *Nautical Almanac* several years ahead of time. Indeed, the "world" with which we have to deal in theoretical or mathematical mechanics is but a mathematical scheme, the function of which it is to imitate, by logical consequences of the properties assigned to it by definition, certain processes of nature as closely as possible. Thus our "dynamical world" may be called a model of reality, and must not be confused with the reality itself.

That this model of reality is constructed solely out of logical conceptions will result from our conclusion that mathematics is based on logic, and on logic alone; that such a model is possible is really surprising on reflection. The need for completing facts of nature in thought was, no doubt, first felt as a *practical* need—the need that arises because we feel it convenient to be able to predict certain kinds of future events. Thus, with a purely mathematical model of the solar system, we can tell, with an approximation which depends upon the completeness of the model, the relative positions of the sun, stars, and planets several years ahead of time; this it is that enables us to publish the *Nautical Almanac,* and makes up to us, in some degree, for our inability "to grasp this sorry scheme of things entire . . . and remould it nearer to the heart's desire."

Now, what is called "mechanics" deals with a very important part of the structure of this model. We spoke of a billiard ball just now. Everybody gets into the way, at an early age, of abstracting from the colour, roughness, and so on, of the ball, and forming for himself the conception of a *sphere*. A sphere can be exactly described; and so can what we call a "square," a "circle" and an "ellipse," in terms of certain conceptions such as those called "point," "distance," "straight line," and so on. Not so easily describable are certain other things, like a person or an emotion. In the world of moving and what we roughly class as *inanimate* objects— that is to say, objects whose behaviour is not perceptibly complicated by the phenomena of what we call "life" and "will"—people have sought from very ancient times, and with increasing success, to discover rules for the motions and rest of given systems of objects (such as a lever or a wedge) under given circumstances (pulls, pressures, and so on). Now, this discovery means: the discovery of an ideal, exactly describable motion which should approximate as nearly as possible to a natural motion or class of motions. Thus Galileo (1564–1642) discovered the approximate law of bodies falling freely, or on an inclined plane, near the earth's surface; and Newton (1642–1727) the still more accurate law of the motions of any number of bodies under any forces.

Let us now try to think clearly of what we mean by such a rule, or, as it is usually called, a "scientific" or "natural law," and why it plays an important part in the arrangement of our knowledge in such a convenient way that we can at once, so to speak, lay our hand on any particular fact the need of which is shown by practical or theoretical circumstances.

For this purpose, we will see how Galileo, in a work published in 1638, attacked the problem of a falling body. Consider a body falling freely to the earth: Galileo tried to find out, not *why* it fell, but *how* it fell—that is to say, in what mathematical form the distance fallen through and the velocity attained depends on the time taken in falling and the space fallen through. Freely falling bodies are followed with more difficulty by the eye the farther they have fallen; their impact on the hand receiving them is, in like measure, sharper; the sound of their striking louder. The velocity accordingly increases with the time elapsed and the space traversed. Thus, the modern inquirer would ask: What function is the number (v) representing the velocity of those (s and t) representing the distance fallen through and the time of falling? Galileo asked, in his primitive way: Is v proportional to s; or again, is v proportional to t? Thus he made assumptions, and then *ascertained by actual trial the correctness or otherwise of these assumptions.*

One of Galileo's assumptions was, thus, that the velocity acquired in the descent is proportional to the time of the descent. That is to say, if a body falls once, and then falls again during twice as long an interval of time as it first fell, it will attain in the second instance double the velocity it acquired in the first. To find by experiment whether or not this assumption accorded with observed facts, as it was difficult to prove by any direct means that the velocity acquired was proportional to the time of descent, but easier to investigate by what law the distance increased with the time, *Galileo deduced from his assumption the relation that obtained between the distance and the time.* This very important deduction he effected as follows.

On the straight line *OA*, let the abscissæ *OE, OC, OG,* and so on, represent in length various lengths of time elapsed from a certain instant represented by *O*, and let the ordinates *EF, CD, GH,* and so on, corresponding to these abscissæ, represent in length the magnitude of the velocities acquired at the time represented by the respective abscissæ.

We observe now that, by our assumption, *O, F, D, H, lie in a straight line OB,* and so: (1) At the instant *C*, at which one-half *OC* of the time of descent *OA* has elapsed, the velocity *CD* is also one-half of the final velocity AB; (2) If E and G are equally distant in opposite directions on OA from C, the velocity GH exceeds the mean velocity CD by the same amount that the velocity EF falls short of it; and for every instant antecedent to C there exists a corresponding one subsequent to C and equally distant from it. Whatever loss, therefore, as compared with uniform motion with half the final velocity, is suffered in the first half of the motion, such loss is made up in the second half. The distance fallen through we may consequently regard as having been uniformly described with half the final velocity.

In symbols, if the number of units of velocity acquired in t units of time is v, and suppose that v is proportional to t, the number s of units of space descended through is proportional to $\frac{1}{2}t^2$. In fact, s is given by $\frac{1}{2}vt$, and, as v is proportional to t, s is proportional to $\frac{1}{2}t^2$.

Now, Galileo verified this relation between s and t experimentally. The motion of free falling was too quick for Galileo to observe accurately with the very imperfect means—such as water-clocks—at his disposal. There were no mechanical clocks at the beginning of the seventeenth century; they were first made possible by the dynamical knowledge of which Galileo laid the foundations. Galileo, then, made the motion slower, so that s and t were big enough to be measured by rather primitive apparatus in which the moving balls ran down grooves in inclined planes. That the spaces traversed by the ball are proportional to the squares of the measures of the times in free descent as well as in motion on an inclined plane, Galileo verified by experimentally proving that a ball which falls through the height of an inclined plane attains the same final velocity as a ball which falls through its length. This experiment was an ingenious one with a pendulum whose string, when half the swing had been accomplished, caught on a fixed nail so placed that the remaining half of the swing was with a shorter string than the other half. This experiment showed that the bob of the pendulum rose, in virtue of the velocity acquired in its descent, *just as high* as it had fallen. This fact is in agreement with our instinctive knowledge of natural events; for if a ball which falls down the length of an inclined plane could attain a greater velocity than one which falls through its height, we should only have to let the body pass with the acquired velocity to another more inclined plane to make it rise to a greater vertical height than that from which it had fallen. Hence we can deduce, from the acceleration on an inclined plane, the acceleration of free descent, for, since the final velocities are the same and $s = \frac{1}{2}vt$, the lengths of the sides of the inclined plane are simply proportional to the times taken by the ball to pass over them.

The motion of falling that Galileo found actually to exist is, accordingly, a motion of which the velocity increases proportionally to the time.

Like Galileo, we have started with the notions familiar to us (through the practical arts, for example), such as that of *velocity*. Let us consider this motion more closely.

If a motion is *uniform* and c feet are travelled over in every second, at the end of t seconds it will have travelled ct feet. Put $ct = s$ for short. Then we call the "velocity" of the moving body the distance traversed in a unit of time so that it is $\dfrac{s}{t}$ units of length per second, the number which is the measure of the distance divided by the number which is the measure of the time elapsed. Galileo, now, attained to the conception of a motion in which the velocity increases proportionally to the time. If we draw a diagram and set off, from the origin O along the x-axis OA, a series of abscissæ which represent the times in length, and erect the corresponding ordinates to represent the velocities, the ends of these ordinates will lie on a line OB, which, in the case of the "uniformly accelerated motion" to which Galileo attained, is *straight*, as we have already seen. But if the ordinates represent *spaces* instead of *velocities*, the straight line OB becomes a curve. We see the distinction between the "curve of spaces" and "the curve of velocities," with times as abscissæ in both cases. If the velocity is uniform, the curve of spaces is a straight line OB drawn from the origin O, and the curve of velocities is a straight line parallel to the x-axis. If the velocity is variable, the curve of spaces is never a straight line; but if the motion is uniformly accelerated, the curve of velocities is a straight line like OB. The relations between the curve of spaces, the curve of velocities, and the areas of such curves AOB are, as we shall see, relations which are at once expressible by the "differential and integral calculus"—indeed, it is mainly because of this important illustration of the calculus that the elementary problems of dynamics have been treated here. And the measurement of velocity in the case where the velocity varies from time to time is an illustration of the formation of the fundamental conception of the differential calculus.

It may be remarked that the finding of the velocity of a particle at a given instant and the finding of a tangent to a curve at a given point are both of them the same kind of problem—the finding of the "differential quotient" of a function. We will now enter into the matter more in detail.

Consider a curve of spaces. If the motion is uniform, the number measuring *any* increment of the distance divided by the number measuring the corresponding increment of the time gives the same value for the measure of the velocity. But if we were to proceed like this where the velocity is variable, we should obtain widely differing values for the velocity. However, the smaller the increment of the time, the more nearly does the bit of the curve of spaces which corresponds to this increment approach straightness, and hence uniformity of increase (or decrease) of s. Thus, if we denote the increment of t by "Δt,"—where "Δ" does not stand

for a number but for the phrase "the increment of,"—and the corresponding increment (or decrement) of s by "Δs," we may define the measure
of average velocity in this element of the motion as $\dfrac{\Delta s}{\Delta t}$. But, however
small Δt is, the line represented by Δs is not, usually at least, quite straight,
and the velocity at the instant t, which, in the language of Leibniz's differential calculus, is defined as the quotient of "infinitely small" increments
and symbolised by $\dfrac{ds}{dt}$, —the Δ's being replaced by d's when we consider
"infinitesimals,"—appears to be only defined approximately. We have met
this difficulty when considering the method of indivisibles, and will meet
it again when considering the infinitesimal calculus, and will only see how
it is overcome when we have become familiar with the conception of a
"limit."

   This new notion of velocity includes that of uniform velocity as a particular case. In fact, the rules of the infinitesimal calculus allow us to
conclude, from the equation $\dfrac{ds}{dt} = a$, where a is some constant, the equation $s = at + b$, where b is another constant. We must remember that all
this was not *expressly* formulated until about fifty years after Galileo had
published his investigations on the motion of falling.

   If we consider the curve of velocities, uniformly accelerated motion
occupies in it exactly the same place as uniform velocity does in the
curve of spaces. If we denote by v the numerical measure of the velocity
at the end of t units of time, the acceleration, in the notation of the differential calculus, is measured by $\dfrac{dv}{dt}$, and the equation $\dfrac{dv}{dt} = h$, where h is
some constant, is the equation of uniformly accelerated motion. In New-
tonian dynamics, we have to consider *variably* accelerated motions, and
this is where the infinitesimal calculus or some practically equivalent
calculus such as Newton's "method of fluxions" becomes so necessary in
theoretical mechanics.

   We will now consider the curve of spaces for uniformly accelerated
motion. On this diagram—the arcs being t and s—we will draw the curve

$$s = \frac{gt^2}{2},$$

where g denotes a constant. Of course, this is the same thing as drawing
the curve $y = \dfrac{gx^2}{2}$ in a plane divided up by the x-axis and the y-axis of

Descartes. This curve is a parabola passing through the origin. An interesting thing about this curve is that it is the curve that would be described by a body projected obliquely near the surface of the earth if the air did not resist, and is very nearly the path of such a projectile in the resisting atmosphere. A free body, according to Galileo's view, always falls towards the earth with a uniform vertical acceleration measured by the above number g. If we project a body vertically upwards with the initial velocity of c units, its velocity at the end of t units of time is $-c + gt$ units, for if the direction downwards (of g) is reckoned positive, the direction upwards (of c) must be reckoned negative. If we project a body horizontally with the velocity of a units, and neglect the resistance of the air, Galileo recognised that it would describe, in the horizontal direction, a distance of at units in t units of time, while *simultaneously* it would fall a distance of $\dfrac{gt^2}{2}$ units. The two motions are to be considered as going on *independently* of each other. Thus also, oblique projection may be considered as compounded of a horizontal and a vertical projection. In all these cases the path of the projectile is a parabola; in the case of the horizontal projection, its equation in x and y co-ordinates is got from the two equations

$$x = at \text{ and } y = \frac{gt^2}{2}, \text{ and is thus } y = \frac{gx^2}{2a^2}.$$

Now, suppose that the velocity is neither uniform nor increases uniformly, but is different and increases at a different rate at different points of time. Then in the curve of velocities, the line OB is no longer straight. *In the former case, the number s was the number of square units in the area of the triangle AOB.* In this case the figure AOB is not a triangle, though we shall find that its area is the s units we seek, although v does not increase uniformly from O to A.

Notice again that if, on OA, we take points C and E very close together, the little arc DF is very nearly straight, and the figure DGF very nearly a rectilinear triangle. Note that we are only trying, in this, to get a first approximation to the value of s, so that, instead of the continuously changing velocities we know—or think we know—from our daily experience, we are considering a fictitious motion in which the velocity increases (or decreases) so as to be the same as that of the motion thought of at a large number of points at minute and equal distances, and between successive points increases (or decreases) uniformly.

Note also that we are assuming (what usually happens with the curves with which we shall have to do) that the arc DF which corresponds to CE becomes as straight as we wish if we take C and E close enough together.

And now let us calculate *s* approximately. Starting from *O*, in the first small interval *OH* the rectilinear triangle *OHK*, where *HK* is the ordinate at *H*, represents approximately the space described. In the next small interval *HL*, where the length of *HL* is equal to that of *OH*, the space described is represented by the rectilinear figure *KHLM*. The rectangle *KL* is the space passed over with the uniform velocity *HK* in time *HL*; and the triangle *KNM* is the space passed over by a motion in which the velocity increases from zero to *MN*. And so on for other intervals beyond *HL*. Thus *s* is ultimately given (approximately) as the number of square units in a polygon which closely approximates to the figure *AOB*.

We must now say a few words about the meaning of the letters in geometrical and mechanical *equations* which, following Descartes, we use instead of the proportions used by Galileo and even many of his contemporaries and followers. It seems better, when beginning mechanics, to think in proportions, but afterwards, for convenience in dealing with the symbolism of mathematical data, it is better to think in equations.

A typical proportion is: Final velocities are to one another as the times; or, in symbols,

$$\text{``}V : V' : : T : T'.\text{''}$$

Here "*V*" (for example) is just short for "the velocity attained at the end of the period of time" (reckoned from some fixed instant) denoted by "*T*," and $V : V'$, and $T : T'$, are just *numbers* (real numbers); and the proportion states the equality of these numbers. Hence the proportion is sometimes written "$V : V' = T : T'$." If, now, *v* is the numerical measure, merely, of *V*, *v'* that of *V'*, and so on, we have $\dfrac{v}{v'} = \dfrac{t}{t'}$ or $vt' = v't$.

In the last equation, the letters *v* and *t* have a mnemonic significance, as reminding us that we started from *velocities* and *times, but we must carefully avoid the idea that we are "multiplying"* (or can do so) *velocities by times;* what we *are* doing is multiplying the numerical measures of them. People who write on geometry and mechanics often say inaccurately, simply for shortness, "Let *s* denote the distance, *t* the time," and so on; whereas, by a tacit convention, small italics are usually employed to denote *numbers*. However, in future, for the sake of shortness, I shall do as the writers referred to, and speak of *v* as "the velocity." Equations in mechanics, such as "$s = \dfrac{gt^2}{2}$." are only possible if the left-hand side is of the same kind as the right-hand side: we cannot equate spaces and times, for example.

Suppose that we have fixed on the unit of length as one inch and the unit of time as one second. As unit of velocity we might choose the velocity with which, say, *a* inches are described uniformly in one second. If we did this, we should express the relation between the *s* units of space passed over by a body with a given velocity (*v* units) in a given time (*t* units) as "*s = avt*"; whereas, if we defined the unit of velocity as the velocity with which the unit of length is travelled over in the unit of time, we should write "*s = vt.*"

Among the units derived from the fundamental units—such as those of length and time—the simplest possible relations are made to hold. Thus, as the unit of area and the unit of volume, the square and the cube of unit sides are respectively used, the unit of velocity is the uniform rate at which unit of length is travelled over in the unit of time, the unit of acceleration is the gain of unit velocity in unit time, and so on.

The derived units depend on the fundamental units, and the *function* which a given derived unit is of its fundamental units is called its "dimensions." Thus the velocity *v* is got by dividing the length *s* by the time *t*. The dimensions of a velocity are written

$$"[V] = \frac{[L]}{[T]},"$$

and those of an acceleration—denoted "*F*"—

$$"[F] = \frac{[V]}{[T]} = \frac{[L]}{[T]^2}."$$

These equations are merely mnemonic; the letters do not mean numbers. The mnemonic character comes out when we wish to pass from one set of units to another. Thus, if we pass to a unit of length *b* times greater and one of time *c* times greater, the acceleration *f* with the old units is related to that (*f'*) with the new units by the equation

$$f'\left(\frac{c^2}{b}\right) = f.$$

As the units become greater, *f'* becomes less; and, since the dimensions of *F* are $\frac{[L]}{[T]^2}$, the factor $\frac{c^2}{b}$ is obviously suggested to us—the symbol "$[T]^2$" suggesting a squaring of the number measuring the time.

From Galileo's work resulted the conclusion that, where there is no change of *velocity in a straight line,* there is no force. The state of a body unacted upon by force is uniform rectilinear motion; and rest in a special case of this motion where the velocity is and remains zero. This "law of inertia" was exactly opposite to the opinion, derived from Aristotle, that force is requisite to keep up a uniform motion, and may be roughly verified by noticing the behaviour of a body projected with a given velocity and moving under little resistance—as a stone moving on a sheet of ice. Newton and his contemporaries saw how important this law was in the explanation of the motion of a planet—say, about the sun. Think of a simple case, and imagine the orbit to be a circle. The planet tends to move along the tangent with uniform velocity, but the attraction of the sun simultaneously draws the planet towards itself, and the result of this continual combination of two motions is the circular orbit. Newton succeeded in calculating the shapes of the orbits for different laws of attraction, and found that, when attraction varies inversely as the square of the distance, the shapes are conic sections, as had been observed in the case of our solar system.

The problem of the solar system appeared, then, in a mathematical dress; various things move about in space, and this motion is completely described if we know the geometrical relations—distances, positions, and angular distances—between these things at some moment, the velocities at this moment, and the accelerations at *every* moment. Of course, if we knew all the positions of all the things at all the instants, our description would be complete; it happens that the *accelerations* are usually simpler to find directly than the positions: thus, in Galileo's case the acceleration was simply constant. Thus, we are given functional relations between these positions and their rates of change. We have to determine the positions from these relations.

It is the business of the "method of fluxions" or the "infinitesimal calculus" to give methods for finding the relations between variables from relations between their rates of change or between them and these rates. This shows the importance of the calculus in such physical questions.

Mathematical physics grew up—perhaps too much so—on the model of theoretical astronomy, its first really extensive conquest. There are signs that mathematical physics is freeing itself from its traditions, but we need not go further into the subject in this place.

Roberval devised a method of tangents which is based on Galileo's conception of the composition of motions. The tangent is the direction of the resultant motion of a point describing the curve. Newton's method, which is to be dealt with in the fifth chapter, is analogous to this, and the idea of velocity is fundamental in his "method of fluxions."

CHAPTER V

THE RISE OF MODERN MATHEMATICS
—THE INFINITESIMAL CALCULUS

IN the third chapter we have seen that the ancient Greeks were some-times occupied with the theoretically exact determination of the areas enclosed by curvilinear figures, and that they used the "method of exhaustion," and, to demonstrate the results which they got, an indirect method. We have seen, too, a "method of indivisibles," which was direct and seemed to gain in brevity and efficiency from a certain lack of correctness in expression and perhaps even a small inexactness in thought. We shall find the same merits and demerits—both, especially the merits, intensified —in the "infinitesimal calculus."

By the side of researches on quadratures and the finding of volumes and centres of gravity developed the methods of drawing tangents to curves. We have begun to deal with this subject in the third chapter: here we shall illustrate the considerations of Fermat (1601–1665) and Barrow (1630–1677)—the intellectual descendants of Kepler—by a simple example.

Let it be proposed to draw a tangent at a given point P in the circumference of a circle of centre O and equation $x^2 + y^2 = 1$. Let us take the circle to be a polygon of a great number of sides; let PQ be one of these sides, and produce it to meet the x-axis at T. Then PT will be the tangent in question. Let the co-ordinates of P be X and Y; those of Q will be $X + e$ and $Y + a$, where e and a are infinitely small increments, positive or negative. From a figure in which the ordinates and abscissæ of P and Q are drawn, so that the ordinate of P is PR, we can see, by a well-known property of triangles, that TR is to RP (or Y) as e is to a. Now, X and Y are related by the equation $X^2 + Y^2 = 1$, and, since Q is also on the locus $x^2 + y^2 = 1$, we have $(X + e)^2 + (Y + a)^2 = 1$. From the two equations in which X and Y occur, we conclude that $2eX + e^2 + 2aY + a^2 = 0$,

$$\text{and hence } -(X + \frac{e}{a}) + Y + \frac{e}{2} + \frac{a}{2} = 0. \text{ But } \frac{e}{a} = \frac{TR}{Y}; \text{ hence } TR = \frac{-Y(Y + a/_2)}{X + e/_2}.$$

Now, a and e may be neglected in comparison with X and Y, and thus we can say that, at any rate *very* nearly, we have $TR = \dfrac{Y^2}{X}$. But this is

exactly right, for, since TP is at right angles to OP, we know that OR is to RP as PR is to RT. Here X and Y are constant, but we can say that the abscissa of the point where the tangent at *any* point (say y) of the circle cuts the x-axis is given by adding $-\dfrac{y^2}{x}$ to x.

Thus, we can find tangents by considering the ratios of infinitesimals to one another. The method obviously applies to other curves besides circles; and Barrow's method and nomenclature leads us straight to the notation and nomenclature of Leibniz. Barrow called the triangle PQS, where S is where a parallel to the x-axis through Q meets PR, the "differential triangle," and Leibniz denoted Barrow's a and e by dy and dx (short for the "differential of y" and "the differential of x," so that "d" does not denote a number but "dx" altogether stands for an "infinitesimal") respectively, and called the collection of rules for working with his signs the "differential calculus."

But before the notation of the differential calculus and the rules of it were discovered by Gottfried Wilhelm von Leibniz (1646–1716), the celebrated German philosopher, statesman, and mathematician, he had invented the notation and found some of the rules of the "integral calculus": thus, he had used the now well-known sign "$\int$" or long "s" as short for "the sum of," when considering the sum of an infinity of infinitesimal elements as we do in the method of indivisibles. Suppose that we propose to determine the area included between a certain curve $y = f(x)$, the x-axis, and two fixed ordinates whose equations are $x = a$ and $x = b$; then, if we make use of the idea and notation of differentials, we notice that the area in question can be written as

$$\text{"}\int y \cdot dx,\text{"}$$

the summation extending from $x = a$ to $x = b$. We will not here further concern ourselves about these boundaries. Notice that in the above expression we have put a dot between the "y" and the "dx": this is to indicate that y is to multiply dx. Hitherto we have used juxtaposition to denote multiplication, but here d is written close to x with another end in view; and it is desirable to emphasise the difference between "d" used in the sense of an adjective and "d" used in the sense of a multiplying number, at least until the student can easily tell the difference by the context. If, then, we imagine the abscissa divided into equal infinitesimal parts, each of length dx, corresponding to the constituents called "points" in the method of indivisibles, $y \cdot dx$ is the area of the little rectangle of sides dx and y which stand at the end of the abscissa x. If, now, instead of extending to $x = b$, the summation extends to the ordinate at the indeterminate or "variable" point x, $y \cdot dx$ becomes a function of x.

Now, if we think what must be the differential of this sum—the infinitesimal increment that it gets when the abscissa of length x, which is one of the boundaries, is increased by dx—we see that it must be $y \cdot dx$. Hence

$$d(\textstyle\int y \cdot dx) = y \cdot dx,$$

and hence the sign of "d" destroys, so to speak, the effect of the sign "$\int$". We also have $\int dx = x$, and find that this summation is the inverse process to differentiation. *Thus the problems of tangents and quadratures are inverses of one another.* This vital discovery seems to have been first made by Barrow without the help of any technical symbolism. The quantity which by its differentiation produces a proposed differential, is called the "integral" of this differential; since we consider it as having been formed by infinitely small continual additions: each of these additions is what we have named the differential of the increasing quantity, it is a fraction of it: and the sum of all these fractions is the entire quantity which we are in search of. For the same reason we call "integrating" or "taking the sum of" a differential the finding the integral of the sum of all the infinitely small successive additions which form the series, the differential of which, properly speaking, is the general term.

It is evident that two variables which constantly remain equal increase the one as much as the other during the same time, and that consequently their differences are equal: and the same holds good even if these two quantities had differed by any quantity whatever when they began to vary; provided that this primitive difference be always the same, their differentials will always be equal.

Reciprocally, it is clear that two variables which receive at each instant infinitely small equal additions must also either remain constantly equal to one another, or always differ by the same quantity—that is, the integrals of two differentials which are equal can only differ from each other by a constant quantity. For the same reason, if any two quantities whatever differ in an infinitely small degree from each other, their differentials will also differ from one another infinitely little: and reciprocally if two differential quantities differ infinitely little from one another, their integrals, putting aside the constant, can also differ but infinitely little one from the other.

Now, some of the rules for differentiation are as follows. If $y = f(x)$, $dy = f(x + dx) - f(x)$, in which higher powers of differentials added to lower ones may be neglected. Thus, if $y = x^2$, then $dy = (x + dx)^2 - x^2 = 2x \cdot dx + (dx)^2 = 2x \cdot dx$. Here it is well to refer back to the treatment of the problem of tangents at the beginning of this chapter. Again, if $y = a \cdot x$, where a is constant, $dy = a \cdot dx$. If $y = x \cdot z$, then

$$dy = (x + dx)(z + dz) - x \cdot z = x \cdot dz + z \cdot dx. \text{ If } y = \frac{x}{z}, \ x = y \cdot z, \text{ so}$$

$$dx = y \cdot dz + z \cdot dy; \text{ hence } dy = \frac{dx - y \cdot dz}{z}. \text{ Since the integral calculus}$$

is the inverse of the differential calculus, we have at once

$$\int 2x \, . \, dx = x^2, \quad \int a \, . \, dx = a \int dx,$$
$$\int x \, . \, dz + \int z \, . \, dx = xz,$$

and so on. More fully, from $d(x^3) = 3x^2 \, . \, dx$, we conclude, not that $\int x^2 \, . \, dx = \frac{1}{3}x^3$, but that $\int x^2 \, . \, dx = \frac{1}{3}x^3 + c$, where "$c$" denotes some constant depending on the fixed value for x from which the integration starts.

Consider a parabola $y^2 = ax$; then $2y \, . \, dy = a \, . \, dx$, or $dx = \dfrac{2y \, . \, dy}{a}$.

Thus the area from the origin to the point x is $\displaystyle\int \frac{2y^2 \, . \, dy}{a} + c$; but $d\dfrac{2y^3}{3a} = \dfrac{2y^2 \, . \, dy}{a}$; thus the area is $\dfrac{2y^3}{3a} + c$, or, since $y^2 = ax$, $\frac{2}{3}x \, . \, y + c$. To determine c when we measure the area from 0 to x, we have the area zero when $x = 0$; hence the above equation gives $c = 0$. This whole result, now quite simple to us, is one of the greatest discoveries of Archimedes.

Let us now make a few short reflections on the infinitesimal calculus. First, the extraordinary power of it in dealing with complicated questions lies in that the question is split up into an infinity of *simpler* ones which can all be dealt with at once, thanks to the wonderfully economical fashion in which the calculus, like analytical geometry, deals with variables. Thus, a *curvilinear* area is split up into *rectangular* elements, all the rectangles are added together at once when it is observed that integral is the inverse of the easily acquired practice of differentiation. We must never lose sight of the fact that, when we differentiate y or integrate $y \, . \, dx$, we are considering, not a particular x or y, but *any* one of an infinity of them. Secondly, we have seen that what in the first place had been regarded but as a simple method of approximation, leads at any rate in certain cases to results perfectly exact. The fact is that the exact results are due to a compensation of errors: the error resulting from the false supposition made, for example, by regarding a curve as a polygon with an infinite number of sides each infinitely small and which when produced is a tangent of the curve, is corrected or compensated for by that which springs from the very processes of the calculus, according to which we retain in differentiation infinitely small quantities of the same order alone. In fact, after having introduced these quantities into the calculation to facilitate the expression of the conditions of the problem, and after having regarded them as absolutely zero in comparison with the proposed quantities, with a view to simplify these equations, in order to banish the errors that they had occasioned, and to obtain a result perfectly exact, there remains but to eliminate these same quantities from the equations where they may still be.

But all this cannot be regarded as a strict proof. There *are* great difficulties in trying to determine what infinitesimals are: at one time they are treated like finite numbers and at another like zeros or as "ghosts of departed quantities," as Bishop Berkeley, the philosopher, called them.

Another difficulty is given by differentials "of higher orders than the first." Let us take up again the considerations of the fourth chapter. We saw that $v = \dfrac{ds}{dt}$, and found that s was got by integration: $s = \int v \,.\, dt$. This is now an immediate inference, since $\dfrac{ds}{dt} \, dt = ds$. Now, let us substitute for v in $\dfrac{dv}{dt}$. Here t is the independent variable, and all of the older mathematicians treated the elements dt as constant—the interval of the independent variable was split up into atoms, so to speak, which themselves were regarded as known, and in terms of which other differentials, ds, dx, dy, were to be determined. Thus

$$\frac{dv}{dt} = \frac{d(^{ds}/_{dt})}{dt} = \frac{^{1}/_{dt} \,.\, d(ds)}{dt} = \frac{d^2s}{dt^2},$$

"d^2s" being written for "$d(ds)$" and "dt^2" for "$(dt)^2$". Thus the acceleration was expressed as "the second differential of the space divided by the square of dt." If $\dfrac{d^2s}{dt^2}$ were constant, say, a, then $\dfrac{d^2s}{dt} = a \,.\, dt$; and, integrating both sides:

$$\frac{ds}{dt} = \int a \,.\, dt = a\!\int\! dt = at + b,$$

where b is a new constant. Integrating again, we have:

$$s = a\!\int\! t \,.\, dt + b\!\int\! dt = \frac{at^2}{2} + bt + c,$$

which is a more general form of Galileo's result. Many complicated problems which show how far-reaching Galileo's principles are were devised by Leibniz and his school.

Thus, the infinitesimal calculus brought about a great advance in our powers of describing nature. And this advance was mainly due to Leibniz's notation; Leibniz himself attributed all of his mathematical discoveries to his improvements in notation. Those who know something of Leibniz's work know how conscious he was of the suggestive and economical value of a good notation. And the fact that we still use and appreciate Leibniz's

"∫" and "*d*," even though our views as to the principles of the calculus are very different from those of Leibniz and his school, is perhaps the best testimony to the importance of this question of notation. This fact that Leibniz's notations have become permanent is the great reason why I have dealt with his work before the analogous and prior work of Newton.

Isaac Newton (1642–1727) undoubtedly arrived at the principles and practice of a method equivalent to the infinitesimal calculus much earlier than Leibniz, and, like Roberval, his conceptions were obtained from the dynamics of Galileo. He considered curves to be described by moving points. If we conceive a moving point as describing a curve, and the curve referred to co-ordinate axes, then the velocity of the moving point can be decomposed into two others parallel to the axes of x and y respectively; these velocities are called the "fluxions" of x and y, and the velocity of the point is the fluxion of the arc. Reciprocally the arc is the "fluent" of the velocity with which it is described. From the given equation of the curve we may seek to determine the relations between the fluxions—and this is equivalent to Leibniz's problem of differentiation;—and reciprocally we may seek the relations between the co-ordinates when we know that between their fluxions, either alone or combined with the co-ordinates themselves. This is equivalent to Leibniz's general problem of integration, and is the problem to which we saw, at the end of the fourth chapter, that theoretical astronomy reduces.

Newton denoted the fluxion of x by "$\dot{x}$," and the fluxion of the fluxion (the acceleration) of $\dot{x}$ by "$\ddot{x}$." It is obvious that this notation becomes awkward when we have to consider fluxions of higher orders; and further, Newton did not indicate by his notation the independent variable considered. Thus "$\dot{y}$" might possibly mean either $\dfrac{dy}{dt}$ or $\dfrac{dy}{dx}$. We have $\dot{x} = \dfrac{dx}{dt}$,

$\ddot{x} = \dfrac{d\dot{x}}{dt} = \dfrac{d^2x}{dt^2}$; but a dot-notation for $\dfrac{d^n x}{dt^n}$ would be clumsy and inconvenient. Newton's notation for the "inverse method of fluxions" was far clumsier even, and far inferior to Leibniz's "∫".

The relations between Newton and Leibniz were at first friendly, and each communicated his discoveries to the other with a certain frankness. Later, a long and acrimonious dispute took place between Newton and Leibniz and their respective partisans. Each accused—unjustly, it seems—the other of plagiarism, and mean suspicions gave rise to meanness of conduct, and this conduct was also helped by what is sometimes called "patriotism." Thus, for considerably more than a century, British mathematicians failed to perceive the great superiority of Leibniz's notation. And thus, while the Swiss mathematicians, James Bernoulli (1654–1705), John Bernoulli (1667–1748), and Leonhard Euler (1707–1783), the French mathematicians d'Alembert (1707–1783), Clairaut (1713–1765), Lagrange (1736–1813), Laplace (1749–1827), Legendre (1752–1833), Fourier (1768–1830), and Poisson (1781–1850), and many other Continental mathematicians, were rapidly [6] extending knowledge by using the infinitesimal calculus in all branches of pure and applied mathematics, in England comparatively little progress was made. In fact, it was not until the beginning of the nineteenth century that there was formed, at Cambridge, a Society to introduce and spread the use of Leibniz's notation among British mathematicians: to establish, as it was said, "the principles of pure *d*-ism in opposition to the *dot*-age of the university."

The difficulties met and not satisfactorily solved by Newton, Leibniz, or their immediate successors, in the principles of the infinitesimal calculus, centre about the conception of a "limit"; and a great part of the meditations of modern mathematicians, such as the Frenchman Cauchy (1789–1857), the Norwegian Abel (1802–1829), and the German Weierstrass (1815–1897), not to speak of many still living, have been devoted to the putting of this conception on a sound logical basis.

We have seen that, if $y = x^2$, $\dfrac{dy}{dx} = 2x$. What we do in forming $\dfrac{dy}{dx}$ is to form $\dfrac{(x + \Delta x)^2 - x^2}{\Delta x}$, which is readily found to be $2x + \Delta x$, and then consider that, as Δx approaches 0 more and more, the above quotient approaches $2x$. We express this by saying that the "limit, as h [Δx] approaches 0," is $2x$. We do not consider Δx as being a fixed "infinitesimal" or as an absolute zero (which would make the above quotient become indeterminate $\dfrac{0}{0}$), nor need we suppose that the quotient *reaches* its limit (the state of Δx being 0). What we need to consider is that "Δx" should represent a variable which can take values differing from 0 by as little as we please. That is to say, if we choose *any* number, however small, there is a value which Δx can take, and which differs from 0 by less than that

[6] It is difficult for a mathematician not to think that the sudden and brilliant dawn on eighteenth-century France of the magnificent and apparently all-embracing physics of Newton and the wonderfully powerful mathematical method of Leibniz inspired scientific men with the belief that the goal of all knowledge was nearly reached and a new era of knowledge quickly striding towards perfection begun; and that this optimism had indirectly much to do in preparing for the French Revolution.

number. As before, when we speak of a "variable" we mean that we are considering a certain *class*. When we speak of a "limit," we are considering a certain *infinite* class. Thus the sequence of an infinity of terms 1, ½, ¼, ⅛, 1/16, and so on, whose law of formation is easily seen, has the limit 0. In this case 0 is such that any number greater than it is greater than some term of the sequence, but 0 itself is not greater than any term of the sequence and is not a term of the sequence. A sequence like 1, $1 + \frac{1}{2}$, $1 + \frac{1}{2} + \frac{1}{4}$, $1 + \frac{1}{2} + \frac{1}{4} + \frac{1}{8}$ . . ., has an analogous *upper* limit 2. A function $f(x)$, as the independent variable x approaches a certain value, like $\dfrac{2x}{x}$ as x approaches 0, may have a value (in this case 2, though *at* 0, $\dfrac{2x}{x}$ is indeterminate). The question of the limits of a function in general is somewhat complicated, but the most important limit is $\dfrac{f(x + \Delta x) - f(x)}{\Delta x}$ as Δx approaches 0; this, if $y = f(x)$, is $\dfrac{dy}{dx}$.

That the infinitesimal calculus, with its rather obscure "infinitesimals"— treated like finite numbers when we write $\dfrac{dy}{dx} dx = dy$ and $\dfrac{1}{dy/dx} = \dfrac{dx}{dy}$, and then, on occasion, neglected—leads so often to correct results is a most remarkable fact, and a fact of which the true explanation only appeared when Cauchy, Gauss (1777–1855), Riemann (1826–1866), and Weierstrass had developed the theory of an extensive and much used class of functions. These functions happen to have properties which make them especially easy to be worked with, and nearly all the functions we habitually use in mathematical physics are of this class. A notable thing is that the complex numbers spoken of in the second chapter *make* this theory to a great extent.

Large tracts of mathematics have, of course, not been mentioned here. Thus, there is an elaborate theory of integer numbers to be referred to in a note to the seventh chapter, and a geometry using the conceptions of the ancient Greeks and methods of modern mathematical thought; and very many men still regard space-perception as something mathematics deals with. We will return to this soon. Again, algebra has developed and branched off; the study of functions in general and in particular has grown; and soon a list of some of the many great men who have helped in all this would not be very useful. Let us now try to resume what we have seen of the development of mathematics along what seem to be its main lines.

In the earliest times men were occupied with particular questions—the properties of particular numbers and geometrical properties of particular figures, together with simple mechanical questions. With the Greeks, a more general study of classes of geometrical figures began. But traces of an earlier exception to this study of particulars are afforded by "algebra." In it and its later form symbols—like our present x and y—took the place of numbers, so that, what is a great advance in economy of thought and other labour, a part of calculation could be done with symbols instead of numbers, so that the one result stated, in a manner analogous to that of Greek geometry, a proposition valid for a whole infinite class of different numbers.

The great revolution in mathematical thought brought about by Descartes in 1637 grew out of the application of this general algebra to geometry by the very natural thought of substituting the numbers expressing the lengths of straight lines for those lines. Thus a point in a plane—for instance—is determined in position by two numbers x and y, or coordinates. Now, as the point in question varies in position, x and y both vary; to every x belongs, in general, one or more y's, and we arrive at the most beautiful idea of a single algebraical equation between x and y representing the whole of a curve—the one "equation of the curve" expressing the general law by which, given any particular x out of an infinity of them, the corresponding y or y's can be found.

The problem of drawing a tangent—the limiting position of a secant, when the two meeting points approach indefinitely close to one another—at any point of a curve came into prominence as a result of Descartes' work, and this, together with the allied conceptions of velocity and acceleration "at an instant," which appeared in Galileo's classical investigation, published in 1638, of the law according to which freely falling bodies move, gave rise at length to the powerful and convenient "infinitesimal calculus" of Leibniz and the "method of fluxions" of Newton. Mathematically, the finding of the tangent at the point of a curve, and finding the velocity of a particle describing this curve when it gets to that point, are identical problems. They are expressed as finding the "differential quotient," or the "fluxion" at the point. It is now known to be very probable that the above two methods, which are theoretically—but not practically—the same, were discovered independently; Newton discovered his first, and Leibniz published his first, in 1684. The finding of the areas of curves and of the shapes of the curves which moving particles describe under given forces showed themselves, in this calculus, as results of the inverse process to that of the direct process which serves to find tangents and the law of attraction to a given point from the datum of the path described by a particle. The direct process is called "differentiation," the inverse process "integration."

Newton's fame is chiefly owing to his application of this method to the solution, which, in its broad outlines, he gave of the problem of the motion of the bodies in the solar system, which includes his discovery of the law according to which all matter gravitates towards—is attracted by—other matter. This was given in his *Principia* of 1687; and for more than a century afterwards mathematicians were occupied in extending and applying the calculus.

Then came more modern work, more and more directed towards the putting of mathematical methods on a sound logical basis, and the separation of mathematical processes from the sense-perception of space with which so much in mathematics grew and grows up. Thus trigonometry took its place by algebra as a study of certain mathematical functions, and it began to appear that the true business of geometry is to supply beautiful and suggestive pictures of abstract—"analytical" or "algebraical" or even "arithmetical," as they are called—processes of mathematics. In the next chapter we shall be concerned with part of the work of logical examination and reconstruction.

CHAPTER VI

MODERN VIEWS OF LIMITS AND NUMBERS

LET us try to form a clear idea of the conception which showed itself to be fundamental in the principles of the infinitesimal calculus, the conception of a *limit*.

Notice that the limit of a sequence is a number which is already defined. We cannot prove that there is a limit to a sequence unless the limit sought is among the numbers already defined. Thus, in the system of "numbers" —here we must refer back to the second chapter—consisting of all fractions (or ratios), we can say that the sequence (where 1 and 2 are written for the ratios $\frac{1}{1}$ and $\frac{2}{1}$) 1, $1 + \frac{1}{2}$, $1 + \frac{1}{2} + \frac{1}{4}$, . . . , has a limit (2), but that the sequence

$$1, \ 1 + \tfrac{4}{10}, \ 1 + \tfrac{4}{10} + \tfrac{1}{100}, \ 1 + \tfrac{4}{10} + \tfrac{1}{100} + \tfrac{4}{1000}, \ . \ . \ . ,$$

$$\text{or } 1 \cdot 4142 \ . \ . \ . ,$$

got by extracting the square root of 2 by the known process of decimal arithmetic, has not. In fact, it can be proved that there is no ratio such that it is a limit for the above sequence. If there were, and it were denoted by "x," we would have $x^2 = 2$. Here we come again to the question of incommensurables and "irrational numbers." The Greeks were quite right in distinguishing so sharply between numbers and magnitudes, and it was a tacit, natural, and unjustified—not, as it happens, incorrect—presupposition that the series of numbers, completed into the series of what are called "real numbers," exactly corresponds to the series of points on a straight line. The series of points which represents the sequence last named seems undoubtedly to possess a limit; this limiting point was assumed to represent some number, and, since it could not represent an integer or a ratio, it was said to represent an "irrational number," $\sqrt{2}$. Another irrational number is that which is represented by the incommensurable ratio of the circumference of a circle to its diameter. This number is denoted by the Greek letter "π," and its value is nearly $3 \cdot 1416$. . . . Of course, the process of approximation by decimals never comes to an end.

The subject of limits forced itself into a very conspicuous place in the seventeenth and eighteenth centuries owing to the use of infinite series as a means of approximate calculation. I shall distinguish what I call "sequences" and "series." A sequence is a collection—finite or infinite—of numbers; a series is a finite or infinite collection of numbers *connected by addition*. Sequences and series can be made to correspond in the following way. To the sequence 1, 2, 3, 4, . . . belongs a series of which the terms are got by subtracting, in order, the terms of the sequence from the ones immediately following them, thus:
$$(2-1) + (3-2) + (4-3) + \ldots = 1 + 1 + 1 + \ldots;$$
and from a series the corresponding sequence can be got by making the sum of the first, the first two, the first three, . . . terms the first, second, third . . . term of the sequence respectively. Thus, to the series $1 + 1 + 1 + \ldots$ corresponds the sequence 1, 2, 3, . . .

Now, if a series has only a finite number of terms, it is possible to find the sum of all the terms; but if the series is unending, we evidently cannot. But in certain cases the corresponding sequence has a limit, and this limit is called by mathematicians, neither unnaturally nor accurately, "the sum to infinity of the series." Thus, the sequence 1, $1 + \frac{1}{2}$, $1 + \frac{1}{2} + \frac{1}{4}$ . . . has the limit 2, and so the sum to infinity of the series $1 + \frac{1}{2} + \frac{1}{4} + \frac{1}{8} + \ldots$ is 2. Of course, all series do not have a sum: thus $1 + 1 + 1 + \ldots$ to infinity has not—the terms of the corresponding sequence increase continually beyond all limits. Notice particularly that the terms of a sequence may increase continually, and yet have a limit—those of the above sequence with limit 2 so increase, but not beyond 2, though they do beyond any number less than 2; also notice that the terms of a sequence may increase beyond all limits even if the terms of the corresponding series continually diminish, remaining positive, towards 0. The series $1 + \frac{1}{2} + \frac{1}{3} + \frac{1}{4} + \frac{1}{5} + \ldots$ is such a series; the terms of the sequence slowly increase beyond all limits, as we see when we reflect that the sums
$$\frac{1}{3} + \frac{1}{4}, \frac{1}{5} + \frac{1}{6} + \frac{1}{7} + \frac{1}{8}, \frac{1}{9} + \ldots + \frac{1}{16}, \ldots$$
are all greater than $\frac{1}{2}$. It is very important to realise the fact illustrated by this example; for it shows that the conditions under which an infinite

series has a sum are by no means as simple as they might appear at first sight.

The logical scrutiny to which, during the last century, the processes and conceptions of mathematics have been subjected, showed very plainly that it was a sheer assumption that such a process as $1 \cdot 4142$ . . ., though *all* its terms are less than 2, for example, has any limit at all. When we replace numbers by points on a straight line, we feel fairly sure that there is one point which behaves to the points representing the above sequence in the same sort of way as 2 to the sequence $1, 1 + \frac{1}{2}, 1 + \frac{1}{2} + \frac{1}{4}, \ldots$ Now, if our system of numbers is to form a *continuum*, as a line seems to our thoughts to be; so that we can affirm that our number system is adequate, when we introduce axes in the manner of analytical geometry, to the description of all the phenomena of change of position which take place in our space,[7] *then* we must have a number $\sqrt{2}$ which is the limit of the sequence $1 \cdot 4142$ . . . if 2 is of the series $1 + \frac{1}{2} + \frac{1}{4} + \ldots$, for to every point of a line must correspond a number which is subject to the same rules of calculation as the ratios or integers. Thus we must, to justify from a logical point of view our procedure in the great mathematical methods, show what irrationals are, and define them *before* we can prove that they are limits. We cannot take a series, whose law is evident, which has no ratio for sum, and yet such that the terms of the corresponding sequence all remain less than some fixed number (such as $1 + \dfrac{1}{1} + \dfrac{1}{1 \cdot 2} + \dfrac{1}{1 \cdot 2 \cdot 3} + \dfrac{1}{1 \cdot 2 \cdot 3 \cdot 4} + \ldots$, when all the terms of the corresponding sequence are less than 3, for example), and then say that it "defines a limit." All we can prove is that *if* such a series has a limit, then, if the terms of its corresponding sequence do not decrease as we read from left to right (as in the preceding example), it cannot have more than one limit.

Some mathematicians have simply *postulated* the irrationals. At the beginning of their discussions they have, tacitly or not, said: "In what follows we will *assume* that there are such things as fill up kinds of gaps in the system of rationals (or ratios)." Such a gap is shown by this. The rationals less than ½ and those greater than ½ form two sets and ½ divides them. The rationals x such that x^2 is greater than 2 and those x's such that x^2 is less than 2 form two analogous sets, but there is only an analogue to the dividing number ½ if we postulate a number $\sqrt{2}$. Thus by

---

[7] The only kind of change dealt with in the science of mechanics is change of position, that is, motion. It does not seem to me to be necessary to adopt the doctrine that the complete description of any physical event is of a mechanical event; for it is possible to assign and calculate with numbers of our number-continuum to other varying characteristics (such as temperature) of the state of a body besides position.

postulation we fill up these subtle gaps in the set of rationals and get a continuous set of real numbers. But we can avoid this postulation if we define "$\sqrt{2}$" as the name of the class of rationals x such that x^2 is less than 2 and "($\frac{1}{2}$)" as the name of the class of rationals x such that x is less than $\frac{1}{2}$. Proceeding thus, we arrive at a set of *classes*, some of which correspond to rationals, as ($\frac{1}{2}$) to $\frac{1}{2}$, but the rest satisfy our need of a set without gaps. There is no reason why we should not say that these classes *are* the *real numbers* which include the irrationals. But we must notice that rationals are never real numbers; $\frac{1}{2}$ is not ($\frac{1}{2}$), though analogous to it. We have much the same state of things as in the second chapter, where 2, $+2$ and $\frac{2}{1}$ were distinguished and then deliberately confused because, with the mathematicians, we felt the importance of analogy in calculation. Here again we identify ($\frac{1}{2}$) with $\frac{1}{2}$, and so on.

Thus, integers, positive and negative "numbers," ratios, and real "numbers" are all different things: real numbers are classes, ratios and positive and negative numbers are relations. Integers, as we shall see, are classes. Very possibly there is a certain arbitrariness about this, but this is unimportant compared with the fact that in modern mathematics we have reduced the definitions of all "numbers" to logical terms. Whether they are classes or relations or propositions or other logical entities is comparatively unimportant.

Integers can be defined as certain classes. Mathematicians like Weierstrass stopped before they got as far as this: they reduced the other numbers of analysis to logical developments out of the conception of integer, and thus freed analysis from any remaining trace of the sway of geometry. But it was obvious that integers had to be defined, if possible, in logical terms. It has long been recognised that two collections consist of the same number of objects if, and only if, these collections can be put in such a relation to one another that to every object of each one belongs one and only one object of the other. We must not think that this implies that we have already the idea of the *number one*. It is true that "one and only one" seems to use this idea. But "the class a has one and only one member" is simply a short way of expressing: "x is a member of a, and if y is also a member of a, then y is identical with x." It is true, also, that we use the idea of the *unity* or the *individuality* of the things considered. But this unity is a property of each individual, while the number 1 is a property of a *class*. If a class of pages of a book is itself, under the name of a "volume," a member of a class of books, the same class of pages has both a number (say 360) and a unity as being itself a member of a class.

The relation spoken of above in which two classes possessing the same number stand to one another does not involve counting. Think of the fingers on your hands. If to every finger of each hand belongs, by some process of correspondence, one and only one—remember the above meaning of this phrase—of the other, they are said to have "the same number." This is a definition of what "the same number" is to mean for us; the word "number" by itself is to have, as yet, no meaning for us; and, to avoid confusion, we had better replace the phrase "have the same number" by the words "are equivalent." Any other word would, of course, do, but this word happens to be fairly suggestive and customary. Now, if the variable u is any class, "the number of u" is defined as short for the phrase: "the class whose members are classes which are similar to u." Thus the number of u is an entity which is purely logical in its nature. Some people might urge that by "number" they mean something different from this, and that is quite possible. All that is maintained by those who agree to the process sketched above is: (1) Classes of the kind described are identical in all known arithmetical properties with the undefined things people call "integer numbers"; (2) It is futile to say: "These classes are not *numbers*," if it is not also said what *numbers* are—that is to say, if "the number of" is not defined in some more satisfactory way. There may be more satisfactory definitions, but this one is a perfectly sound foundation for all mathematics, including the theory not touched upon here of *ordinal numbers* (denoted by "first," "second," . . .) which apply to sets arranged in some order, known at present.

To illustrate (1), think of this. Acording to the above definition 2 is the general idea we call "couple." We say: "Mr. and Mrs. A. are a couple"; our definition would ask us to say in agreement with this: "The class consisting of Mr. and Mrs. A. is a member of the class 2." We define "2" as "the class of classes u such that, if x is a u, u lacking x is a 1"; the definition of "3" follows that of "2"; and so on. In the same way, we see that the class of fingers on your right hand and the class of fingers on your left hand are each of them members of the class 5. It follows that the classes of the fingers are equivalent in the above sense.

Out of the striving of human minds to reproduce conveniently and anticipate the results of experience of geometrical and natural events, mathematics has developed. Its development gave priceless hints to the development of logic, and then it appeared that there is no gap between the science of number and the science of the most general relations of objects of thought. As for geometry and mathematical physics, it becomes possible clearly to separate the logical parts from those parts which formulate the data of our experience.

We have seen that mathematics has often made great strides by sacrificing accuracy to analogy. Let us remember that, though mathematics and logic give the highest forms of certainty within the reach of us, the process of mathematical discovery, which is so often confused with what is discovered, has led through many doubtful analogies and errors arriving from the great help of symbolism in making the difficult easy. Fortunately symbolism can also be used for precise and subtle analysis, so that we can say that it can be made to show up the difficulties in what appears easy and even negligible—like $1 + 1 = 2$. This is what much modern fundamental work does.

CHAPTER VII

THE NATURE OF MATHEMATICS

IN the preceding chapters we have followed the development of certain branches of knowledge which are usually classed together under the name of "mathematical knowledge." These branches of knowledge were never clearly marked off from all other branches of knowledge: thus geometry was sometimes considered as a logical study and sometimes as a natural science—the study of the properties of the space we live in. Still less was there an absolutely clear idea of what it was that this knowledge was about. It had a name—"mathematics"—and few except "practical" men and some philosophers doubted that there was something about which things were known in that kind of knowledge called "mathematical." But what it was did not interest very many people, and there was and is a great tendency to think that the question as to what mathematics is could be answered if we only knew all the facts of the development of our mathematical knowledge. It seems to me that this opinion is, to a great extent, due to an ambiguity of language: one word—"mathematics"—is used both for our knowledge of a certain kind and the thing, if such a thing there be, about which this knowledge is. I have distinguished, and will now explicitly distinguish, between "Mathematics," a collection of truths of which we know something, and "mathematics," our knowledge of Mathematics. Thus, we may speak of "Euclid's mathematics" or "Newton's mathematics," and say truly that mathematics has developed and therefore had history; but Mathematics is eternal and unchanging, and therefore has no history—it does not belong, even in part, to Euclid or Newton or anybody else, but is something which is discovered, in the course of time, by human minds. An analogous distinction can be drawn between "Logic" and "logic." The small initial indicates that we are writing of a psychological process which may lead to Truth; the big initial indicates that we are writing of the entity—the part of Truth—to which this process leads us. The reason why mathematics is important is that Mathematics is not incomprehensible, though it is eternal and unchanging.

Grammatical usage makes us use a capital letter even for "mathematics" in the psychological sense when the word begins a sentence, but in this case I have guarded and will guard against ambiguity.

That particular function of history which I wish here to emphasise will now, I think, appear. In mathematics we gradually learn, by getting to know some thing about mathematics, to know that there is such a thing as Mathematics.

We have, then, glanced at the mathematics of primitive peoples, and have seen that at first discoveries were of isolated properties of abstract things like numbers or geometrical figures, and of abstract relations between concrete things like the relations between the weights and the arms of a lever in equilibrium. These properties were, at first, discovered and applied, of course, with the sole object of the satisfaction of bodily needs. With the ancient Greeks comes a change in point of view which perhaps seems to us, with our defective knowledge, as too abrupt. So far as we know, Greek geometry was, from its very beginning, deductive, general, and studied for its own interest and not for any applications to the concrete world it might have. In Egyptian geometry, if a result was stated as universally true, it was probably only held to be so as a result of induction —the conclusion from a great number of particular instances to a general proposition. Thus, if somebody sees a very large number of officials of a certain railway company, and notices that all of them wear red ties, he might conclude that *all* the officials of that company wear red ties. This might be probably true: it would not be certain: for *certainty* it would be necessary to know that there was some rule according to which all the officials were compelled to wear red ties. Of course, even then the conclusion would not be certain, since these sort of laws may be broken. Laws of *Logic*, however, cannot be broken. These laws are not, as they are sometimes said to be, laws of *thought*; for logic has nothing to do with the way people think, any more than poetry has to do with the food poets must eat to enable them to compose. Somebody might *think* that 2 and 2 make 5: we know, by a process which rests on the laws of Logic, that they make 4.

This is a more satisfactory case of induction: Fermat stated that no integral values of x, y, and z can be found such that $x^n + y^n = z^n$, if n be an integer greater than 2. This theorem has been proved to be true for $n = 3, 4, 5, 7$, and many other numbers, and there is no reason to doubt that it is true. But to this day no general proof of it has been given.[8] This, then, is an example of a mathematical proposition which has been reached and stated as probably true by induction.

Now, in Greek geometry, propositions were stated and proved by the

---

[8] This is an example of the "theory of numbers," the study of the properties of integers, to which the chief contributions, perhaps, have been made by Fermat and Gauss.

laws of Logic—helped, as we now know, by tacit appeals to the conclusions which common sense draws from the pictorial representation in the mind of geometrical figures—about *any* triangles, say, or *some* triangles, and thus not about one or two particular things, but about an *infinity* of them. Thus, consider any two triangles *ABC* and *DEF*. It helps the thinking of most of us to draw pictures of particular triangles, but our conclusions do not hold merely for these triangles. If the sides *BA* and *AC* are equal in length to the sides *ED* and *DF* respectively, and the angle at *A* is equal to the angle at *D*, then *BC* is equal to *EF*. This is proved rather imperfectly in the fourth proposition of the first Book of Euclid's *Elements*.

When we examine into and complete the reasonings of geometricians, we find that the conception of space vanishes, and that we are left with logic alone. Philosophers and mathematicians used to think—and some do now—that, in geometry, we had to do, not with the space of ordinary life in which our houses stand and our friends move about, and which certain quaint people say is "annihilated" by electric telegraphs or motor cars, but an abstract form of the same thing from which all that is personal or material has disappeared, and only things like *distance* and *order* and *position* have remained. Indeed, some have thought that position did not remain; that, in abstract space, a circle, for example, had no position of its own, but only with respect to other things. Obviously, we can only, in practice, give the position of a thing with respect to other things—"relatively" and not "absolutely." These "relativists" denied that position had any properties which could not be practically discovered. Relativism, in a thought-out form, seems quite tenable; in a crude form, it seems like excluding the number 2, as distinguished from classes of two things, from notice as a figment of the brain, because it is not visible or tangible like a poker or a bit of radium or a mutton-chop.

In fact, a perfected geometry reduces to a series of deductions holding not only for figures in space, but for any abstract things. Spatial figures give a striking illustration of some abstract things; and that is the secret of the interest which analytical geometry has. But it is into algebra that we must now look to discover the nature of Mathematics.

We have seen that Egyptian arithmetic was more general than Egyptian geometry: like algebra, by using letters to denote unknown numbers, it began to consider propositions about *any* numbers. In algebra and algebraical geometry this quickly grew, and then it became possible to treat branches of mathematics in a systematic way and make whole classes of problems subject to the uniform and almost mechanical working of one method. Here we must again recall the economical function of science.

At the same time as methods—algebra and analytical geometry and the infinitesimal calculus—grew up from the application of mathematics to natural science, grew up also the new conceptions which influenced the form which mathematics took in the seventeenth, eighteenth, and nineteenth centuries. The ideas of *variable* and *function* became more and more prominent. These ideas were brought in by the conception of motion, and, unaffected by the doubts of the few logicians in the ranks of the mathematicians, remained to fructify mathematics. When mathematicians woke up to the necessity of explaining mathematics logically and finding out what Mathematics is, they found that, in mathematics the striving for generality had led, from very early times, to the use of a method of deduction used but not recognised and distinguished from the method usually used by the Aristotelians. I will try to indicate the nature of these methods, and it will be seen how the ideas of variable and function, in a form which does not depend on that particular kind of variability known as motion, come in.

A *proposition* in logic is the kind of thing which is denoted by such a phrase as: "Socrates was a mortal and the husband of a scold." If—and this is the characteristic of modern logic—we notice that the notions of variable and function (correspondence, relation) which appeared first in a special form in mathematics, are fundamental in all the things which are the objects of our thought, we are led to replace the particular conceptions in a proposition by variables, and thus see more clearly the structure of the proposition. Thus: "x is a y and has the relation R to z, a member of the class u" gives the general form of a multitude of propositions, of which the above is a particular case; the above proposition may be true, but it is not a judgment of logic, but of history or experience. The proposition is false if "Kant" or "Westminster Abbey" is substituted for "Socrates": it is neither if "x," a sign for a variable, is, and then becomes what we call a "propositional function" of x and denote by "ϕx" or "ψx." If more variables are involved, we have the notation "$\phi(x,y)$," and so on.

Relations between propositional functions may be true or false. Thus x is a member of the class a, and a is contained in the class b, together imply that x is a b, is true. Here the *implication* is true, and we do not say that the *functions* are. The kind of implication we use in mathematics is of the form: "If ϕx is true, then ψx is true"; that is, any particular value of x which makes ϕx true also makes ψx true.

From the perception that, when the notions of variable and function are introduced into logic, as their fundamental character necessitates, all mathematical methods and all mathematical conceptions can be defined in purely logical terms, leads us to see that Mathematics is only a part of Logic and is the class of all propositions of the form: $\phi(x,y,z, \ . \ . \ .)$ implies, for all values of the variables, $\psi(x,y,z, \ . \ . \ .)$. The structure of the propositional functions involves only such ideas as are fundamental in logic, like implication, class, relation, the relation of a term to a class of which it is a member, and so on. And, of course, Mathematics depends on the notion of Truth.

When we say that "1 + 1 = 2," we seem to be making a mathematical statement which does not come under the above definition. But the statement is rather mistakenly written: there is, of course, only *one* whole class of unit classes, and the notation "1 + 1" makes it look as if there were two. Remembering that 1 is a class of certain classes, what the above proposition means is: If x and y are members of 1, and x differs from y, then x and y together make up a member of 2.

At last, then, we arrive at seeing that the nature of Mathematics is independent of us personally and of the world outside, and we can feel that our own discoveries and views do not affect the Truth itself, but only the extent to which we or others see it. Some of us discover things in science, but we do not really create anything in science any more than Columbus created America. Common sense certainly leads us astray when we try to use it for the purposes for which it is not particularly adapted, just as we may cut ourselves and not our beards if we try to shave with a carving knife; but it has the merit of finding no difficulty in agreeing with those philosophers who have succeeded in satisfying themselves of the truth and position of Mathematics. Some philosophers have reached the startling conclusion that Truth is made by men, and that Mathematics is created by mathematicians, and that Columbus created America; but common sense, it is refreshing to think, is at any rate above being flattered by philosophical persuasion that it really occupies a place sometimes reserved for an even more sacred Being.

BIBLIOGRAPHY

THE view that science is dominated by the principle of the economy of thought has been, in part,[9] very thoroughly worked out by Ernst Mach (see especially the translation of his *Science of Mechanics*, 5th ed., La Salle, Ill., 1942). On the history of mathematics, we may mention W. W. Rouse Ball's books, *A Primer of the History of Mathematics* (7th ed., 1930), and the fuller *Short Account of the History of Mathematics* (4th ed., 1908, both published in London by Macmillan), and Karl Fink's *Brief History of Mathematics* (Chicago and London, 3rd ed., 1910).

As text-books of mathematics, De Morgan's books on *Arithmetic, Algebra,* and *Trigonometry* are still unsurpassed, and his *Trigonometry and Double Algebra* contains one of the best discussions of complex numbers, for students, that there is. As De Morgan's books are not all easy to get,

[9] *Cf.* above, pp. 5, 11, 13, 15, 16, 42.

the reprints of his *Elementary Illustrations of the Differential and Integral Calculus* and his work *On the Study and Difficulties of Mathematics* (Chicago and London, 1899 and 1902) may be recommended. Where possible, it is best to read the works of the great mathematicians themselves. For elementary books, Lagrange's *Lectures on Elementary Mathematics*, of which a translation has been published at Chicago and London (2nd ed., 1901), is the most perfect specimen.

The questions dealt with in the fourth chapter are more fully discussed in Mach's *Mechanics*. An excellent collection of methods and problems in graphical arithmetic and algebra, and so on, is contained in H. E. Cobb's book on *Elements of Applied Mathematics* (Boston and London: Ginn & Co., 1911).

PART II

Historical and Biographical

The Great Mathematicians

AT the outset of assembling this anthology I decided that I ought to include a biographical history of the subject. This would provide a setting for the other selections, and also serve as a small reference manual for the general reader. It was not easy to find a history which was brief, authoritative, elementary and readable. W. W. R. Ball's *A Primer of the History of Mathematics* is a book of merit but rather old-fashioned. J. W. N. Sullivan's *The History of Mathematics in Europe,* an admirable outline, carries the story only as far as the end of the eighteenth century; I commend this book to your attention. Dirk Struik's *A Concise History of Mathematics* has solid virtues but is a trifle too advanced for my purposes and at times dull. A few French and German books which might have been suitable were not considered because of the labor of translating them.

Turnbull's excellent little volume, a biographical history, turned out to meet the standard in all respects. It is the story of several great mathematicians, "representatives of their day in this venerable science." "I have tried to show," says Professor Turnbull in his preface, "how a mathematician thinks, how his imagination, as well as his reason, leads him to new aspects of the truth. Occasionally it has been necessary to draw a figure or quote a formula—and in such cases the reader who dislikes them may skip, and gather up the thread undismayed a little further on. Yet I hope that he will not too readily turn aside in despair, but will, with the help of the accompanying comment, find something to admire in these elegant tools of the craft." There is overlapping between this survey and the preceding selection, but the two books are complementary, and the reader who enjoys one will derive no less pleasure from the other. Jourdain makes ideas the heroes of his account while Turnbull devotes a good deal of space to lively sketches of the men who made the ideas.

H. W. Turnbull, distinguished for his researches in algebra (determinants, matrices, theory of equations), is Regius professor of mathematics at the University of St. Andrews in Scotland, a Fellow of the Royal Society, and, as demonstrated not only in this volume but in other writings, a gifted simplifier of mathematical ideas.

We think of Euclid as of fine ice; we admire Newton as we admire the Peak of Teneriffe. Even the intensest labors, the most remote triumphs of the abstract intellect, seem to carry us into a region different from our own—to be in a terra incognita *of pure reasoning, to cast a chill on human glory.*   —WALTER BAGEHOT

Many small make a great.   —CHAUCER

Everything of importance has been said before by somebody who did not discover it.   —ALFRED NORTH WHITEHEAD

1   The Great Mathematicians

By HERBERT WESTREN TURNBULL

PREFACE

THE usefulness of mathematics in furthering the sciences is commonly acknowledged: but outside the ranks of the experts there is little inquiry into its nature and purpose as a deliberate human activity. Doubtless this is due to the inevitable drawback that mathematical study is saturated with technicalities from beginning to end. Fully conscious of the difficulties in the undertaking, I have written this little book in the hope that it will help to reveal something of the spirit of mathematics, without unduly burdening the reader with its intricate symbolism. The story is told of several great mathematicians who are representatives of their day in this venerable science. I have tried to show how a mathematician thinks, how his imagination, as well as his reason, leads him to new aspects of the truth. Occasionally it has been necessary to draw a figure or quote a formula—and in such cases the reader who dislikes them may skip, and gather up the thread undismayed a little further on. Yet I hope that he will not too readily turn aside in despair, but will, with the help of the accompanying comment, find something to admire in these elegant tools of the craft.

Naturally in a work of this size the historical account is incomplete: a few references have accordingly been added for further reading. It is pleasant to record my deep obligation to the writers of these and other larger works, and particularly to my college tutor, the late Mr. W. W. Rouse Ball, who first woke my interest in the subject. My sincere thanks are also due to several former and present colleagues in St. Andrews who have made a considerable and illuminating study of mathematics among the Ancients: and to kind friends who have offered many valuable suggestions and criticisms.

In preparing the Second Edition I have had the benefit of suggestions which friends from time to time have submitted. I am grateful for this means of removing minor blemishes, and for making a few additions. In particular, a date list has been added.

PREFACE TO THIRD EDITION

A FEW additions have been made to the earlier chapters and to Chapter VI, which incorporate results of recent discoveries among mathematical inscriptions and manuscripts, particularly those which enlarge our knowledge of the mathematics of Ancient Babylonia and Egypt. I gratefully acknowledge the help derived from reading the *Manual of Greek Mathematics* (1931) by Sir Thomas Heath. It provides a short but masterly account of these developments, for which the scientific world is greatly indebted.

H. W. T.

December, 1940.

PREFACE TO FOURTH EDITION

AT the turn of the half-century it is appropriate to add a postscript to Chapter XI, which brought the story of mathematical development as far as the opening years of the century. What has happened since has followed very directly from the wonderful advances that opened up through the algebraical discoveries of Hamilton, the analytical theories of Weierstrass and the geometrical innovations of Von Staudt, and of their many great contemporaries. One very noteworthy development has been the rise of American mathematics to a place in the front rank, and this has come about with remarkable rapidity and principally through the study of abstract algebra such as was inspired by Peirce, a great American disciple of the Hamiltonian school. Representative of this advance in algebra is Wedderburn who built upon the foundations laid, not only by Peirce, but also by Frobenius in Germany and Cartan in France. Through abandoning the commutative law of multiplication by inventing quaternions, Hamilton had opened the door for the investigation of systems of algebra distinct from the ordinary familiar system. Algebra became algebras just as, through the discovery of non-Euclidean systems, geometry became geometries. This plurality, which had been unsuspected for so long, naturally led to the study of the classification of algebras. It was here that Wedderburn, following a hint dropped by Cartan, attained great success. The matter led to deeper and wider understanding of abstract theory, while at the same time it provided a welcome and fertile medium for the further developments in quantum mechanics. Simultaneously with this abstract approach to algebra a powerful advance was made in the technique of algebraical manipulation through the discoveries of Frobenius, Schur and A. Young in the theory of groups and of their representations and applications.

Similar trends may be seen in arithmetic and analysis where the same plurality is in evidence. Typical of this are the theory of valuation and the recognition of Banach spaces. The axiom of Archimedes (p. 99) is here in jeopardy: which is hardly surprising once the concept of regular equal steps upon a straight line had been broadened by the newer forms of geometry. Arithmetic and analysis were, so to speak, projected and made more abstract. It is remarkable that, with these trends towards generalization in each of the four great branches of pure mathematics, the branches lose something of their distinctive qualities and grow more alike. Whitehead's description of geometry as the science of cross-classification remains profoundly true. The applications of mathematics continue to extend, particularly in logic and in statistics.

H. W. T.

May, 1951.

CONTENTS

DATE LIST

? 18th	Century	B.C.	Ahmes (? 1800–).
6th	"	"	Thales (640–550), Pythagoras (569–500).
5th	"	"	Anaxagoras (500–428), Zeno (495–435), Hippocrates (470–), Democritus (? 470–).
4th	"	"	Archytas (? 400), Plato (429–348), Eudoxus (408–355), Menaechmus (375–325).
3rd	"	"	Euclid (? 330–275), Archimedes (287–212), Apollonius (? 262–200).
2nd	"	"	Hipparchus (? 160–).
1st	"	A.D.	Menelaus (? 100).
2nd	"	"	Ptolemy (? 100–168).
3rd	"	"	Hero (? 250), Pappus (? 300), Diophantus (– 320 ?).
6th	"	"	Arya-Bhata (? 530).
7th	"	"	Brahmagupta (? 640).
12th	"	"	Leonardo of Pisa (1175–1230).
16th	"	"	Scipio Ferro (1465–1526), Tartaglia (1500–1557), Cardan (1501–1576), Copernicus (1473–1543), Vieta (1540–1603), Napier (1550–1617), Galileo (1564–1642), Kepler (1571–1630), Cavalieri (1598–1647).
17th	"	"	Desargues (1593–1662), Descartes (1596–1650), Fermat (1601–1665), Pascal (1623–1662), Wallis (1616–1703), Barrow (1630–1677), Gregory (1638–1675), Newton (1642–1727), Leibniz (1646–1716), Jacob Bernoulli (1654–1705), John Bernoulli (1667–1748).
18th	"	"	Euler (1707–1783), Demoivre (1667–1754), Taylor (1685–1741), Maclaurin (1698–1746), D'Alembert (1717–1783), Lagrange (1736–1813), Laplace (1749–1827), Cauchy (1759–1857).
19th	"	"	Gauss (1777–1855), Von Staudt (1798–1867), Abel (1802–1829), Hamilton (1805–1865), Galois (1811–1832), Riemann (1826–1866), Sylvester (1814–1897), Cayley (1821–1895), Weierstrass (1815–1897), and many others.
20th	"	"	Ramanujan (1887–1920), and many living mathematicians.

CHAPTER I

EARLY BEGINNINGS: THALES, PYTHAGORAS AND THE PYTHAGOREANS

TO-DAY with all our accumulated skill in exact measurements, it is a noteworthy event when lines driven through a mountain meet and make a tunnel. How much more wonderful is it that lines, starting at the corners of a perfect square, should be raised at a certain angle and successfully brought to a point, hundreds of feet aloft! For this, and more, is what is meant by the building of a pyramid: and all this was done by the Egyptians in the remote past, far earlier than the time of Abraham.

Unfortunately we have no actual record to tell us who first discovered enough mathematics to make the building possible. For it is evident that such gigantic erections needed very accurate plans and models. But many general statements of the rise of mathematics in Egypt are to be found in the writings of Herodotus and other Greek travellers. Of a certain king Sesostris, Herodotus says:

'This king divided the land among all Egyptians so as to give each one a quadrangle of equal size and to draw from each his revenues, by imposing a tax to be levied yearly. But everyone from whose part the river tore anything away, had to go to him to notify what had happened; he then sent overseers who had to measure out how much the land had become smaller, in order that the owner might pay on what was left, in proportion to the entire tax imposed. In this way, it appears to me, geometry originated, which passed thence to Hellas.'

Then in the *Phaedrus* Plato remarks:

'At the Egyptian city of Naucratis there was a famous old god whose name was Theuth; the bird which is called the Ibis was sacred to him, and he was the inventor of many arts, such as arithmetic and calculation and geometry and astronomy and draughts and dice, but his great discovery was the use of letters.'

According to Aristotle, mathematics originated because the priestly class in Egypt had the leisure needful for its study; over two thousand years later exact corroboration of this remark was forthcoming, through the discovery of a papyrus, now treasured in the Rhind collection at the British Museum. This curious document, which was written by the priest Ahmes, who lived before 1700 B.C., is called 'directions for knowing all dark things'; and the work proves to be a collection of problems in geometry and arithmetic. It is much concerned with the reduction of fractions such as $2/(2n + 1)$ to a sum of fractions each of whose numerators is unity. Even with our improved notation it is a complicated matter to work through such remarkable examples as:

$$\tfrac{2}{29} = \tfrac{1}{24} + \tfrac{1}{58} + \tfrac{1}{174} + \tfrac{1}{232}.$$

There is considerable evidence that the Egyptians made astonishing progress in the science of exact measurements. They had their land surveyors, who were known as *rope stretchers,* because they used ropes, with knots or marks at equal intervals, to measure their plots of land. By this simple means they were able to construct right angles; for they knew that three ropes, of lengths three, four, and five units respectively, could be formed into a right-angled triangle. This useful fact was not confined to Egypt: it was certainly known in China and elsewhere. But the Egyptian skill in practical geometry went far beyond the construction of right angles: for it included, besides the angles of a square, the angles of other regular figures such as the pentagon, the hexagon and the heptagon.

If we take a pair of compasses, it is very easy to draw a circle and then to cut the circumference into *six* equal parts. The six points of division form a regular hexagon, the figure so well known as the section of the honey cell. It is a much more difficult problem to cut the circumference into *five* equal parts, and a very much more difficult problem to cut it into *seven* equal parts. Yet those who have carefully examined the design of the ancient temples and pyramids of Egypt tell us that these particular figures and angles are there to be seen. Now there are two distinct methods of dealing with geometrical problems—the practical and the theoretical. The Egyptians were champions of the practical, and the Greeks of the theoretical method. For example, as Röber has pointed out, the Egyptians employed a practical rule to determine the angle of a regular heptagon. And although it fell short of theoretical precision, the rule was accurate enough to conceal the error, unless the figure were to be drawn on a grand scale. It would barely be apparent even on a circle of radius 50 feet.

Unquestionably the Egyptians were masters of practical geometry; but whether they knew the theory, the underlying reason for their results, is another matter. Did they know that their right-angled triangle, with sides of lengths three, four and five units, contained an *exact* right angle? Probably they did, and possibly they knew far more. For, as Professor D'Arcy Thompson has suggested, the very *shape* of the Great Pyramid indicates a considerable familiarity with that of the regular pentagon. A certain obscure passage in Herodotus can, by the slightest literal emendation, be made to yield excellent sense. It would imply that the area of each triangular face of the Pyramid is equal to the square of the vertical height; and this accords well with the actual facts. If this is so, the ratios of height, slope and base can be expressed in terms of the 'golden section', or of the radius of a circle to the side of the inscribed decagon. In short, there was already a wealth of geometrical and arithmetical results treasured by the priests of Egypt, before the early Greek travellers became acquainted with mathematics. But it was only after the keen imaginative eye of the Greek fell upon these Egyptian figures that they yielded up their wonderful secrets and disclosed their inner nature.

Among these early travellers was THALES, a rich merchant of Miletus, who lived from about 640 to 550 B.C. As a man of affairs he was highly successful: his duties as merchant took him to many countries, and his native wit enabled him to learn from the novelties which he saw. To his admiring fellow-countrymen of later generations he was known as one of the Seven Sages of Greece, many legends and anecdotes clustering round his name. It is said that Thales was once in charge of some mules, which were burdened with sacks of salt. Whilst crossing a river one of the animals slipped; and the salt consequently dissolving in the water, its load became instantly lighter. Naturally the sagacious beast deliberately submerged itself at the next ford, and continued this trick until Thales hit upon the happy expedient of filling the sack with sponges! This proved an effectual cure. On another occasion, foreseeing an unusually fine crop of olives, Thales took possession of every olive-press in the district; and having made this 'corner', became master of the market and could dictate his own terms. But now, according to one account, as he had *proved* what could be done, his purpose was achieved. Instead of victimizing his buyers, he magnanimously sold the fruit at a price reasonable enough to have horrified the financier of to-day.

Like many another merchant since his time Thales early retired from commerce, but unlike many another he spent his leisure in philosophy and mathematics. He seized on what he had learnt in his travels, particularly from his intercourse with the priests of Egypt; and he was the first to bring out something of the true significance of Egyptian scientific lore. He was both a great mathematician and a great astronomer. Indeed, much of his popular celebrity was due to his successful prediction of a solar eclipse in 585 B.C. Yet it is told of him that in contemplating the stars during an evening walk, he fell into a ditch; whereupon the old woman attending him exclaimed, 'How canst thou know what is doing in the heavens when thou seest not what is at thy feet?'

We live so far from these beginnings of a rational wonder at natural things, that we run the risk of missing the true import of results now so very familiar. There are the well-known propositions that a circle is bisected by any diameter, or that the angles at the base of an isosceles triangle are equal, or that the angle in a semicircle is a right angle, or that the sides about equal angles in similar triangles are proportional. These and other like propositions have been ascribed to Thales. Simple as they are, they mark an epoch. They elevate the endless details of Egyptian mensuration to general truths; and in like manner his astronomical results replace what was little more than the making of star catalogues by a genuine science.

It has been well remarked that in this geometry of Thales we also have the true source of algebra. For the theorem that the diameter bisects a circle is indeed a true equation; and in his experiment conducted, as Plutarch says, 'so simply, without any fuss or instrument' to determine the height of the Great Pyramid by comparing its shadow with that of a vertical stick, we have the notion of equal ratios, or proportion.

The very idea of abstracting all solidity and area from a material shape, such as a square or triangle, and pondering upon it as a pattern of lines, seems to be definitely due to Thales. He also appears to have been the first to suggest the importance of a geometrical *locus*, or curve traced out by a point moving according to a definite law. He is known as the father of Greek mathematics, astronomy and philosophy, for he combined a practical sagacity with genuine wisdom. It was no mean achievement, in his day, to break through the pagan habit of mind which concentrates on particular cults and places. Thales asserted the existence of the abstract and the more general: these, said he, were worthier of deep study than the intuitive or sensible. Here spoke the philosopher. On the other hand he gave to mankind such practical gifts as the correct number of days in the year, and a convenient means of finding by observation the distance of a ship at sea.

Thales summed up his speculations in the philosophical proposition 'All things are water'. And the fact that all things are not water is trivial compared with the importance of his outlook. He saw the field; he asked the right questions; and he initiated the search for underlying law beneath all that is ephemeral and transient.

Thales never forgot the debt that he owed to the priests of Egypt; and when he was an old man he strongly advised his pupil PYTHAGORAS to pay them a visit. Acting upon this advice, Pythagoras travelled and gained a wide experience, which stood him in good stead when at length he settled and gathered round him disciples of his own, and became even more famous than his master. Pythagoras is supposed to have been a native of Samos, belonging like Thales to the Ionian colony of Greeks planted on the western shores and islands of what we now call Asia Minor. He lived from about 584 B.C. to 495 B.C. In 529 B.C. he settled at Crotona, a town of the Dorian colony in South Italy, and there he began to lecture upon philosophy and mathematics. His lecture-room was thronged with enthusiastic hearers of all ranks. Many of the upper classes attended, and even women broke a law which forbade them to attend public meetings, and flocked to hear him. Among the most attentive was Theano, the young and beautiful daughter of his host Milo, whom he married. She wrote a biography of her husband, but unfortunately it is lost.

So remarkable was the influence of this great master that the more attentive of his pupils gradually formed themselves into a society or brotherhood. They were known as the Order of the Pythagoreans, and they were soon exercising a great influence far across the Grecian world. This influence was not so much political as religious. Members of the Society shared everything in common, holding the same philosophical beliefs, engaging in the same pursuits, and binding themselves with an oath not to reveal the secrets and teaching of the school. When, for example, Hippasus perished in a shipwreck, was not his fate due to a broken promise? For he had divulged the secret of the sphere with its twelve pentagons!

A distinctive badge of the brotherhood was the beautiful star pentagram—a fit symbol of the mathematics which the school discovered. It was also the symbol of health. Indeed, the Pythagoreans were specially interested in the study of medicine. Gradually, as the Society spread, teachings once treasured orally were committed to writing. Thereby a copy of a treatise by Philolaus, we are told, ultimately came into the possession of Plato; probably a highly significant event in the history of mathematics.

FIGURE 1

In mathematics the Pythagoreans made very great progress, particularly in the theory of numbers and in the geometry of areas and solids. As it was the generous practice among members of the brotherhood to attribute all credit for each new discovery to Pythagoras himself, we cannot be quite certain about the authorship of each particular theorem. But at any rate in the mathematics which are now to be described, his was the dominating influence.

In thinking of these early philosophers we must remember that open air and sunlight and starry nights formed their surroundings—not our grey mists and fettered sunshine. As Pythagoras was learning his mensuration from the priests of Egypt, he would constantly see the keen lines cast by the shadows of the pillars across the pavements. He trod chequered floors with their arrays of alternately coloured squares. His mind was stirred by interesting geometrical truths learnt from his master Thales, his interest in number would lead him to count the squares, and the sight of the long straight shadow falling obliquely across them would suggest sequences of special squares. It falls maybe across the centre of the first, the fourth, the seventh; the arithmetical progression is suggested. Then again the square is interesting for its *size*. A fragment of more diverse pattern would demonstrate a larger square enclosing one exactly half its size. A

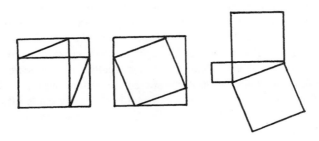

FIGURE 2

FIGURE 2

little imaginative thought would reveal, within the larger, a smaller square placed unsymmetrically, and so would lead to the great theorem which somehow or other was early reached by the brotherhood (and some say by Pythagoras himself), that the square on one side of a right-angled triangle is equal to the sum of the squares on the remaining sides. The above figures (Figure 2) actually suggest the proof, but it is quite possible that several different proofs were found, one being by the use of similar triangles. According to one story, when Pythagoras first discovered this fine result, in his exultation he sacrificed an ox!

Influenced no doubt by these same orderly patterns, he pictured numbers as having characteristic designs. There were the triangular numbers, one, three, six, ten, and so on, ten being the *holy tetractys*, another sym-

bol highly revered by the brotherhood. Also there were the square numbers, each of which could be derived from its predecessor by adding an

L-shaped border. Great importance was attached to this border: it was called a gnomon (γνώμων, carpenter's rule). Then it was recognized that each odd number, three, five, seven, etc., was a gnomon of a square number. For example, seven is the gnomon of the square of three to make the square of four. Pythagoras was also interested in the more abstract natural objects, and he is said to have discovered the wonderful harmonic progressions in the notes of the musical scale, by finding the relation between the length of a string and the pitch of its vibrating note. Thrilled by his discovery, he saw in numbers the element of all things. To him numbers were no mere attributes: three was not that which is common to three cats or three books or three Graces: but numbers were themselves the stuff out of which all objects we see or handle are made—the rational reality. Let us not judge the doctrine too harshly; it was a great advance on the cruder water philosophy of Thales.

So, in geometry, *one* came to be identified with the point; *two* with the line, *three* with the surface, and *four* with the solid. This is a noteworthy disposition that really is more fruitful than the usual allocation in which the line is said to have one, the surface two, and the solid three, dimensions.

More whimsical was the attachment of *seven* to the maiden goddess Athene 'because seven alone within the decade has neither factors nor product'. *Five* suggested marriage, the union of the first even with the first genuine odd number. *One* was further identified with reason; *two* with opinion—a wavering fellow is Two; he does not know his own mind: *four* with justice, steadfast and square. Very fanciful no doubt: but has not Ramanujan, one of the greatest arithmeticians of our own days, been thought to treat the positive integers as his personal friends? In spite of this exuberance the fact remains, as Aristotle sums it up: 'The Pythagoreans first applied themselves to mathematics, a science which they improved; and, penetrated with it, they fancied that the principles of mathematics were the principles of all things'. And a younger contemporary, Eudemus, shrewdly remarked that 'they changed geometry into a liberal science; they diverted arithmetic from the service of commerce'.

To Pythagoras we owe the very word mathematics and its doubly twofold branches:

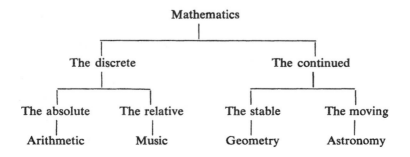

```
                          Mathematics
                              |
            ┌─────────────────┴─────────────────┐
            |                                    |
       The discrete                        The continued
            |                                    |
      ┌─────┴─────┐                        ┌─────┴─────┐
      |           |                        |           |
 The absolute  The relative           The stable   The moving
      |           |                        |           |
  Arithmetic    Music                  Geometry    Astronomy
```

This classification is the origin of the famous Quadrivium of knowledge. In geometry Pythagoras or his followers developed the theory of space-filling figures. The more obvious of these must have been very well known. If we think of each piece in such a figure as a unit, the question arises, can we fill a flat surface with repetitions of these units? It is very likely that this type of inquiry was what first led to the theorem that the three angles of a triangle are together equal to two right angles. The same train of thought also extends naturally to solid geometry, including the conception of regular solids. One of the diagrams (Figure 3) shows six equal triangles filling flat space round their central point. But five such equilateral triangles can likewise be fitted together, to form a blunted bell-tent-shaped figure, spreading from a central vertex: and now their bases form a regular pentagon. Such a figure is no longer flat; it makes a solid angle, the corner, in fact, of a regular icosahedron. The process could be repeated by surrounding each vertex of the original triangles with five triangles. Exactly twenty triangles would be needed, no more and no less, and the

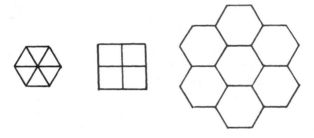

FIGURE 3

result would be the beautiful figure of the icosahedron of twenty triangles surrounding its twelve vertices in circuits of five.

It is remarkable that in solid geometry there are only five such regular figures, and that in the plane there is a very limited number of associations of regular space-filling figures. The three simplest regular solids, including the cube, were known to the Egyptians. But it was given to Pythagoras to discover the remaining two—the dodecahedron with its twelve pentagonal faces, and the icosahedron. Nowadays we so often become acquainted with these regular solids and plane figures only after a long excursion through the intricacies of mensuration and plane geometry that we fail to see their full simplicity and beauty.

Another kind of problem that interested Pythagoras was called the *method of application of areas.* His solution is noteworthy because it provided the geometrical equivalent of solving a quadratic equation in algebra. The main problem consisted in drawing, upon a given straight line, a figure that should be the size of one and the shape of another given figure. In the course of the solution one of three things was bound to happen. The base of the constructed figure would either fit, fall short of, or exceed the length of the given straight line. Pythagoras thought it proper to draw attention to these three possibilities; accordingly he introduced the terms *parabole, ellipsis* and *hyperbole*. Many years later his nomenclature was adopted by Apollonius, the great student of the conic section, because the same threefold characteristics presented themselves in the construction of the curve. And we who follow Apollonius still call the curve the parabola, the ellipse, or the hyperbola, as the case may be. The same threefold classification underlies the signs $=$, $<$, $>$ in arithmetic.

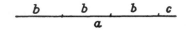

FIGURE 4

Many a time throughout the history of mathematics this classification has proved to be the key to further discoveries.

For example, it is closely connected with the theory of irrational numbers; and this brings us to the greatest achievement of Pythagoras, who is credited with discovering the (ἄλογον) irrational. In other words, he proved that it was not always possible to find a common measure for two given lengths a and b. The practice of measuring one line against another must have been very primitive. Here is a long line a, into which the shorter line b fits three times, with a still shorter piece c left over (Figure 4). To-day we express this by the equation $a = 3b + c$, or more generally by $a = nb + c$. If there is no such remainder c, the line b measures a; and a is called a multiple of b. If, however, there is a remainder c, further subdivision might perhaps account for each length a, b, c without remainder: experiment might show, for instance, that in tenths of inches, $a = 17$, $b = 5$, $c = 2$. At one time it was thought that it was *always* possible to reduce lengths a and b to such multiples of a smaller length. It appeared to be simply a question of patient subdivision, and sooner or later the desired measure would be found. So the required subdivision, in the present example, is found by measuring b with c. For c fits twice into b with a remainder d; and d fits exactly twice into c *without* remainder. Consequently d measures c, and also measures b and also a. This is how the numbers 17, 5 and 2 come to be attached to a, b, and c: namely a contains d 17 times.

Incidentally this shows how naturally the arithmetical progression arises. For although the original subdivisions, and extremity, of the line a occur at distances 5, 10, 15, 17, measured from the left in quarter inches, they occur at distances 2, 7, 12, 17, from the right. These numbers form a typical arithmetical progression, with a rhythmical law of succession that alone would be incentive for a Pythagorean to study them further.

This reduction of the comparison of a line a with a line b to that of the number 17 with 5, or, speaking more technically, this reduction of the ratio $a : b$ to 17 : 5 would have been agreeable to the Pythagorean. It exactly fitted in with his philosophy; for it helped to reduce space and geometry to pure number. Then came the awkward discovery, by Pythagoras himself, that the reduction was not always possible; that something in geometry eluded whole numbers. We do not know exactly how this discovery of the *irrational* took place, although two early examples can be cited. First when a is the diagonal and b is the side of a square, no common measure can be found; nor can it be found in a second example, when a line a is divided in *golden section* into parts b and c. By this is meant that the ratio of a, the whole line, to the part b is equal to the ratio of b to the other part c. Here c may be fitted once into b with remainder d: and then d may be fitted once into c with a remainder e: and so on. It is not hard to prove that such lengths a, b, c, d, . . . form a geometrical progression without end; and the desired common measure is never to be found.

If we prefer algebra to geometry we can verify this as follows. Since it is given that $a = b + c$ and also $a : b :: b : c$, it follows that $a(a - b) = b^2$. This is a quadratic equation for the ratio $a : b$, whose solution gives the result

$$a : b = \sqrt{5} + 1 : 2.$$

The presence of the surd $\sqrt{5}$ indicates the irrational.

The underlying reason why such a problem came to be studied is to be found in the star badge of the brotherhood (p. 83); for every line therein is divided in this golden section. The star has five lines, each cut into three

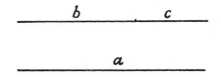

FIGURE 5

parts, the lengths of which can be taken as a, b, a. As for the ratio of the diagonal to the side of a square, Aristotle suggests that the Pythagorean proof of its irrationality was substantially as follows:

If the ratio of diagonal to side is commensurable, let it be $p : q$, where

p and q are whole numbers prime to one another. Then p and q denote the number of equal subdivisions in the diagonal and the side of a square respectively. But since the square on the diagonal is double that on the side, it follows that $p^2 = 2q^2$. Hence p^2 is an even number, and p itself must be even. Therefore p may be taken to be $2r$, p^2 to be $4r^2$, and consequently q^2 to be $2r^2$. This requires q to be even; which is impossible because two numbers p, q, prime to each other cannot both be even. So the original supposition is untenable: no common measure can exist; and the ratio is therefore irrational.

This is an interesting early example of an indirect proof or *reductio ad absurdum*; and as such it is a very important step in the logic of mathematics.

We can now sum up the mathematical accomplishments of these early Greek philosophers. They advanced in geometry far enough to cover roughly our own familiar school course in the subject. They made substantial progress in the theoretical side of arithmetic and algebra. They had a geometrical equivalent for our method of solving quadratic equations; they studied various types of progressions, arithmetical, geometrical and harmonical. In Babylon, Pythagoras is said to have learnt the 'perfect proportion'

$$a : \frac{a+b}{2} :: \frac{2ab}{a+b} : b$$

which involves the arithmetical and harmonical means of two numbers. Indeed, to the Babylonians the Greeks owed many astronomical facts, and the sexagesimal method of counting by sixties in arithmetic. But they lacked our arithmetical notation and such useful abbreviations as are found in the theory of indices. From a present-day standpoint these results may be regarded as elementary: it is otherwise with their discovery of irrational numbers. That will ever rank as a piece of essentially advanced mathematics. As it upset many of the accepted geometrical proofs it came as a 'veritable logical scandal'. Much of the mathematical work in the succeeding era was coloured by the attempt to retrieve the position, and in the end this was triumphantly regained by Eudoxus.

Recent investigations of the Rhind Papyrus, the Moscow Papyrus of the Twelfth Egyptian Dynasty, and the Strassburg Cuneiform texts have greatly added to the prestige of Egyptian and Babylonian mathematics. While no general proof has yet been found among these sources, many remarkable *ad hoc* formulae have come to light, such as the Babylonian solution of complicated quadratic equations dating from 2000 B.C., which O. Neugebauer published in 1929, and an Egyptian approximation to the area of a sphere (equivalent to reckoning $\pi = 256/81$).

CHAPTER II

EUDOXUS AND THE ATHENIAN SCHOOL

A SECOND stage in the history of mathematics occupied the fifth and fourth centuries B.C., and is associated with Athens. For after the wonderful victories at Marathon and Salamis early in the fifth century, when the Greeks defeated the Persians, Athens rose to a position of pre-eminence. The city became not only the political and commercial, but the intellectual centre of the Grecian world. Her philosophers congregated from East and West, many of whom were remarkable mathematicians and astronomers. Perhaps the greatest among these were Hippocrates, Plato, Eudoxus and Menaechmus; and contemporary with the three latter was Archytas the Pythagorean, who lived at Tarentum.

Thales and Pythagoras had laid the foundations of geometry and arithmetic. The Athenian school concentrated upon special aspects of the superstructure; and, whether by accident or design, found themselves embarking upon three great problems: (i) the *duplication of the cube*, or the attempt to find the edge of a cube whose volume is double that of a given cube; (ii) the *trisection of a given angle*; and (iii) the *squaring of a circle*, or, the attempt to find a square whose area is equal to that of a given circle. These problems would naturally present themselves in a systematic study of geometry; while, as years passed and no solutions were forthcoming they would attract increasing attention. Such is their inherent stubbornness that not until the nineteenth century were satisfactory answers to these problems found. Their innocent enunciations are at once an invitation and a paradox. Early attempts to solve them led indirectly to results that at first sight seem to involve greater difficulties than the problems themselves. For example, in trying to square the circle Hippocrates discovered that two moon-shaped figures could be drawn whose areas were together equal to that of a right-angled triangle. This diagram (Figure 6) with its three semicircles described on the respective sides of the triangle illustrates his theorem. One might readily suppose that it would be easier to determine the area of a single circle than that of these lunes, or lunules, as they are called, bounded by pairs of circular arcs. Yet such is not the case.

In this by-product of the main problem Hippocrates gave the first example of a solution in *quadratures*. By this is meant the problem of constructing a rectilinear area equal to an area bounded by one or more curves. The sequel to attempts of this kind was the invention of the integral calculus by Archimedes, who lived in the next century. But his first success in the method was not concerned with the area of a circle, but

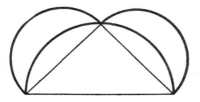

FIGURE 6

with that of a parabola, a curve that had been discovered by Menaechmus in an attempt to duplicate the cube. This shows how very interdependent mathematics had now become with its interplay between branch and branch. All this activity led to the discovery of many other new curves, including the ellipse, the hyperbola, the quadratrix, the conchoid (the shell), the cissoid (the ivy leaf), various spirals, and other curves classed as *loci on surfaces.*

The Greeks now found it useful to adopt a special classification for their problems, calling them *plane, solid* and *linear.* Problems were *plane* if their solution depended only on the use of straight lines and circles. These were of distinctly the Pythagorean type. They were *solid* if they depended upon conic sections: and they were *linear* if in addition they depended upon still more complicated curves. This early classification shows true mathematical insight, because later experience has revealed close algebraic and analytic parallels. For example, the *plane* problem corresponds in algebra to the problem soluble by quadratic equations. The Greeks quite naturally but vainly supposed that the three famous problems above were soluble by plane methods. It is here that they were wrong: for by solid or linear methods the problems were not necessarily insoluble.

One of the first philosophers to bring the new learning from Ionia to Athens was ANAXAGORAS (? 500–428 B.C.), who came from near Smyrna. He is said to have neglected his possessions, which were considerable, in order to devote himself to science, and in reply to the question, what was the object of being born, he remarked: 'The investigation of the sun, moon and heaven.' In Athens he shared the varying fortunes of his friend Pericles, the great statesman, and at one time was imprisoned for impiety. This we know from an ancient record which adds that 'while in prison he wrote (or drew) the squaring of the circle', a brief but interesting allusion to the famous problem. Nor has the geometry of the circle suffered unduly from the captivity of its devotees. Centuries later another great chapter was opened, when the Russians flung Poncelet, an officer serving under Napoleon, into prison, where he discovered the circular points at infinity. Anaxagoras, however, was famous chiefly for his work in astronomy.

HIPPOCRATES [1] was his younger contemporary, who came from Chios to Athens about the middle of the fifth century. A lawsuit originally lured him to the city: for he had lost considerable property in an attack by Athenian pirates near Byzantium. Indeed, the tastes of Athenian citizens were varied: they were not all artists, sculptors, statesmen, dramatists, philosophers, or honest seamen, in spite of the wealth there and then assembled. After enduring their ridicule first at being cheated and then for hoping to recover his money, the simple-minded Hippocrates gave up the quest, and found his solace in mathematics and philosophy.

He made several notable advances. He was the first author who is known to have written an account of elementary mathematics; in particular he devoted his attention to properties of the circle. To-day his actual work survives among the theorems of Euclid, although his original book is lost. His chief result is the proof of the statement that circles are to one another in the ratio of the squares on their diameters. This is equivalent to the discovery of the formula πr^2 for the area of a circle in terms of its radius. It means that a certain number π exists, and is the same for all circles, although his method does not give the actual numerical value of π. It is thought that he reached his conclusions by looking upon a circle as the limiting form of a regular polygon, either inscribed or circumscribed. This was an early instance of the *method of exhaustions*—a particular use of approximation from below and above to a desired limit.

The introduction of the method of exhaustions was an important link in the chain of thought culminating in the work of Eudoxus and Archimedes. It brought the prospect of unravelling the mystery of irrational numbers, that had sorely puzzled the early Pythagoreans, one stage nearer. A second important but perhaps simpler work of Hippocrates was an example of the useful device of reducing one theorem to another. The Pythagoreans already had shown how to find the geometric mean between two magnitudes by a geometrical construction. They merely drew a square equal to a given rectangle. Hippocrates now showed that to duplicate a cube was tantamount to finding *two* such geometric means. Put into more familiar algebraic language, if

$$a : x = x : b, \text{ then } x^2 = ab,$$

and if

$$a : x = x : y = y : 2a$$

then $x^3 = 2a^3$. Consequently if a was the length of the edge of the given cube, x would be that of a cube twice its size. But the statement also shows that x is the first of two geometric means between a and $2a$.

We must, of course, bear in mind that the Greeks had no such convenient algebraic notation as the above. Although they went through the

[1] Not the great physician.

same reasoning and reached the same conclusions as we can, their statements were prolix, and afforded none of the help which we find in these
concise symbols of algebra.

It is supposed that the study of the properties of two such means, x
and y, between given lengths a and b, led to the discovery of the parabola
and hyperbola. As we should say, nowadays, the above continued proportions indicate the equations $x^2 = ay$, and $xy = 2a^2$. These equations
represent a parabola and a hyperbola: taken together they determine a
point of intersection which is the key to the problem. This is an instance
of a *solid* solution for the duplication of the cube. It represents the ripe
experience of the Athenian school; for MENAECHMUS (? 375–325 B.C.),
to whom it is credited, lived a century later than Hippocrates.

Where two lines, straight or curved, cross, is a point: where three surfaces meet is a point. The two walls and the ceiling meeting at the corner
of a room give a convenient example. But two curved walls, meeting a
curved ceiling would also make a corner, and in fact illustrate a truly
ingenious method of dealing with this same problem of the cube. The
author of this geometrical novelty was ARCHYTAS (? 400 B.C.), a contemporary of Menaechmus. This time the problem was reduced to finding
the position of a certain point in space: and the point was located as the
meeting-place of three surfaces. For one surface Archytas chose that generated by a circle revolving about a fixed tangent as axis. Such a surface
can be thought of as a ring, although the hole through the ring is completely stopped up. His other surfaces were more commonplace, a cylinder
and a cone. With this unusual choice of surfaces he succeeded in solving
the problem. When we bear in mind how little was known, in his day,
about solid geometry, this achievement must rank as a gem among mathematical antiquities. Archytas, too, was one of the first to write upon
mechanics, and he is said to have been very skilful in making toys and
models—a wooden dove which could fly, and a rattle which, as Aristotle
says, 'was useful to give to children to occupy them from breaking things
about the house (for the young are incapable of keeping still)'.

Unlike the majority of mathematicians who lived in this Athenian era,
Archytas lived at Tarentum in South Italy. He found time to take a considerable part in the public life of his city, and is known for his enlightened attitude in his treatment of slaves and in the education of children.
He was a Pythagorean, and was also in touch with the philosophers of
Athens, numbering Plato among his friends. He is said upon one occasion
to have used his influence in high quarters to save the life of Plato.

Between Crotona and Tarentum upon the shore of the gulf of Southern
Italy was the city of Elea: and with each of these places we may associate
a great philosopher or mathematician. At Crotona Pythagoras had insti-
tuted his lecture-room; nearly two centuries later Archytas made his

mechanical models at Tarentum. But about midway through the intervening period there lived at Elea the philosopher ZENO. This acutely original thinker played the part of philosophical critic to the mathematicians, and some of his objections to current ideas about motion and the infinitesimal were very subtle indeed. For example, he criticized the infinite geometrical progression by proposing the well-known puzzle of Achilles and the Tortoise. How, asked Zeno, can the swift Achilles overtake the Tortoise if he concedes a handicap? For if Achilles starts at A and the tortoise at B, then when Achilles reaches B the tortoise is at C, and when Achilles reaches C the tortoise is at D. As this description can go on and on, apparently Achilles never overtakes the tortoise. But actually he may do so; and this is a paradox. The point of the inquiry is not *when,* but *how* does Achilles overtake the tortoise.

Somewhat similar questions were asked by DEMOCRITUS, the great philosopher of Thrace, who was a contemporary of Archytas and Plato. Democritus has long been famous as the originator of the atomic theory, a speculation that was immediately developed by Epicurus, and later provided the great theme for the Latin poet Lucretius. It is, however, only quite recently that any mathematical work of Democritus has come to light. This happened in 1906, when Heiberg discovered a lost book of Archimedes entitled the *Method.* We learn from it that Archimedes re-

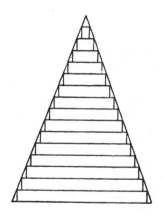

FIGURE 7

garded Democritus as the first mathematician to state correctly the formula for the volume of a cone or a pyramid. Each of these volumes was one-third part of a circumscribing cylinder, or prism, standing on the same base. To reach his conclusions, Democritus thought of these solids as built up of innumerable parallel layers. There would be no difficulty in the case of the cylinder, for each layer would be equal. But for the cone or pyramid the sizes of layer upon layer would taper off to a point. The appended diagram (Figure 7), showing the elevation of a cone or pyramid, illustrates this tapering of the layers, although the picture that Democritus had in mind consisted of very much thinner layers. He was puzzled by their diminishing sizes. 'Are they equal or unequal?' he asked, 'for if they are unequal, they will make the cone irregular as having many indentations, like steps, and unevennesses; but, if they are equal, the sections will be equal, and the cone will appear to have the property of the cylinder and to be made up of equal, not unequal circles, which is very absurd.'

This quotation is striking; for it foreshadows the great constructive work of Archimedes, and, centuries later, that of Cavalieri and Newton. It exhibits the infinitesimal calculus in its infancy. The notion of stratification—that a solid could be thought of as layer upon layer—would occur quite naturally to Democritus, because he was a physicist; it would *not* so readily have occurred to Pythagoras or Plato with their more algebraic turn of mind which attracted them to the pattern or arrangement of things. But here the acute Greek thought is once more restless. No mere rough and ready approximation will satisfy Democritus: there is discrepancy between stratified pyramid and smoothly finished whole. The deep question of the theory of limits is at issue; but how far he foresaw any solution, we do not know.

This brings us to the great arithmetical work at Athens, associated with the names of PLATO (429–348 B.C.) and EUDOXUS (408–355 B.C.).

Among the philosophers of Athens only two were native to the place, Socrates and Plato, master and disciple, both of whom were well read mathematicians. Plato was perhaps an original investigator; but whether this is so or not, he exerted an immense influence on the course that mathematics was to take, by founding and conducting in Athens his famous Academy. Over the entrance of his lecture room his students read the telling inscription, 'Let no one destitute of geometry enter my doors'; and it was his earnest wish to give his pupils the finest possible education. A man, said he, should acquire no mere bundle of knowledge, but be trained to see below the surface of things, seeking rather for the eternal reality and the Good behind it all. For this high endeavour the study of mathematics is essential; and numbers, in particular, must be studied, simply as numbers and not as embodied in anything. They impart a character to nature; for instance, the periods of the heavenly bodies can only be characterized by invoking the use of irrationals.

Originally the Greek word ἀριθμοί, from which we derive 'arithmetic,' meant the natural numbers, although it was at first questioned whether unity was a number; for 'how can unity, the measure, be a number, the thing measured?' But by including irrationals as numbers Plato made a great advance: he was in fact dealing with what we nowadays call the positive real numbers. Zero and negative numbers were proposed at a far later date.

There is grandeur here in the importance which Plato ascribes to arithmetic for forming the mind: and this is matched by his views on geometry, 'the subject which has very ludicrously been called mensuration' ($\gamma\epsilon\omega\mu\epsilon\tau\rho\iota\alpha$ = land measuring) but which is really an art, a more than human miracle in the eyes of those who can appreciate it. In his book, the *Timaeus,* where he dramatically expounds the views of his hero Timaeus, the Pythagorean, reference is made to the five regular solids and to their supposed significance in nature. The speaker tells how that the four elements earth, air, fire and water have characteristic shapes: the cube is appropriated to earth, the octahedron to air, the sharp pyramid or tetrahedron to fire, and the blunter icosahedron to water, while the Creator used the fifth, the dodecahedron, for the Universe itself. Is it sophistry, or else a brilliant foretaste of the molecular theory of our own day? According to Proclus, the late Greek commentator, 'Plato caused mathematics in general and geometry in particular to make a very great advance, by reason of his enthusiasm for them, which of course is obvious from the way he filled his books with mathematical illustrations, and everywhere tries to kindle admiration for these subjects, in those who make a pursuit of philosophy.' It is related that to the question, What does God do? Plato replied, 'God always geometrizes.'

Among his pupils was a young man of Cnidus, named EUDOXUS, who came to Athens in great poverty, and, like many another poor student, had a struggle to maintain himself. To relieve his pocket he lodged down by the sea at the Piraeus, and every day used to trudge the dusty miles to Athens. But his genius for astronomy and mathematics attracted attention and finally brought him to a position of eminence. He travelled and studied in Egypt, Italy and Sicily, meeting Archytas, the geometer, and other men of renown. About 368 B.C., at the age of forty, Eudoxus returned to Athens in company with a considerable following of pupils, about the time when Aristotle, then a lad of seventeen, first crossed the seas to study at the Academy.

In astronomy the great work of Eudoxus was his theory of concentric spheres explaining the strange wanderings of the planets; an admirable surmise that went far to fit the observed facts. Like his successor Ptolemy, who lived many centuries later, and all other astronomers until Kepler, he found in circular motion a satisfactory basis for a complete planetary theory. This was great work; yet it was surpassed by his pure mathematics which touched the zenith of Greek brilliance. For Eudoxus placed the doctrine of irrationals upon a thoroughly sound basis, and so well was his task done that it still continues, fresh as ever, after the great arithmetical reconstructions of Dedekind and Weierstrass during the nineteenth century. An immediate effect of the work was to restore confidence in the geometrical methods of proportion and to complete the proofs of several important theorems. The *method of exhaustions* vaguely underlay the results of Democritus upon the volume of a cone and of Hippocrates on the area of a circle. Thanks to Eudoxus this method was fully explained.

An endeavour will now be made to indicate in a simple way how this great object was achieved. This study of higher arithmetic at Athens was stimulated by the Pythagorean, Theodorus of Cyrene, who is said to have been Plato's teacher. For Theodorus discovered many irrationals, $\sqrt{3}$, $\sqrt{5}$, $\sqrt{6}$, $\sqrt{7}$, $\sqrt{8}$, $\sqrt{10}$, $\sqrt{11}$, $\sqrt{12}$, $\sqrt{13}$, $\sqrt{14}$, $\sqrt{15}$, $\sqrt{17}$, 'at which point,' says Plato, 'for some reason he stopped'. The omissions in the list are obvious: $\sqrt{2}$ had been discovered by Pythagoras through the ratio of diagonal to side of a square, while $\sqrt{4}$, $\sqrt{9}$, $\sqrt{16}$ are of course irrelevant. Now it is one thing to discover the *existence* of an irrational such as $\sqrt{2}$; it is quite another matter to find a way of *approach* to the number. It was this second problem that came prominently to the fore: it provided the arithmetical aspect of the *method of exhaustions* already applied to the circle: and it revealed a wonderful example of ancient arithmetic. We learn the details from a later commentator, Theon of Smyrna.

Unhampered by a decimal notation (which here is a positive hindrance, useful as it is in countless other examples), the Greeks set about their task in the following engaging fashion. To approximate to $\sqrt{2}$ they built a ladder of whole numbers. A brief scrutiny of the ladder shows how the rungs are devised: $1 + 1 = 2$, $1 + 2 = 3$, $2 + 3 = 5$, $2 + 5 = 7$, $5 + 7 = 12$, and so on. Each rung of the ladder consists of two numbers x and y, whose ratio approaches nearer and nearer to the ratio $1:\sqrt{2}$, the further down the ladder it is situated. Again, these numbers x and y,

1	1
2	3
5	7
12	17
29	41
etc.	

at each rung, satisfy the equation

$$y^2 - 2x^2 = \pm 1.$$

The positive and negative signs are taken at alternate rungs, starting with a negative. For example, at the third rung $7^2 - 2 \cdot 5^2 = -1$.

As these successive ratios are alternately less than and greater than all that follow, they nip the elusive limiting ratio $1 : \sqrt{2}$ between two extremes, like the ends of a closing pair of pincers. They approximate *from both sides* to the desired irrational: $\frac{5}{7}$ is a little too large, but $\frac{12}{17}$ is a little too small. Like pendulum swings of an exhausted clock they die down—but they never actually come to rest. Here again, is the Pythagorean notion of *hyperbole* and *ellipsis*; it was regarded as very significant, and was called by the Greeks the 'dyad' of the 'great and small'.

Such a ladder could be constructed for any irrational; and another very pretty instance, which has been pointed out by Professor D'Arcy Thompson, is closely connected with the problem of the Golden Section.

1	1	Here the right member of each rung is the sum of the pair on
1	2	the preceding rung, so that the ladder may be extended with
2	3	the greatest ease. In this case the ratios approximate, again
3	5	by the little more and the little less, to the limit $\sqrt{5} + 1 : 2$.
5	8	It is found that they provide the arithmetical approach to the
etc.		golden section of a line AB, namely when C divides AB so that

FIGURE 8

CB : AC = AC : AB. In fact, AC is roughly ⅗ of the length AB, but more nearly ⅝ of AB; and so on. It is only fair to say that this simplest of all such ladders has not yet been found in the ancient literature, but owing to its intimate connexion with the pentagon, it is difficult to resist the conclusion that the later Pythagoreans were familiar with it. The series 1, 2, 3, 5, 8, . . . was known in mediaeval times to Leonardo of Pisa, surnamed Fibonacci, after whom it is nowadays named.

Let us now combine this ladder-arithmetic with the geometry of a divided line. For example, let a line AB be divided at random by C, into lengths a and b, where AC = a, CB = b. Then the question still remains, what is the *exact* arithmetical meaning of the ratio $a : b$, whether or not this is irrational? The wonderful answer to this question is what has made Eudoxus so famous. Before considering it, let us take as an illustration the strides of two walkers. A tall man A takes a regular stride of length a, and his short friend B takes a stride b. Now suppose that eight strides of A cover the same ground as thirteen of B: in this case the *single* strides of A and B are in the ratio 13 : 8. The repetition of strides, to make them cover a considerable distance, acts as a magnifying glass and helps in the measurement of the *single* strides a and b, one against the other. Here we have the point of view adopted by Eudoxus. Let us, says he in effect, multiply our magnitudes a and b, whose ratio is required, and see what happens.

Let us, he continues, be able to recognize if a and b are equal, and if not, which is greater. Then if a is the greater, let us secondly be able to find multiples $2b$, $3b$, . . . , nb, of the smaller magnitude b; and thirdly, let us always be able to find a multiple nb of b which exceeds a. (The tall man may have seven-league boots and the short man may be Tom Thumb. Sooner or later the dwarf will be able to overtake one stride of his friend!) Few will gainsay the propriety of these mild assumptions: yet their mathematical implications have proved to be very subtle. This third supposition of Eudoxus has been variously credited, but to-day it is known as the *Axiom of Archimedes*.

A definition of equal ratios can now be stated. Let *a, b, c, d* be four given magnitudes, then the ratio *a* : *b* is equal to that of *c* : *d*, if, *whatever* equimultiples *ma, mc* are chosen and whatever equimultiples *nb, nd* are chosen,

$$\text{either } ma > nb, \quad mc > nd, \quad \text{(i)}$$
$$\text{or } ma = nb, \quad mc = nd, \quad \text{(ii)}$$
$$\text{or } ma < nb, \quad mc < nd. \quad \text{(iii)}$$

On this strange threefold statement the whole theory of proportion for geometry and algebra was reared. It is impossible to develop the matter here in any convincing way, but the simplicity of the ingredients in this definition is remarkable enough to merit attention. It has the characteristic threefold pattern already noticed by Pythagoras. As far as ordinary commensurable ratios go, the statement (ii) would suffice; *m* and *n* are whole numbers and the ratios *a* : *b*, *c* : *d* are each equal to the ratio *n* : *m*. But the essence of the new theory lies in (i) and (iii), because (ii) *never* holds for incommensurables—the geometrical equivalent of irrationals in arithmetic. But it is extraordinary that out of these inequalities *equal* ratios emerge.

Lastly it was a stroke of genius when Eudoxus put on record the above Axiom of Archimedes. To continue our illustration, marking time is not striding, and Eudoxus excluded marking time. However small the stride *b* might be, it had a genuine length. Eudoxus simply ruled out the case of a ratio *a* : *b* when either *a* or *b* was zero. Thereby he avoided a trap that Zeno had already set, and into which many a later victim was to fall. So the axiom was a notice-board to warn the unwary. It also had another use; it automatically required *a* and *b* to be magnitudes of the same kind. For if *a* denoted length and *b* weight, no number of ounces could be said to exceed the length of a foot.

The logical triumphs of this great period in Grecian mathematics overshadow important but less spectacular advances which were made in numerical notation and in music. From the earliest times the significance of the numbers five and ten for counting had been recognised in Babylonia, China and Egypt: and in Homer πεμπάζειν 'to five' means to count. Eventually the Greeks systematised their written notation by using the letters of the alphabet to denote definite numbers ($\alpha = 1$, $\beta = 2$, $\gamma = 3$ and so on). A Halicarnassus inscription (circa 450 B.C.) provides perhaps the earliest attested use of this alphabetical numeration.

In music Archytas gave the numerical ratios for the intervals of the

tetrachord on three scales, the enharmonic, the chromatic and the dia-
tonic. He held that sound was due to impact, and that higher tones
correspond to quicker, and lower tones to slower, motion communicated
to the air.

CHAPTER III

ALEXANDRIA: EUCLID, ARCHIMEDES AND APOLLONIUS

TOWARDS the end of the fourth century B.C., the scene of mathematical
learning shifted from Europe to Africa. By an extraordinary sequence of
brilliant victories the young soldier-prince, Alexander of Macedonia, conquered the Grecian world, and conceived the idea of forming a great
empire. But he died at the age of thirty-three (323 B.C.), only two years
after founding the city of Alexandria. He had planned this stronghold
near the mouth of the Nile on a magnificent scale, and the sequel largely
fulfilled his hopes. Geographically it was a convenient meeting-place for
Greek and Jew and Arab. There, what was finest in Greek philosophy
was treasured in great libraries: the mathematics of the ancients was
perfected: the intellectual genius of the Greek came into living touch with
the moral and spiritual genius of the Jew: the Septuagint translation of
Old Testament Scriptures was produced: and in due time it was there
that the great philosophers of the early Christian Church taught and
prospered. In spite of ups and downs the city endured for about six
hundred years, but suffered grievous losses in the wilder times that followed. The end came in A.D. 642, when a great flood of Arab invasion
surged westward, and Alexandria fell into the hands of the Calif Omar.

A great library, reputed to hold 700,000 volumes, was lost or destroyed
in this series of disasters. But happily a remnant of its untold wealth
filtered through to later days when the Arabs, who followed the original
warriors, came to appreciate the spoils upon which they had fallen.
Ptolemy, the successor of Alexander in his African dominions, had
founded this library about 300 B.C. He had in effect established a University; and among the earliest of the teachers was EUCLID. We know
little of his life and character, but he most probably passed his years of
tuition at Athens before accepting the invitation of Ptolemy to settle in
Alexandria. For twenty or thirty years he taught, writing his well-known
Elements and many other works of importance. This teaching bore notable fruit in the achievements of Archimedes and Apollonius, two of the
greatest members of the University.

The picture has been handed down of a genial man of learning, modest
and scrupulously fair, always ready to acknowledge the original work of

others, and conspicuously kind and patient. Some one who had begun to read geometry with Euclid, on learning the first theorem asked, 'What shall I get by learning these things?' Euclid called his slave and said, 'Give him threepence, since he must make gain out of what he learns'. Apparently Euclid made much the same impression as he does to-day. The schoolboy, for whom the base angles of an isosceles triangle 'are forced to be equal, without any nasty proof', is but re-echoing the ancient critic who remarked that two sides of a triangle were greater than the third, as was evident to an ass. And no doubt they told Euclid so.

In the *Elements* Euclid set about writing an exhaustive account of mathematics—a colossal task even in his day. The Work consisted of thirteen books, and the subjects of several books are extremely well known. Books I, II, IV, VI on lines, areas and simple regular plane figures are mostly Pythagorean, while Book III on circles expounds Hippocrates. The lesser known Book V elaborates the work of Eudoxus on proportion, which was needed to justify the properties of similar figures discussed in Book VI. Books VII, VIII and IX are arithmetical, giving an interesting account of the theory of numbers; and again much here is probably Pythagorean. Prime and composite numbers are introduced—a relatively late distinction; so are the earlier G.C.M. and L.C.M. of numbers, the theory of geometrical progressions, and in effect the theorem $a^{m+n} = a^m a^n$, together with a method of summing the progression by a beautiful use of equal ratios. Incidentally Euclid utilized this method to give his *perfect* numbers, such as 6, 28, 496, each of which is equal to the sum of its factors. The collection of perfect numbers still interests the curious; they are far harder to find than the rarest postage stamps. The ninth specimen alone has thirty-seven digits, while a still larger one is $2^{126}(2^{127} - 1)$.

Book X of Euclid places the writer in the forefront among analysts. It is largely concerned with the doctrine of irrational numbers, particularly of the type $\sqrt{(\sqrt{a} \pm \sqrt{b})}$, where a and b are positive integers. Here Euclid elaborates the arithmetical side of the work of Eudoxus, having already settled the geometrical aspect in Books V and VI, and here we duly find the method of exhaustions carefully handled. After Book XI on elementary solid geometry comes the great Book XII, which illustrates the same method of exhaustions by formally proving Hippocrates' theorem for πr^2, the area of a circle. Finally in Book XIII we have the climax to which all this stately procession has been leading. The Greeks were never in a hurry; and it is soothing, in these days of bustle, to contemplate the working of their minds. This very fine book gives and proves the constructions for the five regular solids of Pythagoras, extolled by Plato; and it ends with the dodecahedron, the symbol of the Universe itself.

By this great work Euclid has won the admiration and helped to form the minds of all his successors. To be sure a few logical blemishes occur in his pages, the gleanings of centuries of incessant criticism; but the wonder is that so much has survived unchanged. In point of form he left nothing to be desired, for he first laid down his careful definitions, then his common assumptions or axioms, and then his postulates, before proceeding with the orderly arrangement of their consequences. There were, however, certain gaps and tautologies among these preliminaries of his work: they occur in the geometrical, not in the Eudoxian parts of his books; and it has been one of the objects of latter-day criticism to supply what Euclid left unsaid.

But on one point Euclid was triumphant; in his dealing with parallel lines. For he made no attempt to hide, by a plausible axiom, his inability to prove a certain property of coplanar lines. Most of his other assumptions, or necessary bases of his arguments, were such as would reasonably command general assent. But in the case of parallel lines he started with the following elaborate supposition, called the *Parallel Postulate*:

If a straight line meet two straight lines, so as to make the two interior angles on the same side of it taken together less than two right angles, these straight lines, being continually produced, shall at length meet on that side on which are the angles which are less than two right angles.

By leaving this unproved, and by actually proving its converse, Euclid laid himself open to ridicule and attack. Surely, said the critics, this is no proper assumption; it must be capable of proof. Hundreds of attempts were vainly made to remove this postulate by proving its equivalent; but each so-called proof carried a lurking fallacy. The vindication of Euclid came with the discovery in the nineteenth century of non-Euclidean geometry, when fundamental reasons were found for some such postulate. There is dignity in the way that Euclid left this curious rugged excrescence, like a natural outcrop of rock in the plot of ground that otherwise had been so beautifully smoothed.

Many of his writings have come down to us, dealing with astronomy, music and optics, besides numerous other ways of treating geometry in his *Data* and *Division of Figures*. But his *Book of Fallacies* with its intriguing title, and the *Porisms* are lost; and we only learn of them indirectly through Pappus, another great commentator. It is one of the historical puzzles of mathematics to discover what porisms were, and many geometers, notably Simson in Scotland and Chasles in France, have tried to do so. Very likely they were properties relating to the organic description of figures—a type of geometry that appealed to Newton, Maclaurin, and to workers in projective geometry of our own days. Geometry at Alexandria was in fact a wide subject, and it has even been thought by some that the *porisms* consisted of an analytical method, foreshadowing the co-ordinate geometry of Descartes.

Euclid was followed by ARCHIMEDES of Samos, and APOLLONIUS of Perga. After the incomparable discoveries of Eudoxus, so well consolidated by Euclid, it was now the time for great constructive work to be launched; and here were the giants to do it. Archimedes, one of the greatest of all mathematicians, was the practical man of common sense, the Newton of his day, who brought imaginative skill and insight to bear upon metrical geometry and mechanics, and even invented the integral calculus. Apollonius, one of the greatest of geometers, endowed with an eye to see form and design, followed the lead of Menaechmus, and perfected the geometry of conic sections. They sowed in rich handfuls the seeds of pure mathematics, and in due time the harvest was ingathered by Kepler and Newton.

Little is known of the outward facts in the life of ARCHIMEDES. His father was Phidias the astronomer, and he was possibly related to Hiero II, King of Syracuse, who certainly was his friend. As a youth Archimedes spent some time in Egypt, presumably at Alexandria with the immediate successors of Euclid—perhaps studying with Euclid himself. Then on returning home he settled in Syracuse, where he made his great reputation. In 212 B.C., at the age of seventy-five, he lost his life in the tumult that followed the capture of Syracuse by the Romans. Rome and Carthage were then at grips in the deadly Punic wars, and Sicily with its capital Syracuse lay as a 'No man's land' between them. During the siege of Syracuse by the Romans, Archimedes directed his skill towards the discomfiture of the enemy, so that they learnt to fear the machines and contrivances of this intrepid old Greek. The story is vividly told by Plutarch, how at last Marcellus, the Roman leader, cried out to his men, 'Shall we not make an end of fighting against this geometrical Briareus who uses our ships like cups to ladle water from the sea, drives off our sambuca ignominiously with cudgel-blows, and by the multitude of missiles that he hurls at us all at once, outdoes the hundred-handed giants of mythology!' But all to no purpose, for if the soldiers did but see a piece of rope or wood projecting above the wall, they would cry, 'There it is,' declaring that Archimedes was setting some engine in motion against them, and would turn their backs and run away. Of course the geometrical Briareus attached no importance to these toys; they were but the diversions of geometry at play. Ignoble and sordid, unworthy of written record, was the business of mechanics and every sort of art which was directed to use and profit. Such was the outlook of Archimedes.

He held these views to the end; for even after the fall of the city he was still pondering over mathematics. He had drawn a diagram in the sand on the ground and stood lost in thought, when a soldier struck him down. As Whitehead has remarked:

'The death of Archimedes at the hands of a Roman soldier is symbolical of a world change of the first magnitude. The Romans were a great race, but they were cursed by the sterility which waits upon practicality. They were not dreamers enough to arrive at new points of view, which could give more fundamental control over the forces of nature. No Roman lost his life because he was absorbed in the contemplation of a mathematical diagram.'

Many, but not all, of the wonderful writings of Archimedes still survive. They cover a remarkable mathematical range, and bear the incisive marks of genius. It has already been said that he invented the integral calculus. By this is meant that he gave strict proofs for finding the areas, volumes and centres of gravity of curves and surfaces, circles, spheres, conics, and spirals. By his method of finding a tangent to a spiral he even embarked on what is nowadays called differential geometry. In this work he had to invoke algebraic and trigonometric formulae; here, for example, are typical results:

$$1^2 + 2^2 + 3^2 + \ldots + n^2 = \tfrac{1}{6}n(n+1)(2n+1),$$
$$\sin \frac{\pi}{2n} + \sin \frac{2\pi}{2n} + \ldots + \sin (2n-1) \frac{\pi}{2n} = \cot \frac{\pi}{4n}.$$

This last is the concise present-day statement of a geometrical theorem, arising in his investigation of the value of π, which he gave approximately in various ways, such as

$$3\tfrac{1}{7} > \pi > 3\tfrac{10}{71}.$$

Elsewhere he casually states approximations to $\sqrt{3}$ in the form

$$\tfrac{265}{153} < \sqrt{3} < \tfrac{1351}{780},$$

which is an example of the ladder-arithmetic of the Pythagoreans (p. 97). As these two fractions are respectively equal to

$$\tfrac{1}{3}\left(5 + \frac{1}{5 + \tfrac{1}{10}}\right) \text{ and } \tfrac{1}{3}\left(5 + \cfrac{1}{5 + \cfrac{1}{10 + \tfrac{1}{5}}}\right),$$

it is natural to suppose that Archimedes was familiar with continued fractions, or else with some virtually equivalent device; especially as $\sqrt{3}$ itself is given by further continuation of this last fraction, with denominators 10, 5, 10, 5 in endless succession. The same type of arithmetic occurs elsewhere in his writings, as well as in those of his contemporary, Aristarchus of Samos, a great astronomer who surmised that the earth travels round the sun.

Allusion has already been made to the recent discovery of the *Method* of Archimedes, a book that throws light on the mathematical powers of Democritus. It also reveals Archimedes in a confidential mood, for in it he lifts the veil and tells us how some of his results were reached. He *weighed* his parabola to ascertain the area of a segment, and this experiment suggested the theorem that the parabolic area is two-thirds of the area of a circumscribing parallelogram (Figure 9). He admits the value of such experimental methods for arriving at mathematical truths, which afterwards of course must be rigidly proved.

Indeterminate equations, with more unknowns than given equations, have attracted great interest from the earliest days. For example, there may be *one* equation for *two* unknowns:

$$3x - 2y = 5.$$

Many whole numbers x and y satisfy this equation, but it is often interest-

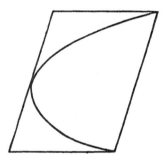

FIGURE 9

ing to discover the simplest numbers to do so. Such problems are closely connected with continued fractions, and perhaps Archimedes was beginning to recognize this. At all events we are told that he set the *Cattle Problem* to his friends in Alexandria.

The problem dealt with eight herds, four of bulls and four of cows, according to the colours, white, black, yellow and dappled. Certain facts were stated; for example, that the dappled bulls exceeded the yellow bulls in multitude by $(\frac{1}{6} + \frac{1}{7})$ of the number of white bulls: and the problem required, for its solution, the exact size of each herd. In other words there were eight unknown numbers to be found, but unfortunately, when turned into algebra, the data of the problem provided only seven equations. One of these equations, typical of all seven, can readily be formed from the facts already cited. If x denotes the number of dappled bulls, y that of the white bulls, and z that of the yellow bulls, it follows that

$$x = (\frac{1}{6} + \frac{1}{7})y + z.$$

From seven such equations for eight unknowns, of which only three x, y, z occur in this particular equation, all the unknowns have to be found.

There are, of course, an infinite number of solutions to seven equations for eight unknowns. The simplest solution of the above equation, taken apart from its context, is $x = 14$, $y = 42$, $z = 1$. As this does not fit in with the other six equations, a more complicated set of numbers for x, y, z must be found. Who would guess that the smallest value of x satisfying all seven such innocent looking equations is a number exceeding 3½ million? In our decimal notation this is a number seven figures long. But Archimedes improved on the problem by stating that 'when the white bulls joined in number with the black, they stood firm, with depth and breadth of equal measurement; and the plains of Thrinakia, far stretching all ways, were filled with their multitude'. Taking this to mean that the total number of black and white bulls was square, an enterprising investigator, fifty years ago, showed that the smallest such herd amounted to a number 200,000 figures long. The plains of Thrinakia would have to be replaced by the Milky Way.

The so-called *Axiom of Archimedes* bears his name probably because of its application on a grand scale, when he showed that the amount of sand in the world was finite. This appears in the *Sand-Reckoner*, a work full of quaint interest, and important for its influence on the arithmeticians of the last century. The opening sentences run as follows:

'There are some, King Gelon, who think that the number of the sand is infinite in multitude: and I mean by the sand not only that which exists about Syracuse and the rest of Sicily but also that which is found in every region whether inhabited or uninhabited. And again, there are some who, without regarding it as infinite, yet think that no number has been named which is great enough to exceed its multitude.'

So far from weeping to see such quantities of sand, Archimedes cheerfully fancies the whole Universe to be stuffed with sandgrains and then proceeds to count them. After a tilt at the astronomer Aristarchus for talking of the ratio of the centre of a sphere to the surface—'it being easy to see that this is impossible, the centre having no magnitude'—he gently puts Aristarchus right and then turns to the problem. First he settles the question, how many grains of sand placed side by side would measure the diameter of a poppy seed. Then, how many poppy seeds would measure a finger breadth. From poppy seed to finger breadth, from finger breadth to stadium, and so on to a span of 10,000 million stadia, he serenely carries out his arithmetical reductions. Mathematically he is developing something more elaborate than the theory of indices: his arithmetic might be called *the theory of indices of indices,* in which he classifies his gigantic numbers by orders and periods. The first order consists of all numbers from 1 to $100,000,000 = 10^8$, and the first period ends with the number $10^{800,000,000}$. This number can be expressed more compactly as $(10^8)^{10^8}$, but in the ordinary decimal notation consists of eight hundred million and one figures. Archimedes advances through further periods of this enormous size, never pausing in his task until the hundred-millionth period is reached.

In conclusion:

'I conceive that these things, King Gelon, will appear incredible to the great majority of people who have not studied mathematics, but that to those who are conversant therewith and have given thought to the question of the distances and sizes of the earth, the sun and moon, and the whole universe, the proof will carry conviction. And it was for this reason that I thought the subject would be not inappropriate for your consideration.'

One cannot pass from the story of Archimedes without reference to his work on statics and hydrostatics, in which he created a new application for mathematics. Like the rest of his writings this was masterly. Finally, in a book now lost, he discussed the semiregular solids, which generalize on the Pythagorean group of five regular solids. When each face of the solid is to be a regular polygon, exactly thirteen and no more forms are possible, as Kepler was one of the first to verify.

The third great mathematician of this period was APOLLONIUS of Perga in Pamphilia (? 262–200 B.C.), who earned the title 'the great geometer'. Little is known of him but that he came as a young man to Alexandria, stayed long, travelled elsewhere, and visited Pergamum, where he met Eudemus, one of the early historians of our science. Apollonius wrote extensively, and many of his books are extant. His prefaces are admirable, showing how perfect was the style of the great mathematicans when they were free from the trammels of technical terminology. He speaks with evident pleasure of some results: 'the most and prettiest of these theorems are new.'

What Euclid did for elementary geometry, Apollonius did for conic sections. He defined these curves as sections of a cone standing on a circular base; but the cone may be *oblique*. He noticed that not only were all sections parallel to the base, circular, but that there was also a secondary set of circular sections.

Although a circle is much easier to study than an ellipse, yet every property of a circle gives rise to a corresponding property of an ellipse. For example, if a circle and tangent are looked at obliquely, what the eye sees is an ellipse and its tangent. This matter of perspective leads on to projective geometry; and in this manner Apollonius simplified his problems. By pure geometry he arrived at the properties of conics which we nowadays express by equations such as

$$\frac{x^2}{a^2} \pm \frac{y^2}{b^2} = 1$$

and $ax^2 + bxy + cy^2 = 1$, and even $\sqrt{ax} + \sqrt{by} = 1$. In the second equation a, b, c denote given multiples of certain squares and a rectangle, the total area being constant. From our analytical geometry of conics he had clearly very little to learn except the notation, which improves on his own. He solved the difficult problem of finding the shortest and longest distances from a given point P to a conic. Such lines cut the curve at right angles and are called *normals*. He found that as many as four normals could be drawn from favourable positions of P, and less from others. This led him to consider a still more complicated curve called the *evolute*, which he fully investigated. He worked with what is virtually an equation of the sixth degree in x and y, or its geometrical equivalent— in its day a wonderful feat. His general problem, [*locus*] *ad tres et quattuor lineas*, will be considered when we turn to the work of Pappus.

Another achievement of Apollonius was his complete solution of a problem about a circle satisfying three conditions. When a circle passes through a given point, *or* touches a given line, *or* touches a given circle, it is said to satisfy one condition. So the problem of Apollonius really involved nine cases, ranging from the description of a circle through three given points to that of a circle touching three given circles. The simplest of these cases were probably quite well known: in fact, one of them occurs in the *Elements* of Euclid.

Apollonius was also a competent arithmetician and astronomer. It is reported that he wrote on *Unordered Irrationals*, and invented a 'quick delivery' method of approximating to the number π. Here, to judge from his title, it looks as if he had begun the theory of uniform convergence.

It may now be wondered what was left for their successors to discover after Archimedes and Apollonius had combed the field? So complete was their work that only a few trivial gaps needed to be filled, such as the addition of a focus to a parabola or a directrix to a conic, properties which Apollonius seems to have overlooked. The next great step could not be taken until algebra was abreast of geometry, and until men like Kepler, Cavalieri and Descartes were endowed with both types of technique.

CHAPTER IV

THE SECOND ALEXANDRIAN SCHOOL: PAPPUS AND DIOPHANTUS

WITH the death of Apollonius the golden age of Greek mathematics came to an end. From the time of Thales there had been almost a continuous chain of outstanding mathematicians. But until about the third century A.D., when Hero, Pappus and Diophantus once more brought fame to Alexandria, there seems to have been no mathematician of pre-eminence. During this interval of about half a millennium the pressure of Roman culture had discouraged Greek mathematics, although a certain interest in mechanics and astronomy was maintained; and the age produced the great astronomer HIPPARCHUS, and two noteworthy commentators, MENELAUS and PTOLEMY. Menelaus lived about the year A.D. 100, and Ptolemy was perhaps fifty years his junior. There is a strange monotony in trying to detail any facts whatsoever about these men—so little is known for certain, beyond their actual writings. The same uncertainty hangs over Hero, Pappus and Diophantus, whose names may be associated together as forming the Second Alexandrian School, because they each appear to have been active about the year A.D. 300. Yet Pappus and Diophantus are surrounded by mystery. Each seems to be a solitary echo of bygone days, in closer touch with Pythagoras and Archimedes than with their contemporaries, or even with each other.

MENELAUS is interesting, more particularly to geometers, because he made a considerable contribution to spherical trigonometry. Many new theorems occur in his writings—new in the sense that no earlier records are known to exist. But it is commonly supposed that most of the results originated with Hipparchus, Apollonius and Euclid. A well-known theorem, dealing with the points in which a straight line drawn across a triangle meets the sides, still bears his name. For some reason, hard to fathom, it is often classed to-day as 'modern geometry', a description which scarcely does justice to its hoary antiquity. The occasion of its appearance in the work of Menelaus is the more significant because he used it to prove a similar theorem for a triangle drawn on a sphere. Menelaus gave several theorems which hold equally well for triangles and other figures, whether they are drawn on a sphere, or on a flat plane. They include a very fundamental theorem known as the *cross ratio* property of a transversal drawn across a pencil of lines. This too is 'modern geometry'. He also gave the celebrated theorem that the angles of a spherical triangle are together greater than two right angles.

PTOLEMY (? 100–168 A.D.), who was a good geometer, will always be remembered for his work in astronomy. He treated this subject with a completeness comparable to that which Euclid achieved in geometry. His compilation is known as the Almagest—a name which is thought to be an Arabic abbreviation of the original Greek title.[2] His work made a strong appeal to the Arabs, who were attracted by the less abstract branches of mathematics; and through the Arabs it ultimately found a footing in mediaeval Europe. In this way a certain planetary theory called the Ptolemaic system became commonly accepted, holding sway for many centuries until it was superseded by the Copernican system. Ptolemy, following the lead

[2] Meaning 'The Great Compilation.'

of Hipparchus, chose one of several competing explanations of planetary motion, and interpreted the facts by an ingenious combination of circular orbits, or epicycles. Fundamental to his theory was the supposition that the Earth is fixed in space: and, if this is granted, his argument follows very adequately. But there were other explanations, such as that of Aristarchus, the friend of Archimedes, who supposed that the Earth travels round the Sun. When, therefore, Copernicus superseded the Ptolemaic theory by his own well-known system, centred on the Sun, he was restoring a far older theory to its rightful place.

HERO of Alexandria was a very practical genius with considerable mathematical powers. It is generally assumed that all the great mathematicians of the Hellenic world were Greek; but it is supposed that Hero was not. He was probably an Egyptian. At any rate there is in his work a strong bias towards the applications and away from the abstractions of mathematics, which is quite in keeping with the national characteristics of Egypt. Yet Hero proved to be a shrewd follower of Archimedes, bringing his mathematics to bear on engineering and surveying. He not only made discoveries in geometry and physics, but is also reputed to have invented a steam engine. One of his most interesting theorems proves that, when light from an object is broken by reflection on mirrors, the path of the ray between object and eye is a minimum. This is an instance of a *principle of least action*, which was formally adopted for optics and dynamics by Hamilton in the last century, and which has recently been incorporated in the work of Einstein. We may, therefore, regard Hero as the pioneer of Relativity (*c.* 250 A.D.).

At the beginning of the fourth century there was a revival of pure mathematics, when something of the Pythagorean enthusiasm for geometry and algebra existed once again in Alexandria under the influence of Pappus and Diophantus. PAPPUS wrote a great commentary called the Collection (συναγωγή); and happily many of his books are preserved. They form a valuable link with still more ancient sources, and particularly with the lost work of Euclid and Apollonius. As an expositor, Pappus rivals Euclid himself, both in completeness of design and wealth of outlook. To discover what Euclid and his followers were about, from reading the Collection, is like trying to follow a masterly game of chess by listening to the comments of an intelligent onlooker who is in full sight of the board.

Pappus was somewhat vain and occasionally unscrupulous, but he had enough sympathy to enter into the spirit of each great epoch. The space-filling figures of Pythagorean geometry made him brood over the marvels of bee-geometry; for God has endowed these sagacious little creatures with a power to construct their honey cells with the *smallest* enclosing surface. How far the bee knows this is not for the mathematician to say, but the fact is perfectly true. Triangular, or square, cells could be crowded together, each holding the same amount of honey as the hexagonal cell; but the hexagonal cell requires least wax. Like the mirrors of Hero, this again suggests *least action* in nature: and Pappus was disclosing another important line of inquiry. He put the question, What is the maximum volume enclosed by a given superficial area? This was perhaps the earliest suggestion of a branch of mathematics called the *calculus of variations*.

Most striking, and in true Archimedean style, is his famous theorem which determines the volume of a surface of revolution. His leading idea may be grasped by first noting that the volume of a straight tube is known if its cross-section A and its length *l* are given. For the volume is the product A.*l*. Pappus generalized this elementary result by considering such a tube to be no longer straight but circular. The cross-section A was taken to be the same at every place; but the length of the tube would need further definition. For example, the length of an inflated bicycle tyre is least if measured round the inner circle in contact with the rim, and is greatest round the outermost circle. This illustration suggests that an *average* or mean length *l* may exist for which the formula A.*l* still gives the volume. Pappus found that for such a circular tube this was so, and he located his average length as that of the circle passing through the centroid of each cross-section A. By centroid is meant that special point of a plane area often called the centre of gravity. As the shape of the section A is immaterial to his result, the theorem is one of the most general conclusions in ancient mathematics. In later years P. Guldin (1577–1643) without even the excuse of an unconscious re-discovery, calmly annexed this theorem, and it has become unjustly associated with his name.

As two further examples of important geometrical work by Pappus, the properties of the two following diagrams may be given. There are no hidden subtleties about the drawing of either figure. In the first (Figure 10),

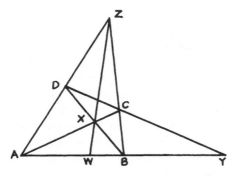

FIGURE 10

A, B, C, D are four points through which various straight lines have been drawn; and these intersect as shown at X, Y, Z. The line joining ZX is produced and cuts AB at W. The interest of this construction lies in the fact that, no matter what the shape of the quadrilateral ABCD may be, the lines AW, AB, AY are in harmonical progression. In the second figure (Figure 11), ABC and DEF are any two straight lines. These trios of points are joined crosswise by the three pairs of lines meeting at X, Y, Z. Then it follows that X, Y, Z are themselves in line. Here the interest lies in the *symmetry* of the result. It has nine lines meeting by threes in nine points: but it also has nine points lying by threes on the nine lines, as the reader may verify. This nice balance between points and lines of a figure is an early instance of reciprocation, or the *principle of duality*, in geometry.

In the parts of geometry which deal with such figures of points and lines, Pappus excelled. He gave a surprisingly full account of kindred properties connected with the quadrilateral, and particularly with a grouping of six points upon a line into three pairs. This so-called *involution* of six points would be effected by erasing the line ZXW in the first of the above figures, and re-drawing it so as to cross the other six lines at random in six distinct points.

In a significant passage of commentary on Apollonius, Pappus throws

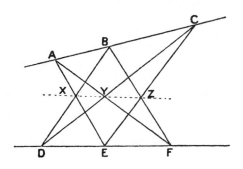

FIGURE 11

light upon what was evidently a very famous problem—the *locus ad tres et quattuor lineas*. It sums up so well the best Greek thought upon conics and it so very nearly inaugurates analytical geometry that it deserves special mention. Apollonius, says Pappus, considered the locus or trace of a roving point P in relation to three or four fixed lines. Suppose P were at a distance x from the first line, y from the second, z from the third, and t from the fourth. Suppose further that these distances were measured in specified directions, but not necessarily at right angles to their several lines. Then, as P moves, the values of x, y, z, t would vary; although it would always be possible to construct a rectangle of area xy, or a solid rectangular block of volume xyz. But as space is three-dimensional there is apparently nothing in geometry corresponding to the product $xyzt$ derived from four lines. On the other hand the ratio $x : y$ of two lines is a *number*, and there is nothing to prevent us from multiplying together as many such ratios as we like. So from the four lines x, y, z, t we can form two ratios $x : y$ and $z : t$, and then multiply them together. This gives $xz : yt$. Now if the resulting ratio is given as a constant, and equal to c, we can write

$$xz/yt = c, \text{ or } xz = cyt.$$

This is a way of stating the Apollonian problem about four lines: it indicates that the rectangle of the distances x, z from P to two of the lines is proportional to that of its distances y, t from the other two. When this happens, as Apollonius proved, P describes a conic. By a slight modification the same scrutiny may be applied to the problem, if three and not four lines are given. Pappus continues his commentary by generalizing the result with any number of lines: but it will be clearest if we confine ourselves to six lines.

If the distances of the point P from six given lines are x, y, z, u, v, w, then we can form them into three ratios $x : y$, $z : u$, $v : w$. If, further, it is given that the product of these three ratios is fixed, then we can write

$$\frac{x}{y} \cdot \frac{z}{u} \cdot \frac{v}{w} = c.$$

Pappus draws the correct conclusion that, when this happens, the point P is constrained to lie upon a certain locus or curve. But after a few more remarks he turns aside as if ashamed of having said something obvious. He had nevertheless again made one of the most general statements in all ancient geometry. He had begun the theory of Higher Plane Curves. For the number of ratios involved in such a constant product defines what is called the *order* of the locus. So a conic is a curve of order two, because it involves two ratios, as is shown in the Apollonian case above. In the simpler case, when only one ratio $x : y$ is utilized, the locus is a straight line. For this reason a straight line is sometimes called a curve of the first order. But Pappus had suggested curves of order higher than the second. These are now called cubics, quartics, quintics, and so on. To be sure, particular cases of cubic and other curves had already been discovered. The ancients had invented them for *ad hoc* purposes of trisecting an angle, and the like: but mathematicians had to wait for Descartes to clinch the matter.

The other great mathematician who brought fame to Alexandria was DIOPHANTUS. He is celebrated for his writings on algebra, and lived at the time of Pappus, or perhaps a little earlier. This we gather from a letter of Psellus, who records that Anatolius, Bishop of Laodicea about A.D. 280, dedicated to Diophantus a concise treatise on the Egyptian method of reckoning. Diophantus was devoted to algebra, as the wording of a Greek epigram indicates, which tells us the scanty record of his life. His boyhood lasted ⅙th of his life; his beard grew after ¹⁄₁₂th more; he married after ⅐th more, and his son was born five years later; the son lived to half his father's age, and the father died four years after his son.

If x was the age when he died, then,

$$\tfrac{1}{6}x + \tfrac{1}{12}x + \tfrac{1}{7}x + 5 + \tfrac{1}{2}x + 4 = x;$$

and Diophantus must have lived to be eighty-four years old.

The chief surviving writings of Diophantus are six of the thirteen books forming the *Arithmetica*, and fragments of his *Polygonal Numbers* and *Porisms*. Twelve hundred years after they were written these books began to attract the attention of scholars in Europe. As Regiomontanus observed in 1463: 'In these old books the very flower of the whole of arithmetic lies hid, the *ars rei et census* which to-day we call by the Arabic name of Algebra.' This work of Diophantus has a twofold importance: he made an essential improvement in mathematical notation, while at the same time he added large instalments to the scope of algebra as it then existed. The full significance of his services to mathematics only became evident with the rise of the early French school in the fifteenth and sixteenth centuries.

The study of notation is interesting, and covers a wider sphere than at first sight may be supposed. For it is the study of symbols; and as words are symbols of thought, it embraces literature itself. Now we may concentrate our attention on the literal symbol as it appears to the eye in a mathematical formula and in a printed sentence; or else on the thing signified, on the sense of the passage, and on the thought behind the symbol. A good notation is a valuable tool; it brings its own fitness and suggestiveness, it is easy to recognize and is comfortable to use. Given this tool and the material to work upon, advance may be expected. In their own language and in their geometrical notation the Greeks were well favoured: and a due succession of triumphs followed. But their arithmetic and algebra only advanced in spite of an unfortunate notation. For the Greeks were hampered by their use of letters α, β, γ for the numbers 1, 2, 3, and this concealed from them the flexibility of ordinary arithmetical calculations. On the other hand, the very excellence of our *decimal notation* has made these operations wellnigh trivial. Before the notation was widely known, even simple addition, without the help of a ball frame, was a task of some skill. The chief merits of this notation are the sign 0 for zero, and the use of one symbol, its meaning being decided by its context, to denote several distinct things, as, for example, the writing of 11 to denote *ten* and *one*. The history of this usage has been traced to a source in Southern India, dating shortly after the time of Diophantus. Thence it spread to the Moslem world and so to mediaeval Europe.

In the previous chapters many algebraic formulæ have occurred. They are, of course, not a literal transcription of the Greek, but are concise symbolic statements of Greek theorems originally given in verbal sentences or in geometrical form. For instance, a^2 has been used instead of 'the square on AB'. The earliest examples of this symbolic algebra occur in the work of Vieta, who lived in the sixteenth century, though it only came into general use about the year A.D. 1650. Until that time the notation of Diophantus had been universally adopted.

An old classification speaks of

Rhetorical Algebra,
Syncopated Algebra,
Symbolic Algebra,

and these names serve to indicate broad lines of development. By the rhetorical is meant algebra expressed in ordinary language. Then syncopations, or abbreviations, similar to our use of H.M.S. for His Majesty's Ship, and the like, became common among the ancients. To Diophantus more than to any other we owe this essential improvement. The third, symbolic algebra, became finally established, once Vieta had invented it, through the influence of Napier, Descartes and Wallis.

A typical expression of symbolic algebra is

$$(250x^2 + 2520) \div (x^4 + 900 - 60x^2):$$

and this serves to indicate the type of complication which Diophantus successfully faced. His syncopations enabled him to write down, and deal with, equations involving this or similar expressions. For $250x^2$ he wrote $\Delta^{\Upsilon}\sigma\nu$: here the letter ν meant 50 and σ, 200, according to the ordinary Greek practice. But the Δ^{Υ} was short for the Greek word meaning *power* (it is our word, dynamic); and power meant the square of the unknown number. Diophantus used the letter ς for the first power of the unknown, and the abbreviation of the word *cube* for the third power. He used no sign for *plus*, but a sort of inverted ψ for *minus*, the letter ι for *equals*, and a special phrase to denote the *division* of one expression by another. It is interesting that his idea of addition and subtraction was 'forthcoming' and 'wanting', and that the Greek word for wanting is related to the Pythagorean term *ellipse*.

Those who have solved quadratic equations will remember the little refrain—'the square of half the co-efficient of x'. It is a quotation from Diophantus, who dealt with such equations very thoroughly. He even ventured on the easier cases of cubic equations. Yet he speaks of 'the impossible solution of the absurd equation $4 = 4x + 20$': such an equation requires a negative solution; and it was not until much later that negative numbers as things in themselves were contemplated. But fractions and alternative roots of quadratic equations presented to him no difficulties.

We need not go far into the puzzles of 'problems leading to simple equations' to convince ourselves of the value of using several letters x, y, z for the unknown quantities. Each different symbol comes like a friendly hand to help in disentangling the skein. As Diophantus attempted such problems with the sole use of his symbol ς he was, so to speak, tying one hand behind his back and successfully working single-handed. This was clearly the chief drawback of his notation. Nevertheless he cleverly solved simultaneous equations such as

$$yz = m\ (y + z),\quad zx = n\ (z + x),\quad xy = p\ (x + y);$$

and it is evident from this instance that he saw the value of *symmetry* in algebra.

All this is valuable for its general influence upon mathematical manipulation: and had the genius of Diophantus taken him no farther, he would still be respected as a competent algebraist. But he attained far greater heights, and his abiding work lies in the Theory of Numbers and of Indeterminate Equations. Examples of these last occurred in the *Cattle Problem* of Archimedes (p. 105) and in the relation $2x^2 - y^2 = 1$ (of p. 97). His name is still attached to simple equations, such as enter the Cattle Problem, although he never appears to have interested himself in them. Instead he was concerned with the more difficult quadratic and higher types, as, for example, the equation

$$x^4 + y^4 + z^4 = u^2.$$

He discovered four whole numbers x, y, z, u for which this statement was true. Centuries later his pages were eagerly read by Fermat, who proved to be a belated but brilliant disciple. 'Why', says Fermat, 'did not Diophantus seek *two* fourth powers such that their sum is square? This problem is, in fact, impossible, as by my method I am able to prove with all rigour.' No doubt Diophantus had experimented far enough with the easier looking equation $x^4 + y^4 = u^2$ to prove that no solution was available.

This brings us to the close of the Hellenic period; and we are now in a position to appreciate the contribution which the Greeks made to mathematics. They virtually sketched the whole design that was to give incessant opportunities for the mathematicians and physicists of later centuries. In some parts of geometry and in the theory of the irrational the picture had been actually completed.

Within a glittering heap of numerical and geometrical puzzles and trifles —the accumulation in Egypt or the East of bygone ages—the Greeks had found order. Their genius had made mathematics and music out of the discord. And now in turn their own work was to appear as a wealth of scattered problems whose interrelations would be seen as parts of a still grander whole. New instruments were to be invented—the decimal notation, the logarithm, the analytical geometry of Descartes, and the magnifying glass. Each in its way has profoundly modified and enriched the mathematics handed down by the Greeks. Such profound changes have been wrought that we have been in some danger of losing a proper perspective in mathematics as a whole. So ingrained to-day is our habit of microscopic scrutiny that we are apt to think that all accuracy is effected by examining the infinitesimal under a glass or by reducing everything to decimals. It is well to remember that, even in the scientific world, this is but a partial method of arriving at exact results. Speaking numerically, multiplication, and not division, was the guiding process of the Greeks. The spacious definition of equal ratios which the astronomer Eudoxus bequeathed was not the work of a man with one eye glued to a micrometer.

CHAPTER V

THE RENAISSANCE: NAPIER AND KEPLER; THE RISE OF ANALYSIS

AFTER the death of Pappus, Greek mathematics and indeed European mathematics lay dormant for about a thousand years. The history of the science passed almost entirely to India and Arabia; and by far the most important event of this long period was the introduction of the Indian decimal notation into Europe. The credit for this innovation is due to Leonardo of Pisa, who was mentioned on p. 98, and certainly ranks as a remarkable mathematician in these barren centuries. From time to time there were others of merit and even of genius; but, judged by the lofty standard of past achievements and of what the future held in store, no one rose supreme. The broad fact remains: Pappus died in the middle of the fourth century, and the next great forward step for Western mathematics was taken in the sixteenth century.

It is still an obscure historical problem to determine whether Indian mathematics is independent of Greek influence. When Alexander conquered Eastern lands he certainly reached India, so that at any rate there was contact between East and West. This took place about 300 B.C., whereas the early mathematical work of India is chiefly attributed to the far later period A.D. 450–650. So in the present state of our knowledge it is safest to assume that considerable independent work was done in India. An unnamed genius invented the decimal notation; he was followed by ĀRYABHATA and BRAHMAGUPTA, who made substantial progress in algebra and trigonometry. Their work brings us to the seventh century, an era marked by the fall of Alexandria and the rise of the Moslem civilization.

The very word *algebra* is part of an Arabic phrase for 'the science of reduction and cancellation', and the digits we habitually use are often called the Arabic notation. These survivals remind us that mathematical knowledge was mediated to western Europe through the Arabs. But it will be clear from what has already been said that the Arabs were in no sense the originators of either algebra or the number notation. The Arabs rendered homage to mathematics; they valued the ancient learning whether it came from Greece or India. They proved apt scholars; and soon they were industriously translating into Arabic such valuable old manuscripts as their forerunners had not destroyed. In practical computation and the making of tables they showed their skill, but they lacked the originality and genius of Greece and India. Great tracts of Diophantine algebra and of geometry left them quite unmoved. For long centuries they were the safe custodians of mathematical science.

Then came the next chapter in the story, when northern Italy and the nations beyond the Alps began to feel their wakening strength. Heart and mind alike were stirred by the great intellectual and spiritual movements of the Renaissance and the Reformation. Once again mathematics was investigated with something of the ancient keenness, and its study was greatly stimulated by the invention of printing. There were centres of learning, in touch with the thriving city life of Venice and Bologna and other celebrated towns of mediaeval Europe. Italy led the way; France, Scotland, Germany and England were soon to follow. The first essential advance beyond Greek and Oriental mathematics was made by SCIPIO FERRO (1465–1526), who picked up the threads where Diophantus left them. Ferro discovered a solution to the cubic equation.

$$x^3 + mx = n;$$

and, as this solved a problem that had baffled the Greeks, it was a remarkable achievement.

Scipio was the son of a paper-maker in Bologna whose house can still be precisely located. He became Reader in mathematics at the University in 1496 and continued in office, except for a few years' interval at Venice, till his death in the year 1526. In those days mathematical discoveries were treasured as family secrets, only to be divulged to a few intimate disciples. So for thirty years this solution was carefully guarded, and it only finally came to light owing to a scientific dispute. Such wranglings were very fashionable: they were the jousts and tournaments of the intellectual world, and mathematical devices, often double-edged, were the weapons. Some protagonists preferred to spar with slighter blades—only drawing their mightiest swords as a last resort. Among them were Tartaglia and Cardan, both very celebrated, and ranking with Scipio as leading figures in this drama of the Cubic Equation. Scipio himself was dragged rather unwillingly into the fray: others relished it.

NICCOLO FONTANA (1500–1557) received the nickname TARTAGLIA because he stammered. When he was quite a little lad he had been almost killed by a wound on the head, which permanently affected his speech. This had occurred in the butchery that followed the capture of Brescia, his native town, by the French. His father, a postal messenger, was amongst the slain, but his mother escaped, and rescued the boy. Although they lived in great poverty, Tartaglia was determined to learn. Lacking the ordinary writing materials, he even used tombstones as slates, and eventually rose to a position of eminence for his undoubted mathematical ability. He emulated Ferro by solving a new type of cubic equation, $x^3 + mx^2 = n$; and when he heard of the original problem, he was led to re-discover Ferro's solution. This is an interesting example of what frequently happens,—the mere knowledge that a certain step had been taken being inducement enough for another to take the same step. Tartaglia was the first to apply mathematics to military problems in artillery.

GIROLAMO CARDAN (1501–1576) was a turbulent man of genius, very unscrupulous, very indiscreet, but of commanding mathematical ability. With strange versatility he was astrologer and philosopher, gambler and algebraist, physician yet father and defender of a murderer, heretic yet receiver of a pension from the Pope. He occupied the Chair of Mathematics at Milan and also practised medicine. In 1552 he visited Scotland at the invitation of John Hamilton, Archbishop of St. Andrews, whom he cured of asthma. He was interested one day to find that Tartaglia held a solution of the cubic equation. Cardan begged to be told the details, and eventually under a pledge of secrecy obtained what he wanted. Then he calmly proceeded to publish it as his own unaided work in the *Ars Magna*, which appeared in 1545. Such a blot on his pages is deplorable because of the admittedly original algebra to be found in the book. He seems to have been equally ungenerous in the treatment of his pupil Ferrari, who was the first to solve a quartic equation. Yet Cardan combined piracy with a measure of honest toil; and he had enough mathematical genius in him to profit by these spoils. He opened up the general theory of the cubic and quartic equations, by discussing how many roots an equation may have. He surmised the need not only for negative but for complex (or imaginary) numbers to effect complete solutions. He also found out the more important relations between the roots.

By these mathematical achievements, so variously conducted, Italy made a substantial advance. It was now possible to state, in an algebraic formula, the solution of the equation

$$ax^4 + bx^3 + cx^2 + dx + e = 0.$$

The matter had proceeded step by step from the simple to the quadratic, the cubic and the quartic equation. Naturally the question of the quintic and higher equations arose, but centuries passed before further light was thrown upon them. About a hundred years ago a young Scandinavian mathematician named Abel found out the truth about these equations. They proved to be insoluble by finite algebraic formulae such as these Italians had used. Cardan, it would seem, had unwittingly brought the algebraic theory of equations to a violent full stop!

Now what was going on at this time elsewhere in Europe? Something very significant in Germany, and a steady preparation for the new learning in France, Flanders and England. Contemporary with Scipio Ferro were three German pioneers, DÜRER, STIFEL and COPERNICUS. Dürer is renowned for his art; Stifel was a considerable writer on algebra; and Copernicus revolutionized astronomy by postulating that the Earth and all the planets revolve around the Sun as centre. About this time, in 1522, the first book on Arithmetic was published in England: it was a fine scholarly production by TONSTALL, who became Bishop of London. In the preface the author explains the reason for his belated interest in arithmetic. Having forgotten what he had learnt as a boy, he realized his disadvantage when certain gold- and silver-smiths tried to cheat him, and he wished to check their transactions.

Half a century later another branch of mathematics came into prominence, when STEVINUS left his mark in work on Statics and Hydrostatics. He was born at Bruges in 1548, and lived in the Low Countries. Then once more the scene shifts to Italy, where GALILEO of Pisa (1564–1642) invented dynamics, by rebuilding the scanty and ill-conceived system which had come down from the time of Aristotle. Galileo showed the importance of experimental evidence as an essential prelude to a theoretical account of moving objects. This was the beginning of physical science—which really lies outside our present scope—and by taking this step Galileo considerably enlarged the possible applications of mathematics. In such applications it was no longer possible for the mathematician to make his discoveries merely by sitting in his study or by taking a walk. He had to face stubborn facts, often very baffling to common sense, but always the outcome of systematic experiments. Two of the first to do this were Galileo and his contemporary, Kepler. Galileo found out the facts of dynamics for himself by dropping pebbles from a leaning tower at Pisa. Kepler took, for the basis of his astronomical speculations, the results of patient observations made by Tycho Brahe, of whom more anon.

The latter half of the sixteenth century also saw the rise of mathematics in France and Scotland. France produced VIETA, and Scotland NAPIER. The work of these two great men reminds us how deep was the influence of Ancient Greece upon the leaders of this mathematical Renaissance. Allusion has already been made to the share which Vieta took in improving the notation of algebra: he also attacked several outstanding problems that had baffled the Greeks, and he made excellent progress. He showed, for example, that the famous problem of trisecting an angle really depended on the solution of a cubic equation. Also he reduced the problem of squaring a circle to that of evaluating the elegant expression:

$$\frac{2}{\pi} = \sqrt{\tfrac{1}{2}} \times \sqrt{(\tfrac{1}{2} + \tfrac{1}{2}\sqrt{\tfrac{1}{2}})} \times \sqrt{(\tfrac{1}{2} + \tfrac{1}{2}\sqrt{(\tfrac{1}{2} + \tfrac{1}{2}\sqrt{\tfrac{1}{2}})})} \times \cdots$$

Here was a considerable novelty—the first actual formula for the time-honoured number π, which Archimedes had located to lie somewhere between $3\tfrac{1}{7}$ and $3^{10}\!/_{71}$. Vieta was also the first to make explicit use of that wonderful principle of duality, or reciprocation, which was hinted at by Pappus. We had an instance in the figure 11 of p. 112. For Vieta pointed out the importance of a polar triangle, obtained from a spherical triangle ABC. He drew three great circular arcs whose poles were respectively A, B, C; and then he formed a second triangle from these arcs. The study of the two triangles jointly turned out to be easier than that of the original triangle by itself.

Perhaps the most remarkable of all these eminent mathematicians was JOHN NAPIER, Baron of Merchiston, who discovered the logarithm. This achievement broke entirely new ground, and it had great consequences, both practical and theoretical. It gave not only a wonderful labour-saving device for arithmetical computation, but it also suggested several leading principles in higher analysis.

John Napier was born in 1550 and died in 1617: he belonged to a noble Scottish family notable for several famous soldiers. His mother was sister of Adam Bothwell, first reformed Bishop of Orkney, who assisted at the marriage of his notorious kinsman, the Earl of Bothwell, to Queen Mary, and who also anointed and crowned the infant King James VI. Scotland was a country where barbarous hospitality, hunting, the military art and keen religious controversy occupied the time and attention of Napier's contemporaries: a country of baronial leaders whose knowledge of arithmetic went little farther than counting on the fingers of their mail-clad hands. It was a strange place for the nurture of this fair spirit who seemed to belong to another world. The boy lost his mother when he was thirteen, and in the same year was sent to the University of St. Andrews, where he matriculated in 'the triumphant college of St. Salvator'. In those days St. Andrews was no home of quiet academic studies: accordingly the Bishop, who always took a kindly interest in the lad, advised a change. 'I pray you, schir,' he wrote to John's father, 'to send your son Jhone to the schuyllis; oyer to France or Flanderis; for he can leyr na guid at hame, nor get na proffeit in this maist perullous worlde.' So abroad he went; but it is probable that he soon returned to Merchiston, his home near Edinburgh, where he was to spend so many years of his serene life.

During the year at St. Andrews his interest was aroused in both arithmetic and theology. The preface to his *Plain Discovery of the Whole Revelation of St. John,* which was published in 1593, contains a reference to his 'tender yeares and barneage in Sanct Androis' where he first was led to devote his talents to the study of the Apocalypse. His book is full of profound but, it is to be feared, fruitless speculations; yet in form it follows the finest examples of Greek mathematical argument, of which he was master, while in sober manner of interpretation it was far ahead of its time. Unlike Cardan, before him, and Kepler, after him, he was innocent of magic and astrology.

Napier acquired a great reputation as an inventor; for with his intellectual gifts he combined a fertile nimbleness in making machines. His constant efforts to fashion easier modes of arithmetical calculation led him to produce a variety of devices. One was a sort of chess-arithmetic where digits moved like rooks and bishops on a board: another survives under the name of Napier's Bones. But what impressed his friends was a piece of artillery of such appalling efficiency that it was able to kill all cattle within the radius of a mile. Napier, horrified, refused to develop this terrifying invention, and it was forgotten.

During his sojourn abroad he eagerly studied the history of the Arabic notation, which he traced to its Indian source. He brooded over the mysteries of arithmetic and in particular over the principle which underlies the number notation. He was interested in reckoning not only, as is customary, in *tens,* but also in *twos.* If the number eleven is written 11, the notation indicates *one* ten and *one.* In the common scale of ten each number is denoted by so many *ones,* so many *tens,* so many *hundreds,* and so on. But Napier also saw the value of a binary scale—in which a number is broken up into parts 1, 2, 4, 8, etc. Thus he speaks with interest of the fact that any number of pounds can be weighed by loading the other scale pan with one or more from among the weights 1 lb., 2 lb., 4 lb., 8 lb., and so on.

When Napier returned to Scotland he wrote down his thoughts on arithmetic and algebra, and many of his writings remain. They are very systematic, showing a curious mixture of theory and practice: the main business is the theory, but now and then comes an illustration that 'would please the mechanicians more than the mathematicians'. Somewhere on his pages the following table appears:

I	II	III	IIII	V	VI	VII	. . .
1	2	4	8	16	32	64	128 . . .

Perhaps the reader thinks that it is simple and obvious; yet in the light of the sequel, it is highly significant. Men were still feeling for a notation of indices, and the full implications of the Arabic decimal notation had hardly yet been grasped. Napier was looking with the eyes of a Greek-trained mathematician upon this notation as upon a new plaything. He saw in the above parallel series of numbers the matching of an arithmetical with a geometrical progression. A happy inspiration made him think of these two progressions as *growing continuously from term to term*. The above table then became to him a sort of slow kinematograph record, implying that things are happening *between* the recorded terms. By the year 1590, or perhaps earlier, he discovered logarithms—the device which replaces multiplication by addition in arithmetic: and his treatment of the matter shows intimate knowledge of the correspondence between arithmetical and geometrical progressions. So clearly did he foresee the practical benefit of logarithms in astronomy and trigonometry, that he deliberately turned aside from his speculations in algebra, and quietly set himself the lifelong task of producing the requisite tables. Twenty-five years later they were published.

Long before the tables appeared, they created a stir abroad. There dwelt on an island of Denmark the famous Tycho Brahe, who reigned in great pomp over his sea-girt domain. It was called Uraniburg—the Castle of the Heavens—and had been given to him by a beneficent monarch, King Frederick II, for the sole purpose of studying astronomy. Here prolonged gazings and much accurate star chronicling proceeded; but the stars in their courses were getting too much for Tycho. Like a voice from another world word came of a portentous arithmetical discovery in Scotland, the *terra incognita*. The Danish astronomer looked for an early publication of the logarithmic tables; but it was long before they were completed. Napier, in fact, was slow but sure. 'Nothing', said he, 'is perfect at birth. I await the judgment and criticism of the learned on this, before unadvisedly publishing the others and exposing them to the detraction of the envious.' The first tables appeared in 1614, and immediately attracted the attention of mathematicians in England and on the Continent—notably BRIGGS and KEPLER. The friendship between Napier and Briggs rapidly grew, but was very soon to be cut short: for in 1617, worn out by his incessant toil, Napier died. One of his last writings records how 'owing to our bodily weakness we leave the actual computation of the new canon to others skilled in this kind of work, more particularly to that very learned scholar, my dear friend, Henry Briggs, public Professor of Geometry in London'.

A picturesque account of their first meeting has been handed down. The original publication had so delighted Briggs that

'he could have no quietness in himself, until he had seen that noble person whose only invention they were. . . . Mr. Briggs appoints a certain day when to meet in Edinburgh; but failing thereof, Merchiston was fearful he would not come. It happened one day as John Marr and the Lord

Napier were speaking of Mr. Briggs: "Ah, John," saith Merchiston, "Mr. Briggs will not now come": at the very instant one knocks at the gate; John Marr hasted down and it proved to be Mr. Briggs, to his great contentment. He brings Mr. Briggs up into My Lord's chamber, where almost one quarter of an hour was spent, each beholding other with admiration before one word was spoken: at last Mr. Briggs began. "My Lord, I have undertaken this long journey purposely to see your person, and to know by what engine of wit or ingenuity you came first to think of this most excellent help unto Astronomy, viz. the Logarithms: but My Lord, being by you found out, I wonder nobody else found it out before, when now being known it appears so easy." '

Exactly: and perhaps this was the highest praise. It is pleasant to record the excellent harmony existing between Napier, Briggs and Kepler. Kepler the same year discovered his third great planetary *canon* which he published in the *Ephemerides* of 1620, a work inscribed to Napier; and there for frontispiece was a telescope of Galileo, the elliptic orbit of a planet, the system of Copernicus, and a female figure with the Napierian logarithm of half the radius of a circle arranged as a glory round her head!

And what was a logarithm? Put into unofficial language it can be explained somewhat as follows. A point G may be conceived as describing a straight line TS with diminishing speed, slowing towards its destination S, in such wise that the speed is always proportional to the distance it has to go. When the point G is at the place *d* its speed is proportional to the distance *d*S. What a problem in dynamics to launch on the world,

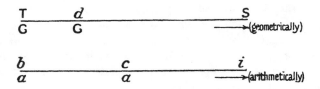

FIGURE 12

before dynamics were even invented! This motion Napier called *decreasing geometrically*. Alongside this, and upon a parallel line *bi,* a point *a* moves off uniformly from its starting position *b*. This Napier called *increasing arithmetically*. The race between the moving points G and *a* is supposed to begin at T and *b*, both starting off at the same speed; and then at any subsequent instant the places reached by G and *a* are recorded. When G has reached *d* let *a* have reached *c*. Then the number measuring the length *bc* is called by Napier the *logarithm* of the number measuring *d*S. In short, the distance *a* has gone is the logarithm of the distance G has to go.

Beginning with this as his definition Napier built up not only the theoretical properties of logarithms, but also his seven-figure tables. The definition is in effect the statement of a differential equation; and his superstructure provides the complete solution. It even suggests a theory of functions on a genuinely arithmetical basis. As this was done before either the theory of indices or the differential calculus had been invented, it was a wonderful performance.

Napier was also a geometer of some imagination. He devised new methods in spherical trigonometry. Particularly beautiful is his treatment of a right-angled spherical triangle as part of a fivefold figure, reminiscent of the Pythagorean symbol.

The story of Napier shows how the time was ripe for logarithms to be invented, and it is scarcely surprising that another should also have discovered them. This was his contemporary BÜRGI, a Swiss watchmaker, who reached his conclusions through the idea of indices, and published his results in 1620. Great credit must also be given to Briggs for the rapid progress he made in fashioning logarithmic tables of all kinds. None but an expert mathematician of considerable originality could have done the work so quickly.

The rapid spread of Napier's logarithms on the Continent was due to the enthusiasm of KEPLER, an astronomer, who was born in 1571 of humble parents near Stuttgart in Würtemberg, and died at Ratisbon in 1630. He was a man of affectionate disposition, abundant energy and methodical habits, with the intuition of true genius and the readiness to look for new relations between familiar things. He combined a love of general principles with the habit of attending to details. To his knowledge of ancient and mediaeval lore which included, in onè comprehensive grasp, the finest Greek-geometry and the extravagances of astrology, he added the new learning of Copernicus and Napier. He learnt of the former in his student days at Tübingen whence at the age of twenty-two he migrated to Gratz in Austria, where he was appointed Professor. There he imprudently married a wealthy widow—a step which brought him no happiness. Within three years of his appointment he became famous through the publication of his *Mysterium*, a work full of fancies and strange theories of the heavens.

Kepler's interest in the stars and planets developed as he corresponded with the great Tycho Brahe at Uraniburg, who held even kings spellbound by his discoveries. When in course of time Brahe lost royal favour and began to wander, he accepted a post at the new observatory near Prague. He even persuaded Kepler, who also was rather unsettled, to become his assistant. This arrangement was made in 1599 at the instigation of Rudolph II, a taciturn monarch much addicted to astrology, who hoped that these two astrological adepts would bring distinction to his kingdom. In this he was disappointed: for collaboration was not a success between these two strong personalities, with their widely different upbringing. Yet the experience was good for Kepler, especially as he also came under the influence of Galileo. It helped to stabilize his wayward genius. When Tycho died in 1601, Kepler succeeded him as astronomer; but his career was dogged by bad luck. He was often unpaid; his wife died;—nor did a second matrimonial venture prove more successful, although he acted with the greatest deliberation: for he carefully analysed and weighed the virtues and defects of several young ladies until he found his desire. It is a warning to all scientists that there *are* matters in life which elude weights and measures. The axiom of Archimedes has its limitations!

Kepler brimmed over with new ideas. Possessed with a feeling for number and music, and imbued through and through with the notions of Pythagoras, he sought for the underlying harmony in the cosmos. Temperamentally he was as ready to listen as to look for a clue to these secrets. Nor was there any current scientific reason to suppose that light would yield more significant results than sound. So he brought all his genius to bear on the problem of the starry universe: and he dreamt of a harmony in arithmetic, geometry and music that would solve its deepest mysteries. Eventually he was able to disclose his great laws of planetary motion, two in 1609, and the third and finest in the *Harmonices Mundi* of 1619.

These laws, which mark an epoch in the history of mathematical science, are as follows:

1. The orbit of each planet is an ellipse, with the sun at a focus.

2. The line joining the planet to the sun sweeps out equal areas in equal times.

3. The square of the period of the planet is proportional to the cube of its mean distance from the sun.

The period in the case of the earth is, of course, a year. So this third law states that a planet situated twice as far from the sun would take nearly three years to perform its orbit, since the cube of two is only a little less than the square of three. This first law itself made a profound change in the scientific outlook upon nature. From ancient times until the days of Copernicus and Tycho Brahe, circular motion had reigned supreme. But the circle was now replaced by the ellipse: and with the discovery that the ellipse was a path actually performed in the heavens and by the earth itself, a beautiful chapter in ancient geometry had unexpectedly become the centre of a practical natural philosophy. In reaching this spectacular result Kepler inevitably pointed out details in the abstract theory that Apollonius had somehow missed—such as the importance of the focus of a conic, and even the existence of a focus for a parabola. Then by a shrewd combination of his new ideas with the original conical properties, Kepler began to see ellipses, parabolas, hyperbolas, circles, and pairs of lines as so many phases of *one* type of curve. To Kepler, starlight, radiating from points unnumbered leagues away, suggested that in geometry parallel lines have a common point at infinity. Kepler therefore not only found out something to interest the astronomer; he made essential progress in geometry. An enthusiastic geometer once lamented that here was a genius spoilt for mathematics by his interest in astronomy!

The second law of Kepler is remarkable as an early example of the infinitesimal calculus. It belongs to the same order of mathematics as the definition that Napier gave for a logarithm. Again we must remember that this calculus, as a formal branch of mathematics, still lay hidden in the future. Yet Kepler made further important contributions by his accurate methods of calculating the size of areas within curved boundaries. His interest in these matters arose partly through reading the ancient work of Archimedes and partly through a wish to improve on the current method of measuring wine casks. Kepler recorded his results in a curious document, which incidentally contained an ingenious number notation based on the Roman system, where subtraction as well as addition is involved. Kepler used symbols analogous to I, V, X, L, but instead of the numbers one, five, ten and fifty he selected one, three, nine, twenty-seven, and so on. In this way he expressed any whole number very economically; for instance,

$$20 = 27 - 9 + 3 - 1.$$

As an algebraist he also touched upon the theory of recurring series and difference relations. He performed prodigies of calculation from the sheer love of handling numbers. The third of his planetary laws, which followed ten years after the other two, was no easy flight of genius: it represented prolonged hard work.

Something may be quoted of the contents in the *Harmonices Mundi* which enshrines this great planetary law. It is typical of the work of this extraordinary man. In it he makes a systematic search into the theory of musical intervals, and their relations to the distances between the planets and the sun: he discusses the significance of the five Platonic regular solids for interplanetary space: he elaborates the properties of the thirteen semi-regular solids of Archimedes: he philosophizes on the place of harmonic and other algebraic progressions in civil life, drawing his illustrations from the dress óf Cyrus as a small boy, and the equity of Roman marriage laws. Few indeed are the great discoverers in science who can rival Kepler in richness of imagery! For Kepler, every planet sang its tune: Venus a monotone, the Earth (in the sol-fa notation) the notes *m*, *f*, *m*, signifying that in this world man may expect but misery and hunger. This gave Kepler an opportunity for a Latin pun—'in hoc nostro domicilio *mi*seriam et *fa*men obtinere'. The italics are his, and in fact the whole book was written in solemn mediaeval Latin. The song of Mercury, in his arpeggio-like orbit, is

$$d \; r \; m \; f \; s \; l \; t \; d' \; r' \; m' \; d' \; s \; m \; d$$

—stated originally of course in the staff notation. As for the comets, surely they must be live things, darting about with will and purpose 'like fishes in the sea'! This frisky skirl of Mercury amid the sober hummings of the other planets, is no idle fancy: it duly records a curious fact, that the orbit of Mercury is more strongly elliptical, and less like a circle, than that of any other planet. It was this very peculiarity of Mercury which provided Einstein with one of his clues leading to the hypothesis of Relativity.

Carlyle, in his *Frederick the Great* (Book III, Chapter XIV) has preserved a delightful picture of John Kepler as he appeared to a contemporary, Sir Henry Wotton, Ambassador to the King of Bohemia.

" 'He hath a little black Tent . . .,' says the Ambassador, 'which he can suddenly set up where he will in a Field; and it is convertible (like a windmill) to all quarters at pleasure; capable of not much more than one man, as I conceive, and perhaps at no great ease; exactly close and dark, —save at one hole, about an inch and a half in the diameter to which he applies a long perspective Trunk, with the convex glass fitted to the said hole, and the concave taken out at the other end . . .' . . . An ingenious person, truly, if there ever was one among Adam's Posterity. Just turned fifty, and ill-off for cash. This glimpse of him, in his little black tent with perspective glasses, while the Thirty-Years War blazes out, is welcome as a date."

CHAPTER VI

DESCARTES AND PASCAL: THE EARLY FRENCH GEOMETERS
AND THEIR CONTEMPORARIES

HITHERTO the mathematicians of outstanding ability, whose names have survived, have been comparatively few; but from the beginning of the seventeenth century the number increased so rapidly that it is quite impossible in a short survey to do justice to all. In France alone there were as many mathematicians of genius as Europe had produced during the preceding millennium. Three names will therefore be singled out to be representatives of their time, Descartes and Pascal from among the French, and Newton from among the English. In this heroic age that followed the performances of Napier and Kepler, mathematics attained a remarkable prestige. The age was mathematical; the habits of mind were mathematical; and its methods were deemed necessary for an exact philosophy, or an exact anything else. It was the era when what is called modern philosophy began; and the pioneers among its philosophers, like the Greek philosophers of old, were expert mathematicians. They were Descartes and Leibniz.

DESCARTES was born of Breton parents near Tours in 1596 and died at Stockholm in 1650. In his youth he was delicate, and until the age of twenty his friends despaired of his life. After receiving the traditional scholastic education of mathematics, physics, logic, rhetoric and ancient languages, at which he was an apt pupil, he declared that he had derived no other benefit from his studies than the conviction of his utter ignorance and profound contempt for the systems of philosophy then in vogue.

'And this is why, as soon as my age permitted me to quit my preceptors,' he says, 'I entirely gave up the study of letters; and resolving to seek no other science than that which I could find in myself or else in the great book of the world, I employed the remainder of my youth in travel, in seeing courts and camps, in frequenting people of diverse humours and conditions, . . . and above all in endeavouring to draw profitable reflection from what I saw. For it seemed to me that I should meet with more truth in the reasonings which each man makes in his own affairs, and which if wrong would be speedily punished by failure, than in those reasonings which the philosopher makes in his study.'

In this frame of mind he led a roving, unsettled life; sometimes serving in the army, sometimes remaining in solitude. At the age of three and twenty, when residing in his winter quarters at Neuberg on the Danube, he conceived the idea of a reformation in philosophy. Thereupon he began his travels, and ten years later retired to Holland to arrange his thoughts into a considered whole. In 1638 he published his *Discourse on Method* and his *Meditations*. An immense sensation was produced by the *Discourse*, which contained important mathematical work. The name of Descartes became known throughout Europe; Princes sought him; and it was only the outbreak of the civil war in England which prevented him from accepting a liberal appointment from Charles I. Instead, he went to Sweden at the invitation of Queen Christina, arriving at Stockholm in 1649, where it was hoped that he would found an Academy of Sciences. Such a replica cf the Platonic School in Athens already existed in Paris. But his health gave way under the severity of the climate, and shortly after his arrival he died.

The work of Descartes changed the face of mathematics: it gave geometry a universality hitherto unattained; and it consolidated a position which made the differential calculus the inevitable discovery of Newton and Leibniz. For Descartes founded *analytical geometry*, and by so doing provided mathematicians with occupation lasting over two hundred years.

Descartes was led to his analytical geometry by systematically fitting algebraic symbols to the still fashionable rhetorical geometry. Examples of this procedure have already been given on p. 14 and elsewhere. Those examples were stated in algebraic formulae in order to convey the sense of the propositions more readily to the reader. Strictly speaking, they were an anachronism before the time of Descartes. His next step concerned the famous Apollonian problem (p. 112) [*locus*] *ad tres et quattuor lineas,* disclosed by Pappus. It will be recalled that a point moves so that the product of its oblique distances from certain given lines is proportional to that of its distances from certain others. Descartes took a step that from one point

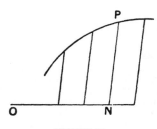

FIGURE 13

of view was simplicity itself—he enlisted the fact that plane geometry is *two*-dimensional. So he expressed everything in the figure in terms of two variable lengths, x and y, together with fixed quantities. This at once gave an algebraic statement for the results of Pappus: it put them into a form now typified by $f(x, y) = 0$, an equation where x and y alone are variable. The fundamental importance of this result lies in the further consequence that such an equation can be looked on as the definition of y in terms of x. It defined y as a function of x: it did geometrically very much what Napier's definition of a logarithm did dynamically. It also gave a new significance to the method of Archimedes for discussing the area of a curve, using an abscissa ON and an ordinate NP: in the notation of Descartes ON became x and NP, y. But, besides this, it linked the wealth of Apollonian geometry with what Archimedes had found; by forging this link Descartes rendered his most valuable service to mathematics.

Although Descartes deserves full credit for this, because he took considerable pains to indicate its significance, he was not alone in the discovery. Among others to reach the same conclusion was FERMAT—another of the great French mathematicians, a man of deeper mathematical imagination than Descartes. But Fermat had a way of hiding his discoveries.

Before indicating some of the principal consequences of this new method in geometry, there are other aspects of the notation which should be mentioned. The letter x has become world-famous: and it was the methodical Descartes who first set the fashion of denoting variables by *x*, *y*, *z* and constants by *a*, *b*, *c*. He also introduced indices to denote continued products of the same factor, a step which completed the improvements in notation originating with Diophantus. The fruitful suggestion of negative and fractional indices followed soon afterwards: it was due to WALLIS, one of our first great English mathematicians. A profound step in classification also was taken when Descartes distinguished between two classes of curves, *geometrical* and *mechanical*, or, as LEIBNIZ preferred to call them, *algebraic* and *transcendental*. By the latter is meant a curve, such as the spiral of Archimedes, whose Cartesian equation has no finite degree.

Apollonius had solved the problem of finding the shortest distance from a given point to a given ellipse, or other conic. Following this lead Des-

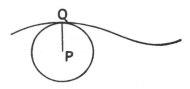

FIGURE 14

cartes addressed himself to the same problem in general: he devised a method of determining the shortest line PQ from a given point P to a given curve. Such a line meets the curve at right angles in the point Q, and is often called the *normal* at Q to the curve. Descartes took a circle with centre P, and arranged that the radius should be just large enough for the circle to reach the curve. The point where it reached the curve gave him Q, the required foot of the normal. His way of getting the proper radius was interesting; it depended on solving a certain equation, two of whose roots were equal. It is hardly appropriate to go into further details

FIGURE 15

here; but the reader who has some familiarity with analytical geometry, and has found the tangent to a circle or conic by the method of equal roots, has really employed the same general principle. Had Descartes been so inclined he could also have used his method for finding a tangent to a curve, i.e. a line PQ touching a given curve at a point Q (Figure 15). This is one of the first problems of the differential calculus; and one of the earliest solutions was found by Fermat and not by Descartes.

Fermat had discovered how to draw the tangent at certain points of a curve, namely at points Q which were, so to speak, at a crest or in the trough of a wave of the curve. They were points at a maximum or minimum distance from a certain standard base line called the axis of x. By so

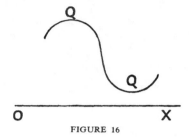

O   X

FIGURE 16

doing, Fermat had followed up a fertile hint, which Kepler had let fall, concerning the behaviour of a variable quantity near its maximum or minimum values.

An interesting curve, still called the Cartesian oval, was discovered by Descartes, and has led to far-reaching research in geometry and analysis. It was found in an endeavour to improve the shape of a lens, so as to condense a pencil of light to an accurate focus. Although a lens of this shape would successfully focus a wide-angled pencil of light, if it issued from a certain particular position, the lens would be otherwise useless. But it has a physical, besides a mathematical, interest: for the principle underlying its construction is identical with that which Hero of Alexandria first noticed in the case of plane mirrors. It is the principle of Least Action, which was ultimately exhibited in a general form by Hamilton.

All this mathematical work was but part of a comprehensive philosophical programme culminating in a theory of vortices, by which Descartes sought to account for the planetary motions. Just as Kepler had thought of comets as live fishes darting through a celestial sea, Descartes imagined the planets as objects swirling in vast eddies. It remained for Newton not only to point out that this theory was incompatible with Kepler's planetary laws, but to propose a truer solution.

In philosophy Descartes made a serious attempt to build up a system in the only way which would appeal to a mathematician——by first framing his axioms and postulates. In doing this he was the true symbol of an age, filled with self-confidence after the triumphs of Copernicus, Napier and Kepler. We cannot but admire the intellectual force of a man who undertook to revise philosophy and achieved so much. Nevertheless he lacked certain gifts that might be thought essential to success in the venture. He was cold, prudent and selfish, and offered a great contrast to his younger contemporary, the mathematician and philosopher, Blaise Pascal.

The analytical geometry of Descartes is a kind of machine: and 'the clatter of the co-ordinate mill', as Study has remarked, may be too insistent. The phenomenal success of this machine in the hands of Newton, Euler and Lagrange almost completely diverted thought from pure geometry. The great geometrical work in France, contemporary with that of Descartes, actually sank into oblivion for about two centuries, until it came into prominence once more, a hundred years ago. Two of the early French geometers were PASCAL and DESARGUES, and their work was the natural continuation of what Kepler had begun in projective geometry. Desargues, who was an engineer and architect residing at Lyons, gave to the ancient geometry of Apollonius its proper geometrical setting. He showed, for example, with grand economy, how to cut conics of different shapes from a single cone, and that a right circular cone. He won the admiration of Chasles, the great French geometer of the nineteenth century, who speaks of Desargues as an artist, but goes on to say that his work

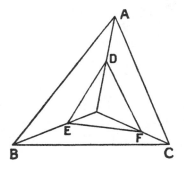

FIGURE 17

bears the stamp of universality uncommon in that of an artist. Desargues had the distinction of finding out one of the most important theorems of geometry, which takes its place, with a theorem of Pappus already quoted, as a fundamental element in the subject. It runs as follows: if two triangles ABC and DEF are such that AD, BE, CF meet in a point, then BC, EF; CA, FD; AB, DE, taken in pairs, meet in three points which are in line (Figure 17). The theorem is remarkable because it is easier to prove if the triangles are not in the same plane. As a rule, solid geometry is more difficult to handle than plane geometry—but not invariably. The perspective outline drawing of a cube on a sheet of paper is a more complicated figure than the actual outline of the solid cube. Desargues began the method of disentangling plane figures by raising them out of the flat into three dimensions. This is a choice method that has only lately borne its finest fruit in the *many-dimensional* geometry of Segre and the Italian school.

The work of Desargues is intimately linked with that of Pascal. Even in the grand century which produced Descartes, Fermat and Desargues, the fourth great French mathematician, BLAISE PASCAL, stands out for the brilliancy of his genius and for his astonishing gifts. He was born at Clermont-Ferrand in Auvergne on 19th June, 1623; and was educated with the greatest care by his father, who was a lawyer and president of the Court of Aids. As it was thought unwise to begin mathematics too early, the boy was put to the study of languages. But his mathematical curiosity was aroused, when he was twelve years old, on being told in reply to a question as to the nature of geometry, that it consisted in constructing exact figures and in studying the relations between the parts. Pascal was doubtless stimulated by the injunction against reading it, for he gave up his playtime to the new study, and before long had actually deduced several leading properties of the triangle. He found out for himself the fact that the angles of a triangle are together equal to two right angles. When his father knew of it, he was so overcome with wonder that he wept for joy, repented, and gave him a copy of Euclid. This, eagerly read and soon mastered, was followed by the conics of Apollonius, and within four years Pascal had written and published an original essay on conic sections, which astounded Descartes. Everything turned on a miracle of a theorem that Pascal called 'L'hexagramme mystique', commonly acknowledged to be the greatest theorem of mediaeval geometry. It states that, if a hexagon is inscribed in a conic, the three points of intersection of pairs of opposite sides always lie on a straight line: and from this proposition he is said to have deduced hundreds of corollaries, the whole being infused with the method of projection. The theorem has had a remarkably rich history, after the two hundred year eclipse, culminating in the enchantments of Segre when he presents it as a cubic locus in space of four dimensions, transfigured yet in its simplest and most inevitable form!

During these years Pascal was fortunate in enjoying the society in Paris of Roberval, Mersenne and other mathematicians of renown, whose regular weekly meetings finally grew into the French Academy. Such a stimulating atmosphere bore fruit after the family removed to Rouen, where at the age of eighteen Pascal amused himself by making his first calculating machine, and six years later he published his *Nouvelles Expériences sur le vide*, containing important experimental results which verified the work of Torricelli upon the barometer. Pascal was, in fact, as capable and original in the practical and experimental sciences as in pure geometry. At Rouen his father was greatly influenced by the Jansenists, a newly formed religious sect who denied certain tenets of Catholic doctrine, and in this atmosphere occurred his son's first conversion. A second conversion took place seven years later, arising from a narrow escape in a carriage accident. Henceforth Pascal led a life of self-denial and charity, rarely equalled and still more rarely surpassed. When one of his friends was condemned for heresy, Pascal undertook a vigorous defence in *A Letter written to a Provincial*, full of scathing irony against the Jesuits. Then the idea came to him to write an Apologia of the Christian Faith, but in 1658 his health, always feeble, gave way; and after some years of suffering borne with noble patience he died at the age of thirty-nine. The notes in which he jotted down his thoughts in preparation for this great project, have been treasured up and published in his *Pensées*, a literary classic.

In Pascal the simplest faith graced the holder of the highest intellectual gifts: and for him mathematics was something to be taken up or laid aside at the will of God. So when in the years of his retirement, as he lay awake suffering, certain mathematical thoughts came to him and the pain disappeared, he took this as a divine token to proceed. The problem which occurred to him concerned a curve called the cycloid, and in eight days he found out its chief properties by a brilliant geometrical argument. This curve may be described by the rotation of a wheel: if the axle is fixed, like that of a flywheel in a machine, a point on the rim describes a circle; but if the wheel rolls along a line, a point on the rim describes a cycloid. Galileo, Descartes and others were interested in the cycloid, but Pascal surpassed them all. To do so he made use of a new tool, the *method of indivisibles* recently invented by the Italian CAVALIERI. Though Pascal threw out a challenge, no one could compete with him: and his work may be regarded as the second chapter in the integral calculus, to which Archimedes had contributed the first.

An account of Pascal, the mathematician, would be incomplete without reference to his algebra, which, in the present-day sense of the word, he practically founded. It arose out of a game of chance that had formed a topic of discussion between Pascal and Fermat. From the debate the notion of mathematical *probability* emerged; this in turn Pascal looked upon as a problem in arrangements or combinations of given things and in counting those arrangements. With characteristic insight he lit upon the proper mechanism for handling the subject. It was the *Arithmetic Triangle*, a device already used by Napier for another purpose, and dating from still earlier times.

1	1	1	1	1	1	Certain numbers are written down in a tri-
1	2	3	4	5		angular table, as shown by the diagram.
1	3	6	10			The table can at any stage be enlarged by
1	4	10				affixing further numbers, one each at the
1	5					right-hand extremities of the rows, with a
1						single 1 added at the bottom of the first

column to start a new row. For example, underneath the 5 of the second row, and alongside the 10 of the third row a new number can be placed. This number is 15, the sum of the 5 and the 10. According to this rule of simple addition each new number is entered in the table. The diagram exhibits a 1 in the top left-hand corner followed by five parallel diagonals, the fifth and last being (1, 5, 10, 10, 5, 1). A sixth, which has not been filled in, would consist of 1, 6, 15, 20, 15, 6, 1, according to the addition rule. Instead of locating an entry, 10 for example, as standing in the fourth row and third column, it is more important to locate it by the *fifth diagonal* and *third* column. Pascal discovered that this gave the number of combinations of *five* things taken *two* at a time; and he found a formula for the general case, when the number stood in the mth diagonal and the $(n + 1)$th column. He stated this correctly to be $(n + 1)(n + 2)(n + 3)$ . . . $(m)/1.2.3. . . . (m - n)$. He also utilized the diagonals for working out the binomial expansion of $(a + b)^m$. For example,

$$(a + b)^5 = a^5 + 5a^4b + 10a^3b^2 + 10a^2b^3 + 5ab^4 + b^5.$$

Numbers and quantities are not always so important for their size or bulk as for their patterns and arrangements. What Pascal did was to bring this notion of pattern, common enough in geometry, to bear upon number itself—a highly significant step in the history of mathematics. By so doing he created higher algebra and prepared the way for Bernoulli, Euler and Cayley. 'Let no one say that I have said nothing new', writes Pascal in his *Pensées*; 'the arrangement of the subject is new. When we play tennis, we both play with the same ball, but one of us places it better.'

FERMAT, who shared with Pascal the beginning of this algebra, is most famous for his theory of numbers. In the margin of a copy of Diophantus he made a habit of scribbling notes of ideas which came into his mind as he read. These notes are unique in their interest and profundity: he seemed to grasp properties of whole numbers by intuition rather than reason. The most celebrated note, which is often called *Fermat's Last Theorem*, has baffled the wit of all his analytical successors: for no one has yet been able to say whether Fermat was right or wrong. The theorem asserts that it is impossible to find whole numbers x, y, z which satisfy the equation

$$x^n + y^n = z^n$$

when n is an integer greater than 2. He adds: 'I have found for this a truly wonderful proof, but the margin is too small to hold it.' The problem has led to a wealth of new methods and new ideas about number; valuable prizes have been offered for a solution; but to-day its quiet challenge still remains unanswered.

Great things were also going on in Italy and England during this early seventeenth century. CAVALIERI of Bologna will always rank as a remarkable geometer who went far in advancing the integral calculus by his *Method of Indivisibles*, following up Kepler's wine-cask geometry. One of his theorems is a gem: upon concentric circles equally spaced apart he drew a spiral of Archimedes whose starting-point was the centre. Then in order to discover its area he re-drew the figure with all the circles straightened out into parallel lines the same distances apart as before. As a result the spiral became a parabola: and 'Unless I am mistaken', he adds, 'this is a new and very beautiful way of describing a parabola.' This is an early example of a transcendental mathematical transformation that not only preserves the area of a sector of the original curve but also the length of its arc.

Another very fine piece of work was done in 1695 by PIETRO MENGOLI, who gave an entirely new setting to the celebrated logarithm, by showing that it was intimately linked with a harmonical progression. His definition and treatment was on true Eudoxian lines and rigorous enough to satisfy the strictest arithmetical disciple of Weierstrass.

It is natural that, in these years succeeding Napier's death, a great deal of attention was bestowed upon the logarithm. Besides the practical business of constructing tables there was the still more interesting theory of logarithms to consider. The stimulus of analytical geometry encouraged several mathematicians to treat the logarithm by the method of co-ordinates. This led to a beautiful result that connected the area between a hyperbola and its asymptote with the logarithm. It was found in 1647 by GREGOIRE DE SAINT VINCENT, of Flanders: but several others turned their attention to the matter, reaching the same general conclusions more or less independently; notably Mercator, Mersennes, Brouncker, Wallis, James Gregory, Newton and Leibniz. (This Mercator was not the maker of geographical maps: he was a mathematician who had lived in the previous century.)

It is not difficult to suggest how this result was attained. A start was made with the geometrical progression whose sum is $1/(1-x)$; namely,

$$\frac{1}{1-x} = 1 + x + x^2 + x^3 + x^4 + \ldots,$$

and a curve was determined whose co-ordinate equation is $y = 1/(1-x)$. This curve is a hyperbola. Next, its area was determined, by following much the same course that Archimedes had taken for the case of the parabola. There was no difficulty in finding a requisite formula, thanks to Napier's original definition of the logarithm. It led to the result

$$\log(1-x) = -x - \frac{x^2}{2} - \frac{x^3}{3} - \frac{x^4}{4} - \ldots,$$

which is called the logarithmic series. As may be seen, it is a union of the geometrical and harmonical progression.

Among the names which have just been given we find one Scot, one Irishman, and two Englishmen. For at last England produced mathematicians of the first rank, and in Gregory Scotland possessed a worthy successor to Napier. It is interesting to give, as typical specimens from the work of these our fellow-countrymen, the following formulae, which may be compared and contrasted with the logarithmic series:

$$\frac{4}{\pi} = \cfrac{1}{1 + \cfrac{1^2}{2 + \cfrac{3^2}{2 + \cfrac{5^2}{2 + \ldots}}}},$$

$$\frac{\pi}{4} = \frac{2 \times 4 \times 4 \times 6 \times 6 \times 8 \times \ldots}{3 \times 3 \times 5 \times 5 \times 7 \times 7 \times \ldots},$$

$$\frac{\pi}{4} = 1 - \tfrac{1}{3} + \tfrac{1}{5} - \tfrac{1}{7} + \ldots$$

The first is due to LORD BROUNCKER, an Irish peer; the second to WALLIS, who was educated in Cambridge and later became Savilian Professor of Mathematics in Oxford. The third was given by LEIBNIZ, but is really a special case of a formula discovered by JAMES GREGORY. Two of these formulae have been slightly altered from their original statements. The reader is not asked to prove, but merely to accept the results! After all, as they stand, they are readily grasped. The row of dots, with which each concludes, signifies that the formula can be carried farther; in fact, they each have something in common with the ladder-arithmetic of Athens (p. 97). They have this in common also with Vieta's formula for π (p. 121); but they improve on it, not only for their greater simplicity, but because each converges, as Plato would have it, by 'the great and small'— each step slightly overshooting the mark. This is not always done when such sequences are used, as in the more ordinary formula

$$\frac{\pi}{4} = \tfrac{1}{4} \text{ of } 3 \cdot 1415926 \ldots = \cdot 785398 \ldots,$$

which approximates from one side only, like the putts of a timid golfer who *never* gives the ball a chance, or like the race of Achilles and the tortoise. Such series need careful handling, as Zeno had broadly hinted;

and Gregory (by framing the notions of convergency and divergency) was the first to provide this.

In the last of these four formulae for $\dfrac{\pi}{4}$, the digits occur at random, and for this reason the statement is of little interest except to the practical mathematician. It is far otherwise with the other three: the *arrangement* of their parts has the inevitability of the highest works of art. It would be a pleasure to hear Pythagoras commenting upon them.

The Gregory family has long been associated with the county of Aberdeen. It had not been distinguished intellectually until John Gregory of Drumoak married Janet Anderson, herself a mathematician and a relative of the Professor of Mathematics in Paris. Many of their descendants have been eminent either as mathematicians or physicians. Chief among them all was their son James, who learnt mathematics from his mother. Unhappily, like Pascal, he died in his prime; but he lived long enough to exhibit his powers. After spending several years in Italy he occupied the Chair of Mathematics in St. Andrews for six years, followed by one year in Edinburgh. Shortly before his death he became blind.

Gregory was a great mathematical analyst, and many of his incidental results are striking. From the study of the logarithm he discovered the binomial theorem, generally and rightly attributed to Newton, who had probably found it out a few years earlier without publishing the result. It was but another case of independent discovery, as were also their invention of the reflecting telescope, and their attainments in the differential and integral calculus. The work of Gregory opened out a broad region of higher trigonometry, algebra and analysis. It is important not merely in detailed theorems but for its general aim, which was to prove that no finite algebraic formula could be found to express the functions that arise in trigonometry and logarithms. In other words, he held that circle-squarers were pursuing a vainer phantom than those who endeavour with rule and compass to trisect an angle. His project was lofty, even if it inevitably failed: it was a brilliant failure in an attempt to disentangle parts of pure mathematics which were only satisfactorily resolved during the nineteenth century.

Some of his greatest work remained in manuscript until the Gregory tercentenary (1938) gave an opportunity to publish it. This included an important general theorem which was later discovered by Brook Taylor (1715). Paper was scarce in 1670 when Gregory used the blank spaces of old letters to record his work. This was the year when BARROW produced his masterpiece, the *Lectiones Geometricae*, in which the foundations of the differential and the integral calculus were truly but geometrically laid.

If it is asked what is the peculiar national contribution made by our country to mathematics, the reply is: the mathematics of interpolation—the mathematical art of reading between the lines. As an illustration let us consider the arithmetical triangle of Pascal, supposing it to be a fragment of an Admiralty chart. The numbers indicate the depth in fathoms at various points on the surface of the sea. Such a chart with these particular readings obviously indicates a submarine valley trending downwards south-east. What the chart does not show is the actual depth at positions intermediate between the readings. Mathematical interpolation is concerned with discovering a formula for the most probable depth consistent with these measured soundings. Certain isolated points are given: what is happening between? Napier, Briggs, Wallis, Gregory and Newton, each in his way gave an answer.

> From gap to gap
> One hangs up a huge curtain so,
> Grandly, nor seeks to have it go
> Foldless and flat against the wall.

Indeed, some faith was needed to believe that there *was* a curtain, and some imagination to see its pattern. For Napier it was the pattern of the logarithm; Wallis wrought a continuous chain out of the isolated exponents x^1, x^2, x^3, . . ., by filling in fractional indices. Newton found out the pattern which fills in the triangle of Pascal; and from this he discovered the binomial theorem in its general form. Briggs suggested and Gregory found an interpolation formula of very wide application, while Newton supplemented it with several other alternatives which have usually been attributed to Stirling, Bessel and Gauss.

CHAPTER VII

ISAAC NEWTON

IN the country near Grantham during a great storm, which occurred about the time of Oliver Cromwell's death, a boy might have been seen amusing himself in a curious fashion. Turning his back to the wind he took a jump, which of course was a long jump. Then he turned his face to the wind and again took a jump, which was not nearly so long as his first. These distances he carefully measured, for this was his way of ascertaining the force of the wind. The boy was Isaac Newton, and he was one day to measure the force, if force it be, that carries a planet in its orbit.

From school at Grantham his friends took him to tend sheep and go regularly to the Grantham market. But as he *would* read mathematics instead of minding his business, it was at last agreed that he should go back to school, and from school to college. At school he lodged with Mr. Clark, apothecary, and in his lodgings spent much time, hammering and knocking. In the room were picture-frames and pictures of his own making, portraits, and drawings of birds and beasts and ships. Somewhere in the house might be seen a clock that was worked by water, and a mill which had a mouse as its miller. The boy made a carriage which could be propelled by the passenger, and a sundial that stood in the yard. To the little ladies of the house he was a very good friend, making tables and chairs for their dolls. His schoolfellows looked up to him as a skilful mechanic. As for his studies, when he first came to school he was somewhat lazy, but a fight that he had one day woke him up, and thereafter he made good progress. This quiet boy had great powers which were yet to be brought out.

In his twentieth year he went to Cambridge, where for more than thirty years he lived at Trinity College. He entered the college as a sizar, that is to say, being too poor to live in the style of other undergraduates he received help from the college. His tutor invited him to join a class reading Kepler's *Optics*. So Newton procured a copy of the book, and soon surprised the tutor by mastering it. Then followed a book on astrology; but this contained something which puzzled him. It was a diagram of the heavens. He found that, in order to understand the diagram, he must first understand geometry. So he bought Euclid's *Elements*, but was disappointed to find it too simple. He called it a 'trifling book' and threw it aside (an act of which he lived to repent). But turning to the work of Descartes he found his match, and by fighting patiently and steadily he won the battle.

After taking his degree Isaac Newton still went on learning all the mathematics and natural philosophy that Cambridge could teach him, and finding out new things for himself, until the Lucasian Professor of Mathematics in the University had become so convinced of the genius of this young man that, incredible as it may seem, he gave up to him his professorship. Isaac Barrow, the master of Newton's college, who thus resigned, was at no time a man to prefer self-interest before honour. He was possessed of great personal courage, and is reputed to have fought with a savage dog in an early morning's walk, and to have defended a ship from pirates. He was a mathematician of no mean powers; and as a divine he gained a lasting reputation.

Newton made three famous discoveries: one was in light, one was in mathematics, and one in astronomy. We are not to suppose that these flashed upon him all at once. They were prepared for by long pondering. 'I keep', said he, 'the subject of my inquiry constantly before me, and wait till the first dawning opens gradually, by little and little, into a full and clear light.' Early in his career he discovered that white light was composed of coloured lights, by breaking up a sunbeam and making the separate beams paint a rainbow ribbon of colours upon a screen. This discovery was occasioned by the imperfection of the lenses in telescopes as they were then made. Newton chose to cure the defect by inventing a reflecting telescope with a mirror to take the place of the principal lens, because he found that mirrors do not suffer from this awkwardness of lenses. It is one of his distinctions, shared with Archimedes and a few other intellectual giants, that his own handiwork was so excellent. In the chapel of his college there is a statue, holding a prism:

> —Newton, with his prism and silent face;
> The marble index of a mind for ever
> Voyaging through strange seas of thought, alone.

In mathematics his most famous discovery was the differential and integral calculus—which he called the method of *Fluxions*: and in astronomy it was the conception and elaboration of universal gravitation. It would be a mistake to suppose that he dealt with these subjects one by one: rather they were linked together, and reinforced each other. Already at the age of twenty-three, when for parts of the years 1665 and 1666 the college was shut down owing to the plague, Newton had thought out, in his quiet country home, the principles of gravitation and, for the better handling of the intense mathematical difficulties which the principles involved, he had worked at the fluxional calculus. In the space of three years after first reading geometry, he had so completely mastered the range of mathematics from Archimedes to Barrow, that he had fitted their wonderful infinitesimal geometry into a systematic discipline. Newton gave to analysis the same universality that Descartes had already given to geometry.

Newton may be said to have fused the points of view adopted by Napier and Descartes into a single whole. Napier thought of points M and N racing along parallel tracks OX and OY, N moving steadily and M at a variable speed. The co-ordinates of Descartes provide a chart of the race in the following way: the lines OX and OY can be placed, no longer parallel, but at right angles to each other, and a curve can be plotted, traced by a point P which is simultaneously abreast of the points N and M. In this way two figures can be drawn, one the Napierian and the other the Cartesian. The figures are symbols of two lines of thought—the kinematical and the geometrical. Newton may never actually have drawn such figures side by side, but he certainly had the two trains of thought. 'I fell by degrees on the method of fluxions,' he remarks: and by *fluxions* he simply meant what we call the simultaneous speeds of the points N and M. Then by seeking to compare the speed of M with that of N, he devised the method which the geometrical figure suggests. 'Fluxions' was his name

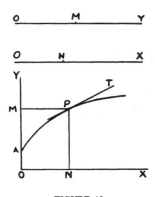

FIGURE 18

for what we call the differential and integral calculus, but he kept the discovery to himself.

In after years LEIBNIZ announced that *he* had found this new mathematical method. Then a quarrel arose between the followers of Newton and the followers of Leibniz, and unhappily it grew into a quarrel between the great men themselves. It is enough to say that the time was ripe for such a discovery: and both Newton and the German philosopher were sufficiently gifted to effect it. Newton was the first to do so, and only brought the trouble unwittingly on his head by refraining from publishing his results. It is also probable that Leibniz was influenced more by Pascal and Barrow than by Newton: and in turn we owe to Leibniz the record of parts of Pascal's work which would otherwise have been lost.

About this time the Royal Society of London was founded by King Charles II. It corresponded with the Academy of Paris, and provided a rendezvous for the leading mathematicians and natural philosophers in the country. Two of the Fellows of this Society were Gregory and Newton, who had become friends through their common interest in the reflecting telescope. Besides maintaining a correspondence, they may actually have met. They were certainly brought into contact with other leading mathematicians and astronomers. Among those who have not already been mentioned in the last chapter were Wren, Hooke, and Halley.

Christopher Wren is now so famous as the builder of St. Paul's Cathedral that we never hear of his scientific fame, though a man of science he was. Hooke, who was in appearance a puny little man, was a hard student, often working till long after midnight, but caring too excessively for his own reputation. When Newton found out anything, Hooke would commonly remark, 'That is just what I found out before.' But he was a great inventor, whose eager speculations stirred people up to think about the questions which Newton was to solve. Halley was an astronomer—a very active man, always travelling about the world to make some addition to his science. Every one has heard of Halley's comet, and to Halley is due the credit of bringing Newton before the world as the discoverer of gravitation.

One day these three friends were talking earnestly together: the subject of their conversation was the whirlpool theory of Descartes, which they felt to be hardly a satisfactory explanation of planetary motion. It did not seem to give a proper explanation of the focal position of the Sun within the elliptic orbit. Instead of imagining the planets to be propelled by a whirling current, they preferred to think that each planet was forcibly attracted by the Sun. 'Supposing', said they, 'the Sun pulls a planet with such and such a force, how ought the planet to go? We want to see clearly that the planet will go in an ellipse. If we can see that, we shall be pretty sure that the Sun *does* pull the planet in the way we supposed.' 'I can answer that,' said Hooke: upon which Wren offered him forty shillings on condition of his producing the answer within a certain time. However, nothing more was heard of Hooke's solution. So at last after several months Halley went to Cambridge, to consult Newton; and, without mentioning the discussion which had taken place in London, he put the question: If a planet were pulled by the Sun with a force which varies inversely as the square of the distance between them, in what sort of a curve ought the planet to go? Newton, to Halley's astonishment and delight answered, 'An ellipse.' 'How do you know that?' 'Why, I have calculated it.' 'Where's the calculation?' Oh, it was somewhere among his papers; he would look for it and send it to Halley. It appeared that Newton had worked all this out long before; and only now in this casual way was the matter made known to the world. Then Halley did a wise thing: he persuaded his retiring friend to develop the entire problem, explaining the whole complicated system of planetary motion. This Newton did; it was a tremendous task, taking two or three years; at the end of which appeared the famous book called *The Mathematical Principles of Natural Philosophy*, or more shortly the *Principia*, one of the supreme achievements of the human mind.

It is impossible to exaggerate the importance of the book, which at once attracted the keenest attention not only in England but throughout Europe. It was a masterpiece alike of mathematics and of natural philosophy. Perhaps the strangest part of the work was not so much the conception that the Sun pulls the planet, but that the planet pulls the Sun—and pulls equally hard! And that the whole Universe is full of falling bodies: and everything pulls everything else—literally everything, down to the minutest speck of dust. When Newton's friends had discussed the effect of the solar attraction upon a planet, they had correctly surmised the requisite force: it was determined by what is called the *law of the inverse square*. Newton had already adopted this law of force in his early conjectures, during the long vacation of 1666, over twenty years before the publication of the *Principia* (1687). That early occasion is also the date to which the well-known apple story may be referred. It is said that the sight of a falling apple set in motion the train of thought, leading Newton to his discovery of universal gravitation. But after working out the mathematical consequences of his theory and finding them to disagree with the observed facts he had tossed his pages aside. Only after many years he became aware of later and more careful calculations of the observations. This time, to his delight, they fitted his mathematical theory, and so Newton was ready with his answer, when Halley paid him the memorable visit.

In the *Principia* Newton demonstrated that, if his rule of gravitation is universally granted, it becomes the key to all celestial motions. Newton could not *prove* that it was the right key, for not all the celestial motions were known at the time, but very nearly all that have since been discovered help to prove that he was right. Even so, there was enough already known to give Newton plenty of trouble. The moon, for instance, that refuses to go round the Earth in an exact ellipse, but has all sorts of fanciful little excursions of her own—the moon was very trying to Isaac Newton.

Newton's great book was written in Latin, and, in order to make it intelligible to current habits of mind, it was couched in the style of Greek geometry. Newton had of course worked the mathematics out by fluxions, but he preferred to launch the main gravitational discovery alone, without further perplexing his readers by the use of a novel method. Outside his Cambridge lecture-room little was known of his other mathematical performances until a much later date. His *Arithmetica Universalis* was published in 1707, and two more important works, on algebra and geometry, appeared about the same time. Newton left his mark on every branch of mathematics which he touched; indeed, there are few parts of the subject which escaped his attention. Allusion has already been made to his work in interpolation and algebra. The power of his methods may be judged from one celebrated theorem which he gave without proof for determining the positions of the roots of an equation. A hundred and fifty years elapsed before Sylvester discovered how to prove his theorem.

The publication of the *Principia* forced Newton to abandon his sheltered life. In 1689 he became a Member of Parliament, and a few years later was appointed Master of the Mint. In 1705 he was knighted by Queen Anne. He died in 1727 at an advanced age, and was buried in Westminster Abbey. Voltaire has recorded his pride at having lived for a time 'in a land where a Professor of Mathematics, only because he was great in his vocation, was buried like a king who had done good to his subjects'. The world at large is often more generous in showing appreciation and gratitude than are mathematicians themselves, who feel, but are slow to exhibit their feelings. It is therefore the more noteworthy that, two hundred years later, in 1927, the English mathematical world made a pilgrimage to Grantham, to signal their respect for the genius of Newton. This alone is enough to indicate that the immense reputation which he always enjoyed was fully deserved.

It is proper to associate, with Newton, the great Dutch natural philosopher HUYGENS (1629–1693), who was in close touch with scientists of England, and did much to stimulate their wonderful advances. His own work in physics is so grand that his mathematics are apt to be overlooked. He contributed many elegant results in the infinitesimal calculus, particularly in its bearings upon mechanical phenomena, the oscillations of a pendulum, the shape of a hanging string, and the like. But he is best known for his undulatory theory of light.

As a mathematical concept this has proved to be a landmark in the history, and it is particularly interesting because it has thrown Newton's universal gravitation into intense relief. To Newton light seemed to be so many tiny particles streaming in luminous lines: to Huygens, on the contrary, light was propagated by waves. The sequel has shown that, of these rival theories, the latter is the more valuable. Not only has it provided a better key to optical puzzles, but it has also answered many purposes in the theory of electricity and magnetism. One by one, the natural phenomena were absorbed in this all-enveloping wave-theory, and gravitation alone remained untouched—a single physical exception. This unwavelike behaviour of gravitation, this action at a distance, sorely perplexed Newton himself, long before these further instances of natural behaviour had made wave motion the correct deportment. As the mystery of gravitation deepened, it became more and more the conscious aim of scientists to explain the contrast: and the matter has only lately been settled by Einstein, who solves the problem by drastically embedding gravitation in the very texture of space and time.

But it would be wrong to suppose that this left the field clear for the wave-theory. Quietly and unobtrusively other interruptions have been congregating, and reasons have once more been urged in favour of Newton's corpuscular theory of light. At the present time there is no clear-cut decision one way or the other: the work of both Newton and Huygens appears to be fulfilled in the Quantum Theory and the Wave Mechanics.

CHAPTER VIII

THE BERNOULLIS AND EULER

THE story of mathematics during the eighteenth century is centred upon Euler, and the scene of action is chiefly laid in Switzerland and Russia. About the time when Napier was experiencing the turmoil of the Reformation, violent persecution of Protestants took place in Antwerp. One of the many refugees, whom Belgium could ill afford to lose, was a certain Jacques Bernoulli, who fled to Frankfort. In 1622 his grandson settled at Basel, and there, on the frontiers of Switzerland, the BERNOULLI family were destined to bring fame to the country of their adoption. As evidence of the power of heredity, or of early home influence, their mathematical record is unparalleled. No less than nine members of the family attained eminence in mathematics or physics, four of whom received signal honours from the Paris Academy of Sciences. Of these nine the two greatest were the brothers JACOB and JOHN, great-grandsons of the fugitive from Antwerp. Jacob was fifth child in the large family, and John, thirteen years his junior, was tenth. Each in turn became Professor of Mathematics at Basel.

The elder brother settled to his distinguished career, as a mathematical analyst, only after considerable experiment and travel. At one time his father had forbidden him to study either mathematics or astronomy, hoping that he would devote himself to theology. But an inborn talent urged his son to spend his life in perfecting what Pascal and Newton had begun. Among his many discoveries, and perhaps the finest of them all, is the equiangular spiral. It is a curve to be found in the tracery of the spider's web, in the shells upon the shore and in the convolutions of the far-away nebulae. Mathematically it is related in geometry to the circle and in analysis to the logarithm. A circle threads its way over the radii by crossing them always at right angles; this spiral also crosses its radii at a constant angle—but the angle is not a right angle. Wonderful are the phœnix-like properties of the curve: let all the mathematical equivalents of burning it and tearing it in pieces be performed—it will but reappear unscathed! To Bernoulli in his old age the curve seemed to be no unworthy symbol of his life and faith; and in accordance with his wishes the spiral was engraved upon his tombstone, and with it the words *Eadem mutata resurgo.*

His younger brother JOHN (1667–1748) followed in his footsteps, continually adding fresh material to the store of analysis, which now included differential equations. His works exhibits a bolder use of negative and imaginary numbers, thereby realizing 'the great emolument' which Napier himself had hoped to bestow on mathematics by 'this ghost of a quantity,' had not his own attention been absorbed by logarithms. His sons Daniel and Nicolas Bernoulli were also very able mathematicians, and it was under their influence at college that Euler discovered his vocation.

LEONARD EULER (1707–1783) was the son of a clergyman who lived in the neighbourhood of Basel. His natural aptitude for mathematics was soon apparent from the eagerness and facility with which he mastered the elements under the tuition of his father. At an early age he was sent to the University of Basel, where he attracted the attention of John Bernoulli. Inspired by such a teacher he rapidly matured, and at the age of seventeen, when he received the degree of Master of Arts, he provoked high applause for a probationary discourse, the subject of which was a Comparison between the Cartesian and Newtonian Systems.

His father earnestly wished him to enter the ministry and directed his son to study theology. But unlike the father of Bernoulli, he abandoned his views when he saw that his son's talents lay in another direction. Leonard was allowed to resume his favourite pursuits and, at the age of nineteen, he transmitted two dissertations to the Paris Academy, one upon the masting of ships, and the other on the philosophy of sound. These essays mark the beginning of his splendid career.

About this time, in consequence of the keen disappointment at failing to attain a vacant professorship in Basel, he resolved to leave his native country. So in 1727, the year when Newton died, Euler set off for St. Petersburg to join his friends, the younger Bernoullis, who had preceded him thither a few years earlier. On the way to Russia, he learnt that Nicolas Bernoulli had fallen a victim to the stern northern climate; and the very day upon which he set foot on Russian soil the Empress Catherine I died—an event which at first threatened the dissolution of the Academy, of which she had laid the foundation. Euler, in dismay, was ready to give up all hope of an intellectual career and to join the Russian navy. But, happily for mathematics, when a change took place in the aspect of public affairs in 1730, Euler obtained the Chair of Natural Philosophy. In 1733 he succeeded his friend Daniel Bernoulli, who wished to retire; and the same year he married Mademoiselle Gsell, a Swiss lady, the daughter of a painter who had been brought to Russia by Peter the Great.

Two years later, Euler gave a signal example of his powers, when in three days he effected the solution of a problem urgently needed by members of the Academy, though deemed insoluble in less than several months' toil. But the strain of the work told upon him, and he lost the sight of an eye. In spite of this calamity he prospered in his studies and discoveries, each step seeming only to invigorate his future exertions. At about the age of thirty he was honoured by the Paris Academy when he received recognition, as also did Daniel Bernoulli and our own country-man Colin Maclaurin, for dissertations upon the flux and reflux of the sea. The work of Maclaurin contained a celebrated theorem upon the equilib-rium of elliptical spheroids; that of Euler brought the hope considerably nearer of solving outstanding problems on the motions of the heavenly bodies.

In the summer of 1741 King Frederick the Great invited Euler to reside in Berlin. This invitation was accepted, and until 1766 Euler lived in Germany. On first arriving he received a royal letter written from the camp at Reichenbach, and he was soon after presented to the queen-mother, who always took a great interest in conversing with illustrious men. Though she tried to put Euler at his ease, she never succeeded in drawing him into any conversation but that of monosyllables. One day when she asked the reason for this, Euler replied, 'Madam, it is because I have just come from a country where every person who speaks is hanged.' It was during his residence in Berlin that Euler wrote a remark-able set of letters, or lessons, on natural philosophy, for the Princess of Anhalt Dessau, who was eager for instruction from so great a teacher. These letters are a model of perspicuous and interesting teaching, and it is noteworthy that Euler should have found time for such detailed ele-mentary work, amid all his other literary interests.

For eleven years his widowed mother lived in Berlin also, receiving assiduous attention from her son, and enjoying the pleasure of seeing him universally esteemed and admired. Euler became intimate in Berlin with M. de Maupertuis, President of the Academy, a Frenchman from Brittany who strongly favoured Newtonian philosophy in preference to Cartesian. His influence was important, as it was exerted at a time when Continental opinion was still reluctant to accept the views of Newton. Maupertuis much impressed Euler with his favourite principle of least action, which Euler used with great effect in his mechanical problems.

It speaks highly for the esteem in which Euler was held that, when in 1760 a Russian army invaded Germany and pillaged a farm belonging to Euler, and the act became known to the general, the loss was immediately made good, and a gift of four thousand florins was added by the Empress Elizabeth when she learnt of the circumstance. In 1766 Euler returned to Petersburg, to spend the remainder of his days, but shortly after his arrival he lost the sight of his other eye. For some time he had been forced to use a slate, upon which in large characters he would make his calculations. Now, however, his pupils and children copied his work, writing the memoirs exactly as Euler dictated them. Magnificent work it was too, astonishing at once for its labour and its originality. He developed an amazing facility for figures, and that rare gift of mentally carrying out far-reaching calculations. It is recorded that on one occasion when two of his pupils, working the sum of a series to seventeen terms, disagreed in their results by one unit at the fiftieth significant figure, an appeal was made to Euler. He went over the calculation in his own mind, and his decision was found to be correct.

In 1771, when a great fire broke out in the town and reached Euler's house, a fellow-countryman from Basel, Peter Grimm, dashed into the flames, discovered the blind man and carried him off on his shoulders into safety. Although books and furniture were all lost, his precious writings were saved. For twelve years more Euler continued his excessive labours, until the day of his death, in the seventy-sixth year of his age.

Like Newton and many others, Euler was a man of parts, who had studied anatomy, chemistry and botany. As is reported of Leibniz, he could repeat the *Aeneid* from beginning to end, and could even remember the first and last lines in every page of the edition which he had been accustomed to use. The power seems to have been the result of his most wonderful concentration, that great constituent of inventive power, to which Newton himself has borne witness, when the senses are locked up in intense meditation, and no external idea can intrude.

Sweetness of disposition, moderation and simplicity of manner were his characteristics. His home was his joy, and he was fond of children. In spite of his affliction he was lively and cheerful, possessed of abundant energy; as his pupil M. Fuss has testified, 'his piety was rational and sincere; his devotion was fervent.'

In an untechnical account it is impossible to do justice to the mathematics of Euler: but while Newton is a national hero, surely Euler is a hero for mathematicians. Newton was the Archimedes and Euler was the Pythagoras. Great was the work of Euler in the problems of physics—but only because their mathematical pattern caught and retained his attention. His delight was to speculate in the realms of pure intellect, and here he reigns a prince of analysts. Not even geometry, not even the study of lines and figures, diverted him: his ultimate and constant aim was the perfection of the calculus and analysis. His ideas ran so naturally in this train, that even in Virgil's poetry he found images which suggested philosophic inquiry, leading on to new mathematical adventures. Adventures they were, which his more wary followers sometimes hailed with delight and occasionally condemned. The full splendour of the early Greek beginnings and the later works of Napier, Newton and Leibniz, was now displayed. Let one small formula be quoted as an epitome of what Euler achieved:

$$e^{i\pi} + 1 = 0.$$

Was it not Felix Klein who remarked that all analysis was centred here? Every symbol has its history—the principal whole numbers 0 and 1; the chief mathematical relations $+$ and $=$; π the discovery of Hippocrates; i the sign for the 'impossible' square root of minus one; and e the base of Napierian logarithms.

CHAPTER IX

MACLAURIN AND LAGRANGE

AMONG the contemporaries of Euler there were many excellent mathematicians in England and France, such as Cotes, Taylor, Demoivre, D'Alembert, Clairaut, Stirling, Maclaurin, and, somewhat later, Ivory, Wilson and Waring. This by no means exhaustive list contains the names of several friends of Newton—notably Cotes, Maclaurin and Demoivre. They were Newton's disciples, and each was partly responsible for making the work of the Master generally accessible. Cotes and Maclaurin were highly gifted geometers: the others of their time were interested in analysis. It was therefore a loss not only to British but to European mathematics that Cotes and Maclaurin should both have died young.

COLIN MACLAURIN (1698–1746), a Highlander from the county of Argyle, was educated at the University of Glasgow. Such was his outstanding ability that, at the age of nineteen, he was elected Professor of Mathematics in Aberdeen. Eight years later, when he acted as deputy Professor in Edinburgh, Newton wrote privately offering to pay part of the salary, as there was difficulty in raising the proper sum. Maclaurin took an active part in opposing the march of the Young Pretender in 1745 at the head of a great Highland army, which overran the country and finally seized Edinburgh. Maclaurin escaped, but the hardships of trench warfare and the subsequent flight to York proved fatal, and in 1746 he died.

Stirred by the brilliant work of Cotes, which luckily came into his hands, Maclaurin wrote a wonderful account of higher geometry. He dealt with the part which is called the *organic description of plane curves,* a subject belonging to Euclid, Pappus, Pascal and Newton. It is the mathematics of rods and bars, constrained by pivots and guiding rails—the abstract replica of valve gears and link motions familiar to the engineer—and it fascinates the geometer who 'likes to see the wheels go round'. Maclaurin carried on what Pascal had begun with, the celebrated mystic hexagram (which at that date still lay hid), and in so doing he reached a result of great generality. It provided a basis for the advances in pure geometry that were made a century later by Chasles, Salmon and Clifford.

In this kind of geometry the Cartesian method of co-ordinates fails to keep pace with the purely geometrical. In it men breathe a rarer air, akin to that in the theory of numbers.

The very success of Maclaurin partakes of the tragic. For there are huge tracts of mathematics where co-ordinates provide the natural medium —where, for any but a supreme master, analysis succeeds and pure geometry leaves one helpless. When Maclaurin wrote his essay on the equilibrium of spinning planets, which gained him the honours of the Paris Academy, he set out on a course wherein few could follow: for the problem was rendered in the purest geometry. When, in addition to this, Maclaurin produced a great geometrical work on fluxions, the scale was so heavily loaded that it diverted England from Continental habits of thought. During the remainder of the century British mathematics were relatively undistinguished, and there was no proper revival until the differential calculus began to be taught in Cambridge, according to the methods of Leibniz—a change which took place about a hundred years ago. This delay was the unhappy legacy of the Newton-Leibniz controversy, which need never have arisen.

The circumstances that prompted Maclaurin to adopt a geometrical style in his book on fluxions, extended beyond his partiality for geometry. Many philosophical influences were at work, and there were logical difficulties to face, which seemed to be insurmountable except by recourse to geometry. The difficulties were focused on the word *infinitesimal*—which Eudoxus had so carefully excluded from the vocabulary of Greek mathematics (the mere fact that it is a Latin, and not a Greek, word is not without its significance; so many of our ordinary mathematical terms have a Greek derivation). By an infinitesimal is meant something, distinguishable from zero, yet which is exceedingly small—so minute indeed that *no* multiple of it can be made into a finite size. It evades the axiom of Archimedes. Practically all analysts, from Kepler onwards, believed in the efficacy of infinitesimals, until Weierstrass taught otherwise. The differential calculus of Leibniz was founded on this belief, and its tremendous success, in the hands of the Bernoullis, Euler and Lagrange, obscured the issue. Men were disinclined to reject a doctrine which worked so brilliantly, and they turned a deaf ear to the philosophers, ancient and modern. In our own country a lively attack on infinitesimals was headed by the Irish philosopher and theologian, Bishop Berkeley. His criticism of the calculus was not lost upon Maclaurin, who was also well versed in Greek mathematics and the careful work of Eudoxus. So Maclaurin made up his mind to put Fluxions upon a sound basis and for this reason threw the work into a geometrical frame. It was his tribute to Newton, the master 'whose caution', said Maclaurin, 'was almost as distinguishing a part of his character as his invention.'

One of the chief admirers of Maclaurin was Lagrange, the great French analyst, whose own work offered a complete contrast to that of the geometer. Maclaurin had dealt in lines and figures—those characters, as Galileo has finely said, in which the great book of the Universe is written. Lagrange, on the contrary, pictured the Universe as an equally rhythmical theme of numbers and equations; and was proud to say, of his masterpiece, the *Mécanique Analytique,* that it contained not a single geometrical diagram. Nevertheless he appreciated the true geometer, declaring that the work of Maclaurin surpassed that of Archimedes himself, while as for Newton, he was 'the greatest genius the world has ever seen—and the most fortunate, for only once can it be given a man to discover the system of the Universe!'

JOSEPH-LOUIS LAGRANGE (1736–1813) came of an illustrious Parisian family which had long connection with Sardinia, and some trace of noble Italian ancestry. He spent his early years in Turin, his active middle life in Berlin, and his closing years in Paris, where he attained his greatest fame. Foolish speculation on the part of his father threw Lagrange, at an early age, upon his own resources, but this change of fortunes proved to be no great calamity, 'for otherwise', he says, 'I might never have discovered my vocation.' At school his boyish interests were Homer and Virgil, and it was not until a memoir of Halley came his way, that the mathematical spark was kindled. Like Newton, but at a still earlier age, he reached to the heart of the matter in an incredibly short space of time. At the age of sixteen he was made Professor of Mathematics in the Royal School of Artillery at Turin, where the diffident lad, possessed of no tricks of oratory and very few words, held the attention of men far older than himself. His winning personality elicited their friendship and enthusiasm. Very soon he was conducting a youthful band of scientists who became the earliest members of the Turin Academy. With a pen in his hand Lagrange was transfigured; and from the first, his writings were elegance itself. He would set to mathematics all the little themes on physical inquiries which his friends brought him, much as Schubert would set to music any stray rhyme that took his fancy.

At the age of nineteen he won fame by solving the so-called isoperimetrical problem, that had puzzled the mathematical world for half a century. He communicated his proof in a letter to Euler, who was immensely interested in the solution, particularly as it agreed with a result that he himself had found. With admirable tact and kindness Euler replied to Lagrange, deliberately withholding his own work, that all the credit might fall on his young friend. Lagrange had indeed not only solved a problem, he had also invented a new method, a new *Calculus of Variations,* which was to be the central subject of his life-work. This calculus belongs to the story of Least Action, which began with the reflecting mirrors of Hero (p. 110) and continued when Descartes pondered over his curiously shaped oval lenses. Lagrange was able to show that the somewhat varied Newtonian postulates of matter and motion fitted in with a broad principle of economy in nature. The principle has led to the still more fruitful results of Hamilton and Maxwell, and it continues today in the work of Einstein and in the latest phases of Wave Mechanics.

Lagrange was ready to appreciate the fine work of others, but he was equally able to detect a weakness. In an early memoir on the mathematics of sound, he pointed out faults even in the work of his revered Newton. Other mathematicians ungrudgingly acknowledged him first as their peer, and later as the greatest living mathematician. After several years of the utmost intellectual effort he succeeded Euler in Berlin. From time to time he was seriously ill from overwork. In Germany King Frederick, who had always admired him, soon grew to like his unassuming manner, and would lecture him for his intemperance in study which threatened to unhinge his mind. The remonstrances must have had some effect, because Lagrange changed his habits and made a programme every night of what was to be read the next day, never exceeding the ration. For twenty years he continued to reside in Prussia, producing work of high distinction that culminated in his *Mécanique Analytique*. This he decided to publish in France, whither it was safely conveyed by one of his friends.

The publication of this masterpiece aroused great interest, which was considerably augmented in 1787 by the arrival in Paris of the celebrated author himself, who had left Germany after the death of King Frederick, as he no longer found a sympathetic atmosphere in the Prussian Court. Mathematicians thronged to meet him and to show him every honour, but they were dismayed to find him distracted, melancholy, and indifferent to his surroundings. Worse still—his taste for mathematics had gone! The years of activity had told; and Lagrange was mathematically worn out. Not once for two whole years did he open his *Mécanique Analytique*: instead, he directed his thoughts elsewhere, to metaphysics, history, religion, philology, medicine, botany, and chemistry. As Serret has said, 'That thoughtful head could only change the objects of its meditations.' Whatever subject he chose to handle, his friends were impressed with the originality of his remarks. His saying that chemistry was 'easy as algebra' vastly astonished them. In those days the first principles of atomic chemistry were keenly canvassed: but it seemed odd to draw a comparison between such palpable things as chemicals, that can be handled and seen, and such abstractions as algebraic symbols.

In this philosophical and unmathematical state of mind Lagrange continued for two years, when suddenly the country was plunged into the Revolution. Many avoided the ordeal by flight abroad, but Lagrange refused to leave. He remained in Paris, wondering as he saw his friends done to death if *his* turn was coming, and surprised at his good fortune in surviving. France has reason to be glad that he was not cut down as was his friend Lavoisier, the great chemist; for in later years mathematical skill once again returned to him, and he produced many gems of algebra and analysis.

One mathematical effect of the Revolution was the adoption of the metric system, in which the subdivision of money, weights and measures is strictly based on the number ten. When someone objected to this number, naturally preferring twelve, because it has more factors, Lagrange unexpectedly remarked, what a pity it was that the number eleven had not been chosen as base, because it was prime. The M.C.C. appears to be one of the few official bodies who have followed this hint, by thinking systematically in terms of such a unit!

For music he had a liking. He said it isolated him and helped him to think, as it interrupted general conversation. 'For three bars I listen to it; thereafter I distinguish nothing, but give myself up to my thoughts. In this way I have solved many a difficult problem.' He was twice married: first when he lived in Berlin, where he lost his wife after a long illness, in which he nursed her devotedly. Then again in Paris he married Mlle. Lemonnier, daughter of a celebrated astronomer. Happy in his home life, simple and almost austere in his tastes, he spent his quiet fruitful years, till he died in 1813 at the age of seventy-six.

Lagrange is one of the great mathematicians of all time, not only for the abundance and originality of his work but for the beauty and propriety of his writings. They possess the grandeur and ease of the ancient geometers, and Hamilton has described the *Mécanique Analytique* as 'a scientific poem'. He was equally at home rivalling Fermat in the theory of numbers and Newton in analytical mechanics. Much of the contemporary and later work of Laplace, Legendre, Monge, Fourier and Cauchy, was the outcome of his inspiration. Lagrange sketched the broad design; it was left to others to fill in the finished picture. One must turn to the historians of mathematics to learn how fully and completely this was done. The breadth of the canvas attracted men of widely different interests. Nothing could afford a greater contrast to the mind of Lagrange than that of Laplace, the other great contributor to natural philosophy, whose most notable work was the *Mécanique Céleste*. To Laplace mathematics were the accidents and natural phenomena the substance—a point of view exactly opposite to that of Lagrange. To Laplace mathematics were tools, and they were handled with extraordinary skill, but any makeshift of a proof would do, provided that the problem was solved. It remained for the nineteenth century to show the faultiness of this naïve attitude. The instinct of the Greeks was yet to be justified.

CHAPTER X

GAUSS AND HAMILTON: THE NINETEENTH CENTURY

THE nineteenth century, which links the work of Lagrange with that of our own day, is perhaps the most brilliant era in the long history of mathematics. The subject assumed a grandeur in which all that was great in Greek mathematics was fully recovered; geometry once again came into its own, analysis further broadened its scope, and the outlets for its applications were ever enlarging. The century was marked in three noteworthy ways: there was deeper insight into the familiar properties of number; there was positive discovery of new processes of calculation, which, in the quaint words of Sylvester, ushered in 'the reign of Algebra the Second'; and there was also a philosophy of mathematics. During these years England once again rivalled mathematical France, and Germany and Italy rose to positions of scientific importance; while pre-eminent over all was the genius of one man, a mathematician worthy of a place of honour in the supreme rank with Archimedes and Newton.

CARL FRIEDRICH GAUSS was born in 1777 at Brunswick, and died in 1855, aged seventy-eight. He was the son of a bricklayer, and it was the wish of his father that he should be a bricklayer too. But at a very early age it was clear that the boy had unusual talents. Unlike Newton and Lagrange he showed the precocity of Pascal and Mozart. It is said that Mozart wrote a minuet at the age of four, while Gauss pointed out to his father an error in an account when he was three. At school his cleverness attracted attention, and eventually be came known to the Duke of Brunswick himself, who took an interest in the lad. In spite of parental protest the Duke sent him for a few years to the Collegium Carolinum and in 1795 to Göttingen. Still undecided whether to pursue mathematics or philology, Gauss now came under the influence of Kaestner—'that first of geometers among poets, and first of poets among geometers', as the pupil was proud to remark. In the course of his college career Gauss became known for his marvellous intuition in higher arithmetic. 'Mathematics, the Queen of the Sciences, and Arithmetic, the Queen of Mathematics', he would say: and mathematics became the main study of his life.

The next nine years were spent at Brunswick, varied by occasional travels, in the course of which he first met his friend Pfaff, who alone in Germany was a mathematician approximating to his calibre. After declining the offer of a Chair at the Academy in St. Petersburg, Gauss was appointed in 1807 to be first director of the new observatory at Göttingen, and there he lived a studious and simple life, happy in his surroundings, and blessed with good health, until shortly before his death. Once, in

1828, he visited Berlin, and once, in 1854, he made a pilgrimage to be present at the opening of the railway from Hanover to Göttingen. He saw his first railway engine in 1836, but except for these quiet adventures, it is said that until the last year of his life he never slept under any other roof than that of his own observatory!

His simple and direct character made a profound impression upon his pupils, who, seated round a table and not allowed to take notes, would listen with delight to the animated address of the master. Vivid accounts have been handed down of the chief figure in the group as he stood before his pupils 'with clear bright eyes, the right eyebrow raised higher than the left (for was he not an astronomer?), with a forehead high and wide, overhung with grey locks, and a countenance whose variations were expressive of the great mind within.'

Like Euler, Lagrange and Laplace, Gauss wrote voluminously, but with a difference. Euler never condensed his work; he revelled in the richness of his ideas. Lagrange had the easy style of a poet; that of Laplace was jerky and difficult to read. Gauss governed his writings with austerity, cutting away all but the essential results, after taking endless trouble to fill in the details. His pages stimulate but they demand great patience of the reader.

Gauss made an early reputation by his work in the theory of numbers. This was but one of his many mathematical activities, and, apart from all that followed, it would have placed him in the front rank. Like Fermat, he manifested that baffling genius which leaps—one knows not how—to the true conclusion, leaving the long-drawn-out deductive proof for others to formulate. A typical example is provided by the *Prime Number Theorem* which has taken a century to prove. Prime numbers were studied by Euclid, and continue to be an eternal source of interest to mathematicians. They are the numbers, such as 2, 3, 5, 7, 11, that cannot be broken up into factors. They are infinitely numerous, as Euclid himself was aware, and they occur, scattered through the orderly scale of numbers, with an irregularity that at once teases and captivates the mathematician. The question is naturally suggested: *How often, or how rarely, do prime numbers occur on the average?* Or, put in another way, What is the chance that a specified number is prime? In some form or other this problem was known to Gauss; and here is his innocent-looking answer:

$$\text{``Primzahlen unter } a \ (= \infty \,)$$
$$\frac{a}{la}\text{.''}$$

It means that when a is a very large number, the result of dividing a by its logarithm gives a good approximation to the total number of primes less than a: and the larger a is, the more precise is the result. Whether

Gauss proved his statement is not known: the quotation is taken from the back page of a copy of Schulze's Table of Logarithms which came into his possession when he was fourteen. Probably he recorded his note a few years later. Even if we recall the history of the logarithm and its diverse relations with so many remote parts of mathematics, this latest example is not a little strange. The contents of a book of logarithms, with its thickly crowded tables of decimal fractions, appears to be foreign indeed to the delicate distribution of primes among the whole numbers.

An actual proof of this theorem was only given as recently as thirty years ago by Hadamard and de la Vallée Poussin. It is an example of a new and very abstract part of the subject, now called the analytical theory of numbers. This is one of the striking developments in the present century, that has been notably advanced in Germany by Landau and in our country by Hardy and Littlewood.

Ever since the time of Gauss, mathematics has increased so extensively that no individual could hope to master the whole. Gauss was the last complete mathematician, and of him it can truly be said that he adorned every branch in the science. The beginnings of nearly all his discoveries are to be found in the youthful notes that he jotted down in a diary, unmethodically kept for several years, which has happily been preserved. The diary reveals pioneer facts in higher trigonometry, a subject generally known as Elliptic Functions: it also contains certain aspects of non-Euclidean geometry.

There is no doubt that Gauss was led to take an interest in geometry through Kaestner, his master, who himself had written on the fundamentals of the subject. Another contemporary influence was that of Legendre, whose book, the *Éléments de géometrie*, had appeared in 1794. These authors were interested in a problem that had often been discussed, notably by Wallis in England, and Saccheri, an Italian monk of the early eighteenth century. It concerned the parallel postulate of Euclid—that curious rugged feature in the smooth logic of the ancients, the removal of which seemed so very desirable. Gauss was perhaps the first to offer a satisfactory explanation of the anomaly: and his diary shows how early in his career this occurred. But, like Newton he was a cautious man, particularly when he handled strange and disconcerting novelties. For some years he kept the matter to himself, until he found that others were thinking of the same things. Among his college friends was a Hungarian, W. Bolyai, with whom he still corresponded; and in 1804 Bolyai wrote him a letter bearing on this theory of parallels. The interest spread, and out of it grew a branch of geometry called hyperbolic geometry. This branch of the subject is now always associated with the names of Gauss and his two friends the Bolyai's, father and son, and of Lobatchewski, a Russian, who wrote some twenty years later. It is another case of several independent discoveries on one theme all taking place in the same era.

Hyperbolic geometry was not merely a novelty; it was a revolution. In a very practical way it ran counter to Euclid, and in a still more practical way it ran counter to the current opinions of what Euclid was

FIGURE 19

supposed to teach. Euclid declared, for instance, that the sum of the three angles of a triangle is equal to two right angles. He also·declared that the sum of two adjacent angles, made by cross lines, is equal to two right angles. Both of these properties were implied, as he showed, in his fundamental axioms and postulates. But according to Gauss and Bolyai, while the declaration about the cross lines is true, that about the triangle is not: in fact they fashioned a triangle for which the sum of the angles is *less* than two right angles. Then, by nice balance, a little later, RIEMANN and others did the same for a triangle in which the sum is *greater* than two right angles. They called theirs elliptic geometry: it is the geometry that sailors know so well who voyage in direct courses over curved oceans of the globe. Less, equal and greater: three contradictory statements. These give rise to three bodies of geometrical doctrine, elliptic, parabolic and hyperbolic, the parabolic being that of Euclid. Here was the making of a first-class battle, not between opposing scientific camps holding relatively vague conflicting hypotheses, but in the very stronghold of logical argument—the realm which every one had taken for granted as settled and secure. The battle was fought, and it came to an end. As a three-cornered contest all sides lost in this sense, that any partisan for one of the views, saying *this* is true and both the others are false, would find himself pursuing a wild-goose chase. Instead of holding sovereign sway, each of these three is found to subserve a more fundamental whole. The programme of last century was designed to unravel the essential from the unessential, to isolate and underline right reasons for each geometrical fact, stripped of all misleading lumber. If, for example, we speak of the straight line AB, in the direction AB, we are not speaking redundantly, but of two different things, straightness and direction. Admittedly this is very puzzling, but it is none the less a fact.

An illustration of these abstract ideas is very literally ready to hand.

Every one knows that it is easy to fix a small piece of plaster to the back of the hand, but quite difficult to fix it over a knuckle or between two knuckles. In these cases the plaster has to be crumpled or stretched, to make it adhere. This has a mathematical explanation. The back of the hand presents a surface agreeable to Euclidean geometry, but the knuckles and the gaps do not. The knuckles illustrate Riemann, and the gaps, Gauss. In the gap a triangular piece of plaster would have its angles crumpled and therefore *less* than two right angles; it would need to be elastic, and expand, in order to fit over a knuckle. There is nothing very difficult in apprehending these ideas, because the surface of the hand is two-dimensional. When, however, the same notions are applied to space itself, as inevitably they were, the mind refuses readily to accommodate itself to the effort. A necessary prelude was the study of the sticking-plaster type of geometry, and this was done by Gauss. He developed the theory of surfaces, with special attention to their curvature and the conditions for one surface to fit another. It is said that he laid aside several questions which he treated analytically, and hoped to apply to them geometrical methods in some future state of existence, when his conceptions of space should have become amplified and extended.

RIEMANN, one of his many celebrated pupils, partly fulfilled this aspiration of Gauss. He certainly improved analysis out of all knowledge by his ingenious geometrical interpretation for the theory of functions. Also, in a few pages of epoch-making dissertation, he not only contemplated geometry for space of any dimensions—a surmise that he shared with Cayley—but showed that the earlier three types of geometry were particular instances of a still more general geometry. If geometry is likened to the surface of a sea, then these three correspond to the surface in a calm; that of Riemann corresponds to the surface in a calm or storm. His thesis was the necessary prelude to that of Einstein.

Another great pupil of Gauss was an Irish Rugby schoolboy of the days of Arnold, HENRY J. S. SMITH by name. He became Savilian Professor of Mathematics at Oxford, and handed on the tradition of Gauss in the theory of numbers. Even among mathematicians, the highly original work of Smith on the borderland of arithmetic and algebra is hardly as well known as it ought to be: for he originated certain important developments which brought fame to others, notably Weierstrass, Frobenius and Kronecker.

Smith owed much to his talented mother, who was early left a widow, schooling her children with abundant leisure in comparative isolation, to grow up like the young Brontës in a world of their own. He became an excellent linguist, in doubt at first whether to follow mathematics or classics, and it is said of him that no British mathematician ever came nearer to the spirit of the ancient Greek philosophers. As a boy he had something of the wisdom of a man, and to the day of his death he retained the simplicity and high spirits of a boy.

Ireland has produced many great mathematicians: and another was SALMON, who did so much to reconcile the geometry of Pascal and Descartes, and whose books have been an education in themselves. Later, when he became a distinguished theologian, he showed the same power and lucidity in his theological writings that had marked his mathematics. But the greatest figure of all was WILLIAM ROWAN HAMILTON, who made two splendid discoveries, an early one in optics, on the Principle of Least Action, and later the Quaternions in algebra. He was born in 1805, and educated at Trinity College, Dublin, where at the age of twenty-one he became Professor of Astronomy, continuing to hold the office until 1865, the year of his death. He was a poet, and a friend of Wordsworth and Coleridge, and between these three passed a highly interesting correspondence, dealing with philosophy, science, and literature.

As a child, Hamilton astonished every one by his early powers. At three he could read English; at four he was thoroughly interested in geography and had begun to read Latin, Greek and Hebrew; before he was ten he had slaked his thirst for Oriental languages by forming an intimate acquaintance with Sanscrit, and grounding himself in Persian, Arabic, Chaldee, Syriac and sundry Indian dialects. Italian and French were imbibed as a matter of course, and he was ready to give vent to his feelings in extemporized Latin. Taking to this monumental programme with ease and diligence, he was for all that a vigorous child, as ready to romp and run and swim as any other small boy.

In his seventeenth year he began to think for himself upon the subject of optics, and worked out his great principle of the *Characteristic Function* which, four years later, he presented to the Irish Academy in a thesis entitled an *Account of a Theory of Systems of Rays*. This youthful production was a work of capital importance in natural philosophy, as may be gathered from the sequel. Along with certain work in electro-magnetism by Clerk Maxwell, it shares the hard-won distinction of triumphantly surviving the latter-day revolution caused by the theory of Relativity.

Owing to its importance in the history of mathematics a quotation from Hamilton's thesis may not be inappropriate. After noticing how others—and particularly Malus, an officer who served under Napoleon—had invoked the principle of least action in studying rays of light, he says:

'A certain quantity which in one physical theory is the *action* and in another the *time*, expended by light in going from any first to any second point, is found to be less than if the light had gone in any other than its actual path. . . . The mathematical novelty of my method consists in considering this quantity as a function . . . and *in reducing all researches respecting optical systems of rays to the study of this single function*: a reduction which presents mathematical optics under an entirely novel view, and one analogous (as it appears to me) to the aspect under which Descartes presented the application of algebra to geometry.'

So light navigates space, as sailors navigate an ocean, by seeking the direct path. It is therefore hardly surprising that the work of Gauss and Hamilton should have eventually merged in a broader mathematical harmony. It needed but one step more—to devise a means of applying these ideas to space of *more* than the three ordinary dimensions—for the theory of Relativity to come into being. This essential step was taken by Christoffel, who worked with Riemannian geometry, and to-day the grandly sounding *World Function* of Hilbert is none other than the Characteristic Function of the youthful Hamilton, rehabilitated in four dimensions. It was a stroke of genius when Einstein found in this exceedingly elaborate geometry the very medium needed to cope with actual physical phenomena.

Hamilton also had high views on algebra, which he designated 'the science of pure time', and in making his discovery of *quaternions* brought into being an entirely new method of computation. Though his quaternions behaved very like numbers, they were not numbers; for they broke the commutative law. By this is meant the law under which it is asserted that $2 \times 3 = 3 \times 2$ or $a \times b = b \times a$ for ordinary numbers. As this law had gently been appropriated, at each stage, for all the new types of number, fractional, negative, irrational and indeed complex—the finishing touches of Gauss and Cauchy being, so to say, hardly dry—the mathematical world was lulled into slumber; little expecting something explosive to occur in *this* quarter. Nevertheless an explosion came—two explosions, in fact. One was fired by Hamilton and the other in Germany by GRASSMANN. For each discovered independently the need, in geometry or dynamics, of algebraic symbols whose behaviour was exemplary, judged by all accepted numerical standards—except the commutative law. For such symbols, the products ij and ji differed. If, says Hamilton, $ij = k$ then $ji = -k$.

Hamilton made this discovery in 1843 at the age of thirty-eight. It came like a flash, to relieve an intellectual need that had haunted him for fifteen years. Already Möbius had invented a sort of geometrical weighing machine which he called the *Barycentric Calculus*, whereby, not merely numbers, but points and forces could be *added* together. Out of this grew the notion of vectors, a name given to cover diverse physical phenomena such as forces and velocities. Hamilton called his vectors triplets, because forces act in three dimensions, and in course of time he was anxious to find a way for their multiplication. His home circle became interested in this puzzle. Every morning, on coming down to breakfast, one of his little boys used to ask, 'Well, Papa, can you multiply triplets?' Whereto he was obliged to answer with a sad shake of the head, 'No, I can only *add* and subtract them.' But one day, so he relates, he was walking with his wife beside the Royal canal on his way to a meeting in Dublin of the Academy. Although she talked with him now and then, yet an undercurrent of thought was going on in his mind, which at last gave a result. It came in a very tangible form, and it at once suggested to him many a long year of purposeful work upon an important theme. He could not resist the impulse to cut with a knife, on the stone of Brougham Bridge, as they passed it, the fundamental formulae

$$i^2 = j^2 = k^2 = ijk = -1,$$

indicative of the *Quaternions* which gave the solution of the problem.

This instinctive feeling that the discovery was important was well founded. Hamilton and Grassmann provided the first examples of a vast range in mathematics to which the same *algebra* has been appropriated. Arithmetic, algebra, analysis and geometry—these are the ingredients of Mathematics, each with its influence on the others, yet each with its own peculiar flavour. It was one of the characteristics of last century to emphasize the peculiarities, so that now we have a much clearer notion of their several significances. The subjects have been present from the beginnings of the science, and Eudoxus, Pythagoras, Archimedes, and Apollonius were forerunners of these several branches: Eudoxus with his interest in pure number, Pythagoras for his patterns and arrangements of things, Archimedes for his speculations about the infinite, and Apollonius for his projections of lines and curves.

CHAPTER XI

MORE RECENT DEVELOPMENTS

THE discovery of quaternions by Sir William Hamilton was a signal for the revival of mathematics throughout the country. Henceforward through the nineteenth century not only Ireland, but England and Scotland were once again represented in the foremost rank. A very prominent group of English mathematicians included Boole, Cayley and Sylvester, all of whom made important contributions to the new algebra inaugurated by Hamilton. Boole, who came originally from Lincolnshire but spent many of his most active years in Ireland, made an important discovery, not unlike that of the quaternions. He found that it was possible to apply algebraic symbols to logic, a step that went far towards clarifying our fundamental ideas in both logic and mathematics. He was also the pioneer in the algebraic theory of invariants, for in 1841 he discovered the first specimen of such a function. Out of this grew the work of Cayley and Sylvester, two of the greatest British mathematicians of the century. Cayley brought mathematical glory to Cambridge, second only to that of Newton, and the fertility of his suggestions, in geometry and algebra, continues to influence the whole range that is now studied at home and abroad. To this versatility Cayley added a Gauss-like care and industry.

Cayley was a sage, but his friend, the fiery, enthusiastic Sylvester, was a poet. At one time they both resided in London, where they were studying for a legal profession, in days before a happy turn of fortune drew them into their truer avocations. Many an important addition to higher algebra was made, as they strolled round the Law Courts eagerly discussing invariants. Subsequently Cayley returned to his old University, becoming Sadlerian Professor of Mathematics in Cambridge. For a time Sylvester lived in America, fostering an algebraic tradition, the fruit of which is one of the features of twentieth-century mathematics. The ground had already been prepared by PEIRCE of Harvard, who played the rôle, in America, of Hamilton and Grassmann in Europe.

It is noteworthy that in 1925 another example occurred of an abstract mathematical theory providing the mechanism for a new physical development. This took place when Heisenberg found, in the algebra of Hamilton as generalized by Cayley, the key to his new mechanics. To-day the subject is known as the Wave Mechanics, and it has received several different treatments. In one of these, which is due to Schrödinger, the Characteristic Function of Hamilton reappears, as a natural vehicle to explain the heart-beats of the atom.

Our ancestors in the Middle Ages received a shock when it was found that the surface of the earth, boundless as it appeared to them, was limited and could be circumnavigated. A similar shock was felt, in the narrower world of mathematics, when in 1868 it was found that a certain set of algebraic expressions, or invariants, that appeared to be endless, was finite. The credit of this startling discovery rests with GORDAN of Erlangen, a small university town of South Germany—a place already famous for the geometer, Von Staudt, and his able successors. This theorem of Gordan led HILBERT, a few years later, to formulate his *Basis Theorem*, of extraordinarily wide application, which can be regarded as giving a sort of algebraic blessing to the quaternions of Hamilton and Grassmann, as well as to large tracts of actual arithmetic. Gordan's own proof had involved a long piece of mathematical induction, marvellously handled. That of Hilbert, on the contrary, was short and depended on such general principles that it drew from Gordan the comment, 'This is theology, not mathematics!'

The geometer VON STAUDT shares with Grassmann a belated fame, for their contemporaries quite failed to realize the profundity and originality of their work. Von Staudt belongs to a distinguished group of geometers, among whom was found once again the spirit of Pappus, Pascal and Desargues. His name is singled out for special mention because he alone in the group considered, and successfully dealt with, fundamental principles. In abandoning the geometry of Euclid and adopting one or other of the non-Euclidean systems, it is difficult to avoid a feeling of being cast hopelessly adrift on a sea of chance. But the work of Von Staudt, as much as that of any man, has rendered a sober discussion of non-Euclidean geometry possible; for he has revealed the solid foundations common to all these types of geometry. His 'unpretentious but imperishable little volume' has led us to apprehend what is fundamental and what is not. For example, the first concern in geometry is whether three points A, B, C are in line or not, quite apart from whether B lies betwen A and C, or how far apart they are: these are doubtless important, but *secondary*, considerations. This is good theory; but after all it is also good common sense. It is easier on a level plain to get two distant posts in line with the eye, than to discover which of them is nearer, or exactly how far off they are. The really remarkable thing about it all is that Pappus and Desargues had actually hit upon the fundamental theorems of geometry (pp. 112, 133) in spite of using proofs involving unnecessary assumptions. It is just as if they had drawn their figures in red ink and had thought that they would not be valid in any other colour. Nineteenth-century mathematics was largely concerned with getting rid of that red ink.

The developments of Analysis indicate the same general princples at work. Analysis is the branch of mathematics dealing with the infinite, the unbounded—either the immeasurably large or the immeasurably small. After Gauss, prodigious advances were made. Perhaps the most important of these was that of WEIERSTRASS and the Berlin school, which finally settled the Newton-Leibniz controversy by a return to the ancient methods of Eudoxus. In geometry, Pappus and Desargues had given wrong reasons for the right results, and the same thing frequently happened in the calculus. It may be recalled that Zeno had cured the Greeks of any such loose reasoning and that his criticisms had lead to the epoch-making work of Eudoxus. But never since the Renaissance had this been properly assimilated, although Wallis, Newton and Maclaurin came nearest to doing so. The task of reconciliation was undertaken by Weierstrass, and also by DEDEKIND of Göttingen. They said, in effect, that analysis has to do with number—not with geometry; therefore let a strictly arithmetical account be given. In this they succeeded, both for the calculus and for the theory of irrational numbers; and two of their principal means, in attaining this end, were the definitions of irrationals due to Eudoxus, and those of limits given by Wallis and Newton.

Euler and his contemporaries had provided an armoury of analytical weapons: they were now rendered keen and refined. Many a famous old problem consequently fell before the cunning onslaughts of the analysts. One of the most spectacular results was that of LINDEMANN, who proved that the irrational number π satisfies no algebraic equation with integer coefficients. This settled the matter of squaring the circle, once and for all.

With such a piling up of analytical armaments it might be thought that all simplicity had forsaken mathematics. At the close of the century nothing could have seemed more desirable than the rise of a genius who could dispense with all these elaborations, and yet find something new to say. Very dramatically this took place in India, and the career of SRINIVASA RAMANUJAN has marked a new epoch. India has from time to time possessed mathematicians of great power: they may be traced through the ages back to the later Greek period. But judged by absolute standards of greatness, among all mathematicians of the East, the genius of Ramanujan appears to be supreme. He was born at Erode, a town not far from Madras, in 1887 and died in 1920. At school his extraordinary powers seem to have been recognized, but owing to his weakness in English he failed to matriculate at the University of Madras. He therefore proceeded to work out mathematics for himself, deriving what help he could from Carr's *Synopsis of Pure Mathematics*. After several years' work in the Harbour Board office at Madras he became known to someone sufficiently interested in the contents of his mystifying notebooks to put him in touch with the mathematical experts. Notes that were locally unintelligible received immediate recognition in Cambridge as the work of a self-taught genius. An invitation to visit Cambridge was accepted, and in due time Ramanujan became a Fellow of Trinity College and of the Royal Society, in token of his conspicuous merit. Unhappily the residence in England destroyed his health, and the year after his return to India he died.

Difficult as it is to form a judgment of almost contemporary work, there can be no question that here was an exceptional mathematician. In spite of all the disadvantages of his mathematical upbringing, with its scanty supply of material, he attained a command of certain branches in analysis and the theory of numbers that placed him in the very front rank, even before he was discovered by the West. Sylvester once took Huxley to task for thinking that 'mathematics is the study which knows nothing of observation, nothing of experiment, nothing of induction, nothing of causation'. Such a description completely misrepresents mathematics, which in its making unceasingly calls forth the highest efforts of imagination and invention. As for induction, there is no more wonderful instance of its use than that of Ramanujan. One of his problems happened to have been solved independently by Landau; yet, as Hardy has remarked, Ramanujan 'had none of Landau's weapons at his command; he had never seen a French or German book. It is sufficiently marvellous that he should have even dreamt of problems such as these, problems which have taken the finest mathematicians of Europe a hundred years to solve, and of which the solution is incomplete to the present day.' Nevertheless there were curious blind spots in his work where it went definitely wrong. Traversing many of the roads taken by Wallis, the Bernoullis and Euler, who occasionally blundered on to the wrong path in their enterprising adventures, Ramanujan repeated in his own short career the experiences of three centuries. Yet 'with his memory, his patience, and his power of calculation, he combined a power of generalization, a feeling for form, and a capacity for rapid modification of his hypotheses, that was often really startling, and made him, in his own peculiar field, without a rival in his day.'

Perhaps his greatest monument is a theorem that he discovered jointly with Hardy, dealing with the partitions of a number n. The theorem determines the number of ways in which n can be expressed as a sum of smaller whole numbers: and once more the simplicity of the enunciation completely disguises the profound difficulty of the quest. The theorem was a genuine example of collaboration, involving characteristic leaps in the dark, that border on the miraculous, followed by searching applications of Western mathematical analysis. As Littlewood has said: 'We owe the theorem to the happy collaboration of two men of quite unlike gifts, in which each contributed the best, most characteristic, and most fortunate work that was in him. Ramanujan's genius did have this one opportunity worthy of it.'

There were still other very general concepts that were brought into mathematics during the latter half of last century, particularly the theory of groups, which dates from the student days in Paris of two friends, SOPHUS LIE and FELIX KLEIN, of Scandinavia and Germany; and the theory of assemblages created by GEORG CANTOR, a Dane. Both of these ideas have had enormous influence on the trend of recent and contemporary thought; it is enough to say here that they exhibit in diverse ways the more mathematical side of that philosophical search into the principles of our subject which has marked the most recent stage of its history. For mathematics had now reached a state in which it was possible to do for the whole what Euclid tried to do for geometry, by disclosing the underlying axioms or primitive propositions, as Peano called them: and the most patient investigation has been made—notably in our own land by Whitehead and Russell—first of the subject-matter itself and next of the very ideas that govern the subject-matter. As all this was conceived on a sublimely universal scale, it is hardly remarkable that certain paradoxes have come to light. How to face these paradoxes is an urgent problem, and there are at the present time two or three different schools of thought employed upon this. The school associated with Brouwer of Holland has adopted a most drastic policy. They trace the presence of paradoxes to the use of indirect proofs, or more precisely to what is called in logic the law of the excluded middle. To this they object, very much as others have earlier objected to the Parallel Postulate of Euclid; indeed, it may be a symptom of the advent of a higher synthesis in arithmetic and analysis, just as the earlier was in geometry. As nothing less than the whole edifice from Eudoxus to Cantor is at stake, little wonder that these views cause a stir in the mathematical world. 'Of what use,' said Kronecker to Lindemann, 'is your beautiful investigation regarding π? Why study such problems, since irrational numbers are non-existent?' So back we are once more at a logical scandal such as troubled the Greeks. The Greeks survived and conquered it, and so shall we. At any rate, it is all a sign of the eternal freshness of mathematics.

*   *   *   *   *

The story has now been told of a few among many whose admirable genius has composed the lofty themes which go to form our present-day heritage. Agelong has been the noble toil that has called forth a simplicity and steadfastness of purpose in all its greatest exponents. And if this little book perhaps may bring to some, whose acquaintance with mathematics is full of toil and drudgery, a knowledge of those great spirits who have found in it an inspiration and delight, the story has not been told in vain. There is a largeness about mathematics that transcends race and time: mathematics may humbly help in the market-place, but it also reaches to the stars. To one, mathematics is a game (but what a game!) and to another it is the handmaiden of theology. The greatest mathematics has the simplicity and inevitableness of supreme poetry and music, standing on the borderland of all that is wonderful in Science, and all that is beautiful in Art. Mathematics transfigures the fortuitous concourse of atoms into the tracery of the finger of God.

The Rhind Papyrus

THE oldest mathematical documents in existence are two Egyptian papyrus rolls dating from around the Twelfth Dynasty (2000–1788 B.C.). The earlier of the scrolls, the Golenischev—both are named after their former owners—reposes in Moscow; the other, the Rhind papyrus, is in the British Museum. These remarkable texts make evident what has not always been acknowledged, namely, that the Egyptians possessed a good deal of arithmetic and geometric knowledge. Their methods were clumsy and they were incapable of grand generalizations—a preëminent ability of the Greeks. Yet it is nonsense to depreciate the real skill and imagination exhibited in these texts, to belittle the contribution made by the Egyptians to mathematical thought. Egyptian mathematics was precocious, as George Sarton has remarked; [1] its major achievements came early. It was also arrested in its development; after a short period of vigorous growth it made little further progress. The static character of Egyptian culture, the blight that fell upon Egyptian science around the middle of the second millennium, has often been emphasized but never adequately explained. Religious and political factors undoubtedly played a part in turning a dynamic society into one of stone.

The Rhind papyrus is described in the article which follows; the Golenischev deserves a note here. A scroll of the same length (544 cm.) as the Rhind, but only one quarter as wide (8 cm.), the Moscow papyrus is a collection of twenty-five problems rather than a treatise. The method of solving these problems agrees with rules given in the Rhind papyrus. [2] One of the problems indicates that the Egyptians may have known the formula for the volume of a truncated pyramid, $V = (^h/_3)\ (a^2 + ab + b^2)$, where a and b are the lengths of the sides of the square (the base of the pyramid) and h is the height. [3] Sarton calls this solution the "masterpiece" of their geometry. It was indeed an impressive step forward, unsurpassed in three more millennia of Egyptian mathematics. [4]

---

[1] George Sarton, *A History of Science*, Cambridge (Mass.), 1952, p. 40.

[2] O. Neugebauer, *The Exact Sciences in Antiquity*, Princeton, 1952, p. 78.

[3] W. Struve, *Mathematischer Papyrus des Staatlichen Museums der Schönen Kunste in Moskau*, Berlin, 1930.

[4] For an interesting survey of the beginnings of geometry, including the Babylonian, Egyptian, Indian, Chinese and Japanese contributions, see Julian Lowell Coolidge, *A History of Geometrical Methods*, Oxford, 1940, pp. 1–23. Another problem in the Moscow Papyrus which, as Coolidge mentions, has excited scholars, is that of finding the area of a basket, in connection with which exercise the Egyptians gave the excellent approximation, $\pi = (^{16}/_9)^2$. See also B. L. Van der Waerden, *Science Awakening*, Groningen, 1954.

2 The Rhind Papyrus

By JAMES R. NEWMAN

IN the winter of 1858 a young Scottish antiquary named A. Henry Rhind,
sojourning in Egypt for his health, purchased at Luxor a rather large
papyrus said to have been found in the ruins of a small ancient building
at Thebes. Rhind died of tuberculosis five years later, and his papyrus
was acquired by the British Museum. The document was not intact; evidently it had originally been a roll nearly 18 feet long and 13 inches high,
but it was broken into two parts, with certain portions missing. By one
of those curious chances that sometimes occur in archaeology, several
fragments of the missing section turned up half a century later in the
deposits of the New York Historical Society. They had been obtained,
along with a noted medical papyrus, by the collector Edwin Smith. The
fragments cleared up some points essential for understanding the whole
work.

The scroll was a practical handbook of Egyptian mathematics, written
about 1700 B.C. Soon after its discovery several scholars satisfied them-
selves that it was an antiquity of first importance, no less, as D'Arcy
Thompson later said, than "one of the ancient monuments of learning."
It remains to this day our principal source of knowledge as to how the
Egyptians counted, reckoned and measured.

The Rhind was indited by a scribe named A'h-mosè (another, more
sonorous form of his name is Aāh-mes) under a certain Hyksos king who
reigned "somewhere between 1788 and 1580 B.C." A'h-mosè, a modest
man, introduces his script with the notice that he copied the text "in
likeness of writings of old made in the time of the King of Upper [and
Lower] Egypt, [Ne-ma] 'et-Rê'." The older document to which he refers
dates back to the 12th Dynasty, 1849–1801 B.C. But there the trail ends,
for one cannot tell whether the writing from which A'h-mosè copied was
itself a copy of an even earlier work. Nor is it clear for what sort of
audience the papyrus was intended, which is to say we do not know
whether "it was a great work or a minor one, a compendium for the
scholar, a manual for the clerk, or even a lesson book for the schoolboy."

The Egyptians, it has been said, made no great contributions to mathe-

matical knowledge. They were practical men, not much given to speculative or abstract inquiries. Dreamers, as Thompson suggests, were rare
among tnem, and mathematics is nourished by dreamers—as it nourishes
them. Egyptian mathematics nonetheless is not a subject whose importance
the historian or student of cultural development can afford to disparage.
And the Rhind Papyrus, though elementary, is a respectable mathematical
accomplishment, proffering problems some of which the average intelligent man of the modern world—38 centuries more intelligent, perhaps,
than A'h-mosè—would have trouble solving.

Scholars disagree as to A'h-mosè's mathematical competence. There are
mistakes in his manuscript, and it is hard to say whether he put them

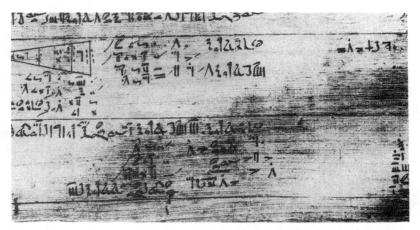

The papyrus was originally a roll 13 inches high and almost 18 feet long. This photograph shows
a small section of it about 4 inches high and 10 inches wide. Hieratic script reads from right to
left and top to bottom.

there or copied them from the older document. But he wrote a "fine bold
hand" in hieratic, a cursive form of hieroglyphic; altogether it seems
unlikely that he was merely an ignorant copyist.

It would be misleading to describe the Rhind as a treatise. It is a collection of mathematical exercises and practical examples, worked out in a
syncopated, sometimes cryptic style. The first section presents a table of
the division of 2 by odd numbers—from $\frac{2}{3}$ to $\frac{2}{101}$. This conversion was
necessary because the Egyptians could operate only with unit fractions
and had therefore to reduce all others to this form. With the exception
of $\frac{2}{3}$, for which the Egyptians had a special symbol, every fraction had
to be expressed as the sum of a series of fractions having 1 as the numera-
tor. For example, the fraction $\frac{3}{4}$ was written as $\frac{1}{2}$, $\frac{1}{4}$ (note they did not
use the plus sign), and $\frac{2}{61}$ was expressed as $\frac{1}{40}$, $\frac{1}{244}$, $\frac{1}{488}$, $\frac{1}{610}$.

It is remarkable that the Egyptians, who attained so much skill in their arithmetic manipulations, were unable to devise a fresh notation and less cumbersome methods. We are forced to realize how little we understand the circumstances of cultural advance: why societies move—or is it perhaps jump—from one orbit to another of intellectual energy, why the science of Egypt "ran its course on narrow lines" and adhered so rigidly to its clumsy rules. Unit fractions continued in use, side by side with improved methods, even among Greek mathematicians. Archimedes, for instance, wrote ½, ¼ for ¾, and Hero, ½, $\frac{1}{17}$, $\frac{1}{34}$, $\frac{1}{51}$ for $\frac{31}{51}$. Indeed, as late as the 17th century certain Russian documents are said to have expressed $\frac{1}{96}$ as a "half-half-half-half-half-third."

The Rhind Papyrus contains some 85 problems, exhibiting the use of fractions, the solution of simple equations and progressions, the mensuration of areas and volumes. The problems enable us to form a pretty clear notion of what the Egyptians were able to do with numbers. Their arithmetic was essentially additive, meaning that they reduced multiplication and division, as children and electronic computers do, to repeated additions and subtractions. The only multiplier they used, with rare exceptions, was 2. They did larger multiplications by successive duplications. Multiplying 19 by 6, for example, the Egyptians would double 19, double the result and add the two products, thus:

	1	19
\	2	38
\	4	76
Total	6	114

The symbol \ is used to designate the sub-multipliers that add up to the total multiplier, in this case 6. The problem 23 times 27 would, in the Rhind, look like this:

\	1	27
\	2	54
\	4	108
	8	216
\	16	432
Total	23	621

In division the doubling process had to be combined with the use of fractions. One of the problems in the papyrus is "the making of loaves 9 for man 10," meaning the division of 9 loaves among 10 men. This problem is not carried out without pain. Recall that except for ⅔ the Egyptians had to reduce all fractions to sums of fractions with the numerator 1. The Rhind explains:

"The doing as it occurs: Make thou the multiplication ⅔ ⅕ $\frac{1}{30}$ times 10.

$$
\begin{array}{llll}
1 & \frac{2}{3} & \frac{1}{5} & \frac{1}{30} \\
\diagdown\ 2 & 1\frac{2}{3} & \frac{1}{10} & \frac{1}{30} \\
4 & 3\frac{1}{2} & \frac{1}{10} & \\
\diagdown\ 8 & 7\frac{1}{5} & &
\end{array}
$$

Total loaves 9; it, this is."

In other words, if one adds the fractions obtained by the indicated multiplications $(2 + 8 = 10)$, he arrives at 9. The reader understandably, may find the demonstration baffling. For one thing, the actual working of the problem is not given. If 10 men are to share 9 loaves, each man, says A'h-mosè, is to get $\frac{2}{3}$, $\frac{1}{5}$, $\frac{1}{30}$ (*i.e.*, $\frac{27}{30}$) times 10 loaves; but we have no idea how the figure for each share was arrived at. The answer to the problem ($\frac{27}{30}$, or $\frac{9}{10}$) is given first and then verified, not explained. It may be, in truth, that the author had nothing to explain, that the problem was solved by trial and error—as, it has been suggested, the Egyptians solved all their mathematical problems.

An often discussed problem in the Rhind is: "Loaves 100 for man 5, $\frac{1}{7}$ of the 3 above to man 2 those below. What is the difference of share?" Freely translated this reads: "Divide 100 loaves among 5 men in such a way that the shares received shall be in arithmetical progression and that $\frac{1}{7}$ of the sum of the largest three shares shall be equal to the sum of the smallest two. What is the difference of the shares?" This is not as easy to answer as its predecessors, especially when no algebraic symbols or processes are used. The Egyptian method was that of "false position"—a mixture of trial and error and arithmetic proportion. Let us look at the solution in some detail:

"Do it thus: Make the difference of the shares $5\frac{1}{2}$. Then the amounts that the five men receive will be 23, $17\frac{1}{2}$, 12, $6\frac{1}{2}$, 1: total 60."

Now the assumed difference $5\frac{1}{2}$, as we shall see, turns out to be correct. It is the key to the solution. But how did the author come to this disingenuously "assumed" figure? Probably by trial and error. Arnold Buffum Chace, in his definitive study *The Rhind Papyrus*—from which I have borrowed shamelessly—proposes the following ingenious reconstruction of the operation:

Suppose, as a starter, that the difference between the shares were 1. Then the terms of the progression would be 1, 2, 3, 4, 5; the sum of the smallest two would be 3, and $\frac{1}{7}$ of the largest three shares would be $1\frac{5}{7}$ ($1\frac{1}{2}$, $\frac{1}{7}$, $\frac{1}{14}$ Egyptian style). The difference between the two groups (3 minus $1\frac{5}{7}$) would be $1\frac{2}{7}$, or $1\frac{1}{4}$, $\frac{1}{28}$. Next, trying 2 as the difference between the successive shares, the progression would be 1, 3, 5, 7, 9. The sum of the two smallest terms would be 4; $\frac{1}{7}$ of the three largest terms would be 3, and the difference between the two sides, 1. The experimenter might then begin to notice that for each increase of 1 in the assumed common difference, the inequality between the two sides was reduced by

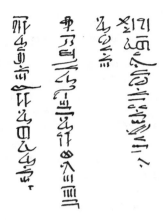

Part of title page of the papyrus is reproduced in facsimile. Here the hieratic script reads from top to bottom and right to left. It has been translated: "Accurate reckoning of entering into things, knowledge of existing things all, mysteries . . . secrets all. Now was copied book this in year 33, month four of the inundation season [under the majesty of the] King of [Upper and] Lower Egypt, 'A-user-Rê', endowed with life, in likeness of writings of old made in the time of the King of Upper [and Lower] Egypt, [Ne-ma] 'et-Rê'. Lo the scribe A'h-mosè writes copy this."

¼, ⅟₂₈. Very well: to make the two sides equal, apparently he must multiply his increase 1 by as many times as ¼, ⅟₂₈ is contained in 1¼, ⅟₂₈. That figure is 4½. Added to the first assumed difference, 1, it gives 5½ as the true common difference. "This process of reasoning is exactly in accordance with Egyptian methods," remarks Chace.

Having found the common difference, one must now determine whether the progression fulfills the second requirement of the problem: namely, that the number of loaves shall total 100. In other words, multiply the progression whose sum is 60 (see above) by a factor to convert it into 100; the factor, of course, is 1⅔. This the papyrus does: "As many times as is necessary to multiply 60 to make 100, so many times must these terms be multiplied to make the true series." (Here we see the essence of the method of false position.) When multiplied by 1⅔, 23 becomes 38⅓, and the other shares, similarly, become 29⅙, 20, 10⅚ and 1⅔. Thus one arrives at the prescribed division of the 100 loaves among 5 men.

The author of the papyrus computes the areas of triangles, trapezoids and rectangles and the volumes of cylinders and prisms, and of course the area of a circle. His geometrical results are even more impressive than his arithmetic solutions, though his methods, as far as one can tell, are quite unrelated to the discipline today called geometry. "A cylindrical granary of 9 diameter and height 6. What is the amount of grain that goes into it?" In solving this problem a rule is used for determining the area of a circle which comes to Area = $(\%d)^2$, where d denotes the diameter. Matching this against the modern formula, Area = πr^2, gives a value for π of 3.16—a very close approximation to the correct value. The Rhind Papyrus gives the area of a triangle as ½ the base times the length of a line which may be the altitude of the triangle, but, on the other hand—Egyptologists are not sure—may be its side. In an isosceles triangle, tall and with a narrow base, the error resulting from using the side instead of the altitude in computing area would make little difference. The three triangle problems in the Rhind Papyrus involve triangles of this type, but it is clear that the author had only the haziest notion of what triangles were like. What he was thinking of was (as one expert conjectures) "a piece of land, of a certain width at one end and coming to a point, or at least narrower at the other end."

Egyptian geometry makes a very respectable impression if one considers the information derived not only from the Rhind but also from another Egyptian document known as the Moscow Papyrus and from lesser sources. Its attainments, besides those already mentioned, include the correct determination of the area of a hemisphere (some scholars, however, dispute this) and the formula for the volume of a truncated pyramid, $V = (^h/_3)\ (a^2 + ab + b^2)$, where a and b are the lengths of the sides of the square and h is the height.

I should like to give one more example taken from the Rhind Papyrus, something by way of a historical oddity. Chace offers the following translation of the hard-to-translate Problem 79:

"Sum the geometrical progression of five terms, of which the first term is 7 and the multiplier 7.

"The sum according to the rule. Multiply 2801 by 7.

\ 1	2801
\ 2	5602
\ 4	11204
Total	19607

"The sum by addition

houses	7
cats	49
mice	343
spelt (wheat)	2401
hekat (half a peck)	16807
Total	19607"

This catalogue of miscellany provides a strange little prod to fancy. It has been interpreted thus: In each of 7 houses are 7 cats; each cat kills 7 mice; each mouse would have eaten 7 ears of spelt; each ear of spelt would have produced 7 hekat of grain. Query: How much grain is saved

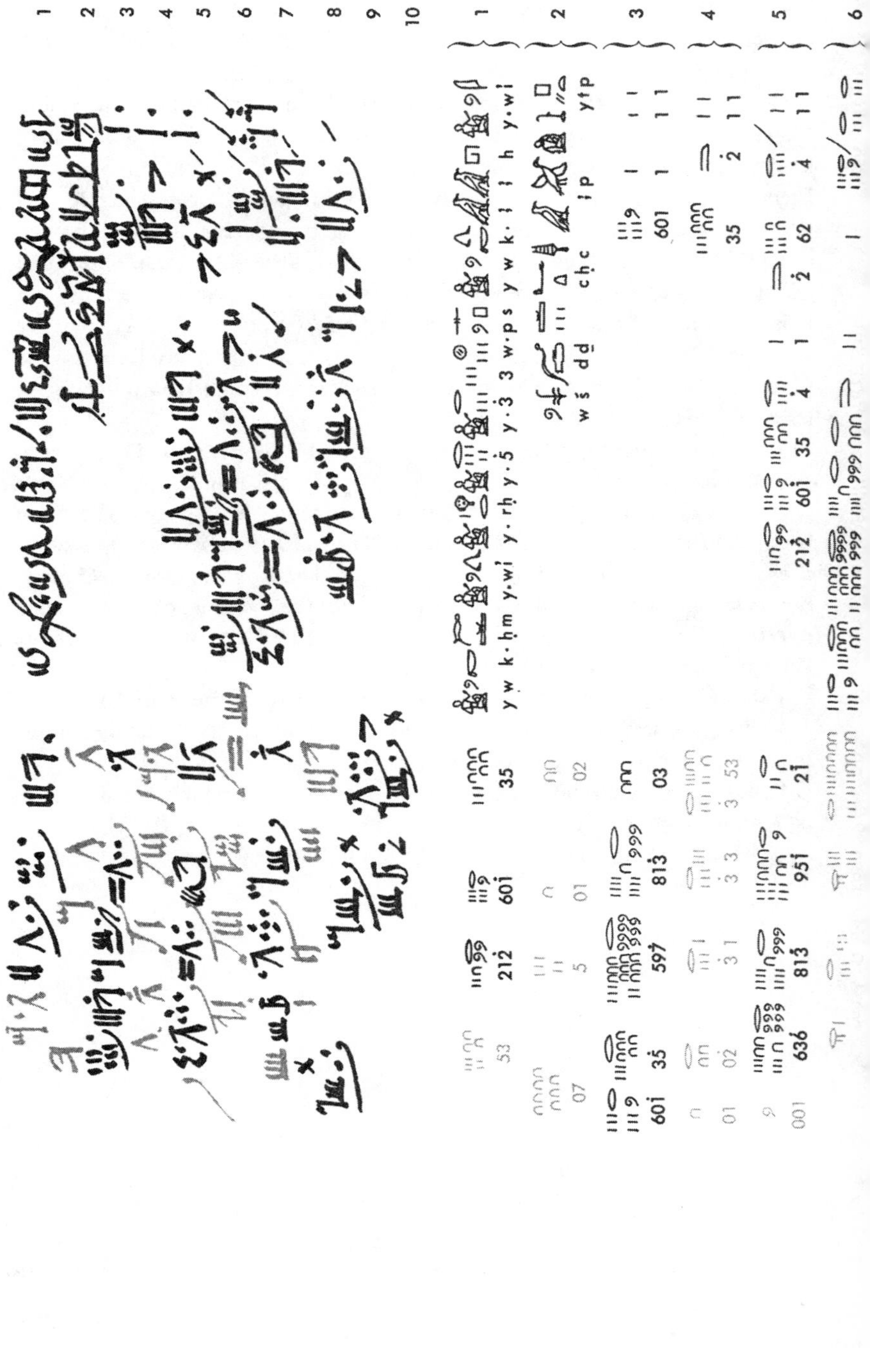

iw·y hꜣ·kwy sp·w 3   3 · y   ⅓ · y   5 · y   ḥr·y   iw·y mḥ·kwy   ply   pꜣ   'ḥ'   dd   šw

Go down I times 3, ⅓ of me, ⅕ of me is added to me; return I, filled am I. What is the quantity saying it?

1	53	106	212
2	30	318	795
3	12	159	318	636
5	20	265	530	1060

53	106	212		
20	10	5		35
30	318	795	53	106	70
35 3	33	13	20	10	
12	159	318	636		100
88 3	6 3	3 3	13		
20	265	530	1060		80
53	4	2	1	4	265

2			530
4			265
4			265
dmd			1060
Total			

Problem 36 of the papyrus begins: "Go down I times 3, ⅓ of me, ⅕ of me is added to me; return I, filled am I. What is the quantity saying it?" The problem is then solved by the Egyptian method. On these pages is a facsimile of the problem as it appears in the papyrus. The hieratic script reads from right to left. The characters are reproduced in gray and black (the original papyrus was written in red and black). In the middle of the page is a rendering in hieroglyphic script, which also reads from right to left. Beneath each line of hieroglyphs is a phonetic translation. The numbers are given in Arabic with the Egyptian notation. Each line of hieroglyphs and its translation is numbered to correspond to a line of the hieratic. At the bottom of the page the phonetic and numerical translation has been reversed to read from left to right. Beneath each phonetic expression is its English translation. A dot above a number indicates that it is a fraction with a numerator of one. Two dots above a 3 represent ⅔, the only Egyptian fraction with a numerator of more than one. Readers who have the desire to trace the entire solution are cautioned that the scribe made several mistakes in the various translations.

by the 7 houses' cats? (The author confounds us by not only giving the hekats of grain saved but by adding together the entire heterogeneous lot.) Observe the resemblance of this ancient puzzle to the 18th-century Mother Goose rhyme:

"As I was going to St. Ives
I met a man with seven wives.
Every wife had seven sacks,
Every sack had seven cats,
Every cat had seven kits.
Kits, cats, sacks and wives,
How many were there going to St. Ives?"

(To this question, unlike the question in the papyrus, the correct answer is "one" or "none," depending on how it is interpreted.)

A considerable difference of opinion exists among students of ancient science as to the caliber of Egyptian mathematics. I am not impressed with the contention based partly on comparison with the achievements of other ancient peoples, partly on the wisdom of hindsight, that the Egyptian contribution was negligible, that Egyptian mathematics was consistently primitive and clumsy. The Rhind Papyrus, though it demonstrates the inability of the Egyptians to generalize and their penchant for clinging to cumbersome calculating processes, proves that they were remarkably pertinacious in solving everyday problems of arithmetic and mensuration, that they were not devoid of imagination in contriving algebraic puzzles, and that they were uncommonly skillful in making do with the awkward methods they employed.

It seems to me that a sound appraisal of Egyptian mathematics depends upon a much broader and deeper understanding of human culture than either Egyptologists or historians of science are wont to recognize. As to the question how Egyptian mathematics compares with Babylonian or Mesopotamian or Greek mathematics, the answer is comparatively easy and comparatively unimportant. What is more to the point is to understand why the Egyptians produced their particular kind of mathematics, to what extent it offers a culture clue, how it can be related to their social and political institutions, to their religious beliefs, their economic practices, their habits of daily living. It is only in these terms that their mathematics can be judged fairly.

COMMENTARY ON
ARCHIMEDES

M ANY of Archimedes' wonderful writings have survived, but of his life only fragments are known. Heracleides wrote a biography but it has been lost and one must turn to various ancient sources of uneven reliability for such particulars as remain. The Byzantine grammarian Tzetzes records that Archimedes died at the age of seventy-five; since he perished at the fall of Syracuse, it is inferred that he was born about 287 B.C. From the Greek historian Diodorus one learns that he studied mathematics at Alexandria; from Pappus, that he wrote a book on mechanics, *On Sphere-making;* from Cicero, that he constructed a sphere which imitated the motion of the sun, the moon and the planets (Cicero claims to have seen this miniature planetarium); from Lucian, that he set the Roman ships on fire by an arrangement of concave mirrors or burning glasses; from Ptolemy, that he made many astronomical observations; from the Roman philosopher Macrobius, that he discovered the distances of the planets; from Vitruvius' history of architecture, that he once ran naked through the streets shouting "Eureka"—the occasion being familiar to every schoolboy. From Pappus, again, comes the no less famous tale that after Archimedes had solved the problem: "To move a given weight by a given force," he declared triumphantly: "Give me a place to stand on, and I can move the earth."

Plutarch tells the dramatic story of Archimedes' death: it was a violent conclusion to what had apparently been a quiet and contemplative life. The story appears as an aside in the biography of Marcellus, a Roman general who sacked Syracuse after a two-year siege (the resistance was much prolonged, it is said, by the military machines Archimedes designed for defending the city). Observe the curious working of history: Archimedes' death is familiar because of Plutarch's interest in Marcellus; of Marcellus it is generally remembered only that one of his soldiers murdered Archimedes.

There is an astonishing imagination, even in the science of mathematics.
. . . We repeat, there was far more imagination in the head of Archimedes
than in that of Homer. —VOLTAIRE

3 Archimedes

By PLUTARCH, VITRUVIUS, TZETZES

PLUTARCH, *"Marcellus"*

MARCELLUS, incensed by injuries done him by Hippocrates, commander of the Syracusans, (who, to give proof of his good affection to the Carthaginians, and to acquire the tyranny to himself, had killed a number of Romans at Leontini,) besieged and took by force the city of Leontini; yet violated none of the townsmen; only deserters, as many as he took, he subjected to the punishment of the rods and axe. But Hippocrates, sending a report to Syracuse, that Marcellus had put all the adult population to the sword, and then coming upon the Syracusans, who had risen in tumult upon that false report, made himself master of the city. Upon this Marcellus moved with his whole army to Syracuse, and, encamping near the wall, sent ambassadors into the city to relate to the Syracusans the truth of what had been done in Leontini. When these could not prevail by treaty, the whole power being now in the hands of Hippocrates, he proceeded to attack the city both by land and by sea. The land forces were conducted by Appius: Marcellus, with sixty galleys, each with five rows of oars, furnished with all sorts of arms and missiles, and a huge bridge of planks laid upon eight ships chained together, upon which was carried the engine to cast stones and darts, assaulted the walls, relying on the abundance and magnificence of his preparations, and on his own previous glory; all which, however, were, it would seem, but trifles for Archimedes and his machines.

These machines he had designed and contrived, not as matters of any importance, but as mere amusements in geometry; in compliance with King Hiero's desire and request, some little time before, that he should reduce to practice some part of his admirable speculations in science, and by accommodating the theoretic truth to sensation and ordinary use, bring it more within the appreciation of people in general. Eudoxus and Archytas had been the first originators of this far-famed and highly prized art of mechanics, which they employed as an elegant illustration of geometrical truths, and as a means of sustaining experimentally, to the satisfaction of the senses, conclusions too intricate for proof by words and diagrams. As, for example, to solve the problem, so often required in constructing geometrical figures, given the two extremes, to find the two mean lines of a proportion, both these mathematicians had recourse to the aid of instruments, adapting to their purpose certain curves and sections of lines.[1] But what with Plato's indignation at it, and his invectives against it as the mere corruption and annihilation of the one good of geometry,—which was thus shamefully turning its back upon the unembodied objects of pure intelligence to recur to sensation, and to ask help (not to be obtained without base subservience and depravation) from matter; so it was that mechanics came to be separated from geometry, and, repudiated and neglected by philosophers, took its place as a military art. Archimedes, who was a kinsman and friend of King Hiero, wrote to him that with a given force it was possible to move any given weight; and emboldened, as it is said, by the strength of the proof, he averred that, if there were another world and he could go to it, he would move this one. Hiero was amazed and besought him to give a practical demonstration of the problem and show some great object moved by a small force; he thereupon chose a three-masted merchantman among the king's ships which had been hauled ashore with great labour by a large band of men, and after putting on board many men and the usual cargo, sitting some distance away and without any special effort, he pulled gently with his hand at the end of a compound pulley [2] and drew the vessel smoothly and evenly towards himself as though she were running along the surface of the water. Astonished at this, and understanding the power of his art, the king persuaded Archimedes to construct for him engines to be used in every type of siege warfare, some defensive and some offensive; he had not himself used these engines because he spent the greater part of his life remote from war and amid the rites of peace, but now his apparatus proved of great advantage to the Syracusans, and with the apparatus its inventor.[3]

Accordingly, when the Romans attacked them from two elements, the

---

[1] The *mesolabes* or *mesolabium*, was the name by which this instrument was commonly known.

[2] πολυσπάστος. Galen, *in Hipp. De Artic.* iv. 47 uses the same word. Tzetzes speaks of a triple-pulley device (τῇ τρισπάστω μηχανῇ) in the same connexion, and Oribasius, *Coll. med.* xlix. 22 mentions the τρίσπαστος as an invention of Archimedes; he says that it was so called because it had three ropes, but Vitruvius says it was thus named because it had three wheels. Athenaeus v. 207 a-b says that a *helix* was used. Heath, *The Works of Archimedes*, p. xx, suggests that the vessel, once started, was kept in motion by the system of pulleys, but the first impulse was given by a machine similar to the κοχλίας described by Pappus viii. ed. Hultsch 1066, 1108 ff., in which a cog-wheel with oblique teeth moves on a cylindrical helix turned by a handle.

[3] Similar stories of Archimedes' part in the defence are told by Polybius viii. 5. 3–5 and Livy xxiv. 34.

Syracusans were struck dumb with fear, thinking that nothing would avail against such violence and power. But Archimedes began to work his engines and hurled against the land forces all sorts of missiles and huge masses of stones, which came down with incredible noise and speed; nothing at all could ward off their weight, but they knocked down in heaps those who stood in the way and threw the ranks into disorder. Furthermore, beams were suddenly thrown over the ships from the walls, and some of the ships were sent to the bottom by means of weights fixed to the beams and plunging down from above; others were drawn up by iron claws, or crane-like beaks, attached to the prow and were plunged down on their sterns, or were twisted round and turned about by means of ropes within the city, and dashed against the cliffs set by Nature under the wall and against the rocks, with great destruction of the crews, who were crushed to pieces. Often there was the fearful sight of a ship lifted out of the sea into mid-air and whirled about as it hung there, until the men had been thrown out and shot in all directions, when it would fall empty upon the walls or slip from the grip that had held it. As for the engine which Marcellus was bringing up from the platform of ships, and which was called *sambuca* from some resemblance in its shape to the musical instrument,[4] while it was still some distance away as it was being carried to the wall a stone ten talents in weight was discharged at it, and after this a second and a third; some of these, falling upon it with a great crash and sending up a wave, crushed the base of the engine, shook the framework and dislodged it from the barrier, so that Marcellus in perplexity sailed away in his ships and passed the word to his land forces to retire.

In a council of war it was decided to approach the walls, if they could, while it was still night; for they thought that the ropes used by Archimedes, since they gave a powerful impetus, would send the missiles over their heads and would fail in their object at close quarters since there was no space for the cast. But Archimedes, it seems, had long ago prepared for such a contingency engines adapted to all distances and missiles of short range, and through openings in the wall, small in size but many and continuous, short-ranged engines called scorpions could be trained on objects close at hand without being seen by the enemy.

When, therefore, the Romans approached the walls, thinking to escape notice, once again they were met by the impact of many missiles; stones fell down on them almost perpendicularly, the wall shot out arrows at them from all points, and they withdrew to the rear. Here again, when they were drawn up some distance away, missiles flew forth and caught

---

[4] The σαμβύκη was a triangular musical instrument with four strings. Polybius (viii. 6) states that Marcellus had eight quinqueremes in pairs locked together, and on each pair a "sambuca" had been erected; it served as a penthouse for raising soldiers on to the battlements.

them as they were retiring, and caused much destruction among them; many of the ships, also, were dashed together and they could not retaliate upon the enemy. For Archimedes had made the greater part of his engines under the wall, and the Romans seemed to be fighting against the gods, inasmuch as countless evils were poured upon them from an unseen source.

Nevertheless Marcellus escaped, and, twitting his artificers and craftsmen, he said: "Shall we not cease fighting against this geometrical Briareus, who uses our ships like cups to ladle water from the sea, who has whipped our *sambuca* and driven it off in disgrace, and who outdoes all the hundred-handed monsters of fable in hurling so many missiles against us all at once?" For in reality all the other Syracusans were only a body for Archimedes' apparatus, and his the one soul moving and turning everything: all other weapons lay idle, and the city then used his alone, both for offence and for defence. In the end the Romans became so filled with fear that, if they saw a little piece of rope or of wood projecting over the wall, they cried, "There it is, Archimedes is training some engine upon us," and fled; seeing this Marcellus abandoned all fighting and assault, and for the future relied on a long siege.

Yet Archimedes possessed so lofty a spirit, so profound a soul, and such a wealth of scientific inquiry, that although he had acquired through his inventions a name and reputation for divine rather than human intelligence, he would not deign to leave behind a single writing on such subjects. Regarding the business of mechanics and every utilitarian art as ignoble and vulgar, he gave his zealous devotion only to those subjects whose elegance and subtlety are untrammelled by the necessities of life; these subjects, he held, cannot be compared with any others; in them the subject-matter vies with the demonstration, the former possessing strength and beauty, the latter precision and surpassing power; for it is not possible to find in geometry more difficult and weighty questions treated in simpler and purer terms. Some attribute this to the natural endowments of the man, others think it was the result of exceeding labour that everything done by him appeared to have been done without labour and with ease. For although by his own efforts no one could discover the proof, yet as soon as he learns it, he takes credit that he could have discovered it: so smooth and rapid is the path by which he leads to the conclusion. For these reasons there is no need to disbelieve the stories told about him— how, continually bewitched by some familiar siren dwelling with him, he forgot his food and neglected the care of his body; and how, when he was dragged by main force, as often happened, to the place for bathing and anointing, he would draw geometrical figures in the hearths, and draw lines with his finger in the oil with which his body was anointed, being overcome by great pleasure and in truth inspired of the Muses. And though he made many elegant discoveries, he is said to have besought his friends and kinsmen to place on his grave after his death a cylinder enclosing a sphere, with an inscription giving the proportion by which the including solid exceeds the included.[5]

Such was Archimedes, who now showed himself, and, so far as lay in him, the city also, invincible. While the siege continued, Marcellus took Megara, one of the earliest founded of the Greek cities in Sicily, and capturing also the camp of Hippocrates at Acilæ, killed above eight thousand men, having attacked them whilst they were engaged in forming their fortifications. He overran a great part of Sicily; gained over many towns from the Carthaginians, and overcame all that dared to encounter him. As the siege went on, one Damippus, a Lacedæmonian, putting to sea in a ship from Syracuse, was taken. When the Syracusans much desired to redeem this man, and there were many meetings and treaties about the matter betwixt them and Marcellus, he had opportunity to notice a tower into which a body of men might be secretly introduced, as the wall near to it was not difficult to surmount, and it was itself carelessly guarded. Coming often thither, and entertaining conferences about the release of Damippus, he had pretty well calculated the height of the tower, and got ladders prepared. The Syracusans celebrated a feast to Diana; this juncture of time, when they were given up entirely to wine and sport, Marcellus laid hold of, and, before the citizens perceived it, not only possessed himself of the tower, but, before the break of day, filled the wall around with soldiers, and made his way into the Hexapylum. The Syracusans now beginning to stir, and to be alarmed at the tumult, he ordered the trumpets everywhere to sound, and thus frightened them all into flight, as if all parts of the city were already won, though the most fortified, and the fairest, and most ample quarter was still ungained. It is called Acradina, and was divided by a wall from the outer city, one part of which they call Neapolis, the other Tycha. Possessing himself of these, Marcellus, about break of day, entered through the Hexapylum, all his officers congratulating him. But looking down from the higher places upon the beautiful and spacious city below, he is said to have wept much, commiserating the calamity that hung over it, when his thoughts represented to him, how dismal and foul the face of the city would in a few hours be, when plundered and sacked by the soldiers. For among the officers of his army there was not one man that durst deny the plunder of the city to the soldiers' demands; nay, many were instant that it should be set on fire and laid level to the ground: but this Marcellus would not listen to. Yet he granted, but with great unwillingness

[5] Cicero, when quaestor in Sicily, found this tomb overgrown with vegetation, but still bearing the cylinder with the sphere, and he restored it (*Tusc. Disp.* v. 64–66).

and reluctance, that the money and slaves should be made prey; giving orders, at the same time, that none should violate any free person, nor kill, misuse, or make a slave of any of the Syracusans. Though he had used this moderation, he still esteemed the condition of that city to be pitiable, and, even amidst the congratulations and joy, showed his strong feelings of sympathy and commiseration at seeing all the riches accumulated during a long felicity, now dissipated in an hour. For it is related, that no less prey and plunder was taken here, than afterward in Carthage. For not long after, they obtained also the plunder of the other parts of the city, which were taken by treachery; leaving nothing untouched but the king's money, which was brought into the public treasury. But what specially grieved Marcellus was the death of Archimedes. For it chanced that he was alone, examining a diagram closely; and having fixed both his mind and his eyes on the object of his inquiry, he perceived neither the inroad of the Romans nor the taking of the city. Suddenly a soldier came up to him and bade him follow to Marcellus, but he would not go until he had finished the problem and worked it out to the demonstration. Thereupon the soldier became enraged, drew his sword and dispatched him. Others, however, say that the Roman came upon him with drawn sword intending to kill him at once, and that Archimedes, on seeing him, besought and entreated him to wait a little while so that he might not leave the question unfinished and only partly investigated; but the soldier did not understand and slew him. There is also a third story, that as he was carrying to Marcellus some of his mathematical instruments, such as sundials, spheres and angles adjusted to the apparent size of the sun, some soldiers fell in with him and, under the impression that he carried treasure in the box, killed him. What is, however, agreed is that Marcellus was distressed, and turned away from the slayer as from a polluted person, and sought out the relatives of Archimedes to do them honour.

Vitruvius, *On Architecture*

ARCHIMEDES made many wonderful discoveries of different kinds, but of all these that which I shall now explain seems to exhibit a boundless ingenuity. When Hiero was greatly exalted in the royal power at Syracuse, in return for the success of his policy he determined to set up in a certain shrine a golden crown as a votive offering to the immortal gods. He let out the work for a stipulated payment, and weighed out the exact amount of gold for the contractor. At the appointed time the contractor brought his work skilfully executed for the king's approval, and he seemed to have fulfilled exactly the requirement about the weight of the crown. Later information was given that gold had been removed and an equal weight of silver added in the making of the crown. Hiero was indignant at this disrespect for himself, and, being unable to discover any means by which he might unmask the fraud, he asked Archimedes to give it his attention. While Archimedes was turning the problem over, he chanced to come to the place of bathing, and there, as he was sitting down in the tub, he noticed that the amount of water which flowed over the tub was equal to the amount by which his body was immersed. This indicated to him a means of solving the problem, and he did not delay, but in his joy leapt out of the tub and, rushing naked towards his home, he cried out with a loud voice that he had found what he sought. For as he ran he repeatedly shouted in Greek, *heureka, heureka.*

Then, following up his discovery, he is said to have made two masses of the same weight as the crown, the one of gold and the other of silver. When he had so done, he filled a large vessel right up to the brim with water, into which he dropped the silver mass. The amount by which it was immersed in the vessel was the amount of water which overflowed. Taking out the mass, he poured back the amount by which the water had been depleted, measuring it with a pint pot, so that as before the water was made level with the brim. In this way he found what weight of silver answered to a certain measure of water.

When he had made this test, in like manner he dropped the golden mass into the full vessel. Taking it out again, for the same reason he added a measured quantity of water, and found that the deficiency of water was not the same, but less; and the amount by which it was less corresponded with the excess of a mass of silver, having the same weight, over a mass of gold. After filling the vessel again, he then dropped the crown itself into the water, and found that more water overflowed in the case of the crown than in the case of the golden mass of identical weight; and so, from the fact that more water was needed to make up the deficiency in the case of the crown than in the case of the mass, he calculated and detected the mixture of silver with the gold and the contractor's fraud stood revealed.[6]

[6] The method may be thus expressed analytically.

Let w be the weight of the crown, and let it be made up of a weight w_1 of gold and a weight w_2 of silver, so that $w = w_1 + w_2$.

Let the crown displace a volume v of water.

Let the weight w of gold displace a volume v_1 of water; then a weight w_1 of gold displaces a volume $\dfrac{w_1}{w} \cdot v_1$ of water.

Let the weight w of silver displace a volume v_2 of water; then a weight w_2 of silver displaces a volume $\dfrac{w_2}{w} \cdot v_2$ of water.

It follows that $v = \dfrac{w_1}{w} \cdot v_1 + \dfrac{w_2}{w} \cdot v_2$

Tzetzes, *Book of Histories* [7]

ARCHIMEDES the wise, the famous maker of engines, was a Syracusan by race, and worked at geometry till old age, surviving five-and-seventy-years [8]; he reduced to his service many mechanical powers, and with his triple-pulley device, using only his left hand, he drew a vessel of fifty thousand medimni burden. Once, when Marcellus, the Roman general, was assaulting Syracuse by land and sea, first by his engines he drew up some merchant-vessels, lifted them up against the wall of Syracuse, and sent them in a heap again to the bottom, crews and all. When Marcellus had withdrawn his ships a little distance, the old man gave all the Syracusans power to lift stones large enough to load a waggon and, hurling them one after the other, to sink the ships. When Marcellus withdrew them a bow-shot, the old man constructed a kind of hexagonal mirror, and at an interval proportionate to the size of the mirror he set similar small mirrors with four edges, moved by links and by a form of hinge, and made it the centre of the sun's beams—its noon-tide beam, whether in summer or in mid-winter. Afterwards, when the beams were reflected in the mirror, a fearful kindling of fire was raised in the ships, and at the distance of a bow-shot he turned them into ashes. In this way did the old man prevail over Marcellus with his weapons. In his Doric dialect, and in its Syracusan variant, he declared: "If I have somewhere to stand, I will move the whole earth with my *charistion*." [9]

$$= \frac{w_1 v_1 + w_2 v_2}{w_1 + w_2},$$

so that
$$\frac{w_1}{w_2} = \frac{v_2 - v}{v - v_1}.$$

[7] The lines which follow are an example of the "political" verse which prevailed in Byzantine times. The name is given to verse composed by accent instead of quantity, with an accent on the last syllable but one, especially an iambic verse of fifteen syllables. The twelfth-century Byzantine pedant, John Tzetzes, preserved in his *Book of Histories* a great treasure of literary, historical, theological and scientific detail, but it needs to be used with caution. The work is often called the *Chiliades* from its arbitrary division by its first edition (N. Gerbel, 1546) into books of 1000 lines each—it actually contains 12,674 lines.

[8] As he perished in the sack of Syracuse in 212 B.C., he was therefore born about 287 B.C.

[9] The instrument is otherwise mentioned by Simplicius (in *Aristot. Phys.*, ed. Diels 1110. 2–5) and it is implied that it was used for weighing. As Tzetzes in another place writes of a triple-pulley device in the same connexion, it may be presumed to have been of this nature.

COMMENTARY ON
Greek Mathematics

FROM Ivor Thomas' admirable source book in the Loeb Classical Library,[1] I have chosen a number of examples of the work of Greek mathematicians, fragments of a tale of "one of the most stupendous achievements in the history of human thought." The selections vary from brief comments and anecdotes to complete proofs, in original form, of some of the beautiful theorems on which all subsequent mathematical science is based. Turnbull's little book contains (pp. 75–168) a lucid survey of Greek mathematics; here are reproduced a few of the classical achievements which he discusses. Among the items below are Euclid's proof of the Pythagorean theorem; Plato's comments in the *Republic* on the philosophy of mathematics; Euclid's demonstration of the method of exhaustion; Archimedes on the principle of the lever, on the prodigious cattle problem and on the displacement of fluids by solids—whereby he may have discovered the proportions of gold and silver in King Hiero's crown; Eratosthenes on measuring the size of the earth; Diophantus' algebraic epitaph, inscribed on his tombstone; Pappus on isoperimetric figures and on the "geometrical forethought" of bees. I hope these will enable the reader to sense the excitement of great inventions and to imbibe ideas and the method of deduction from some of the foremost of ancient thinkers.

[1] *Selections Illustrating the History of Greek Mathematics*, 2 vols., Cambridge, Mass., 1939.

Except the blind forces of Nature, nothing moves in this world which is not Greek in its origin. —SIR HENRY JAMES SUMNER MAINE

Grammarian, rhetorician, geometer, painter, trainer, soothsayer, rope-dancer, physician, wizard—he knows everything. Bid the hungry Greekling go to heaven! He'll go. —JUVENAL

4 Greek Mathematics

By IVOR THOMAS

PYTHAGORAS

(DIOGENES LAERTIUS)

HE [Pythagoras] it was who brought geometry to perfection, after Moeris had first discovered the beginnings of the elements of that science, as Anticleides says in the second book of his *History of Alexander*. He adds that Pythagoras specially applied himself to the arithmetical aspect of geometry and he discovered the musical intervals on the monochord; nor did he neglect even medicine. Apollodorus the calculator says that he sacrificed a hecatomb on finding that the square on the hypotenuse of the right-angled triangle is equal to the squares on the sides containing the right angle. And there is an epigram as follows:

As when Pythagoras the famous figure found,
For which a sacrifice renowned he brought.

(PROCLUS, *on Euclid*)

Whatsoever offers a more profitable field of research and contributes to the whole of philosophy, we shall make the starting-point of further inquiry, therein imitating the Pythagoreans, among whom there was prevalent this motto, "A figure and a platform, not a figure and sixpence," by which they implied that the geometry deserving study is that which, at each theorem, sets up a platform for further ascent and lifts the soul on high, instead of allowing it to descend among sensible objects and so fulfil the common needs of mortal men and in this lower aim neglect conversion to things above.

(PLUTARCH, *The Epicurean Life*)

Pythagoras sacrificed an ox in virtue of his proposition, as Apollodorus says—

As when Pythagoras the famous figure found
For which the noble sacrifice he brought

whether it was the theorem that the square on the hypotenuse is equal to the squares on the sides containing the right angle, or the problem about the application of the area.

(PLUTARCH, *Convivial Questions*)

Among the most geometrical theorems, or rather problems, is this— given two figures, to apply a third equal to the one and similar to the other; it was in virtue of this discovery they say Pythagoras sacrificed. This is unquestionably more subtle and elegant than the theorem which he proved that the square on the hypotenuse is equal to the squares on the sides about the right angle.

SUM OF THE ANGLES OF A TRIANGLE
(PROCLUS, *on Euclid*)

Eudemus the Peripatetic ascribes to the Pythagoreans the discovery of this theorem, that any triangle has its internal angles equal to two right angles. He says they proved the theorem in question after this fashion. Let

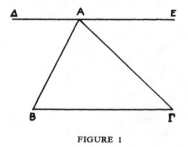

FIGURE 1

ABΓ be a triangle, and through A let ΔE be drawn parallel to BΓ. Now since BΓ, ΔE are parallel, and the alternate angles are equal, the angle ΔAB is equal to the angle ABΓ, and EAΓ is equal to AΓB. Let BAΓ be added to both. Then the angles ΔAB, BAΓ, ΓAE, that is, the angles ΔAB, BAE, that is, two right angles, are equal to the three angles of the triangle. Therefore the three angles of the triangle are equal to two right angles.

"PYTHAGORAS'S THEOREM"
(EUCLID, *Elements* i. 47)

In right-angled triangles the square on the side subtending the right angle is equal to the squares on the sides containing the right angle.

Let ABΓ be a right-angled triangle having the angle BAΓ right; I say that the square on BΓ is equal to the squares on BA, AΓ.

For let there be described on BΓ the square BΔEΓ, and on BA, AΓ the squares HB, ΘΓ [Eucl. i. 46], and through A let AΛ be drawn parallel to

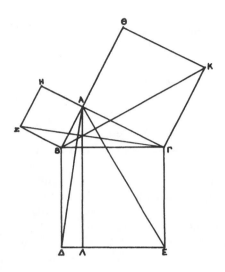

FIGURE 2

either BΔ or ΓE, and let AΔ, ZΓ be joined.[1] Then, since each of the angles
BAΓ, BAH is right, it follows that with a straight line BA and at the point
A on it, two straight lines AΓ, AH, not lying on the same side, make the
adjacent angles equal to two right angles; therefore ΓA is in a straight line
with AH [Eucl. i. 14]. For the same reasons BA is also in a straight line
with AΘ. And since the angle ΔBΓ is equal to the angle ZBA, for each is
right, let the angle ABΓ be added to each; the whole angle ΔBA is therefore

---

[1] In this famous "windmill" figure, the lines AΔ, BK, ΓZ meet in a point. Euclid
has no need to mention this fact, but it was proved by Heron.

If AΔ, the perpendicular from A, meets BΓ in M, as in the detached portion of the
figure here reproduced, the triangles MBA, MAΓ are similar to the triangle ABΓ and
to one another. It follows from Eucl. *Elem.* vi. 4 and 17 (which do not depend on
i. 47) that

$$BA^2 = BM . BΓ, \text{ and } AΓ^2 = ΓM . BΓ.$$
Therefore $$BA^2 + AΓ^2 = BΓ (BM + ΓM) = BΓ^2.$$

The theory of proportion developed in Euclid's sixth book therefore offers a simple
method of proving "Pythagoras's Theorem." This proof, moreover, is of the same type
as Eucl. *Elem.* i. 47 inasmuch as it is based on the equality of the square on BΓ to
the sum of two rectangles. This has suggested that Pythagoras proved the theorem by
means of his inadequate theory of proportion, which applied only to commensurable
magnitudes. When the incommensurable was discovered, it became necessary to find a
new proof independent of proportions. Euclid therefore recast Pythagoras's invali-
dated proof in the form here given so as to get it into the first book in accordance
with his general plan of the *Elements*.

For other methods by which the theorem can be proved, the complete evidence
bearing on its reputed discovery by Pythagoras, and the history of the theorem in
Egypt, Babylonia, and India, see Heath, *The Thirteen Books of Euclid's Elements*, i.,
pp. 351–366, *A Manual of Greek Mathematics*, pp. 95–100.

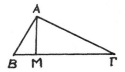

equal to the whole angle ZBΓ. And since ΔB is equal to BΓ, and ZB to BA,
the two [sides] ΔB, BA are equal to the two [sides] BΓ, ZB respectively;
and the angle ΔBA is equal to the angle ZBΓ. The base AΔ is therefore
equal to the base ZΓ, and the triangle ABΔ is equal to the triangle ZBΓ
[Eucl. i. 4]. Now the parallelogram BΛ is double the triangle ABΔ, for they
have the same base BΔ and are in the same parallels BΔ, AΛ [Eucl. i. 41].
And the square HB is double the triangle ZBΓ, for they have the same base
ZB and are in the same parallels ZB, HΓ. Therefore the parallelogram BΛ
is equal to the square HB. Similarly, if AE, BK are joined, it can also be
proved that the parallelogram ΓΛ is equal to the square ΘΓ. Therefore the
whole square BΔEΓ is equal to the two squares HB, ΘΓ. And the square
BΔEΓ is described on BΓ, while the squares HB, ΘΓ are described on BA,
AΓ. Therefore the square on the side BΓ is equal to the squares on the
sides BA, AΓ.

Therefore in right-angled triangles the square on the side subtending
the right angle is equal to the squares on the sides containing the right
angle; which was to be proved.

PLATO

(ARISTOXENUS, *Elements of Harmony*)

IT is perhaps well to go through in advance the nature of our inquiry,
so that, knowing beforehand the road along which we have to travel, we
may have an easier journey, because we will know at what stage we are
in, nor shall we harbour to ourselves a false conception of our subject.
Such was the condition, as Aristotle often used to tell, of most of the
audience who attended Plato's lecture on the Good. Every one went there
expecting that he would be put in the way of getting one or other of the
things accounted good in human life, such as riches or health or strength
or, in fine, any extraordinary gift of fortune. But when they found that
Plato's arguments were of mathematics and numbers and geometry and
astronomy and that in the end he declared the One to be the Good, they

were altogether taken by surprise. The result was that some of them scoffed at the thing, while others found great fault with it.

PHILOSOPHY OF MATHEMATICS
(PLATO, *Republic* vi. 510)

I think you know that those who deal with geometrics and calculations and such matters take for granted the odd and the even, figures, three kinds of angles and other things cognate to these in each field of inquiry; assuming these things to be known, they make them hypotheses, and henceforth regard it as unnecessary to give any explanation of them either to themselves or to others, treating them as if they were manifest to all; setting out from these hypotheses, they go at once through the remainder of the argument until they arrive with perfect consistency at the goal to which their inquiry was directed.

Yes, he said, I am aware of that.

Therefore I think you also know that although they use visible figures and argue about them, they are not thinking about these figures but of those things which the figures represent; thus it is the square in itself and the diameter in itself which are the matter of their arguments, not that which they draw; similarly, when they model or draw objects, which may themselves have images in shadows or in water, they use them in turn as images, endeavouring to see those absolute objects which cannot be seen otherwise than by thought.

EUCLID

(STOBAEUS, *Extracts*)

SOMEONE who had begun to read geometry with Euclid, when he had learnt the first theorem asked Euclid, "But what advantage shall I get by learning these things?" Euclid called his slave and said, "Give him threepence, since he must needs make profit out of what he learns."

METHOD OF EXHAUSTION
(EUCLID, *Elements* xii, 2 [2])

Circles are to one another as the squares on the diameters.

Let ABΓΔ, EZHΘ be circles, and BΔ, ZΘ their diameters; I say that, as the circle ABΓΔ is to the circle EZHΘ, so is the square on BΔ to the square on ZΘ.

For if the circle ABΓΔ is not to the circle EZHΘ as the square on BΔ to the square on ZΘ, then the square on BΔ will be to the square on ZΘ as

[2] Eudemus attributed the discovery of this important theorem to Hippocrates. Unfortunately we do not know how Hippocrates proved it.

the circle ABΓΔ is to some area either less than the circle EZHΘ or greater. Let it first be in that ratio to a lesser area Σ. And let the square EZHΘ be inscribed in the circle EZHΘ; then the inscribed square is greater than the half of the circle EZHΘ, inasmuch as, if through the points E, Z, H, Θ we draw tangents to the circle, the square EZHΘ is half the square circumscribed about the circle, and the circle is less than the circumscribed square; so that the inscribed square EZHΘ is greater than the half of the circle EZHΘ. Let the circumferences EZ, ZH, HΘ, ΘE be bisected at the points K, Λ, M, N, and let EK, KZ, ZΛ, ΛH, HM, MΘ, ΘN, NE be joined;

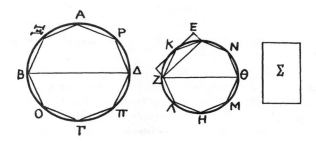

FIGURE 4

therefore each of the triangles EKZ, ZΛH, HMΘ, ΘNE is greater than the half of the segment of the circle about it, inasmuch as, if through the points K, Λ, M, N we draw tangents to the circle and complete the parallelograms on the straight lines EZ, ZH, HΘ, ΘE, each of the triangles EKZ, ZΛH, HMΘ, ΘNE will be half of the parallelogram about it, while the segment about it is less than the parallelogram; so that each of the triangles EKZ, ZΛH, HMΘ, ΘNE is greater than the half of the segment of the circle about it. Thus, by bisecting the remaining circumferences and joining straight lines, and doing this continually, we shall leave some segments of the circle which will be less than the excess by which the circle EZHΘ exceeds the area Σ. For it was proved in the first theorem of the tenth book that, if two unequal magnitudes be set out, and if from the greater there be subtracted a magnitude greater than its half, and from the remainder a magnitude greater than its half, and so on continually, there will be left some magnitude which is less than the lesser magnitude set out. Let such segments be then left, and let the segments of the circle EZHΘ on EK, KZ, ZΛ, ΛH, HM, MΘ, ΘN, NE be less than the excess by which the circle EZHΘ exceeds the area Σ. Therefore the remainder, the polygon EKZΛHMΘN, is greater than the area Σ. Let there be inscribed, also, in the circle ABΓΔ the polygon AΞBOΓΠΔP similar to the polygon EKZΛHMΘN; therefore as the square on BΔ is to the square on ZΘ, so is the polygon AΞBOΓΠΔP to the polygon EKZΛHMΘN. But as the square on BΔ is to the square on ZΘ, so is the circle ΑΒΓΔ to the area Σ; therefore also as the circle ΑΒΓΔ is to the area Σ, so is the polygon ΑΞΒΟΓΠΔΡ to the polygon ΕΚΖΛΗΜΘΝ; therefore, alternately, as the circle ΑΒΓΔ is to the polygon in it, so is the area Σ to the polygon ΕΚΖΛΗΜΘΝ. Now the circle ΑΒΓΔ is greater than the polygon in it; therefore the area Σ also is greater than the polygon ΕΚΖΛΗΜΘΝ. But it is also less; which is impossible. Therefore it is not true that, as is the square on ΒΔ to the square on ΖΘ, so is the circle ΑΒΓΔ to some area less than the circle ΕΖΗΘ. Similarly we shall prove that neither is it true that, as the square on ΖΘ is to the square on ΒΔ, so is the circle ΕΖΗΘ to some area less than the circle ΑΒΓΔ.

I say now that neither is the circle ΑΒΓΔ towards some area greater than the circle ΕΖΗΘ as the square on ΒΔ is to the square on ΖΘ.

For, if possible, let it be in that ratio to some greater area Σ. Therefore, inversely, as the square on ΖΘ is to the square on ΔΒ, so is the area Σ to the circle ΑΒΓΔ. But as the area Σ is to the circle ΑΒΓΔ, so is the circle ΕΖΗΘ to some area less than the circle ΑΒΓΔ; therefore also, as the square on ΖΘ is to the square on ΒΔ, so is the circle ΕΖΗΘ to some area less than the circle ΑΒΓΔ; which was proved impossible. Therefore it is not true that, as the square on ΒΔ is to the square on ΖΘ, so is the circle ΑΒΓΔ to some area greater than the circle ΕΖΗΘ. And it was proved not to be in that relation to a less area; therefore as the square on ΒΔ is to the square on ΖΘ, so is the circle ΑΒΓΔ to the circle ΕΖΗΘ.

Therefore circles are to one another as the squares on the diameters; which was to be proved.

ARCHIMEDES

PRINCIPLE OF THE LEVER
(ARCHIMEDES, *On Plane Equilibriums*)

Proposition 6

COMMENSURABLE magnitudes balance at distances reciprocally proportional to their weights.

Let A, B be commensurable magnitudes with centres [of gravity] A, B, and let ΕΔ be any distance, and let

$$A : B = \Delta\Gamma : \Gamma E;$$

it is required to prove that the centre of gravity of the magnitude composed of both A, B is Γ.

Since $A : B = \Delta\Gamma : \Gamma E,$

and A is commensurate with B, therefore ΓΔ is commensurate with ΓΕ, that is, a straight line with a straight line [Eucl. x. 11]; so that EΓ, ΓΔ have a common measure. Let it be N, and let ΔH, ΔK be each equal to EΓ, and let EΛ be equal to ΔΓ. Then since ΔH = ΓE, it follows that ΔΓ = EH; so that ΛEE = H. Therefore ΛH = 2ΔΓ and HK = 2ΓE; so that N measures both ΛH and HK, since it measures their halves [Eucl. x. 12]. And since

$$A : B = ΔΓ : ΓE,$$

while $$ΔΓ : ΓE = ΛH : HK—$$

for each is double of the other—

therefore $$A : B = ΛH : HK.$$

Now let Z be the same part of A as N is of ΔH;

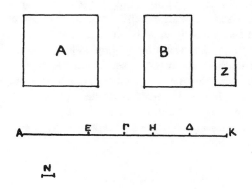

FIGURE 5

then $$ΔH : N = A : Z.$$ [Eucl. v., Def. 5
And $$KH : ΛH = B : A;$$ [Eucl. v. 7, coroll.
therefore, *ex aequo*,

$$KH : N = B : Z;$$ [Eucl. v. 22

therefore Z is the same part of B as N is of KH. Now A was proved to be a multiple of Z; therefore Z is a common measure of A, B. Therefore, if ΛH is divided into segments equal to N and A into segments equal to Z, the segments in ΛH equal in magnitude to N will be equal in number to the segments of A equal to Z. It follows that, if there be placed on each of the segments in ΛH a magnitude equal to Z, having its centre of gravity at the middle of the segment, the sum of the magnitudes will be equal to Λ, and the centre of gravity of the figure compounded of them all will be E; for they are even in number, and the numbers on either side of E will be equal because ΛE = HE.

Similarly it may be proved that, if a magnitude equal to Z be placed on each of the segments [equal to N] in KH, having its centre of gravity at the middle of the segment, the sum of the magnitudes will be equal to B, and the centre of gravity of the figure compounded of them all will be Δ. Therefore A may be regarded as placed at E, and B at Δ. But they will be a set of magnitudes lying on a straight line, equal one to another, with their centres of gravity at equal intervals, and even in number; it is therefore clear that the centre of gravity of the magnitude compounded of them all is the point of bisection of the line containing the centres [of gravity] of the middle magnitudes. And since ΛE = ΓΔ and EΓ = ΔK, therefore ΛΓ = ΓK; so that the centre of gravity of the magnitude compounded of them all is the point Γ. Therefore if A is placed at E and B at Δ, they will balance about Γ.

INDETERMINATE ANALYSIS: THE CATTLE PROBLEM
(ARCHIMEDES (?), *Cattle Problem* [3])

A problem which Archimedes solved in epigrams, and which he communicated to students of such matters at Alexandria in a letter to Eratosthenes of Cyrene.

If thou art diligent and wise, O stranger, compute the number of cattle of the Sun, who once upon a time grazed on the fields of the Thrinacian isle of Sicily, divided into four herds of different colours, one milk white, another a glossy black, the third yellow and the last dappled. In each herd were bulls, mighty in number according to these proportions: Understand, stranger, that the white bulls were equal to a half and a third of the black together with the whole of the yellow, while the black were equal to the fourth part of the dappled and a fifth, together with, once more, the whole of the yellow. Observe further that the remaining bulls, the dappled, were equal to a sixth part of the white and a seventh, together with all the yellow. These were the proportions of the cows: The white were precisely equal to the third part and a fourth of the whole herd of the black; while the black were equal to the fourth part once more of the dappled and with it a fifth part, when all, including the bulls, went to pasture together. Now the dappled in four parts [4] were equal in number to a fifth part and a sixth of the yellow herd. Finally the yellow were in number equal to a sixth part and a seventh of the white herd. If thou canst accurately tell, O stranger, the number of cattle of the Sun, giving separately the number of well-fed

---

[3] It is unlikely that the epigram itself, first edited by G. E. Lessing in 1773, is the work of Archimedes, but there is ample evidence from antiquity that he studied the actual problem. The most important papers bearing on the subject are A. Amthor, *Zeitschrift für Math. u. Physik* (Hist.-litt. Abtheilung), xxv. (1880), pp. 153–171, supplementing an article by B. Krumbiegel (pp. 121–136) on the authenticity of the problem. See also T. L. Heath, *The Works of Archimedes*, pp. 319–326.

[4] *i.e.* a fifth and a sixth both of the males and of the females.

bulls and again the number of females according to each colour, thou wouldst not be called unskilled or ignorant of numbers, but not yet shalt thou be numbered among the wise. But come, understand also all these conditions regarding the cows of the Sun. When the white bulls mingled their number with the black, they stood firm, equal in depth and breadth,[5] and the plains of Thrinacia, stretching far in all ways, were filled with their multitude. Again, when the yellow and the dappled bulls were gathered into one herd they stood in such a manner that their number, beginning from one, grew slowly greater till it completed a triangular figure, there being no bulls of other colours in their midst nor none of them lacking. If thou art able, O stranger, to find out all these things and gather them together in your mind, giving all the relations, thou shalt depart crowned with glory and and knowing that thou hast been adjudged perfect in this species of wisdom.[6]

[5] At a first glance this would appear to mean that the sum of the number of white and black bulls is a square, but this makes the solution of the problem intolerably difficult. There is, however, an easier interpretation. If the bulls are packed together so as to form a square figure, their number need not be a square, since each bull is longer than it is broad. The simplified condition is that the sum of the number of white and black bulls shall be a rectangle.

[6] If

X, x are the numbers of white bulls and cows respectively,
Y, y "   "   " black   "   "   "
Z, z "   "   " yellow   "   "   "
W, w "   "   " dappled   "   "   "

the first part of the epigram states that

(*a*)
$$X = (\tfrac{1}{2} + \tfrac{1}{3}) Y + Z \qquad (1)$$
$$Y = (\tfrac{1}{4} + \tfrac{1}{5}) W + Z \qquad (2)$$
$$W = (\tfrac{1}{6} + \tfrac{1}{7}) X + Z \qquad (3)$$

(*b*)
$$x = (\tfrac{1}{3} + \tfrac{1}{4})(Y + y) \qquad (4)$$
$$y = (\tfrac{1}{4} + \tfrac{1}{5})(W + w) \qquad (5)$$
$$w = (\tfrac{1}{5} + \tfrac{1}{6})(Z + z) \qquad (6)$$
$$z = (\tfrac{1}{6} + \tfrac{1}{7})(X + x) \qquad (7)$$

The second part of the epigram states that

$$X + Y = \text{a rectangular number} \qquad (8)$$
$$Z + W = \text{a triangular number} \qquad (9)$$

This was solved by J. F. Wurm, and the solution is given by A. Amthor, *Zeitschrift für Math. u. Physik.* (*Hist.-litt. Abtheilung*), xxv. (1880), pp. 153–171, and by Heath, *The Works of Archimedes*, pp. 319–326. For reasons of space, only the results can be noted here.

Equations (1) to (7) give the following as the values of the unknowns in terms of an unknown integer n:

$X = 10366482n$	$x = 7206360n$
$Y = 7460514n$	$y = 4893246n$
$Z = 4149387n$	$z = 5439213n$
$W = 7358060n$	$w = 3515820n.$

We have now to find a value of n such that equation (9) is also satisfied—equation (8) will then be simultaneously satisfied. Equation (9) means that

$$Z + W = \frac{p(p + 1)}{2},$$

SOLID IMMERSED IN A FLUID

(ARCHIMEDES, *On Floating Bodies*)

Solids heavier than a fluid will, if placed in the fluid, sink to the bottom, and they will be lighter [if weighed] in the fluid by the weight of a volume of the fluid equal to the volume of the solid.[7]

That they will sink to the bottom is manifest; for the parts of the fluid under them are under greater pressure than the parts lying evenly with them, since it is postulated that the solid is heavier than water; that they will be lighter, as aforesaid will be [thus] proved.

Let A be any magnitude heavier than the fluid, let the weight of the magnitude A be B + Γ, and let the weight of fluid having the same volume as A be B. It is required to prove that in the fluid the magnitude A will have a weight equal to Γ.

FIGURE 6

where p is some positive integer, or

$$(4149387 + 7358060)n = \frac{p(p+1)}{2},$$

i.e.

$$2471 . 4657n = \frac{p(p+1)}{2}.$$

This is found to be satisfied by $n = 3^3 . 4349,$ and the final solution is

$X = 1217263415886$	$x = 846192410280$
$Y = 876035935422$	$y = 574579625058$
$Z = 487233469701$	$z = 638688708099$
$W = 864005479380$	$w = 412838131860$

and the total is 5916837175686.

If equation (8) is taken to be that $X + Y = $ a square number, the solution is much more arduous; Amthor found that in this case,

$$W = 1598 \; \langle \overline{206541} \rangle,$$

where $\langle \overline{206541} \rangle$ means that there are 206541 more digits to follow, and the whole number of cattle $= 7766 \; \langle \overline{206541} \rangle$. Merely to write out the eight numbers, Amthor calculates, would require a volume of 660 pages, so we may reasonably doubt whether the problem was really framed in this more difficult form, or, if it were, whether Archimedes solved it.

[7] Or, as we should say, "lighter by the weight of fluid displaced."

For let there be taken any magnitude Δ lighter than the same volume of the fluid such that the weight of the magnitude Δ is equal to the weight B, while the weight of the fluid having the same volume as the magnitude Δ is equal to the weight B + Γ. Then if we combine the magnitudes A, Δ, the combined magnitude will be equal to the weight of the same volume of the fluid; for the weight of the combined magnitudes is equal to the weight (B + Γ) + B, while the weight of the fluid having the same volume as both the magnitudes is equal to the same weight. Therefore, if the [combined] magnitudes are placed in the fluid, they will balance the fluid, and will move neither upwards nor downwards; for this reason the magnitude A will move downwards, and will be subject to the same force as that by which the magnitude Δ is thrust upwards, and since Δ is lighter than the fluid it will be thrust upwards by a force equal to the weight Γ; for it has been proved that when solid magnitudes lighter than the fluid are forcibly immersed in the fluid, they will be thrust upwards by a force equal to the difference in weight between the magnitude and an equal volume of the fluid [Prop. 6]. But the fluid having the same volume as Δ is heavier than the magnitude Δ by the weight Γ; it is therefore plain that the magnitude A will be borne upwards by a force equal to Γ.[8]

APOLLONIUS OF PERGA

CONSTRUCTION OF THE SECTIONS

(APOLLONIUS, *Conics*)

Proposition 7 [9]

IF a cone be cut by a plane through the axis, and if it be also cut by another plane cutting the plane containing the base of the cone in a straight line perpendicular to the base of the axial triangle,[10] or to the base produced, a section will be made on the surface of the cone by the cutting plane, and straight lines drawn in it parallel to the straight line perpendicu-

---

[8] This proposition suggests a method, alternative to that given by Vitruvius, whereby Archimedes may have discovered the proportions of gold and silver in King Hiero's crown.

Let w be the weight of the crown, and let w_1 and w_2 be the weights of gold and silver in it respectively, so that $w = w_1 + w_2$.

Take a weight w of gold and weigh it in a fluid, and let the loss of weight be P_1. Then the loss of weight when a weight w_1 of gold is weighed in the fluid, and consequently the weight of fluid displaced, will be $\dfrac{w_1}{w} \cdot P_1$.

Now take a weight of w of silver and weigh it in the fluid, and let the loss of weight be P_2. Then the loss of weight when a weight w_2 of silver is weighed in the fluid, and consequently the weight of fluid displaced, will be $\dfrac{w_2}{w} \cdot P_2$.

lar to the base of the axial triangle will meet the common section of the cutting plane and the axial triangle and, if produced to the other part of the section, will be bisected by it; if the cone be right, the straight line in the base will be perpendicular to the common section of the cutting plane and the axial triangle; but if it be scalene, it will not in general be perpendicular, but only when the plane through the axis is perpendicular to the base of the cone.

Let there be a cone whose vertex is the point A and whose base is the circle BΓ, and let it be cut by a plane through the axis, and let the section so made be the triangle ABΓ. Now let it be cut by another plane cutting the plane containing the circle BΓ in a straight line ΔE which is either perpendicular to BΓ or to BΓ produced, and let the section made on the surface of the cone be ΔZE [11]; then the common section of the cutting plane

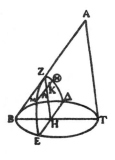

FIGURE 7

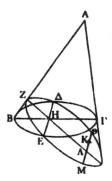

FIGURE 8

Finally, weigh the crown itself in the fluid, and let the loss of weight, and consequently the weight of fluid displaced, be P.

It follows that $\dfrac{w_1}{w} \cdot P_1 + \dfrac{w_2}{w} \cdot P_2 = P,$

whence $\dfrac{w_1}{w_2} = \dfrac{P_2 - P}{P - P_1}.$

[9] This proposition defines a conic section in the most general way with reference to any diameter. It is only much later in the work that the principal axes are introduced as diameters at right angles to their ordinates. The proposition is an excellent example of the generality of Apollonius's methods.

Apollonius followed rigorously the Euclidean form of proof. In consequence his general enunciations are extremely long and often can be made tolerable in an English rendering only by splitting them up; but, though Apollonius seems to have taken a malicious pleasure in their length, they are formed on a perfect logical pattern without a superfluous word.

[10] Lit. "the triangle through the axis."

[11] This applies only to the first two of the figures given in the MSS. (i.e. figures 7, 8, above).

and of the triangle ABΓ is ZH. Let any point Θ be taken on ΔZE, and through Θ let ΘK be drawn parallel to ΔE. I say that ΘK intersects ZH and, if produced to the other part of the section ΔZE, it will be bisected by the straight line ZH.

For since the cone, whose vertex is the point A and base the circle BΓ, is cut by a plane through the axis and the section so made is the triangle ABΓ, and there has been taken any point Θ on the surface, not being on a side of the triangle ABΓ, and ΔH is perpendicular to BΓ, therefore the straight line drawn through Θ parallel to ΔH, that is ΘK, will meet the triangle ABΓ and, if produced to the other part of the surface, will be bisected by the triangle. Therefore, since the straight line drawn through Θ parallel to ΔE meets the triangle ABΓ and is in the plane containing the section ΔZE, it will fall upon the common section of the cutting plane and the triangle ABΓ. But the common section of those planes is ZH; therefore the straight line drawn through Θ parallel to ΔE will meet ZH; and if it be

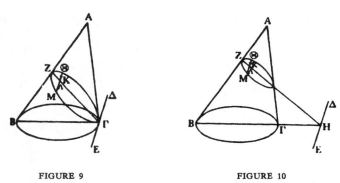

FIGURE 9   FIGURE 10

produced to the other part of the section ΔZE it will be bisected by the straight line ZH.

Now the cone is right, or the axial triangle ABΓ is perpendicular to the circle BΓ, or neither.

First, let the cone be right; then the triangle ABΓ will be perpendicular to the circle BΓ [Eucl. xi. 18]. Then since the plane ABΓ is perpendicular to the plane BΓ, and ΔE is drawn in one of the planes perpendicular to their common section BΓ, therefore ΔE is perpendicular to the triangle ABΓ [Eucl. xi. Def. 4]; and therefore it is perpendicular to all the straight lines in the triangle ABΓ which meet it [Eucl. xi. Def. 3]. Therefore it is perpendicular to ZH.

Now let the cone be not right. Then, if the axial triangle is perpendicular to the circle BΓ, we may similarly show that ΔE is perpendicular to ZH. Now let the axial triangle ABΓ be not perpendicular to the circle BΓ. I say that neither is ΔE perpendicular to ZH. For if it is possible, let it be; now it is also perpendicular to ΒΓ; therefore ΔΕ is perpendicular to both ΒΓ, ΖΗ. And therefore it is perpendicular to the plane through ΒΓ, ΖΗ [Eucl. xi. 4]. But the plane through ΒΓ, ΗΖ is ΑΒΓ; and therefore ΔΕ is perpendicular to the triangle ΑΒΓ. Therefore all the planes through it are perpendicular to the triangle ΑΒΓ [Eucl. xi. 18]. But one of the planes through ΔΕ is the circle ΒΓ; therefore the circle ΒΓ is perpendicular to the triangle ΑΒΓ. Therefore the triangle ΑΒΓ is perpendicular to the circle ΒΓ; which is contrary to hypothesis. Therefore ΔΕ is not perpendicular to ΖΗ.

Corollary

From this it is clear that ΖΗ is a diameter of the section ΔΖΕ, inasmuch as it bisects the straight lines drawn parallel to the given straight line ΔΕ, and also that parallels can be bisected by the diameter ΖΗ without being perpendicular to it.

FUNDAMENTAL PROPERTIES

(APOLLONIUS, *Conics*)

Proposition 11

Let a cone be cut by a plane through the axis, and let it be also cut by another plane cutting the base of the cone in a straight line perpendicular to the base of the axial triangle, and further let the diameter of the section be parallel to one side of the axial triangle; then if any straight line be drawn from the section of the cone parallel to the common section of the cutting plane and the base of the cone as far as the diameter of the section, its square will be equal to the rectangle bounded by the intercept made by it on the diameter in the direction of the vertex of the section and a certain other straight line; this straight line will bear the same ratio to the intercept between the angle of the cone and the vertex of the segment as the square on the base of the axial triangle bears to the rectangle bounded by the remaining two sides of the triangle; and let such a section be called a parabola.

For let there be a cone whose vertex is the point Α and whose base is the circle ΒΓ, and let it be cut by a plane through the axis, and let the section so made be the triangle ΑΒΓ, and let it be cut by another plane cutting the base of the cone in the straight line ΔΕ perpendicular to ΒΓ, and let the section so made on the surface of the cone be ΔΖΕ, and let ΖΗ, the diameter of the section, be parallel to ΑΓ, one side of the axial triangle and from the point Ζ let ΖΘ be drawn perpendicular to ΖΗ, and let ΒΓ² : ΒΑ . ΑΓ = ΖΘ : ΖΑ, and let any point Κ be taken at random on the section, and through Κ let ΚΛ be drawn parallel to ΔΕ. I say that ΚΛ² = ΘΖ . ΖΛ.

For let ΜΝ be drawn through Λ parallel to ΒΓ; but ΚΛ is parallel to ΔΕ;

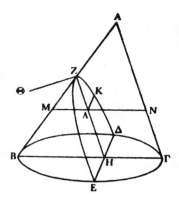

therefore the plane through KΛ, MN is parallel to the plane through BΓ, ΔE [Eucl. xi. 15], that is to the base of the cone. Therefore the plane through KΛ, MN is a circle, whose diameter is MN [Prop. 4]. And KΛ is perpendicular to MN, since ΔE is perpendicular to BΓ [Eucl. xi. 10];

therefore $MΛ . ΛN = KΛ^2$.
And since $BΓ^2 : BA . AΓ = ΘZ : ZA$,
while $BΓ^2 : BA . AΓ = (BΓ : ΓA)(BΓ : BA)$,
therefore $ΘZ : ZA = (BΓ : ΓA)(ΓB : BA)$.
But $BΓ : ΓA = MN : NA$
                                     $= MΛ : ΛZ$,   [Eucl. vi. 4
and $BΓ : BA = MN : MA$
                                     $= ΛM : MZ$ [*ibid.*
                                     $= NΛ : ZA$.   [Eucl. vi. 2
Therefore $ΘZ : ZA = (MΛ : ΛZ)(NΛ : ZA)$.
But $(MΛ : ΛZ)(ΛN : ZA) = MΛ . ΛN : ΛZ . ZA$.
Therefore $ΘZ : ZA = MΛ . ΛN : ΛZ . ZA$.
But $ΘZ : ZA = ΘZ . ZΛ : ΛZ . ZA$,
by taking a common height ZΛ;
therefore $MΛ . ΛN : ΛZ . ZA = ΘZ . ZΛ : ΛZ . ZA$.
Therefore $MΛ . ΛN = ΘZ . ZΛ$.   [Eucl. v. 9
But $MΛ . ΛN = KΛ^2$;
and therefore $KΛ^2 = ΘZ . ZΛ$.

Let such a section be called a *parabola,* and let ΘZ be called the *parameter of the ordinates* to the diameter ZH, and let it also be called the *erect side* (*latus rectum*).[12]

───────────

[12] A *parabola* (παραβολή) because the square on the ordinate KΛ is *applied* (παραβαλεῖν) to the parameter ΘZ in the form of the rectangle ΘZ . ZΛ, and is

ERATOSTHENES

MEASUREMENT OF THE EARTH

(CLEOMEDES,[13] *On the Circular Motion of the Heaven Bodies*)

SUCH then is Posidonius's method of investigating the size of the earth, but Eratosthenes' method depends on a geometrical argument, and gives the impression of being more obscure. What he says will, however, become clear if the following assumptions are made. Let us suppose, in this case also, first that Syene and Alexandria lie under the same meridian circle; secondly, that the distance between the two cities is 5000 stades; and thirdly, that the rays sent down from different parts of the sun upon different parts of the earth are parallel; for the geometers proceed on this assumption. Fourthly, let us assume that, as is proved by the geometers, straight lines falling on parallel straight lines make the alternate angles equal, and fifthly, that the arcs subtended by equal angles are similar, that is, have the same proportion and the same ratio to their proper circles— this also being proved by the geometers. For whenever arcs of circles are subtended by equal angles, if any one of these is (say) one-tenth of its proper circle, all the remaining arcs will be tenth parts of their proper circles.

Anyone who has mastered these facts will have no difficulty in understanding the method of Eratosthenes, which is as follows. Syene and Alexandria, he asserts, are under the same meridian. Since meridian circles are great circles in the universe, the circles on the earth which lie under them are necessarily great circles also. Therefore, of whatever size this method shows the circle on the earth through Syene and Alexandria to be, this will be the size of the great circle on the earth. He then asserts, as is indeed the case, that Syene lies under the summer tropic. Therefore, whenever the sun, being in the Crab at the summer solstice, is exactly in the middle of the heavens, the pointers of the sundials necessarily throw no shadows, the sun being in the exact vertical line above them; and this is said to be true over a space 300 stades in diameter. But in Alexandria at the same hour the pointers of the sundials throw shadows, because this city lies farther to the north than Syene. As the two cities lie under the

---

exactly equal to this rectangle. It was Apollonius's most distinctive achievement to have based his treatment of the conic sections on the Pythagorean theory of the *application of areas* ($\pi\alpha\rho\alpha\beta o\lambda\acute{\eta}\ \tau\hat{\omega}\nu\ \chi\omega\rho\acute{\iota}\omega\nu$).

[13] Cleomedes probably wrote about the middle of the first century B.C. His handbook *De motu circulari corporum caelestium* is largely based on Posidonius.

same meridian great circle, if we draw an arc from the extremity of the
shadow of the pointer to the base of the pointer of the sundial in Alexandria, the arc will be a segment of a great circle in the bowl of the sundial,
since the bowl lies under the great circle. If then we conceive straight
lines produced in order from each of the pointers through the earth, they
will meet at the centre of the earth. Now since the sundial at Syene is
vertically under the sun, if we conceive a straight line drawn from the
sun to the top of the pointer of the sundial, the line stretching from
the sun to the centre of the earth will be one straight line. If now we
conceive another straight line drawn upwards from the extremity of the
shadow of the pointer of the sundial in Alexandria, through the top of
the pointer to the sun, this straight line and the aforesaid straight line will
be parallel, being straight lines drawn through from different parts of the
sun to different parts of the earth. Now on these parallel straight lines
there falls the straight line drawn from the centre of the earth to the
pointer at Alexandria, so that it makes the alternate angles equal; one of
these is formed at the centre of the earth by the intersection of the straight
lines drawn from the sundials to the centre of the earth; the other is at
the intersection of the top of the pointer in Alexandria and the straight
line drawn from the extremity of its shadow to the sun through the point
where it meets the pointer. Now this latter angle subtends the arc carried
round from the extremity of the shadow of the pointer to its base, while
the angle at the centre of the earth subtends the arc stretching from Syene
to Alexandria. But the arcs are similar since they are subtended by equal
angles. Whatever ratio, therefore, the arc in the bowl of the sundial has

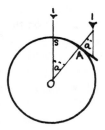

FIGURE 12

to its proper circle, the arc reaching from Syene to Alexandria has the
same ratio. But the arc in the bowl is found to be the fiftieth part of its
proper circle. Therefore the distance from Syene to Alexandria must necessarily be a fiftieth part of the great circle of the earth. And this distance

is 5000 stades. Therefore the whole great circle is 250000 stades. Such is the method of Eratosthenes.[14]

DIOPHANTUS

(*Palatine Anthology* [15] xiv)

THIS tomb holds Diophantus. Ah, what a marvel! And the tomb tells scientifically the measure of his life. God vouchsafed that he should be a boy for the sixth part of his life; when a twelfth was added, his cheeks acquired a beard; He kindled for him the light of marriage after a seventh, and in the fifth year after his marriage He granted him a son. Alas! late-begotten and miserable child, when he had reached the measure of half his father's life, the chill grave took him. After consoling his grief by this science of numbers for four years, he reached the end of his life.[16]

REVIVAL OF GEOMETRY: PAPPUS

ISOPERIMETRIC FIGURES [17]

(PAPPUS, *Collection*)

THOUGH God has given to men, most excellent Megethion, the best and most perfect understanding of wisdom and mathematics, He has allotted

[14] The figure on p. 206 will help to elucidate Cleomedes. S is Syene and A Alexandria; the centre of the earth is O. The sun's rays at the two places are represented by the broken straight lines. If a be the angle made by the sun's rays with the pointer of the sundial at Alexandria (OA produced), the angle SOA is also equal to a, or one-fiftieth of four right angles. The arc SA is known to be 5000 stades and it follows that the whole circumference of the earth must be 250000 stades.

[15] There are in the *Anthology* 46 epigrams which are algebraical problems. Most of them were collected by Metrodorus, a grammarian who lived about A.D. 500, but their origin is obviously much earlier and many belong to a type described by Plato and the scholiast to the Charmides.

Problems in indeterminate analysis solved before the time of Diophantus include the Pythagorean and Platonic methods of finding numbers representing the sides of right-angled triangles, the methods (also Pythagorean) of finding "side- and diameter-numbers," Archimedes' *Cattle Problem* and Heron's problems.

[16] If x was his age at death, then

$$\tfrac{1}{6}x + \tfrac{1}{12}x + \tfrac{1}{7}x + 5 + \tfrac{1}{2}x + 4 = x,$$

whence $\qquad\qquad\qquad x = 84.$

[17] The whole of Book v. in Pappus's *Collection* is devoted to isoperimetry. The first section follows closely the exposition of Zenodorus as given by Theon, except that Pappus includes the proposition that *of all circular segments having the same circumference the semicircle is the greatest.* The second section compares the volumes of solids whose surfaces are equal, and is followed by a digression, on the semi-regular solids discovered by Archimedes. After some propositions on the lines of Archimedes' *De sph. et cyl.,* Pappus finally proves that *of regular solids having equal surfaces, that is greatest which has most faces.*

The introduction, here cited, on the sagacity of bees is rightly praised by Heath as an example of the good style of the Greek mathematicians when freed from the restraints of technical language.

a partial share to some of the unreasoning creatures as well. To men, as being endowed with reason, He granted that they should do everything in the light of reason and demonstration, but to the other unreasoning creatures He gave only this gift, that each of them should, in accordance with a certain natural forethought, obtain so much as is needful for supporting life. This instinct may be observed to exist in many other species of creatures, but it is specially marked among bees. Their good order and their obedience to the queens who rule in their commonwealths are truly admirable, but much more admirable still is their emulation, their cleanliness in the gathering of honey, and the forethought and domestic care they give to its protection. Believing themselves, no doubt, to be entrusted with the task of bringing from the gods to the more cultured part of mankind a share of ambrosia in this form, they do not think it proper to pour it carelessly into earth or wood or any other unseemly and irregular material, but, collecting the fairest parts of the sweetest flowers growing on the earth, from them they prepare for the reception of the honey the vessels called honeycombs, [with cells] all equal, similar and adjacent, and hexagonal in form.

That they have contrived this in accordance with a certain geometrical forethought we may thus infer. They would necessarily think that the figures must all be adjacent one to another and have their sides common, in order that nothing else might fall into the interstices and so defile their work. Now there are only three rectilineal figures which would satisfy the condition, I mean regular figures which are equilateral and equiangular, inasmuch as irregular figures would be displeasing to the bees. For equilateral triangles and squares and hexagons can lie adjacent to one another and have their sides in common without irregular interstices. For the space about the same point can be filled by six equilateral triangles and six angles, of which each is $\frac{2}{3}$. right angle, or by four squares and four right angles, or by three hexagons and three angles of a hexagon, of which each is $1\frac{1}{3}$. right angle. But three pentagons would not suffice to fill the space about the same point, and four would be more than sufficient; for three angles of the pentagon are less than four right angles (inasmuch as each angle is $1\frac{1}{5}$. right angle), and four angles are greater than four right angles. Nor can three heptagons be placed about the same point so as to have their sides adjacent to each other; for three angles of a heptagon are greater than four right angles (inasmuch as each is $1\frac{3}{7}$. right angle). And the same argument can be applied even more to polygons with a greater number of angles. There being, then, three figures capable by themselves of filling up the space around the same point, the triangle, the square and the hexagon, the bees in their wisdom chose for their work that which has the most angles, perceiving that it would hold more honey than either of the two others.

Bees, then, know just this fact which is useful to them, that the hexagon is greater than the square and the triangle and will hold more honey for the same expenditure of material in constructing each. But we, claiming a greater share in wisdom than the bees, will investigate a somewhat wider problem, namely that, *of all equilateral and equiangular plane figures having an equal perimeter, that which has the greater number of angles is always greater, and the greatest of them all is the circle having its perimeter equal to them.*

COMMENTARY ON
ROBERT RECORDE

R OBERT RECORDE, born in Wales in 1510, taught mathematics at
Oxford and Cambridge, got his M.D. degree at the latter university
in 1545, became physician to Edward VI and Queen Mary, served for a
time as "Comptroller of Mines and Monies" in Ireland and died in the
King's Bench Prison, Southwark, where he was confined for debt—some
historians hint at a darker offense—in 1558. Recorde published a number
of mathematical works, chiefly in the then not uncommon form of dialogue between master and scholar. The most popular and influential of
these treatises were *The Ground of Artes* (1540), and *The Whetstone of
Witte* (1557), "containying extraction of Rootes: The *Cossike* practise,
with the rule of *Equation:* and the woorkes of Surde Numbers."

England lagged behind Italy and France in the publication of mathematical books.[1] By the close of the fifteenth century Italy alone had
printed more than 200 treatises on mathematics, whilst it was not until
1522 that the first book dealing exclusively with arithmetic [2]—the "erudite
but dull" *De Arte Supputandi* of Cuthbert Tonstall—appeared in Great
Britain. *The Ground of Artes* (the selections below were taken from its
edition of 1646) was an immensely successful commercial arithmetic. It
went through at least eighteen editions in the sixteenth century and a
dozen more in the seventeenth. Its influence, and that of *The Whetstone
of Witte,* is at least partly explained by the economic development—manufacture and commerce—of England in the reign of Elizabeth. "Never
(says D. E. Smith) was there a better opportunity for a commercial arithmetic, and never was the opportunity more successfully met."

Recorde's book discusses operations with arabic numerals as well as
computation with counters, proportion, the "golden rule" of three, fractions and "allegation" and "contains the usual commercial topics which
European countries north of the Alps had derived from Italy." It is hard
to understand how anyone could learn from a dialogue delivered in the
"formal" language of this primer. On second thought, it is perhaps just as
remarkable that any of us profited from the average arithmetic texts
visited on pupils as recently as twenty-five years ago. Recorde, whose
work was "entail'd upon the People, ratified and sign'd by the approbation

---

[1] For the source of these data see David Eugene Smith, *Rara Arithmetica*, Boston,
1908; also, the same author's standard *History of Mathematics*, Boston and New
York, 1923.

[2] The first printed English book containing a reference to arithmetic is an anony-
mous translation from the French, *The Mirrow of the World or Thymage of the
same*, issued in 1480 from the Caxton press. Chapter 10 of this book begins: "And
after of Arsmetrike and whereof it proceedeth." See Smith, *Rara Arithmetica*.

of Time," [3] was at least modest in his claims for it. He wrote in his preface: "And if any man obiect, that other books haue bene written of Arithmetike alreadie so sufficiently, that I needed not now to put penne to the booke, except I will codemne other mens writings: to them I answere. That as I codemne no mans diligence, so I know that no man can satisfie euery man, and therefore like as many do esteeme greatly other bookes, so I doubt not but some will like this my booke aboue any other English Arithmetike hitherto written, & namely such as shal lacke instructers, for whose sake I haue plain-ly set forth the exaples, as no book (that I haue seene) hath hitherto: which thing shall be great ease to the rude readers." [4]

[3] A comment by Thomas Willsford in his 1662 edition of *The Ground of Artes*.
[4] From the 1594 edition; see Smith, *Rara Arithmetica*, p. 216.

Someone who had begun to read geometry with Euclid, when he had learned the first proposition, asked Euclid, "But what shall I get by learning these things?" whereupon Euclid called his slave and said "Give him three-pence since he must make gain out of what he learns." —STOBAEUS

There still remain three studies suitable for free man. Arithmetic is one of them. —PLATO

5   The Declaration of the Profit of Arithmeticke

By ROBERT RECORDE

TO THE LOVING READERS,

THE PREFACE OF MR. ROBERT RECORD

SORE oft times have I lamented with my self the unfortunate condition of England, seeing so many great Clerks to arise in sundry other parts of the world, and so few to appear in this our Nation: whereas for pregnancy of naturall wit (I think) few Nations do excell Englishmen: But I cannot impute the cause to any other thing, then to be contempt, or misregard of learning. For as Englishmen are inferiour to no men in mother wit, so they passe all men in vain pleasures, to which they may attain with great pain and labour: and are as slack to any never so great commodity; if there hang of it any painfull study or travelsome labour.

Howbeit, yet all men are not of that sort, though the most part be, the more pity it is: but of them that are so glad, not onely with painfull study, and studious pain to attain learning, but also with as great study and pain to communicate their learning to other, and make all England (if it might be) partakers of the same; the most part are such, that unneath they can support their own necessary charges, so that they are not able to bear any charges in doing of that good, that else they desire to do.

But a greater cause of lamentation is this, that when learned men have taken pains to do things for the aid of the unlearned, scarce they shall be allowed for their wel-doing, but derided and scorned, and so utterly discouraged to take in hand any like enterprise again.

The following is "The declaration of the profit of Arithmeticke" and constitutes the first ten pages of the text. It may be said to represent the influence of this text upon establishing for a long period what educators at present speak of as "the objectives" of elementary arithmetic.

A DIALOGUE BETWEEN THE MASTER AND THE SCHOLAR: TEACHING
THE ART AND USE OF ARITHMETICK WITH PEN.
THE SCHOLAR SPEAKETH.

*SIR, such is your authority in mine estimation, that I am content to con-
sent to your saying, and to receive it as truth, though I see none other
reason that doth lead me thereunto: whereas else in mine own conceit it
appeareth but vain, to bestow any time privately in learning of that thing,
that every childe may, and doth learn at all times and hours, when he
doth any thing himself alone, and much more when he talketh or reason-
eth with others.*

Master.   Lo, this is the fashion and chance of all them that seek to
defend their blinde ignorance, that when they think they have made strong
reason for themselves, then have they proved quite contrary. For if numbring be so common (as you grant it to be) that no man can do anything
alone, and much lesse talk or bargain with other, but he shall still have to
do with number: this proveth not number to be contemptible and vile, but
rather right excellent and of high reputation, sith it is the ground of all
mens affairs, in that without it no tale can be told, no communication
without it can be continued, no bargaining without it can duely be ended,
or no businesse that man hath, justly completed. These commodities, if
there were none other, are sufficient to approve the worthinesse of number.
But there are other innumerable, farre passing all these, which declare
number to exceed all praise. Wherefore in all great works are Clerks so
much desired? Wherefore are Auditors so richly fed? What causeth Geometricians so highly to be enhaunsed? Why are Astronomers so greatly
advanced? Because that by number such things they finde, which else
would farre excell mans minde.

Scholar.   Verily, sir, if it bee so, that these men by numbring, their
cunning do attain, at whose great works most men do wonder, then I see
well I was much deceived, and numbring is a more cunning thing then
I took it to be.

Master.   If number were so vile a thing as you did esteem it, then
need it not to be used so much in mens communication. Exclude number,
and answer to this question: How many years old are you?

Scholar.   Mum.

Master.   How many dayes in a weeke? How many weeks in a year?
What lands hath your Father? How many men doth hee keep? How long
is it since you came from him to me?

Scholar.   Mum.

Master.   So that if number want, you answer all by Mummes: How
many miles to London?

Scholar.   A poak full of plums.

Master. Why, thus you may see, what rule number beareth, and that if number bee lacking it maketh men dumb, so that to most questions they must answer Mum.

Scholar. This is the cause, sir, that I judged it so vile, because it is so common in talking every while: Nor plenty is not dainty, as the common saying is.

Master. No, nor store is no sore, perceive you this? The more common that the thing is, being needfully required, the better is the thing, and the more to be desired. But in numbring, as some of it is light and plain, so the most part is difficult, and not easie to attain. The easier part serveth all men in common, and the other requireth some learning. Wherefore as without numbring a man can do almost nothing, so with the help of it, you may attain to all things.

Scholar. Yes, sir, why then it were best to learn the Art of numbring, first of all other learning, and then a man need learn no more, if all other come with it.

Master. Nay not so: but if it be first learned, then shall a man be able (I mean) to learn, perceive, and attain to other Sciences; which without it he could never get.

Scholar. I perceive by your former words, that Astronomy and Geometry depend much on the help of numbring: but that other Sciences, as Musick, Physick, Law, Grammer, and such like, have any help of Arithmetick, I perceive not.

Master. I may perceive your great Clerk-linesse by the ordering of your Sciences: but I will let that passe now, because it toucheth not the matter that I intend, and I will shew you how Arithmetick doth profit in all these somewhat grosly, according to your small understanding, omitting other reasons more substantiall.

First (as you reckon them) Musick hath not onely great help of Arithmetic, but is made, and hath his perfectnesse of it: for all Musick standeth by number and proportion: And in Physick, beside the calculation of criticall dayes, with other things, which I omit, how can any man judge the pulse rightly, that is ignorant of the proportion of numbers?

And so for the Law, it is plain, that the man that is ignorant of Arithmetick, is neither meet to be a Judge, neither an Advocate, nor yet a Proctor. For how can hee well understand another mans cause, appertaining to distribution of goods, or other debts, or of summes of money, if he be ignorant of Arithmetick? This oftentimes causeth right to bee hindered, when the Judge either delighteth not to hear of a matter that hee perceiveth not, or cannot judge for lack of understanding: this commeth by ignorance of Arithmetick.

Now, as for Grammer, me thinketh you would not doubt in what it needeth number, sith you have learned that Nouns of all sorts, Pronouns,

Verbs, and Participles are distinct diversly by numbers: besides the variety of Nouns of Numbers, and Adverbs. And if you take away number from Grammer, then is all the quantity of Syllables lost. And many other ways doth number help Grammer. Whereby were all kindes of Meeters found and made? was it not by number?

But how needfull Arithmetick is to all parts of Philosophy, they may soon see, that do read either Aristotle, Plato, or any other Philosophers writings. For all their examples almost, and their probations, depend of Arithmetick. It is the saying of Aristotle, that hee that is ignorant of Arithmetick, is meet for no Science. And Plato his Master wrote a little sentence over his Schoolhouse door, Let none enter in hither (quoth he) that is ignorant of Geometry. Seeing hee would have all his Scholars expert in Geometry, much rather he would the same in Arithmetick, without which Geometry cannot stand.

And how needfull Arithmetick is to Divinity, it appeareth, seeing so many Doctors gather so great mysteries out of number, and so much do write of it. And if I should go about to write all the commodities of Arithmetick in civill acts, as in governance of Common-weales in time of peace, and in due provision & order of Armies, in time of war, for numbering of the Host, summing of their wages, provision of victuals, viewing of Artillery, with other Armour; beside the cunningest point of all, for casting of ground, for encamping of men, with such other like: And how many wayes also Arithmetick is conducible for all private Weales, of Lords and all Possessioners, of Merchants, and all other occupiers, and generally for all estates of men, besides Auditors, Treasurers, Receivers, Stewards, Bailiffes, and such like, whose Offices without Arithmetick are nothing: If I should (I say) particularly repeat all such commodities of the noble Science of Arithmetick, it were enough to make a very great book.

Scholar. No, no sir, you shall not need: For I doubt not, but this, that you have said, were enough to perswade any man to think this Art to be right excellent and good, and so necessary for man, that (as I think now) so much as a man lacketh of it, so much hee lacketh of his sense and wit.

Master. What, are you so farre changed since, by hearing these few commodities in generall: by likelihood you would be farre changed if you knew all the particular Commodities.

Scholar. I beseech you Sir, reserve those Commodities that rest yet behinde unto their place more convenient: and if yee will bee so good as to utter at this time this excellent treasure, so that I may be somewhat inriched thereby, if ever I shall be able, I will requite your pain.

Master. I am very glad of your request, and will do it speedily, sith that to learn it you bee so ready.

Scholar.   And I to your authority my wit do subdue; whatsoever you say, I take it for true.

Master.   That is too much; and meet for no man to bee beleeved in all things, without shewing of reason. Though I might of my Scholar some credence require, yet except I shew reason, I do it not desire. But now sith you are so earnestly set this Art to attaine, best it is to omit no time, lest some other passion coole this great heat, and then you leave off before you see the end.

Scholar.   Though many there bee so unconstant of mind, that flitter and turn with every winde, which often begin, and never come to the end, I am none of this sort, as I trust you partly know. For by my good will what I once begin, till I have it fully ended, I would never blin.

Master.   So have I found you hitherto indeed, and I trust you will increase rather then go back. For, better it were never to assay, then to shrink and flie in the mid way: But I trust you will not do so; therefore tell mee briefly: What call you the Science that you desire so greatly.

Scholar: Why sir, you know.

Master.   That maketh no matter, I would hear whether you know, and therefore I ask you. For great rebuke it were to have studied a Science, and yet cannot tell how it is named.

Scholar.   Some call it Arsemetrick, and some Augrime.

Master.   And what do these names betoken?

Scholar.   That, if it please you, of you would I learn.

Master.   Both names are corruptly written: Arsemetrick for Arithmetick, as the Greeks call it, and Augrime for Algorisme, as the Arabians found it: which both betoken the Science of Numbring: for Arithmos in Greek is called Number: and of it commeth Arithmetick, the Art of Numbring. So that Arithmetick is a Science or Art teaching the manner and use of Numbring: This Art may be wrought diversly, with Pen or with Counters. But I will first shew you the working with the Pen, and then the other in order.

Scholar.   This I will remember. But how many things are to bee learned to attain this Art fully?

Master.   There are reckoned commonly seven parts or works of it. Numeration, Addition, Subtraction, Multiplication, Division, Progression, and Extraction of roots: to these some men adde Duplication, Triplation, and Mediation. But as for these three last they are contained under the other seven. For Duplication, and Triplation are contained under Multiplication; as it shall appear in their place: And Mediation is contained under Division, as I will declare in his place also.

Scholar.   Yet then there remain the first seven kinds of Numbring.

Master.   So there doth: Howbeit if I shall speak exactly of the parts of Numbring, I must make but five of them: for Progression is a compound operation of Addition, Multiplication and Division. And so is the Extractions of roots. But it is no harme to name them as kindes severall, seeing they appear to have some severall working. For it forceth not so much to contend for the number of them, as for the due knowledge and practising of them.

Scholar. Then you will that I shall name them as seven kindes distinct. But now I desire you to instruct mee in the use of each of them.

Master. So I will, but it must be done in order: for you may not learn the last so soon as the first, but you must learn them in that order, as I did rehearse them, if you will learn them speedily, and well.

Scholar. Even as you please. Then to begin; Numeration is the first in order: what shall I do with it?

Master. First, you must know what the thing is, and then after learn the use of the same.

KEPLER and LODGE

J OHANN KEPLER (1571–1630) is usually considered the founder of physical astronomy. Copernicus conceived the heliocentric theory—reviving Pythagorean beliefs—and worked it out in his famous book *De Revolutionibus Orbium Coelestium*; Tycho Brahe invented and improved astronomical instruments, and by his wonderful skill in observation introduced undreamed-of accuracy into celestial measurements; Galileo contributed the telescope, the discovery of new stars and nebulae, the support and diffusion of Copernican ideas in his brilliant writings. Kepler ranks foremost as the mathematician of the sky.

Though a devout Lutheran, Kepler was also inclined toward Pythagorean mysticism. He was number-intoxicated, a variety of religious experience not restricted to the weak-minded. Fortunately he not only loved numbers but knew how to handle them. His fanatical search for simple mathematical harmonies in the physical universe produced some silly ideas but also three great laws. The first was that the planets move in ellipses with the sun in one focus. Before he made this discovery it was believed that the planets, being perfect creations of God, followed the most perfect of orbits, namely circles. The second law was that the line joining sun and planet (the radius vector) sweeps out equal areas in equal times. The third law was published in 1618, nine years after the other two. It connected the times and distances of the planets: "The square of the time of revolution of each planet is proportional to the cube of its mean distance from the sun." The discovery of this beautifully simple relationship threw Kepler into a picturesque but justifiable rapture: ". . . What sixteen years ago, I urged as a thing to be sought . . . for which I have devoted the best part of my life . . . at length I have brought to light, and recognized its truth beyond my most sanguine expectations. . . . Nothing holds me; I will indulge my sacred fury; I will triumph over mankind by the honest confession that I have stolen the golden vases of the Egyptians to build up a tabernacle for my God far away from the confines of Egypt."

The story of Kepler's life and work is briefly and attractively told in the selection which follows. The teller is Sir Oliver Lodge, distinguished for his physical researches and famous as a student of psychical phenomena. Lodge was born in England in 1851 and died in 1940. His father was a prosperous business man who decided when Oliver was fourteen that he should leave school and learn the routine of a pottery-material supply agency. This important work kept him busy for the next eight years.

Thanks to a maternal aunt, who insisted that he be permitted to visit her occasionally in London, he was able to attend classes in chemistry and geology, and was fortunate enough to hear John Tyndall's lectures on heat. After he had won a scholarship, his father grudgingly allowed him to complete his education at the Royal College of Science. While a graduate student, Lodge did research in electricity, on thermal conductivity and on the foundations of mechanics. Later, when appointed to the chair of physics at Liverpool, he performed important experiments in connection with the ether concept, to which he was devoted. He also investigated electromagnetic radiation and contributed to the development of wireless. It has been said that his work was the "major influence in making spark telegraphy possible." [1] Lodge won many honors and held high posts— for example, Principal of Birmingham University, 1900–1919 and President of the British Association, 1913. The public, however, came to know of him less because of these attainments than because of his psychical research and his pronounced opinions on the survival of the mind after death. He was a pioneer practitioner of thought transference, and attracted world-wide attention by his inquiries into the powers of the famous mediums "Mrs. Piper" and the unlikely "Eusapia Palladino," a lady "telekinesis" expert. (Miss Palladino could move things by thinking about them, at least so Sir Oliver believed. Unfortunately, she was caught at one séance assisting her thoughts with her hands.) Lodge was a fine scientist, a most estimable man, and an able expositor; in his spiritualistic beliefs he was inclined to be something of a goose. The essay on Kepler is from a book anyone interested in science will enjoy reading. It is called *Pioneers of Science,* and is based on a course of astronomy lectures Lodge gave at University College, Liverpool, in 1887.

[1] Article on Lodge, *Dictionary of National Biography;* supplemental volume 1931–40. See also *The Proceedings of the* (London) *Physical Society,* Vol. 53; *Obituary Notices of Fellows of the Royal Society,* Vol. 3.

Our Souls, whose faculties can comprehend
The wondrous Architecture of the world:
And measure every wand'ring planet's course,
Still climbing after knowledge infinite.

—CHRISTOPHER MARLOWE *(Conquests of Tamburlaine)*

The die is cast; I have written my book; it will be read either in the present
age or by posterity, it matters not which; it may well await a reader, since
God has waited six thousand years for an interpreter of his words.

—JOHANN KEPLER

6   Johann Kepler

By SIR OLIVER LODGE

KEPLER AND THE LAWS OF PLANETARY MOTION

IT is difficult to imagine a stronger contrast between two men engaged in the same branch of science than exists between Tycho Brahé, the subject of the last lecture, and Kepler, our subject on the present occasion.

The one, rich, noble, vigorous, passionate, strong in mechanical ingenuity and experimental skill, but not above the average in theoretical and mathematical power.

The other, poor, sickly, devoid of experimental gifts, and unfitted by nature for accurate observation, but strong almost beyond competition in speculative subtlety and innate mathematical perception.

The one is the complement of the other; and from the fact of their following each other so closely arose the most surprising benefits to science.

The outward life of Kepler is to a large extent a mere record of poverty and misfortune. I shall only sketch in its broad features, so that we may have more time to attend to his work.

He was born (so his biographer assures us) in longitude 29° 7', latitude 48° 54', on the 21st of December, 1571. His parents seem to have been of fair condition, but by reason, it is said, of his becoming surety for a friend, the father lost all his slender income, and was reduced to keeping a tavern. Young John Kepler was thereupon taken from school, and employed as pot-boy between the ages of nine and twelve. He was a sickly lad, subject to violent illnesses from the cradle, so that his life was frequently despaired of. Ultimately he was sent to a monastic school and thence to the University of Tübingen, where he graduated second on the list. Meanwhile home affairs had gone to rack and ruin. His father abandoned the home, and later died abroad. The mother quarrelled with all her relations, including her son John; who was therefore glad to get away as soon as possible.

All his connection with astronomy up to this time had been the hearing the Copernican theory expounded in University lectures, and defending it in a college debating society.

An astronomical lectureship at Graz happening to offer itself, he was urged to take it, and agreed to do so, though stipulating that it should not debar him from some more brilliant profession when there was a chance.

For astronomy in those days seems to have ranked as a minor science, like mineralogy or meteorology now. It had little of the special dignity with which the labours of Kepler himself were destined so greatly to aid in endowing it.

Well, he speedily became a thorough Copernican, and as he had a most singularly restless and inquisitive mind, full of appreciation of everything relating to number and magnitude—was a born speculator and thinker just as Mozart was a born musician, or Bidder a born calculator—he was agitated by questions such as these: Why are there exactly six planets? Is there any connection between their orbital distances, or between their orbits and the times of describing them? These things tormented him, and he thought about them day and night. It is characteristic of the spirit of the times—this questioning why there should be six planets. Nowadays, we should simply record the fact and look out for a seventh. Then, some occult property of the number six was groped for, such as that it was equal to $1 + 2 + 3$ and likewise equal to $1 \times 2 \times 3$, and so on. Many fine reasons had been given for the seven planets of the Ptolemaic system, but for the six planets of the Copernican system the reasons were not so cogent.

Again, with respect to their successive distances from the sun, some law would seem to regulate their distance, but it was not known. (Parenthetically I may remark that it is not known even now: a crude empirical statement known as Bode's law is all that has been discovered.)[1]

[1] [Write down the series 0, 3, 6, 12, 24, 48 &c.; add 4 to each, and divide by 10. The resulting series gives the approximate mean distances of the planets from the sun in astronomical units.

Planet	Bode Distance	Actual Mean Distance
Mercury	0.4	0.39
Venus	0.7	0.72
Earth	1.0	1.00
Mars	1.6	1.52
—	2.8	—
Jupiter	5.2	5.20
Saturn	10.0	9.53
Uranus	19.6	19.19
Neptune	38.8	30.07
Pluto	77.2	39.5

This is the law discovered by the German astronomer Johann Elert Bode (1747–1826) in 1772. Its failure in the case of Neptune and Pluto, planets found after

Once more, the further the planet the slower it moved; there seemed to be some law connecting speed and distance. This also Kepler made continual attempts to discover.

One of his ideas concerning the law of the successive distances was based on the inscription of a triangle in a circle. If you inscribe in a circle a large number of equilateral triangles, they envelop another circle bearing a definite ratio to the first: these might do for the orbits of two planets

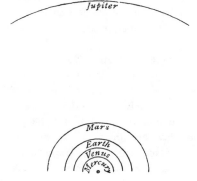

FIGURE 1—Orbits of some of the planets drawn to scale: showing the gap between Mars and
Jupiter.

(see Figure 2). Then try inscribing and circumscribing squares, hexagons, and other figures, and see if the circles thus defined would correspond to the several planetary orbits. But they would not give any satisfactory result. Brooding over this disappointment, the idea of trying solid figures suddenly strikes him. "What have plane figures to do with the celestial orbits?" he cries out; "inscribe the regular solids." And then— brilliant idea—he remembers that there are but five. Euclid had shown that there could be only five regular solids.[2] The number evidently corresponds to the gaps between the six planets. The reason of there being only six seems to be attained. This coincidence assures him he is on the right track, and with great enthusiasm and hope he "represents the earth's orbit by a sphere as the norm and measure of all"; round it he circumscribes a dodecahedron, and puts another sphere round that, which is approximately the orbit of Mars; round that, again, a tetrahedron, the

Bode's time, has led most astronomers to conclude that the "law" is a "purely empirical relationship, more in the realm of coincidence than an actual physical law." Besides not being a law, it was not in fact discovered by Bode but by the German mathematician J. D. Titius (1729–1796). Ed.]

[2] The proof is easy, and ought to occur in books on solid geometry. By a "regular" solid is meant one with all its faces, edges, angles, &c., absolutely alike: it is of these perfectly symmetrical bodies that there are only five. Crystalline forms are very numerous.

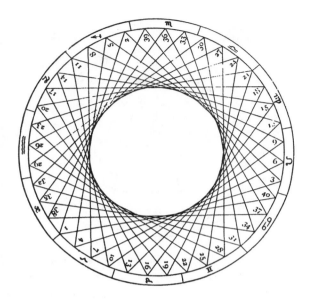

FIGURE 2—Many-sided polygon or approximate circle enveloped by straight lines, as for instance by a number of equilateral triangles. No internal circle has been drawn.

corners of which mark the sphere of the orbit of Jupiter; round that sphere, again, he places a cube, which roughly gives the orbit of Saturn.

On the other hand, he inscribes in the sphere of the earth's orbit an icosahedron; and inside the sphere determined by that, an octahedron; which figures he takes to inclose the sphere of Venus and of Mercury respectively.

The imagined discovery is purely fictitious and accidental. First of all, eight planets are now known; and secondly, their real distances agree only very approximately with Kepler's hypothesis.

Nevertheless, the idea gave him great delight. He says:—"The intense pleasure I have received from this discovery can never be told in words. I regretted no more the time wasted; I tired of no labour; I shunned no toil of reckoning, days and nights spent in calculation, until I could see whether my hypothesis would agree with the orbits of Copernicus, or whether my joy was to vanish into air."

He then went on to speculate as to the cause of the planets' motion. The old idea was that they were carried round by angels or celestial intelligences. Kepler tried to establish some propelling force emanating from the sun, like the spokes of a windmill.

This first book of his brought him into notice, and served as an introduction to Tycho and to Galileo.

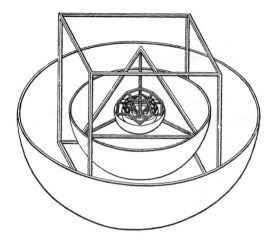

FIGURE 3—Frameworks with inscribed and circumscribed spheres, representing the five regular
solids distributed as Kepler supposed them to be among the planetary orbits.

Tycho Brahé was at this time at Prague under the patronage of the
Emperor Rudolph; and as he was known to have by far the best planetary
observations of any man living, Kepler wrote to him to know if he might
come and examine them so as to perfect his theory.

Tycho immediately replied, "Come, not as a stranger, but as a very
welcome friend; come and share in my observations with such instruments
as I have with me, and as a dearly beloved associate." After this visit,
Tycho wrote again, offering him the post of mathematical assistant, which
after hesitation was accepted. Part of the hesitation Kepler expresses by
saying that "for observations his sight was dull, and for mechanical operations his hand was awkward. He suffered much from weak eyes, and
dare not expose himself to night air." In all this he was, of course, the
antipodes of Tycho, but in mathematical skill he was greatly his superior.

On his way to Prague he was seized with one of his periodical illnesses,
and all his means were exhausted by the time he could set forward again,
so that he had to apply for help to Tycho.

It is clear, indeed, that for some time now he subsisted entirely on the
bounty of Tycho, and he expresses himself most deeply grateful for all
the kindness he received from that noble and distinguished man, the head
of the scientific world at that date.

To illustrate Tycho's kindness and generosity, I must read you a letter
written to him by Kepler. It seems that Kepler, on one of his absences
from Prague, driven half mad with poverty and trouble, fell foul of Tycho,
whom he thought to be behaving badly in money matters to him and his
family, and wrote him a violent letter full of reproaches and insults.

Tycho's secretary replied quietly enough, pointing out the groundlessness and ingratitude of the accusation.

Kepler repents instantly, and replies:—

"MOST NOBLE TYCHO," (these are the words of his letter), "how shall I enumerate or rightly estimate your benefits conferred on me? For two months you have liberally and gratuitously maintained me, and my whole family; you have provided for all my wishes; you have done me every possible kindness; you have communicated to me everything you hold most dear; no one, by word or deed, has intentionally injured me in anything; in short, not to your children, your wife, or yourself have you shown more indulgence than to me. This being so, as I am anxious to put on record, I cannot reflect without consternation that I should have been so given up by God to my own intemperance as to shut my eyes on all these benefits; that, instead of modest and respectful gratitude, I should indulge for three weeks in continual moroseness towards all your family, in headlong passion and the utmost insolence towards yourself, who possess so many claims on my veneration, from your noble family, your extraordinary learning, and distinguished reputation. Whatever I have said or written against the person, the fame, the honour, and the learning of your excellency; or whatever, in any other way, I have injuriously spoken or written (if they admit no other more favourable interpretation), as, to my grief, I have spoken and written many things, and more than I can remember; all and everything I recant, and freely and honestly declare and profess to be groundless, false, and incapable of proof."

Tycho accepted the apology thus heartily rendered, and the temporary breach was permanently healed.

In 1601, Kepler was appointed "Imperial mathematician," to assist Tycho in his calculations.

The Emperor Rudolph did a good piece of work in thus maintaining these two eminent men, but it is quite clear that it was as astrologers that he valued them; and all he cared for in the planetary motions was limited to their supposed effect on his own and his kingdom's destiny. He seems to have been politically a weak and superstitious prince, who was letting his kingdom get into hopeless confusion, and entangling himself in all manner of political complications. While Bohemia suffered, however, the world has benefited at his hands; and the tables upon which Tycho was now engaged are well called the Rudolphine tables.

These tables of planetary motion Tycho had always regarded as the main work of his life; but he died before they were finished, and on his death-bed he intrusted the completion of them to Kepler, who loyally undertook their charge.

The Imperial funds were by this time, however, so taxed by wars and other difficulties that the tables could only be proceeded with very slowly, a staff of calculators being out of the question. In fact, Kepler could not get even his own salary paid: he got orders, and promises, and drafts on estates for it; but when the time came for them to be honoured they were worthless, and he had no power to enforce his claims.

So everything but brooding had to be abandoned as too expensive, and he proceeded to study optics. He gave a very accurate explanation of the action of the human eye, and made many hypotheses, some of them shrewd and close to the mark, concerning the law of refraction of light in dense media: but though several minor points of interest turned up, nothing of the first magnitude came out of this long research.

The true law of refraction was discovered some years after by a Dutch professor, Willebrod Snell, and by Descartes.

We must now devote a little time to the main work of Kepler's life. All the time he had been at Prague he had been making a severe study of the motion of the planet Mars, analyzing minutely Tycho's books of observations, in order to find out, if possible, the true theory of his motion. Aristotle had taught that circular motion was the only perfect and natural motion, and that the heavenly bodies therefore necessarily moved in circles.

So firmly had this idea become rooted in men's minds, that no one ever seems to have contemplated the possibility of its being false or meaningless.

When Hipparchus and others found that, as a matter of fact, the planets did *not* revolve in simple circles, they did not try other curves, as we should at once do now, but they tried combinations of circles. The small circle carried by a bigger one was called an Epicycle. The carrying circle was called the Deferent. If for any reason the earth had to be placed out of the centre, the main planetary orbit was called an Excentric, and so on.

But although the planetary paths might be roughly represented by a combination of circles, their speeds could not, on the hypothesis of uniform motion in each circle round the earth as a fixed body. Hence was introduced the idea of an Equant, *i.e.*, an arbitrary point, not the earth, about which the speed might be uniform. Copernicus, by making the sun the centre, had been able to simplify a good deal of this, and to abolish the equant.

But now that Kepler had the accurate observations of Tycho to refer to, he found immense difficulty in obtaining the true positions of the planets for long together on any such theory.

He specially attacked the motion of the planet Mars, because that was sufficiently rapid in its changes for a considerable collection of data to have accumulated with respect to it. He tried all manner of circular orbits for the earth and for Mars, placing them in all sorts of aspects with respect to the sun. The problem to be solved was to choose such an orbit and such a law of speed, for both the earth and Mars, that a line joining them, produced out to the stars, should always mark correctly the apparent position of Mars as seen from the earth. He had to arrange the size of the orbits that suited best, then the positions of their centres, both being supposed excentric with respect to the sun; but he could not get any such arrangement to work with uniform motion about the sun. So he reintroduced the equant, and thus had another variable at his disposal—in fact, two, for he had an equant for the earth and another for Mars, getting a pattern of the kind suggested in Figure 4.

The equants might divide the line in any arbitrary ratio. All sorts of combinations had to be tried, the relative positions of the earth and Mars to be worked out for each, and compared with Tycho's recorded observations. It was easy to get them to agree for a short time, but sooner or later a discrepancy showed itself.

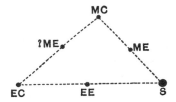

FIGURE 4—S represents the sun; EC, the centre of the earth's orbit, to be placed as best suited; MC, the same for Mars; EE, the earth's equant, or point about which the earth uniformly revolved (i.e., the point determining the law of speed about the sun), likewise to be placed anywhere, but supposed to be in the line joining S to EC; ME, the same thing for Mars; with ? ME for an alternative hypothesis that perhaps Mars' equant was on line joining EC with MC.

I need not say that all these attempts and gropings, thus briefly summarized, entailed enormous labour, and required not only great pertinacity, but a most singularly constituted mind, that could thus continue groping in the dark without a possible ray of theory to illuminate its search. Grope he did, however, with unexampled diligence.

At length he hit upon a point that seemed nearly right. He thought he had found the truth; but no, before long the position of the planet, as calculated, and as recorded by Tycho, differed by eight minutes of arc, or about one-eighth of a degree. Could the observation be wrong by this small amount? No, he had known Tycho, and knew that he was never wrong eight minutes in an observation.

So he set out the whole weary way again, and said that with those eight minutes he would yet find out the law of the universe. He proceeded to see if by making the planet librate, or the plane of its orbit tilt up and down, anything could be done. He was rewarded by finding that at any rate the plane of the orbit did not tilt up and down: it was fixed, and this was a simplification on Copernicus's theory. It is not an absolute fixture, but the changes are very small.

At last he thought of giving up the idea of *uniform* circular motion, and of trying *varying* circular motion, say inversely as its distance from the

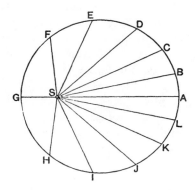

FIGURE 5—Excentric circle supposed to be divided into equal areas. The sun, S, being placed at
    a selected point, it was possible to represent the varying speed of a planet by saying
    that it moved from A to B, from B to C, and so on, in equal times.

sun. To simplify calculation, he divided the orbit into triangles, and tried
if making the triangles equal would do. A great piece of luck, they did
beautifully: the rate of description of areas (not arcs) is uniform. Over
this discovery he greatly rejoices. He feels as though he had been carrying
on a war against the planet and had triumphed; but his gratulation was
premature. Before long fresh little errors appeared, and grew in importance. Thus he announces it himself:—

"While thus triumphing over Mars, and preparing for him, as for one
already vanquished, tabular prisons and equated excentric fetters, it is
buzzed here and there that the victory is vain, and that the war is raging
anew as violently as before. For the enemy left at home a despised cap-
tive has burst all the chains of the equations, and broken forth from the
prisons of the tables."

Still, a part of the truth had been gained, and was not to be abandoned
any more. The law of speed was fixed: that which is now known as his
second law. But what about the shape of the orbit—Was it after all possible that Aristotle, and every philosopher since Aristotle, had been wrong?
that circular motion was not the perfect and natural motion, but that
planets might move in some other closed curve?

Suppose he tried an oval. Well, there are a great variety of ovals, and
several were tried: with the result that they could be made to answer
better than a circle, but still were not right.

Now, however, the geometrical and mathematical difficulties of calculation, which before had been tedious and oppressive, threatened to become overwhelming; and it is with a rising sense of despondency that
Kepler sees his six years' unremitting labour leading deeper and deeper
into complication.

One most disheartening circumstance appeared, viz. that when he made the circuit oval his law of equable description of areas broke down. That seemed to require the circular orbit, and yet no circular orbit was quite accurate.

While thinking and pondering for weeks and months over this new dilemma and complication of difficulties, till his brain reeled, an accidental ray of light broke upon him in a way not now intelligible, or barely intelligible. Half the extreme breadth intercepted between the circle and oval was $^{429}\!/_{100,000}$ of the radius, and he remembered that the "optical inequality" of Mars was also about $^{429}\!/_{100,000}$. This coincidence, in his own words, woke him out of sleep; and for some reason or other impelled him instantly to try making the planet *oscillate in the diameter of its epicycle instead of revolve round it*—a singular idea, but Copernicus had had a similar one to explain the motions of Mercury.

Away he started through his calculations again. A long course of work night and day was rewarded by finding that he was now able to hit off the

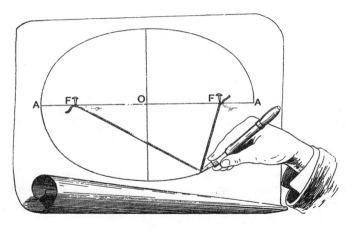

FIGURE 6—Mode of drawing an ellipse. The two pins F are the foci.

motions better than before; but what a singularly complicated motion it was. Could it be expressed no more simply? Yes, the curve so described by the planet is a comparatively simple one: it is a special kind of oval— the ellipse. Strange that he had not thought of it before. It was a famous curve, for the Greek geometers had studied it as one of the sections of a cone, but it was not so well known in Kepler's time. The fact that the planets move in it has raised it to the first importance, and it is familiar enough to us now. But did it satisfy the law of speed? Could the rate of description of areas be uniform with it? Well, he tried the ellipse, and to his inexpressible delight he found that it did satisfy the condition of equable description of areas, if the sun was in one focus. So, moving the planet in a selected ellipse, with the sun in one focus, at a speed given by the equable area description, its position agreed with Tycho's observations within the limits of the error of experiment. Mars was finally conquered, and remains in his prison-house to this day. The orbit was found.

In a paroxysm of delight Kepler celebrates his victory by a triumphant figure, sketched actually on his geometrical diagram—the diagram which proves that the law of equable description of areas can hold good with an ellipse. Below is a tracing of it.

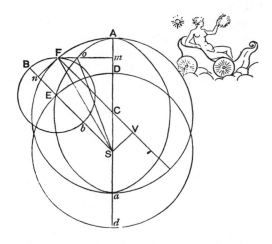

FIGURE 7

Such is a crude and bald sketch of the steps by which Kepler rose to his great generalizations—the two laws which have immortalized his name.

All the complications of epicycle, equant, deferent, excentric, and the like, were swept at once away, and an orbit of striking and beautiful properties substituted. Well might he be called, as he was, "the legislator," or law interpreter, "of the heavens."

He concludes his book on the motions of Mars with a half comic appeal to the Emperor to provide him with the sinews of war for an attack on Mars's relations—father Jupiter, brother Mercury, and the rest—but the death of his unhappy patron in 1612 put an end to all these schemes, and reduced Kepler to the utmost misery. While at Prague his salary was in continual arrear, and it was with difficulty that he could provide sustenance for his family. He had been there eleven years, but they had been hard years of poverty, and he could leave without regret were it not that he should have to leave Tycho's instruments and observations behind him. While he was hesitating what best to do, and reduced to the verge of despair, his wife, who had long been suffering from low spirits and despondency, and his three children, were taken ill; one of the sons died of small-pox, and the wife eleven days after of low fever and epilepsy. No money could be got at Prague, so after a short time he accepted a pro-

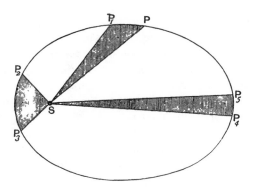

FIGURE 8—If S is the sun, a planet or comet moves from P to P₁, from P₂ to P₃, and from P₄ to P₅ in the same time; if the shaded areas are equal.

fessorship at Linz, and withdrew with his two quite young remaining children.

He provided for himself now partly by publishing a prophesying almanack, a sort of Zadkiel arrangement—a thing which he despised, but the support of which he could not afford to do without. He is continually attacking and throwing sarcasm at astrology, but it was the only thing for which people would pay him, and on it after a fashion he lived. We do not find that his circumstances were ever prosperous, and though 8,000 crowns were due to him from Bohemia he could not manage to get them paid.

About this time occurred a singular interruption to his work. His old mother, of whose fierce temper something has already been indicated, had been engaged in a law-suit for some years near their old home in Würtemberg. A change of judge having in process of time occurred, the defendant saw his way to turn the tables on the old lady by accusing her of sorcery. She was sent to prison, and condemned to the torture, with the usual intelligent idea of extracting a "voluntary" confession. Kepler had to hurry from Linz to interpose. He succeeded in saving her from the torture, but she remained in prison for a year or so. Her spirit, however, was unbroken, for no sooner was she released than she commenced a fresh action against her accuser. But fresh trouble was averted by the death of the poor old dame at the age of nearly eighty.

This narration renders the unflagging energy shown by her son in his mathematical wrestlings less surprising.

Interspersed with these domestic troubles, and with harassing and unsuccessful attempts to get his rights, he still brooded over his old problem of some possible connection between the distances of the planets from the sun and their times of revolution, *i.e.* the length of their years.

It might well have been that there was no connection, that it was purely imaginary, like his old idea of the law of the successive distances of the planets, and like so many others of the guesses and fancies which he entertained and spent his energies in probing. But fortunately this time there was a connection, and he lived to have the joy of discovering it.

The connection is this, that if one compares the distance of the different planets from the sun with the length of time they take to go round him, the cube of the respective distances is proportional to the square of the corresponding times. In other words, the ratio of r^3 to T^2 for every planet is the same. Or, again, the length of a planet's year depends on the $\frac{3}{2}$th power of its distance from the sun. Or, once more, the speed of each planet in its orbit is as the inverse square-root of its distance from the sun. The product of the distance into the square of the speed is the same for each planet.

This fact (however stated) is called Kepler's third law. It welds the planets together, and shows them to be one system. His rapture on detecting the law was unbounded, and he breaks out into an exulting rhapsody:—

"What I prophesied two-and-twenty years ago, as soon as I discovered the five solids among the heavenly orbits—what I firmly believed long before I had seen Ptolemy's *Harmonics*—what I had promised my friends in the title of this book, which I named before I was sure of my discovery —what sixteen years ago, I urged as a thing to be sought—that for which I joined Tycho Brahé, for which I settled in Prague, for which I have devoted the best part of my life to astronomical contemplations, at length I have brought to light, and recognized its truth beyond my most sanguine expectations. It is not eighteen months since I got the first glimpse of light, three months since the dawn, very few days since the unveiled sun, most admirable to gaze upon, burst upon me. Nothing holds me; I will indulge my sacred fury; I will triumph over mankind by the honest confession that I have stolen the golden vases of the Egyptians to build up a tabernacle for my God far away from the confines of Egypt. If you forgive me, I rejoice; if you are angry, I can bear it; the die is cast, the book is written, to be read either now or by posterity, I care not which; it may well wait a century for a reader, as God has waited six thousand years for an observer."

Soon after this great work his third book appeared: it was an epitome of the Copernican theory, a clear and fairly popular exposition of it, which had the honour of being at once suppressed and placed on the list of books prohibited by the Church, side by side with the work of Copernicus himself, *De Revolutionibus Orbium Coelestium*.

This honour, however, gave Kepler no satisfaction—it rather occasioned him dismay, especially as it deprived him of all pecuniary benefit, and made it almost impossible for him to get a publisher to undertake another book.

Still he worked on at the Rudolphine tables of Tycho, and ultimately, with some small help from Vienna, completed them; but he could not get the means to print them. He applied to the Court till he was sick of applying: they lay idle four years. At last he determined to pay for the type himself. What he paid it with, God knows, but he did pay it, and he did bring out the tables, and so was faithful to the behest of his friend.

This great publication marks an era in astronomy. They were the first really accurate tables which navigators ever possessed; they were the precursors of our present *Nautical Almanack.*

After this, the Grand Duke of Tuscany sent Kepler a golden chain, which is interesting inasmuch as it must really have come from Galileo, who was in high favour at the Italian Court at this time.

Once more Kepler made a determined attempt to get his arrears of salary paid, and rescue himself and family from their bitter poverty. He travelled to Prague on purpose, attended the imperial meeting, and pleaded his own cause, but it was all fruitless; and exhausted by the journey, weakened by over-study, and disheartened by the failure, he caught a fever, and died in his fifty-ninth year. His body was buried at Ratisbon, and a century ago a proposal was made to erect a marble monument to his memory, but nothing was done. It matters little one way or the other whether Germany, having almost refused him bread during his life, should, a century and a half after his death, offer him a stone.

The contiguity of the lives of Kepler and Tycho furnishes a moral too obvious to need pointing out. What Kepler might have achieved had he been relieved of those ghastly struggles for subsistence one cannot tell, but this much is clear, that had Tycho been subjected to the same misfortune, instead of being born rich and being assisted by generous and enlightened patrons, he could have accomplished very little. His instruments, his observatory—the tools by which he did his work—would have been impossible for him. Frederick and Sophia of Denmark, and Rudolph of Bohemia, are therefore to be remembered as co-workers with him.

Kepler, with his ill-health and inferior physical energy, was unable to command the like advantages. Much, nevertheless, he did; more one cannot but feel he might have done had he been properly helped. Besides, the world would have been free from the reproach of accepting the fruits of his bright genius while condemning the worker to a life of misery, relieved only by the beauty of his own thoughts and the ecstasy awakened in him by the harmony and precision of Nature.

Concerning the method of Kepler, the mode by which he made his discoveries, we must remember that he gives us an account of all the steps, unsuccessful as well as successful, by which he travelled. He maps out his route like a traveller. In fact he compares himself to Columbus or Magellan, voyaging into unknown lands, and recording his wandering route. This being remembered, it will be found that his methods do not differ so utterly from those used by other philosophers in like case. His imagination was perhaps more luxuriant and was allowed freer play than most men's, but it was nevertheless always controlled by rigid examination and comparison of hypotheses with fact.

Brewster says of him:—"Ardent, restless, burning to distinguish himself by discovery, he attempted everything; and once having obtained a glimpse of a clue, no labour was too hard in following or verifying it. A few of his attempts succeeded—a multitude failed. Those which failed seem to us now fanciful, those which succeeded appear to us sublime. But his methods were the same. When in search of what really existed he sometimes found it; when in pursuit of a chimæra he could not but fail; but in either case he displayed the same great qualities, and that obstinate perseverance which must conquer all difficulties except those really insurmountable."

To realize what he did for astronomy, it is necessary for us now to consider some science still in its infancy. Astronomy is so clear and so thoroughly explored now, that it is difficult to put oneself into a contemporary attitude. But take some other science still barely developed: meteorology, for instance. The science of the weather, the succession of winds and rain, sunshine and frost, clouds and fog, is now very much in the condition of astronomy before Kepler.

We have passed through the stage of ascribing atmospheric disturbances—thunderstorms, cyclones, earthquakes, and the like—to supernatural agency; we have had our Copernican era: not perhaps brought about by a single individual, but still achieved. Something of the laws of cyclone and anticyclone are known, and rude weather predictions across the Atlantic are roughly possible. Barometers and thermometers and anemometers, and all their tribe, represent the astronomical instruments in the island of Huen; and our numerous meteorological observatories, with their continual record of events, represent the work of Tycho Brahé.

Observation is heaped on observation; tables are compiled; volumes are filled with data; the hours of sunshine are recorded, the fall of rain, the moisture in the air, the kind of clouds, the temperature—millions of facts; but where is the Kepler to study and brood over them? Where is the man to spend his life in evolving the beginnings of law and order from the midst of all this chaos?

Perhaps as a man he may not come, but his era will come. Through this stage the science must pass, ere it is ready for the commanding intellect of a Newton.

But what a work it will be for the man, whoever he be that undertakes it—a fearful monotonous grind of calculation, hypothesis, hypothesis, calculation, a desperate and groping endeavour to reconcile theories with facts.

A life of such labour, crowned by three brilliant discoveries, the world owes (and too late recognizes its obligation) to the harshly treated German genius, Kepler.

COMMENTARY ON
DESCARTES and Analytical Geometry

RENÉ DESCARTES (1596–1650) came from a noble family settled since the fourteenth century in southern Touraine. His father was a councilor of the parliament of Brittany and reasonably well off. Descartes inherited from him enough money to support a life of study and travel. He was educated, from 1604 to 1612, at the Jesuit college at La Flèche, where he not only received a good training in the humanities and mathematics but was treated with exquisite consideration. He was allowed by Father Charlet, the rector and evidently a sensible man, to lie in bed in the morning—"a habit which he maintained all his life, and which he regarded as above all conducive to intellectual profit and comfort." [1] After leaving school, Descartes took a turn at social life in Paris, but he found this tiresome and soon shut himself up in lodgings in the Faubourg Saint Germain to study mathematics. In 1621 he entered the army of Prince Maurice of Nassau. Descartes said that in his youth he liked war, attributing this taste "to a certain animal heat in his liver, which cooled down in the course of time." [2] Whether or not this is true, being a soldier, though occasionally dangerous, was not an arduous occupation. It provided leisure for meditation as well as an opportunity to see something of other countries. In 1619 Descartes enlisted in the Bavarian Army and during the winter of that year had a major philosophical inspiration which he reports in the *Discourse on Method*. Bertrand Russell gives a witty description of the circumstances: "The weather being cold, he got into a stove [3] in the morning, and stayed there all day meditating; by his own account, his philosophy was half finished when he came out, but this need not be accepted too literally. Socrates used to meditate all day in the snow, but Descartes' mind only worked when he was warm." [4]

After a few years of fighting, interrupted by trips to Italy and Paris, Descartes retired to Holland where for twenty years, 1629 to 1649, he spent his time on science and philosophy. The freedom of thought permitted him in his work was characteristic of Holland in the seventeenth century. It was only slightly marred by the interference of "Protestant bigots," for the French ambassador and the Prince of Orange protected

---

[1] J. P. Mahaffy, *Descartes*, Philadelphia, 1881, p. 12.

[2] Mahaffy, *op. cit.*, p. 20.

[3] Descartes *says* it was a stove (*poêle*), but the authors of the standard translation, Elizabeth S. Haldane and G. R. T. Ross, write "shut up alone in a stove-heated room." With Russell, I prefer "stove."

[4] Bertrand Russell, *A History of Western Philosophy*, New York, 1945, p. 558.

him from the malevolence of his assailants. At one time, booksellers were forbidden to print or sell any of his works and he was even haled before the magistrates to answer charges by the theologians of Utrecht and Leyden that he was an atheist, a "vagabond" and a profligate; but the excitement passed without serious consequences. Nevertheless, when Queen Christina of Sweden invited him in 1649 to adorn her court, he had had enough of controversy and of Holland and he accepted her offer. It is said that a "presentiment" of death came over him as he prepared for the journey. Stockholm, when he arrived there in October, was a cold and disagreeable spot and the Queen, a yearner for wisdom, shattered the unhappy man's lifelong routine by requiring him to instruct her daily at five in the morning. Descartes withstood the weather and the sovereign for only four months; on February 1 he came down with pneumonia and ten days later he died.

Descartes has been called the father of modern philosophy, perhaps because he attempted to build a new system of thought from the ground up, emphasized the use of logic and scientific method, and was "profoundly affected in his outlook by the new physics and astronomy." Undoubtedly he had great influence on philosophy; in recent years, however, critics have depreciated his originality by pointing out how much he owed to the scholastics. His contribution to mathematics was of enormous importance. He is usually considered the inventor of analytical geometry, but this notion is "historically inadequate" because the subject did not spring fullarmed from Descartes' head.[5] The study of curves by means of their equations, defined as the "essence" of analytic geometry, was known to the Greeks and "was the basis of their study of the conic sections." [6] Menaechmus, the tutor of Alexander the Great, is reputed to have made this discovery. Among Descartes' other predecessors were the French theologian Nicole Oresme, whose system of "latitudes and longitudes" roughly foreshadowed "the use of co-ordinates in the graphical representation of arbitrary functions," [7] and François Viète, the sixteenth-century counselor to the King of France, whose improvements in notation substantially facilitated the development of algebra. The most formidable claimant to Descartes' title of inventor of analytic geometry was his famous contemporary, Pierre Fermat, who enriched every branch of mathematics known in his time, founded the modern theory of numbers and significantly advanced the study of probability. The truth seems to be that each man, simultaneously and independently, carried the subject far beyond where any

[5] For an interesting, popular and authoritative account of the development of this branch of mathematics see Carl B. Boyer, "The Invention of Analytic Geometry," *Scientific American*, January 1949. A more advanced and comprehensive, but no less readable survey appears in J. L. Coolidge's excellent treatise, *A History of Geometrical Methods*, Oxford, 1940, pp. 117 *et seq.*

[6] See J. L. Coolidge, cited in the preceding note, p. 119.

[7] Boyer, *op. cit.*, p. 42.

earlier mathematician had taken it. Thus was enacted a prelude to the classic example of simultaneous discovery, Newton and Leibniz at work on the calculus. Fermat may have preceded Descartes in stating problems of maxima and minima; but Descartes went far past Fermat in the use of symbols, in "arithmetizing" analytic geometry, in extending it to equations of higher degree. The fixing of a point's position in the plane by assigning two numbers, co-ordinates, giving its distance from two lines perpendicular to each other, was entirely Descartes' invention.[8] Fermat's little treatise of eight folio pages appeared in 1679, half a century after it had been composed; Descartes' *La Géométrie* was published as an appendix to the *Discourse* in 1673. He made other contributions to mathematics but this was by far the most important. He did not, it should be added, dream up the idea of co-ordinates while lying in bed at La Flèche. This agreeable fable (which I myself have on other occasions repeated) was the concoction of one Daniel Lipstorpius, a Lübeck professor who wrote a life of Descartes in the style of Parson Weems. I will not go so far as to say that men cannot have great thoughts in bed; only that the sixteen-year-old Descartes, having studied mathematics for a few months, was not up to conceiving the co-ordinates during a morning reverie.

The following selection is a facsimile, with translation, of the first eight pages of the first edition of *La Géométrie*. It is written in a "contemptuous vein." Descartes was evidently more interested in showing what he knew than in instructing beginners. He concludes the work with an ironic paragraph: "I hope that posterity will judge me kindly, not only as to the things which I have explained, but also as to those which I have intentionally omitted so as to leave to others the pleasure of discovery."

[8] Coolidge, *op. cit.*, p. 126; Boyer, *op. cit.*, p. 43.

"He (Descartes) contemptuously rejected the idea that the only curves we should consider were plane and solid loci, that is to say lines, circles and conics, and maintained that we were at liberty to make use of any smooth curve which has a recognizable equation. . . . He also introduced exponents, except in the case of the square, and other algebraic simplifications, developing an algebra that was infinitely more manageable than that of Vieta and Fermat." Coolidge, *op. cit.*, p. 126.

Des minimes de Paris

LA

GEOMETRIE.

LIVRE PREMIER.

Des problefmes qu'on peut conftruire fans
y employer que des cercles & des
lignes droites.

Ous les Problefmes de Geometrie fe
peuuent facilement reduire a tels termes,
qu'il n'eft befoin par aprés que de connoiftre la longeur de quelques lignes droites,
pour les conftruire.

Et comme toute l'Arithmetique n'eft compofée, que
de quatre ou cinq operations, qui font l'Addition, la
Souftraction, la Multiplication, la Diuifion, & l'Extraction des racines, qu'on peut prendre pour vne efpece
de Diuifion : Ainfi n'at'on autre chofe a faire en Geometrie touchant les lignes qu'on cherche, pour les pre-
parer a eftre connuës, que leur en adioufter d'autres , ou
en ofter, Oubien en ayant vne, que ie nommeray l'vnité·
pour la rapporter d'autant mieux aux nombres , & qui
peut ordinairement eftre prife a difcretion, puis en ayant
encore deux autres, en trouuer vne quatriefme, qui foit
à l'vne de ces deux, comme l'autre eft a l'vnité, ce qui eft
le mefme que la Multiplication ; oubien en trouuer vne
quatriefme, qui foit a l'vne de ces deux , comme l'vnité

Commēt le calcul d'Ari- thmeti- que fe rapporte aux ope- rations de Geome- trie.

P p   eft

It is impossible not to feel stirred at the thought of the emotions of men at certain historic moments of adventure and discovery—Columbus when he first saw the Western shore, Pizarro when he stared at the Pacific Ocean, Franklin when the electric spark came from the string of his kite, Galileo when he first turned his telescope to the heavens. Such moments are also granted to students in the abstract regions of thought, and high among them must be placed the morning when Descartes lay in bed and invented the method of co-ordinate geometry. —ALFRED NORTH WHITEHEAD

There are some men who are counted great because they represent the actuality of their own age, and mirror it as it is. Such an one was Voltaire, of whom it was epigrammatically said: "he expressed everybody's thoughts better than anybody." But there are other men who attain greatness because they embody the potentiality of their own day and magically reflect the future. They express the thoughts which will be everybody's two or three centuries after them. Such an one was Descartes. —THOMAS HUXLEY

7   The Geometry

By RENE DESCARTES

BOOK I

PROBLEMS THE CONSTRUCTION OF WHICH REQUIRES ONLY
STRAIGHT LINES AND CIRCLES

ANY problem in geometry can easily be reduced to such terms that a knowledge of the lengths of certain straight lines is sufficient for its construction. Just as arithmetic consists of only four or five operations, namely, addition, subtraction, multiplication, division and the extraction of roots, which may be considered a kind of division, so in geometry, to find required lines it is merely necessary to add or subtract other lines;

298   LA GEOMETRIE.

eſt a l'autre, ce qui eſt le meſme que la Diuiſion; ou enfin trouuer vne, ou deux, ou pluſieurs moyennes proportion-nelles entre l'vnité, & quelque autre ligne; ce qui eſt le meſme que tirer la racine quarrée, on cubique, &c. Et ie ne craindray pas d'introduire ces termes d'Arithmeti-que en la Geometrie, affin de me rendre plus intel-ligibile.

La Multi-
plication.

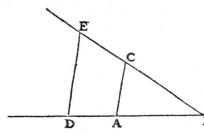

Soit par exemple A B l'vnité, & qu'il fail-le multiplier B D par B C, ie n'ay qu'a ioindre les poins A & C, puis ti-rer D E parallele a C A, & B E eſt le produit de cete Multiplication.

La Divi-
ſion.

Oubien s'il faut diuiſer B E par B D, ayant ioint les poins E & D, ie tire A C parallele a D E, & B C eſt le produit de cete diuiſion.

l'Extra-
ſtion dela
racine
quarrée.

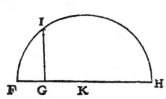

Ou s'il faut tirer la racine quarrée de G H, ie luy ad-iouſte en ligne droite F G, qui eſt l'vnité, & diuiſant F H en deux parties eſgales au point K, du centre K ie tire le cercle F I H, puis eſleuant du point G vne ligne droite iuſques à I, à angles droits ſur F H, c'eſt G I la racine cherchée. Ie ne dis rien icy de la racine cubique, ny des autres, à cauſe que i en parleray plus commodement cy aprés.

Commēt
on peut

Mais ſouuent on n'a pas beſoin de tracer ainſi ces li-gne

or else, taking one line which I shall call unity in order to relate it as closely as possible to numbers,[1] and which can in general be chosen arbitrarily, and having given two other lines, to find a fourth line which shall be to one of the given lines as the other is to unity (which is the same as multiplication); or, again, to find a fourth line which is to one of the given lines as unity is to the other (which is equivalent to division); or, finally, to find one, two, or several mean proportionals between unity and some other line (which is the same as extracting the square root, cube root, etc., of the given line).[2] And I shall not hesitate to introduce these arithmetical terms into geometry, for the sake of greater clearness.

For example, let AB be taken as unity, and let it be required to multiply BD by BC. I have only to join the points A and C, and draw DE parallel to CA; then BE is the product of BD and BC.

If it be required to divide BE by BD, I join E and D, and draw AC parallel to DE; then BC is the result of the division.

If the square root of GH is desired, I add, along the same straight line, FG equal to unity; then, bisecting FH at K, I describe the circle FIH about K as a center, and draw from G a perpendicular and extend it to I, and GI is the required root. I do not speak here of cube root, or other roots, since I shall speak more conveniently of them later.

Often it is not necessary thus to draw the lines on paper, but it is sufficient to designate each by a single letter. Thus, to add the lines BD and GH, I call one a and the other b, and write $a + b$. Then $a - b$ will indicate that b is subtracted from a; ab that a is multiplied by b; $\dfrac{a}{b}$ that a is divided by b; aa or a^2 that a is multiplied by itself; a^3 that this result is

---

[1] In general, the translation runs page for page with the facing original. On account of figures and footnotes, however, this plan is occasionally varied, but not in such a way as to cause the reader any serious inconvenience.

[2] While in arithmetic the only exact roots obtainable are those of perfect powers, in geometry a length can be found which will represent exactly the square root of a given line, even though this line be not commensurable with unity. Of other roots, Descartes speaks later.

gnes fur le papier, & il fuffift de les defigner par quelques
lettres, chafcune par vne feule. Comme pour adioufter
la ligne B D a G H, ie nomme l'vne a & l'autre b, & efcris
$a + b$; Et $a - b$, pour fouftraire b d' a; Et $a b$, pour les multiplier l'vne par l'autre; Et $\frac{a}{b}$, pour diuifer a par b; Et $a a$,
ou a^2, pour multiplier a par foy mefme; Et a^3, pour le
multiplier encore vne fois par a, & ainfi a l'infini; Et
$\sqrt{a^2 + b^2}$, pour tirer la racine quarrée d' $a^2 + b^2$; Et
$\sqrt{C. a^3 - b^3 + a b b}$, pour tirer la racine cubique d' $a^3 - b^3$
$+ a b b$, & ainfi des autres.

Où il eft a remarquer que par a^2 ou b^3 ou femblables,
ie ne conçoy ordinairement que des lignes toutes fimples, encore que pour me feruir des noms vfités en l'Algebre, ie les nomme des quarrés ou des cubes, &c.

Il eft auffy a remarquer que toutes les parties d'vne
mefme ligne, fe doiuent ordinairement exprimer par autant de dimenfions l'vne que l'autre, lorfque l'vnité n'eft
point déterminée en la queftion, comme icy a^3 en con-
tient autant qu' $a b b$ ou b^3 dont fe compofe la ligne que
i'ay nommée $\sqrt{C. a^3 - b^3 + a b b}$: mais que ce n'eft
pas de mefme lorfque l'vnité eft déterminée, a caufo
qu'elle peut eftre foufentendue par tout ou il y a trop ou
trop peu de dimenfions : comme s'il faut tirer la racine
cubique de $a a b b - b$, il faut penfer que la quantité
$a a b b$ eft diuifée vne fois par l'vnité, & que l'autre quantité b eft multipliée deux fois par la mefme.

P p 2   Au

multiplied by a, and so on, indefinitely.[3] Again, if I wish to extract the square root of $a^2 + b^2$, I write $\sqrt{a^2 + b^2}$; if I wish to extract the cube root of $a^3 - b^3 + ab^2$, I write $\sqrt[3]{a^3 - b^3 + ab^2}$, and similarly for other roots.[4] Here it must be observed that by a^2, b^3, and similar expressions, I ordinarily mean only simple lines, which, however, I name squares, cubes, etc., so that I may make use of the terms employed in algebra.[5]

It should also be noted that all parts of a single line should always be expressed by the same number of dimensions, provided unity is not determined by the conditions of the problem. Thus, a^3 contains as many dimensions as ab^2 or b^3, these being the component parts of the line which I have called $\sqrt[3]{a^3 - b^3 + ab^2}$. It is not, however, the same thing when unity is determined, because unity can always be understood, even where there are too many or too few dimensions; thus, if it be required to extract the cube root of $a^2b^2 - b$, we must consider the quantity a^2b^2 divided once by unity, and the quantity b multiplied twice by unity.[6]

Finally, so that we may be sure to remember the names of these lines, a separate list should always be made as often as names are assigned or changed. For example, we may write, $AB = 1$, that is AB is equal to 1;[7] $GH = a$, $BD = b$, and so on.

---

[3] Descartes uses a^3, a^4, a^5, a^6, and so on, to represent the respective powers of a, but he uses both aa and a^2 without distinction. For example, he often has $aabb$, but he also uses $\dfrac{3a^2}{4b^2}$.

[4] Descartes writes: $\sqrt{C.a^3 - b^3 + abb}$.

[5] At the time this was written, a^2 was commonly considered to mean the surface of a square whose side is a, and b^3 to mean the volume of a cube whose side is b; while b^4, b^5, . . . were unintelligible as geometric forms. Descartes here says that a^2 does not have this meaning, but means the line obtained by constructing a third proportional to 1 and a, and so on.

[6] Descartes seems to say that each term must be of the third degree, and that therefore we must conceive of both a^2b^2 and b as reduced to the proper dimension.

[7] Descartes writes, $AB \propto 1$. He seems to have been the first to use this symbol. Among the few writers who followed him, was Hudde (1633–1704). It is very commonly supposed that $\propto$ is a ligature representing the first two letters (or dipththong) of "æquare." See, for example, M. Aubry's note in W. W. R. Ball's *Recréations Mathématiques et Problèmes des Temps Anciens et Modernes,* French edition, Paris, 1909, Part III, p. 164.

La Geometrie.

Au reſte affin de ne pas manquer a ſe ſouuenir des
noms de ces lignes, il en faut touſiours faire vn regiſtre
ſeparé , à meſure qu'on les poſe ou qu'on les change,
eſcriuant par exemple.

A B ∽ 1 ,   c'eſt a dire, A B eſgal à 1.

G H ∽ *a*

B D ∽ *b* , ℰc.

Commēt
il faut ve-
nir aux
Equatiōs
qui ſer-
uent a re-
ſoudre les
probleſ-
mes.
Ainſi voulant reſoudre quelque probleſme, on doit d'a-
bord le conſiderer comme deſia fait, & donner des noms
a toutes les lignes, qui ſemblent neceſſaires pour le con-
ſtruire, auſſy bien a celles qui ſont inconnuës , qu'aux
autres. Puis ſans conſiderer aucune difference entre ces
lignes connuës, & inconnuës , on doit parcourir la diffi-
culté, ſelon l'ordre qui monſtre le plus naturellement
de tous en qu'elle ſorte elles dependent mutuellement
les vnes des autres, iuſques a ce qu'on ait trouué moyen
d'exprimer vne meſme quantité en deux façons: ce qui
ſe nomme vne Equation; car les termes de l'vne de ces
deux façons ſont eſgaux a ceux de l'autre.   Et on doit
trouuer autant de telles Equations, qu'on a ſuppoſé de lignes, qui eſtoient inconnuës. Oubien s'il ne s'en trouue
pas tant, & que nonobſtant on n'omette rien de ce qui eſt
deſiré en la queſtion, cela teſmoigne qu'elle n'eſt pas entierement determinée. Et lors on peut prendre a diſcre-
tion des lignes connuës , pour toutes les inconnuës auſ-
qu'elles ne correſpond aucune Equation. Aprés cela s'il
en reſte encore pluſieurs , il ſe faut ſeruir par ordre de
chaſcune des Equations qui reſtent auſſy , ſoit en la con-
ſiderant toute ſeule, ſoit en la comparant auec les autres,
pour expliquer chaſcune de ces lignes inconnuës; & faire
                                                         ainſi

If, then, we wish to solve any problem, we first suppose the solution already effected,[8] and give names to all the lines that seem needful for its construction,—to those that are unknown as well as to those that are known. Then, making no distinction between known and unknown lines, we must unravel the difficulty in any way that shows most naturally the relations between these lines, until we find it possible to express a single quantity in two ways.[9] This will constitute an equation, since the terms of one of these two expressions are together equal to the terms of the other.

We must find as many such equations as there are supposed to be unknown lines; but if, after considering everything involved, so many cannot be found, it is evident that the question is not entirely determined. In such a case we may choose arbitrarily lines of known length for each unknown line to which there corresponds no equation.

If there are several equations, we must use each in order, either considering it alone or comparing it with the others, so as to obtain a value for each of the unknown lines; and so we must combine them until there remains a single unknown line [10] which is equal to some known line, or whose square, cube, fourth power, fifth power, sixth power, etc., is equal to the sum or difference of two or more quantities, one of which is known,

[8] This plan, as is well known, goes back to Plato. It appears in the work of Pappus as follows: "In analysis we suppose that which is required to be already obtained, and consider its connections and antecedents, going back until we reach either something already known (given in the hypothesis), or else some fundamental principle (axiom or postulate) of mathematics." *Pappi Alexandrini Collectiones quae supersunt e libris manu scriptis edidit Latina interpellatione et commentariis instruxit Fredericus Hultsch*, Berlin, 1876–1878; vol. II, p. 635 (hereafter referred to as Pappus). See also Commandinus, *Pappi Alexandrini Mathematicae Collectiones*, Bologna, 1588, with later editions.

Pappus of Alexandria was a Greek mathematician who lived about 300 A.D. His most important work is a mathematical treatise in eight books, of which the first and part of the second are lost. This was made known to modern scholars by Commandinus. The work exerted a happy influence on the revival of geometry in the seventeenth century. Pappus was not himself a mathematician of the first rank, but he preserved for the world many extracts or analyses of lost works, and by his commentaries added to their interest.

[9] That is, we must solve the resulting simultaneous equations.

[10] That is, a line represented by x, x^2, x^3, x^4, ...

Livre Premier.   301

ainſi en les demeſlant, qu'il n'en demeure qu'vne ſeule,
eſgale a quelque autre, qui ſoit connuë, oubien dont le
quarré, ou le cube, ou le quarré de quarré, ou le ſurſoli-
de, ou le quarré de cube, &c. ſoit eſgal a ce, qui ſe produiſt par l'addition, ou ſouſtraction de deux ou pluſieurs
autres quantités, dont l'vne ſoit connuë, & les autres
ſoient compoſées de quelques moyennes proportion‐
nelles entre l'vnité, & ce quarré, ou cube, ou quarré de
quarré, &c. multipliées par d'autres connuës. Ce que i'e-
ſcris en cete ſorte.

$$z \infty \; b. \text{ ou}$$

$$z^2 \infty -- a \; z + b b. \text{ ou}$$

$$z^3 \infty + a \; z^2 + b b z -- c^3. \text{ ou}$$

$$z^4 \infty \; a \; z^3 \; -- c^3 z + d^4. \text{ &c.}$$

C'eſt a dire, z, que ie prens pour la quantité inconnuë,
eſt eſgale a b, ou le quarré de z eſt eſgal au quarré de b
moins a multiplié par z. ou le cube de z eſt eſgal à a
multiplié par le quarre de z plus le quarré de b multiplié
par z moins le cube de c. & ainſi des autres.

Et on peut touſiours reduire ainſi toutes les quantités
inconnuës à vne ſeule, lorſque le Problefme ſe peut con‐
ſtruire par des cercles & des lignes droites, ou auſſy par
des ſections coniques, ou meſme par quelque autre ligne
qui ne ſoit que d'vn ou deux degrés plus compoſée. Mais,
ie ne m'areſte point a expliquer cecy plus en detail, a
cauſe que ie vous oſterois le plaiſir de l'apprendre de
vous meſme, & l'vtilité de cultiuer voſtre eſprit en vous
y exerceant, qui eſt a mon auis la principale, qu'on puiſſe

Pp 3   tirer

while the others consist of mean proportionals between unity and this square, or cube, or fourth power, etc., multiplied by other known lines. I may express this as follows:

$$z = b,$$
$$\text{or } z^2 = -az + b^2,$$
$$\text{or } z^3 = az^2 + b^2z - c^3,$$
$$\text{or } z^4 = az^3 - c^3z + d^4, \text{ etc.}$$

That is, z, which I take for the unknown quantity, is equal to b; or, the square of z is equal to the square of b diminished by a multiplied by z; or, the cube of z is equal to a multiplied by the square of z, plus the square of b multiplied by z, diminished by the cube of c; and similarly for the others.

Thus, all the unknown quantities can be expressed in terms of a single quantity,[11] whenever the problem can be constructed by means of circles and straight lines, or by conic sections, or even by some other curve of degree not greater than the third or fourth.[12]

But I shall not stop to explain this in more detail, because I should deprive you of the pleasure of mastering it yourself, as well as of the advantage of training your mind by working over it, which is in my opinion the principal benefit to be derived from this science. Because, I find nothing here so difficult that it cannot be worked out by any one at all familiar with ordinary geometry and with algebra, who will consider carefully all that is set forth in this treatise.[13]

[11] See line 20 on the opposite page.

[12] Literally, "Only one or two degrees greater."

[13] In the Introduction to the 1637 edition of *La Géométrie*, Descartes made the following remark: "In my previous writings I have tried to make my meaning clear to everybody; but I doubt if this treatise will be read by anyone not familiar with the books on geometry, and so I have thought it superfluous to repeat demonstrations contained in them." See *Oeuvres de Descartes*, edited by Charles Adam and Paul Tannery, Paris, 1897–1910, vol. VI, p. 368. In a letter written to Mersenne in 1637 Descartes says: "I do not enjoy speaking in praise of myself, but since few people can understand my geometry, and since you wish me to give you my opinion of it, I think it well to say that it is all I could hope for, and that in *La Dioptrique* and *Les Météores*, I have only tried to persuade people that my method is better than

302   LA GEOMETRIE.

tirer de cete science. Auſſy que ie n y remarque rien de
ſi difficile, que ceux qui ſeront vn peu verſés en la Geometrie commune, & en l'Algebre, & qui prendront gar-
de a tout ce qui eſt en ce traité, ne puiſſent trouuer.

C'eſt pourquoy ie me contenteray icy de vous auer-
tir, que pourvû qu'en demeſlant ces Equations on ne
manque point a ſe ſeruir de toutes les diuiſions, qui ſeront poſſibles, on aura infalliblement les plus ſimples
termes, auſquels la queſtion puiſſe eſtre reduite.

Quels
ſont les
probleſ-
mes plans Et que ſi elle peut eſtre reſolue par la Geometrie ordinaire, c'eſt a dire, en ne ſe ſeruant que de lignes droites
& circulaires tracées ſur vne ſuperficie plate, lorſque la
derniere Equation aura eſté entierement démeſlée, il n'y
reſtera tout au plus qu'vn quarré inconnu, eſgal a ce qui
ſe produiſt de l'Addition, ou ſouſtraction de ſa racine
multipliée par quelque quantité connue, & de quelque
autre quantité auſſy connue

Com-
ment ils
ſe reſol-
uent. Et lors cete racine, ou ligne inconnue ſe trouue ayſe-
ment. Car ſi i'ay par exemple

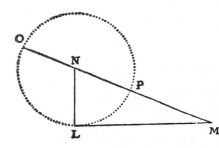

$$z \infty\, a\,z + bb$$

ie fais le triangle rectangle N L M, dont le co-
ſté L M eſt eſgal à b ra-
cine quarrée de la quan-
tité connue b b, & l'au-
tre L N eſt ½ a, la moi-
tié de l'autre quantité
connue, qui eſtoit multipliée par z que ie ſuppoſe eſtre la
ligne inconnue. puis prolongeant M N la baze de ce triangle,

I shall therefore content myself with the statement that if the student, in solving these equations, does not fail to make use of division wherever possible, he will surely reach the simplest terms to which the problem can be reduced.

And if it can be solved by ordinary geometry, that is, by the use of straight lines and circles traced on a plane surface, when the last equation shall have been entirely solved there will remain at most only the square of an unknown quantity, equal to the product of its root by some known quantity, increased or diminished by some other quantity also known.[14] Then this root or unknown line can easily be found. For example, if I have $z^2 = az + b^2$,[15] I construct a right triangle NLM with one side LM,

the ordinary one. I have proved this in my geometry, for in the beginning I have solved a question which, according to Pappus, could not be solved by any of the ancient geometers.

"Moreover, what I have given in the second book on the nature and properties of curved lines, and the method of examining them, is, it seems to me, as far beyond the treatment in the ordinary geometry, as the rhetoric of Cicero is beyond the a, b, c of children. . . .

"As to the suggestion that what I have written could easily have been gotten from Vieta, the very fact that my treatise is hard to understand is due to my attempt to put nothing in it that I believed to be known either by him or by any one else. . . . I begin the rules of my algebra with what Vieta wrote at the very end of his book, *De emendatione aequationum.* . . . Thus, I begin where he left off." *Oeuvres de Descartes, publiées par Victor Cousin,* Paris, 1824, Vol. VI, p. 294.

In another letter to Mersenne, written April 20, 1646, Descartes writes as follows: "I have omitted a number of things that might have made it (the geometry) clearer, but I did this intentionally, and would not have it otherwise. The only suggestions that have been made concerning changes in it are in regard to rendering it clearer to readers, but most of these are so malicious that I am completely disgusted with them." Cousin, Vol. IX, p. 553.

In a letter to the Princess Elizabeth, Descartes says: "In the solution of a geometrical problem I take care, as far as possible, to use as lines of reference parallel lines or lines at right angles; and I use no theorems except those which assert that the sides of similar triangles are proportional, and that in a right triangle the square of the hypotenuse is equal to the sum of the squares of the sides. I do not hesitate to introduce several unknown quantities, so as to reduce the question to such terms that it shall depend only on these two theorems." Cousin, Vol. IX, p. 143.

[14] That is, an expression of the form $z^2 = az \pm b$. "Esgal a ce qui se produit de l'Addition, ou soustraction de sa racine multiplée par quelque quantité connue, & de quelque autre quantité aussy connue," as it appears in line 14, opposite page.

[15] Descartes proposes to show how a quadratic may be solved geometrically.

angle, iufques a O, en forte qu'N O foit efgale a N L,
la toute O M eft z la ligne cherchée· Et elle s'exprime
en cete forte

$$z \infty \tfrac{1}{2} a + \sqrt{\tfrac{1}{4} a a + b b}.$$

Que fi iay $yy \infty -- ay + bb$, & qu'y foit la quantité
qu'il faut trouuer , ie fais le mefme triangle rectangle
N L M, & de fa baze M N i'ofte N P efgale a N L, & le
refte P M eft y la racine cherchée. De façon que iay
$y \infty -\tfrac{1}{2} a + \sqrt{\tfrac{1}{4} a a + b b}$. Et tout de mefme fi i'a-
uois $x^4 \infty -- a x^2 + b^2$. P M feroit x^2. & i'aurois
$x \infty \sqrt{-\tfrac{1}{2} a + \sqrt{\tfrac{1}{4} a a + b b}}$: & ainfi des autres.

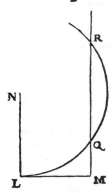

Enfin fi i'ay

$$z \infty a z -- b b:$$

ie fais N L efgale à $\tfrac{1}{2} a$, & L M
efgale à b côme deuãt, puis, au lieu
de ioindre les poins M N , ie tire
M Q R parallele a L N. & du centre N par L ayant defcrit vn cercle qui la couppe aux poins Q &
R, la ligne cherchée z eft M Q,
oubiẽ M R, car en ce cas elle s'ex-
prime en deux façons, a fçauoir $z \infty \tfrac{1}{2} a + \sqrt{\tfrac{1}{4} a a -- b b}$,
& $z \infty \tfrac{1}{2} a -- \sqrt{\tfrac{1}{4} a a -- b b}$.

Et fi le cercle, qui ayant fon centre au point N , paffe
par le point L, ne couppe ny ne touche la ligne droite
M Q R, il n'y a aucune racine en l'Equation, de façon
qu'on peut affurer que la conftruction du problefme
propofé eft impoffible.

Au

equal to b, the square root of the known quantity b^2, and the other side, LN, equal to $\frac{1}{2}a$, that is, to half the other known quantity which was multiplied by z, which I supposed to be the unknown line. Then prolonging MN, the hypotenuse [16] of this triangle, to O, so that NO is equal to NL, the whole line OM is the required line z. This is expressed in the following way: [17]

$$z = \tfrac{1}{2}a + \sqrt{\tfrac{1}{4}a^2 + b^2}.$$

But if I have $y^2 = -ay + b^2$, where y is the quantity whose value is desired, I construct the same right triangle NLM, and on the hypotenuse MN lay off NP equal to NL and the remainder PM is y, the desired root. Thus I have

$$y = -\tfrac{1}{2}a + \sqrt{\tfrac{1}{4}a^2 + b^2}.$$

In the same way, if I had

$$x^4 = -ax^2 + b^2,$$

PM would be x^2 and I should have

$$x = \sqrt{-\tfrac{1}{2}a + \sqrt{\tfrac{1}{4}a^2 + b^2}},$$

and so for other cases.

Finally, if I have $z^2 = az - b^2$, I make NL equal to $\frac{1}{2}a$ and LM equal to b as before: then, instead of joining the points M and N, I draw MQR parallel to LN, and with N as a center describe a circle through L cutting MQR in the points Q and R; then z, the line sought, is either MQ or MR, for in this case it can be expressed in two ways, namely: [18]

$$z = \tfrac{1}{2}a + \sqrt{\tfrac{1}{4}a^2 - b^2},$$

and

$$z = \tfrac{1}{2}a - \sqrt{\tfrac{1}{4}a^2 - b^2}.$$

[16] Descartes says "prolongeant MN la baze de ce triangle," because the hypotenuse was commonly taken as the base in earlier times.

[17] From the figure OM . PM $= \overline{\text{LM}}^2$. If OM $= z$, PM $= z - a$, and since LM $= b$, we have $z(z-a) = b^2$ or $z^2 = az + b^2$. Again, MN $= \sqrt{\tfrac{1}{4}a^2 + b^2}$, whence OM $= z =$ ON + MN $= \tfrac{1}{2} + \sqrt{\tfrac{1}{4}a^2 + b^2}$. Descartes ignores the second root, which is negative.

[18] Since $\overline{\text{MR}}/\text{MQ} = \text{LM}^2$, then if R $= z$, we have MQ $= a - z$, and so
$$z(a - z) = b^2 \text{ or } z^2 = az - b^2.$$

La Geometrie.

Au refte ces mefmes racines fe peuuent trouuer par
vne infinité d'autres moyens , & i'ay feulement veulu
mettre ceux cy, comme fort fimples, affin de faire voir
qu'on peut conftruire tous les Problefmes de la Geometrie ordinaire, fans faire autre chofe que le peu qui eft
compris dans les quatre figures que i'ay expliquées. Ce
que ie ne croy pas que les anciens ayent remarqué. car
autrement ils n'euffent pas pris la peine d'en efcrire tant
de gros liures, ou le feul ordre de leurs propofitions nous
fait connoiftre qu'ils n'ont point eu la vraye methode
pour les trouuer toutes, mais qu'ils ont feulement ramaf-
fé celles qu'ils ont rencontrées.

Exemple
tiré de
Pappus.   Et on le peut voir auffy fort clairement de ce que Pappus a mis au commencement de fon feptiefme liure , ou
aprés s'eftre arefté quelque tems a denombrer tout ce
qui auoit efté efcrit en Geometrie par ceux qui l'auoient
precedé, il parle enfin d vne queftion , qu'il dit que ny
Euclide, ny Apollonius, ny aucun autre n'auoient fceu
entierement refoudre. & voycy fes mots.

Ie cite
plutoft la
verfion la-   *Quem autem dicit (Apollonius) in tertio libro locum ad*
tine que le *tres, & quatuor lineas ab Euclide perfeſtum non effe , neque*
texte grec *ipſe perficere poterat , neque aliquis alius . fed neque pau-*
affin que *lulum quid addere iis , quæ Euclides ſcripfit, per ea tantum*
chafcun *conica , quæ ufque ad Euclidis tempora præmonſtrata*
l'entende
plus ayſe- *ſunt, &c.*
ment.
Et vn peu aprés il explique ainfi qu'elle eft cete queftion.

At locus ad tres, & quatuor lineas , in quo (Apollonius)
magnifice ſe iaſtat, & oſtentat, nulla habita gratia ei , qui
prius ſcripſerat , eſt hujuſmodi. Si poſitione datis tribus
                                                      rectis

And if the circle described about N and passing through L neither cuts nor touches the line MQR, the equation has no root, so that we may say that the construction of the problem is impossible.

These same roots can be found by many other methods. I have given these very simple ones to show that it is possible to construct all the problems of ordinary geometry by doing no more than the little covered in the four figures that I have explained.[19] This is one thing which I believe the ancient mathematicians did not observe, for otherwise they would not have put so much labor into writing so many books in which the very sequence of the propositions shows that they did not have a sure method of finding all,[20] but rather gathered together those propositions on which they had happened by accident.

If, instead of this, MQ $= z$, then MR $= a - z$, and again, $z^2 = az - b^2$. Furthermore, letting O be the mid-point of QR,

$$MQ = OM - OQ = \tfrac{1}{2}a - \sqrt{\tfrac{1}{4}a^2 - b^2},$$

and

$$MR = MO + OR = \tfrac{1}{2}a + \sqrt{\tfrac{1}{4}a^2 - b^2}.$$

Descartes here gives both roots, since both are positive. If MR is tangent to the circle, that is, if $b = \tfrac{1}{2}a$, the roots will be equal; while if $b > \tfrac{1}{2}a$, the line MR will not meet the circle and both roots will be imaginary. Also, since RM . QM $= \overline{\text{LM}}^2$, $z_1 z_2 = b^2$, and RM + QM $= z_1 + z_2 = a$.

[19] It will be seen that Descartes considers only three types of the quadratic equation in z, namely, $z^2 + az - b^2 = 0$, $z^2 - az - b^2 = 0$, and $z^2 - az + b^2 = 0$. It thus appears that he has not been able to free himself from the old traditions to the extent of generalizing the meaning of the coefficients, — as negative and fractional as well as positive. He does not consider the type $z^2 + az + b^2 = 0$, because it has no positive roots.

[20] "Qu'ils n'ont point eu la vraye methode pour les trouuer toutes."

COMMENTARY ON
ISAAC NEWTON

O N Christmas Day, 1642, the year Galileo died, there was born in the Manor House of Woolsthorpe-by-Colsterworth a male infant so tiny that, as his mother told him in later years, he might have been put into a quart mug, and so frail that he had to wear "a bolster around his neck to support his head." This unfortunate creature was entered in the parish register as "Isaac, sonne of Isaac and Hanna Newton." There is no record that the wise men honored the occasion, yet this child was to alter the thought and habit of the world.

The Royal Society of London, over which Newton presided for almost a quarter century, planned to celebrate the tercentenary of his birth in 1942. Strangely, this was to be the first international event in Newton's honor since one during his lifetime at which he was elected a foreign associate of the French Academy of Sciences. Postponed because of the war, the celebration was finally held at London and Cambridge in July of 1946. With representatives of thirty-five nations attending, it was an international gathering such as had rarely been convoked even before passports, iron curtains and the congealing effects of "security" persuaded many a traveler to stay home.

The addresses delivered at celebrations are rarely worth remembering. This occasion proved an exception. A high standard was maintained in lectures by (among others) the English physicist E. N. da Costa Andrade, by the mathematician H. W. Turnbull (whose little biographical book appears elsewhere in this volume), by Niels Bohr of Denmark ("Newton's Principles and Modern Atomic Mechanics"), by the French mathematician Jacques Hadamard, by Lord Keynes (who had died and whose paper was read by his brother, Geoffrey Keynes). I have selected two of the lectures. The first, by Andrade, is a lucid survey of Newton's vast achievement; the second is a sensitive, brilliant delineation of Newton the man by Lord Keynes. I reproduce a fragment of Keynes' eloquence lest you be tempted, having read the comprehensive lecture that precedes it, to pass over his wonderful appreciation. "Newton was not the first of the age of reason. He was the last of the magicians, the last of the Babylonians and Sumerians, the last great mind which looked out on the visible and intellectual world with the same eyes as those who began to build our intellectual inheritance rather less than 10,000 years ago."

Newton did not shew the cause of the apple falling, but he shewed a similitude between the apple and the stars.

—SIR D'ARCY WENTWORTH THOMPSON

Great Nature's well-set clock in pieces took.   —WILLIAM COWPER

Where the statue stood
Of Newton, with his prism and silent face,
The marble index of a mind forever
Voyaging through strange seas of thought alone.

—WILLIAM WORDSWORTH

8   Isaac Newton

By E. N. DA C. ANDRADE

FROM time to time in the history of mankind a man arises who is of universal significance, whose work changes the current of human thought or of human experience, so that all that comes after him bears evidence of his spirit. Such a man was Shakespeare, such a man was Beethoven, such a man was Newton, and, of the three, his kingdom is the most widespread. The poet's full greatness is for those to whom his native language is familiar—he can be translated, but in translation his glory is diminished. The musician also expresses himself in an idiom that is limited—the music of the West means little to the East. To-day, natural science is the one universal learning; the language of science is understood by the initiated in every quarter of the globe, and the great leaders of science are reverenced by thoughtful men wherever the pursuit of knowledge is deemed desirable. In his lifetime the achievements of Newton were celebrated throughout civilized Europe: to-day we have in this room learned men from the five continents, who are gathered together to do honour to his name.

Isaac Newton was not one of those precocious figures, like Blaise Pascal and Evariste Galois and William Rowan Hamilton, who seemed singled out for greatness at a schoolboy's age. Apparently he did well in his later years at school, but the little that is recorded of his childhood is derived from schoolmates of his who were questioned in their old age, when their recollections may have been coloured by Newton's fame. He certainly was fond of making things with his hands and of copying pictures and passages from books. The verses attributed to him were taken by him from the *Eikon Basilike*, where they appear under a picture of Charles I which he copied: the extensive notes on drawing and painting which he made as a schoolboy were all taken from John Bate's *Mysteries of Art and Nature*, which also contains a description and picture of a wooden waterclock which corresponds closely to the waterclock which Stukeley tells us that Newton made. He likewise made sundials, but while there is fair evidence that he was clever and ingenious with his hands I do not think it can be claimed that, as a boy, he showed any greater signs of genius than the ordinary run of mechanically minded boys.

He did well enough at school, and badly enough as a farmer, to be sent to Cambridge in 1661, at the age of eighteen, probably with the intention that he should eventually enter the church, which was the best way for an educated man with influence to make his ways in those days. Trinity College could be counted upon to supply the influence. There is little known about his first two years at the University, but, proceeding, we cannot celebrate Newton's birth without bestowing a pious word of praise on Isaac Barrow, an excellent mathematician who was Lucasian Professor at the time. Newton was about twenty-one when he came under his influence, and it was Barrow who first recognized the young man's genius, encouraged his mathematics and directed his attention to optics. By 1669 Barrow had such faith in Newton that, when he published his *Lectiones Opticae*, he turned to Newton for help.

Newton took his degree at Cambridge early in 1665. In the autumn of that year the great plague, which was raging in London, caused the University to close, and Newton went back to live at the isolated little house at Woolsthorpe where he was born in 1642. Here he spent most of his time until the spring of 1667, when the University reopened and he returned, being then twenty-four. Newton is throughout his life an enigmatic figure, but nothing is more extraordinary than his development in the period from 1663 to this spring of 1667, that is, during the Woolsthorpe period and the time at Cambridge immediately preceding it. Newton was always so secretive that we cannot say definitely that at the earlier date he had done nothing extraordinary, but we can say that there is no evidence that he had done so. By the time of his return to Cambridge it is tolerably certain that he had already firmly laid the foundations of his work in the three great fields with which his name is for ever associated— the calculus, the nature of white light, and universal gravitation and its consequences. The binomial theorem, which was fundamental for his early mathematical work, he discovered about 1664. Fontenelle justly observes that we may apply to Newton what Lucan said of the Nile, whose source was unknown to the ancient—that it has not been permitted to mankind to see the Nile feeble and taking its rise as a tiny stream.

I cannot here forbear to quote Newton's own words about this golden period of his achievement, although they must be familiar to many of you. They are taken from a memorandum in the Portsmouth Collection, probably written when he was about seventy-three years old:

In the same year [1666] I began to think of gravity extending to the orb of the moon, and having found out how to estimate the force with which a globe revolving within a sphere presses the surface of the sphere, from Kepler's rule of the periodical times of the planets being in a sesquialterate proportion of their distances from the centres of their orbs [sesquialterate means one and a half times, or, as we say, the square of the years are as the cubes of the orbits] I deduced that the forces which keep the planets in their orbs must be reciprocally as the squares of their distances from the centres about which they revolve: and thereby compared the force requisite to keep the moon in her orb with the force of gravity at the surface of the earth, and found them answer pretty nearly. All this was in the two plague years of 1665 and 1666, for in those days I was in the prime of my age for invention, and minded mathematics and philosophy more than at any time since.

In Newton's own words, then, even when he was writing the *Principia* he was not bending his mind so intensely to science as he did in the short period at Woolsthorpe which ended when he was twenty-four. In those great two years he had to the full two priceless gifts which no one enjoys to-day, full leisure and quiet. Leisure and quiet do not produce a Newton, but without them even a Newton is unlikely to bring to ripeness the fruits of his genius.

It is hard for us to realize the powers of abstraction necessary for that great mind to break away from all previous theories of the celestial motions, and to formulate his new mechanics of the universe. Action at a distance, gravity extending into the remote depths of space are conceptions so familiar to us that we are apt to forget their difficulties. Einstein's innovations were less revolutionary to his time than Newton's were to his. Remember that the ruling scheme of the universe was Descartes's great theory of vortices, vortices of subtle particles which swept the planets along. Subtle particles were a familiar conception: the picture which the vortices offered was easy to apprehend and had behind it the overwhelming authority of Descartes, which was so great that it continued in France until long after Newton's death. Descartes's scheme was a pictorial fancy: Newton's was an abstract mathematical machinery. Even great men of science found it hard to understand. Jean Bernoulli, who died long after Newton, never accepted it, but remained a Cartesian to his death.

There are two stories about the work at Woolsthorpe to which I may perhaps allude The one is that the fall of an apple from a tree there led Newton to the view that the earth was pulling at the apple. There was, however, nothing radically new about the conception that the earth exercised an attractive force on bodies near its surface. What really happened, according to Stukeley's report of a conversation with Newton in his old age, was that, when he was thinking of what pull could hold the moon in its path, the fall of the apple put it into his head that it might be the same gravitational pull, suitably diminished by distance, as acted on the apple.

The other story is that, finding a small discrepancy between the force required to keep the moon in her orbit and the gravitational pull, calculated from that at the earth's surface by the application of the inverse square law, Newton put the work aside, and that the cause of this discrepancy was that Newton took a wrong value of the earth's radius. I think we may take it that this was not the cause. To find the pull on the moon from that on bodies at the earth's surface it is necessary—it is essential— to show that the gravitational pull of a spherical earth is the same as it would be if the whole mass were concentrated at the centre. This is by no means obvious: in fact, it is true only for an inverse square law of attraction and for no other. We know that it gave Newton some trouble to prove this point, which is the subject of a Proposition in the *Principia*. This seems to be the reason that, having found that 'it answered pretty nearly', Newton turned to other things. An essential link in the argument was missing, and this, it seems, was not produced by him until 1685.

No special pleading, however, is required to account for the fact that Newton published no account of this work on gravitation at the time. He never felt any strong desire to bring his work before the world, and later, as we shall see, had a positive aversion from doing so. In particular, he had a horror of controversy. In any case, all that he did to record his great mathematical discoveries, including the generalized binomial theorem and the new calculus of fluxions and inverse fluxions, as he called the differential and integral calculus, was to write a manuscript account under the title *De Analysi per Aequationes Numero Terminorum Infinitas*, which he gave to Barrow in 1669. This was not published until 1711. It contained an account of the work done about four years earlier: Newton produced it to Barrow only when he was told of Mercator's published calculation, in 1668, of the area of a hyperbola. We shall, no doubt, hear more of this in the following lectures on Newton's mathematical work.

In 1669 Barrow resigned his chair of mathematics, the Lucasian chair, to Newton. In the same year he published his lectures on light, in which he thanks Newton warmly for suggestions and for revising the proof sheets. There is, however, in this book no suggestion of any of Newton's great discoveries in optics, although from his own pen we know that he had already made many of them. No, there is little need to seek special reasons why Newton withheld publication in any particular case. In mathematics, in particular, he never published anything except under strong persuasion.

I am not attempting any systematic history of Newton's life—I am more concerned to try to put the man and his achievements before you— but it is necessary if we are to understand Newton to refer to the happenings of the next few years. In the very year 1669 he started to lecture at Cambridge on optics, describing his discoveries: written copies of these lectures were deposited by him, at the time that they were read, in the archives of the University, and were published in 1729, some sixty years after their delivery, under the title *Lectiones Opticae*, quite a different work from the famous *Opticks*. The *Lectiones* were in Latin: an English translation of the first part only was published in 1728. When I was in Russia last year M. Vavilov told me that he was producing a Russian translation of the whole work. Apparently his lectures did not attract large audiences—his amanuensis, Humphrey Newton, referring to 'when he read in the schools' (or lectured, as we now say) writes 'when he had no auditors he commonly returned . . .'. We know practically nothing of his life at that time, but it seems certain that in 1668, after having tried to grind non-spherical lenses, he had convinced himself that chromatic aberration would prevent the construction of a satisfactory refracting telescope and had made a reflecting telescope. This was seen and discussed in Cambridge: in 1671 he constructed a second reflecting telescope and, in response to urgent requests, sent it to the Royal Society. He was then twenty-nine, and had accomplished a body of scientific work such as no one before or since had done at that age, but, although Barrow was convinced of his exceptional genius, he was unknown to the scientific world at large. He had published nothing, and certainly felt no urge to publish. If I insist upon this it is because it affords an important clue to his character and to subsequent disputes about priority. The telescope aroused the greatest interest at the Royal Society and Newton wrote to Oldenburg, the Secretary,

At the reading of your letter I was surprised to see so much care taken about securing an invention to me, of which I have hitherto had so little value. And therefore since the Royal Society is pleased to think it worth the patronizing, I must acknowledge it deserves much more of them for that, than of me, who, had not the communication of it been desired, might have let it still remain in private as it hath already done some years. . . . I am very sensible of the honour done me by the Bishop of Sarum [Seth Ward, himself a famous mathematician] in proposing me candidate, and which I hope will be further conferred upon me by my election into the Society. And if so, I shall endeavour to testify my gratitude by communicating what my poor and solitary endeavours can effect towards the promoting your philosophical designs.

Stimulated by the interest of the Royal Society, Newton communicated his first paper to be published. Optics was, perhaps, Newton's favourite study—his *Opticks* was the last considerable scientific work that he produced, and his most daring speculations were put forward in the Queries appended to this work, to which he added in subsequent editions. At any rate, except, perhaps, for a few words which he wrote in the *Principia* after having shown that Descartes's vortices could not possibly give a true account of the solar system, it is the only subject on which he ever betrayed any enthusiasm or warmth of feeling. On 18 January 1671/2, he wrote to Oldenburg that he would send the Royal Society an account of a philosophical discovery 'being in my judgement the oddest, if not the most considerable detection, which hath hitherto been made in the operation of nature'. The paper in question is dated 19 February.

In it Newton describes how he formed the spectrum by refracting with a prism a beam of sunlight from a circular hole, a very careful series of experiments leading up to the famous *experimentum crucis*. In this experiment, having formed a spectrum he isolated a blue beam by a hole in a screen and then refracted it with a second prism. He found that the blue beam remained blue and was refracted more than was a red beam similarly isolated, or in his own words, 'Light consists of Rays differently refrangible'. The spectrum is formed because white light contains mixed in it monochromatic—or in Newton's term, homogeneal—lights of all kinds, which are merely separated out because they are differently refrangible.

This was in the most direct opposition to the views of his time, which were that white light was homogeneal, and that coloured bodies changed its nature in various ways, which I have not time to describe. It is not always easy to follow what Newton's predecessors meant, as they used the language of the Aristotelian philosophy. Grimaldi and others asked: 'Was light a substance or an accident?'—they always wanted to know what light was. Newton's design was, as he later said in his *Optics*, 'Not to explain the Properties of Light by Hypotheses but to propose and prove them by Reason and Experiment'. I shall later say a word as to what Newton meant by hypotheses.

In this first paper Newton points out that, with the object glass of a telescope such as was used in his time, dispersion causes insuperable indistinctness of focus, and, having proved that there was no dispersion by reflexion, he then made his reflecting telescope. Remember that he made his own alloy, and himself cast his mirror, and polished it, and that in all these matters he made great advances. He never published any details on the composition and casting of the metal for his mirrors, but after his death a paper was found giving particulars. His alloy was 12 parts of copper, 4 of tin and 1 of white arsenic, or much like modern speculum metal. The description in his *Opticks* of this work on chromatic aberration, although it was not published until 1704, derives from this time.

This paper of Newton's was so revolutionary in its time that it led to much misunderstanding and also to attacks on the unknown young man from those of established reputations. Ignatius Pardies, Franciscus Linus, Gascoigne and Lucas offered objections, mostly trivial; Newton replied in detail, and convinced Pardies, a temperate and courteous critic. Linus was a troublesome pedant, who also attacked Boyle. The trouble with Lucas, which centred on the length of the spectrum, may have been due to different dispersive powers of the glasses used by the two observers.

The effect on Newton was, however, deplorable. In his own words to Leibnitz a few years later: 'I was so persecuted with discussions arising from the publication of my theory of light, that I blamed my own imprudence for parting with so substantial a blessing as my quiet to run after a shadow.'

The most troublesome and acutest critics were, however, Huygens and Hooke. Newton always admired Huygens—in his old age he told Pemberton that he thought Huygens the most elegant of any mathematical writer of modern times, and the most just imitator of the ancients, which Newton considered the highest praise. Huygens, however, was chiefly concerned about questions that Newton did not attempt to answer: even if Newton was right, he said, 'there would still remain the great difficulty of explaining, by mechanical principles, in what consists this diversity of colour'. Hooke was concerned to defend his own views that white light was a pulse which was disturbed by the refraction in some way that he did not make clear. Both critics were themselves experimenters of the highest class, and could not but express admiration of Newton's experiments. The criticisms were dealt with by Newton, who was at some pains to explain what he was after. These details need not concern us, but it is necessary to refer to Hooke, as his opposition had a great influence on Newton's life. If there had been a wise and temperate influence, a man who appreciated them both and was concerned to reconcile them, all might have been well, for both men thought highly of one another at bottom. Newton had studied Hooke's *Micrographia* carefully, and quoted it in his early work, although he unjustifiably omitted all reference to it in the *Opticks*. Instead of a reconciling influence there was Oldenburg, who hated Hooke and made it his business to promote dissension between him and Newton. In a very temperate letter to Newton, Hooke, no doubt with Oldenburg in view, spoke of 'two hard-to-yield contenders . . . put together by the ears by other's hands and incentives'. Newton says in a letter to Oldenburg about this time: 'Pray present my service to Mr Hooke, for I suppose there is nothing but misapprehension in what has lately happened.' I do not believe for a moment that Oldenburg did convey this compliment: as More says, Newton was continually being urged and nagged by Oldenburg to justify himself. Hooke died in 1703 and Newton did not publish his *Opticks* until 1704, although practically all the work that it described had been completed much earlier. It seems that he waited for Hooke's death to avoid criticism. To me the misunderstandings between these two great Englishmen is one of the saddest features of the scientific history of the time.

Newton sent a further important communication on light to the Royal Society in 1675. In this he gives a theory of light, ostensibly to satisfy those critics who wanted a machinery by which the experimental effects might be produced. His words are very characteristic of his attitude to speculative theory. 'Because I have observed the heads of some great virtuosos to run much upon hypotheses . . . for this reason I have here thought fit to send you a description of the circumstances of this hypothesis, as much tending to the illustration of the papers I herewith send you.' Here he supposes space to be filled with a subtle ether, the forerunner of the ethers that ruled in the nineteenth century. He would not have light a wave motion, because of sharp shadows, although later he described a series of acute experiments on diffraction phenomena, first discovered by Grimaldi. Light he considered to be 'possibly' a stream of small swift corpuscles of various sizes, red being the largest and violet the smallest: when these corpuscles strike a boundary between two media, such as a glass-air surface, they put the ether, which is denser in the glass, into a vibrating motion, which travels faster than the ray and, striking any second surface, puts it in a state to reflect or transmit according to the phase. The same alternation of transmission and reflexion takes places at the surface first struck, owing to the vibrations of the denser ether. This is Newton's theory of 'fits'; fits of easy reflexion and easy transmission. It can easily be seen how he could account for interference effects—Newton's rings and soap-bubble colours—on this basis. Newton's rings had been observed by Hooke much earlier, but Newton showed that a much greater number could be observed in monochromatic light and explained in detail the colours obtained with white light. Diffraction he accounted for later by condensation of the ether at solid boundaries.

It is unjust to Newton to call his a pure corpuscular theory and to contrast it with Huygens's so-called wave theory. Huygens's theory was nothing like a modern wave theory. In his famous 'construction' the only part of the wavelet that was effective was the pole, the point where it touched the envelope. The particular waves, or wavelets, were incapable of causing light. He thus gave his wavelets point or particle properties. Again, the pulses followed one another quite irregularly. He had no conception of a transverse wave. Newton's theory had many points of resemblance to the most recent theory, although, be sure, I am not claiming that he really anticipated modern developments. Nevertheless, the coexisting particles and waves and the supervelocity of the phase wave that accompanied the particles cannot but make us think of the developments of to-day. His explanation of interference phenomena was essentially a wave explanation. As Michelson, something of an authority in these matters, says: 'It is true that Newton's explanation of the "colours of thin films" is no longer accepted but the fact remains that he did actually measure the quantity which is now designated as the wave length and showed that every spectral colour is characterized by a definite wave length.' Michelson was writing before wave mechanics came in: he might have said more if he had written later.

In his *Opticks*, one of the supreme productions of the human mind, which, as I have said, did not appear until he was sixty-one years old, Newton sets down with beautiful precision and elegance the substances of these early papers, and much beside. The proof that a precise refrangibility is an unalterable characteristic of monochromatic light, and that sunlight is a mixture of lights of all colours; the explanation of the rainbow; the revolutionary discussion of natural colours, with the clear distinction between addition and subtraction colours; the explanation of colour mixture; the investigation of the colours of thin films, such as those displayed by soap bubbles and by Newton's rings, with the explanation to which I have referred; some acute experiments on diffraction—these things are some of the treasures contained in the *Opticks*. There are appended to the book certain 'Queries' which contain some of Newton's most daring speculations, some of them probably made during his old age, since they appear in the second edition of 1717 and not in the first of 1704. One of them concerns the phenomena exhibited by Iceland spar—what we now call polarization phenomena—to explain which he has to endow the particles with 'sides'. That is, he assumes them to be different in different directions transverse to their direction of travel. Thus, in explaining interference colours he introduced, as I have pointed out, an element of periodicity and practically found a wavelength—in explaining polarization he assumed a transverse property. It is strange that he came so near to transverse waves. He cited sharp shadows as proving that light was not a simple wave motion, but his own experiments on diffraction showed him that shadows were not always sharp. Strange, not that he missed the wave theory of light, if he had been so wedded to his corpuscles as he is usually represented to be, but strange that he had himself supplied all the elements of a wave theory and just failed to weld them together. In any case, we can observe nothing about the nature of light in empty space: we only know of its interaction with matter, and in considering this interaction Newton introduced the wave properties.

As an example of the close-packed and original detail of the book, I may point out that in some twelve pages Newton compares quantitatively the lack of precise focus due to chromatic and to spherical aberration, discusses the circle of least confusion and the relative brilliance of the different colours, and points out that the errors due to spherical aberration are as the cube of the aperture, while those due to chromatic aberration are directly proportional to the aperture. He describes carefully how to grind metallic mirrors; his description is precise and would hold as good technique to-day with only slight modification. He gives, for instance, what I believe to be the first account of the use of pitch for optical polishing, in the modern manner. Mr. Frank Twyman tells me that he has never found any earlier mention of the use of pitch. He describes how to make a reflecting telescope, in which he uses a right-angled prism as a reflector, which I believe to be the first mention of this device. All this in twelve pages. In the second edition he adds: 'If the Theory of making Telescopes could at length be fully brought into practice, yet there would be certain Bounds beyond which Telescopes could not perform. For the Air through which we look upon the Stars, is in a perpetual Tremor.' And after some discussion he concludes: 'The only remedy is a most serene and quiet Air, such as may perhaps be found on the tops of the highest Mountains above the grosser Clouds.' I would also add that Newton quite clearly realized the advantage of a slit over a hole in his spectral work, and discusses the point, and also the use of a triangular aperture, which, he points out, gives a pure spectrum on one side of the strip and an overlapping spectrum on the other.

Let us now consider another aspect of Newton's work and for the purpose return to 1675. Round about this time Newton had done the bulk of his optical work and had on several occasions expressed a great distaste for science, and especially for publication. I could give many examples, but I will content myself with one of the most notable, when he wrote to Hooke in 1679: 'But yet my affection to philosophy being worn out, so that I am almost as little concerned about it as one tradesman uses to be about another man's trade or a countryman about learning, I must acknowledge myself averse from spending that time in writing about it which I think I can spend otherwise more to my own content and the good of others: and I hope neither you nor any body else will blame for this averseness.' So far he had published nothing about mechanics or gravity, but he expressed this distaste for science at a time when he was soon to be spurred, cajoled, importuned into writing his greatest work, the *Principia*. I say 'greatest work', for such everyone must admit it to be, but in reading the *Opticks* I still feel that Newton wrote it with more creative joy, that the beautiful presentation represents him at the height of his pleasure in shaping.

The story of the events that led to the writing of the *Principia* has often been told. Hooke, a genius second to Newton but to few others, had come to the firm conclusion that the motion of the planets could be explained on the basis of an inverse square law of attraction, but could not prove it. The matter had been discussed by him with Christopher Wren and with Halley. Newton was by now known as a mathematician, and Halley visited him at Cambridge to ask his opinion in the matter. Newton told the inquirer that he had already proved that the path of a body under a central attraction acting according to the inverse square law would be an ellipse; Jones says that he had done so in 1676–7. Later he sent Halley two different proofs. This incident apparently roused Newton from his distaste for science and revived the old fire. After writing a little treatise *De Motu*

Corporum, Newton, at the instigation of Halley, who undertook to see the book printed at his own charge, began to write the *Principia*. There were the usual troubles. Hooke, who was in the irksome position of having been convinced of the truth of the inverse square law—although much later than Newton's unpublished work at Woolsthorpe—and having published his conviction, without being able to support it with a proof, claimed that Newton had the notion from him, though apparently all that he wanted was some mention in the preface of Newton's book. Newton, irritated beyond bounds at Hooke's contention, wrote that he would suppress the third book of the *Principia*, which is the crown of the work and contains the celestial mechanics. 'Philosophy is such an impertinently litigious Lady, that a man had as good be engaged in lawsuits, as have to do with her. I found it so formerly, and now I am no sooner come near her again, but she gives me warning.' Halley persuaded him not to mutilate the work. The *Principia* eventually came forth in 1687. The composition of this book, which changed the face of science, had taken about eighteen months.

The book is not an easy one to read, either now or then, a fact which Newton recognized. Writing to Gilbert Clarke in the year of its publication he called it 'a hard book', and in the work itself he tells the reader that even a good mathematician may find many of the propositions difficult, and advises him as to what he can skip in the preceding parts if he wants to read the third book. The mathematical machinery is on the lines of strict classical geometry—Pemberton, who saw much of the Master in his old age, since he edited the third edition of the *Principia*, reports that Newton always expressed great admiration for the geometers of ancient Greece and censured himself for not following them more closely than he did. Many moderns have suggested that he obtained his results by other methods and then threw them into geometrical form: this may be so, but it may equally be argued that he was so familiar with the methods of classical geometry that he could use them with a facility unknown to-day. In any case, Whewell has expressed the situation with great force:

> Nobody since Newton has been able to use geometrical methods to the same extent for the like purposes; and as we read the *Principia* we feel as when we are in an ancient armoury where the weapons are of gigantic size; and as we look at them we marvel what manner of man he was who could use as a weapon what we can scarcely lift as a burden.

In the first book Newton enunciates his laws of motion, with due acknowledgements to Galileo, and lays down his mechanical foundations, including the composition of forces, clearly formulated for the first time. He proceeds with extreme caution where fundamentals are concerned, and typically refuses to discuss underlying causes not susceptible to observation and experiment. For example, he says, 'For I here design only to give a Mathematical notion of those forces, without considering their Physical causes and seats'; and a little later, he says again,

> Wherefore, the reader is not to imagine that by those words, I any where take upon me to define the kind, or the manner of any Action, the causes or the physical reason thereof, or that I attribute Forces, in a true and Physical sense, to certain centres (which are only Mathematical points); when at any time I happen to speak of centres as attracting, or as endued with attractive powers.[1]

Compare this with his words in a letter to Bentley some six years after the first publication of the *Principia*: 'You sometimes speak of gravity as essential and inherent to matter. Pray do not ascribe that notion to me; for the cause of gravity is what I do not pretend to know, and therefore would take some time to consider of it.' This and many other passages, both in the *Principia* and the *Opticks*, make clear what Newton meant when he wrote 'Hypotheses non fingo'. That he made hypotheses, in the modern sense, is abundantly clear by the fact that certain passages in the *Principia* are boldly headed 'Hypothesis'—he was prepared to make a quantitative assumption from which exact laws could be worked out and compared with observation. He would not, however, willingly speculate beyond the limits where quantitative confirmation could be sought from Nature. The sense of the famous words is better given by his remark to Conduitt in his eighty-third year, 'I do not deal in conjectures', or in his first sentence of the *Opticks*, 'My design in this Book is not to explain the Properties of Light by Hypotheses, but to propose and prove them by Reason and Experiments'. Very characteristic is the way in which he introduces certain optical theories: 'They who are averse from assenting to any new Discoveries, but such as they can explain by an Hypothesis, may for the present suppose. . . .' If I seem to labour the point, and to have left the *Principia* to wander in other fields, it is because the spirit was a new one in science, and because it animated the whole of Newton's work. It is particularly clearly expressed in the famous *Rules of Reasoning in Philosophy* which usher in the third book of the *Principia*.

To return to the first book, in it the laws of simple orbits, in particular Kepler's laws, are deduced from the inverse square law, and it is demonstrated that, if every point attracts every other point with an inverse square law, the attractive mass of a homogeneous sphere may be considered as concentrated at the centre. The vast generalization that every mass point attracts every other mass point with an inverse square law is typical of Newton's genius. It is clear that he was brought to it by considering the particular case of the moon's motion. He solved the problem of the motion of two bodies mutually gravitating. He worked out experimentally the laws of impact of two bodies. In general, this book of the *Principia* is the first

---

[1] Andrew Motte's translation, as with the other passages quoted from the *Principia*.

text-book of theoretical mechanics, written in the modern spirit, and it is practically all original.

The second book is devoted to motion in a resisting medium and is the first treatment of the motion of real fluids. He considers at length the motion of bodies of various shapes through a resisting liquid, taking the resistance to vary as the square of the velocity. Among other problems propounded is that of the solid of revolution of least resistance, 'which I conceive may be of use in the building of ships'. This difficult problem has been discussed very learnedly by A. R. Forsyth. Newton gives no details as to how he reached his conclusions, but it is clear from this, and from his solution of the brachistochrone problem, that he must have invented the general methods of the calculus of variations. Incidentally, he was also familiar with the calculus of finite differences. Particularly important is the mathematical treatment of wave motion, the first ever given. Newton strongly insists upon the way in which a wave passing through a hole spreads out—or is diffracted, as we now say—which was his chief reason for rejecting a direct wave theory of light. He deduces the fundamental law that the velocity is expressed by the square root of the elasticity divided by the density. In this book also is laid down the law for the behaviour of real fluids from which the term 'Newtonian viscosity' is derived. The whole book constitutes the first text-book of mathematical physics, and of hydrodynamics in particular, and it is embellished by an account of many experiments, carried out with the highest degree of precision and skill.

The third book of the *Principia* is the crown of the work. Like the third movement of a supreme symphony it opens with a recapitulation of previous themes and short statement of the new theme.

> In the preceding books I have laid down the principles of philosophy; principles not philosophical but mathematical. . . . These principles are, the laws and conditions of certain motions, and powers or forces, which chiefly have respect to philosophy. But lest they should have appeared of themselves dry and barren, I have illustrated them here and there, with some philosophical scholiums, giving an account of such things . . . as the density and the resistance of bodies, spaces void of all bodies, and the motion of light and sounds. It remains, that from the same principles, I now demonstrate the frame of the System of the World.

In this third book he establishes the movements of the satellites round their planets and of the planets round the sun on the basis of universal gravitation; he also shows how to find the masses of the planets in terms of the earth's mass. The density of the earth he estimated at between five and six times that of water, the accepted figure to-day being almost exactly 5·5. From this he calculated the masses of the sun and of planets which have satellites, a feat which Adam Smith considered to be 'above the reach of human reason and experience'. But he went much further than this. He accounted quantitatively for the flattened figure of the earth and calculated the ellipticity to be $\frac{1}{230}$, as against the most recent figure of $\frac{1}{297}$. It was of La Condamine's measurement of the arc at the equator after Newton's death, which verified Newton's calculation, that Voltaire wrote to La Condamine:

> Vous avez trouvé par de long ennuis
> Ce que Newton trouva sans sortir de chez lui.

He showed that while the gravitational pull on a sphere acts as if concentrated at the centre, the pull on a spheroid does not, and from this he calculated the conical motion of the earth's axis known as the precession of the equinoxes, which Sir George Airy held to be his most amazing achievement. Incidentally, he discussed the variation of gravitational acceleration over the earth's surface. He worked out the main irregularities of the moon's motion due to the pull of the sun. He laid the foundation of all sound work on the theory of the tides. He was the first to establish the orbits of comets and to show that they, too, were moving under the sun's attraction, so that their return could be calculated. These are a few of the astonishing feats recorded in the third book, feats which would have affirmed Newton as a sublime genius if he had never done anything else.

Throughout the *Principia* there is clear evidence of Newton's wonderful skill as an experimenter. As a comparatively small matter I may cite his proof that gravitational mass and inertial mass are the same, a trifle that is often overlooked. He took two exactly similar spherical wooden boxes as pendulum bobs, so as to make the air resistance the same, filled them with different substances—gold, silver, lead, glass, sand, salt, wood, water, wheat—and showed that their time periods were equal. His pendulum work on damping in different circumstances is extraordinary for the attention to essential detail and for its discrimination 'being not very sollicitous for an accurate calculus, in an experiment that was not very accurate'. His optical work is not his only experimental triumph.

Let me now in a few words run through the remaining events in Newton's life. After the composition of the *Principia*, Newton, possibly exhausted by the strain of producing this prodigious work, appears to have abandoned science for a time. There were many distractions. His friend Henry More died. James II came into conflict with the authorities at Cambridge, and Newton became active in the defence of the University. I take from a letter which he wrote on this occasion a sentence that does him credit as a man: 'An honest courage in these matters will secure all, having law on our sides.' In 1688 James fled the country and Newton was elected M.P. for Cambridge in the so-called Convention Parliament. This brought him much to London. In 1689 his mother died: he was an affectionate son and sat up all night with her during her last illness. Whether as a result of these shocks and changes or no, he seems to have been much disturbed about this time, and to have sought an administrative post. He was despondent and appears to have been mainly employed on theological matters. In 1693 he fell into a profound melancholy, with long periods of sleeplessness, during which he thought that all his friends had turned against him. Apparently this attack lasted only some months, and it is a gross exaggeration to refer to this as 'Newton's madness'. Newton's general state certainly alarmed his friends, but the derangement was not more than an exaggeration of his usual suspicious nature and it was not permanent. The story of his little dog Diamond having knocked over a candle and burned papers, the loss of which drove him to madness, is an absurd story first made current in 1780—apparently Newton never had a dog, there is no satisfactory evidence of a really serious fire of Newton's manuscripts, and about the time in question it seems clear that Newton was in one of his periods of distaste for science. He conducted some correspondence with Flamsteed about the moon's motion in 1694, but in 1696 came an event which changed his whole mode of life. He was made Warden of the Mint, from which office he passed to be Master, the chief post, in 1699.

This post was no sinecure, for Newton's friend Charles Montague, later Lord Halifax, was putting through a great recoinage scheme and Newton was actively involved in it. He took his work very seriously and rendered valuable service to his country. There have been many regrets that such a mind should be distracted from science by an administrative post. It seems clear, however, that Newton wanted such a post and that, while he retained his prodigious powers, he no longer felt inclined to devote his main energies to science. He published his *Opticks,* he was interested in the later editions of the *Principia,* but he entrusted the bringing out of the second to Roger Cotes and of the third to Henry Pemberton, although he supplied certain material. I have already referred to the queries appended to the *Opticks,* which remind me somewhat of Beethoven's last quartets. The Master's main work done, he seems reaching forward tentatively to new regions which he will not have time to explore.

Newton in his old age occupied a unique position. His scientific reputation was unrivalled throughout the learned world. From 1703 to his death he held undisputed sway in the Royal Society as President. He was a national figure: when in 1705 Queen Anne knighted him she awarded an honour never before, I believe, conferred for services to science. For, strangely enough, it was not as Master of the Mint that he was knighted: Conduitt tells us that the Queen, 'the Minerva of her age', thought it a happiness to have lived at the same time as, and to have known, so great a man. He held an honourable and lucrative post. His niece, who lived on terms of close friendship with Montague—about the exact nature of which friendship there has been unseemly speculation—was a toast of the town and secured him, should he wish it, the entry into fashionable society. When he died in 1727 the greatest honours were accorded to him: his body lay in state in the Jerusalem Chamber, his pall was supported by the Lord High Chancellor, two dukes and three earls—which meant something in those days—and the place allotted for his monument had been previously refused to the greatest of our nobility. It was a great occasion— the first and the last time that national honours of this kind have been accorded to a man of science or, I believe, to any figure in the world of thought, learning, or art in England. But then those were extraordinary times.

Newton was by no means a perfect character, and it is doing science no service to pretend that he was. He was easily irritated, as a man who had wearied himself with such prodigious and concentrated effort might well be. Whiston, who had the greatest admiration for his powers, said of him that 'He was of the most fearful, cautious, and suspicious temper, that ever I knew', but it may with truth be said that Whiston had quarrelled with him. So had Flamsteed—and with some cause—who said that he was 'insidious, ambitious, and excessively covetous of praise, and impatient of contradiction'. This is undoubtedly the exaggeration of one who himself was of no angelic temperament, but John Locke was a firm friend, a man of admirable character, and he wrote 'he is a nice'—that is, difficult and over-precise—'man to deal with, and a little too apt to raise in himself suspicions where there is no ground'. This evident suspicious element in his nature, which also was a characteristic of Flamsteed and Hooke, is his worst blemish. At the same time, he was kind and helpful to young men, and many authentic stories are told of his generosity. In any case, such imperfections of character are not inconsistent with high performance, even in spiritual matters, as any one who has studied the Church Fathers must admit.

One aspect of this strangeness, this almost morbid sensitiveness, was his abnormal dread of controversy, to which I have already alluded. 'There is nothing I desire to avoid in matters of philosophy more than contention, nor any kind of contention more than one in print', he wrote to Hooke, who, judging by his writings, was not averse from printed dispute. Even stronger are his words to Oldenburg: 'I see I have made myself a slave to philosophy, but if I get free of Mr. Linus's business, I will resolutely bid adieu to it eternally, excepting what I do for my private satisfaction, or leave to come out after me; for I see a man must either resolve to put out nothing new, or to become a slave to defend it.' Phrases such as 'and signify, but not from me' and 'pray keep this letter private to yourself' are common: when he gave Barrow permission to send some of his early mathematical work to John Collins he stipulated that his name should be withheld. He intended to publish his Optical Lectures in 1671, but was deterred by horror of controversy. Not one of his mathematical writings was voluntarily given to the world by himself.

At the same time, when not worried or irritated he was modest about his achievements. Some of his phrases, as when he says that if he is in London he might possibly supply a vacant week or two at the Royal Society 'with something by me, but that's not worth mentioning', may have been the language of compliment, but when he said that if he had seen further, it was by standing on the shoulders of giants, it sounds like genuine humility, against which, however, we must set that he would never make acknowledgements to anyone, like Hooke, with whom he had quarrelled, even when they were plainly due. His aims were so high, the problems which he wished to solve so general and so difficult, his scope so wide that I, for one, am sure that he was sincere when, shortly before his death, he said: 'I do not know what I may appear to the world; but to myself·I seem to have been only like a boy, playing on the seashore, and diverting myself, in now and then finding a smoother pebble or a prettier shell than ordinary, while the great ocean of truth lay all undiscovered before me.' Evidence can be cited for the view that Newton was most modest or most overweening: the truth is that he was a very complex character. Further, he could with perfect consistency be modest about his performance in respect to his aims and completely confident in it when viewed from the standpoint of his contemporaries. Many able men will tolerate self-criticism, but revolt against the criticism of men puffed up by place or exalted by supporters and by the fashionable schools.

No estimate of Newton would be balanced without some reference to the mystical element in his nature. This raises the question of his work in chemistry. Newton devoted probably as much time and effort to alchemy and chemistry, which were one study in his time, as he did to the physical sciences. His library was well stocked with the standard alchemical and mystical books, such as Agrippa *De Occulta Philosophia,* Birrius *De Transmutatione Metallorum, Fame and Confession of the Rosie Cross,* Geber *The Philosopher's Stone,* Kerkringius *Currus Triumphalis Basilii Valentini,* Libavius *Alchymia,* eight books by Raymond Lully, five by Maier, four by Paracelsus, *The Marrow of Alchemy, The Musaeum Hermeticum* and so on, mostly copiously annotated. In most of these books the mystical element is prominent. In many of them the results of experiment were expressed in the allegorical language ridiculed by Goethe in *Faust*:

> The Lion Red, bold wooer, bolder mate,
> In tepid bath was to the Lily married,
> And both were then by open fire-flame straight
> From one bride-chamber to another harried.

> Thus in due time the Youthful Queen, inside
> The glass retort, in motley colours hovered:
> This was the medicine; the patients died,
> And no one thought of asking who recovered.

Man's spiritual life, death and resurrection, were paralleled in the chemical changes of the material world. The lines of thought, the spirit and the language of these toilers in the early chemical laboratories have passed, and it is hard without prolonged study to disentangle meanings, if any, veiled in allegory, tinged with prophecy, coloured with religious belief and clouded with charlatanism in many cases. There is correspondence between Newton and Boyle about *a* mercury that grows hot with gold—not every mercury obtained by extraction will do this, says Boyle, so that he did not mean mercury as we know it. What did he mean? Newton, no doubt, was deeply interested in chemical operations—what was he seeking? Humphrey Newton, his amanuensis, tells us of the period 1685–90, during which the *Principia* was written, that, especially at spring and the fall of the leaf, he used to employ periods of about six weeks in his elaboratory, the fire scarce going out either by day or night. 'What his aim might be', says this faithful, but not overbright, assistant, 'I was not able to penetrate, but his pains, his diligence at these times made me think he aimed at something beyond the reach of human art and industry.'

There are some extraordinary passages in Newton's writings on this subject. I give you two about the transmutation of metals, from a letter written in 1676, a period when he was mainly concerned with chemical operations. I cannot hope to convince the sceptical that Newton had some power of prophecy or special vision, had some inkling of atomic power, but I do say that they do not read to me as if all that he meant was that the manufacture of gold would upset world trade—'Because the way by which mercury may be so impregnated, has been thought fit to be concealed by others that have known it, and therefore may possibly be an inlet to something more noble, not to be communicated without immense danger to the world, if there should be any verity in the Hermetic writers', and a little further on 'there being other things beside the *transmutation of metals* (if those *great pretenders* brag not)'—the word *pretender* had no offensive sense in those days, any more than *professor* has now—'which none but they understand'. In pondering what these passages may import, consider the no greater reticence with which he speaks of his optical discoveries in the letter of 23 February 1668/9.

He published nothing on chemistry except a short but very significant note *De Natura Acidorum,* which I should like to have time to discuss, and certain 'queries' appended to the *Opticks,* which must have been the fruit of mature consideration, since they do not appear in the first edition of 1704. In the last query he sketches a theory of chemistry in terms of attractive forces which seems to me to be an immense advance on the old picture of hooked atoms, a conception 'which is begging the question', says Newton.

Have not small Particles of Bodies certain Powers, Virtues or Forces, by which they act at a distance, not only upon the Rays of Light for reflecting refracting and inflecting them, but also upon one another, for producing a great part of the Phænomena of Nature. . . . How these Attractions may be perform'd I do not here consider. . . . The Attractions of Gravity, Magnetism and Electricity, reach to very sensible distances, and so have been observed by vulgar Eyes, and there may be others which reach to so small distances as hitherto escape Observation, and perhaps electrical Attraction may reach to such small distances, even without being excited by Friction.

This, surely, is a distinct foreshadowing of modern chemical theory. We have small particles—atoms and molecules—endowed with inherent attractive forces of an electrical nature, by the agency of which they act on light and combine with one another. What I have quoted is but a small part of this 'query', which I should discuss at length did time permit. In the query before it he asks whether gross bodies and light are not convertible into one another. Surely here we have a great seer as well as the greatest man of science who ever lived.

Newton's chemistry, then, was not all alchemy, but there are some half million words of manuscript on alchemical subjects in the Portsmouth Collection which have never been digested, let alone published. How much of this matter is copied out of books I cannot find: what there is of worth among these writings no one knows. It may be that Newton never found the great truth for which he was seeking, it may be that much of it is of little value, like his *Chronology of the Ancient Kingdoms Amended*. That it is exclusively mystical I do not believe—that there is a mystical element seems certain. I hope that one day some profound student—no one less will suffice—will study this mass of papers, among which there may well be matter of prime interest. We need not seek for any special reason why he never published the results of his chemical investigations—it was only by chance that the *Principia* was published.

The mystical element in Newton is abundantly clear. He was a close student of the mystic Jacob Boehme, from whose works he copied large extracts. Strange passages occur in Newton's letters, such as 'but it is plain to me by the fountain that I draw it from, though I will not undertake to prove it to others'. Whiston says: 'Sir Isaac, in mathematics, could sometimes see almost by intuition, even without demonstration. . . . And when he did but propose conjectures in natural philosophy, he almost always knew them to be true at the same time', yet adds caustic comment on Newton's *Chronology* which shows him to have been anything but a blind worshipper. As an example of his mathematical prescience I may cite his rule for the discovery of imaginary roots of equations, which was not finally proved until by Sylvester in 1865. The psychology of inspiration is always difficult—the psychology of unique genius is beyond us. If I may, without arrogance, I will use myself the words of Emmanuel Kant: 'I am not aware that anybody has ever perceived in me an inclination to the marvellous, or a weakness tending to credulity': nevertheless, I feel that Newton derived his knowledge by something more like a direct contact with the unknown sources that surround us, with the world of mystery, than has been vouchsafed to any other man of science. A mixture of mysticism and natural science is not unexampled—Swedenborg has important achievements in geology, physiology and engineering to his credit. In any case, it is not honest to neglect half of a man's intellectual life because to do so makes the other half easier to explain.

I do not propose to deal with Newton's religious views, although, if you want a picture of the complete man, you must remember that theological questions occupied much of his attention. Archbishop Tenison said to him, 'You know more divinity than all of us put together', and Locke wrote 'divinity too, and his great knowledge in the Scriptures, wherein I know few his equals'. He is supposed to have been unsound on the Trinity, in fact to have been tainted with the doctrine of Arianism. The works of the Church Fathers were prominent in his library. His two books the *Chronology of Ancient Kingdoms Amended* and *Observations upon the Prophecies of Daniel and the Apocalypse of St John* probably cost him as much effort as the *Principia*. There were over 1,300,000 words in manuscript on theology in the Portsmouth papers, according to my estimate from the catalogue.

Of his immense and very valuable work as Master of the Mint I will also forbear to speak. But I must insist that these activities emphasize the point that he spent a comparatively small part of his long life, probably only ten years or so in all, on the work that has made his name famous throughout the civilized world. For long periods he was indifferent to science—we might even justify the expression that he had a distaste for it —but he never lost his powers. In 1696, shortly after his first appointment to the Mint, when he received one afternoon Bernoulli's problem, set as a challenge to 'the acutest mathematicians in the world', he solved it before going to bed. The problem, I may remind you, was that of the brachistochrone, or line of quickest descent, and required for its solution the calculus of variations. Later, in 1716, when Leibnitz set a problem 'for the purpose of feeling the pulse of the English analysts', he likewise solved it in a few hours. He took an active part in the production of the second edition of the *Principia,* which appeared in 1713, and showed his wonted powers. It is clear that he had what we now call a nervous breakdown in 1693, showing to an exaggerated degree the irritability and suspicion which often characterized his actions, but it was nothing like madness and he made a complete recovery. Nothing could be easier to refute than

Biot's contention that after this breakdown his intellect was permanently impaired.

At our celebrations to-day, at which so many great and learned men are gathered to do honour to Newton's name, I see in imagination shadowy figures, each bearing his tribute. I see the austere form of Leibnitz, reluctant, perhaps in view of the great strife that subsequently arose about the invention of the calculus, I see him, whom we made a Fellow of our Society two hundred and seventy-three years ago, repeating what he wrote: 'Taking mathematics from the beginning of the world to the time when Newton lived, what he has done is much the better part.' I see a small band of the greatest men of science that France has produced—Laplace, who wrote that 'The *Principia* is pre-eminent above any other production of human genius': Lagrange, who frequently asserted that Newton was the greatest genius that ever existed: Biot, who said of the monumental inscription in Westminster Abbey, which runs 'Let mortals rejoice that such and so great an ornament of the human race has existed', that, though true in speaking of Newton, it can be applied to no one else: Arago, declaring 'The efforts of the great philosopher were always superhuman: the questions which he did not solve were incapable of solution in his time'. Gauss is here, who uses the term *clarus* for some scholars, and *clarissimus* for others, but applies *summus* to Newton and to Newton alone. Boltzmann and Ernst Mach come forward to represent their generation of mathematical physicists, Boltzmann declaring that the *Principia* is the first and greatest work ever written on theoretical physics, Mach saying in 1901: 'All that has been accomplished in mechanics since his day has been a deductive, formal and mathematical development of mechanics on the basis of Newton's laws.' And lest it be supposed that the coming of relativity has lessened the awe and admiration with which the great leaders of scientific thought have always regarded him, let me call on Eddington, our much lamented brother, to repeat his words 'To suppose that Newton's great scientific reputation is tossing up and down on these latter-day revolutions is to confuse science with omniscience'; let me take from Einstein's lips his words on Newton: 'Nature to him was an open book, whose letters he could read without effort. In one person he combined the experimenter, the theorist, the mechanic and, not least, the artist in expression.' I think, however, that Einstein is perhaps incorrect in using the words 'without effort'. Newton's own words are eloquent on this point. He told one inquirer that he made his discoveries 'By always thinking unto them', and, in a more communicative mood than usual, said: 'I keep the subject constantly before me and wait till the first dawnings open little by little into the full light.' I would rather say that Newton was capable of greater sustained mental effort than any man, before or since.

I have endeavoured within the limits of time which custom and your endurance impose to give some account of the life, but more particularly of the achievements, of this extraordinary man. If any of you think that I have claimed too much for him, turn, I beg you, not so much to what has been written of him but to what he himself wrote and consider his works against the background of his time, which was, nevertheless, a time as fertile in discovery and as filled with great men as any of which record is made. Consider them, too, in the light of subsequent advances. Ponder the scope of his performance—the invention of the calculus, the establishment of the fundamental features of physical optics and the explanation of celestial mechanics and terrestrial perturbations and tides in terms of the laws of motion and of universal gravitation. Remark the clarity with which he laid down the principles which have since governed fertile scientific work. Remember that he was as supreme in experiment as he was in theory, and, when he chose, as in the *Opticks*, unrivalled as an expositor.

Stand back a little, and range the times in order. The face of science changes, theories fail and rise again transformed. Achievement such as Newton's is not lessened by the great advances of the subsequent centuries: it survives in undiminished strength and beauty the strange and formidable shaping of our own times. From being the preoccupation of a few curious spirits science has grown to be a universal study, on the fruits of which peace among people and the prosperity of nations depend, but the great principles enunciated by Newton and their orderly development by him remain as the foundations of the discipline and as a shining example of the exalted power of the human mind.

I should like to know if any man could have laughed if he had seen Sir Isaac Newton rolling in the mud. —SIDNEY SMITH

If we evolved a race of Isaac Newtons, that would not be progress. For the price Newton had to pay for being a supreme intellect was that he was incapable of friendship, love, fatherhood, and many other desirable things. As a man he was a failure; as a monster he was superb.—ALDOUS HUXLEY

9 Newton, the Man

By JOHN MAYNARD KEYNES

IT is with some diffidence that I try to speak to you in his own home of Newton *as he was himself.* I have long been a student of the records and had the intention to put my impressions into writing to be ready for Christmas Day 1942, the tercentenary of his birth. The war has deprived me both of leisure to treat adequately so great a theme and of opportunity to consult my library and my papers and to verify my impressions. So if the brief study which I shall lay before you to-day is more perfunctory than it should be, I hope you will excuse me.

One other preliminary matter. I believe that Newton was different from the conventional picture of him. But I do not believe he was less great. He was less ordinary, more extraordinary, than the nineteenth century cared to make him out. Geniuses *are* very peculiar. Let no one here suppose that my object to-day is to lessen, by describing, Cambridge's greatest son. I am trying rather to see him as his own friends and contemporaries saw him. And they without exception regarded him as one of the greatest of men.

In the eighteenth century and since, Newton came to be thought of as the first and greatest of the modern age of scientists, a rationalist, one who taught us to think on the lines of cold and untinctured reason.

I do not see him in this light. I do not think that any one who has pored over the contents of that box which he packed up when he finally left Cambridge in 1696 and which, though partly dispersed, have come down to us, can see him like that. Newton was not the first of the age of reason. He was the last of the magicians, the last of the Babylonians and Sumerians, the last great mind which looked out on the visible and intellectual world with the same eyes as those who began to build our intellectual inheritance rather less than 10,000 years ago. Isaac Newton, a posthumous child born with no father on Christmas Day, 1642, was the last wonder-child to whom the Magi could do sincere and appropriate homage.

Had there been time, I should have liked to read to you the contemporary record of the child Newton. For, though it is well known to his biographers, it has never been published *in extenso,* without comment, just as it stands. Here, indeed, is the makings of a legend of the young magician, a most joyous picture of the opening mind of genius free from the uneasiness, the melancholy and nervous agitation of the young man and student.

For in vulgar modern terms Newton was profoundly neurotic of a not unfamiliar type, but—I should say from the records—a most extreme example. His deepest instincts were occult, esoteric, semantic—with profound shrinking from the world, a paralyzing fear of exposing his thoughts, his beliefs, his discoveries in all nakedness to the inspection and criticism of the world. 'Of the most fearful, cautious and suspicious temper that I ever knew', said Whiston, his successor in the Lucasian Chair. The too well-known conflicts and ignoble quarrels with Hooke, Flamsteed, Leibnitz are only too clear an evidence of this. Like all his type he was wholly aloof from women. He parted with and published nothing except under the extreme pressure of friends. Until the second phase of his life, he was a wrapt, consecrated solitary, pursuing his studies by intense introspection with a mental endurance perhaps never equalled.

I believe that the clue to his mind is to be found in his unusual powers of continuous concentrated introspection. A case can be made out, as it also can with Descartes, for regarding him as an accomplished experimentalist. Nothing can be more charming than the tales of his mechanical contrivances when he was a boy. There are his telescopes and his optical experiments. These were essential accomplishments, part of his unequalled all-round technique, but not, I am sure, his *peculiar* gift, especially amongst his contemporaries. His peculiar gift was the power of holding continuously in his mind a purely mental problem until he had seen straight through it. I fancy his pre-eminence is due to his muscles of intuition being the strongest and most enduring with which a man has ever been gifted. Anyone who has ever attempted pure scientific or philosophical thought knows how one can hold a problem momentarily in one's mind and apply all one's powers of concentration to piercing through it, and how it will dissolve and escape and you find that what you are surveying is a blank. I believe that Newton could hold a problem in his mind for hours and days and weeks until it surrendered to him its secret. Then being a supreme mathematical technician he could dress it up, how you will, for purposes of exposition, but it was his intuition which was pre-eminently extraordinary—'so happy in his conjectures', said de Morgan, 'as to seem to know more than he could possibly have any means of proving'. The proofs, for what they are worth, were, as I have said, dressed up afterwards—they were not the instrument of discovery.

There is the story of how he informed Halley of one of his most fundamental discoveries of planetary motion. 'Yes,' replied Halley, 'but how do you know that? Have you proved it?' Newton was taken aback—'Why, I've known it for years', he replied. 'If you'll give me a few days, I'll certainly find you a proof of it'—as in due course he did.

Again, there is some evidence that Newton in preparing the *Principia* was held up almost to the last moment by lack of proof that you could treat a solid sphere as though all its mass was concentrated at the centre, and only hit on the proof a year before publication. But this was a truth which he had known for certain and had always assumed for many years.

Certainly there can be no doubt that the peculiar geometrical form in which the exposition of the *Principia* is dressed up bears no resemblance at all to the mental processes by which Newton actually arrived at his conclusions.

His experiments were always, I suspect, a means, not of discovery, but always of verifying what he knew already.

Why do I call him a magician? Because he looked on the whole universe and all that is in it *as a riddle*, as a secret which could be read by applying pure thought to certain evidence, certain mystic clues which God had laid about the world to allow a sort of philosopher's treasure hunt to the esoteric brotherhood. He believed that these clues were to be found partly in the evidence of the heavens and in the constitution of elements (and that is what gives the false suggestion of his being an experimental natural philosopher), but also partly in certain papers and traditions handed down by the brethren in an unbroken chain back to the original cryptic revelation in Babylonia. He regarded the universe as a cryptogram set by the Almighty—just as he himself wrapt the discovery of the calculus in a cryptogram when he communicated with Leibnitz. By pure thought, by concentration of mind, the riddle, he believed, would be revealed to the initiate.

He *did* read the riddle of the heavens. And he believed that by the same powers of his introspective imagination he would read the riddle of the Godhead, the riddle of past and future events divinely fore-ordained, the riddle of the elements and their constitution from an original undifferentiated first matter, the riddle of health and of immortality. All would be revealed to him if only he could persevere to the end, uninterrupted, by himself, no one coming into the room, reading, copying, testing—all by himself, no interruption for God's sake, no disclosure, no discordant breakings in or criticism, with fear and shrinking as he assailed these half-ordained, half-forbidden things, creeping back into the bosom of the Godhead as into his mother's womb. 'Voyaging through strange seas of thought *alone*', not as Charles Lamb 'a fellow who believed nothing unless it was as clear as the three sides of a triangle'.

And so he continued for some twenty-five years. In 1687, when he was forty-five years old, the *Principia* was published.

Here in Trinity it is right that I should give you an account of how he lived amongst you during these years of his greatest achievement. The east end of the Chapel projects farther eastwards than the Great Gate. In the second half of the seventeenth century there was a walled garden in the free space between Trinity Street and the building which joins the Great Gate to the Chapel. The south wall ran out from the turret of the Gate to a distance overlapping the Chapel by at least the width of the present pavement. Thus the garden was of modest but reasonable size. This was Newton's garden. He had the Fellow's set of rooms between the Porter's Lodge and the Chapel—that, I suppose, now occupied by Professor Broad. The garden was reached by a stairway which was attached to a veranda raised on wooden pillars projecting into the garden from the range of buildings. At the top of this stairway stood his telescope—not to be confused with the observatory erected on the top of the Great Gate during Newton's lifetime (but after he had left Cambridge) for the use of Roger Cotes and Newton's successor, Whiston. This wooden erection was, I think, demolished by Whewell in 1856 and replaced by the stone bay of Professor Broad's bedroom. At the Chapel end of the garden was a small two-storied building, also of wood, which was his elaboratory. When he decided to prepare the *Principia* for publication he engaged a young kinsman, Humphrey Newton, to act as his amanuensis (the MS. of the *Principia*, as it went to the press, is clearly in the hand of Humphrey). Humphrey remained with him for five years—from 1684 to 1689. When Newton died Humphrey's son-in-law Conduitt wrote to him for his reminiscences, and among the papers I have is Humphrey's reply.

During these twenty-five years of intense study mathematics and astronomy were only a part, and perhaps not the most absorbing, of his occupations. Our record of these is almost wholly confined to the papers which he kept and put in his box when he left Trinity for London.

Let me give some brief indications of their subject. They are enormously voluminous—I should say that upwards of 1,000,000 words in his handwriting still survive. They have, beyond doubt, no substantial value whatever except as a fascinating sidelight on the mind of our greatest genius.

Let me not exaggerate through reaction against the other Newton myth which has been so sedulously created for the last two hundred years. There was extreme method in his madness. All his unpublished works on esoteric and theological matters are marked by careful learning, accurate method and extreme sobriety of statement. They are just as *sane* as the *Principia*, if their whole matter and purpose were not magical. They were nearly all composed during the same twenty-five years of his mathematical studies. They fall into several groups.

Very early in life Newton abandoned orthodox belief in the Trinity.

At this time the Socinians were an important Arian sect amongst intellectual circles. It may be that Newton fell under Socinian influences, but I think not. He was rather a Judaic monotheist of the school of Maimonides. He arrived at this conclusion, not on so-to-speak rational or sceptical grounds, but entirely on the interpretation of ancient authority. He was persuaded that the revealed documents give no support to the Trinitarian doctrines which were due to late falsifications. The revealed God was one God.

But this was a dreadful secret which Newton was at desperate pains to conceal all his life. It was the reason why he refused Holy Orders, and therefore had to obtain a special dispensation to hold his Fellowship and Lucasian Chair and could not be Master of Trinity. Even the Toleration Act of 1689 excepted anti-Trinitarians. Some rumours there were, but not at the dangerous dates when he was a young Fellow of Trinity. In the main the secret died with him. But it was revealed in many writings in his big box. After his death Bishop Horsley was asked to inspect the box with a view to publication. He saw the contents with horror and slammed the lid. A hundred years later Sir David Brewster looked into the box. He covered up the traces with carefully selected extracts and some straight fibbing. His latest biographer, Mr More, has been more candid. Newton's extensive anti-Trinitarian pamphlets are, in my judgement, the most interesting of his unpublished papers. Apart from his more serious affirmation of belief, I have a completed pamphlet showing up what Newton thought of the extreme dishonesty and falsification of records for which St Athanasius was responsible, in particular for his putting about the false calumny that Arius died in a privy. The victory of the Trinitarians in England in the latter half of the seventeenth century was not only as complete, but also as extraordinary, as St Athanasius's original triumph. There is good reason for thinking that Locke was a Unitarian. I have seen it argued that Milton was. It is a blot on Newton's record that he did not murmur a word when Whiston, his successor in the Lucasian Chair, was thrown out of his professorship and out of the University for publicly avowing opinions which Newton himself had secretly held for upwards of fifty years past.

That he held this heresy was a further aggravation of his silence and secrecy and inwardness of disposition.

Another large section is concerned with all branches of apocalyptic writings from which he sought to deduce the secret truths of the Universe —the measurements of Solomon's Temple, the Book of David, the Book of Revelations, an enormous volume of work of which some part was published in his later days. Along with this are hundreds of pages of Church History and the like, designed to discover the truth of tradition.

A large section, judging by the handwriting amongst the earliest, relates to alchemy—transmutation, the philosopher's stone, the elixir of life. The scope and character of these papers have been hushed up, or at least minimized, by nearly all those who have inspected them. About 1650 there was a considerable group in London, round the publisher Cooper, who during the next twenty years revived interest not only in the English alchemists of the fifteenth century, but also in translations of the medieval and post-medieval alchemists.

There is an unusual number of manuscripts of the early English alchemists in the libraries of Cambridge. It may be that there was some continuous esoteric tradition within the University which sprang into activity again in the twenty years from 1650 to 1670. At any rate, Newton was clearly an unbridled addict. It is this with which he was occupied 'about 6 weeks at spring and 6 at the fall when the fire in the elaboratory scarcely went out' at the very years when he was composing the *Principia* —and about this he told Humphrey Newton not a word. Moreover, he was almost entirely concerned, not in serious experiment, but in trying to read the riddle of tradition, to find meaning in cryptic verses, to imitate the alleged but largely imaginary experiments of the initiates of past centuries. Newton has left behind him a vast mass of records of these studies. I believe that the greater part are translations and copies made by him of existing books and manuscripts. But there are also extensive records of experiments. I have glanced through a great quantity of this— at least 100,000 words, I should say. It is utterly impossible to deny that it is wholly magical and wholly devoid of scientific value; and also impossible not to admit that Newton devoted years of work to it. Some time it might be interesting, but not useful, for some student better equipped and more idle than I to work out Newton's exact relationship to the tradition and MSS. of his time.

In these mixed and extraordinary studies, with one foot in the Middle Ages and one foot treading a path for modern science, Newton spent the first phase of his life, the period of life in Trinity when he did all his real work. Now let me pass to the second phase.

After the publication of the *Principia* there is a complete change in his habit and way of life. I believe that his friends, above all Halifax, came to the conclusion that he must be rooted out of the life he was leading at Trinity which must soon lead to decay of mind and health. Broadly speaking, of his own motion or under persuasion, he abandons his studies. He takes up University business, represents the University in Parliament; his friends are busy trying to get a dignified and remunerative job for him —the Provostship of King's, the Mastership of Charterhouse, the Controllership of the Mint.

Newton could not be Master of Trinity because he was a Unitarian and so not in Holy Orders. He was rejected as Provost of King's for the more prosaic reason that he was not an Etonian. Newton took this rejection very ill and prepared a long legalistic brief, which I possess, giving reasons why it was not unlawful for him to be accepted as Provost. But, as ill-luck had it, Newton's nomination for the Provostship came at the moment when King's had decided to fight against the right of Crown nomination, a struggle in which the College was successful.

Newton was well qualified for any of these offices. It must not be inferred from his introspection, his absent-mindedness, his secrecy and his solitude that he lacked aptitude for affairs when he chose to exercise it. There are many records to prove his very great capacity. Read, for example, his correspondence with Dr Covell, the Vice-Chancellor when, as the University's representative in Parliament, he had to deal with the delicate question of the oaths after the revolution of 1688. With Pepys and Lowndes he became one of the greatest and most efficient of our civil servants. He was a very successful investor of funds, surmounting the crisis of the South Sea Bubble, and died a rich man. He possessed in exceptional degree almost every kind of intellectual aptitude—lawyer, historian, theologian, not less than mathematician, physicist, astronomer.

And when the turn of his life came and he put his books of magic back into the box, it was easy for him to drop the seventeenth century behind him and to evolve into the eighteenth-century figure which is the traditional Newton.

Nevertheless, the move on the part of his friends to change his life came almost too late. In 1689 his mother, to whom he was deeply attached, died. Somewhere about his fiftieth birthday on Christmas Day 1692, he suffered what we should now term a severe nervous breakdown. Melancholia, sleeplessness, fears of persecution—he writes to Pepys and to Locke and no doubt to others letters which lead them to think that his mind is deranged. He lost, in his own words, the 'former consistency of his mind'. He never again concentrated after the old fashion or did any fresh work. The breakdown probably lasted nearly two years, and from it emerged, slightly 'gaga', but still, no doubt, with one of the most powerful minds of England, the Sir Isaac Newton of tradition.

In 1696 his friends were finally successful in digging him out of Cambridge, and for more than another twenty years he reigned in London as the most famous man of his age, of Europe, and—as his powers gradually waned and his affability increased—perhaps of all time, so it seemed to his contemporaries.

He set up house with his niece Catharine Barton, who was beyond reasonable doubt the mistress of his old and loyal friend Charles Montague, Earl of Halifax and Chancellor of the Exchequer, who had been one of Newton's intimate friends when he was an undergraduate at Trinity. Catharine was reputed to be one of the most brilliant and charming women in the London of Congreve, Swift and Pope. She is celebrated, not least for the broadness of her stories, in Swift's *Journal to Stella*. Newton puts on rather too much weight for his moderate height. 'When he rode in his coach one arm would be out of his coach on one side and the other on the other.' His pink face, beneath a mass of snow-white hair, which 'when his peruke was off was a venerable sight', is increasingly both benevolent and majestic. One night in Trinity after Hall he is knighted by Queen Anne. For nearly twenty-four years he reigns as President of the Royal Society. He becomes one of the principal sights of London for all visiting intellectual foreigners, whom he entertains handsomely. He liked to have clever young men about him to edit new editions of the *Principia*—and sometimes merely plausible ones as in the case of Facio de Duillier.

Magic was quite forgotten. He has become the Sage and Monarch of the Age of Reason. The Sir Isaac Newton of orthodox tradition—the eighteenth-century Sir Isaac, so remote from the child magician born in the first half of the seventeenth century—was being built up. Voltaire returning from his trip to London was able to report of Sir Isaac—' 'twas his peculiar felicity, not only to be born in a country of liberty, but in an Age when all scholastic impertinences were banished from the World. Reason alone was cultivated and Mankind cou'd only be his Pupil, not his Enemy.' Newton, whose secret heresies and scholastic superstitions it had been the study of a lifetime to conceal!

But he never concentrated, never recovered 'the former consistency of his mind'. 'He spoke very little in company.' 'He had something rather languid in his look and manner.'

And he looked very seldom, I expect, into the chest where, when he left Cambridge, he had packed all the evidences of what had occupied and so absorbed his intense and flaming spirit in his rooms and his garden and his elaboratory between the Great Gate and Chapel.

But he did not destroy them. They remained in the box to shock profoundly any eighteenth- or nineteenth-century prying eyes. They became the possession of Catharine Barton and then of her daughter, the Countess of Portsmouth. So Newton's chest, with many hundreds of thousands of words of his unpublished writings, came to contain the 'Portsmouth Papers'.

In 1888 the mathematical portion was given to the University Library at Cambridge. They have been indexed, but they have never been edited. The rest, a very large collection, were dispersed in the auction room in 1936 by Catharine Barton's descendant, the present Lord Lymington. Disturbed by this impiety, I managed gradually to reassemble about half of them, including nearly the whole of the biographical portion, that is, the 'Conduitt Papers', in order to bring them to Cambridge which I hope they will never leave. The greater part of the rest were snatched out of my reach by a syndicate which hoped to sell them at a high price, probably in America, on the occasion of the recent tercentenary.

As one broods over these queer collections, it seems easier to understand—with an understanding which is not, I hope, distorted in the other direction—this strange spirit, who was tempted by the Devil to believe at the time when within these walls he was solving so much, that he could reach *all* the secrets of God and Nature by the pure power of mind—Copernicus and Faustus in one.

COMMENTARY ON
BISHOP BERKELEY and
Infinitesimals

O NE of the great polemics of philosophy is a tract called *The Analyst*, issued in 1734 by the famous Irish metaphysician Bishop Berkeley. The object of the attack was the new calculus, especially the concept of the "fixed infinitesimal," as set forth by Isaac Newton in the *Principia*, in an appendix to the *Opticks*, and in other writings. Berkeley, though not a mathematician, made a number of extremely effective points in exposing the weak and confused conceptual foundations of the subject. Since he was an acute thinker and a brilliant writer, his arguments provoked controversy among mathematicians and led to the clarification of central ideas underlying the new system of analysis. *The Analyst* has been described as marking "a turning point in the history of mathematical thought in Great Britain." [1]

To appreciate the significance of Berkeley's tract the reader should refer to the discussions, in an earlier part of this volume, of the invention of the calculus (pp. 53–62) and of the work of Newton and Leibniz (pp. 140–146 and 255–276). *The Analyst*, a book of 104 pages, is addressed "to an infidel mathematician." The gentleman so designated is generally supposed to have been Newton's friend, the astronomer Edmund Halley.[2] Halley financed the publication of the *Principia* and helped to prepare it for the press. It is said that he also persuaded a friend of Berkeley's of the "inconceivability of the doctrines of Christianity"; the Bishop thereupon set out to demonstrate that the great innovation of fluxions was neither clearer nor more securely grounded than the tenets of theology.[3] He did not in this work "deny the utility of the new devices nor the validity of the results obtained. He merely asserted, with some show of justice, that mathematicians had given no legitimate arguments for their procedure, having used inductive instead of deductive reasoning." [4] It was not difficult to make ridiculous the concept of "evanescent terms" (the "ghosts of departed quantities," Berkeley called them); to expose the absurdity of the infinitesimal, a quantity greater than zero, yet so small that *no* multiple of it attains a measurable size; to suggest that the calculus, though based upon false and contradictory notions, yielded correct results by a "compensation of errors" ("by virtue of a twofold mistake, you arrive, though

[1] Florian Cajori, *A History of the Conception of Limits and Fluxions in Great Britain/From Newton to Woodhouse;* Chicago, 1919, p. 89.
[2] For further biographical details about Halley see p. 1418.
[3] Florian Cajori, *A History of Mathematics;* New York, second edition, 1919, p. 218.
[4] Carl B. Boyer, *The Concepts of the Calculus;* New York, 1939, p. 225.

not at science, yet at the truth"). But it is difficult to understand why Berkeley should have thought he could restore faith in religion by proving that mathematicians were often as muddleheaded as theologians. Whatever his purposes and whether or not he was sincere—both points have been disputed—Berkeley threw light on contradictions and thus promoted sounder definitions of crucial concepts. Among the writers who helped achieve this objective were the Englishman Benjamin Robins [5] and the able Scotch mathematician Colin Maclaurin.[6] The rigorous formulation of the calculus based upon the limit concept (and wholly banishing the infinitesimal) waited upon the labors of the French mathematician Augustin Cauchy in the first half of the nineteenth century. It was brought to "logical exactitude" in the second half of the century by the noted German analyst Karl Weierstrass, who "constructed a purely formal arithmetic basis for analysis, quite independent of all geometric intuition." [7]

The excerpt from *The Analyst* is worth reading for its literary quality, if for no other reason. Berkeley (1685–1753) was a writer and a philosopher of exceptional force, and a formidable controversialist. To be sure, his doctrines are perhaps less well remembered than the attempt of Samuel Johnson to refute them by kicking at a stone. The bishop based his philosophy of immaterialism on the argument that the "absolute existence" of sensible things is a meaningless phrase, since the term "existence" when applied to sensible things necessarily implies a relation to perception.[8] To critics and scoffers he replied (in his *A Defense of Free-Thinking in Mathematics*, a response to an attack on *The Analyst*), "My aim is truth; my reasons I have given. Confute them if you can, but think not to overbear me with either authorities or harsh words." Berkeley was a many-sided man, practical as well as prophetic. His full life included an ill-fated project to found a university in Bermuda, a two-and-a-half-year sojourn in America during which he made generous gifts to Harvard and Yale, an active career in Irish politics and a successful one in the church, a passionate advocacy of tarwater as a universal medicament, high achievement as a critic and essayist, extensive philosophical writings, and at least one serious poem (it contains the famous line, "Westward the course of empire takes its way," often ascribed to Rudyard Kipling).

[5] Benjamin Robins, *A Discourse Concerning the Nature and Certainty of Sir Isaac Newton's Method of Fluxions and of Prime and Ultimate Ratios*, London, 1735; also other papers in his *Mathematical Tracts*, London, 1761.

[6] Colin Maclaurin, *A Treatise of Fluxions*, 2 vols., Edinburgh, 1742.

[7] Carl Boyer, *op. cit.*, p. 284. Boyer's careful study is recommended to anyone with an appetite for more details about the long and fascinating record of this branch of mathematical thought.

[8] "It is interesting to notice that just as the arch-materialist Hobbes, being unable to conceive of lines without thickness, denied them to geometry, so also Berkeley, the extreme idealist, wishes to exclude from mathematics the 'inconceivable' idea of instantaneous velocity. This is in keeping with Berkeley's early sensationalism which led him to think of geometry as an applied science dealing with finite magnitudes which are composed of indivisible 'minima sensibilia.'" Carl Boyer, *op. cit.*, p. 227.

10   The Analyst

By BISHOP BERKELEY

A DISCOURSE ADDRESSED TO AN INFIDEL MATHEMATICIAN

THOUGH I am a stranger to your person, yet I am not, Sir, a stranger to the reputation you have acquired in that branch of learning which hath been your peculiar study; nor to the authority that you therefore assume in things foreign to your profession; nor to the abuse that you, and too many more of the like character, are known to make of such undue authority, to the misleading of unwary persons in matters of the highest concernment, and whereof your mathematical knowledge can by no means qualify you to be a competent judge . . .

Whereas then it is supposed that you apprehend more distinctly, consider more closely, infer more justly, and conclude more accurately than other men, and that you are therefore less religious because more judicious, I shall claim the privilege of a Freethinker; and take the liberty to inquire into the object, principles, and method of demonstration admitted by the mathematicians of the present age, with the same freedom that you presume to treat the principles and mysteries of Religion; to the end that all men may see what right you have to lead, or what encouragement others have to follow you . . .

The Method of Fluxions is the general key by help whereof the modern mathematicians unlock the secrets of Geometry, and consequently of Nature. And, as it is that which hath enabled them so remarkably to outgo the ancients in discovering theorems and solving problems, the exercise and application thereof is become the main if not the sole employment of all those who in this age pass for profound geometers. But whether this method be clear or obscure, consistent or repugnant, demonstrative or precarious, as I shall inquire with the utmost impartiality, so I submit my inquiry to your own judgment, and that of every candid reader.—Lines are supposed to be generated [1] by the motion of points, planes by the motion of lines, and solids by the motion of planes. And whereas quantities generated in equal times are greater or lesser according to the greater or lesser velocity wherewith they increase and are generated, a method hath been found to determine quantities from the velocities of their generating motions. And such velocities are called fluxions: and the quantities generated are called flowing quantities. These fluxions are said to be nearly as the increments of the flowing quantities, generated in the least equal particles of time; and to be accurately in the first proportion of the nascent, or in the last of the evanescent increments. Sometimes, instead of velocities, the momentaneous increments or decrements of undetermined flowing quantities are considered, under the appellation of moments.

By moments we are not to understand finite particles. These are said not to be moments, but quantities generated from moments, which last are only the nascent principles of finite quantities. It is said that the minutest errors are not to be neglected in mathematics: that the fluxions are celerities, not proportional to the finite increments, though ever so small; but only to the moments or nascent increments, whereof the proportion alone, and not the magnitude, is considered. And of the aforesaid fluxions there be other fluxions, which fluxions of fluxions are called second fluxions. And the fluxions of these second fluxions are called third fluxions: and so on, fourth, fifth, sixth, etc., *ad infinitum*. Now, as our Sense is strained and puzzled with the perception of objects extremely minute, even so the Imagination, which faculty derives from sense, is very much strained and puzzled to frame clear ideas of the least particles of time, or the least increments generated therein: and much more so to comprehend the moments, or those increments of the flowing quantities in *statu nascenti*, in their very first origin or beginning to exist, before they become finite particles. And it seems still more difficult to conceive the abstracted velocities of such nascent imperfect entities. But the velocities of the velocities—the second, third, four, and fifth velocities, etc.—exceed, if I mistake not, all human understanding. The further the mind analyseth and pursueth these fugitive ideas the more it is lost and bewildered; the objects, at first fleeting and minute, soon vanishing out of sight. Certainly, in any sense, a second or third fluxion seems an obscure Mystery. The incipient celerity of an incipient celerity, the nascent augment of a nascent augment, *i. e.*, of a thing which hath no magnitude—take it in what light you please, the clear conception of it will, if I mistake not, be found impossible; whether it be so or no I appeal to the trial of every thinking reader.

[1] *Introd. ad Quadraturam Curvarum.*

And if a second fluxion be inconceivable, what are we to think of third, fourth, fifth fluxions, and so on without end? . . .

All these points, I say, are supposed and believed by certain rigorous exactors of evidence in religion, men who pretend to believe no further than they can see. That men who have been conversant only about clear points should with difficulty admit obscure ones might not seem altogether unaccountable. But he who can digest a second or third fluxion, a second or third difference, need not, methinks, be squeamish about any point in divinity . . .

Nothing is easier than to devise expressions or notations for fluxions and infinitesimals of the first, second, third, fourth, and subsequent orders, proceeding in the same regular form without end or limit $\dot{x}. \ddot{x}. \dddot{x}. \ddddot{x}.$ etc. or $dx. ddx. dddx. ddddx.$ etc. These expressions, indeed, are clear and distinct, and the mind finds no difficulty in conceiving them to be continued beyond any assignable bounds. But if we remove the veil and look underneath, if, laying aside the expressions, we set ourselves attentively to consider the things themselves which are supposed to be expressed or marked thereby, we shall discover much emptiness, darkness, and confusion; nay, if I mistake not, direct impossibilities and contradictions. Whether this be the case or no, every thinking reader is entreated to examine and judge for himself . . .

This is given for demonstration.[2] Suppose the product or rectangle AB increased by continual motion: and that the momentaneous increments of the sides A and B are a and b. When the sides A and B were deficient, or lesser by one half of their moments, the rectangle was

$$\overline{A - \tfrac{1}{2}a} \times \overline{B - \tfrac{1}{2}b}, \; i. \, e., \; AB - \tfrac{1}{2}aB - \tfrac{1}{2}bA + \tfrac{1}{4}ab.$$

And as soon as the sides A and B are increased by the other two halves of their moments, the rectangle becomes

$$\overline{A + \tfrac{1}{2}a} \times \overline{B + \tfrac{1}{2}b} \text{ or } AB + \tfrac{1}{2}aB + \tfrac{1}{2}bA + \tfrac{1}{4}ab.$$

From the latter rectangle subduct the former, and the remaining difference will be $aB + bA$. Therefore the increment of the rectangle generated by the entire increments a and b is $aB + bA$. Q. E. D. But it is plain that the direct and true method to obtain the moment or increment of the rectangle AB, is to take the sides as increased by their whole increments, and so multiply them together, $A + a$ by $B + b$, the product whereof $AB + aB + bA + ab$ is the augmented rectangle; whence, if we subduct AB the remainder $aB + bA + ab$ will be the true increment of the rectangle, exceeding that which was obtained by the former illegitimate and indirect method by the quantity ab. And this holds universally by the quantities a and b be what they will, big or little, finite or infinitesi-

[2] *Philosophiae Naturalis Principia Mathematica,* Lib. II, lem. 2.

mal, increments, moments, or velocities. Nor will it avail to say that *ab* is a quantity exceedingly small: since we are told that *in rebus mathematicis errores quam minimi non sunt contemnendi* [3] . . .

But, as there seems to have been some inward scruple or consciousness of defect in the foregoing demonstration, and as this finding the fluxion of a given power is a point of primary importance, it hath therefore been judged proper to demonstrate the same in a different manner, independent of the foregoing demonstration. But whether this method be more legitimate and conclusive than the former, I proceed now to examine; and in order thereto shall premise the following lemma:—"If, with a view to demonstrate any proposition, a certain point is supposed, by virtue of which certain other points are attained; and such supposed point be itself afterwards destroyed or rejected by a contrary supposition; in that case, all the other points attained thereby, and consequent thereupon, must also be destroyed and rejected, so as from thenceforward to be no more supposed or applied in the demonstration." [4] This is so plain as to need no proof.

Now, the other method of obtaining a rule to find the fluxion of any power is as follows. Let the quantity x flow uniformly, and be it proposed to find the fluxion of x^n. In the same time that x by flowing becomes $x + o$, the power x^n becomes $\overline{x + o}|^n$, *i. e.*, by the method of infinite series

$$x^n + nox^{n-1} + \frac{nn - n}{2} oox^{n-2} + \&c.,$$

and the increments

$$o \text{ and } nox^{n-1} + \frac{nn - n}{2} oox^{n-2} + \&c.$$

are one to another as

$$1 \text{ to } nx^{n-1} + \frac{nn - n}{2} ox^{n-2} + \&c.$$

---

[3] *Introd. ad Quadraturam Curvarum.*

[4] [Berkeley's *lemma* was rejected as invalid by James Jurin and some other mathematical writers. The first mathematician to acknowledge openly the validity of Berkeley's *lemma* was Robert Woodhouse in his *Principles of Analytical Calculation*, Cambridge, 1803, p. XII. Instructive, in this connection, is a passage in A. N. Whitehead's *Introduction to Mathematics*, New York and London, 1911, p. 227. Whitehead does not mention Berkeley's *lemma* and probably did not have it in mind. Nevertheless, Whitehead advances an argument which is essentially the equivalent of Berkeley's, though expressed in different terms. When discussing the difference-quotient $\frac{(x + h)^2 - x^2}{h}$, Whitehead says: "In reading over the Newtonian method of statement, it is tempting to seek simplicity by saying that $2x + b$ is $2x$, when b is zero. *But this will not do; for it thereby abolishes the interval from x to $x + b$*, over which the average increase was calculated. The problem is, how to keep an interval of length b over which to calculate the average increase, and at the same time to treat b as if it were zero. Newton did this by the conception of a limit, and we now proceed to give Weierstrass's explanation of its real meaning."]

[The above note, and the two which follow are taken from Florian Cajori's edition of this excerpt in D. E. Smith, *A Source Book in Mathematics*, New York, 1929. ED.]

Let now the increments vanish, and their last proportion will be 1 to nx^{n-1}. But it should seem that this reasoning is not fair or conclusive. For when it is said, let the increments vanish, *i. e.*, let the increments be nothing, or let there be no increments, the former supposition that the increments were something, or that there were increments, is destroyed, and yet a consequence of that supposition, *i. e.*, an expression got by virtue thereof, is retained. Which, by the foregoing lemma, is a false way of reasoning. Certainly when we suppose the increments to vanish, we must suppose their proportions, their expressions, and everything else derived from the supposition of their existence, to vanish with them . . .

I have no controversy about your conclusions, but only about your logic and method: how you demonstrate? what objects you are conversant with, and whether you conceive them clearly? what principles you proceed upon; how sound they may be; and how you apply them? . . .

Now, I observe, in the first place, that the conclusion comes out right, not because the rejected square of dy was infinitely small, but because this error was compensated by another contrary and equal error [5] . . .

The great author of the method of fluxions felt this difficulty, and therefore he gave in to those nice abstractions and geometrical metaphysics without which he saw nothing could be done on the received principles: and what in the way of demonstration he hath done with them the reader will judge. It must, indeed, be acknowledged that he used fluxions, like the scaffold of a building, as things to be laid aside or got rid of as soon as finite lines were found proportional to them. But then these finite exponents are found by the help of fluxions. Whatever therefore is got by such exponents and proportions is to be ascribed to fluxions: which must therefore be previously understood. And what are these fluxions? The velocities of evanescent increments. And what are these same evanescent increments? They are neither finite quantities, nor quantities infinitely small, nor yet nothing. May we not call them the ghosts of departed quantities . . . ?

You may possibly hope to evade the force of all that hath been said, and to screen false principles and inconsistent reasonings, by a general pretence that these objections and remarks are *metaphysical*. But this is a vain pretence. For the plain sense and truth of what is advanced in the foregoing remarks, I appeal to the understanding of every unprejudiced intelligent reader . . .

And, to the end that you may more clearly comprehend the force and design of the foregoing remarks, and pursue them still farther in your own meditations, I shall subjoin the following Queries:—

[5] [Berkeley explains that the calculus of Leibniz leads from false principles to correct results by a "Compensation of errors." The same explanation was advanced again later by Maclaurin, Lagrange, and, independently, by L. N. M. Carnot in his *Réflexions sur la métaphysique du calcul infinitésimal*, 1797.]

Query 1. Whether the object of geometry be not the proportions of assignable extensions? And whether there be any need of considering quantities either infinitely great or infinitely small? . . .

Qu. 4. Whether men may properly be said to proceed in a scientific method, without clearly conceiving the object they are conversant about, the end proposed, and the method by which it is pursued? . . .

Qu. 8. Whether the notions of absolute time, absolute place, and absolute motion be not most abstractely metaphysical? Whether it be possible for us to measure, compute, or know them? . . .

Qu. 16. Whether certain maxims do not pass current among analysts which are shocking to good sense? And whether the common assumption, that a finite quantity divided by nothing is infinite, be not of this number? [6] . . .

Qu. 31. Where there are no increments, whether there can be any *ratio* of increments? Whether nothings can be considered as proportional to real quantities? Or whether to talk of their proportions be not to talk nonsense? Also in what sense we are to understand the proportion of a surface to a line, of an area to an ordinate? And whether species or numbers, though properly expressing quantities which are not homogeneous, may yet be said to express their proportion to each other? . . .

Qu. 54. Whether the same things which are now done by infinites may not be done by finite quantities? And whether this would not be a great relief to the imaginations and understandings of mathematical men? . . .

Qu. 63. Whether such mathematicians as cry out against mysteries have ever examined their own principles?

Qu. 64. Whether mathematicians, who are so delicate in religious points, are strictly scrupulous in their own science? Whether they do not submit to authority, take things upon trust, and believe points inconceivable? Whether they have not *their* mysteries, and what is more, their repugnances and contradictions? . . .

---

[6] [The earliest exclusion of division by zero in ordinary elementary algebra, on the ground of its being a procedure that is inadmissible according to reasoning based on the fundamental assumptions of this algebra, was made in 1828, by Martin Ohm, in his *Versuch eines vollkommen consequenten Systems der Mathematik*, Vol. I, p. 112. In 1872, Robert Grassmann took the same position. But not until about 1881 was the necessity of excluding division by zero explained in elementary school books on algebra.]

COMMENTARY ON
GAUSS

G AUSS is often referred to as the Prince of Mathematics, a not very helpful designation. No one would dispute that Archimedes, Gauss and Newton are in a special class among mathematicians. Each of the three is a tremendous, incredible figure towering above even the most eminent of his contemporaries, and the attempt to assign rank among them is silly. The royal image is further confused by calling arithmetic the Queen of Mathematics—for which Gauss himself was responsible—and mathematics the Queen of the Sciences. Where this leaves the Prince is not wholly clear. Gauss had a long, productive and interesting life, but a full-scale account of it has not yet been written, not even by a German pedant.[1] Mathematicians apparently frighten biographers, and Gauss is a formidable subject. He made so many outstanding contributions to mathematics, mathematical physics and other applied branches of the science that a book describing his work would need to be longer, so Bell conjectures, than a similar treatise on Newton. Considering the material that had to be covered, the chapter Bell devotes to Gauss in his *Men of Mathematics* is a skillful summary. It is a creative essay from which the average reader can gain a sound appreciation of Gauss's role in the development of mathematics and scientific thought. The chapter is reproduced below.

Eric Temple Bell was born in 1883 in Aberdeen, Scotland. He studied at the University of London, came to the United States at the beginning of the century, got his Ph.D. in mathematics from Columbia University in 1912, taught at the University of Washington, and since 1926 has been professor of mathematics at the California Institute of Technology. Dr. Bell, now an American citizen, is a former president of the Mathematical Association of America and a former vice-president of the American Association for the Advancement of Science. He has won many honors for mathematical research and is a member of the National Academy of Science. I count at least twenty-one books by Bell, besides numerous mathematical articles. His books include several items of science fiction which are simply painful, a study of numerology, an able historical survey for experts, *The Development of Mathematics*, and his biographical collection, *Men of Mathematics*. Bell is a lively, stimulating writer, inoffensively crotchety and opinionated, with a good sense of historical circumstance, a fine impatience with humbug, a sound grasp of the entire mathematical scene, and a gift for clear and orderly explanation.

[1] An authoritative German source is Heinrich Mack, *C. F. Gauss und Die Seinen*, E. Applehaus und Comp., Brunswick, 1927.

As yet a child, nor yet a fool to fame,
I lisped in numbers, for the numbers came.

—ALEXANDER POPE

11 The Prince of Mathematicians

By ERIC TEMPLE BELL

The further elaboration and development of systematic arithmetic, like
nearly everything else which the mathematics of our [nineteenth] century
has produced in the way of original scientific ideas, is knit to Gauss.

—LEOPOLD KRONECKER

ARCHIMEDES, Newton, and Gauss, these three, are in a class by themselves among the great mathematicians, and it is not for ordinary mortals
to attempt to range them in order of merit. All three started tidal waves
in both pure and applied mathematics: Archimedes esteemed his pure
mathematics more highly than its applications; Newton appears to have
found the chief justification for his mathematical inventions in the scientific uses to which he put them, while Gauss declared that it was all one to
him whether he worked on the pure or the applied side. Nevertheless Gauss
crowned the higher arithmetic, in his day the least practical of mathematical studies, the Queen of all.

The lineage of Gauss, Prince of Mathematicians, was anything but
royal. The son of poor parents, he was born in a miserable cottage at
Brunswick (Braunschweig), Germany, on April 30, 1777. His paternal
grandfather was a poor peasant. In 1740 this grandfather settled in Brunswick, where he drudged out a meager existence as a gardener. The second
of his three sons, Gerhard Diederich, born in 1744, became the father of
Gauss. Beyond that unique honor Gerhard's life of hard labor as a gardener, canal tender, and bricklayer was without distinction of any kind.

The picture we get of Gauss' father is that of an upright, scrupulously
honest, uncouth man whose harshness to his sons sometimes bordered on
brutality. His speech was rough and his hand heavy. Honesty and persistence gradually won him some measure of comfort, but his circumstances
were never easy. It is not surprising that such a man did everything in his
power to thwart his young son and prevent him from acquiring an education suited to his abilities. Had the father prevailed, the gifted boy would
have followed one of the family trades, and it was only by a series of
happy accidents that Gauss was saved from becoming a gardener or a
bricklayer. As a child he was respectful and obedient, and although he

never criticized his poor father in later life, he made it plain that he had never felt any real affection for him. Gerhard died in 1806. By that time the son he had done his best to discourage had accomplished immortal work.

On his mother's side Gauss was indeed fortunate. Dorothea Benz's father was a stonecutter who died at the age of thirty of tuberculosis, the result of unsanitary working conditions in his trade, leaving two children, Dorothea and her younger brother Friederich.

Here the line of descent of Gauss' genius becomes evident. Condemned by economic disabilities to the trade of weaving, Friederich was a highly intelligent, genial man whose keen and restless mind foraged for itself in fields far from his livelihood. In his trade Friederich quickly made a reputation as a weaver of the finest damasks, an art which he mastered wholly by himself. Finding a kindred mind in his sister's child, the clever uncle Friederich sharpened his wits on those of the young genius and did what he could to rouse the boy's quick logic by his own quizzical observations and somewhat mocking philosophy of life.

Friederich knew what he was doing; Gauss at the time probably did not. But Gauss had a photographic memory which retained the impressions of his infancy and childhood unblurred to his dying day. Looking back as a grown man on what Friederich had done for him, and remembering the prolific mind which a premature death had robbed of its chance of fruition, Gauss lamented that "a born genius was lost in him."

Dorothea moved to Brunswick in 1769. At the age of thirty four (in 1776) she married Gauss' father. The following year her son was born. His full baptismal name was Johann Friederich Carl Gauss. In later life he signed his masterpieces simply Carl Friedrich Gauss. If a great genius was lost in Friederich Benz his name survives in that of his grateful nephew.

Gauss' mother was a forthright woman of strong character, sharp intellect, and humorous good sense. Her son was her pride from the day of his birth to her own death at the age of ninety seven. When the "wonder child" of two, whose astounding intelligence impressed all who watched his phenomenal development as something not of this earth, maintained and even surpassed the promise of his infancy as he grew to boyhood, Dorothea Gauss took her boy's part and defeated her obstinate husband in his campaign to keep his son as ignorant as himself.

Dorothea hoped and expected great things of her son. That she may sometimes have doubted whether her dreams were to be realized is shown by her hesitant questioning of those in a position to judge her son's abilities. Thus, when Gauss was nineteen, she asked his mathematical friend Wolfgang Bolyai whether Gauss would ever amount to anything.

When Bolyai exclaimed "The greatest mathematician in Europe!" she burst into tears.

The last twenty two years of her life were spent in her son's house, and for the last four she was totally blind. Gauss himself cared little if anything for fame; his triumphs were his mother's life.[1] There was always the completest understanding between them, and Gauss repaid her courageous protection of his early years by giving her a serene old age. When she went blind he would allow no one but himself to wait on her, and he nursed her in her long last illness. She died on April 19, 1839.

Of the many accidents which might have robbed Archimedes and Newton of their mathematical peer, Gauss himself recalled one from his earliest childhood. A spring freshet had filled the canal which ran by the family cottage to overflowing. Playing near the water, Gauss was swept in and nearly drowned. But for the lucky chance that a laborer happened to be about his life would have ended then and there.

In all the history of mathematics there is nothing approaching the precocity of Gauss as a child. It is not known when Archimedes first gave evidence of genius. Newton's earliest manifestations of the highest mathematical talent may well have passed unnoticed. Although it seems incredible, Gauss showed his caliber before he was three years old.

One Saturday Gerhard Gauss was making out the weekly payroll for the laborers under his charge, unaware that his young son was following the proceedings with critical attention. Coming to the end of his long computations, Gerhard was startled to hear the little boy pipe up, "Father, the reckoning is wrong, it should be . . ." A check of the account showed that the figure named by Gauss was correct.

Before this the boy had teased the pronunciations of the letters of the alphabet out of his parents and their friends and had taught himself to read. Nobody had shown him anything about arithmetic, although presumably he had picked up the meanings of the digits 1, 2, . . . along with the alphabet. In later life he loved to joke that he knew how to reckon before he could talk. A prodigious power for involved mental calculations remained with him all his life.

Shortly after his seventh birthday Gauss entered his first school, a squalid relic of the Middle Ages run by a virile brute, one Büttner, whose idea of teaching the hundred or so boys in his charge was to thrash them into such a state of terrified stupidity that they forgot their own names.

---

[1] The legend of Gauss' relations to his parents has still to be authenticated. Although, as will be seen later, the *mother* stood by her son, the *father* opposed him; and, as was customary *then* (usually, also, *now*), in a German household, the *father* had the last word.—I allude later to legends from living persons who had known members of the Gauss family, particularly in respect to Gauss' treatment of his sons. These allusions refer to first-hand evidence; but I do not vouch for them, as the people were very old.

More of the good old days for which sentimental reactionaries long. It was in this hell-hole that Gauss found his fortune.

Nothing extraordinary happened during the first two years. Then, in his tenth year, Gauss was admitted to the class in arithmetic. As it was the beginning class none of the boys had ever heard of an arithmetical progression. It was easy then for the heroic Büttner to give out a long problem in addition whose answer he could find by a formula in a few seconds. The problem was of the following sort, $81297 + 81495 + 81693 + \ldots + 100899$, where the step from one number to the next is the same all along (here 198), and a given number of terms (here 100) are to be added.

It was the custom of the school for the boy who first got the answer to lay his slate on the table; the next laid his slate on top of the first, and so on. Büttner had barely finished stating the problem when Gauss flung his slate on the table: "There it lies," he said—*"Ligget se' "* in his pleasant dialect. Then, for the ensuing hour, while the other boys toiled, he sat with his hands folded, favored now and then by a sarcastic glance from Büttner, who imagined the youngest pupil in the class was just another blockhead. At the end of the period Büttner looked over the slates. On Gauss' slate there appeared but a single number. To the end of his days Gauss loved to tell how the one number he had written was the correct answer and how all the others were wrong. Gauss had not been shown the trick for doing such problems rapidly. It is very ordinary once it is known, but for a boy of ten to find it instantaneously by himself is not so ordinary.

This opened the door through which Gauss passed on to immortality. Büttner was so astonished at what the boy of ten had done without instruction that he promptly redeemed himself and to at least one of his pupils became a humane teacher. Out of his own pocket he paid for the best textbook on arithmetic obtainable and presented it to Gauss. The boy flashed through the book. "He is beyond me," Büttner said; "I can teach him nothing more."

By himself Büttner could probably not have done much for the young genius. But by a lucky chance the schoolmaster had an assistant, Johann Martin Bartels (1769–1836), a young man with a passion for mathematics, whose duty it was to help the beginners in writing and cut their quill pens for them. Between the assistant of seventeen and the pupil of ten there sprang up a warm friendship which lasted out Bartels' life. They studied together, helping one another over difficulties and amplifying the proofs in their common textbook on algebra and the rudiments of analysis.

Out of this early work developed one of the dominating interests of Gauss' career. He quickly mastered the binomial theorem,

$$(1 + x)^n = 1 + \frac{n}{1}x + \frac{n(n-1)}{1 \times 2}x^2 + \frac{n(n-1)(n-2)}{1 \times 2 \times 3}x^3 + \ldots,$$

in which n is not necessarily a positive integer, but may be any number. If n is not a positive integer, the series on the right is *infinite* (nonterminating), and in order to state when this series is actually equal to $(1 + x)^n$, it is mandatory to investigate what restrictions must be imposed upon x and n in order that the infinite series shall *converge to a definite, finite limit.* Thus, if $x = -2$, and $n = -1$, we get the absurdity that $(1 - 2)^{-1}$, which is $(-1)^{-1}$ or $1/(-1)$, or finally -1, is equal to $1 + 2 + 2^2 + 2^3 + \ldots$ and so on *ad infinitum*; that is, -1 is equal to the "infinite number" $1 + 2 + 4 + 8 + \ldots$, which is nonsense.

Before young Gauss asked himself whether infinite series *converge* and really do enable us to calculate the mathematical expressions (functions) they are used to represent, the older analysts had not seriously troubled themselves to explain the mysteries (and nonsense) arising from an uncritical use of infinite processes. Gauss' early encounter with the binomial theorem inspired him to some of his greatest work and he became the first of the "rigorists." A *proof* of the binomial theorem when n is not an integer greater than zero is even today beyond the range of an elementary textbook. Dissatisfied with what he and Bartels found in their book, Gauss made a proof. This initiated him to mathematical analysis. The very essence of analysis is the correct use of infinite processes.

The work thus well begun was to change the whole aspect of mathematics. Newton, Leibniz, Euler, Lagrange, Laplace—all great analysts for their times—had practically no conception of what is now acceptable as a proof involving infinite processes. The first to see clearly that a "proof" which may lead to absurdities like "minus 1 equals infinity" is no proof at all, was Gauss. Even if in *some* cases a formula gives consistent results, it has no place in mathematics until the precise conditions under which it will continue to yield consistency have been determined.

The rigor which Gauss imposed on analysis gradually overshadowed the whole of mathematics, both in his own habits and in those of his contemporaries—Abel, Cauchy—and his successors—Weierstrass, Dedekind, and mathematics after Gauss became a totally different thing from the mathematics of Newton, Euler, and Lagrange.

In the constructive sense Gauss was a revolutionist. Before his schooling was over the same critical spirit which left him dissatisfied with the binomial theorem had caused him to question the demonstrations of elementary geometry. At the age of twelve he was already looking askance at the foundations of Euclidean geometry; by sixteen he had caught his first glimpse of a geometry other than Euclid's. A year later he had begun a searching criticism of the proofs in the theory of numbers which had satisfied his predecessors and had set himself the extraordinarily difficult task of filling up the gaps and *completing* what had been only half done. Arithmetic, the field of his earliest triumphs, became his favorite study and the locus of his masterpiece. To his sure feeling for what constitutes proof Gauss added a prolific mathematical inventiveness that has never been surpassed. The combination was unbeatable.

Bartels did more for Gauss than to induct him into the mysteries of algebra. The young teacher was acquainted with some of the influential men of Brunswick. He now made it his business to interest these men in his find. They in turn, favorably impressed by the obvious genius of Gauss, brought him to the attention of Carl Wilhelm Ferdinand, Duke of Brunswick.

The Duke received Gauss for the first time in 1791. Gauss was then fourteen. The boy's modesty and awkward shyness won the heart of the generous Duke. Gauss left with the assurance that his education would be continued. The following year (February, 1792) Gauss matriculated at the Collegium Carolinum in Brunswick. The Duke paid the bills and he continued to pay them till Gauss' education was finished.

Before entering the Caroline College at the age of fifteen, Gauss had made great headway in the classical languages by private study and help from older friends, thus precipitating a crisis in his career. To his crassly practical father the study of ancient languages was the height of folly. Dorothea Gauss put up a fight for her boy, won, and the Duke subsidized a two-years' course at the Gymnasium. There Gauss' lightning mastery of the classics astonished teachers and students alike.

Gauss himself was strongly attracted to philological studies, but fortunately for science he was presently to find a more compelling attraction in mathematics. On entering college Gauss was already master of the supple Latin in which many of his greatest works are written. It is an ever-to-be-regretted calamity that even the example of Gauss was powerless against the tides of bigoted nationalism which swept over Europe after the French Revolution and the downfall of Napoleon. Instead of the easy Latin which sufficed for Euler and Gauss, and which any student can master in a few weeks, scientific workers must now acquire a reading knowledge of two or three languages in addition to their own. Gauss resisted as long as he could, but even he had to submit when his astronomical friends in Germany pressed him to write some of his astronomical works in German.

Gauss studied at the Caroline College for three years, during which he mastered the more important works of Euler, Lagrange and, above all, Newton's *Principia*. The highest praise one great man can get is from another in his own class. Gauss never lowered the estimate which as a boy of seventeen he had formed of Newton. Others—Euler, Laplace, Lagrange, Legendre—appear in the flowing Latin of Gauss with the complimentary *clarissimus*; Newton is *summus*.

While still at the college Gauss had begun those researches in the higher arithmetic which were to make him immortal. His prodigious powers of calculation now came into play. Going directly to the numbers themselves he experimented with them, discovering by induction recondite general theorems whose proofs were to cost even him an effort. In this way he rediscovered "the gem of arithmetic," *"theorema aureum,"* which Euler also had come upon inductively, which is known as the law of quadratic reciprocity, and which he was to be the first to prove. (Legendre's attempted proof slurs over a crux.)

The whole investigation originated in a simple question which many beginners in arithmetic ask themselves: How many digits are there in the period of a repeating decimal? To get some light on the problem Gauss calculated the decimal representations of all the fractions $1/n$ for $n = 1$ to 1000. He did not find the treasure he was seeking, but something infinitely greater—the law of quadratic reciprocity. As this is quite simply stated we shall describe it, introducing at the same time one of the revolutionary improvements in arithmetical nomenclature and notation which Gauss invented, that of *congruence*. All numbers in what follows are integers (common whole numbers).

If the *difference* $(a - b$ or $b - a)$ of two numbers a, b is exactly divisible by the number m, we say that a, b are *congruent* with respect to the modulus m, or simply *congruent modulo m*, and we symbolize this by writing $a \equiv b \pmod{m}$. Thus $100 \equiv 2 \pmod 7$, $35 \equiv 2 \pmod{11}$.

The advantage of this scheme is that it recalls the way we write algebraic equations, traps the somewhat elusive notion of arithmetical divisibility in a compact notation, and suggests that we try to carry over to arithmetic (which is much harder than algebra) some of the manipulations that lead to interesting results in algebra. For example we can "add" equations, and we find that congruences also can be "added," provided the modulus is the same in all, to give other congruences.

Let x denote an unknown number, r and m given numbers, of which r is not divisible by m. Is there a number x such that

$$x^2 \equiv r \pmod{m}?$$

If there is, r is called a *quadratic residue of m*, if not, a *quadratic non-residue of m*.

If r *is* a quadratic residue of m, then it must be possible to find at least one x whose square when divided by m leaves the remainder r; if r is a quadratic non-residue of m, then there is no x whose square when divided by m leaves the remainder r. These are immediate consequences of the preceding definitions.

To illustrate: is 13 a quadratic residue of 17? If so, it must be possible to solve the *congruence*

$$x^2 \equiv 13 \ (\text{mod } 17)$$

Trying 1, 2, 3, . . . , we find that $x = 8, 25, 42, 59, \ldots$ are solutions ($8^2 = 64 = 3 \times 17 + 13$; $25^2 = 625 = 36 \times 17 + 13$; etc.,) so that 13 *is* a quadratic residue of 17. But there is *no* solution of $x^2 \equiv 5 \ (\text{mod } 17)$, so 5 is a quadratic non-residue of 17.

It is now natural to ask what are the quadratic residues and non-residues of a given m? Namely, given m in $x^2 \equiv r \ (\text{mod } m)$, what numbers r can appear and what numbers r cannot appear as x runs through all the numbers 1, 2, 3, . . . ?

Without much difficulty it can be shown that it is sufficient to answer the question when both r and m are restricted to be primes. So we restate the problem: If p is a *given* prime, what primes q will make the congruence $x^2 \equiv q \ (\text{mod } p)$ solvable? This is asking altogether too much in the present state of arithmetic. However, the situation is not utterly hopeless.

There is a beautiful "reciprocity" between the *pair* of congruences

$$x^2 \equiv q \ (\text{mod } p), \ x^2 \equiv p \ (\text{mod } q),$$

in which *both* of p, q are *primes*: *both* congruences are *solvable*, or *both* are *unsolvable, unless both* of p, q leave the remainder 3 when divided by 4, in which case *one* of the congruences *is* solvable and *the other* is *not*. This is the law of quadratic reciprocity.

It was not easy to prove. In fact it baffled Euler and Legendre. Gauss gave the first proof at the age of nineteen. As this reciprocity is of fundamental importance in the higher arithmetic and in many advanced parts of algebra, Gauss turned it over and over in his mind for many years, seeking to find its taproot, until in all he had given six distinct proofs, one of which depends upon the straightedge and compass construction of regular polygons.

A numerical illustration will illuminate the statement of the law. First, take $p = 5$, $q = 13$. Since both of 5, 13 leave the remainder 1 on division by 4, *both* of $x^2 \equiv 13 \ (\text{mod } 5)$, $x^2 \equiv 5 \ (\text{mod } 13)$ must be *solvable*, or *neither* is solvable. The latter is the case for this pair. For $p = 13$, $q = 17$, both of which leave the remainder 1 on division by 4, we get $x^2 \equiv 17$ (mod 13), $x^2 \equiv 13 \ (\text{mod } 17)$, and *both*, or *neither* again must be solvable. The former is the case here: the first congruence has the solutions $x = 2, 15, 28, \ldots$; the second has the solutions $x = 8, 25, 42, \ldots$. There remains to be tested only the case when *both* of p, q leave the remainder 3 on division by 4. Take $p = 11$, $q = 19$. Then, according to the law, *precisely one* of $x^2 \equiv 19$ (mod 11), $x^2 \equiv 11$ (mod 19) must be solvable. The first congruence has no solution; the second has the solutions, 7, 26, 45, ...

The mere discovery of such a law was a notable achievement. That it was first proved by a boy of nineteen will suggest to anyone who tries to prove it that Gauss was more than merely competent in mathematics.

When Gauss left the Caroline College in October, 1795 at the age of eighteen to enter the University of Göttingen he was still undecided whether to follow mathematics or philology as his life work. He had already invented (when he was eighteen) the method of "least squares," which today is indispensable in geodetic surveying, in the reduction of observations and indeed in all work where the "most probable" value of anything that is measured is to be inferred from a large number of measurements. (The most probable value is furnished by making the sum of the squares of the "residuals"—roughly, divergences from assumed exactness—a minimum.) Gauss shares this honor with Legendre who published the method independently in 1806. This work was the beginning of Gauss' interest in the theory of errors of observation. The Gaussian law of normal distribution of errors and its accompanying bell-shaped curve is familiar today to all who handle statistics, from high-minded intelligence testers to unscrupulous market manipulators.

March 30, 1796, marks the turning point in Gauss' career. On that day, exactly a month before his twentieth year opened, Gauss definitely decided in favor of mathematics. The study of languages was to remain a lifelong hobby, but philology lost Gauss forever on that memorable day in March.

The regular polygon of seventeen sides was the die whose lucky fall induced Gauss to cross his Rubicon.[2] The same day Gauss began to keep his

---

[2] "Before leaving 'Fermat's numbers' $2^n + 1$ we shall glance ahead to the last decade of the eighteenth century where these mysterious numbers were partly responsible for one of the two or three most important events in all the long history of mathematics. For some time a young man in his eighteenth year had been hesitating —according to the tradition—whether to devote his superb talents to mathematics or to philology. He was equally gifted in both. What decided him was a beautiful discovery in connection with a simple problem in elementary geometry familiar to every schoolboy.

"A *regular* polygon of n sides has all its n sides equal and all its n angles equal. The ancient Greeks early found out how to construct regular polygons of 3, 4, 5, 6, 8, 10 and 15 sides by the use of straightedge and compass alone, and it is an easy matter, with the same implements, to construct from a regular polygon having a given number of sides another regular polygon having twice that number of sides. The next step then would be to seek straightedge and compass constructions for regular polygons of 7, 9, 11, 13, . . . sides. Many sought, but failed to find, because such constructions are impossible, only they did not know it. After an interval of over 2200 years the young man hesitating between mathematics and philology took the next step—a long one—forward.

"As has been indicated it is sufficient to consider only polygons having an *odd* number of sides. The young man proved that a straightedge and compass construction of a regular polygon having an odd number of sides is possible when, and only when, that number is either a *prime* Fermat number (that is a prime of the form $2^n + 1$), or scientific diary (*Notizen-journal*). It is one of most precious documents in the history of mathematics. The first entry records his great discovery.

The diary came into scientific circulation only in 1898, forty three years after the death of Gauss, when the Royal Society of Göttingen borrowed it from a grandson of Gauss for critical study. It consists of nineteen small octavo pages and contains 146 extremely brief statements of discoveries or results of calculations, the last of which is dated July 9, 1814. A facsimile reproduction was published in 1917 in the tenth volume (part 1) of Gauss' collected works, together with an exhaustive analysis of its contents by several expert editors. Not all of Gauss' discoveries in the prolific period from 1796 to 1814 by any means are noted. But many of those that are jotted down suffice to establish Gauss' priority in fields— elliptic functions, for instance—where some of his contemporaries refused to believe he had preceded them. (Recall that Gauss was born in 1777.)

Things were buried for years or decades in this diary that would have made half a dozen great reputations had they been published promptly. Some were never made public during Gauss' lifetime, and he never claimed in anything he himself printed to have anticipated others when they caught up with him. But the record stands. He did anticipate some who doubted the word of his friends. These anticipations were not mere trivialities. Some of them became major fields of nineteenth century mathematics.

A few of the entries indicate that the diary was a strictly private affair of its author's. Thus for July 10, 1796, there is the entry

$$\text{EYPHKA! } \text{num} = \Delta + \Delta + \Delta.$$

Translated, this echoes Archimedes' exultant "Eureka!" and states that every positive integer is the sum of three triangular numbers—such a number is one of the sequence 0, 1, 3, 6, 10, 15, . . . where each (after 0) is of the form $\frac{1}{2}n(n+1)$, n being any positive integer. Another way of saying the same thing is that every number of the form $8n + 3$ is a sum of three odd squares: $3 = 1^2 + 1^2 + 1^2$; $11 = 1^2 + 1^2 + 3^2$; $19 = 1^2 + 3^2 + 3^2$, etc. It is not easy to prove this from scratch.

Less intelligible is the cryptic entry for October 11, 1796, "Vicimus GEGAN." What dragon had Gauss conquered this time? Or what giant had he overcome on April 8, 1799, when he boxes REV. GALEN up in

---

is made up by multiplying together *different* Fermat primes. Thus the construction is possible for 3, 5, or 15 sides as the Greeks knew, but not for 7, 9, 11 or 13 sides, and is also possible for 17 or 257 or 65537 or—for what the next prime in the Fermat sequence 3, 5, 17, 257, 65537, . . . may be, *if there is one*—nobody yet (1936) knows—and the construction is also possible for 3×17, or $5 \times 257 \times 65537$ sides, and so on. It was this discovery, announced on June 1, 1796, but made on March 30th, which induced the young man to choose mathematics instead of philology as his life work. His name was Gauss."

[From the chapter on Fermat in Bell's *Men of Mathematics*.]

a neat rectangle? Although the meaning of these is lost forever the remaining 144 are for the most part clear enough. One in particular is of the first importance: the entry for March 19, 1797, shows that Gauss had already discovered the double periodicity of certain elliptic functions. He was then not quite twenty. Again, a later entry shows that Gauss had recognized the double periodicity in the general case. This discovery of itself, had he published it, would have made him famous. But he never published it.

Why did Gauss hold back the great things he discovered? This is easier to explain than his genius—if we accept his own simple statements, which will be reported presently. A more romantic version is the story told by W. W. R. Ball in his well-known history of mathematics. According to this, Gauss submitted his first masterpiece, the *Disquisitiones Arithmeticae*, to the French Academy of Sciences, only to have it rejected with a sneer. The undeserved humiliation hurt Gauss so deeply that he resolved thenceforth to publish only what anyone would admit was above criticism in both matter and form. There is nothing in this defamatory legend. It was disproved once for all in 1935, when the officers of the French Academy ascertained by an exhaustive search of the permanent records that the *Disquisitiones* was never even submitted to the Academy, much less rejected.

Speaking for himself Gauss said that he undertook his scientific works only in response to the deepest promptings of his nature, and it was a wholly secondary consideration to him whether they were ever published for the instruction of others. Another statement which Gauss once made to a friend explains both his diary and his slowness in publication. He declared that such an overwhelming horde of new ideas stormed his mind before he was twenty that he could hardly control them and had time to record but a small fraction. The diary contains only the final brief statements of the outcome of elaborate investigations, some of which occupied him for weeks. Contemplating as a youth the close, unbreakable chains of synthetic proofs in which Archimedes and Newton had tamed their inspirations, Gauss resolved to follow their great example and leave after him only finished works of art, severely perfect, to which nothing could be added and from which nothing could be taken away without disfiguring the whole. The work itself must stand forth, complete, simple, and convincing, with no trace remaining of the labor by which it had been achieved. A cathedral is not a cathedral, he said, till the last scaffolding is down and out of sight. Working with this ideal before him, Gauss preferred to polish one masterpiece several times rather than to publish the broad outlines of many as he might easily have done. His seal, a tree with but few fruits, bore the motto *Pauca sed matura* (Few, but ripe).

The fruits of this striving after perfection were indeed ripe but not always easily digestible. All traces of the steps by which the goal had been attained having been obliterated, it was not easy for the followers of Gauss to rediscover the road he had travelled. Consequently some of his works had to wait for highly gifted interpreters before mathematicians in general could understand them, see their significance for unsolved problems, and go ahead. His own contemporaries begged him to relax his frigid perfection so that mathematics might advance more rapidly, but Gauss never relaxed. Not till long after his death was it known how much of nineteenth-century mathematics Gauss had foreseen and anticipated before the year 1800. Had he divulged what he knew it is quite possible that mathematics would now be half a century or more ahead of where it is. Abel and Jacobi could have begun where Gauss left off, instead of expending much of their finest effort rediscovering things Gauss knew before they were born, and the creators of non-Euclidean geometry could have turned their genius to other things.

Of himself Gauss said that he was "all mathematician." This does him an injustice unless it is remembered that "mathematician" in his day included also what would now be termed a mathematical physicist. Indeed his second motto [3]

> *Thou, nature, art my goddess; to thy laws*
> *My services are bound . . . ,*

truly sums up his life of devotion to mathematics and the physical sciences of his time. The "all mathematician" aspect of him is to be understood only in the sense that he did not scatter his magnificent endowment broadcast over all fields where he might have reaped abundantly, as he blamed Leibniz for doing, but cultivated his greatest gift to perfection.

The three years (October, 1795–September, 1798) at the University of Göttingen were the most prolific in Gauss' life. Owing to the generosity of the Duke Ferdinand the young man did not have to worry about finances. He lost himself in his work, making but few friends. One of these, Wolfgang Bolyai, "the rarest spirit I ever knew," as Gauss described him, was to become a friend for life. The course of this friendship and its importance in the history of non-Euclidean geometry is too long to be told here; Wolfgang's son Johann was to retrace practically the same path that Gauss had followed to the creation of a non-Euclidean geometry, in entire ignorance that his father's old friend had anticipated him. The ideas which had overwhelmed Gauss since his seventeenth year were now caught—partly—and reduced to order. Since 1795 he had been meditating a great work on the theory of numbers. This now took definite shape, and by 1798 the *Disquisitiones Arithmeticae* (Arithmetical Researches) was practically completed.

[3] Shakespeare's *King Lear*, Act I, Scene II, 1–2, with the essential change of "laws" for "law."

To acquaint himself with what had already been done in the higher arithmetic and to make sure that he gave due credit to his predecessors, Gauss went to the University of Helmstedt, where there was a good mathematical library, in September, 1798. There he found that his fame had preceded him. He was cordially welcomed by the librarian and the professor of mathematics, Johann Friedrich Pfaff (1765–1825), in whose house he roomed. Gauss and Pfaff became warm friends, although the Pfaff family saw but little of their guest. Pfaff evidently thought it his duty to see that his hard-working young friend took some exercise, for he and Gauss strolled together in the evenings, talking mathematics. As Gauss was not only modest but reticent about his own work, Pfaff probably did not learn as much as he might have. Gauss admired the professor tremendously (he was then the best-known mathematician in Germany), not only for his excellent mathematics, but for his simple, open character. All his life there was but one type of man for whom Gauss felt aversion and contempt, the pretender to knowledge who will not admit his mistakes when he knows he is wrong.

Gauss spent the autumn of 1798 (he was then twenty one) in Brunswick, with occasional trips to Helmstedt, putting the finishing touches to the *Disquisitiones*. He had hoped for early publication, but the book was held up in the press owing to a Leipzig publisher's difficulties till September, 1801. In gratitude for all that Ferdinand had done for him, Gauss dedicated his book to the Duke—*"Serenissimo Principi ac Domino Carolo Guilielmo Ferdinando."*

If ever a generous patron deserved the homage of his protégé, Ferdinand deserved that of Gauss. When the young genius was worried ill about his future after leaving Göttingen—he tried unsuccessfully to get pupils—the Duke came to his rescue, paid for the printing of his doctoral dissertation (University of Helmstedt, 1799), and granted him a modest pension which would enable him to continue his scientific work unhampered by poverty. *"Your* kindness," Gauss says in his dedication, "freed me from all other responsibilities and enabled me to assume this exclusively."

Before describing the *Disquisitiones* we shall glance at the dissertation for which Gauss was awarded his doctor's degree *in absentia* by the University of Helmstedt in 1799: *Demonstratio nova theorematis omnem functionem algebraicam rationalem integram unius variabilis in factores reales primi vel secundi gradus revolvi posse* (A New Proof that Every Rational Integral Function of One Variable Can Be Resolved into Real Factors of the First or Second Degree).

There is only one thing wrong with this landmark in algebra. The first two words in the title would imply that Gauss had merely added a *new* proof to others already known. He should have omitted "nova." His was the *first* proof. (This assertion will be qualified later.) Some before him had published what they supposed were proofs of this theorem—usually called the fundamental theorem of algebra—but none had attained a proof. With his uncompromising demand for logical and mathematical rigor Gauss insisted upon a *proof,* and gave the first. Another, equivalent, statement of the theorem says that every algebraic equation in one unknown has a root, an assertion which beginners often take for granted as being true without having the remotest conception of what it means.

If a lunatic scribbles a jumble of mathematical symbols it does not follow that the writing means anything merely because to the inexpert eye it is indistinguishable from higher mathematics. It is just as doubtful whether the assertion that every algebraic equation has a root means anything until we say *what sort* of a root the equation has. Vaguely, we feel that a *number* will "satisfy" the equation but that half a pound of butter will not.

Gauss made this feeling precise by proving that all the roots of any algebraic equation are "numbers" of the form $a + bi$, where a, b are real numbers (the numbers that correspond to the distances, positive, zero, or negative, measured from a fixed point O on a given straight line, as on the x-axis in Descartes' geometry), and i is the square root of -1. The new sort of "number" $a + bi$ is called *complex.*

Incidentally, Gauss was one of the first to give a coherent account of

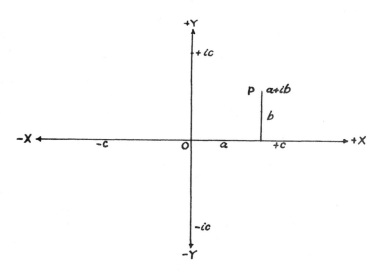

FIGURE 1

complex numbers and to interpret them as labelling the points of a plane, as is done today in elementary textbooks on algebra.

The Cartesian coordinates of P are (a,b); the point P is also labelled $a + bi$. Thus to every point of the plane corresponds precisely one complex number; the numbers corresponding to the points on XOX' are "real," those on YOY' "pure imaginary" (they are all of the type ic, where c is a real number).

The word "imaginary" is the great algebraical calamity, but it is too well established for mathematicians to eradicate. It should never have been used. Books on elementary algebra give a simple interpretation of imaginary numbers in terms of rotations. Thus if we interpret the multiplication $i \times c$, where c is real, as a rotation about O of the segment Oc through one right angle, Oc is rotated onto OY; another multiplication by i, namely $i \times i \times c$, rotates Oc through another right angle, and hence the total effect is to rotate Oc through two right angles, so that $+Oc$ becomes $-Oc$. As an operation, multiplication by $i \times i$ has the same effect as multiplication by -1; multiplication by i has the same effect as a rotation through a right angle, and these interpretations (as we have just seen) are consistent. If we like we may now write $i \times i = -1$, in operations, or $i^2 = -1$; so that the operation of rotation through a right angle is symbolized by $\sqrt{-1}$.

All this of course proves nothing. It is not meant to prove anything. *There is nothing to be proved*; we *assign* to the symbols and operations of algebra *any meanings whatever* that will lead to consistency. Although the *interpretation* by means of rotations *proves* nothing, it may suggest that there is no occasion for anyone to muddle himself into a state of mystic wonderment over nothing about the grossly misnamed "imaginaries." For further details we must refer to almost any schoolbook on elementary algebra.

Gauss thought the theorem that every algebraic equation has a root in the sense just explained so important that he gave four distinct proofs, the last when he was seventy years old. Today some would transfer the theorem from algebra (which restricts itself to processes that can be carried through in a finite number of steps) to analysis. Even Gauss *assumed* that the graph of a polynomial is a continuous curve and that if the polynomial is of odd degree the graph must cross the axis at least once. To any beginner in algebra this is obvious. But today it is *not obvious* without proof, and attempts to prove it again lead to the difficulties connected with continuity and the infinite. The roots of so simple an equation as $x^2 - 2 = 0$ cannot be computed exactly in any finite number of steps. We proceed now to the *Disquisitiones Arithmeticae*.

The *Disquisitiones* was the first of Gauss' masterpieces and by some considered his greatest. It was his farewell to pure mathematics as an exclusive interest. After its publication in 1801 (Gauss was then twenty four), he broadened his activity to include astronomy, geodesy, and electromagnetism in both their mathematical and practical aspects. But arithmetic was his first love, and he regretted in later life that he had never found the time to write the second volume he had planned as a young man. The book is in seven "sections." There was to have been an eighth, but this was omitted to keep down the cost of printing.

The opening sentence of the preface describes the general scope of the book. "The researches contained in this work appertain to that part of mathematics which is concerned with integral numbers, also fractions, surds [irrationals] being always excluded."

The first three sections treat the theory of congruences and give in particular an exhaustive discussion of the binomial *congruence* $x^n \equiv A$ (mod p), where the given integers n, A are arbitrary and p is prime; the unknown integer is x. This beautiful *arithmetical* theory has many resemblances to the corresponding *algebraic* theory of the binomial *equation* $x^n = A$, but in its peculiarly arithmetical parts is incomparably richer and more difficult than the algebra which offers no analogies to the arithmetic.

In the fourth section Gauss develops the theory of quadratic residues. Here is found the first published *proof* of the law of quadratic reciprocity. The proof is by an amazing application of mathematical induction and is as tough a specimen of that ingenious logic as will be found anywhere.

With the fifth section the theory of *binary quadratic forms* from the arithmetical point of view enters, to be accompanied presently by a discussion of *ternary* quadratic forms which are found to be necessary for the completion of the binary theory. The law of quadratic reciprocity plays a fundamental part in these difficult enterprises. For the first forms named the general problem is to discuss the solution in integers x, y of the indeterminate equation

$$ax^2 + 2bxy + cy^2 = m,$$

where a, b, c, m are any given integers; for the second, the integer solutions x, y, z of

$$ax^2 + 2bxy + cy^2 + 2dxz + 2eyz + fz^2 = m,$$

where a, b, c, d, e, f, m, are any given intergers, are the subject of investigation. An easy-looking but hard question in this field is to impose necessary and sufficient restrictions upon a, c, f, m which will ensure the existence of a solution in integers x, y, z of the indeterminate equation

$$ax^2 + cy^2 + fz^2 = m.$$

The sixth section applies the preceding theory to various special cases, for example the integer solutions x, y of $mx^2 + ny^2 = A$, where m, n, A are any given integers.

In the seventh and last section, which many consider the crown of the work, Gauss applies the preceding developments, particularly the theory of binomial congruences, to a wonderful discussion of the algebraic equation $x^n = 1$, where n is any given integer, weaving together arithmetic, algebra, and geometry into one perfect pattern. The equation $x^n = 1$ is the *algebraic* formulation of the geometric problem to construct a regular polygon of n sides, or to divide the circumference of a circle into n equal parts (consult any secondary text book on algebra or trigonometry); the *arithmetical congruence* $x^m \equiv 1 \pmod{p}$, where m, p are given integers, and p is prime, is the thread which runs through the algebra and the geometry and gives the pattern its simple meaning. This flawless work of art is accessible to any student who has had the usual algebra offered in school, but the *Disquisitiones* is not recommended for beginners (Gauss' concise presentation has been reworked by later writers into a more readily assimilable form).

Many parts of all this had been done otherwise before—by Fermat, Euler, Lagrange, Legendre and others; but Gauss treated the whole from his individual point of view, added much of his own, and deduced the isolated results of his predecessors from his general formulations and solutions of the relevant problems. For example, Fermat's beautiful result that every prime of the form $4n + 1$ is a sum of two squares, and is such a sum in only one way, which Fermat proved by his difficult method of "infinite descent," falls out naturally from Gauss' general discussion of binary quadratic forms.

"The *Disquisitiones Arithmeticae* have passed into history," Gauss said in his old age, and he was right. A new direction was given to the higher arithmetic with the publication of the *Disquisitiones,* and the theory of numbers, which in the seventeenth and eighteenth centuries had been a miscellaneous aggregation of disconnected special results, assumed coherence and rose to the dignity of a mathematical science on a par with algebra, analysis, and geometry.

The work itself has been called a "book of seven seals." It is hard reading, even for experts, but the treasures it contains (and partly conceals) in its concise, synthetic demonstrations are now available to all who wish to share them, largely the result of the labors of Gauss' friend and disciple Peter Gustav Lejeune Dirichlet (1805–1859), who first broke the seven seals.

Competent judges recognized the masterpiece for what it was immediately. Legendre [4] at first may have been inclined to think that Gauss had done him but scant justice. But in the preface to the second edition of his

---

[4] Adrien-Marie Legendre (1752–1833). Considerations of space preclude an account of his life; much of his best work was absorbed or circumvented by younger mathematicians.

own treatise on the theory of numbers (1808), which in large part was superseded by the *Disquisitiones,* he is enthusiastic. Lagrange also praised unstintedly. Writing to Gauss on May 31, 1804 he says "Your *Disquisitiones* have raised you at once to the rank of the first mathematicians, and I regard the last section as containing the most beautiful analytical discovery that has been made for a long time. . . . Believe, sir, that no one applauds your success more sincerely than I."

Hampered by the classic perfection of its style the *Disquisitiones* was somewhat slow of assimilation, and when finally gifted young men began studying the work deeply they were unable to purchase copies, owing to the bankruptcy of a bookseller. Even Eisenstein, Gauss' favorite disciple, never owned a copy. Dirichlet was more fortunate. His copy accompanied him on all his travels, and he slept with it under his pillow. Before going to bed he would struggle with some tough paragraph in the hope—frequently fulfilled—that he would wake up in the night to find that a re-reading made everything clear. To Dirichlet is due the marvellous theorem that every arithmetical progression

$$a, a + b, a + 2b, a + 3b, a + 4b, \ . \ . \ . \ ,$$

in which a, b are integers with no common divisor greater than 1, contains an infinity of primes. This was proved by analysis, in itself a miracle, for the theorem concerns integers, whereas analysis deals with the *continuous,* the *non-integral.*

Dirichlet did much more in mathematics than his amplification of the *Disquisitiones,* but we shall not have space to discuss his life. Neither shall we have space (unfortunately) for Eisenstein, one of the brilliant young men of the early nineteenth century who died before their time and, what is incomprehensible to most mathematicians, as the man of whom Gauss is reported to have said, "There have been but three epoch-making mathematicians, Archimedes, Newton, and Eisenstein." If Gauss ever did say this (it is impossible to check) it deserves attention merely because he said it, and he was a man who did not speak hastily.

Before leaving this field of Gauss' activities we may ask why he never tackled Fermat's Last Theorem. He gives the answer himself. The Paris Academy in 1816 proposed the proof (or disproof) of the theorem as its prize problem for the period 1816–18. Writing from Bremen on March 7, 1816, Olbers tries to entice Gauss into competing: "It seems right to me, dear Gauss, that you should get busy about this."

But "dear Gauss" resisted the tempter. Replying two weeks later he states his opinion of Fermat's Last Theorem. "I am very much obliged for your news concerning the Paris prize. But I confess that Fermat's Theorem as an isolated proposition has very little interest for me, because

I could easily lay down a multitude of such propositions, which one could neither prove nor dispose of."

Gauss goes on to say that the question has induced him to recall some of his old ideas for a great extension of the higher arithmetic. This doubtless refers to the theory of algebraic numbers (described in later chapters) which Kummer, Dedekind, and Kronecker were to develop independently. But the theory Gauss has in mind is one of those things, he declares, where it is impossible to foresee what progress shall be made toward a distant goal that is only dimly seen through the darkness. For success in such a difficult search one's lucky star must be in the ascendency, and Gauss' circumstances are now such that, what with his numerous distracting occupations, he is unable to give himself up to such meditations, as he did "in the fortunate years 1796–1798 when I shaped the main points of the *Disquisitiones Arithmeticae*. Still I am convinced that if I am as lucky as I dare hope, and if I succeed in taking some of the principal steps in that theory, then Fermat's Theorem will appear as only one of the least interesting corollaries."

Probably all mathematicians today regret that Gauss was deflected from his march through the darkness by "a couple of clods of dirt which we call planets"—his own words—which shone out unexpectedly in the night sky and led him astray. Lesser mathematicians than Gauss—Laplace for instance—might have done all that Gauss did in computing the orbits of Ceres and Pallas, even if the problem was of a sort which Newton said belonged to the most difficult in mathematical astronomy. But the brilliant success of Gauss in these matters brought him instant recognition as the first mathematician in Europe and thereby won him a comfortable position where he could work in comparative peace; so perhaps those wretched lumps of dirt were after all his lucky stars.

The second great stage in Gauss' career began on the first day of the nineteenth century, also a red-letter day in the histories of philosophy and astronomy. Since 1781 when Sir William Herschel (1738–1822) discovered the planet Uranus, thus bringing the number of planets then known up to the philosophically satisfying seven, astronomers had been diligently searching the heavens for further members of the Sun's family, whose existence was to be expected, according to Bode's law, between the orbits of Mars and Jupiter. The search was fruitless till Giuseppe Piazzi (1746–1826) of Palermo, on the opening day of the nineteenth century, observed what he at first mistook for a small comet approaching the Sun, but which was presently recognized as a new planet—later named Ceres, the first of the swarm of minor planets known today.

By one of the most ironic verdicts ever delivered in the agelong litigation of fact versus speculation, the discovery of Ceres coincided with the publication by the famous philosopher Georg Wilhelm Friedrich Hegel (1770–1831) of a sarcastic attack on astronomers for presuming to search for an eighth planet. Would they but pay some attention to philosophy, Hegel asserted, they must see immediately that there can be precisely seven planets, no more, no less. Their search therefore was a stupid waste of time. Doubtless this slight lapse on Hegel's part has been satisfactorily explained by his disciples, but they have not yet talked away the hundreds of minor planets which mock his Jovian ban.

It will be of interest here to quote what Gauss thought of philosophers who busy themselves with scientific matters they have not understood. This holds in particular for philosophers who peck at the foundations of mathematics without having first sharpened their dull beaks on some hard mathematics. Conversely, it suggests why Bertrand A. W. Russell (1872–   ), Alfred North Whitehead (1861–1947) and David Hilbert (1862–1943) in our own times have made outstanding contributions to the philosophy of mathematics: these men are mathematicians.

Writing to his friend Schumacher on November 1, 1844, Gauss says "You see the same sort of thing [mathematical incompetence] in the contemporary philosophers Schelling, Hegel, Nees von Essenbeck, and their followers; don't they make your hair stand on end with their definitions? Read in the history of ancient philosophy what the big men of that day—Plato and others (I except Aristotle)—gave in the way of explanations. But even with Kant himself it is often not much better; in my opinion his distinction between analytic and synthetic propositions is one of those things that either run out in a triviality or are false." When he wrote this (1844) Gauss had long been in full possession of non-Euclidean geometry, itself a sufficient refutation of some of the things Kant said about "space" and geometry, and he may have been unduly scornful.

It must not be inferred from this isolated example concerning purely mathematical technicalities that Gauss had no appreciation of philosophy. He had. All philosophical advances had a great charm for him, although he often disapproved of the means by which they had been attained. "There are problems," he said once, "to whose solution I would attach an infinitely greater importance than to those of mathematics, for example touching ethics, or our relation to God, or concerning our destiny and our future; but their solution lies wholly beyond us and completely outside the province of science."

Ceres was a disaster for mathematics. To understand why she was taken with such devastating seriousness by Gauss we must remember that the colossal figure of Newton—dead for more than seventy years—still overshadowed mathematics in 1801. The "great" mathematicians of the time were those who, like Laplace, toiled to complete the Newtonian edifice of celestial mechanics. Mathematics was still confused with mathematical physics—such as it was then—and mathematical astronomy. The vision of mathematics as an autonomous science which Archimedes saw in the third century before Christ had been lost sight of in the blaze of Newton's splendor, and it was not until the youthful Gauss again caught the vision that mathematics was acknowledged as a science whose first duty is to itself. But that insignificant clod of dirt, the minor planet Ceres, seduced his unparalleled intellect when he was twenty four years of age, just as he was getting well into his stride in those untravelled wildernesses which were to become the empire of modern mathematics.

Ceres was not alone to blame. The magnificent gift for mental arithmetic whose empirical discoveries had given mathematics the *Disquisitiones Arithmeticae* also played a fatal part in the tragedy. His friends and his father, too, were impatient with the young Gauss for not finding some lucrative position now that the Duke had educated him and, having no conception of the nature of the work which made the young man a silent recluse, thought him deranged. Here now at the dawn of the new century, the opportunity which Gauss had lacked was thrust at him.

A new planet had been discovered in a position which made it extraordinarily difficult of observation. To compute an orbit from the meager data available was a task which might have exercised Laplace himself. Newton had declared that such problems are among the most difficult in mathematical astronomy. The mere arithmetic necessary to establish an orbit with accuracy sufficient to ensure that Ceres on her whirl round the sun should not be lost to telescopes might well deter an electrically-driven calculating machine even today; but to the young man whose inhuman memory enabled him to dispense with a table of logarithms when he was hard pressed or too lazy to reach for one, all this endless arithmetic —*logistica*, not *arithmetica*—was the sport of an infant.

Why not indulge his dear vice, calculate as he had never calculated before, produce the difficult orbit to the sincere delight and wonderment of the dictators of mathematical fashion and thus make it possible, a year hence, for patient astronomers to rediscover Ceres in the place where the Newtonian law of gravitation decreed that she *must* be found—*if* the law were indeed a law of nature? Why not do all this, turn his back on the insubstantial vision of Archimedes and forget his own unsurpassed discoveries which lay waiting for development in his diary? Why not, in short, be popular? The Duke's generosity, always ungrudged, had nevertheless wounded the young man's pride in its most secret place; honor, recognition, acceptance as a "great" mathematician in the fashion of the time with its probable sequel of financial independence—all these were now within his easy reach. Gauss, the mathematical god of all time, stretched forth his hand and plucked the Dead Sea fruits of a cheap fame in his own young generation.

For nearly twenty years the sublime dreams whose fugitive glimpses the boyish Gauss had pictured with unrestrained joy in his diary lay cold and all but forgotten. Ceres was rediscovered, precisely where the marvellously ingenious and detailed calculations of the young Gauss had predicted she must be found. Pallas, Vesta, and Juno, insignificant sister planets of the diminutive Ceres were quickly picked up by prying telescopes defying Hegel, and their orbits, too, were found to conform to the inspired calculations of Gauss. Computations which would have taken Euler three days to perform—one such is sometimes said to have blinded him—were now the simple exercises of a few laborious hours. Gauss had prescribed the *method,* the routine. The major part of his own time for nearly twenty years was devoted to astronomical calculations.

But even such deadening work as this could not sterilize the creative genius of a Gauss. In 1809 he published his second masterpiece, *Theoria motus corporum coelestium in sectionibus conicis solem ambientium* (Theory of the Motion of the Heavenly Bodies Revolving round the Sun in Conic Sections), in which an exhaustive discussion of the determination of planetary and cometary orbits from observational data, including the difficult analysis of perturbations, lays down the law which for many years is to dominate computational and practical astronomy. It was great work, but not as great as Gauss was easily capable of had he developed the hints lying neglected in his diary. No essentially new discovery was added to *mathematics* by the *Theoria motus.*

Recognition came with spectacular promptness after the rediscovery of Ceres. Laplace hailed the young mathematician at once as an equal and presently as a superior. Some time later when the Baron Alexander von Humboldt (1769–1859), the famous traveller and amateur of the sciences, asked Laplace who was the greatest mathematician in Germany, Laplace replied "Pfaff." "But what about Gauss?" the astonished von Humboldt asked, as he was backing Gauss for the position of director at the Göttingen observatory. "Oh," said Laplace, "Gauss is the greatest mathematician in the world."

The decade following the Ceres episode was rich in both happiness and sorrow for Gauss. He was not without detractors even at that early stage of his career. Eminent men who had the ear of the polite public ridiculed the young man of twenty four for wasting his time on so useless a pastime as the computation of a minor planet's orbit. Ceres might be the goddess of the fields, but it was obvious to the merry wits that no corn grown on the new planet would ever find its way into the Brunswick market of a Saturday afternoon. No doubt they were right, but they also ridiculed him in the same way thirty years later when he laid the foundations of the mathematical theory of electromagnetism and invented the electric telegraph. Gauss let them enjoy their jests. He never replied publicly, but in private expressed his regret that men of honor and priests of science could stultify themselves by being so petty. In the meantime he went on with his work, grateful for the honors the learned societies of Europe showered on him but not going out of his way to invite them.

The Duke of Brunswick increased the young man's pension and made it possible for him to marry (October 9, 1805) at the age of twenty eight. The lady was Johanne Osthof of Brunswick. Writing to his old university friend, Wolfgang Bolyai, three days after he became engaged, Gauss expresses his unbelievable happiness. "Life stands still before me like an eternal spring with new and brilliant colors."

Three children were born of this marriage: Joseph, Minna, and Louis, the first of whom is said to have inherited his father's gift for mental calculations. Johanne died on October 11, 1809, after the birth of Louis, leaving her young husband desolate. His eternal spring turned to winter. Although he married again the following year (August 4, 1810) for the sake of his young children it was long before Gauss could speak without emotion of his first wife. By the second wife, Minna Waldeck, who had been a close friend of the first, he had two sons and a daughter.

According to gossip Gauss did not get on well with his sons, except possibly the gifted Joseph who never gave his father any trouble. Two are said to have run away from home and gone to the United States. As one of these sons is said to have left numerous descendants still living in America, it is impossible to say anything further here, except that one of the American sons became a prosperous merchant in St. Louis in the days of the river boats; both first were farmers in Missouri. With his daughters Gauss was always happy. An exactly contrary legend (vouched for forty years ago by old people whose memories of the Gauss family might be considered trustworthy) to that about the sons asserts that Gauss was never anything but kind to his boys, some of whom were rather wild and caused their distracted father endless anxiety. One would think that the memory of his own father would have made Gauss sympathetic with his sons.

In 1808 Gauss lost his father. Two years previously he had suffered an even severer loss in the death of his benefactor under tragic circumstances.

The Duke Ferdinand was not only an enlightened patron of learning and a kindly ruler but a first-rate soldier as well who had won the warm praise of Frederick the Great for his bravery and military brilliance in the Seven Years' War (1756–1763).

At the age of seventy Ferdinand was put in command of the Prussian forces in a desperate attempt to halt the French under Napoleon, after the Duke's mission to St. Petersburg in an effort to enlist the aid of Russia for Germany had failed. The battle of Austerlitz (December 2, 1805) was already history and Prussia found itself forsaken in the face of overwhelming odds. Ferdinand faced the French on their march toward the Saale at Auerstedt and Jena, was disastrously defeated and himself mortally wounded. He turned homeward.

Napoleon the Great here steps on the stage in person at his potbellied greatest. At the time of Ferdinand's defeat Napoleon was quartered at Halle. A deputation from Brunswick waited on the victorious Emperor of all the French to implore his generosity for the brave old man he had defeated. Would the mighty Emperor stretch a point of military etiquette and let his broken enemy die in peace by his own fireside? The Duke, they assured him, was no longer dangerous. He was dying.

It was the wrong time of the month and Napoleon was enjoying one of his womanish tantrums. He not only refused but did so with quite vulgar and unnecessary brutality. Revealing the true measure of himself as a man, Napoleon pointed his refusal with a superfluous vilification of his honorable opponent and a hysterical ridicule of the dying man's abilities as a soldier. There was nothing for the humiliated deputation to do but to try to save their gentle ruler from the disgrace of a death in prison. It does not seem surprising that these same Germans some nine years later fought like methodical devils at Waterloo and helped to topple the Emperor of the French into the ditch.

Gauss at the time was living in Brunswick. His house was on the main highway. One morning in late autumn he saw a hospital wagon hastening by. In it lay the dying Duke on his flight to Altona. With an emotion too deep for words Gauss saw the man who had been more than his own father to him hurried away to die in hiding like a hounded criminal. He said nothing then and but little afterwards, but his friends noticed that his reserve deepened and his always serious nature became more serious. Like Descartes in his earlier years Gauss had a horror of death, and all his life the passing of a close friend chilled him with a quiet, oppressive dread. Gauss was too vital to die or to witness death. The Duke died in his father's house in Altona on November 10, 1806.

His generous patron dead, it became necessary for Gauss to find some reliable livelihood to support his family. There was no difficulty about this as the young mathematician's fame had now spread to the farthest corners of Europe. St. Petersburg had been angling for him as the logical successor of Euler who had never been worthily replaced after his death in 1783. In 1807 a definite and flattering offer was tendered Gauss. Alexander von Humboldt and other influential friends, reluctant to see Germany lose the greatest mathematician in the world, bestirred themselves, and Gauss was appointed director of the Göttingen Observatory with the privilege—and duty, when necessary—of lecturing on mathematics to university students.

Gauss no doubt might have obtained a professorship of mathematics but he preferred the observatory as it offered better prospects for uninterrupted research. Although it may be too strong to say that Gauss hated teaching, the instruction of ordinary students gave him no pleasure, and it was only when a real mathematician sought him out that Gauss, sitting at a table with his students, let himself go and disclosed the secrets of his methods in his perfectly prepared lessons. But such incentives were regrettably rare and for the moşt part the students who took up Gauss' priceless time had better have been doing something other than mathematics. Writing in 1810 to his intimate friend the astronomer and mathematician Friedrich Wilhelm Bessel (1784–1846), Gauss says "This winter I am giving two courses of lectures to three students, of whom one is only moderately prepared, the other less than moderately, and the third lacks both preparation and ability. Such are the burdens of a mathematical calling."

The salary which Göttingen could afford to pay Gauss at the time—the French were then busy pillaging Germany in the interests of good government for the Germans by the French—was modest but sufficient for the simple needs of Gauss and his family. Luxury never attracted the Prince of Mathematicians whose life had been unaffectedly dedicated to science long before he was twenty. As his friend Sartorius von Waltershausen writes, "As he was in his youth, so he remained through his old age to his dying day, the unaffectedly simple Gauss. A small study, a little work table with a green cover, a standing-desk painted white, a narrow sopha and, after his seventieth year, an arm chair, a shaded lamp, an unheated bedroom, plain food, a dressing gown and a velvet cap, these were so becomingly all his needs."

If Gauss was simple and thrifty the French invaders of Germany in 1807 were simpler and thriftier. To govern Germany according to their ideas the victors of Auerstedt and Jena fined the losers for more than the traffic would bear. As professor and astronomer at Göttingen Gauss was rated by the extortionists to be good for an involuntary contribution of 2,000 francs to the Napoleonic war chest. This exorbitant sum was quite beyond Gauss' ability to pay.

Presently Gauss got a letter from his astronomical friend Olbers enclosing the amount of the fine and expressing indignation that a scholar should be subjected to such petty extortion. Thanking his generous friend for his sympathy, Gauss declined the money and sent it back at once to the donor.

Not all the French were as thrifty as Napoleon. Shortly after returning

Olbers' money Gauss received a friendly little note from Laplace telling
him that the famous French mathematician had paid the 2,000-franc fine
for the greatest mathematician in the world and had considered it an
honor to be able to lift this unmerited burden from his friend's shoulders.
As Laplace had paid the fine in Paris, Gauss was unable to return him
the money. Nevertheless he declined to accept Laplace's help. An unexpected (and unsolicited) windfall was presently to enable him to repay
Laplace with interest at the current market rate. Word must have got
about that Gauss disdained charity. The next attempt to help him
succeeded. An admirer in Frankfurt sent 1,000 guilders anonymously.
As Gauss could not trace the sender he was forced to accept the
gift.

The death of his friend Ferdinand, the wretched state of Germany
under French looting, financial straits, and the loss of his first wife all
did their part toward upsetting Gauss' health and making his life miserable in his early thirties. Nor did a constitutional predisposition to hypo-
chondria, aggravated by incessant overwork, help matters. His unhappiness
was never shared with his friends, to whom he is always the serene correspondent, but is confided—only once—to a private mathematical manuscript. After his appointment to the directorship at Göttingen in 1807
Gauss returned occasionally for three years to one of the great things
noted in his diary. In a manuscript on elliptic functions purely scientific
matters are suddenly interrupted by the finely pencilled words "Death
were dearer to me than such a life." His work became his drug.

The years 1811–12 (Gauss was thirty four in 1811) were brighter.
With a wife again to care for his young children Gauss began to have
some peace. Then, almost exactly a year after his second marriage, the
great comet of 1811, first observed by Gauss deep in the evening twilight
of August 22, blazed up unannounced. Here was a worthy foe to test the
weapons Gauss had invented to subjugate the minor planets.

His weapons proved adequate. While the superstitious peoples of
Europe, following the blazing spectacle with awestruck eyes as the comet
unlimbered its flaming scimitar in its approach to the Sun, saw in the
fiery blade a sharp warning from Heaven that the King of Kings was
wroth with Napoleon and weary of the ruthless tyrant, Gauss had the
satisfaction of seeing the comet follow the path he had quickly calculated
for it to the last decimal. The following year the credulous also saw their
own prediction verified in the burning of Moscow and the destruction of
Napoleon's Grand Army on the icy plains of Russia.

This is one of those rare instances where the popular explanation fits
the facts and leads to more important consequences than the scientific.
Napoleon himself had a basely credulous mind—he relied on "hunches,"
reconciled his wholesale slaughters with a childlike faith in a beneficent,

inscrutable Providence, and believed himself a Man of Destiny. It is not impossible that the celestial spectacle of a harmless comet flaunting its gorgeous tail across the sky left its impress on the subconscious mind of a man like Napoleon and fuddled his judgment. The almost superstitious reverence of such a man for mathematics and mathematicians is no great credit to either, although it has been frequently cited as one of the main justifications for both.

Beyond a rather crass appreciation of the value of mathematics in military affairs, where its utility is obvious even to a blind idiot, Napoleon had no conception of what mathematics as practised by masters like his contemporaries, Lagrange, Laplace, and Gauss, is all about. A quick student of trivial, elementary mathematics at school, Napoleon turned to other things too early to certify his promise and, mathematically, never grew up. Although it seems incredible that a man of Napoleon's demonstrated capacity could so grossly underestimate the difficulties of matters beyond his comprehension as to patronize Laplace, it is a fact that he had the ludicrous audacity to assure the author of the *Mécanique céleste* that he would read the book the *first free month* he could find. Newton and Gauss might have been equal to the task; Napoleon no doubt could have turned the pages in his month without greatly tiring himself.

It is a satisfaction to record that Gauss was too proud to prostitute mathematics to Napoleon the Great by appealing to the Emperor's vanity and begging him in the name of his notorious respect for all things mathematical to remit the 2,000-franc fine, as some of Gauss' mistaken friends urged him to do. Napoleon would probably have been flattered to exercise his clemency. But Gauss could not forget Ferdinand's death, and he felt that both he and the mathematics he worshipped were better off without the condescension of a Napoleon.

No sharper contrast between the mathematician and the military genius can be found than that afforded by their respective attitudes to a broken enemy. We have seen how Napoleon treated Ferdinand. When Napoleon fell Gauss did not exult. Calmly and with a detached interest he read everything he could find about Napoleon's life and did his best to understand the workings of a mind like Napoleon's. The effort even gave him considerable amusement. Gauss had a keen sense of humor, and the blunt realism which he had inherited from his hard-working peasant ancestors also made it easy for him to smile at heroics.

The year 1811 might have been a landmark in mathematics comparable to 1801—the year in which the *Disquisitiones Arithmeticae* appeared—had Gauss made public a discovery he confided to Bessel. Having thoroughly understood complex numbers and their geometrical representation as points on the plane of analytic geometry, Gauss proposed himself the problem of investigating what are today called *analytic functions* of such numbers.

The complex number $x + iy$, where i denotes $\sqrt{-1}$, represents the point (x,y). For brevity $x + iy$ will be denoted by the single letter z. As x, y independently take on real values in any prescribed continuous manner, the point z wanders about over the plane, obviously not at random but in a manner determined by that in which x, y assume their values. Any expression containing z, such as z^2, or $1/z$, etc., which takes on a *single* definite value when a value is assigned to z, is called a *uniform function* of z. We shall denote such a function by $f(z)$. Thus if $f(z)$ is the particular function z^2, so that here $f(z) = (x + iy)^2 = x^2 + 2ixy + i^2y^2, = x^2 - y^2 + 2ixy$ (because $i^2 = -1$), it is clear that when any value is assigned to z, namely to $x + iy$, for example $x = 2$, $y = 3$, so that $z = 2 + 3i$, precisely one value of this $f(z)$ is thereby determined; here, for $z = 2 + 3i$ we get $z^2 = -5 + 12i$.

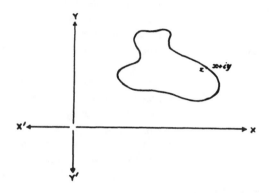

FIGURE 2

Not all uniform functions $f(z)$ are studied in the theory of functions of a complex variable; the *monogenic* functions are singled out for exhaustive discussion. The reason for this will be stated after we have described what "monogenic" means.

Let z move to another position, say to z'. The function $f(z)$ takes on another value, $f(z')$, obtained by substituting z' for z. The *difference* $f(z') - f(z)$ of the new and old values of the function is now divided by the difference of the new and old values of the variable, thus $[f(z') - f(z)]/(z' - z)$, and, precisely as is done in calculating the slope of a graph to find the derivative of the function the graph represents, we here let z' approach z indefinitely, so that $f(z')$ approaches $f(z)$ simultaneously. But here a remarkable new phenomenon appears.

There is not here a unique way in which z' can move into coincidence with z, for z' may wander about all over the plane of complex numbers by any of an infinity of different paths before coming into coincidence with z. We should not expect the limiting value of $[f(z') - f(z)]/(z' - z)$ when z' coincides with z to be *the same* for *all* of these paths, and in general it is *not*. But *if* $f(z)$ is such that the limiting value just described *is* the same for *all* paths by which z' moves into coincidence with z, then $f(z)$ is said to be monogenic at z (or at the point representing z). *Uniformity* (previously described) and *monogenicity* are distinguishing features of *analytic* functions of a complex variable.

Some idea of the importance of analytic functions can be inferred from the fact that vast tracts of the theories of fluid motion (also of mathematical electricity and representation by maps which do not distort angles) are naturally handled by the theory of *analytic* functions of a complex variable. Suppose such a function $f(z)$ is separated into its "real" part (that which does not contain the "imaginary unit" i) and its "imaginary" part, say $f(z) = U + iV$. For the special analytic function z^2 we have $U = x^2 - y^2$, $V = 2xy$. Imagine a film of fluid streaming over a plane. If the motion of the fluid is without vortices, a stream line of the motion is obtainable from *some* analytic function $f(z)$ by plotting the curve $U = a$, in which a is any real number, and likewise the equipotential lines are obtainable from $V = b$ (b any real number). Letting a, b range, we thus get a complete picture of the motion for as large an area as we wish. For a given situation, say that of a fluid streaming around an obstacle, the hard part of the problem is to find what analytic function to choose, and the whole matter has been gone at largely backwards: the simple analytic functions have been investigated and the physical problems which they fit have been sought. Curiously enough, many of these artificially prepared problems have proven of the greatest service in aerodynamics and other practical applications of the theory of fluid motion.

The theory of analytic functions of a complex variable was one of the greatest fields of mathematical triumphs in the nineteenth century. Gauss in his letter to Bessel states what amounts to the fundamental theorem in this vast theory, but he hid it away to be rediscovered by Cauchy and later Weierstrass. As this is a landmark in the history of mathematical analysis we shall briefly describe it, omitting all refinements that would be demanded in an exact formulation.

Imagine the complex variable z tracing out a closed curve of finite length without loops or kinks. We have an intuitive notion of what we mean by the "length" of a piece of this curve.

Mark n points $P_1, P_2, \ldots, P_n$ on the curve so that each of the pieces $P_1P_2, P_2P_3, P_3P_4, \ldots, P_nP_1$ is not greater than some preassigned finite length l. On each of these pieces choose a point, not at either end of the piece; form the value of $f(z)$ for the value of z corresponding to the point, and multiply this value by the length of the piece in which the point lies.

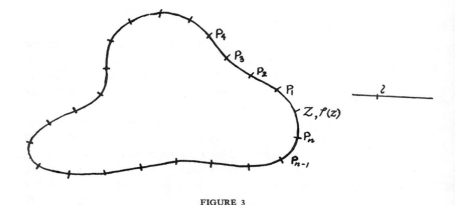

FIGURE 3

Do the like for *all* the pieces, and add the results. Finally take the limiting value of this sum as the number of pieces is indefinitely increased. This gives the *"line integral"* of $f(z)$ for the curve.

When will this line integral be zero? In order that the line integral shall be zero it is sufficient that $f(z)$ be *analytic* (uniform and monogenic) at every point z on the curve and inside the curve.

Such is the great theorem which Gauss communicated to Bessel in 1811 and which, with another theorem of a similar kind, in the hands of Cauchy who rediscovered it independently, was to yield many of the important results of analysis as corollaries.

Astronomy did not absorb the whole of Gauss' prodigious energies in his middle thirties. The year 1812, which saw Napoleon's Grand Army fighting a desperate rear-guard action across the frozen plains, witnessed the publication of another great work by Gauss, that on the *hypergeometric series*

$$1 + \frac{ab}{c}x + \frac{a(a+1)b(b+1)x^2}{c(c+1)1\times 2} + \dots,$$

the dots meaning that the series continues indefinitely according to the law indicated; the next term is

$$\frac{a(a+1)(a+2)b(b+1)(b+2)}{c(c+1)(c+2)} \frac{x^3}{1\times 2\times 3}.$$

This memoir is another landmark. As has already been noted Gauss was the first of the modern rigorists. In this work he determined the restrictions that must be imposed on the numbers a, b, c, x in order that the series shall converge (in the sense explained earlier in this chapter). The series itself was no mere textbook exercise that may be investigated to gain skill in analytical manipulations and then be forgotten. It includes as special cases—obtained by assigning specific values to one or more of *a, b, c, x*—many of the most important series in analysis, for example those by which logarithms, the trigonometric functions, and several of the functions that turn up repeatedly in Newtonian astronomy and mathematical physics are calculated and tabulated; the general binomial theorem also is a special case. By disposing of this series in its general form Gauss slew a multitude at one smash. From this work developed many applications to the differential equations of physics in the nineteenth century.

The choice of such an investigation for a serious effort is characteristic of Gauss. He never published trivialities. When he put out anything it was not only finished in itself but was also so crammed with ideas that his successors were enabled to apply what Gauss had invented to new problems. Although limitations of space forbid discussion of the many instances of this fundamental character of Gauss' contributions to pure mathematics, one cannot be passed over in even the briefest sketch: the work on the law of biquadratic reciprocity. The importance of this was that it gave a new and totally unforeseen direction to the higher arithmetic.

Having disposed of *quadratic* (second degree) reciprocity, it was natural for Gauss to consider the general question of binomial congruences of any degree. If *m* is a given integer not divisible by the prime *p*, and if *n* is a given positive integer, and if further an integer *x* can be found such that $x^n \equiv m$ (mod *p*), *m* is called an *n-ic residue* of *p*; when $n = 4$, *m* is a *biquadratic residue* of *p*.

The case of *quadratic* binomial congruences ($n = 2$) suggests but little to do when *n* exceeds 2. One of the matters Gauss was to have included in the discarded eighth section (or possibly, as he told Sophie Germain, in the projected but unachieved second volume) of the *Disquisitiones Arithmeticae* was a discussion of these higher congruences and a search for the corresponding laws of reciprocity, namely the interconnections (as to solvability or non-solvability) of the pair $x^n \equiv p$ (mod *q*), $x^n \equiv q$ (mod *p*), where *p, q* are rational primes. In particular the cases $n = 3$, $n = 4$ were to have been investigated.

The memoir of 1825 breaks new ground with all the boldness of the great pioneers. After many false starts which led to intolerable complexity Gauss discovered the "natural" way to the heart of his problem. The *rational* integers 1, 2, 3, . . . are *not* those appropriate to the statement of the law of *biquadratic* reciprocity, as they are for *quadratic*; a totally new species of *integers* must be invented. These are called the *Gaussian complex integers* and are all those complex numbers of the form $a + bi$ in which *a, b* are *rational integers* and *i* denotes $\sqrt{-1}$.

To state the law of biquadratic reciprocity an exhaustive preliminary discussion of the laws of arithmetical divisibility for such *complex integers* is necessary. Gauss gave this, thereby inaugurating the theory of algebraic numbers—that which he probably had in mind when he gave his estimate of Fermat's Last Theorem. For *cubic* reciprocity ($n = 3$) he also found the right way in a similar manner. His work on this was found in his posthumous papers.

The significance of this great advance will become clearer when we follow the careers of Kummer and Dedekind. For the moment it is sufficient to say that Gauss' favorite disciple, Eisenstein, disposed of cubic reciprocity. He further discovered an astonishing connection between the law of biquadratic reciprocity and certain parts of the theory of elliptic functions, in which Gauss had travelled far but had refrained disclosing what he found.

Gaussian complex *integers* are of course a subclass of *all* complex *numbers*, and it might be thought that the *algebraic* theory of *all* the numbers would yield the *arithmetical* theory of the included *integers* as a trivial detail. Such is by no means the case. Compared to the arithmetical theory the algebraic is childishly easy. Perhaps a reason why this should be so is suggested by the *rational numbers* (numbers of the form a/b, where a, b are rational integers). We can *always* divide one rational number by another and get *another* rational number: a/b divided by c/d yields the rational number ad/bc. But a rational *integer* divided by another rational integer is not always another rational integer: 7 divided by 8 gives ⅞. Hence if we must restrict ourselves to *integers*, the case of interest for the theory of numbers, we have tied our hands and hobbled our feet before we start. This is one of the reasons why the higher arithmetic is harder than algebra, higher or elementary.

Equally significant advances in geometry and the applications of mathematics to geodesy, the Newtonian theory of attraction, and electromagnetism were also to be made by Gauss. How was it possible for one man to accomplish this colossal mass of work of the highest order? With characteristic modesty Gauss declared that "If others would but reflect on mathematical truths as deeply and as continuously as I have, they would make my discoveries." Possibly. Gauss' explanation recalls Newton's. Asked how he had made discoveries in astronomy surpassing those of all his predecessors, Newton replied, "By always thinking about them." This may have been plain to Newton; it is not to ordinary mortals.

Part of the riddle of Gauss is answered by his *involuntary* preoccupation with mathematical ideas—which itself of course demands explanation. As a young man Gauss would be "seized" by mathematics. Conversing with friends he would suddenly go silent, overwhelmed by thoughts beyond his control, and stand staring rigidly oblivious of his surroundings.

Later he controlled his thoughts—or they lost their control over him—and he consciously directed all his energies to the solution of a difficulty till he succeeded. A problem once grasped was never released till he had conquered it, although several might be in the foreground of his attention simultaneously.

In one such instance (referring to the *Disquisitiones*, page 636) he relates how for four years scarcely a week passed that he did not spend some time trying to settle whether a certain sign should be plus or minus. The solution finally came of itself in a flash. But to imagine that it would have blazed out of itself like a new star without the "wasted" hours is to miss the point entirely. Often after spending days or weeks fruitlessly over some research Gauss would find on resuming work after a sleepless night that the obscurity had vanished and the whole solution shone clear in his mind. The capacity for intense and prolonged concentration was part of his secret.

In this ability to forget himself in the world of his own thoughts Gauss resembles both Archimedes and Newton. In two further respects he also measures up to them, his gifts for precise observation and a scientific inventiveness which enabled him to devise the instruments necessary for his scientific researches. To Gauss geodesy owes the invention of the heliotrope, an ingenious device by which signals could be transmitted practically instantaneously by means of reflected light. For its time the heliotrope was a long step forward. The astronomical instruments he used also received notable improvements at Gauss' hands. For use in his fundamental researches in electromagnetism Gauss invented the bifilar magnetometer. And as a final example of his mechanical ingenuity it may be recalled that Gauss in 1833 invented the electric telegraph and that he and his fellow worker Wilhelm Weber (1804–1891) used it as a matter of course in sending messages. The combination of mathematical genius with first-rate experimental ability is one of the rarest in all science.

Gauss himself cared but little for the possible practical uses of his inventions. Like Archimedes he preferred mathematics to all the kingdoms of the earth; others might gather the tangible fruits of his labors. But Weber, his collaborator in electromagnetic researches, saw clearly what the puny little telegraph of Göttingen meant for civilization. The railway, we recall, was just coming into its own in the early 1830's. "When the globe is covered with a net of railroads and telegraph wires," Weber prophesied in 1835, "this net will render services comparable to those of the nervous system in the human body, partly as a means of transport, partly as a means for the propagation of ideas and sensations with the speed of lightning."

The admiration of Gauss for Newton has already been noted. Knowing the tremendous efforts some of his own masterpieces had cost him, Gauss had a true appreciation of the long preparation and incessant meditation that went into Newton's greatest work. The story of Newton and the falling apple roused Gauss' indignation. "Silly!" he exclaimed. "Believe the story if you like, but the truth of the matter is this. A stupid, officious man asked Newton how he discovered the law of gravitation. Seeing that he had to deal with a child in intellect, and wanting to get rid of the bore, Newton answered that an apple fell and hit him on the nose. The man went away fully satisfied and completely enlightened."

The apple story has its echo in our own times. When teased as to what led him to his theory of the gravitational field Einstein replied that he asked a workman who had fallen off a building, to land unhurt on a pile of straw, whether he noticed the tug of the "force" of gravity on the way down. On being told that no force had tugged, Einstein immediately saw that "gravitation" in a sufficiently small region of space-time can be replaced by an acceleration of the observer's (the falling workman's) reference system. This story, if true, is also probably all rot. What gave Einstein his idea was the hard labor he expended for several years mastering the tensor calculus of two Italian mathematicians, Ricci and Levi-Civita, themselves disciples of Riemann and Christoffel, both of whom in their turn had been inspired by the geometrical work of Gauss.

Commenting on Archimedes, for whom he also had a boundless admiration, Gauss remarked that he could not understand how Archimedes failed to invent the decimal system of numeration or its equivalent (with some base other than 10). The thoroughly un-Greek work of Archimedes in devising a scheme for writing and dealing with numbers far beyond the capacity of the Greek symbolism had—according to Gauss—put the decimal notation with its all-important principle of place-value ($325 = 3 \times 10^2 + 2 \times 10 + 5$) in Archimedes' hands. This oversight Gauss regarded as the greatest calamity in the history of science. "To what heights would science now be raised if Archimedes had made that discovery!" he exclaimed, thinking of his own masses of arithmetical and astronomical calculations which would have been impossible, even to him, without the decimal notation. Having a full appreciation of the significance for all science of improved methods of computation, Gauss slaved over his own calculations till pages of figures were reduced to a few lines which could be taken in almost at a glance. He himself did much of his calculating mentally; the improvements were intended for those less gifted than himself.

Unlike Newton in his later years, Gauss was never attracted by the rewards of public office, although his keen interest and sagacity in all matters pertaining to the sciences of statistics, insurance, and "political arithmetic" would have made him a good minister of finance. Till his last illness he found complete satisfaction in his science and his simple recreations. Wide reading in the literatures of Europe and the classics of antiquity, a critical interest in world politics, and the mastery of foreign languages and new sciences (including botany and mineralogy) were his hobbies.

English literature especially attracted him, although its darker aspect as in Shakespeare's tragedies was too much for the great mathematician's acute sensitiveness to all forms of suffering, and he tried to pick his way through the happier masterpieces. The novels of Sir Walter Scott (who was a contemporary of Gauss) were read eagerly as they came out, but the unhappy ending of *Kenilworth* made Gauss wretched for days and he regretted having read the story. One slip of Sir Walter's tickled the mathematical astronomer into delighted laughter, "the moon rises broad in the northwest," and he went about for days correcting all the copies he could find. Historical works in English, particularly Gibbon's *Decline and Fall of the Roman Empire* and Macaulay's *History of England* gave him special pleasure.

For his meteoric young contemporary Lord Byron, Gauss had almost an aversion. Byron's posturing, his reiterated world-weariness, his affected misanthropy, and his romantic good looks had captivated the sentimental Germans even more completely than they did the stolid British who—at least the older males—thought Byron somewhat of a silly ass. Gauss saw through Byron's histrionics and disliked him. No man who guzzled good brandy and pretty women as assiduously as Byron did could be so very weary of the world as the naughty young poet with the flashing eye and the shaking hand pretended to be.

In the literature of his own country Gauss' tastes were somewhat unusual for an intellectual German. Jean Paul was his favorite German poet; Goethe and Schiller, whose lives partly overlapped his own, he did not esteem very highly. Goethe, he said, was unsatisfying. Being completely at variance with Schiller's philosophical tenets, Gauss disliked his poetry. He called *Resignation* a blasphemous, corrupt poem and wrote "Mephistopheles!" on the margin of his copy.

The facility with which he mastered languages in his youth stayed with Gauss all his life. Languages were rather more to him than a hobby. To test the plasticity of his mind as he grew older he would deliberately acquire a new language. The exercise, he believed, helped to keep his mind young. At the age of sixty two he began an intensive study of Russian without assistance from anyone. Within two years he was reading Russian prose and poetical works fluently, and carrying on his correspondence with scientific friends in St. Petersburg wholly in Russian. In the opinion of Russians who visited him in Göttingen he also spoke the language perfectly. Russian literature he put on a par with English for the pleasure it gave him. He also tried Sanskrit but disliked it.

His third hobby, world politics, absorbed an hour or so of his time every day. Visiting the literary museum regularly, he kept abreast of events by reading all the newspapers to which the museum subscribed, from the London *Times* to the Göttingen local news.

In politics the intellectual aristocrat Gauss was conservative through and through, but in no sense reactionary. His times were turbulent, both in his own country and abroad. Mob rule and acts of political violence roused in him—as his friend Von Waltershausen reports—"an indescribable horror." The Paris revolt of 1848 filled him with dismay.

The son of poor parents himself, familiar from infancy with the intelligence and morality of "the masses," Gauss remembered what he had observed, and his opinion of the intelligence, morality, and political acumen of "the people"—taken in the mass, as demagogues find and take them— was extremely low. *"Mundus vult decepi"* he believed a true saying.

This disbelief in the innate morality, integrity, and intelligence of Rousseau's "natural man" when massed into a mob or when deliberating in cabinets, parliaments, congresses, and senates, was no doubt partly inspired by Gauss' intimate knowledge, as a man of science, of what "the natural man" did to the scientists of France in the early days of the French Revolution. It may be true, as the revolutionists declared, that "the people have no need of science," but such a declaration to a man of Gauss' temperament was a challenge. Accepting the challenge, Gauss in his turn expressed his acid contempt for all "leaders of the people" who lead the people into turmoil for their own profit. As he aged he saw peace and simple contentment as the only good things in any country. Should civil war break out in Germany, he said, he would as soon be dead. Foreign conquest in the grand Napoleonic manner he looked upon as an incomprehensible madness.

These conservative sentiments were not the nostalgia of a reactionary who bids the world defy the laws of celestial mechanics and stand still in the heavens of a dead and unchanging past. Gauss believed in reforms —when they were intelligent. And if brains are not to judge when reforms are intelligent and when they are not, what organ of the human body is? Gauss had brains enough to see where the ambitions of some of the great statesmen of his own reforming generation were taking Europe. The spectacle did not inspire his confidence.

His more progressive friends ascribed Gauss' conservatism to the closeness with which he stuck to his work. This may have had something to do with it. For the last twenty seven years of his life Gauss slept away from his observatory only once, when he attended a scientific meeting in Berlin to please Alexander von Humboldt who wished to show him off. But a man does not always have to be flying about all over the map to see what is going on. Brains and the ability to read newspapers (even when they lie) and government reports (especially when they lie) are sometimes better than any amount of sightseeing and hotel lobby gossip. Gauss stayed at home, read, disbelieved most of what he read, thought, and arrived at the truth.

Another source of Gauss' strength was his scientific serenity and his freedom from personal ambition. All his ambition was for the advancement of mathematics. When rivals doubted his assertion that he had anticipated them—not stated boastfully, but as a fact germane to the matter in hand—Gauss did not exhibit his diary to prove his priority but let his statement stand on its own merits.

Legendre was the most outspoken of these doubters. One experience made him Gauss' enemy for life. In the *Theoria motus* Gauss had referred to his early discovery of the method of least squares. Legendre published the method in 1806, before Gauss. With great indignation he wrote to Gauss practically accusing him of dishonesty and complaining that Gauss, so rich in discoveries, might have had the decency not to appropriate the method of least squares, which Legendre regarded as his own ewe lamb. Laplace entered the quarrel. Whether he believed the assurances of Gauss that Legendre had indeed been anticipated by ten years or more, he does not say, but he retains his usual suavity. Gauss apparently disdained to argue the matter further. But in a letter to a friend he indicates the evidence which might have ended the dispute then and there had Gauss not been "too proud to fight." "I communicated the whole matter to Olbers in 1802," he says, and if Legendre had been inclined to doubt this he could have asked Olbers, who had the manuscript.

The dispute was most unfortunate for the subsequent development of mathematics, as Legendre passed on his unjustified suspicions to Jacobi and so prevented that dazzling young developer of the theory of elliptic functions from coming to cordial terms with Gauss. The misunderstanding was all the more regrettable because Legendre was a man of the highest character and scrupulously fair himself. It was his fate to be surpassed by more imaginative mathematicians than himself in the fields where most of his long and laborious life was spent in toil which younger men—Gauss, Abel, and Jacobi—showed to have been superfluous. At every step Gauss strode far ahead of Legendre. Yet when Legendre accused him of unfair dealing Gauss felt that he himself had been left in the lurch. Writing to Schumacher (July 30, 1806), he complains that "It seems to be my fate to concur in nearly all my theoretical works with Legendre. So it is in the higher arithmetic, in the researches in transcendental functions connected with the rectification [the process for finding the length of an arc of a curve] of the ellipse, in the foundations of geometry and now again here [in the method of least squares, which] . . . is also used in Legendre's work and indeed right gallantly carried through."

With the detailed publication of Gauss' posthumous papers and much of his correspondence in recent years all these old disputes have been settled once for all in favor of Gauss. There remains another score on which he has been criticized, his lack of cordiality in welcoming the great work of others, particularly of younger men. When Cauchy began publishing his brilliant discoveries in the theory of functions of a complex variable, Gauss ignored them. No word of praise or encouragement came from the Prince of Mathematicians to the young Frenchman. Well, why should it have come? Gauss himself (as we have seen) had reached the heart of the matter years before Cauchy started. A memoir on the theory was to have been one of Gauss' masterpieces. Again, when Hamilton's work on quaternions came to his attention in 1852, three years before his death, Gauss said nothing. Why should he have said anything? The crux of the matter lay buried in his notes of more than thirty years before. He held his peace and made no claim for priority. As in his anticipations of the theory of functions of a complex variable, elliptic functions, and non-Euclidean geometry, Gauss was content to have done the work.

The gist of quaternions is the algebra which does for rotations in space of three dimensions what the algebra of complex numbers does for rotations in a plane. But in quaternions (Gauss called them mutations) one of the fundamental rules of algebra breaks down: it is no longer true that $a \times b = b \times a$, and it is impossible to make an algebra of rotations in three dimensions in which this rule *is* preserved. Hamilton, one of the great mathematical geniuses of the nineteenth century, records with Irish exuberance how he struggled for fifteen years to invent a consistent algebra to do what was required until a happy inspiration gave him the clue that $a \times b$ is not equal to $b \times a$ in the algebra he was seeking. Gauss does not state how long it took him to reach the goal; he merely records his success in a few pages of algebra that leave no mathematics to the imagination.

If Gauss was somewhat cool in his printed expressions of appreciation he was cordial enough in his correspondence and in his scientific relations with those who sought him out in a spirit of disinterested inquiry. One of his scientific friendships is of more than mathematical interest as it shows the liberality of Gauss' views regarding women scientific workers. His broadmindedness in this respect would have been remarkable for any man of his generation; for a German it was almost without precedent.

The lady in question was Mademoiselle Sophie Germain (1776–1831) — just a year older than Gauss. She and Gauss never met, and she died (in Paris) before the University of Göttingen could confer the honorary doctor's degree which Gauss recommended to the faculty. By a curious coincidence we shall see the most celebrated woman mathematician of the nineteenth century, another Sophie, getting her degree from the same liberal University many years later after Berlin had refused her on account of her sex. Sophie appears to be a lucky name in mathematics for women —provided they affiliate with broad-minded teachers. The leading woman mathematician of our own times, Emmy Noether (1882–1935) also came from Göttingen.[5]

Sophie Germain's scientific interests embraced acoustics, the mathematical theory of elasticity, and the higher arithmetic, in all of which she did notable work. One contribution in particular to the study of Fermat's Last Theorem led in 1908 to a considerable advance in this direction by the American mathematician Leonard Eugene Dickson (1874–   ).

Entranced by the *Disquisitiones Arithmeticae*, Sophie wrote to Gauss some of her own arithmetical observations. Fearing that Gauss might be prejudiced against a woman mathematician, she assumed a man's name. Gauss formed a high opinion of the talented correspondent whom he addressed in excellent French as "Mr. Leblanc."

Leblanc dropped her—or his—disguise when she was forced to divulge her true name to Gauss on the occasion of her having done him a good turn with the French infesting Hanover. Writing on April 30, 1807, Gauss thanks his correspondent for her intervention on his behalf with the French General Pernety and deplores the war. Continuing, he pays her a high compliment and expresses something of his own love for the theory of numbers. As the latter is particularly of interest we shall quote from this letter which shows Gauss in one of his cordially human moods.

"But how describe to you my admiration and astonishment at seeing my esteemed correspondent Mr. Leblanc metamorphose himself into this illustrious personage [Sophie Germain] who gives such a brilliant example of what I would find it difficult to believe. A taste for the abstract sciences in general and above all the mysteries of numbers is excessively rare: one is not astonished at it; the enchanting charms of this sublime science reveal themselves only to those who have the courage to go deeply into it. But when a person of the sex which, according to our customs and prejudices, must encounter infinitely more difficulties than men to familiarize herself with these thorny researches, succeeds nevertheless in surmounting these obstacles and penetrating the most obscure parts of them, then without doubt she must have the noblest courage, quite extraordinary talents and a superior genius. Indeed nothing could prove to me in so flattering and less equivocal manner that the attractions of this science, which has enriched my life with so many joys, are not chimerical, as the predilection with which *you* have honored it." He then goes on to discuss mathematics

[5] "Came from" is right. When the sagacious Nazis expelled Fräulein Noether from Germany because she was a Jewess, Bryn Mawr College, Pennsylvania, took her in. She was the most creative abstract algebraist in the world. In less than a week of the new German enlightenment, Göttingen lost the liberality which Gauss cherished and which he strove all his life to maintain.

with her. A delightful touch is the date at the end of the letter: "Bronsvic ce 30 Avril 1807 jour de ma naissance—Brunswick, this 30th of April, 1807, my birthday."

That Gauss was not merely being polite to a young woman admirer is shown by a letter of July 21, 1807 to his friend Olbers. ". . . Lagrange is warmly interested in astronomy and the higher arithmetic; the two test-theorems (for what primes 2 is a cubic or a biquadratic residue), which I also communicated to him some time ago, he considers 'among the most beautiful things and the most difficult to prove.' But Sophie Germain has sent me the proofs of these; I have not yet been able to go through them, but I believe they are good; at least she had attacked the matter from the right side, only somewhat more diffusely than would be necessary. . . ." The theorems to which Gauss refers are those stating for what odd primes p each of the congruences $x^3 \equiv 2 \pmod{p}$, $x^4 \equiv 2 \pmod{p}$ is solvable.

It would take a long book (possibly a longer one than would be required for Newton) to describe all the outstanding contributions of Gauss to mathematics, both pure and applied. Here we can only refer to some of the more important works that have not already been mentioned, and we shall select those which have added new techniques to mathematics or which rounded off outstanding problems. As a rough but convenient table of dates (from that adopted by the editors of Gauss' works) we summarize the principal fields of Gauss' interests after 1800 as follows: 1800–1820, astronomy; 1820–1830, geodesy, the theories of surfaces, and conformal mapping; 1830–1840, mathematical physics, particularly electromagnetism, terrestrial magnetism, and the theory of attraction according to the Newtonian law; 1841–1855, analysis situs, and the geometry associated with functions of a complex variable.

During the period 1821–1848 Gauss was scientific adviser to the Hanoverian (Göttingen was then under the government of Hanover) and Danish governments in an extensive geodetic survey. Gauss threw himself into the work. His method of least squares and his skill in devising schemes for handling masses of numerical data had full scope but, more importantly, the problems arising in the precise survey of a portion of the earth's surface undoubtedly suggested deeper and more general problems connected with all curved surfaces. These researches were to beget the mathematics of relativity. The subject was not new: several of Gauss' predecessors, notably Euler, Lagrange, and Monge, had investigated geometry on certain types of curved surfaces, but it remained for Gauss to attack the problem in all its generality, and from his investigations the first great period of *differential geometry* developed.

Differential geometry may be roughly described as the study of properties of curves, surfaces, etc., in the immediate neighborhood of a point, so that higher powers than the second of distances can be neglected. Inspired by this work, Riemann in 1854 produced his classic dissertation on the hypotheses which lie at the foundations of geometry, which, in its turn, began the second great period in differential geometry, that which is today of use in mathematical physics, particularly in the theory of general relativity.

Three of the problems which Gauss considered in his work on surfaces suggested general theories of mathematical and scientific importance: the measurement of *curvature*, the theory of *conformal representation* (or mapping), and the *applicability* of surfaces.

The unnecessarily mystical motion of a "curved" space-time, which is a purely mathematical extension of familiar, visualizable curvature to a "space" described by four coordinates instead of two, was a natural development of Gauss' work on curved surfaces. One of his definitions will illustrate the reasonableness of all. The problem is to devise some precise means for describing how the "curvature" of a surface varies from point to point of the surface; the description must satisfy our intuitive feeling for what "more curved" and "less curved" signify.

The total curvature of any part of a surface bounded by an unlooped closed curve C is defined as follows. The *normal* to a surface at a given

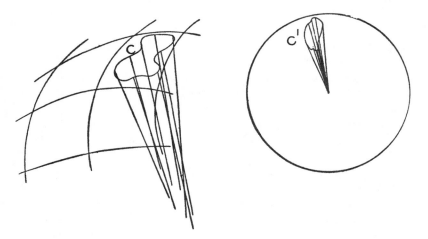

FIGURE 4

point is that straight line passing through the point which is perpendicular to the plane which touches the surface at the given point. At each point of C there is a normal to the surface. Imagine all these normals drawn. Now, from the center of a sphere (which may be anywhere with reference to the surface being considered), whose radius is equal to the unit length, imagine all the radii drawn which are parallel to the normals to C. These radii will cut out a curve, say C', on the sphere of unit radius. The *area* of that part of the spherical surface which is enclosed by C' is defined to be the *total curvature* of the part of the given surface which is enclosed by C. A little visualization will show that this definition accords with common notions as required.

Another fundamental idea exploited by Gauss in his study of surfaces was that of *parametric representation*.

It requires *two* coordinates to specify a particular point on a plane. Likewise on the surface of a sphere, or on a spheroid like the Earth: the coordinates in this case may be thought of as latitude and longitude. This illustrates what is meant by a *two-dimensional manifold*. Generally: if *precisely* n numbers are both necessary and sufficient to specify (individualize) each particular member of a class of things (points, sounds, colors, lines, etc.), the class is said to be an *n-dimensional manifold*. In such specifications it is agreed that only certain characteristics of the members of the class shall be assigned numbers. Thus if we consider only the pitch of sounds, we have a one-dimensional manifold, because one number, the frequency of the vibration corresponding to the sound, suffices to determine the pitch; if we add loudness—measured on some convenient scale —sounds are now a two-dimensional manifold, and so on. If now we regard a *surface* as being made up of *points*, we see that it is a *two-dimensional manifold* (of points). Using the language of geometry we find it convenient to speak of *any* two-dimensional manifold as a "surface," and to apply to the manifold the reasoning of geometry—in the hope of finding something interesting.

The foregoing considerations lead to the parametric representation of surfaces. In Descartes' geometry *one* equation between *three* coordinates represents a surface. Say the coordinates (Cartesian) are x, y, z. Instead of using a single equation connecting x, y, z to represent the surface, we now seek *three*:

$$x = f(u,v), \; y = g(u,v), \; z = h(u,v),$$

where $f(u,v)$, $g(u,v)$, $h(u,v)$ are such functions (expressions) of the new variables u, v that when these variables are eliminated (got rid of—"put over the threshold," literally) there results between x, y, z the equation of of the surface. The elimination is possible, because *two* of the equations can be used to solve for the *two* unknowns u, v; the results can then be substituted in the third. For example, if

$$x = u + v, \; y = u - v, \; z = uv,$$

we get $u = \frac{1}{2}(x + y)$, $v = \frac{1}{2}(x - y)$ from the first two, and hence $4z = x^2 - y^2$ from the third. Now as the variables u, v independently run through any prescribed set of numbers, the functions f, g, h will take on numerical values and x, y, z will move on the surface whose equations are the three written above. The variables u, v are called the *parameters* for the surface, and the three equations $x = f(u,v)$, $y = g(u,v)$, $z = h(u,v)$ their parametric equations. This method of representing surfaces has great advantages over the Cartesian when applied to the study of curvature and other properties of surfaces which vary rapidly from point to point.

Notice that the parametric representation is *intrinsic*; it refers to the surface itself for its coordinates, and not to an extrinsic, or extraneous, set of axes, not connected with the surface, as is the case in Descartes' method. Observe also that the *two* parameters u, v immediately show up the two-dimensionality of the surface. Latitude and longitude on the earth are instances of these intrinsic, "natural" coordinates; it would be most awkward to have to do all our navigation with reference to three mutually perpendicular axes drawn through the center of the Earth, as would be required for Cartesian sailing.

Another advantage of the method is its easy generalization to a space of any number of dimensions. It suffices to increase the number of parameters and proceed as before. When we come to Riemann we shall see how these simple ideas led naturally to a generalization of the metric geometry of Pythagoras and Euclid. The foundations of this generalization were laid down by Gauss, but their importance for mathematics and physical science was not fully appreciated till our own century.

Geodetic researches also suggested to Gauss the development of another powerful method in geometry, that of conformal mapping. Before a map can be drawn, say of Greenland, it is necessary to determine what is to be preserved. Are distances to be distorted, as they are on Mercator's projection, till Greenland assumes an exaggerated importance in comparison with North America? Or are distances to be preserved, so that one inch on the map, measured anywhere along the reference lines (say those for latitude and longitude) shall always correspond to the same distance measured on the surface of the earth? If so, one kind of mapping is demanded, and this kind will not preserve some other feature that we may wish to preserve; for example, if two roads on the earth intersect at a certain angle, the lines representing these roads on the map will intersect at a different angle. That kind of mapping which *preserves angles* is called conformal. In such mapping the theory of analytic functions of a complex variable, described earlier, is the most useful tool.

The whole subject of conformal mapping is of constant use in mathematical physics and its applications, for example in electrostatics, hydrodynamics and its offspring aerodynamics, in the last of which it plays a part in the theory of the airfoil.

Another field of geometry which Gauss cultivated with his usual thoroughness and success was that of the applicability of surfaces, in which it is required to determine what surfaces can be bent onto a given surface without stretching or tearing. Here again the methods Gauss invented were general and of wide utility.

To other departments of science Gauss contributed fundamental researches, for example in the mathematical theories of electromagnetism, including terrestrial magnetism, capillarity, the attraction of ellipsoids (the planets are special kinds of ellipsoids) where the law of attraction is the Newtonian, and dioptrics, especially concerning systems of lenses. The last gave him an opportunity to apply some of the purely abstract technique (continued fractions) he had developed as a young man to satisfy his curiosity in the theory of numbers.

Gauss not only mathematicized sublimely about all these things; he used his hands and his eyes, and was an extremely accurate observer. Many of the specific theorems he discovered, particularly in his researches on electromagnetism and the theory of attraction, have become part of the indispensable stock in trade of all who work seriously in physical science. For many years Gauss, aided by his friend Weber, sought a satisfying theory for all electromagnetic phenomena. Failing to find one that he considered satisfactory he abandoned his attempt. Had he found Clerk Maxwell's (1831–1879) equations of the electromagnetic field he might have been satisfied.

To conclude this long but still far from complete list of the great things that earned Gauss the undisputed title of Prince of Mathematicians we must allude to a subject on which he published nothing beyond a passing mention in his dissertation of 1799, but which he predicted would become one of the chief concerns of mathematics—*analysis situs*. A technical definition of what this means is impossible here (it requires the notion of a *continuous group*), but some hint of the type of problem with which the subject deals can be gathered from a simple instance. Any sort of a knot is tied in a string, and the ends of the string are then tied together. A "simple" knot is easily distinguishable by eye from a "complicated" one, but how are we to give an exact, *mathematical* specification of the difference between the two? And how are we to classify knots mathematically? Although he published nothing on this, Gauss had made a beginning, as was discovered in his posthumous papers. Another type of problem in this subject is to determine the least number of cuts on a given surface which will enable us to flatten the surface out on a plane. For a conical surface one cut suffices; for an anchor ring, two; for a sphere, no finite number of cuts suffices if no stretching is permitted.

These examples may suggest that the whole subject is trivial. But if it had been, Gauss would not have attached the extraordinary importance to it that he did. His prediction of its fundamental character has been fulfilled in our own generation. Today a vigorous school (including many

Americans—J. W. Alexander, S. Lefschetz, O. Veblen, among others) is finding that analysis situs, or the "geometry of position" as it used sometimes to be called, has far-reaching ramifications in both geometry and analysis. What a pity it seems to us now that Gauss could not have stolen a year or two from Ceres to organize his thoughts on this vast theory which was to become the dream of his old age and a reality of our own young age.

His last years were full of honor, but he was not as happy as he had earned the right to be. As powerful of mind and as prolifically inventive as he had ever been, Gauss was not eager for rest when the first symptoms of his last illness appeared some months before his death.

A narrow escape from a violent death had made him more reserved than ever, and he could not bring himself to speak of the sudden passing of a friend. For the first time in more than twenty years he had left Göttingen on June 16, 1854, to see the railway under construction between his town and Cassel. Gauss had always taken a keen interest in the construction and operation of railroads; now he would see one being built. The horses bolted; he was thrown from his carriage, unhurt, but badly shocked. He recovered, and had the pleasure of witnessing the opening ceremonies when the railway reached Göttingen on July 31, 1854. It was his last day of comfort.

With the opening of the new year he began to suffer greatly from an enlarged heart and shortness of breath, and symptoms of dropsy appeared. Nevertheless he worked when he could, although his hand cramped and his beautifully clear writing broke at last. The last letter he wrote was to Sir David Brewster on the discovery of the electric telegraph.

Fully conscious almost to the end he died peacefully, after a severe struggle to live, early on the morning of February 23, 1855, in his seventy eighth year. He lives everywhere in mathematics.

COMMENTARY ON
CAYLEY and SYLVESTER

ARTHUR CAYLEY (1821–1895) and James Joseph Sylvester (1814–1897) were among the giants of the Victorian era. The two men differed widely in background and temperament, and even in attitude toward their common interest. Yet they became close friends and each, as Bell points out, inspired the other to some of his best work. Their most celebrated contributions to mathematics were in algebra, especially in the theory of invariants and matrices. Their inventions, it may justly be asserted, revolutionized physical science as well as mathematical thought.

Cayley was a calm and precise man, born in Surrey and descended from an ancient Yorkshire family. His life ran evenly and successfully. Sylvester, born of orthodox Jewish parents in London, was brilliant, quick-tempered and restless, filled with immense enthusiasms and an insatiable appetite for knowledge. Cayley spent fourteen years at the bar, an experience which, considering where his real talent lay, was time largely wasted. In his youth Sylvester served as professor of mathematics at the University of Virginia. One day a young member of the chivalry whose classroom recitation he had criticized prepared an ambush and fell upon Sylvester with a heavy walking stick. He speared the student with a sword cane; the damage was slight, but the professor found it advisable to leave his post and take "the earliest possible passage for England." Sylvester once remarked that Cayley had been much more fortunate than himself: "that they both lived as bachelors in London, but that Cayley had married and settled down to a quiet and peaceful life at Cambridge; whereas he had never married, and had been fighting the world all his days." [1] This is a fair summary of their lives.

The selection following, another chapter from E. T. Bell's *Men of Mathematics*, presents an excellent account of the careers of these two great British mathematicians and of their work. It is no small achievement in a few pages to explain invariant theory to the layman and also to give lively and convincing portraits of Cayley and Sylvester. Bell is without a peer in this field of literature.

[1] Alexander Macfarlane, *Lectures on Ten British Mathematicians*, New York, 1916, p. 66.

12 Invariant Twins, Cayley and Sylvester

By ERIC TEMPLE BELL

The theory of Invariants sprang into existence under the strong hand of Cayley, but that it emerged finally a complete work of art, for the admiration of future generations of mathematicians, was largely owing to the flashes of inspiration with which Sylvester's intellect illuminated it.
—P. A. MACMAHON

"IT is difficult to give an idea of the vast extent of modern mathematics. The word 'extent' is not the right one: I mean extent crowded with beautiful detail—not an extent of mere uniformity such as an objectless plain, but of a tract of beautiful country seen at first in the distance, but which will bear to be rambled through and studied in every detail of hillside and valley, stream, rock, wood, and flower. But, as for every thing else, so for a mathematical theory—beauty can be perceived but not explained."

These words from Cayley's presidential address in 1883 to the British Association for the Advancement of Science might well be applied to his own colossal output. For prolific inventiveness Euler, Cauchy, and Cayley are in a class by themselves, with Poincaré (who died younger than any of the others) a far second. This applies only to the bulk of these men's work; its quality is another matter, to be judged partly by the frequency with which the ideas originated by these giants recur in mathematical research, partly by mere personal opinion, and partly by national prejudice.

Cayley's remarks about the vast extent of modern mathematics suggest that we confine our attention to some of those features of his own work which introduced distinctly new and far-reaching ideas. The work on which his greatest fame rests is in the theory of invariants and what grew naturally out of that vast theory of which he, brilliantly sustained by his friend Sylvester, was the originator and unsurpassed developer. The concept of invariance is of great importance for modern physics, particularly in the theory of relativity, but this is not its chief claim to attention. Physical theories are notoriously subject to revision and rejection; the theory of invariance as a permanent addition to pure mathematical thought appears to rest on firmer ground.

Another of the ideas originated by Cayley, that of the geometry of "higher space" (space of n dimensions) is likewise of present scientific significance but of incomparably greater importance as pure mathematics. Similarly for the theory of matrices, again an invention of Cayley's. In non-Euclidean geometry Cayley prepared the way for Klein's splendid discovery that the geometry of Euclid and the non-Euclidean geometries of Lobatchewsky and Riemann are, all three, merely different aspects of a more general kind of geometry which includes them as special cases. The nature of these contributions of Cayley's will be briefly indicated after we have sketched his life and that of his friend Sylvester.

The lives of Cayley and Sylvester should be written simultaneously, if that were possible. Each is a perfect foil to the other, and the life of each, in large measure, supplies what is lacking in that of the other. Cayley's life was serene; Sylvester, as he himself bitterly remarks, spent much of his spirit and energy "fighting the world." Sylvester's thought was at times as turbulent as a millrace; Cayley's was always strong, steady, and un-ruffled. Only rarely did Cayley permit himself the printed expression of anything less severe than a precise mathematical statement—the simile quoted at the beginning of this chapter is one of the rare exceptions; Sylvester could hardly talk about mathematics without at once becoming almost orientally poetic, and his unquenchable enthusiasm frequently caused him to go off half-cocked. Yet these two became close friends and inspired one another to some of the best work that either of them did, for example in the theories of invariants and matrices (described later).

With two such temperaments it is not surprising that the course of friendship did not always run smoothly. Sylvester was frequently on the point of exploding; Cayley sat serenely on the safety valve, confident that his excitable friend would presently cool down, when he would calmly resume whatever they had been discussing as if Sylvester had never blown off, while Sylvester for his part ignored his hotheaded indiscretion—till he got himself all steamed up for another. In many ways this strangely congenial pair were like a honeymoon couple, except that one party to the friendship never lost his temper. Although Sylvester was Cayley's senior by seven years, we shall begin with Cayley. Sylvester's life breaks naturally into the calm stream of Cayley's like a jagged rock in the middle of a deep river.

Arthur Cayley was born on August 16, 1821, at Richmond, Surrey, the second son of his parents, then residing temporarily in England. On his father's side Cayley traced his descent back to the days of the Norman Conquest (1066) and even before, to a baronial estate in Normandy. The family was a talented one which, like the Darwin family, should provide much suggestive material for students of heredity. His mother was Maria Antonia Doughty, by some said to have been of Russian origin. Cayley's father was an English merchant engaged in the Russian trade; Arthur was born during one of the periodical visits of his parents to England.

In 1829, when Arthur was eight, the merchant retired, to live thenceforth in England. Arthur was sent to a private school at Blackheath and later, at the age of fourteen, to King's College School in London. His mathematical genius showed itself very early. The first manifestations of superior talent were like those of Gauss; young Cayley developed an amazing skill in long numerical calculations which he undertook for amusement. On beginning the formal study of mathematics he quickly outstripped the rest of the school. Presently he was in a class by himself, as he was later when he went up to the University, and his teachers agreed that the boy was a born mathematician who should make mathematics his career. In grateful contrast to Galois' teachers, Cayley's recognized his ability from the beginning and gave him every encouragement. At first the retired merchant objected strongly to his son's becoming a mathematician but finally, won over by the Principal of the school, gave his consent, his blessing, and his money. He decided to send his son to Cambridge.

Cayley began his university career at the age of seventeen at Trinity College, Cambridge. Among his fellow students he passed as "a mere mathematician" with a queer passion for novel-reading. Cayley was indeed a lifelong devotee of the somewhat stilted fiction, now considered classical, which charmed readers of the 1840's and '50's. Scott appears to have been his favorite, with Jane Austen a close second. Later he read Thackeray and disliked him; Dickens he could never bring himself to read. Byron's tales in verse excited his admiration, although his somewhat puritanical Victorian taste rebelled at the best of the lot and he never made the acquaintance of that diverting scapegrace Don Juan. Shakespeare's plays, especially the comedies, were a perpetual delight to him. On the more solid—or stodgier—side he read and reread Grote's interminable *History of Greece* and Macaulay's rhetorical *History of England*. Classical Greek, acquired at school, remained a reading-language for him all his life; French he read and wrote as easily as English, and his knowledge of German and Italian gave him plenty to read after he had exhausted the Victorian classics (or they had exhausted him). The enjoyment of solid fiction was only one of his diversions; others will be noted as we go.

By the end of his third year at Cambridge Cayley was so far in front of the rest in mathematics that the head examiner drew a line under his name, putting the young man in a class by himself "above the first." In 1842, at the age of twenty one, Cayley was senior wrangler in the mathematical tripos, and in the same year he was placed first in the yet more difficult test for the Smith's prize.

Under an excellent plan Cayley was now in line for a fellowship which would enable him to do as he pleased for a few years. He was elected Fellow of Trinity and assistant tutor for a period of three years. His appointment might have been renewed had he cared to take holy orders, but although Cayley was an orthodox Church of England Christian he could not quite stomach the thought of becoming a parson to hang onto his job or to obtain a better one—as many did, without disturbing either their faith or their conscience.

His duties were light almost to the point of nonexistence. He took a few pupils, but not enough to hurt either himself or his work. Making the best possible use of his liberty he continued the mathematical researches which he had begun as an undergraduate. Like Abel, Galois, and many others who have risen high in mathematics, Cayley went to the masters for his inspiration. His first work, published in 1841 when he was an undergraduate of twenty, grew out of his study of Lagrange and Laplace.

With nothing to do but what he wanted to do after taking his degree Cayley published eight papers the first year, four the second, and thirteen the third. These early papers by the young man who was not yet twenty five when the last of them appeared map out much of the work that is to occupy him for the next fifty years. Already he has begun the study of geometry of n dimensions (which he originated), the theory of invariants, the enumerative geometry of plane curves, and his distinctive contributions to the theory of elliptic functions.

During this extremely fruitful period he was no mere grind. In 1843, when he was twenty two, and occasionally thereafter till he left Cambridge at the age of twenty five, he escaped to the Continent for delightful vacations of tramping, mountaineering, and water-color sketching. Although he was slight and frail in appearance he was tough and wiry, and often after a long night spent in tramping over hilly country, would turn up as fresh as the dew for breakfast and ready to put in a few hours at his mathematics. During his first trip he visited Switzerland and did a lot of mountaineering. Thus began another lifelong passion. His description of the "extent of modern mathematics" is no mere academic exercise by a professor who had never climbed a mountain or rambled lovingly over a tract of beautiful country, but the accurate simile of a man who had known nature intimately at first hand.

During the last four months of his first vacation abroad he became acquainted with northern Italy. There began two further interests which were to solace him for the rest of his life: an understanding appreciation of architecture and a love of good painting. He himself delighted in water-colors, in which he showed marked talent. With his love of good literature, travel, painting, and architecture, and with his deep understanding of natural beauty, he had plenty to keep him from degenerating into the "mere mathematician" of conventional literature—written, for the most part, by people who may indeed have known some pedantic college professor of mathematics, but who never in their lives saw a real mathematician in the flesh.

In 1846, when he was twenty five, Cayley left Cambridge. No position as a mathematician was open to him unless possibly he could square his conscience to the formality of "holy orders." As a mathematician Cayley felt no doubt that it would be easier to square the circle. Anyhow, he left. The law, which with the India Civil Service has absorbed much of England's most promising intellectual capital at one time or another, now attracted Cayley. It is somewhat astonishing to see how many of England's leading barristers and judges in the nineteenth century were high wranglers in the Cambridge tripos, but it does not follow, as some have claimed, that a mathematical training is a good preparation for the law. What seems less doubtful is that it may be a social imbecility to put a young man of Cayley's demonstrated mathematical genius to drawing up wills, transfers, and leases.

Following the usual custom of those looking toward an English legal career of the more gentlemanly grade (that is, above the trade of solicitor), Cayley entered Lincoln's Inn to prepare himself for the Bar. After three years as a pupil of a Mr. Christie, Cayley was called to the Bar in 1849. He was then twenty eight. On being admitted to the Bar, Cayley made a wise resolve not to let the law run off with his brains. Determined not to rot, he rejected more business than he accepted. For fourteen mortal years he stuck it, making an ample living and deliberately turning away the opportunity to smother himself in money and the somewhat blathery sort of renown that comes to prominent barristers, in order that he might earn enough, but no more than enough, to enable him to get on with his work.

His patience under the deadening routine of dreary legal business was exemplary, almost saintly, and his reputation in his branch of the profession (conveyancing) rose steadily. It is even recorded that his name has passed into one of the law books in connection with an exemplary piece of legal work he did. But it is extremely gratifying to record also that Cayley was no milk-and-water saint but a normal human being who could, when the occasion called for it, lose his temper. Once he and his friend Sylvester were animatedly discussing some point in the theory of invariants in Cayley's office when the boy entered and handed Cayley a large batch of legal papers for his perusal. A glance at what was in his hands brought him down to earth with a jolt. The prospect of spending days straightening out some petty muddle to save a few pounds to some comfortable client's already plethoric income was too much for the man with real brains in his head. With an exclamation of disgust and a contemptuous reference to the "wretched rubbish" in his hands, he hurled the stuff to the floor and went on talking mathematics. This, apparently, is the only instance on record when Cayley lost his temper. Cayley got out of the law at the first opportunity—after fourteen years of it. But during his period of servitude he had published between two and three hundred mathematical papers, many of which are now classic.

As Sylvester entered Cayley's life during the legal phase we shall introduce him here.

James Joseph—to give him first the name with which he was born—was the youngest of several brothers and sisters, and was born of Jewish parents on September 3, 1814, in London. Very little is known of his childhood, as Sylvester appears to have been reticent about his early years. His eldest brother emigrated to the United States, where he took the name of Sylvester, an example followed by the rest of the family. But why an orthodox Jew should have decorated himself with a name favored by Christian popes hostile to Jews is a mystery. Possibly that eldest brother had a sense of humor; anyhow, plain James Joseph, son of Abraham Joseph, became henceforth and forevermore James Joseph Sylvester.

Like Cayley's, Sylvester's mathematical genius showed itself early. Between the ages of six and fourteen he attended private schools. The last five months of his fourteenth year were spent at the University of London, where he studied under De Morgan. In a paper written in 1840 with the somewhat mystical title *On the Derivation of Coexistence*, Sylvester says "I am indebted for this term [recurrents] to Professor De Morgan, whose pupil I may boast to have been."

In 1829, at the age of fifteen, Sylvester entered the Royal Institution at Liverpool, where he stayed less than two years. At the end of his first year he won the prize in mathematics. By this time he was so far ahead of his fellow students in mathematics that he was placed in a special class by himself. While at the Royal Institution he also won another prize. This is of particular interest as it establishes the first contact of Sylvester with the United States of America where some of the happiest—also some of the most wretched—days of his life were to be spent. The American brother, by profession an actuary, had suggested to the Directors of the Lotteries Contractors of the United States that they submit a difficult problem in arrangements to young Sylvester. The budding mathematician's solution was complete and practically most satisfying to the Directors, who gave Sylvester a prize of five hundred dollars for his efforts.

The years at Liverpool were far from happy. Always courageous and open, Sylvester made no bones about his Jewish faith, but proudly proclaimed it in the face of more than petty persecution at the hands of the sturdy young barbarians at the Institution who humorously called themselves Christians. But there is a limit to what one lone peacock can stand from a pack of dull jays, and Sylvester finally fled to Dublin with only a few shillings in his pocket. Luckily he was recognized on the street by a distant relative who took him in, straightened him out, and paid his way back to Liverpool.

Here we note another curious coincidence: Dublin, or at least one of its citizens, accorded the religious refugee from Liverpool decent human treatment on his first visit; on his second, some eleven years later, Trinity College, Dublin, granted him the academic degrees (B.A. and M.A.) which his own alma mater, Cambridge University, had refused him because he could not, being a Jew, subscribe to that remarkable compost of nonsensical statements known as the Thirty-Nine Articles prescribed by the Church of England as the minimum of religious belief permissible to a rational mind. It may be added here however that when English higher education finally unclutched itself from the stranglehold of the dead hand of the Church in 1871 Sylvester was promptly given his degrees *honoris causa*. And it should be remarked that in this as in other difficulties Sylvester was no meek, long-suffering martyr. He was full of strength and courage, both physical and moral, and he knew how to put up a devil of a fight to get justice for himself—and frequently did. He was in fact a born fighter with the untamed courage of a lion.

In 1831, when he was just over seventeen, Sylvester entered St. John's College, Cambridge. Owing to severe illnesses his university career was interrupted, and he did not take the mathematical tripos till 1837. He was placed second. The man who beat him was never heard of again as a mathematician. Not being a Christian, Sylvester was ineligible to compete for the Smith's prizes.

In the breadth of his intellectual interests Sylvester resembles Cayley. Physically the two men were nothing alike. Cayley, though wiry and full of physical endurance as we have seen, was frail in appearance and shy and retiring in manner. Sylvester, short and stocky, with a magnificent head set firmly above broad shoulders, gave the impression of tremendous strength and vitality, and indeed he had both. One of his students said he might have posed for the portrait of Hereward the Wake in Charles Kingsley's novel of the same name. As to interests outside of mathematics, Sylvester was much less restricted and far more liberal than Cayley. His knowledge of the Greek and Latin classics in the originals was broad and exact, and he retained his love of them right up to his last illness. Many of his papers are enlivened by quotations from these classics. The quotations are always singularly apt and really do illuminate the matter in hand.

The same may be said for his allusions from other literatures. It might amuse some literary scholar to go through the four volumes of the collected *Mathematical Papers* and reconstruct Sylvester's wide range of reading from the credited quotations and the curious hints thrown out without explicit reference. In addition to the English and classical literatures he was well acquainted with the French, German, and Italian in the originals. His interest in language and literary form was keen and penetrating. To him is due most of the graphic terminology of the theory of invariants. Commenting on his extensive coinage of new mathematical terms from the mint of Greek and Latin, Sylvester referred to himself as the "mathematical Adam."

On the literary side it is quite possible that had he not been a very great mathematician he might have been something a little better than a merely passable poet. Verse, and the "laws" of its construction, fascinated him all his life. On his own account he left much verse (some of which has been published), a sheaf of it in the form of sonnets. The subject matter of his verse is sometimes rather apt to raise a smile, but he frequently showed that he understood what poetry is. Another interest on the artistic side was music, in which he was an accomplished amateur. It is said that he once took singing lessons from Gounod and that he used to entertain workingmen's gatherings with his songs. He was prouder of his "high C" than he was of his invariants.

One of the many marked differences between Cayley and Sylvester may be noted here: Cayley was an omnivorous reader of other mathematicians' work; Sylvester found it intolerably irksome to attempt to master what others had done. Once, in later life, he engaged a young man to teach him something about elliptic functions as he wished to apply them to the theory of numbers (in particular to the theory of partitions, which deals with the number of ways a given number can be made up by adding together numbers of a given kind, say all odd, or some odd and some even). After about the third lesson Sylvester had abandoned his attempt to learn and was lecturing to the young man on his own latest discoveries in algebra. But Cayley seemed to know everything, even about subjects in which he seldom worked, and his advice as a referee was sought by authors and editors from all over Europe. Cayley never forgot anything he had seen; Sylvester had difficulty in remembering his own inventions and once even disputed that a certain theorem of his own could possibly be true. Even comparatively trivial things that every working mathematician knows were sources of perpetual wonder and delight to Sylvester. As a consequence almost any field of mathematics offered an enchanting world for discovery to Sylvester, while Cayley glanced serenely over it all, saw what he wanted, took it, and went on to something fresh.

In 1838, at the age of twenty four, Sylvester got his first regular job, that of Professor of Natural Philosophy (science in general, physics in particular) at University College, London, where his old teacher De Morgan was one of his colleagues. Although he had studied chemistry at Cambridge, and retained a lifelong interest in it, Sylvester found the teaching of science thoroughly uncongenial and, after about two years, abandoned it. In the meantime he had been elected a Fellow of the Royal Society at the unusually early age of twenty five. Sylvester's mathematical merits were so conspicuous that they could not escape recognition, but they did not help him into a suitable position.

At this point in his career Sylvester set out on one of the most singular misadventures of his life. Depending upon how we look at it, this mishap is silly, ludicrous, or tragic. Sanguine and filled with his usual enthusiasm, he crossed the Atlantic to become Professor of Mathematics at the University of Virginia in 1841—the year in which Boole published his discovery of invariants.

Sylvester endured the University only about three months. The refusal of the University authorities to discipline a young gentleman who had insulted him caused the professor to resign. For over a year after this disastrous experience Sylvester tried vainly to secure a suitable position, soliciting—unsuccessfully—both Harvard and Columbia Universities. Failing, he returned to England.

Sylvester's experiences in America gave him his fill of teaching for the next ten years. On returning to London he became an energetic actuary for a life insurance company. Such work for a creative mathematician is poisonous drudgery, and Sylvester almost ceased to be a mathematician. However, he kept alive by taking a few private pupils, one of whom was to leave a name that is known and revered in every country of the world today. This was in the early 1850's, the "potatoes, prunes, and prisms" era of female propriety when young women were not supposed to think of much beyond dabbling in paints and piety. So it is rather surprising to find that Sylvester's most distinguished pupil was a young woman, Florence Nightingale, the first human being to get some decency and cleanliness into military hospitals—over the outraged protests of bull-headed military officialdom. Sylvester at the time was in his late thirties, Miss Nightingale six years younger than her teacher. Sylvester escaped from his makeshift ways of earning a living in the same year (1854) that Miss Nightingale went out to the Crimean War.

Before this however he had taken another false step that landed him nowhere. In 1846, at the age of thirty two, he entered the Inner Temple (where he coyly refers to himself as "a dove nestling among hawks") to prepare for a legal career, and in 1850 was called to the Bar. Thus he and Cayley came together at last.

Cayley was twenty nine, Sylvester thirty six at the time; both were out of the real jobs to which nature had called them. Lecturing at Oxford thirty five years later Sylvester paid grateful tribute to "Cayley, who, though younger than myself is my spiritual progenitor—who first opened my eyes and purged them of dross so that they could see and accept the higher mysteries of our common Mathematical faith." In 1852, shortly after their acquaintance began, Sylvester refers to "Mr. Cayley, who habitually discourses pearls and rubies." Mr. Cayley for his part frequently mentions Mr. Sylvester, but always in cold blood, as it were. Sylvester's earliest outburst of gratitude in print occurs in a paper of 1851 where he says, "The theorem above enunciated [it is his relation between the minor determinants of linearly equivalent quadratic forms] was in part suggested in the course of a conversation with Mr. Cayley (to whom I am indebted for my restoration to the enjoyment of mathematical life). . . ."

Perhaps Sylvester overstated the case, but there was a lot in what he said. If he did not exactly rise from the dead he at least got a new pair of lungs: from the hour of his meeting with Cayley he breathed and lived mathematics to the end of his days. The two friends used to tramp round the Courts of Lincoln's Inn discussing the theory of invariants which both of them were creating and later, when Sylvester moved away, they continued their mathematical rambles, meeting about halfway between their respective lodgings. Both were bachelors at the time.

The theory of algebraic invariants from which the various extensions of the concept of invariance have grown naturally originated in an extremely simple observation. The earliest instance of the idea appears in Lagrange, from whom it passed into the arithmetical works of Gauss. But neither of these men noticed that the simple but remarkable algebraical phenomenon before them was the germ of a vast theory. Nor does Boole seem to have fully realized what he had found when he carried on and greatly extended the work of Lagrange. Except for one slight tiff, Sylvester was always just and generous to Boole in the matter of priority, and Cayley, of course, was always fair.

The simple observation mentioned above can be understood by anyone who has ever seen a quadratic equation solved, and is merely this. A necessary and sufficient condition that the equation $ax^2 + 2bx + c = 0$ shall have two equal roots is that $b^2 - ac$ shall be zero. Let us replace the variable x by its value in terms of y obtained by the transformation $y = (px + q)/(rx + s)$. Thus x is to be replaced by the result of solving this for x, namely $x = (q - sy)/(ry - p)$. This transforms the given equation into another in y; say the new equation is $Ay^2 + 2By + C = 0$. Carrying out the algebra we find that the new coefficients A, B, C are expressed in terms of the old a, b, c as follows,

$$A = as^2 - 2bsr + cr^2,$$
$$B = -aqs + b(qr + sp) - cpr,$$
$$C = aq^2 - 2bpq + cp^2.$$

From these it is easy to show (by brute-force reductions, if necessary, although there is a simpler way of reasoning the result out, without actually calculating A, B, C) that

$$B^2 - AC = (ps - qr)^2 (b^2 - ac).$$

Now $b^2 - ac$ is called the discriminant of the quadratic equation in x; hence the discriminant of the quadratic in y is $B^2 - AC$, and it has been shown that *the discriminant of the transformed equation is equal to the discriminant of the original equation, times the factor* $(ps - qr)^2$ *which depends only upon the coefficients* p, q, r, s *in the transformation* $y = (px + q)/(rx + s)$ *by means of which x was expressed in terms of y.*

Boole was the first (in 1841) to observe something worth looking at in this particular trifle. Every algebraic equation has a discriminant, that is, a certain expression (such as $b^2 - ac$ for the quadratic) which is equal to zero if, and only if, two or more roots of the equation are equal. Boole first asked, does the discriminant of every equation when its x is replaced by the related y (as was done for the quadratic) come back unchanged except for a factor depending only on the coefficients of the transformation? He found that this was true. Next he asked whether there might not be expressions other than discriminants constructed from the coefficients having this same property of *invariance* under *transformation*. He found two such for the general equation of the fourth degree. Then another man, the brilliant young German mathematician, F. M. G. Eisenstein (1823–1852) following up a result of Boole's, in 1844, discovered that certain expressions involving *both the coefficients and the x* of the original equations exhibit the same sort of invariance: the original coefficients and the original x pass into the transformed coefficients and y (as for the quadratic), and the expressions in question constructed from the originals differ from those constructed from the transforms only by a factor which depends solely on the coefficients of the transformation.

Neither Boole nor Eisenstein had any *general* method for finding such *invariant* expressions. At this point Cayley entered the field in 1845 with his pathbreaking memoir, *On the Theory of Linear Transformations*. At the time he was twenty four. He set himself the problem of finding uniform methods which would give him *all* the invariant expressions of the kind described. To avoid lengthy explanations the problem has been stated in terms of equations; actually it was attacked otherwise, but this is of no importance here.

As this question of invariance is fundamental in modern scientific thought we shall give three further illustrations of what it means, none of which involves any symbols or algebra. Imagine any figure consisting of intersecting straight lines and curves drawn on a sheet of paper. Crumple the paper in any way you please without tearing it, and try to think what is the most obvious property of the figure that is the same before and after crumpling. Do the same for any figure drawn on a sheet of rubber, stretching but not tearing the rubber in any complicated manner dictated by whim. In this case it is obvious that sizes of areas and angles, and lengths of lines, have *not* remained "invariant." By suitably stretching the rubber the straight lines may be distorted into curves of almost any tortuosity you like, and at the same time the original curves—or at least some of them—may be transformed into straight lines. Yet *something* about the whole figure has remained unchanged; its very simplicity and obviousness might well cause it to be overlooked. This is the order of the points on any one of the lines of the figure which mark the places where other lines intersect the given one. Thus, if moving the pencil along a given line from *A* to *C*, we had to pass over the point *B* on the line before the figure was distorted, we shall have to pass over *B* in going from *A* to *C* after distortion. The *order* (as described) is an *invariant* under the particular *transformations* which crumpled the sheet of paper into a crinkly ball, say, or which stretched the sheet of rubber.

This illustration may seem trivial, but anyone who has read a nonmathematical description of the intersections of "world-lines" in general relativity, and who recalls that an intersection of two such lines marks a physical *"point-event,"* will see that what we have been discussing is of the same stuff as one of our pictures of the physical universe. The mathematical machinery powerful enough to handle such complicated "transformations" and actually to produce the invariants was the creation of many workers, including Riemann, Christoffel, Ricci, Levi-Civita, Lie, and Einstein—all names well known to readers of popular accounts of relativity; the whole vast program was originated by the early workers in the theory of algebraic invariants, of which Cayley and Sylvester were the true founders.

As a second example, imagine a knot to be looped in a string whose ends are then tied together. Pulling at the knot, and running it along the string, we distort it into any number of "shapes." What remains "invariant," what is "conserved," under all these distortions which, in this case, are our transformations? Obviously neither the shape nor the size of the knot is invariant. But the "style" of the knot itself is invariant; in a sense that need not be elaborated, it is the *same sort* of a knot whatever we do to the string provided we do not untie its ends. Again, in the older physics, energy was "conserved"; the total amount of energy in the universe was assumed to be an invariant, the same under all transformations from one form, such as electrical energy, into others, such as heat and light.

Our third illustration of invariance need be little more than an allusion to physical science. An observer fixes his "position" in space and time with reference to three mutually perpendicular axes and a standard timepiece. Another observer, moving relatively to the first, wishes to describe the same physical event that the first describes. He also has his space-time reference system; his movement relatively to the first observer can be expressed as a transformation of his own coordinates (or of the other observer's). The descriptions given by the two may or may not differ in mathematical form, according to the particular kind of transformation concerned. If their descriptions do differ, the difference is not, obviously, inherent in the physical event they are both observing, but in their reference systems and the transformation. The problem then arises to formulate only those mathematical expressions of natural phenomena which shall be independent, mathematically, of any *particular* reference system and therefore be expressed by all observers in the same form. This is equivalent to finding the invariants of the transformation which expresses the most general shift in "space-time" of one reference system with respect to any other. Thus the problem of finding the mathematical expressions for the intrinsic laws of nature is replaced by an attackable one in the theory of invariants.

In 1863 Cambridge University established a new professorship of mathematics (the Sadlerian) and offered the post to Cayley, who promptly accepted. The same year, at the age of forty two, he married Susan Moline. Although he made less money as a professor of mathematics than he had at the law, Cayley did not regret the change. Some years later the affairs of the University were reorganized and Cayley's salary was raised. His duties also were increased from one course of lectures during one term to two. His life was now devoted almost entirely to mathematical research and university administration. In the latter his sound business training, even temper, impersonal judgment, and legal experience proved invaluable. He never had a great deal to say, but what he said was usually accepted as final, for he never gave an opinion without having reasoned the matter through. His marriage and home life were happy; he had two children, a son and a daughter. As he gradually aged his mind remained as vigorous as ever and his nature became, if anything, gentler. No harsh judgment uttered in his presence was allowed to pass without a quiet protest. To younger men and beginners in mathematical careers he was always generous with his help, encouragement, and sound advice.

During his professorship the higher education of women was a hotly contested issue. Cayley threw all his quiet, persuasive influence on the side of civilization and largely through his efforts women were at last admitted as students (in their own nunneries of course) to the monkish seclusion of medieval Cambridge.

While Cayley was serenely mathematicizing at Cambridge his friend Sylvester was still fighting the world. Sylvester never married. In 1854, at the age of forty, he applied for the professorship of mathematics at the Royal Military Academy, Woolwich. He did not get it. Nor did he get another position for which he applied at Gresham College, London. His trial lecture was too good for the governing board. However, the successful Woolwich candidate died the following year and Sylvester was appointed. Among his not too generous emoluments was the right of pasturage on the common. As Sylvester kept neither horse, cow, nor goat, and did not eat grass himself, it is difficult to see what particular benefit he got out of this inestimable boon.

Sylvester held the position at Woolwich for sixteen years, till he was forcibly retired as "superannuated" in 1870 at the age of fifty six. He was still full of vigor but could do nothing against the hidebound officialdom against him. Much of his great work was still in the future, but his superiors took it for granted that a man of his age must be through.

Another aspect of his forced retirement roused all his fighting instincts. To put the matter plainly, the authorities attempted to swindle Sylvester out of part of the pension which was legitimately his. Sylvester did not take it lying down. To their chagrin the would-be gyppers learned that they were not browbeating some meek old professor but a man who could give them a little better than he took. They came through with the full pension.

While all these disagreeable things were happening in his material affairs Sylvester had no cause to complain on the scientific side. Honors frequently came his way, among them one of those most highly prized by scientific men, foreign correspondent of the French Academy of Sciences. Sylvester was elected in 1863 to the vacancy in the section of geometry caused by the death of Steiner.

After his retirement from Woolwich Sylvester lived in London, versifying, reading the classics, playing chess, and enjoying himself generally, but not doing much mathematics. In 1870 he published his pamphlet, *The Laws of Verse*, by which he set great store. Then, in 1876, he suddenly came to mathematical life again at the age of sixty two. The "old" man was simply inextinguishable.

The Johns Hopkins University had been founded at Baltimore in 1875 under the brilliant leadership of President Gilman. Gilman had been advised to start off with an outstanding classicist and the best mathematician he could afford as the nucleus of his faculty. All the rest would follow, he was told, and it did. Sylvester at last got a job where he might do practically as he pleased and in which he could do himself justice. In 1876 he again crossed the Atlantic and took up his professorship at Johns Hopkins. His salary was generous for those days, five thousand dollars a year. In accepting the call Sylvester made one curious stipulation; his salary was "to be paid in gold." Perhaps he was thinking of Woolwich, which gave him the equivalent of $2750.00 (plus pasturage), and wished to be sure that this time he really got what was coming to him, pension or no pension.

The years from 1876 to 1883 spent at Johns Hopkins were probably the happiest and most tranquil Sylvester had thus far known. Although he did not have to "fight the world" any longer he did not recline on his honors and go to sleep. Forty years seemed to fall from his shoulders and he became a vigorous young man again, blazing with enthusiasm and scintillating with new ideas. He was deeply grateful for the opportunity Johns Hopkins gave him to begin his second mathematical career at the age of sixty three, and he was not backward in expressing his gratitude publicly, in his address at the Commemoration Day Exercises of 1877.

In this Address he outlined what he hoped to do (he did it) in his lectures and researches.

"These are things called Algebraical Forms. Professor Cayley calls them Quantics. [Examples: $ax^2 + 2bxy + cy^2$, $ax^3 + 3bx^2y + 3cxy^2 + dy^3$; the numerical coefficients 1, 2, 1 in the first, 1, 3, 3, 1 in the second, are binomial coefficients, as in the third and fourth lines of Pascal's triangle; the next in order would be $x^4 + 4x^3y + 6x^2y^2 + 4xy^3 + y^4$.] They are not, properly speaking, Geometrical Forms, although capable, to some extent, of being embodied in them, but rather schemes of process, or of operations for forming, for calling into existence, as it were, Algebraic quantities.

"To every such Quantic is associated an infinite variety of other forms that may be regarded as engendered from and floating, like an atmosphere, around it—but infinite as were these derived existences, these emanations from the parent form, it is found that they admit of being obtained by composition, by mixture, so to say, of a certain limited number of fundamental forms, standard rays, as they might be termed in the Algebraic Spectrum of the Quantic to which they belong. And, as it is a leading pursuit of the Physicists of the present day [1877, and even today] to ascertain the fixed lines in the spectrum of every chemical substance, so it is the aim and object of a great school of mathematicians to make out the fundamental derived forms, the *Covariants* [that kind of 'invariant' expression, already described, which involves *both* the variables *and* the coefficients of the form or quantic] and *Invariants*, as they are called, of these Quantics."

To mathematical readers it will be evident that Sylvester is here giving a very beautiful analogy for the fundamental system and the syzygies for a given form; the nonmathematical reader may be recommended to reread the passage to catch the spirit of the algebra Sylvester is talking about, as the analogy is really a close one and as fine an example of "popularized" mathematics as one is likely to find in a year's marching.

In a footnote Sylvester presently remarks "I have at present a class of from eight to ten students attending my lectures on the Modern Higher Algebra. One of them, a young engineer, engaged from eight in the morning to six at night in the duties of his office, with an interval of an hour and a half for his dinner or lectures, has furnished me with the best proof, and the best expressed, I have ever seen of what I call [a certain theorem]. . . ." Sylvester's enthusiasm—he was past sixty—was that of a prophet inspiring others to see the promised land which he had discovered or was about to discover. Here was teaching at its best, at the only level, in fact, which justifies advanced teaching at all.

He had complimentary things to say (in footnotes) about the country of his adoption: ". . . I believe there is no nation in the world where ability with character counts for so much, and the mere possession of wealth (in spite of all that we hear about the Almighty dollar), for so little as in America. . . ."

He also tells how his dormant mathematical instincts were again aroused to full creative power. "But for the persistence of a student of this University [Johns Hopkins] in urging upon me his desire to study with me the modern Algebra, I should never have been led into this investigation. . . . He stuck with perfect respectfulness, but with invincible pertinacity, to his point. He would have the New Algebra (Heaven knows where he had heard about it, for it is almost unknown on this continent), that or nothing. I was obliged to yield, and what was the consequence? In trying to throw light on an obscure explanation in our text-book, my brain took fire, I plunged with requickened zeal into a subject which I had for years abandoned, and found food for thoughts which have engaged my attention for a considerable time past, and will probably occupy all my powers of contemplation advantageously for several months to come."

Almost any public speech or longer paper of Sylvester's contains much that is quotable *about* mathematics in addition to technicalities. A refreshing anthology for beginners and even for seasoned mathematicians could be gathered from the pages of his collected works. Probably no other mathematician has so transparently revealed his personality through his writings as has Sylvester. He liked meeting people and infecting them with his own contagious enthusiasm for mathematics. Thus he says, truly in his own case, "So long as a man remains a gregarious and sociable being, he cannot cut himself off from the gratification of the instinct of imparting what he is learning, of propagating through others the ideas and impressions seething in his own brain, without stunting and atrophying his moral nature and drying up the surest sources of his future intellectual replenishment."

As a pendant to Cayley's description of the extent of modern mathematics, we may hang Sylvester's beside it. "I should be sorry to suppose that I was to be left for long in sole possession of so vast a field as is occupied by modern mathematics. Mathematics is not a book confined within a cover and bound between brazen clasps, whose contents it needs only patience to ransack; it is not a mine, whose treasures may take long to reduce into possession, but which fill only a limited number of veins and lodes; it is not a soil, whose fertility can be exhausted by the yield of successive harvests; it is not a continent or an ocean, whose area can be mapped out and its contour defined: it is limitless as that space which it finds too narrow for its aspirations; its possibilities are as infinite as the worlds which are forever crowding in and multiplying upon the astronomer's gaze; it is as incapable of being restricted within assigned boundaries or being reduced to definitions of permanent validity, as the consciousness, the life, which seems to slumber in each monad, in every atom of matter, in each leaf and bud and cell, and is forever ready to burst forth into new forms of vegetable and animal existence."

In 1878 the *American Journal of Mathematics* was founded by Sylvester and placed under his editorship by Johns Hopkins University. The *Journal* gave mathematics in the United States a tremendous urge in the right direction—research. Today it is still flourishing mathematically but hard pressed financially.

Two years later occurred one of the classic incidents in Sylvester's career. We tell it in the words of Dr. Fabian Franklin, Sylvester's successor in the chair of mathematics at Johns Hopkins for a few years and later editor of the Baltimore *American*, who was an eye (and ear) witness.

"He [Sylvester] made some excellent translations from Horace and from German poets, besides writing a number of pieces of original verse. The tours de force in the way of rhyming, which he performed while in Baltimore, were designed to illustrate the theories of versification of which he gives illustrations in his little book called 'The Laws of Verse.' The reading of the Rosalind poem at the Peabody Institute was the occasion of an amusing exhibition of absence of mind. The poem consisted of no less than four hundred lines, all rhyming with the name Rosalind (the long and short sound of the *i* both being allowed). The audience quite filled the hall, and expected to find much interest or amusement in listening to this unique experiment in verse. But Professor Sylvester had found it necessary to write a large number of explanatory footnotes, and he announced that in order not to interrupt the poem he would read the footnotes in a body first. Nearly every footnote suggested some additional extempore remark, and the reader was so interested in each one that he was not in the least aware of the flight of time, or of the amusement of the audience. When he had dispatched the last of the notes, he looked up at the clock, and was horrified to find that he had kept the audience an hour and a half before beginning to read the poem they had come to hear. The astonishment on his face was answered by a burst of good-humored laughter from the audience; and then, after begging all his hearers to feel at perfect liberty to leave if they had engagements, he read the Rosalind poem."

Doctor Franklin's estimate of his teacher sums the man up admirably: "Sylvester was quick-tempered and impatient, but generous, charitable and tender-hearted. He was always extremely appreciative of the work of others and gave the warmest recognition to any talent or ability displayed by his pupils. He was capable of flying into a passion on slight provocation, but he did not harbor resentment, and was always glad to forget the cause of quarrel at the earliest opportunity."

Before taking up the thread of Cayley's life where it crossed Sylvester's again, we shall let the author of *Rosalind* describe how he made one of his most beautiful discoveries, that of what are called "canonical forms." [This means merely the reduction of a given "quantic" to a "standard" form. For example $ax^2 + 2bxy + cy^2$ can be expressed as the sum of two squares, say $X^2 + Y^2$; $ax^5 + 5bx^4y + 10cx^3y^2 + 10dx^2y^3 + 5exy^4 + fy^5$ can be expressed as a sum of three fifth powers, $X^5 + Y^5 + Z^5$.]

"I discovered and developed the whole theory of canonical binary forms for odd degrees, and, so far as yet made out, for even degrees [1] too, at one sitting, with a decanter of port wine to sustain nature's flagging energies, in a back office in Lincoln's Inn Fields. The work was done, and well done, but at the usual cost of racking thought—a brain on fire, and feet feeling, or feelingless, as if plunged in an ice-pail. *That night we slept no more.*" Experts agree that the symptoms are unmistakable. But it must have been ripe port, to judge by what Sylvester got out of the decanter.

Cayley and Sylvester came together again professionally when Cayley accepted an invitation to lecture at Johns Hopkins for half a year in

---

[1] This part of the theory was developed many years later by E. K. Wakeford (1894–1916), who lost his life in the World War. "Now thanked be God who matched us with this hour." (Rupert Brooke.)

1881–82. He chose Abelian functions, in which he was researching at the time, as his topic, and the 67-year-old Sylvester faithfully attended every lecture of his famous friend. Sylvester had still several prolific years ahead of him, Cayley not quite so many.

We shall now briefly describe three of Cayley's outstanding contributions to mathematics in addition to his work on the theory of algebraic invariants. It has already been mentioned that he invented the theory of matrices, the geometry of space of n dimensions, and that one of his ideas in geometry threw a new light (in Klein's hands) on non-Euclidean geometry. We shall begin with the last because it is the hardest.

Desargues, Pascal, Poncelet, and others had created *projective* geometry, in which the object is to discover those properties of figures which are invariant under projection. Measurements—sizes of angles, lengths of lines—and theorems which depend upon measurement, as for example the Pythagorean proposition that the square on the longest side of a right triangle is equal to the sum of the squares on the other two sides, are not projective but *metrical*, and are not handled by *ordinary* projective geometry. It was one of Cayley's greatest achievements in geometry to transcend the barrier which, before he leapt it, had separated projective from metrical properties of figures. From his higher point of view metrical geometry also became projective, and the great power and flexibility of projective methods were shown to be applicable, by the introduction of "imaginary" elements (for instance points whose coordinates involve $\sqrt{-1}$) to metrical properties. Anyone who has done any analytic geometry will recall that two circles intersect in four points, two of which are always "imaginary." (There are cases of apparent exception, for example concentric circles, but this is close enough for our purpose.) The fundamental notions in metrical geometry are the distance between two points and the angle between two lines. Replacing the concept of distance by another, also involving "imaginary" elements, Cayley provided the means for unifying Euclidean geometry and the common non-Euclidean geometries into one comprehensive theory. Without the use of some algebra it is not feasible to give an intelligible account of how this may be done; it is sufficient for our purpose to have noted Cayley's main advance of uniting projective and metrical geometry with its cognate unification of the other geometries just mentioned.

The matter of n-dimensional geometry when Cayley first put it out was much more mysterious than it seems to us today, accustomed as we are to the special case of four dimensions (space-time) in relativity. It is still sometimes said that a four-dimensional geometry is inconceivable to human beings. This is a superstition which was exploded long ago by Plücker; it is easy to put four-dimensional figures on a flat sheet of paper, and so far as *geometry* is concerned the *whole* of a four-dimensional "space" can be easily imagined. Consider first a rather unconventional three-dimensional space: *all* the *circles* that may be drawn in a *plane*. This "all" is a three-dimensional "space" for the simple reason that it takes *precisely three numbers*, or *three coordinates*, to individualize any one of the swarm of circles, namely *two* to fix the position of the center with reference to any arbitrarily given pair of axes, and *one* to give the length of the radius.

If the reader now wishes to visualize a four-dimensional space he may think of *straight lines*, instead of *points*, as the *elements* out of which our common "solid" space is built. Instead of our familiar solid space looking like an agglomeration of infinitely fine birdshot it now resembles a cosmic haystack of infinitely thin, infinitely long straight straws. That it is indeed four-dimensional in *straight lines* can be seen easily if we convince our-selves (as we may do) that *precisely four numbers* are necessary and sufficient to individualize a particular straw in our haystack. The "dimen-sionality" of a "space" can be anything we choose to make it, provided we suitably select the elements (points, circles, lines, etc.) out of which we build it. Of course if we take *points* as the elements out of which our space is to be constructed, nobody outside of a lunatic asylum has yet succeeded in visualizing a space of more than three dimensions.

Modern physics is fast teaching some to shed their belief in a myste-rious "absolute space" over and above the mathematical "spaces"—like Euclid's, for example—that were *constructed* by geometers to correlate their physical experiences. Geometry today is largely a matter of analysis, but the old terminology of "points," "lines," "distances," and so on, is helpful in suggesting interesting things to do with our sets of coordinates. But it does not follow that these particular things are the most useful that might be done in analysis; it may turn out some day that all of them are comparative trivialities by more significant things which we, hidebound in outworn traditions, continue to do merely because we lack imagination.

If there is any mysterious virtue in talking about situations which arise in analysis as if we were back with Archimedes drawing diagrams in the dust, it has yet to be revealed. Pictures after all may be suitable only for very young children; Lagrange dispensed entirely with such infantile aids when he composed his analytical mechanics. Our propensity to "geome-trize" our analysis may only be evidence that we have not yet grown up. Newton himself, it is known, first got his marvellous results analytically and re-clothed them in the demonstrations of an Apollonius partly because he knew that the multitude—mathematicians less gifted than himself—would believe a theorem true only if it were accompanied by a pretty picture and a stilted Euclidean demonstration, partly because he himself still lingered by preference in the pre-Cartesian twilight of geometry.

The last of Cayley's great inventions which we have selected for mention is that of matrices and their algebra in its broad outline. The subject originated in a memoir of 1858 and grew directly out of simple observations on the way in which the transformations (linear) of the theory of algebraic invariants are combined. Glancing back at what was said on discriminants and their invariance we note the transformation (the arrow is here read "is replaced by") $y \to \dfrac{px + q}{rx + s}$. Suppose we have two such transformations,

$$y \to \frac{px + q}{rx + s}, \quad x \to \frac{Pz + Q}{Rz + S},$$

the second of which is to be applied to the x in the first. We get

$$y \to \frac{(pP + qR)z + (pQ + qS)}{(rP + sR)z + (rQ + sS)}.$$

Attending only to the coefficients in the three transformations we write them in square arrays, thus

$$\left\| \begin{matrix} p & q \\ r & s \end{matrix} \right\|, \quad \left\| \begin{matrix} P & Q \\ R & S \end{matrix} \right\|, \quad \left\| \begin{matrix} pP + qR & pQ + qS \\ rP + sR & rQ + sS \end{matrix} \right\|,$$

and see that the result of performing the first two transformations successively could have been written down by the following rule of "multiplication,"

$$\left\| \begin{matrix} p & q \\ r & s \end{matrix} \right\| \times \left\| \begin{matrix} P & Q \\ R & S \end{matrix} \right\| = \left\| \begin{matrix} pP + qR & pQ + qS \\ rP + sR & rQ + sS \end{matrix} \right\|,$$

where the *rows* of the array on the right are obtained, in an obvious way, by applying the *rows* of the first array on the left onto the columns of the second. Such arrays (of any number of rows and columns) are called *matrices*. Their algebra follows from a few simple postulates, of which we need cite only the following. The matrices

$$\left\| \begin{matrix} a & b \\ c & d \end{matrix} \right\| \quad \text{and} \quad \left\| \begin{matrix} A & B \\ C & D \end{matrix} \right\| \quad \text{are } equal$$

(by definition) when, and only when, $a = A$, $b = B$, $c = C$, $d = D$. The *sum* of the two matrices just written is the matrix $\left\| \begin{matrix} a + A & b + B \\ c + C & d + D \end{matrix} \right\|$. The result of multiplying $\left\| \begin{matrix} a & b \\ c & d \end{matrix} \right\|$ by m (any *number*) is the matrix $\left\| \begin{matrix} ma & mb \\ mc & md \end{matrix} \right\|$. The rule for "multiplying," $\times$, (or "compounding") matrices is as exemplified for $\left\| \begin{matrix} p & q \\ r & s \end{matrix} \right\|$, $\left\| \begin{matrix} P & Q \\ R & S \end{matrix} \right\|$ above.

A distinctive feature of these rules is that multiplication is *not commutative*, except for *special* kinds of matrices. For example, by the rule we

$$\text{get } \left\| \begin{matrix} P & Q \\ R & S \end{matrix} \right\| \times \left\| \begin{matrix} p & q \\ r & s \end{matrix} \right\| = \left\| \begin{matrix} Pp + Qr & Pq + Qs \\ Rp + Sr & Rq + Ss \end{matrix} \right\| ,$$

and the matrix on the right is not equal to that which arises from the multiplication

$$\left\| \begin{matrix} p & q \\ r & s \end{matrix} \right\| \times \left\| \begin{matrix} P & Q \\ R & S \end{matrix} \right\| .$$

All this detail, particularly the last, has been given to illustrate a phenomenon of frequent occurrence in the history of mathematics: the necessary mathematical tools for scientific applications have often been invented decades before the science to which the mathematics is the key was imagined. The bizarre rule of "multiplication" for matrices, by which we get different results according to the order in which we do the multiplication (unlike common algebra where $x \times y$ is always equal to $y \times x$), seems about as far from anything of scientific or practical use as anything could possibly be. Yet sixty seven years after Cayley invented it, Heisenberg in 1925 recognized in the algebra of matrices exactly the tool which he needed for his revolutionary work in quantum mechanics.

Cayley continued in creative activity up to the week of his death, which occurred after a long and painful illness, borne with resignation and unflinching courage, on January 26, 1895. To quote the closing sentences of Forsyth's biography: "But he was more than a mathematician. With a singleness of aim, which Wordsworth would have chosen for his 'Happy Warrior,' he persevered to the last in his nobly lived ideal. His life had a significant influence on those who knew him [Forsyth was a pupil of Cayley and became his successor at Cambridge]: they admired his character as much as they respected his genius: and they felt that, at his death, a great man had passed from the world."

Much of what Cayley did has passed into the main current of mathematics, and it is probable that much more in his massive *Collected Mathematical Papers* (thirteen large quarto volumes of about 600 pages each, comprising 966 papers) will suggest profitable forays to adventurous mathematicians for generations to come. At present the fashion is away from the fields of Cayley's greatest interest, and the same may be said for Sylvester; but mathematics has a habit of returning to its old problems to sweep them up into more inclusive syntheses.

In 1883 Henry John Stephen Smith, the brilliant Irish specialist in the theory of numbers and Savilian Professor of Geometry in Oxford University, died in his scientific prime at the age of fifty seven. Oxford invited the aged Sylvester, then in his seventieth year, to take the vacant chair. Sylvester accepted, much to the regret of his innumerable friends in America. But he felt homesick for his native land which had treated him none too generously; possibly also it gave him a certain satisfaction to feel that "the stone which the builders rejected, the same is become the head of the corner."

The amazing old man arrived in Oxford to take up his duties with a brand-new mathematical theory ("Reciprocants"—differential invariants) to spring on his advanced students. Any praise or just recognition always seemed to inspire Sylvester to outdo himself. Although he had been partly anticipated in his latest work by the French mathematician Georges Halphen, he stamped it with his peculiar genius and enlivened it with his ineffaceable individuality.

The inaugural lecture, delivered on December 12, 1885, at Oxford when Sylvester was seventy one, has all the fire and enthusiasm of his early years, perhaps more, because he now felt secure and knew that he was recognized at last by that snobbish world which had fought him. Two extracts will give some idea of the style of the whole.

"The theory I am about to expound, or whose birth I am about to announce, stands to this ['the great theory of Invariants'] in the relation not of a younger sister, but of a brother, who, though of later birth, on the principle that the masculine is more worthy than the feminine, or at all events, according to the regulations of the Salic law, is entitled to take precedence over his elder sister, and exercise supreme sway over their united realms."

Commenting on the unaccountable absence of a term in a certain algebraic expression he waxes lyric.

"Still, in the case before us, this unexpected absence of a member of the family, whose appearance might have been looked for, made an impression on my mind, and even went to the extent of acting on my emotions. I began to think of it as a sort of lost Pleiad in an Algebraical Constellation, and in the end, brooding over the subject, my feelings found vent, or sought relief, in a rhymed effusion, a *jeu de sottise*, which, not without some apprehension of appearing singular or extravagant, I will venture to rehearse. It will at least serve as an interlude, and give some relief to the strain upon your attention before I proceed to make my final remarks on the general theory.

To a Missing Member
OF A FAMILY OF TERMS IN AN ALGEBRAICAL FORMULA

Lone and discarded one! divorced by fate,
From thy wished-for fellows—whither art flown?
Where lingerest thou in thy bereaved estate,
Like some lost star or buried meteor stone?
Thou mindst me much of that presumptuous one
Who loth, aught less than greatest, to be great,

> *From Heaven's immensity fell headlong down*
> *To live forlorn, self-centred, desolate:*
> *Or who, new Heraklid, hard exile bore,*
> *Now buoyed by hope, now stretched on rack of fear,*
> *Till throned Astraea, wafting to his ear*
> *Words of dim portent through the Atlantic roar,*
> *Bade him 'the sanctuary of the Muse revere*
> *And strew with flame the dust of Isis' shore.'*

Having refreshed ourselves and bathed the tips of our fingers in the Pierian spring, let us turn back for a few brief moments to a light banquet of the reason, and entertain ourselves as a sort of after-course with some general reflections arising naturally out of the previous matter of my discourse."

If the Pierian spring was the old boy's finger bowl at this astonishing feast of reason, it is a safe bet that the faithful decanter of port was never very far from his elbow.

Sylvester's sense of the kinship of mathematics to the finer arts found frequent expression in his writings. Thus, in a paper on Newton's rule for the discovery of imaginary roots of algebraic equations, he asks in a footnote "May not Music be described as the Mathematic of sense, Mathematic as Music of the reason? Thus the musician *feels* Mathematic, the mathematician *thinks* Music——Music the dream, Mathematic the working life——each to receive its consummation from the other when the human intelligence, elevated to its perfect type, shall shine forth glorified in some future Mozart-Dirichlet or Beethoven-Gauss——a union already not indistinctly foreshadowed in the genius and labors of a Helmholtz!"

Sylvester loved life, even when he was forced to fight it, and if ever a man got the best that is in life out of it, he did. He gloried in the fact that the great mathematicians, except for what may be classed as avoidable or accidental deaths, have been long-lived and vigorous of mind to their dying days. In his presidential address to the British Association in 1869 he called the honor roll of some of the greatest mathematicians of the past and gave their ages at death to bear out his thesis that ". . . there is no study in the world which brings into more harmonious action all the faculties of the mind than [mathematics], . . . or, like this, seems to raise them, by successive steps of initiation, to higher and higher states of conscious intellectual being. . . . The mathematician lives long and lives young; the wings of the soul do not early drop off, nor do its pores become clogged with the early particles blown from the dusty highways of vulgar life."

Sylvester was a living example of his own philosophy. But even he at last began to bow to time. In 1893——he was then seventy nine——his eyesight began to fail, and he became sad and discouraged because he could no longer lecture with his old enthusiasm. The following year he asked to be relieved of the more onerous duties of his professorship, and retired to live, lonely and dejected, in London or at Tunbridge Wells. All his brothers and sisters had long since died, and he had outlived most of his dearest friends.

But even now he was not through. His mind was still vigorous, although he himself felt that the keen edge of his inventiveness was dulled forever. Late in 1896, in the eighty second year of his age, he found a new enthusiasm in a field which had always fascinated him, and he blazed up again over the theory of compound partitions and Goldbach's conjecture that every even number is the sum of two primes.

He had not much longer. While working at his mathematics in his London rooms early in March, 1897, he suffered a paralytic stroke which destroyed his power of speech. He died on March 15, 1897, at the age of eighty three. His life can be summed up in his own words, "I really love my subject."

COMMENTARY ON
SRINIVASA RAMANUJAN

I HAVE here set down from the scanty materials available, a brief account of the poor Indian boy who became, as one eminent authority has written, "quite the most extraordinary mathematician of our time." Srinivasa Ramanujan died in India of tuberculosis on April 26, 1920, at the age of thirty-three. He was a mathematician whom only first-class mathematicians can follow and it is not surprising, therefore, that he attracted little attention outside his profession. But his work has left a memorable imprint on mathematical thought.

Two points provide the background for this sketch. The first is that, despite a very limited formal education, Ramanujan was already a brilliant mathematician when he came to England to study in 1914. On the foundation of a borrowed volume, Carr's *Synopsis of Pure Mathematics*, he had built "an outstanding edifice of analytical knowledge and discovery." [1] It was the only book on higher mathematics to which Ramanujan had access, and the nature of his achievement becomes clear when one examines Carr's text. While a work of "some real scholarship and enthusiasm and with a style and individuality of its own," it was in fact no more than a synopsis of some 6,000 theorems of algebra, trigonometry, calculus and analytical geometry with proofs "which are often little more than cross references." [2] In general, the mathematical knowledge contained in Carr's book went no further than the 1860s. Yet in areas that interested him, Ramanujan was abreast, and often ahead, of contemporary mathematical knowledge when he arrived in England. Thus in a mighty sweep he had succeeded in re-creating in his field through his own unaided powers, a rich half-century of European mathematics. One may doubt that so prodigious a feat had ever before been accomplished in the history of thought.

The second noteworthy point is that Ramanujan was a particular kind of mathematician. He was not as versatile as Gauss or Poincaré. He was not a geometer; he cared nothing for mathematical physics, let alone the possible "usefulness" of his mathematical work to other disciplines. Instead, Ramanujan's intuition was much at ease in the bewildering inter-

---

[1] The book was *A Synopsis of Elementary Results in Pure and Applied Mathematics*, by George Shoobridge Carr, a Cambridge mathematician. It was published in two volumes, 1880 and 1886. Carr was a private coach in London and came to Cambridge as an undergraduate when he was nearly forty. The book is "substantially a summary of Carr's coaching notes. . . . He is now completely forgotten, even in his own college, except in so far as Ramanujan has kept his name alive, but he must have been in some ways rather a remarkable man." G. H. Hardy, *Ramanujan—Twelve Lectures Suggested by His Life and Work*; Cambridge, 1940, p. 3.

[2] G. H. Hardy, *op. cit.*, p. 3.

stices of the number system. Numbers, as will appear, were his friends. In the simplest array of digits he detected wonderful properties: congruences, symmetries and relationships which had escaped the notice of even the outstandingly gifted theoreticians. The modern theory of numbers is at once one of the richest, most elusive and most difficult branches of mathematics. Some of its principal theorems, while self-evident and childishly simple in statement, defy repeated and strenuous efforts to prove them. A good example is Goldbach's Theorem, which states that every even number is the sum of two prime numbers. Any fool, as one noted mathematician remarked, might have thought of it; it is altogether obvious, and no even number has been found which does not obey it. Yet no proof of its validity for *every* even number has yet been adduced. It was in dealing with such problems that Ramanujan showed his remarkable powers.

The late G. H. Hardy, a leading mathematician of his time (see pp. 2024–2026), was professionally and personally closest to Ramanujan during his fruitful five years in England. I have taken from Hardy's well-known obituary of Ramanujan,[3] and from his notable course of Ramanujan lectures at Harvard [4] the bulk of the material in the selection following; the rest comes from a brief biographical sketch by P. V. Seshu Aiyar and R. Ramachandra Rao to be found in Ramanujan's *Collected Works*.[5] My contribution has been merely to copy, paraphrase and select. Some of the material is understandable only to the professional mathematician. There is enough, I think, of general interest to justify bringing before the common reader even this inadequate notice of a true genius.

[3] *Proceedings of the London Mathematical Society* (2), XIX (1921), pp. XL–LVIII. Reprinted in *Collected Papers of Srinivasa Ramanujan*, edited by G. H. Hardy, P. V. Seshu Aiyar and B. M. Wilson, Cambridge, 1927, pp. XXI–XXXVI.
[4] G. H. Hardy, *Ramanujan—Twelve Lectures Suggested by His Life and Work*; Cambridge, 1940.
[5] See footnote 3, above.

I have often admired the mystical way of Pythagoras and the secret magic of numbers.
—SIR THOMAS BROWNE

13 Srinivasa Ramanujan

By JAMES R. NEWMAN

SRINIVASA RAMANUJAN AIYANGAR, according to his biographer Seshu Aiyar, was a member of a Brahman family in somewhat poor circumstances in the Tanjore district of the Madras presidency. His father was an accountant to a cloth merchant at Kumbakonam, while his mother, a woman of "strong common sense," was the daughter of a Brahman petty official in the Munsiff's (or legal judge's) Court at Erode. For some time after her marriage she had no children, "but her father prayed to the famous goddess Namagiri, in the neighboring town of Namakkal, to bless his daughter with offspring. Shortly afterwards, her eldest child, the mathematician Ramanujan, was born on 22nd December 1887."

He first went to school at five and was transferred before he was seven to the Town High School at Kumbakonam, where he held a scholarship. His extraordinary powers appear to have been recognized almost immediately. He was quiet and meditative and had an extraordinary memory. He delighted in entertaining his friends with theorems and formulae, with the recitation of complete lists of Sanskrit roots and with repeating the values of *pi* and the square root of two to any number of decimal places.

When he was 15 and in the sixth form at school, a friend of his secured for him the loan of Carr's *Synopsis of Pure Mathematics* from the library of the local Government College. Through the new world thus opened to him Ramanujan ranged with delight. It was this book that awakened his genius. He set himself at once to establishing its formulae. As he was without the aid of other books, each solution was for him a piece of original research. He first devised methods for constructing magic squares. Then he branched off to geometry, where he took up the squaring of the circle and went so far as to get a result for the length of the equatorial circumference of the earth which differed from the true length by only a few feet. Finding the scope of geometry limited, he turned his attention to algebra. Ramanujan used to say that the goddess of Namakkal inspired him with the formulae in dreams. It is a remarkable fact that, on rising from bed, he would frequently note down results and verify them, though he was not always able to supply a rigorous proof. This pattern repeated itself throughout his life.

He passed his matriculation examination to the Government College at

Kumbakonam at 16, and secured the "Junior Subrahmanyam Scholarship." Owing to weakness in English—for he gave no thought to anything but mathematics—he failed in his next examination and lost his scholarship. He then left Kumbakonam, first for Vizagapatam and then for Madras. Here he presented himself for the "First Examination in Arts" in December 1906, but failed and never tried again. For the next few years he continued his independent work in mathematics. In 1909 he was married and it became necessary for him to find some permanent employment. In the course of his search for work he was given a letter of recommendation to a true lover of mathematics, Diwan Bahadur R. Ramachandra Rao, who was then Collector at Nelore, a small town 80 miles north of Madras. Ramanchandra Rao had already seen one of the two fat notebooks kept by Ramanujan into which he crammed his wonderful ideas. His first interview with Ramanujan is best described in his own words.

"Several years ago, a nephew of mine perfectly innocent of mathematical knowledge said to me, 'Uncle, I have a visitor who talks of mathematics; I do not understand him; can you see if there is anything in his talk?' And in the plenitude of my mathematical wisdom, I condescended to permit Ramanujan to walk into my presence. A short uncouth figure, stout, unshaved, not overclean, with one conspicuous feature—shining eyes—walked in with a frayed notebook under his arm. He was miserably poor. He had run away from Kumbakonam to get leisure in Madras to pursue his studies. He never craved for any distinction. He wanted leisure; in other words, that simple food should be provided for him without exertion on his part and that he should be allowed to dream on.

"He opened his book and began to explain some of his discoveries. I saw quite at once that there was something out of the way; but my knowledge did not permit me to judge whether he talked sense or nonsense. Suspending judgment, I asked him to come over again, and he did. And then he had gauged my ignorance and showed me some of his simpler results. These transcended existing books and I had no doubt that he was a remarkable man. Then, step by step, he led me to elliptic integrals and hypergeometric series and at last his theory of divergent series not yet announced to the world converted me. I asked him what he wanted. He said he wanted a pittance to live on so that he might pursue his researches."

Ramachandra Rao undertook to pay Ramanujan's expenses for a time. After a while, other attempts to obtain a scholarship having failed and Ramanujan being unwilling to be supported by anyone for any length of time, he accepted a small appointment in the office of the Madras Port Trust.

But he never slackened his work in mathematics. His earliest contribution was published in the *Journal of the Indian Mathematical Society* in

1911, when Ramanujan was 23. His first long article was on "Some Properties of Bernoulli's Numbers" and was published in the same year. In 1912 he contributed two more notes to the same journal and also several questions for solution.

By this time Ramachandra Rao had induced a Mr. Griffith of the Madras Engineering College to take an interest in Ramanujan, and Griffith spoke to Sir Francis Spring, the chairman of the Madras Port Trust, where Ramanujan was employed. From that time on it became easy to secure recognition of his work. Upon the suggestion of Seshu Aiyar and others, Ramanujan began a correspondence with G. H. Hardy, then Fellow of Trinity College, Cambridge. His first letter to Hardy, dated January 16, 1913, which his friends helped him put in English, follows: ˙

"DEAR SIR,

"I beg to introduce myself to you as a clerk in the Accounts Department of the Port Trust Office at Madras on a salary of only £20 per annum. I am now about 23 years of age. [*He was actually 25—Ed.*] I have had no University education but I have undergone the ordinary school course. After leaving school I have been employing the spare time at my disposal to work at Mathematics. I have not trodden through the conventional regular course which is followed in a University course, but I am striking out a new path for myself. I have made a special investigation of divergent series in general and the results I get are termed by the local mathematicians as 'startling'. . . .

"I would request you to go through the enclosed papers. Being poor, if you are convinced that there is anything of value I would like to have my theorems published. I have not given the actual investigations nor the expressions that I get but I have indicated the lines on which I proceed. Being inexperienced I would very highly value any advice you give me. Requesting to be excused for the trouble I give you.

"I remain, Dear Sir, Yours truly,

"S. RAMANUJAN."

To the letter were attached about 120 theorems, of which the 15 here presented were part of a group selected by Hardy as "fairly representative." Hardy commented on these:

"I should like you to begin by trying to reconstruct the immediate re-

$$(1.1) \quad 1 - \frac{3!}{(1!2!)^3}x^2 + \frac{6!}{(2!4!)^3}x^4 - \ldots$$

$$= \left(1 + \frac{x}{(1!)^3} + \frac{x^2}{(2!)^3} + \ldots\right)\left(1 - \frac{x}{(1!)^3} + \frac{x^2}{(2!)^3} - \ldots\right).$$

$$(1.2) \qquad 1 - 5\left(\frac{1}{2}\right)^3 + 9\left(\frac{1.3}{2.4}\right)^3 - 13\left(\frac{1.3.5}{2.4.6}\right)^3 + \ldots = \frac{2}{\pi}.$$

$$(1.3)\; 1 + 9\left(\frac{1}{4}\right)^4 + 17\left(\frac{1.5}{4.8}\right)^4 + 25\left(\frac{1.5.9}{4.8.12}\right)^4 + \ldots = \frac{2^{3/2}}{\pi^{1/2}\{\Gamma(\tfrac{3}{4})\}^2}.$$

$$(1.4)\; 1 - 5\left(\frac{1}{2}\right)^5 + 9\left(\frac{1.3}{2.4}\right)^5 - 13\left(\frac{1.3.5}{2.4.6}\right)^5 + \ldots = \frac{2}{\{\Gamma(\tfrac{3}{4})\}^4}.$$

$$(1.5) \quad \int_0^\infty \frac{1 + \left(\dfrac{x}{b+1}\right)^2}{1 + \left(\dfrac{x}{a}\right)^2} \cdot \frac{1 + \left(\dfrac{x}{b+2}\right)^2}{1 + \left(\dfrac{x}{a+1}\right)^2} \ldots dx$$

$$= \tfrac{1}{2}\pi^{1/2}\, \frac{\Gamma(a+\tfrac{1}{2})\,\Gamma(b+1)\,\Gamma(b-a+\tfrac{1}{2})}{\Gamma(a)\,\Gamma(b+\tfrac{1}{2})\,\Gamma(b-a+1)}.$$

$$(1.6) \int_0^\infty \frac{dx}{(1+x^2)(1+r^2x^2)(1+r^4x^2)\ldots}$$

$$= \frac{\pi}{2(1 + r + r^3 + r^6 + r^{10} + \ldots)}.$$

(1.7) If $\alpha\beta = \pi^2$, then

$$\alpha^{-1/4}\left(1 + 4\alpha \int_0^\infty \frac{xe^{-\alpha x^2}}{e^{2\pi x} - 1}\, dx\right) = \beta^{-1/4}\left(1 + 4\beta \int_0^\infty \frac{xe^{-\beta x^2}}{e^{2\pi x} - 1}\, dx\right).$$

$$(1.8) \qquad \int_0^a e^{-x^2} dx = \tfrac{1}{2}\pi^{1/2} - \frac{e^{-a^2}}{2a+}\frac{1}{a+}\frac{2}{2a+}\frac{3}{a+}\frac{4}{2a+\ldots}.$$

$$(1.9) \quad 4\int_0^\infty \frac{xe^{-x\sqrt{5}}}{\cosh x}\, dx = \frac{1}{1+}\frac{1^2}{1+}\frac{1^2}{1+}\frac{2^2}{1+}\frac{2^2}{1+}\frac{3^2}{1+}\frac{3^2}{1+\ldots}.$$

(1.10) If $u = \dfrac{x}{1+}\dfrac{x^5}{1+}\dfrac{x^{10}}{1+}\dfrac{x^{15}}{1+\ldots}, \quad v = \dfrac{x^{1/5}}{1+}\dfrac{x}{1+}\dfrac{x^2}{1+}\dfrac{x^3}{1+\ldots},$

then

$$v^5 = u\, \frac{1 - 2u + 4u^2 - 3u^3 + u^4}{1 + 3u + 4u^2 + 2u^3 + u^4}.$$

(1.11) $\dfrac{1}{1+}\dfrac{e^{-2\pi}}{1+}\dfrac{e^{-4\pi}}{1+\dots} = \left\{ \sqrt{\left(\dfrac{5+\sqrt{5}}{2} \right)} - \dfrac{\sqrt{5}+1}{2} \right\} e^{\frac{2}{5}\pi}.$

(1.12) $\dfrac{1}{1+}\dfrac{e^{-2\pi\sqrt{5}}}{1+}\dfrac{e^{-4\pi\sqrt{5}}}{1+\dots} =$

$$\left[\dfrac{\sqrt{5}}{1+\sqrt[5]{\left\{ 5^{3/4}\left(\dfrac{\sqrt{5}-1}{2} \right)^{5/2} - 1 \right\}}} - \dfrac{\sqrt{5}+1}{2} \right] e^{2\pi/\sqrt{5}}.$$

(1.13) If $F(k) = 1 + \left(\dfrac{1}{2} \right)^2 k + \left(\dfrac{1.3}{2.4} \right)^2 k^2 + \dots$ and

$$F(1-k) = \sqrt{(210)}F(k), \text{ then}$$

$$k = (\sqrt{2}-1)^4(2-\sqrt{3})^2(\sqrt{7}-\sqrt{6})^4(8-3\sqrt{7})^2(\sqrt{10}-3)^4$$
$$\times (4-\sqrt{15})^4(\sqrt{15}-\sqrt{14})^2(6-\sqrt{35})^2.$$

(1.14) The coefficient of x^n in $(1 - 2x + 2x^4 - 2x^9 + \dots)^{-1}$ is the integer nearest to

$$\frac{1}{4n}\left(\cosh \pi\sqrt{n} - \frac{\sinh \pi\sqrt{n}}{\pi\sqrt{n}} \right).$$

(1.15) The number of numbers between A and x which are either squares or sums of two squares is

$$K \int_{A}^{x} \frac{dt}{\sqrt{(\log t)}} + \theta(x),$$

where $K = 0\cdot 764\dots$ and $\theta(x)$ is very small compared with the previous integral.

actions of an ordinary professional mathematician who receives a letter like this from an unknown Hindu clerk.

"The first question was whether I could recognise anything. I had proved things rather like (1.7) myself, and seemed vaguely familiar with (1.8). Actually (1.8) is classical; it is a formula of Laplace first proved properly by Jacobi; and (1.9) occurs in a paper published by Rogers in 1907. I thought that, as an expert in definite integrals, I could probably prove (1.5) and (1.6), and did so, though with a good deal more trouble than I had expected. . . .

"The series formulae (1.1)–(1.4) I found much more intriguing, and it soon became obvious that Ramanujan must possess much more general theorems and was keeping a great deal up his sleeve. The second is a formula of Bauer well known in the theory of Legendre series, but the others are much harder than they look. . . .

"The formulae (1.10)–(1.13) are on a different level and obviously both difficult and deep. An expert in elliptic functions can see at once that (1.13) is derived somehow from the theory of 'complex multiplication,' but (1.10)–(1.12) defeated me completely; I had never seen anything in the least like them before. A single look at them is enough to show that they could only be written down by a mathematician of the highest class. They must be true because, if they were not true, no one would have had the imagination to invent them. Finally . . . the writer must be completely honest, because great mathematicians are commoner than thieves or humbugs of such incredible skill. . . .

"While Ramanujan had numerous brilliant successes, his work on prime numbers and on all the allied problems of the theory was definitely wrong. This may be said to have been his one great failure. And yet I am not sure that, in some ways, his failure was not more wonderful than any of his triumphs. . . ."

Ramanujan's notation of one mathematical term in this area, wrote Hardy, "was first obtained by Landau in 1908. Ramanujan had none of Landau's weapons at his command; he had never seen a French or German book; his knowledge even of English was insufficient to qualify for a degree. It is sufficiently marvellous that he should have even dreamt of problems such as these, problems which it had taken the finest mathematicians in Europe a hundred years to solve, and of which the solution is incomplete to the present day."

At last, in May of 1913, as the result of the help of many friends, Ramanujan was relieved of his clerical post in the Madras Port Trust and given a special scholarship. Hardy had made efforts from the first to bring Ramanujan to Cambridge. The way seemed to be open, but Ramanujan refused at first to go because of caste prejudice and lack of his mother's consent.

"This consent," wrote Hardy, "was at last got very easily in an unexpected manner. For one morning his mother announced that she had had a dream on the previous night, in which she saw her son seated in a big hall amidst a group of Europeans, and that the goddess Namagiri had commanded her not to stand in the way of her son fulfilling his life's purpose."

When Ramanujan finally came, he had a scholarship from Madras of £250, of which £50 was allotted to the support of his family in India, and an allowance of £60 from Trinity.

"There was one great puzzle," Hardy observes of Ramanujan. "What was to be done in the way of teaching him modern mathematics? The limitations of his knowledge were as startling as its profundity. Here was a man who could work out modular equations, and theorems of complex multiplication, to orders unheard of, whose mastery of continued fractions was, on the formal side at any rate, beyond that of any mathematician in the world, who had found for himself the functional equation of the Zeta-function and the dominant terms of many of the most famous problems in the analytic theory of numbers; and he had never heard of a doubly periodic function or of Cauchy's theorem, and had indeed but the vaguest idea of what a function of a complex variable was. His ideas as to what constituted a mathematical proof were of the most shadowy description. All his results, new or old, right or wrong, had been arrived at by a process of mingled argument, intuition, and induction, of which he was entirely unable to give any coherent account.

"It was impossible to ask such a man to submit to systematic instruction, to try to learn mathematics from the beginning once more. I was afraid too that, if I insisted unduly on matters which Ramanujan found irksome, I might destroy his confidence or break the spell of his inspiration. On the other hand there were things of which it was impossible that he should remain in ignorance. Some of his results were wrong, and in particular those which concerned the distribution of primes, to which he attached the greatest importance. It was impossible to allow him to go through life supposing that all the zeros of the Zeta-function were real. So I had to try to teach him, and in a measure I succeeded, though obviously I learnt from him much more than he learnt from me. . . .

"I should add a word here about Ramanujan's interests outside mathematics. Like his mathematics, they shewed the strangest contrasts. He had very little interest, I should say, in literature as such, or in art, though he could tell good literature from bad. On the other hand, he was a keen philosopher, of what appeared, to followers of the modern Cambridge school, a rather nebulous kind, and an ardent politician, of a pacifist and ultraradical type. He adhered, with a severity most unusual in Indians resident in England, to the religious observances of his caste; but his religion was a matter of observance and not of intellectual conviction, and I remember well his telling me (much to my surprise) that all religions seemed to him more or less equally true. Alike in literature, philosophy, and mathematics, he had a passion for what was unexpected, strange, and odd; he had quite a small library of books by circle-squarers and other cranks . . . He was a vegetarian in the strictest sense—this proved a terrible difficulty later when he fell ill—and all the time he was in Cambridge he cooked all his food himself, and never cooked it without first changing into pyjamas. . . .

"It was in the spring of 1917 that Ramanujan first appeared to be unwell. He went to a Nursing Home at Cambridge in the early summer, and was never out of bed for any length of time again. He was in sanatoria at Wells, at Matlock, and in London, and it was not until the autumn of 1918 that he shewed any decided symptom of improvement. He had then resumed active work, stimulated perhaps by his election to the Royal Society, and some of his most beautiful theorems were discovered about this time. His election to a Trinity Fellowship was a further encouragement; and each of those famous societies may well congratulate themselves that they recognized his claims before it was too late."

Early in 1919 Ramanujan went home to India, where he died in the following year.

For an evaluation of Ramanujan's method and work in mathematics we must again quote from Hardy. "I have often been asked whether Ramanujan had any special secret; whether his methods differed in kind from those of other mathematicians; whether there was anything really abnormal in his mode of thought. I cannot answer these questions with any confidence or conviction; but I do not believe it. My belief is that all mathematicians think, at bottom, in the same kind of way, and that Ramanujan was no exception. He had, of course, an extraordinary memory. He could remember the idiosyncrasies of numbers in an almost uncanny way. It was Mr. Littlewood (I believe) who remarked that 'every positive integer was one of his personal friends.' I remember once going to see him when he was lying ill at Putney. I had ridden in taxi-cab No. 1729, and remarked that the number seemed to me rather a dull one, and that I hoped it was not an unfavourable omen. 'No,' he replied, 'it is a very interesting number; it is the smallest number expressible as a sum of two cubes in two different ways.' I asked him, naturally, whether he knew the answer to the corresponding problem for fourth powers; and he replied, after a moment's thought, that he could see no obvious example, and thought that the first such number must be very large. His memory, and his powers of calculation, were very unusual, but they could not reasonably be called 'abnormal.' If he had to multiply two large numbers, he multiplied them in the ordinary way; he could do it with unusual rapidity and accuracy, but not more rapidly or more accurately than any mathematician who is naturally quick and has the habit of computation.

"It was his insight into algebraical formulae, transformations of infinite series, and so forth, that was most amazing. On this side most certainly I have never met his equal, and I can compare him only with Euler or Jacobi. He worked, far more than the majority of modern mathematicians, by induction from numerical examples; all of his congruence properties of partitions, for example, were discovered in this way. But with his memory, his patience, and his power of calculation, he combined a power of generalisation, a feeling for form, and a capacity for rapid modification of his hypotheses, that were often really startling, and made him, in his own field, without a rival in his day.

"It is often said that it is much more difficult now for a mathematician to be original than it was in the great days when the foundations of modern analysis were laid; and no doubt in a measure it is true. Opinions may differ as to the importance of Ramanujan's work, the kind of standard by which it should be judged, and the influence which it is likely to have on the mathematics of the future. It has not the simplicity and the inevitableness of the very greatest work; it would be greater if it were less strange. One gift it has which no one can deny—profound and invincible originality. He would probably have been a greater mathematician if he had been caught and tamed a little in his youth; he would have discovered more that was new, and that, no doubt, of greater importance. On the other hand he would have been less of a Ramanujan, and more of a European professor and the loss might have been greater than the gain."

COMMENTARY ON
BERTRAND RUSSELL

[Another article by Bertrand Russell appears on pages 537–543.]

BERTRAND ARTHUR WILLIAM RUSSELL was born at Trelleck, in Monmouthshire, on May 18, 1872. His father was Viscount Amberley, son of Lord John Russell who introduced the first Reform Bill, which became law in 1832; his mother was Kate Stanley, daughter of Baron Stanley of Alderley. Both parents died before Russell was four years old and he was brought up by his grandmother, Countess Russell, "a more powerful influence upon my general outlook than anyone else."

Russell was tutored at home until he was eighteen; he then went to Cambridge where he concentrated on mathematics and philosophy. At Trinity College he joined a circle of what he called "clever" young men, which included the philosophers J. M. E. McTaggart and G. E. Moore, the three brothers Trevelyan and the essayist Lowes Dickinson. Moore, with an intellect "as deeply passionate as Spinoza's," fulfilled Russell's "ideal of genius" and greatly influenced his philosophical opinions. Alfred North Whitehead, one of his teachers, also played an important part in Russell's "gradual transition from a student to an independent writer." [1] The older man guided the younger in his studies, criticized his work "somewhat severely though quite justly" and showed him much personal kindness. Russell recalls that his first contact with Whitehead, "or rather with his father" was in 1877. "I had been told that the earth is round, but trusting to the evidence of the senses, I refused to believe it. The vicar of the parish, who happened to be Whitehead's father, was called in to persuade me. Clerical authority so far prevailed as to make me think an experimental test worth while, and I started to dig a hole in the hopes of emerging at the antipodes. When they told me this was useless, my doubts revived." [2]

Immediately after graduating, Russell spent several years traveling abroad, visited America in 1896, and in 1898 returned to Trinity as a lecturer and fellow. Among his earliest writings were *German Social Democracy* (1896), *An Essay on the Foundations of Geometry* (1897) and

[1] Bertrand Russell, "Portraits from Memory, I: Alfred North Whitehead," *Harper's Magazine*, December, 1952.
[2] *Ibid.*

his admirable monograph *A Critical Exposition of the Philosophy of Leibniz* (1900).[3]

In 1900, "the most important year in my intellectual life," Russell went with Whitehead to the International Congress of Philosophy in Paris where he heard Peano tell of his inventions in symbolic logic. This experience impelled him to prolonged investigations, the principal fruits of which were *The Principles of Mathematics* (1903) and the great work on which he collaborated with Whitehead, *Principia Mathematica* (1910–1913).[4]

Russell says he grew up in an atmosphere of politics and that, although his interests in the subject were secondary, they were "very strong." He was active in the Fabian Society and was a close friend of Sidney and Beatrice Webb. He thought of standing for Parliament but lost his chance to become a Liberal party candidate when he declined to conceal his agnosticism. Always outspoken, he got into trouble during the First World War by his energetic support of the No-Conscription Fellowship. He was fined and dismissed from his college post. In 1918 he was imprisoned for several months because he had written a pamphlet accusing the American Army of "intimidating strikes at home." [5] While in jail he wrote his *Introduction to Mathematical Philosophy*, first published in 1919. The book laid a heavy burden on the Governor of the prison who, though unable to comprehend it, was required to read the manuscript for possible seditious tendencies. It is from this work that I have taken one of the selections below, "Definition of Number."

In 1920 Russell made a trip to Russia, where he met Lenin, Trotsky and Gorki. The outcome of this visit was *The Practice and Theory of Bolshevism* (1920), in which Russell praised fundamental ideas of communism, but warned that the "present holders of power are evil men . . . [and] that there is no depth of cruelty, perfidy or brutality from which [they] will shrink when they feel themselves threatened." After a year spent in China (1920–1921), whose people he loved, Russell returned to teaching and lecturing in England and the United States. His professional reputation was already established by his contributions to the foundations of mathematics, but in the twenties and thirties his literary output brought his ideas to the attention of a very wide audience. Between 1920 and 1940, he published two dozen books and more than 200 articles in journals and

[3] For a bibliography of Russell's writings up to 1944, see Paul Arthur Schilpp, ed., *The Philosophy of Bertrand Russell* (The Library of Living Philosophers, Vol. V), Evanston and Chicago, 1944.

[4] For further discussion of the *Principia* see selections by Nagel (p. 1878), Lewis and Langford (p. 1859), and the introduction to the selection by Whitehead (pp. 395–401).

[5] This was regarded as "likely to prejudice His Majesty's relations with the United States of America." See also H. W. Leggett, *Bertrand Russell, O.M.*, New York, 1950, p. 27.

general magazines on mathematical, philosophical, scientific, political and social subjects. His writings combine profundity with wit, trenchant thinking with literary excellence, honesty and clarity with kindliness and wisdom. No other contemporary philosopher has enjoyed a comparable popular appeal. Among his principal works are *Mysticism and Logic* (1918), a brilliant collection of essays, one of which is here reproduced; *The Analysis of Mind* (1921), lectures on philosophy and psychology given in London and Peking; *The Prospects of Industrial Civilization* (1923), a study of socialism written in collaboration with his second wife, Dora Russell; *The ABC of Relativity* (1925), a stimulating but not in all respects successful popularization of Einstein's theory; *The Analysis of Matter* (1927), an examination of the new concepts of physics; *An Outline of Philosophy* (1927), perhaps the best of modern introductions to philosophical thought; *Marriage and Morals* (1929); *The Scientific Outlook* (1931), a delightfully written survey of scientific knowledge and method and of the relation of science to society; *Education and the Social Order* (1932); *Freedom and Organization* (1934), a historical account of the development of political theory; *Power* (1938), regarded as one of the most important of modern analyses of the theory of the state; *An Inquiry into Meaning and Truth* (1940), the William James Lectures at Harvard University; *History of Western Philosophy* (1945), a wonderfully readable survey; *Human Knowledge* (1948), an examination of the relation "between individual experience and the general body of scientific knowledge."

Russell has been several times married. His first wife was Alys Pearsall Smith, sister of Logan Pearsall Smith and member of an American Quaker family which had settled in England. They were divorced in 1921, but had separated much earlier, after seventeen years of married life. That same year he married Dora Winifred Black by whom he had a son and a daughter. It was mainly because of the problem of educating their children that the Russells decided to start a school on "novel and progressive" lines. The school at Beacon Hill, near Petersfield, England, was unsuccessful, neither Russell nor his wife being practical administrators. After a few years Russell withdrew from participation in its affairs; Dora Russell, from whom he was divorced in 1935, continued to run it until 1939. His third marriage took place in 1936, with Patricia Helen Spence, and after a divorce, Russell married for the fourth time in 1951.

Russell has made many visits to the United States, his longest stay being from 1938 to 1944. During this period he lectured at the University of Chicago, at the University of California in Los Angeles, at Harvard and at the Barnes Foundation at Merion, Pennsylvania. His appointment to a professorship of philosophy at the College of the City of New York precipitated a ludicrous controversy. A woman brought suit to have the appointment annulled on the ground of Russell's "advocacy of free love." [6] Judge John McGeehan of the New York Supreme Court thereupon made himself immortal by revoking the appointment, which he described as an "attempt to establish a chair of indecency."

In 1944 Russell was again appointed to a fellowship at Trinity College, Cambridge. With almost unabated energy he continued to write, lecture and express his vigorous views on a wide variety of subjects. Having received almost every distinction his own country could confer, Russell was awarded the Nobel Prize in Literature in 1950.

Russell was eighty-two and in excellent health when these lines were written in 1954. On his eightieth birthday he offered typical advice on longevity. He recommended a "habit of hilarious olympian controversy," keeping busy, and avoiding every kind of excess—except smoking. ("Until the age of forty-two I was a teetotaler. But for the last sixty years I have smoked incessantly, stopping only to eat and sleep.") Conceding that his advice was neither very instructive nor edifying, he pointed out that he had never done anything "on the ground that it was good for health . . . I am convinced that so long as you are healthy, it is unnecessary to think about health." "I should like," he said, "to live another ten years provided there is not another world war meanwhile. If there is, there will be something to be said for being dead."

                    *   *   *   *   *

These lines are little more than a chronology of Russell's life. This is not the place to appraise his work, nor do I feel qualified for the task. Like many others of my generation I have learned so much from Russell that I cannot, either in agreement or dissent, achieve the necessary disinterest. As to further biographical details, I have included a superb sketch by Russell about himself, entitled "My Mental Development." It offers a comment of one of the greatest of living writers on one of the greatest of living philosophers. The reader will, I hope, accept this as a satisfactory substitute for the account I might have written.

[6] Leggett, *op. cit.*, p. 47.

Every man who rises above the common level has received two educations: the first from his teachers; the second, more personal and important, from himself.
                                                                    —EDWARD GIBBON

I 'spect I grow'd. Don't think nobody ever made me.
            —HARRIET BEECHER STOWE (*Topsy in Uncle Tom's Cabin*)

14   My Mental Development

By BERTRAND RUSSELL

MY mother having died when I was two years old, and my father when I was three, I was brought up in the house of my grandfather, Lord John Russell, afterwards Earl Russell. Of my parents, Lord and Lady Amberley, I was told almost nothing—so little that I vaguely sensed a dark mystery. It was not until I was twenty-one that I came to know the main outlines of my parents' lives and opinions. I then found, with a sense of bewilderment, that I had gone through almost exactly the same mental and emotional development as my father had.

It was expected of my father that he should take to a political career, which was traditional in the Russell family. He was willing, and was for a short time in Parliament (1867–68); but he had not the temperament or the opinions that would have made political success possible. At the age of twenty-one he decided that he was not a Christian, and refused to go to Church on Christmas Day. He became a disciple, and afterwards a friend, of John Stuart Mill, who, as I discovered some years ago, was (so far as is possible in a non-religious sense) my godfather. My parents accepted Mill's opinions, not only such as were comparatively popular, but also those that still shocked public sentiment, such as women's suffrage and birth control. During the general election of 1868, at which my father was a candidate, it was discovered that, at a private meeting of a small society, he had said that birth control was a matter for the medical profession to consider. This let loose a campaign of vilification and slander. A Catholic Bishop declared that he advocated infanticide; he was called in print a "filthy foul-mouthed rake"; on election day, cartoons were exhibited accusing him of immorality, altering his name to "Vice-count Amberley," and accusing him of advocating "The French and American system." [1] By these means he was defeated. The student of comparative sociology may be interested in the similarities between rural England in 1868 and urban New York in 1940. The available documents are collected

[1] My parents, when in America, had studied such experiments as the Oneida community. They were therefore accused of attempting to corrupt the purity of English family life by introducing un-English transatlantic vices.

in *The Amberley Papers*, by my wife and myself. As the reader of this book will see, my father was shy, studious, and ultra-conscientious— perhaps a prig, but the very opposite of a rake.

My father did not give up hope of returning to politics, but never obtained another constituency, and devoted himself to writing a big book, *Analysis of Religious Belief*, which was published after his death. He could not, in any case, have succeeded in politics, because of his very exceptional intellectual integrity; he was always willing to admit the weak points on his own side and the strong points on that of his opponents. Moreover his health was always bad, and he suffered from a consequent lack of physical vigour.

My mother shared my father's opinions, and shocked the 'sixties by addressing meetings in favour of equality for women. She refused to use the phrase "women's rights," because, as a good utilitarian, she rejected the doctrine of natural rights.

My father wished my brother and me to be brought up as free thinkers, and appointed two free thinkers as our guardians. The Court of Chancery, however, at the request of my grandparents, set aside the will, and I enjoyed the benefits of a Christian upbringing.

In 1876, when after my father's death, I was brought to the house of my grandparents, my grandfather was eighty-three and had become very feeble. I remember him sometimes being wheeled about out-of-doors in a bath-chair, sometimes in his room reading Hansard (the official report of debates in Parliament). He was invariably kind to me, and seemed never to object to childish noise. But he was too old to influence me directly. He died in 1878, and my knowledge of him came through his widow, my grandmother, who revered his memory. She was a more powerful influence upon my general outlook than any one else, although, from adolescence onward, I disagreed with very many of her opinions.

My grandmother was a Scotch Presbyterian, of the border family of the Elliots. Her maternal grandfather suffered obloquy for declaring, on the basis of the thickness of the lava on the slopes of Etna, that the world must have been created before b.c. 4004. One of her great-grandfathers was Robertson, the historian·of Charles V.

She was a Puritan, with the moral rigidity of the Covenanters, despising comfort, indifferent to food, hating wine, and regarding tobacco as sinful. Although she had lived her whole life in the great world until my grandfather's retirement in 1866, she was completely unworldly. She had that indifference to money which is only possible to those who have always had enough of it. She wished her children and grandchildren to live useful and virtuous lives, but had no desire that they should achieve what others would regard as success, or that they should marry "well." She had the Protestant belief in private judgment and the supremacy of the individual conscience. On my twelfth birthday she gave me a Bible (which I still possess), and wrote her favourite texts on the fly-leaf. One of them was "Thou shalt not follow a multitude to do evil"; another, "Be strong, and of a good courage; be not afraid, neither be Thou dismayed; for the Lord Thy God is with thee withersoever thou goest." These texts have profoundly influenced my life, and still seemed to retain some meaning after I had ceased to believe in God.

At the age of seventy, my grandmother became a Unitarian; at the same time, she supported Home Rule for Ireland, and made friends with Irish Members of Parliament, who were being publicly accused of complicity in murder. This shocked people more than now seems imaginable. She was passionately opposed to imperialism, and taught me to think ill of the Afghan and Zulu wars, which occurred when I was about seven. Concerning the occupation of Egypt, however, she said little, as it was due to Mr. Gladstone, whom she admired. I remember an argument I had with my German governess, who said that the English, having once gone into Egypt, would never come out, whatever they might promise, whereas I maintained, with much patriotic passion, that the English never broke promises. That was sixty years ago, and they are there still.

My grandfather, seen through the eyes of his widow, made it seem imperative and natural to do something important for the good of mankind. I was told of his introducing the Reform Bill in 1832. Shortly before he died, a delegation of eminent nonconformists assembled to cheer him, and I was told that fifty years earlier he had been one of the leaders in removing their political disabilities. In his sitting-room there was a statue of Italy, presented to my grandfather by the Italian Government, with an inscription: "A Lord John Russell, L'Italia Riconoscente"; I naturally wished to know what this meant, and learnt, in consequence, the whole saga of Garibaldi and Italian unity. Such things stimulated my ambition to live to some purpose.

My grandfather's library, which became my schoolroom, stimulated me in a different way. There were books of history, some of them very old; I remember in particular a sixteenth-century Guicciardini. There were three huge folio volumes called *L'Art de vérifier les dates*. They were too heavy for me to move, and I speculated as to their contents; I imagined something like the tables for finding Easter in the Prayer-Book. At last I became old enough to lift one of the volumes out of the shelf, and I found, to my disgust, that the only "art" involved was that of looking up the date in the book. Then there were *The Annals of Ireland* by the Four Masters, in which I read about the men who went to Ireland before the Flood and were drowned in it; I wondered how the Four Masters knew about them, and read no further. There were also more ordinary books, such as Machiavelli and Gibbon and Swift, and a book in four volumes that I never opened: *The Works of Andrew Marvell Esq. M. P.* It was not till I grew up that I discovered Marvell was a poet rather than a politician. I was not supposed to read any of these books; otherwise I should probably not have read any of them. The net result of them was to stimulate my interest in history. No doubt my interest was increased by the fact that my family had been prominent in English history since the early sixteenth century. I was taught English history as the record of a struggle against the King for constitutional liberty. William Lord Russell, who was executed under Charles II, was held up for special admiration, and the inference was encouraged that rebellion is often praiseworthy.

A great event in my life, at the age of eleven, was the beginning of Euclid, which was still the accepted textbook of geometry. When I had got over my disappointment in finding that he began with axioms, which had to be accepted without proof, I found great delight in him. Throughout the rest of my boyhood, mathematics absorbed a very large part of my interest. This interest was complex: partly mere pleasure in discovering that I possessed a certain kind of skill, partly delight in the power of deductive reasoning, partly the restfulness of mathematical certainty; but more than any of these (while I was still a boy) the belief that nature operates according to mathematical laws, and that human actions, like planetary motions, could be calculated if we had sufficient skill. By the time I was fifteen, I had arrived at a theory very similar to that of the Cartesians. The movements of living bodies, I felt convinced, were wholly regulated by the laws of dynamics; therefore free will must be an illusion. But, since I accepted consciousness as an indubitable datum, I could not accept materialism, though I had a certain hankering after it on account of its intellectual simplicity and its rejection of "nonsense." I still believed in God, because the First-Cause argument seemed irrefutable.

Until I went to Cambridge at the age of eighteen, my life was a very solitary one. I was brought up at home, by German nurses, German and Swiss governesses, and finally by English tutors; I saw little of other children, and when I did they were not important to me. At fourteen or fifteen I became passionately interested in religion, and set to work to examine successively the arguments for free will, immortality, and God. For a few months I had an agnostic tutor with whom I could talk about these problems, but he was sent away, presumably because he was thought to be undermining my faith. Except during these months, I kept my thoughts to myself, writing them out in a journal in Greek letters to prevent others from reading them. I was suffering the unhappiness natural to lonely adolescence, and I attributed my unhappiness to loss of religious belief. For three years I thought about religion, with a determination not to let my thoughts be influenced by my desires. I discarded first free will, then immortality; I believed in God until I was just eighteen, when I found in

Mill's *Autobiography* the sentence: "My father taught me that the question 'Who made me'? cannot be answered, since it immediately suggests the further question 'Who made God'?" In that moment I decided that the First-Cause argument is fallacious.

During these years I read widely, but as my reading was not directed, much of it was futile. I read much bad poetry, especially Tennyson and Byron; at last, at the age of seventeen, I came upon Shelley, whom no one had told me about. He remained for many years the man I loved most among great men of the past. I read a great deal of Carlyle, and admired *Past and Present*, but not *Sartor Resartus*. "The Everlasting Yea" seemed to me sentimental nonsense. The man with whom I most nearly agreed was Mill. His *Political Economy, Liberty,* and *Subjection of Women* influenced me profoundly. I made elaborate notes on the whole of his *Logic,* but could not accept his theory that mathematical propositions are empirical generalizations, though I did not know what else they could be.

All this was before I went to Cambridge. Except during the three months when I had the agnostic tutor mentioned above, I found no one to speak to about my thoughts. At home I concealed my religious doubts. Once I said that I was a utilitarian, but was met with such a blast of ridicule that I never again spoke of my opinions at home.

Cambridge opened to me a new world of infinite delight. For the first time I found that, when I uttered my thoughts, they seemed to be accepted as worth considering. Whitehead, who had examined me for entrance scholarships, had mentioned me to various people a year or two senior to me, with the result that within a week I met a number who became my life-long friends. Whitehead, who was already a Fellow and Lecturer, was amazingly kind, but was too much my senior to be a close personal friend until some years later. I found a group of contemporaries, who were able, rather earnest, hard-working, but interested in many things outside their academic work—poetry, philosophy, politics, ethics, indeed the whole world of mental adventure. We used to stay up discussing till very late on Saturday nights, meet for a late breakfast on Sunday, and then go for an all-day walk. Able young men had not yet adopted the pose of cynical superiority which came in some years later, and was first made fashionable in Cambridge by Lytton Strachey. The world seemed hopeful and solid; we all felt convinced that nineteenth-century progress would continue, and that we ourselves should be able to contribute something of value. For those who have been young since 1914 it must be difficult to imagine the happiness of those days.

Among my friends at Cambridge were McTaggart, the Hegelian philosopher; Lowes Dickinson, whose gentle charm made him loved by all who

knew him; Charles Sanger, a brilliant mathematician at College, afterwards a barrister, known in legal circles as the editor of Jarman on Wills;
two brothers, Crompton and Theodore Llewelyn Davies, sons of a Broad
Church clergyman most widely known as one of "Davies and Vaughan,"
who translated Plato's *Republic*. These two brothers were the youngest
and ablest of a family of seven, all remarkably able; they had also a quite
unusual capacity for friendship, a deep desire to be of use to the world,
and unrivalled wit. Theodore, the younger of the two, was still in the
earlier stages of a brilliant career in the government service when he was
drowned in a bathing accident. I have never known any two men so
deeply loved by so many friends. Among those of whom I saw most were
the three brothers Trevelyan, great-nephews of Macaulay. Of these the
oldest became a Labour politician and resigned from the Labour Government because it was not sufficiently socialistic; the second became a poet
and published, among other things, an admirable translation of Lucretius;
the third, George, achieved fame as an historian. Somewhat junior to me
was G. E. Moore, who later had a great influence upon my philosophy.

The set in which I lived was very much influenced by McTaggart,
whose wit recommended his Hegelian philosophy. He taught me to consider British empiricism "crude," and I was willing to believe that Hegel
(and in a lesser degree Kant) had a profundity not to be found in Locke,
Berkeley, and Hume, or in my former pope, Mill. My first three years at
Cambridge, I was too busy with mathematics to read Kant or Hegel, but
in my fourth year I concentrated on philosophy. My teachers were Henry
Sidgwick, James Ward, and G. F. Stout. Sidgwick represented the British
point of view, which I believed myself to have seen through; I therefore
thought less of him at that time than I did later. Ward, for whom I had
a very great personal affection, set forth a Kantian system, and introduced me to Lotze and Sigwart. Stout, at that time, thought very highly
of Bradley; when *Appearance and Reality* was published, he said it had
done as much as is humanly possible in ontology. He and McTaggart
between them caused me to become a Hegelian; I remember the precise
moment, one day in 1894, as I was walking along Trinity Lane, when I
saw in a flash (or thought I saw) that the ontological argument is valid.
I had gone out to buy a tin of tobacco; on my way back, I suddenly threw
it up in the air, and exclaimed as I caught it: "Great Scott, the ontological
argument is sound." I read Bradley at this time with avidity, and admired
him more than any other recent philosopher.

After leaving Cambridge in 1894, I spent a good deal of time in foreign
countries. For some months in 1894, I was honorary attaché at the British
Embassy in Paris, where I had to copy out long dispatches atttempting to
persuade the French Government that a lobster is not a fish, to which the

French Government would reply that it was a fish in 1713, at the time of the Treaty of Utrecht. I had no desire for a diplomatic career, and left the Embassy in December, 1894. I then married, and spent most of 1895 in Berlin, studying economics and German Social Democracy. The Ambassador's wife being a cousin of mine, my wife and I were invited to dinner at the Embassy; but she mentioned that we had gone to a Socialist meeting, and after this the Embassy closed its doors to us. My wife was a Philadelphia Quaker, and in 1896 we spent three months in America. The first place we visited was Walt Whitman's house in Camden, N.J.; she had known him well, and I greatly admired him. These travels were useful in curing me of a certain Cambridge provincialism; in particular, I came to know the work of Weierstrass, whom my Cambridge teachers had never mentioned. After these travels, we settled down in a workman's cottage in Sussex, to which we added a fairly large work-room. I had at that time enough money to live simply without earning, and I was therefore able to devote all my time to philosophy and mathematics, except the evenings, when we read history aloud.

In the years from 1894 to 1898, I believed in the possibility of proving by metaphysics various things about the universe that religious feeling made me think important. I decided that, if I had sufficient ability, I would devote my life to philosophy. My fellowship dissertation, on the foundations of geometry, was praised by Ward and Whitehead; if it had not been, I should have taken up economics, at which I had been working in Berlin. I remembered a spring morning when I walked in the Tiergarten, and planned to write a series of books in the philosophy of the sciences, growing gradually more concrete as I passed from mathematics to biology; I thought I would also write a series of books on social and political questions, growing gradually more abstract, At last I would achieve a Hegelian synthesis in an encyclopaedic work dealing equally with theory and practice. The scheme was inspired by Hegel, and yet something of it survived the change in my philosophy. The moment had had a certain importance: I can still, in memory, feel the squelching of melting snow beneath my feet, and smell the damp earth that promised the end of winter.

During 1898, various things caused me to abandon both Kant and Hegel. I read Hegel's *Greater Logic*, and thought, as I still do, that all he says about mathematics is muddle-headed nonsense. I came to disbelieve Bradley's arguments against relations, and to distrust the logical bases of monism. I disliked the subjectivity of the "Transcendental Aesthetic." But these motives would have operated more slowly than they did, but for the influence of G. E. Moore. He also had had a Hegelian period, but it was briefer than mine. He took the lead in rebellion, and I followed, with a sense of emancipation. Bradley argued that everything common sense believes in is mere appearance; we reverted to the opposite extreme, and thought that *everything* is real that common sense, uninfluenced by philosophy or theology, supposes real. With a sense of escaping from prison, we allowed ourselves to think that grass is green, that the sun and stars would exist if no one was aware of them, and also that there is a pluralistic timeless world of Platonic ideas. The world, which had been thin and logical, suddenly became rich and varied and solid. Mathematics could be *quite* true, and not merely a stage in dialectic. Something of this point of view appeared in my *Philosophy of Leibniz*. This book owed its origin to chance. McTaggart, who would, in the normal course, have lectured on Leibniz at Cambridge in 1898, wished to visit his family in New Zealand, and I was asked to take his place for this course. For me, the accident was a fortunate one.

The most important year in my intellectual life was the year 1900, and the most important event in this year was my visit to the International Congress of Philosophy in Paris. Ever since I had begun Euclid at the age of eleven, I had been troubled about the foundations of mathematics; when, later, I came to read philosophy, I found Kant and the empiricists equally unsatisfactory. I did not like the synthetic *a priori*, but yet arithmetic did not seem to consist of empirical generalizations. In Paris in 1900, I was impressed by the fact that, in all discussions, Peano and his pupils had a precision which was not possessed by others. I therefore asked him to give me his works, which he did. As soon as I had mastered his notation, I saw that it extended the region of mathematical precision backwards towards regions which had been given over to philosophical vagueness. Basing myself on him, I invented a notation for relations. Whitehead, fortunately, agreed as to the importance of the method, and in a very short time we worked out together such matters as the definitions of series, cardinals, and ordinals, and the reduction of arithmetic to logic. For nearly a year, we had a rapid series of quick successes. Much of the work had already been done by Frege, but at first we did not know this. The work that ultimately became my contribution to *Principia Mathematica* presented itself to me, at first, as a parenthesis in the refutation of Kant.

In June 1901, this period of honeymoon delight came to an end. Cantor had a proof that there is no greatest cardinal; in applying this proof to the universal class, I was led to the contradiction about classes that are not members of themselves. It soon became clear that this is only one of an infinite class of contradictions. I wrote to Frege, who replied with the utmost gravity that *"die Arithmetik ist ins Schwanken geraten."* At first, I hoped the matter was trivial and could be easily cleared up; but early hopes were succeeded by something very near to despair. Throughout 1903 and 1904, I pursued will-o'-the-wisps and made no progress. At last, in the spring of 1905, a different problem, which proved soluble, gave the first glimmer of hope. The problem was that of descriptions, and its solution suggested a new technique.

Scholastic realism was a metaphysical theory, but every metaphysical theory has a technical counterpart. I had been a realist in the scholastic or Platonic sense; I had thought that cardinal integers, for instance, have a timeless being. When integers were reduced to classes of classes, this being was transferred to classes. Meinong, whose work interested me, applied the arguments of realism to descriptive phrases. Everyone agrees that "the golden mountain does not exist" is a true proposition. But it has, apparently, a subject, "the golden mountain," and if this subject did not designate some object, the proposition would seem to be meaningless. Meinong inferred that there is a golden mountain, which is golden and a mountain, but does not exist. He even thought that the existent golden mountain is existent, but does not exist. This did not satisfy me, and the desire to avoid Meinong's unduly populous realm of being led me to the theory of descriptions. What was of importance in this theory was the discovery that, in analysing a significant sentence, one must not assume that each separate word or phrase has significance on its own account. "The golden mountain" can be part of a significant sentence, but is not significant in isolation. It soon appeared that class-symbols could be treated like descriptions, i.e., as non-significant parts of significant sentences. This made it possible to see, in a general way, how a solution of the contradictions might be possible. The particular solution offered in *Principia Mathematica* had various defects, but at any rate it showed that the logician is not presented with a complete *impasse*.

The theory of descriptions, and the attempt to solve the contradictions, had led me to pay attention to the problem of meaning and significance. The definition of "meaning" as applied to words and "significance" as applied to sentences is a complex problem, which I tried to deal with in *The Analysis of Mind* (1921) and *An Inquiry into Meaning and Truth* (1940). It is a problem that takes one into psychology and even physiology. The more I have thought about it, the less convinced I have become of the complete independence of logic. Seeing that logic is a much more advanced and exact science than psychology, it is clearly desirable, as far as possible, to delimit the problems that can be dealt with by logical methods. It is here that I have found Occam's razor useful.

Occam's razor, in its original form, was metaphysical: it was a principle of parsimony as regards "entities." I still thought of it in this way while *Principia Mathematica* was being written. In Plato, cardinal integers are timeless entities; they are equally so in Frege's *Grundgesetze der Arithmetik*. The definition of cardinals as classes of classes, and the discovery that class-symbols could be "incomplete symbols," persuaded me that cardinals as entities are unnecessary. But what had really been demonstrated was something quite independent of metaphysics, which is best stated in terms of "minimum vocabularies." I mean by a "minimum vocabulary" one in which no word can be defined in terms of the others. All definitions are theoretically superfluous, and therefore the whole of any science can be expressed by means of a minimum vocabulary for that science. Peano reduced the special vocabulary of arithmetic to three terms; Frege and *Principia Mathematica* maintained that even these are unnecessary, and that a minimum vocabulary for mathematics is the same as for logic. This problem is a purely technical one, and is capable of a precise solution.

There is need, however, of great caution in drawing inferences from minimum vocabularies. In the first place, there are usually, if not always, a number of different minimum vocabularies for a given subject-matter; for example, in the theory of truth-functions we may take "not-p or not-q" or "not-p and not-q" as undefined, and there is no reason to prefer one choice to the other. Then again there is often a question as to whether what seems to be a definition is not really an empirical proposition. Suppose, for instance, I define "red" as "those visual sensations which are caused by wave-lengths of such and such a range of frequencies." If we take this as what the word "red" means, no proposition containing the word can have been known before the undulatory theory of light was known and wave-lengths could be measured; and yet the word "red" was used before these discoveries had been made. This makes it clear that in all every-day statements containing the word "red" this word does not have the meaning assigned to it in the above definition. Consider the question: "Can everything that we know about colours be known to a blind man?" With the above definition, the answer is yes; with a definition derived from every-day experience, the answer is no. This problem shows how the new logic, like the Aristotelian, can lead to a narrow scholasticism.

Nevertheless, there is one kind of inference which, I think, can be drawn from the study of minimum vocabularies. Take, as one of the most important examples, the traditional problem of universals. It seems fairly certain that no vocabulary can dispense wholly with words that are more or less of the sort called "universals." These words, it is true, need never occur as nouns; they may occur only as adjectives or verbs. Probably we could be content with one such word, the word "similar," and we should never need the word "similarity." But the fact that we need the word "similar" indicates some fact about the world, and not only about language. What fact it indicates about the world, I do not know.

Another illustration of the uses of minimum vocabularies is as regards historical events. To express history, we must have a means of speaking of something which has only happened once, like the death of Caesar. An undue absorption in logic, which is not concerned with history, may cause this need to be overlooked. Spatio-temporal relativity has made it more difficult to satisfy this need than it was in a Newtonian universe, where points and instants supplied particularity.

Thus, broadly speaking, minimum vocabularies are more instructive when they show a certain kind of term to be indispensable than when they show the opposite.

In some respects, my published work, outside mathematical logic, does not at all completely represent my beliefs or my general outlook. Theory of knowledge, with which I have been largely concerned, has a certain essential subjectivity; it asks "how do *I* know what I know?" and starts inevitably from personal experience. Its data are egocentric, and so are the earlier stages of its argumentation. I have not, so far, got beyond the earlier stages, and have therefore seemed more subjective in outlook than in fact I am. I am not a solipsist, nor an idealist; I believe (though without good grounds) in the world of physics as well as in the world of psychology. But it seems clear that whatever is not experienced must, if known, be known by inference. I find that the fear of solipsism has prevented philosophers from facing this problem, and that either the necessary principles of inference have been left vague, or else the distinction between what is known by experience and what is known by inference has been denied. If I ever have the leisure to undertake another serious investigation of a philosophical problem, I shall attempt to analyse the inferences from experience to the world of physics, assuming them capable of validity, and seeking to discover what principles of inference, if true, would make them valid. Whether these principles, when discovered, are accepted as true, is a matter of temperament; what should not be a matter of temperament should be the proof that acceptance of them is necessary if solipsism is to be rejected.

I come now to what I have attempted to do in connection with social questions. I grew up in an atmosphere of politics, and was expected by my elders to take up a political career. Philosophy, however, interested me more than politics, and when it appeared that I had some aptitude for it, I decided to make it my main work. This pained my grandmother, who alluded to my investigation of the foundations of geometry as "the life you have been leading," and said in shocked tones: "O Bertie, I hear you are writing *another* book." My political interests, though secondary, nevertheless, remained very strong. In 1895, when in Berlin, I made a study of German Social Democracy, which I liked as being opposed to the Kaiser, and disliked as (at that time) embodying Marxist orthodoxy. For a time, under the influence of Sidney Webb, I became an imperialist, and even supported the Boer War. This point of view, however, I abandoned completely in 1901; from that time onwards, I felt an intense dislike of the

use of force in human relations, though I always admitted that it is sometimes necessary. When Joseph Chamberlain, in 1903, turned against free
trade, I wrote and spoke against him, my objections to his proposals being
those of an internationalist. I took an active part in the agitation for
Women's Suffrage. In 1910, *Principia Mathematica* being practically finished, I wished to stand for Parliament, and should have done so if the
Selection Committee had not been shocked to discover that I was a free
thinker.

The first world war gave a new direction to my interests. The war,
and the problem of preventing future wars, absorbed me, and the books
that I wrote on this and cognate subjects caused me to become known to
a wider public. During the war I had hoped that the peace would embody
a rational determination to avoid future great wars; this hope was destroyed by the Versailles Treaty. Many of my friends saw hope in Soviet
Russia, but when I went there in 1920 I found nothing that I could like
or admire. I was then invited to China, where I spent nearly a year. I
loved the Chinese, but it was obvious that the resistance to hostile mili-
tarisms must destroy much of what was best in their civilization. They
seemed to have no alternative except to be conquered or to adopt many
of the vices of their enemies. But China did one thing for me that the
East is apt to do for Europeans who study it with sensitive sympathy:
it taught me to think in long stretches of time, and not to be reduced to
despair by the badness of the present. Throughout the increasing gloom
of the past twenty years, this habit has helped to make the world less
unendurable than it would otherwise have been.

In the years after my return from China, the birth of my two older children caused me to become interested in early education, to which, for
some time, I devoted most of my energy. I have been supposed to be an
advocate of complete liberty in schools, but this, like the view that I am
an anarchist, is a mistake. I think a certain amount of force is indispensable, in education as in government; but I also think that methods can
be found which will greatly diminish the necessary amount of force. This
problem has both political and private aspects. As a rule, children or
adults who are happy are likely to have fewer destructive passions, and
therefore to need less restraint, than those who are unhappy. But I do not
think that children can be made happy by being deprived of guidance,
nor do I think that a sense of social obligation can be fostered if complete
idleness is permitted. The question of discipline in childhood, like all
other practical questions, is one of degree. Profound unhappiness and
instinctive frustration is apt to produce a deep grudge against the world,
issuing, sometimes by a very roundabout road, in cruelty and violence.
The psychological and social problems involved first occupied my attention during the war of 1914–18; I was especially struck by the fact that,

at first, most people seemed to enjoy the war. Clearly this was due to a variety of social ills, some of which were educational. But while individual parents can do much for their individual children, large-scale educational reform must depend upon the state, and therefore upon prior political and economic reforms. The world, however, was moving more and more in the direction of war and dictatorship, and I saw nothing useful that I could do in practical matters. I therefore increasingly reverted to philosophy, and to history in relation to ideas.

History has always interested me more than anything else except philosophy and mathematics. I have never been able to accept any general schema of historical development, such as that of Hegel or that of Marx. Nevertheless, general trends can be studied, and the study is profitable in relation to the present. I found much help in understanding the nineteenth century from studying the effect of liberal ideas in the period from 1814 to 1914.[2] The two types of liberalism, the rational and the romantic, represented by Bentham and Rousseau respectively, have continued, ever since, their relations of alternate alliance and conflict.

The relation of philosophy to social conditions has usually been ignored by professional philosophers. Marxists are interested in philosophy as an *effect*, but do not recognize it as a *cause*. Yet plainly every important philosophy is both. Plato is in part an effect of the victory of Sparta in the Peloponnesian war, and is also in part among the causes of Christian theology. To treat him only in the former aspect is to make the growth of the medieval church inexplicable. I am at present writing a history of western philosophy from Thales to the present day, in which every important system is treated equally as an effect and as a cause of social conditions.

My intellectual journeys have been, in some respects, disappointing. When I was young I hoped to find religious satisfaction in philosophy; even after I had abandoned Hegel, the eternal Platonic world gave me something non-human to admire. I thought of mathematics with reverence, and suffered when Wittgenstein led me to regard it as nothing but tautologies. I have always ardently desired to find some justification for the emotions inspired by certain things that seemed to stand outside human life and to deserve feelings of awe. I am thinking in part of very obvious things, such as the starry heavens and a stormy sea on a rocky coast; in part of the vastness of the scientific universe, both in space and time, as compared to the life of mankind; in part of the edifice of impersonal truth, especially truth which, like that of mathematics, does not merely describe the world that happens to exist. Those who attempt to make a religion of humanism, which recognizes nothing greater than man, do not satisfy my emotions. And yet I am unable to believe that, in the

[2] *Freedom and Organization, 1814–1914* (1934).

world as known, there is anything that I can value outside human beings, and, to a much lesser extent, animals. Not the starry heavens, but their effects on human percipients, have excellence; to admire the universe for its size is slavish and absurd; impersonal non-human truth appears to be a delusion. And so my intellect goes with the humanists, though my emotions violently rebel. In this respect, the "consolations of philosophy" are not for me.

In more purely intellectual ways, on the contrary, I have found as much satisfaction in philosophy as any one could reasonably have expected. Many matters which, when I was young, baffled me by the vagueness of all that had been said about them, are now amenable to an exact technique, which makes possible the kind of progress that is customary in science. Where definite knowledge is unattainable, it is sometimes possible to prove that it is unattainable, and it is usually possible to formulate a variety of exact hypotheses, all compatible with the existing evidence. Those philosophers who have adopted the methods derived from logical analysis can argue with each other, not in the old aimless way, but coöperatively, so that both sides can concur as to the outcome. All this is new during my lifetime; the pioneer was Frege, but he remained solitary until his old age. This extension of the sphere of reason to new provinces is something that I value very highly. Philosophic rationality may be choked in the shocks of war and the welter of new persecuting superstitions, but one may hope that it will not be lost utterly or for more than a few centuries. In this respect, my philosophic life has been a happy one.

COMMENTARY ON
ALFRED NORTH WHITEHEAD

A LFRED NORTH WHITEHEAD was an elevated and lovable man, a commanding figure of mathematics and logic, and perhaps the outstanding philosopher of recent decades. Yet as a philosopher he was a failure, judged by his own standards. "Philosophy," he said, "is either self-evident, or it is not philosophy." Whitehead's system is grand and deep and influential but it is certainly not self-evident. It is in fact so abominably obscure in many of its parts that few philosophers claim completely to understand it. Nevertheless, none deny its importance and that in itself is a tribute to Whitehead's position in contemporary thought.

Whitehead was born in 1861 at Ramsgate, England.[1] His father, the Reverend Alfred Whitehead, was then headmaster of a private school, but later took up clerical duties. "My father," Whitehead once remarked, "was not intellectual but he possessed personality." He took an active part in local affairs and was well liked. Archbishop Tait and Sir Moses Montefiore were both his close friends; when the Baptist minister in the parish was dying, the elder Whitehead read the Bible to him. Whitehead was permanently influenced by the religious, educational and social interests that pervaded his home; also by its physical surroundings, which were not only beautiful but rich in archaeological remains. There were Roman forts, Norman churches, the shores of Ebbes Fleet where the Saxons and Augustine landed; Canterbury Cathedral was sixteen miles away. At the age of fifteen he was sent to the Sherborne school, founded in the eighth century. Alfred the Great is said to have been one of its pupils, and "we worked under the sound of the Abbey bells, brought from the Field of the Cloth of Gold by Henry VIII." [2] The relics of English history imbued Whitehead with deep feeling for the past, with a secure sense of belonging to the community, of sharing the past with those about him.

At Sherborne, Whitehead studied Greek and Latin, mathematics, science, Wordsworth, Shelley and Scriptures. "We did not want to explain

---

[1] The material in this sketch is taken from the following sources, among others: Sir Edmund T. Whittaker, "Alfred North Whitehead (1861–1947)," *Obituary Notices of the Royal Society*, 1948, pp. 281–296 (numerous unkeyed quotations); C. D. Broad, "Obituary of Alfred North Whitehead," *Mind*, April, 1948, pp. 139–145; Ernest Nagel, "Obituary of Alfred North Whitehead," *The Nation*, Feb. 14, 1948; Bertrand Russell, "Whitehead and Principia Mathematica," *Mind*, April, 1948, pp. 137–138; Bertrand Russell, "Portraits From Memory: I: Alfred North Whitehead," *Harper's Magazine*, Dec., 1952, pp. 50–52; Victor Lowe, Charles Hartshorne and A. H. Johnson, *Whitehead and the Modern World*, Boston, 1950; Paul Arthur Schilpp, *The Philosophy of Alfred North Whitehead* (in particular Whitehead's "Autobiographical Notes," pp. 3–14), Evanston and Chicago, 1941; A. H. Johnson, *Whitehead's Theory of Reality*, Boston, 1952.

[2] Paul Schilpp, "Autobiographical Notes," *op. cit.*, p. 5.

the origin of anything. We wanted to read about people like ourselves, and to imbibe their ideals." [3] Besides imbibing ideals he played cricket and football ("very enjoyable but taking time"), rose to be "Captain of Games" and "Head of the School." On the strength of his own statement one accepts the remarkable fact that he was happy at an English public school.

In 1880 Whitehead entered Trinity College, Cambridge, where he remained as a student and Fellow until 1910. The emphasis of his education was on mathematics, in which he distinguished himself, but no less important was the "free discussion" among friends. These conversations, as he remembered vividly half a century later, were like "a daily Platonic dialogue." Elsewhere in these pages I have mentioned his membership in the Cambridge society known as the "Apostles." His recollection fastens on many conversations with D'Arcy Thompson, James Stephen, Lowes Dickinson, Henry Head, Frederic Maitland, Henry Sidgwick, Arthur Verrall, "and casual judges, or scientists, or members of Parliament who had come up to Cambridge for the weekend." The Platonic education was "a wonderful influence," but "very limited in its application to life." [4]

Whitehead's first book, published in 1898, was *A Treatise on Universal Algebra, with Applications*. It was a development of the profound idea of a highly general geometry set forth by the great but much-neglected German mathematician, Hermann Grassmann, in his book *Ausdehnungslehre*. *Universal Algebra* deals with the Theory of Matrices and other noncommutative algebras, also with non-Euclidean geometry; but its major concern is the algebra of symbolic logic. This subject attracted Whitehead's for-that-time unorthodox tastes in mathematics, "his philosophic urge to grasp the nature of mathematics in its widest aspects." Whittaker remarks on the prophetic instinct that impelled Whitehead to take up certain out-of-the-way branches of mathematics which have since played a great part in the interpretation of nature. Matrix theory was turned to invaluable use by Heisenberg, Born and Jordan in their system of quantum mechanics; non-Euclidean geometry has found important applications in cosmological investigations; quaternions, the "original" noncommutative algebra, have provided a powerful tool for treating problems raised by the Special Theory of Relativity. [5]

Universal Algebra was widely acclaimed; within five years of its appearance Whitehead was elected to the Royal Society. His reputation as a teacher and original investigator attracted the best students, among them Bertrand Russell who became "a specially attached disciple." In 1900 Whitehead and Russell went to a philosophical congress in Paris where

[3] A. N. Whitehead, "The Education of an Englishman," *The Atlantic Monthly*, Vol. 138, p. 195.
[4] Paul Schilpp, "Autobiographical Notes," *op. cit.*, p. 8.
[5] Whittaker, *op. cit.*, p. 282.

they heard an account of the work of the great Italian logician, Giuseppe Peano, who had recently invented a new system of symbols for use in symbolic logic. The logical algebra of Boole had only a limited repertory of symbols representing such words as "and," "or," "not." Peano enormously increased the range of symbolic logic by introducing symbols to represent other logical notions: "is contained in," "the aggregate of all x's such that," "there exists," "is a," "the only," etc. These ideograms "represent the constitutive elements of all other notions in logic, just as the chemical atoms are the constitutive elements of all substances in chemistry; and they are capable of replacing ordinary language completely for the purposes of any deductive theory." [6] Whitehead and Russell recognized the possibilities opened by Peano's achievement. In their epochmaking *Principia Mathematica* (for further discussion see selections by Nagel and Tarski, pp. 1878 and 1901) they attempted, by developing his methods, to bring order to the sadly disarrayed study of the foundations of mathematics. In particular they sought to prove that mathematics is a part of logic.[7] In this they were partially but brilliantly successful. Their work is acknowledged to be the "greatest single contribution" to logic since Aristotle; it is also the most remarkable of modern collaborations.

The work on the *Principia* lasted for ten years. Russell describes the association as completely harmonious. Each man gave richly of his mathematical and philosophic gifts; each stimulated the other. Whitehead was the more patient and careful of the two; Russell often found ways of simplifying the presentation. "Neither of us alone," says Russell, "could have written the book; even together, and with the alleviation brought by mutual discussion, the effort was so severe that at the end we both turned aside from mathematical logic with a kind of nausea. It was, I suppose, inevitable that we should turn aside in different directions, so that collaboration was no longer possible." [8] Whitehead ascribes the separation to divergences of "our fundamental points of view—philosophic and sociological . . . and so with different interests our collaboration came to a natural end." [9]

In 1910 Whitehead resigned the lectureship at Trinity and moved to London. For a time he taught at the University of London and in 1914

[6] Whittaker, *op. cit.*, p. 283.
[7] "In 1900 [Russell writes] we went together to the International Congress of Philosophy in Paris, where I was impressed by Peano. I saw that methods analogous to his would clarify the logic of relations, and I was led to the definitions of cardinal, ordinal, rational, and real numbers which are given in *The Principles of Mathematics*. Very soon Whitehead became interested. The project of deducing mathematics from logic appealed to him, and to my great joy he agreed to collaborate. I knew that my mathematical capacity was not equal to accomplishing this task unaided. Moreover, in June 1901 I came upon the contradiction about classes which are not members of themselves, and from that time on a large part of my time was occupied with attempts to avoid contradictions." Russell, *Mind, op. cit.*, p. 137.
[8] *Ibid.*, p. 138.
[9] Schilpp, *op. cit.*, p. 5.

accepted a professorship of applied mathematics at Imperial College of Science and Technology in Kensington. His life was a busy one, divided among many academic and administrative activities. He wrote the famous *Introduction to Mathematics*, a classic of popularization, served as Dean of the Faculty of Science in the University, as Chairman of the Academic Council and as an active member of many other educational committees. His interest in philosophy was not crowded out by these duties. On the contrary, his experience in practical affairs stimulated his thoughts and brought out, as he says, "latent capabilities." His philosophical writings began toward the end of the war. In 1919 he published *An Enquiry Concerning the Principles of Natural Knowledge* and in 1920 a volume of Tarner Lectures on *The Concept of Nature*. The earlier volume records the great sorrow of his life; it bears the dedication to his son "Eric Alfred Whitehead, Royal Flying Corps: Killed in action over the Forêt de Gobain, March 13, 1918."

Whitehead was invited in 1924 to become professor of philosophy at Harvard. There he served until his retirement in 1937. This period is marked by the publications expressing most fully his philosophical system. "In the course of private discussion he once remarked: 'From twenty on I was interested in philosophy, religion, logic, and history. Harvard gave me a chance to express myself.' " [10] The first fruit of this opportunity was his Lowell Lectures (1925) published under the title *Science and the Modern World*. Whitehead described this volume, from which the extract below has been taken, as a study of the effect of the development of science during the past three centuries on certain aspects of Western culture. It is the most widely influential of his writings, admired not only by professional philosophers but by the general reader. John Dewey proclaimed its restatement of the present relations of science, philosophy and the issues of life "an unforgettable intellectual experience."

In *Process and Reality* (1929), the Gifford Lectures given at the University of Edinburgh in 1927 and 1928, Whitehead presented a complete and systematic exposition of his philosophy of the organism. The British philosopher C. D. Broad describes the work as "one of the most difficult philosophical books that exist," vying in this respect with the works of Plotinus and Hegel.[11] And that in substance is the judgment of most

[10] A. H. Johnson, *The Wit and Wisdom of Alfred North Whitehead*, Boston, 1947, p. 8.
[11] C. D. Broad, *op. cit.*, p. 144. "It is often desperately difficult to understand what it is that Whitehead is asserting. When one is fairly sure of this, it is often equally hard to discover what he considers to be the reason for asserting it; for he seems often to be 'not *arguing* but just *telling* you.' And, finally, when one thinks that one knows what he is asserting and what he is alleging as the ground for it, one often fails to see how the latter proves or makes probable the former. . . . Still, from my knowledge of Whitehead and of those of his writings which I think I can understand, and from the occasional gleams and glimpses which have been vouchsafed to me in struggling with *Process and Reality*, I feel fairly certain that there is something important concealed beneath the portentous verbiage of the Gifford Lectures."

persons who have tried to follow the book, honestly thinking it through and not merely joining those enthusiastic admirers who "wonder with a foolish face of praise." In *Adventures in Ideas* (1933) Whitehead further elaborated his vision of the universe. The style of this book, though somewhat more involved, resembled that of *Science and the Modern World*, which is to say that many of the adventures were open to any intelligent reader. Of all his writings, Whitehead looked on these two with the greatest personal satisfaction.[12]

It is almost impossible to summarize Whitehead's philosophy in a few sentences. He started out, as we have seen, as a mathematician with strong philosophical inclinations. These are reflected in the breadth and depth of his approach and in the objectives of *Universal Algebra* and the *Principia*. In the second period of his thought he turned to a critique of the prevailing concepts of mathematical physics. He rejected, in Whittaker's words, the commonly accepted starting point of physics: "namely that space and time provide, so to speak, a stage on which ponderable bodies, aether and electricity maintain an unending performance. In place of this he put forward the doctrine that the ultimate components of reality are *events*." Every event, or "spatio-temporal happening," includes other events and is itself included in larger, more comprehensive aggregates. Time and space enter into this scheme of the external world as "derivative concepts . . . abstractions which express relations between events." The notion of an *instant* of time or of a *point* of space are "not truly primitive notions" but are obtained by Whitehead's famous method of extensive abstraction.[13] One of the classical problems Whitehead sought to solve by this intricate method was that of the familiar dualism (or as he called it, "bifurcation") between mind and nature, according to which there is an outside "real" world which is knowable to us only through the imperfect medium of sensations.[14] In the opinion of many philosophers, Whitehead managed to bridge the allegedly unbridgeable gap between what we perceive as a table and the "real thing" itself which produces a table-image in the brain.[15] "Natural philosophy," he said,

[12] F. S. C. Northrop and Mason Gross, *Alfred North Whitehead, An Anthology*, New York, 1953, p. 749.

[13] "The idea of this method is, to consider a set of events each of which includes all the subsequent members of the sequence, like a nest of boxes each slightly larger than the one next inside it: if the events ultimately diminish indefinitely (or, in more precise language, if no event is extended over by every event of the set), then they may be regarded as defining a limit: points of space and instants of time may be defined in this way. . . ." Whittaker, *op. cit.*, pp. 286–287.

[14] The problem was "to overcome the familiar dualism between a world of scientific objects, supposed to be knowable only as remote causes of sensations, and a world of sense data, supposed to be private and mind-dependent." Broad, *op. cit.*, p. 142.

[15] "It is one of Whitehead's signal achievements to have recognized that the radical bifurcation of man and nature associated with what he called 'scientific materialism,' is not necessitated by physical science. He was able to show that the concepts of physics, far from describing 'a real world' with which the genuine reality of things

"should never ask what is in the mind and what is in nature": "there is but one nature, namely the nature which is before us in perpetual knowledge."

Whitehead's philosophy of the organism, the third and crowning phase of his speculations, was an extension of his basic ideas to Nature as a whole. It was a cosmology in which rationalism and religion were blended. The universe is not unintelligible, not an "arbitrary mystery" whose ultimate essence must remain forever concealed from us; on the other hand it is not merely a machine. It is a universe of process, of becoming and perishing; these constitute, he said, "the creative advance of the world." In this endless flux of becoming all entities are seen as related, purpose exhibits itself in a kind of "selective activity," and what we call life and "feeling" are in some sense immanent in everything actual.

Whitehead's philosophic cosmology, as Ernest Nagel has observed, "is a vision which is articulated with amazing virtuosity but which rests less upon detailed argument than upon direct insight." [16] Like other visions, it is not always easily shared by others. Its animism seems to reintroduce a brand of hocus-pocus which rational thought has long sought to extirpate; its mathematical elements are forbidding and its romantic strains confusing. Moreover it is couched in an involved idiom made more involved by the strange, unlovely words—concrescence, prehension, appetition, etc.—that Whitehead coined. Yet even those who are most strongly repelled by the philosophy of organism recognize in its formulation the many profound observations, the honesty and daring imagination of a truly wise human being.

As a person Whitehead was well loved. He was simple in manner, unfailingly gentle and kind. He was passionately devoted to his wife and children. Russell describes him as an "extraordinarily perfect" teacher who "would elicit from a pupil the best of which a pupil was capable. He was never repressive, or sarcastic, or any of the things that inferior teachers like to be." [17] He was a modest man and "his most extreme boast was that he did try to have the qualities of his defects." [18] He often spoke of the "many shortcomings" of his published work, venturing only the mitigation: "Philosophy is an attempt to express the infinity of the

---

apprehended in immediate experience is incompatible, are abstractions referring to qualities and processes discernible in such experience. The task that Whitehead envisioned for philosophy was therefore that of supplying a critique of abstractions in a world that contains much more than the skeletal patterns of change made explicit by natural science. But he not only proposed this task. He also developed a powerful intellectual technique for carrying it out, and used it to illumine the significance of a number of basic scientific concepts." Nagel, *op. cit.*, p. 187.

[16] Nagel, *op. cit.*, p. 188.

[17] Russell, *Harper's, op. cit.*, p. 52.

[18] *Ibid.*, p. 51.

universe in terms of the limitations of language." [19] Everyone who knew him speaks of his delightful humor. He never minded, says Russell, telling stories against himself, and in all ways he was intensely human—given to anxieties, "distressing soliloquies," and self-reproach; "certainly not that inhuman monster 'the rational man.'" Russell illustrates his self-discipline and capacity for concentration in this vivid anecdote: "One hot summer day when I was staying with him at Grantchester, our friend Crompton Davies arrived and I took him into the garden to say how-do-you-do to his host. Whitehead was sitting writing mathematics. Davies and I stood in front of him at a distance of no more than a yard and watched him covering page after page with symbols. He never saw us, and after a time we went away with a feeling of awe." [20]

Whitehead died at Cambridge, Massachusetts, on December 30, 1947, at the age of eighty-six.

[19] Schilpp, "Autobiographical Notes," *op. cit.*, p. 14.
[20] Russell, *Harper's, op. cit.*, p. 51.

Thought is only a flash between two long nights, but this flash is everything.
—HENRI POINCARÉ

15 Mathematics as an Element in the History of Thought

By ALFRED NORTH WHITEHEAD

THE science of pure mathematics, in its modern developments, may claim to be the most original creation of the human spirit. Another claimant for this position is music. But we will put aside all rivals, and consider the ground on which such a claim can be made for mathematics. The originality of mathematics consists in the fact that in mathematical science connections between things are exhibited which, apart from the agency of human reason, are extremely unobvious. Thus the ideas, now in the minds of contemporary mathematicians, lie very remote from any notions which can be immediately derived by perception through the senses; unless indeed it be perception stimulated and guided by antecedent mathematical knowledge. This is the thesis which I proceed to exemplify.

Suppose we project our imagination backwards through many thousands of years, and endeavour to realise the simple-mindedness of even the greatest intellects in those early societies. Abstract ideas which to us are immediately obvious must have been, for them, matters only of the most dim apprehension. For example take the question of number. We think of the number 'five' as applying to appropriate groups of any entities whatsoever—to five fishes, five children, five apples, five days. Thus in considering the relations of the number 'five' to the number 'three,' we are thinking of two groups of things, one with five members and the other with three members. But we are entirely abstracting from any consideration of any particular entities, or even of any particular sorts of entities, which go to make up the membership of either of the two groups. We are merely thinking of those relationships between those two groups which are entirely independent of the individual essences of any of the members of either group. This is a very remarkable feat of abstraction; and it must have taken ages for the human race to rise to it. During a long period, groups of fishes will have been compared to each other in respect to their multiplicity, and groups of days to each other. But the first man who noticed the analogy between a group of seven fishes and a group of seven days made a notable advance in the history of thought. He was the first man who entertained a concept belonging to the science

402

of pure mathematics. At that moment it must have been impossible for him to divine the complexity and subtlety of these abstract mathematical ideas which were waiting for discovery. Nor could he have guessed that these notions would exert a widespread fascination in each succeeding generation. There is an erroneous literary tradition which represents the love of mathematics as a monomania confined to a few eccentrics in each generation. But be this as it may, it would have been impossible to anticipate the pleasure derivable from a type of abstract thinking which had no counterpart in the then-existing society. Thirdly, the tremendous future effect of mathematical knowledge on the lives of men, on their daily avocations, on their habitual thoughts, on the organization of society, must have been even more completely shrouded from the foresight of those early thinkers. Even now there is a very wavering grasp of the true position of mathematics as an element in the history of thought. I will not go so far as to say that to construct a history of thought without profound study of the mathematical ideas of successive epochs is like omitting Hamlet from the play which is named after him. That would be claiming too much. But it is certainly analogous to cutting out the part of Ophelia. This simile is singularly exact. For Ophelia is quite essential to the play, she is very charming—and a little mad. Let us grant that the pursuit of mathematics is a divine madness of the human spirit, a refuge from the goading urgency of contingent happenings.

When we think of mathematics, we have in our mind a science devoted to the exploration of number, quantity, geometry, and in modern times also including investigation into yet more abstract concepts of order, and into analogous types of purely logical relations. The point of mathematics is that in it we have always got rid of the particular instance, and even of any particular sorts of entities. So that for example, no mathematical truths apply merely to fish, or merely to stones, or merely to colours. So long as you are dealing with pure mathematics, you are in the realm of complete and absolute abstraction. All you assert is, that reason insists on the admission that, if any entities whatever have any relations which satisfy such-and-such purely abstract conditions, then they must have other relations which satisfy other purely abstract conditions.

Mathematics is thought moving in the sphere of complete abstraction from any particular instance of what it is talking about. So far is this view of mathematics from being obvious, that we can easily assure ourselves that it is not, even now, generally understood. For example, it is habitually thought that the certainty of mathematics is a reason for the certainty of our geometrical knowledge of the space of the physical universe. This is a delusion which has vitiated much philosophy in the past, and some philosophy in the present. The question of geometry is a test case of some urgency. There are certain alternative sets of purely abstract conditions possible for the relationship of groups of unspecified entities, which I will call *geometrical conditions*. I give them this name because of their general analogy to those conditions, which we believe to hold respecting the particular geometrical relations of things observed by us in our direct perception of nature. So far as our observations are concerned, we are not quite accurate enough to be certain of the exact conditions regulating the things we come across in nature. But we can by a slight stretch of hypothesis identify these observed conditions with some one set of the purely abstract geometrical conditions. In doing so, we make a particular determination of the group of unspecified entities which are the *relata* in the abstract science. In the pure mathematics of geometrical relationships, we say that, if *any* group entities enjoy *any* relationships among its members satisfying *this* set of abstract geometrical conditions, then such-and-such additional abstract conditions must also hold for such relationships. But when we come to physical space, we say that some definitely observed group of physical entities enjoys some definitely observed relationships among its members which do satisfy this above-mentioned set of abstract geometrical conditions. We thence conclude that the additional relationships which we concluded to hold in *any* such case, must therefore hold in *this particular* case.

The certainty of mathematics depends upon its complete abstract generality. But we can have no *a priori* certainty that we are right in believing that the observed entities in the concrete universe form a particular instance of what falls under our general reasoning. To take another example from arithmetic. It is a general abstract truth of pure mathematics that any group of forty entities can be subdivided into two groups of twenty entities. We are therefore justified in concluding that a particular group of apples which we believe to contain forty members can be subdivided into two groups of apples of which each contain twenty members. But there always remains the possibility that we have miscounted the big group; so that, when we come in practice to subdivide it, we shall find that one of the two heaps has an apple too few or an apple too many.

Accordingly, in criticising an argument based upon the application of mathematics to particular matters of fact there are always three processes to be kept perfectly distinct in our minds. We must first scan the purely mathematical reasoning to make sure that there are no mere slips in it— no casual illogicalities due to mental failure. Any mathematician knows from bitter experience that in first elaborating a train of reasoning, it is very easy to commit a slight error which yet makes all the difference. But when a piece of mathematics has been revised, and has been before the expert world for some time, the chance of a casual error is almost negligible. The next process is to make quite certain of all the abstract conditions which have been presupposed to hold. This is the determination of the abstract premises from which the mathematical reasoning proceeds. This is a matter of considerable difficulty. In the past quite remarkable oversights have been made, and have been accepted by generations of the greatest mathematicians. The chief danger is that of oversight, namely, tacitly to introduce some condition, which it is natural for us to presuppose, but which in fact need not always be holding. There is another opposite oversight in this connection which does not lead to error, but only to lack of simplification. It is very easy to think that more postulated conditions are required than is in fact the case. In other words, we may think that some abstract postulate is necessary which is in fact capable of being proved from the other postulates that we have already on hand. The only effects of this excess of abstract postulates are to diminish our aesthetic pleasure in the mathematical reasoning, and to give us more trouble when we come to the third process of criticism.

This third process of criticism is that of verifying that our abstract postulates hold for the particular case in question. It is in respect to this process of verification for the particular case that all the trouble arises. In some simple instances, such as the counting of forty apples, we can with a little care arrive at practical certainty. But in general, with more complex instances, complete certainty is unattainable. Volumes, libraries of volumes, have been written on the subject. It is the battle ground of rival philosophers. There are two distinct questions involved. There are particular definite things observed, and we have to make sure that the relations between these things really do obey certain definite exact abstract conditions. There is great room for error here. The exact observational methods of science are all contrivances for limiting these erroneous conclusions as to direct matters of fact. But another question arises. The things directly observed are, almost always, only samples. We want to conclude that the abstract conditions, which hold for the samples, also hold for all other entities which, for some reason or other, appear to us to be of the same sort. This process of reasoning from the sample to the whole species is Induction. The theory of Induction is the despair of philosophy—and yet all our activities are based upon it. Anyhow, in criticising a mathematical conclusion as to a particular matter of fact, the real difficulties consist in finding out the abstract assumptions involved, and in estimating the evidence for their applicability to the particular case in hand.

It often happens, therefore, that in criticising a learned book of applied mathematics, or a memoir, one's whole trouble is with the first chapter, or even with the first page. For it is there, at the very outset, where the author will probably be found to slip in his assumptions. Farther, the trouble is not with what the author does say, but with what he does not say. Also it is not with what he knows he has assumed, but with what he has unconsciously assumed. We do not doubt the author's honesty. It is his perspicacity which we are criticising. Each generation criticises the unconscious assumptions made by its parents. It may assent to them, but it brings them out in the open.

The history of the development of language illustrates this point. It is a history of the progressive analysis of ideas. Latin and Greek were inflected languages. This means that they express an unanalysed complex of ideas by the mere modification of a word; whereas in English, for example, we use prepositions and auxiliary verbs to drag into the open the whole bundle of ideas involved. For certain forms of literary art— though not always—the compact absorption of auxiliary ideas into the main word may be an advantage. But in a language such as English there is the overwhelming gain in explicitness. This increased explicitness is a more complete exhibition of the various abstractions involved in the complex idea which is the meaning of the sentence.

By comparison with language, we can now see what is the function in thought which is performed by pure mathematics. It is a resolute attempt to go the whole way in the direction of complete analysis, so as to separate the elements of mere matter of fact from the purely abstract conditions which they exemplify.

The habit of such analysis enlightens every act of the functioning of the human mind. It first (by isolating it) emphasizes the direct aesthetic appreciation of the content of experience. This direct appreciation means an apprehension of what this experience is in itself in its own particular essence, including its immediate concrete values. This is a question of direct experience, dependent upon sensitive subtlety. There is then the abstraction of the particular entities involved, viewed in themselves, and as apart from that particular occasion of experience in which we are then apprehending them. Lastly there is the further apprehension of the absolutely general conditions satisfied by the particular relations of those entities as in that experience. These conditions gain their generality from the fact that they are expressible without reference to those particular relations or to those particular relata which occur in that particular occasion of experience. They are conditions which might hold for an indefinite variety of other occasions, involving other entities and other relations between them. Thus these conditions are perfectly general because they refer to no particular occasion, and to no particular entities (such as green, or blue, or trees) which enter into a variety of occasions, and to no particular relationships between such entities.

There is, however, a limitation to be made to the generality of mathematics; it is a qualification which applies equally to all general statements. No statement, except one, can be made respecting any remote occasion which enters into no relationship with the immediate occasion so as to form a constitutive element of the essence of that immediate occasion. By the 'immediate occasion' I mean that occasion which involves as an ingredient the individual act of judgment in question. The one excepted statement is:—If anything out of relationship, then complete ignorance as to it. Here by 'ignorance,' I mean *ignorance*; accordingly no advice can be given as to how to expect it, or to treat it, in 'practice' or in any other way. Either we know something of the remote occasion by the cognition which is itself an element of the immediate occasion, or we know nothing. Accordingly the full universe, disclosed for every variety of experience, is a universe in which every detail enters into its proper relationship with the immediate occasion. The generality of mathematics is the most complete generality consistent with the community of occasions which constitutes our metaphysical situation.

It is further to be noticed that the particular entities require these general conditions for their ingression into any occasions; but the same general conditions may be required by many types of particular entities. This fact, that the general conditions transcend any one set of particular entities, is the ground for the entry into mathematics, and into mathematical logic, of the notion of the 'variable.' It is by the employment of this notion that general conditions are investigated without any specification of particular entities. This irrelevance of the particular entities has not been generally understood: for example, the shape-iness of shapes, *e.g.*, circularity and sphericity and cubicality as in actual experience, do not enter into the geometrical reasoning.

The exercise of logical reason is always concerned with these absolutely general conditions. In its broadest sense, the discovery of mathematics is the discovery that the totality of these general abstract conditions, which are concurrently applicable to the relationships among the entities of any one concrete occasion, are themselves inter-connected in the manner of a pattern with a key to it. This pattern of relationships among general abstract conditions is imposed alike on external reality, and on our abstract representations of it, by the general necessity that every thing must be just its own individual self, with its own individual way of differing from everything else. This is nothing else than the necessity of abstract logic, which is the presupposition involved in the very fact of inter-related existence as disclosed in each immediate occasion of experience.

The key to the patterns means this fact:—that from a select set of those general conditions, exemplified in any one and the same occasion, a pattern involving an infinite variety of other such conditions, also exemplified in the same occasion, can be developed by the pure exercise of abstract logic. Any such select set is called the set of postulates, or premises, from which the reasoning proceeds. The reasoning is nothing else than the exhibition of the whole pattern of general conditions involved in the pattern derived from the selected postulates.

The harmony of the logical reason, which divines the complete pattern as involved in the postulates, is the most general aesthetic property arising from the mere fact of concurrent existence in the unity of one occasion. Wherever there is a unity of occasion there is thereby established an aesthetic relationship between the general conditions involved in that occasion. This aesthetic relationship is that which is divined in the exercise of rationality. Whatever falls within that relationship is thereby exemplified in that occasion, whatever falls without that relationship is thereby excluded from exemplification in that occasion. The complete pattern of general conditions, thus exemplified, is determined by any one of many select sets of these conditions. These key sets are sets of equivalent postulates. This reasonable harmony of being, which is required for the unity of a complex occasion, together with the completeness of the realisation (in that occasion) of all that is involved in its logical harmony, is the primary article of metaphysical doctrine. It means that for things to be together involves that they are reasonably together. This means that thought can penetrate into every occasion of fact, so that by comprehending its key conditions, the whole complex of its pattern of conditions lies open before it. It comes to this:—provided we know something which is perfectly general about the elements in any occasion, we can then know an indefinite number of other equally general concepts which must also be exemplified in that same occasion. The logical harmony involved in the unity of an occasion is both exclusive and inclusive. The occasion must exclude the inharmonious, and it must include the harmonious.

Pythagoras was the first man who had any grasp of the full sweep of this general principle. He lived in the sixth century before Christ. Our knowledge of him is fragmentary. But we know some points which establish his greatness in the history of thought. He insisted on the importance of the utmost generality in reasoning, and he divined the importance of number as an aid to the construction of any representation of the conditions involved in the order of nature. We know also that he studied geometry, and discovered the general proof of the remarkable theorem about right-angled triangles. The formation of the Pythagorean Brotherhood, and the mysterious rumours as to its rites and its influence, afford some evidence that Pythagoras divined, however dimly, the possible importance of mathematics in the formation of science. On the side of philosophy he started a discussion which has agitated thinkers ever since. He asked, 'What is the status of mathematical entities, such as numbers for example, in the realm of things?' The number 'two,' for example, is in some sense exempt from the flux of time and the necessity of position in space. Yet it is involved in the real world. The same considerations apply to geometrical notions—to circular shape, for example. Pythagoras is said to have taught that the mathematical entities, such as numbers and shapes, were the ultimate stuff out of which the real entities of our perceptual experience are constructed. As thus baldly stated, the idea seems crude, and indeed silly. But undoubtedly, he had hit upon a philosophical notion of considerable importance; a notion which has a long history, and which has moved the minds of men, and has even entered into Christian theology. About a thousand years separate the Athanasian Creed from Pythagoras, and about two thousand four hundred years separate Pythagoras from Hegel. Yet for all these distances in time, the importance of definite number in the constitution of the Divine Nature, and the concept of the real world as exhibiting the evolution of an idea, can both be traced back to the train of thought set going by Pythagoras.

The importance of an individual thinker owes something to chance. For it depends upon the fate of his ideas in the minds of his successors. In this respect Pythagoras was fortunate. His philosophical speculations reach us through the mind of Plato. The Platonic world of ideas is the refined, revised form of the Pythagorean doctrine that number lies at the base of the real world. Owing to the Greek mode of representing numbers by patterns of dots, the notions of number and of geometrical configuration are less separated than with us. Also Pythagoras, without doubt, included the shape-iness of shape, which is an impure mathematical entity. So to-day, when Einstein and his followers proclaim that physical facts, such as gravitation, are to be construed as exhibitions of local peculiarities of spatio-temporal properties, they are following the pure Pythagorean tradition. In a sense, Plato and Pythagoras stand nearer to modern physical science than does Aristotle. The two former were mathematicians, whereas Aristotle was the son of a doctor, though of course he was not thereby ignorant of mathematics. The practical counsel to be derived from Pythagoras, is to measure, and thus to express quality in terms of numerically determined quantity. But the biological sciences, then and till our own time, have been overwhelmingly classificatory. Accordingly, Aristotle by his Logic throws the emphasis on classification. The popularity of Aristotelian Logic retarded the advance of physical science throughout the Middle Ages. If only the schoolmen had measured instead of classifying, how much they might have learnt!

Classification is a halfway house between the immediate concreteness of the individual thing and the complete abstraction of mathematical notions. The species take account of the specific character, and the genera of the generic character. But in the procedure of relating mathematical notions to the facts of nature, by counting, by measurement, and by geometrical relations, and by types of order, the rational contemplation is lifted from the incomplete abstractions involved in definite species and genera, to the complete abstractions of mathematics. Classification is necessary. But unless you can progress from classification to mathematics, your reasoning will not take you very far.

Between the epoch which stretches from Pythagoras to Plato and the epoch comprised in the seventeenth century of the modern world nearly two thousand years elapsed. In this long interval mathematics had made immense strides. Geometry had gained the study of conic sections and trigonometry; the method of exhaustion had almost anticipated the integral calculus; and above all the Arabic arithmetical notation and algebra had been contributed by Asiatic thought. But the progress was on technical lines. Mathematics, as a formative element in the development of philosophy, never, during this long period, recovered from its deposition at the hands of Aristotle. Some of the old ideas derived from the Pythagorean-Platonic epoch lingered on, and can be traced among the Platonic influences which shaped the first period of evolution of Christian theology. But philosophy received no fresh inspiration from the steady advance of mathematical science. In the seventeenth century the influence of Aristotle was at its lowest, and mathematics recovered the importance of its earlier period. It was an age of great physicists and great philosophers; and the physicists and philosophers were alike mathematicians. The exception of John Locke should be made; although he was greatly influenced by the Newtonian circle of the Royal Society. In the age of Galileo, Descartes, Spinoza, Newton, and Leibniz, mathematics was an influence of the first magnitude in the formation of philosophic ideas. But the mathematics, which now emerged into prominence, was a very different science from the mathematics of the earlier epoch. It had gained in generality, and had started upon its almost incredible modern career of piling subtlety of generalisation upon subtlety of generalisation; and of finding, with each growth of complexity, some new application, either to physical science, or to philosophic thought. The Arabic notation had equipped the science with almost perfect technical efficiency in the manipulation of numbers. This relief from a struggle with arithmetical details (as instanced, for example, in the Egyptian arithmetic of B. C. 1600) gave room for a development which had already been faintly anticipated in later Greek mathematics. Algebra now came upon the scene, and algebra is a generalisation of arithmetic. In the same way as the notion of number abstracted from reference to any one particular set of entities, so in algebra abstraction is made from the notion of any particular numbers. Just as the number '5' refers impartially to any group of five entities, so in algebra the letters are used to refer impartially to any number, with the proviso that each letter is to refer to the same number throughout the same context of its employment.

This usage was first employed in equations, which are methods of asking complicated arithmetical questions. In this connection, the letters representing numbers were termed 'unknowns.' But equations soon suggested a new idea, that, namely, of a function of one or more general symbols, these symbols being letters representing any numbers. In this employment the algebraic letters are called the 'arguments' of the function, or sometimes they are called the 'variables.' Then, for instance, if an angle is represented by an algebraical letter, as standing for its numerical measure in terms of a given unit, Trigonometry is absorbed into this new algebra. Algebra thus develops into the general science of analysis in which we consider the properties of various functions of undetermined arguments. Finally the particular functions, such as the trigonometrical functions, and the logarithmic functions, and the algebraic functions, are generalised into the idea of 'any function.' Too large a generalisation leads to mere barrenness. It is the large generalisation, limited by a happy particularity, which is the fruitful conception. For instance the idea of any *continuous* function, whereby the limitation of continuity is introduced, is the fruitful idea which has led to most of the important applications. This rise of algebraic analysis was concurrent with Descartes' discovery of analytical geometry, and then with the invention of the infinitesimal calculus by Newton and Leibniz. Truly, Pythagoras, if he could have foreseen the issue of the train of thought which he had set going would have felt himself fully justified in his brotherhood with its excitement of mysterious rites.

The point which I now want to make is that this dominance of the idea of functionality in the abstract sphere of mathematics found itself reflected in the order of nature under the guise of mathematically expressed laws of nature. Apart from this progress of mathematics, the seventeenth century developments of science would have been impossible. Mathematics supplied the background of imaginative thought with which the men of science approached the observation of nature. Galileo produced formulae, Descartes produced formulae, Huyghens produced formulae, Newton produced formulae.

As a particular example of the effect of the abstract development of mathematics upon the science of those times, consider the notion of periodicity. The general recurrences of things are very obvious in our ordinary experience. Days recur, lunar phases recur, the seasons of the year recur, rotating bodies recur to their old positions, beats of the heart recur, breathing recurs. On every side, we are met by recurrence. Apart from recurrence, knowledge would be impossible; for nothing could be referred to our past experience. Also, apart from some regularity of recurrence, measurement would be impossible. In our experience, as we gain the idea of exactness, recurrence is fundamental.

In the sixteenth and seventeenth centuries, the theory of periodicity took a fundamental place in science. Kepler divined a law connecting the major axes of the planetary orbits with the periods in which the planets respectively described their orbits: Galileo observed the periodic vibrations of pendulums: Newton explained sound as being due to the disturbance of air by the passage through it of periodic waves of condensation and rarefaction: Huyghens explained light as being due to the transverse waves of vibration of a subtle ether: Mersenne connected the period of the vibration of a violin string with its density, tension, and length. The birth of modern physics depended upon the application of the abstract idea of periodicity to a variety of concrete instances. But this would have been impossible, unless mathematicians had already worked out in the abstract the various abstract ideas which cluster round the notions of periodicity. The science of trigonometry arose from that of the relations of the angles of a right-angled triangle, to the ratios between the sides and hypotenuse of the triangle. Then, under the influence of the newly discovered mathematical science of the analysis of functions, it broadened out into the study of the simple abstract periodic functions which these ratios exemplify. Thus trigonometry became completely abstract; and in thus becoming abstract, it became useful. It illuminated the underlying analogy between sets of utterly diverse physical phenomena; and at the same time it supplied the weapons by which any one such set could have its various features analysed and related to each other.[1]

Nothing is more impressive than the fact that as mathematics withdrew increasingly into the upper regions of ever greater extremes of abstract thought, it returned back to earth with a corresponding growth of importance for the analysis of concrete fact. The history of the seventeenth century science reads as though it were some vivid dream of Plato or Pythagoras. In this characteristic the seventeenth century was only the forerunner of its successors.

The paradox is now fully established that the utmost abstractions are the true weapons with which to control our thought of concrete fact. As the result of the prominence of mathematicians in the seventeenth century, the eighteenth century was mathematically minded, more especially where French influence predominated. An exception must be made of the English empiricism derived from Locke. Outside France, Newton's direct influence on philosophy is best seen in Kant, and not in Hume.

In the nineteenth century, the general influence of mathematics waned. The romantic movement in literature, and the idealistic movement in philosophy were not the products of mathematical minds. Also, even in

[1] For a more detailed consideration of the nature and function of pure mathematics *cf.* my *Introduction to Mathematics.*

science, the growth of geology, of zoölogy, and of the biological sciences generally, was in each case entirely disconnected from any reference to mathematics. The chief scientific excitement of the century was the Darwinian theory of evolution. Accordingly, mathematicians were in the background so far as the general thought of that age was concerned. But this does not mean that mathematics was being neglected, or even that it was uninfluential. During the nineteenth century pure mathematics made almost as much progress as during all the preceding centuries from Pythagoras onwards. Of course progress was easier, because the technique had been perfected. But allowing for that, the change in mathematics between the years 1800 and 1900 is very remarkable. If we add in the previous hundred years, and take the two centuries preceding the present time, one is almost tempted to date the foundation of mathematics somewhere in the last quarter of the seventeenth century. The period of the discovery of the elements stretches from Pythagoras to Descartes, Newton, and Leibniz, and the developed science has been created during the last two hundred and fifty years. This is not a boast as to the superior genius of the modern world; for it is harder to discover the elements than to develop the science.

Throughout the nineteenth century, the influence of the science was its influence on dynamics and physics, and thence derivatively on engineering and chemistry. It is difficult to overrate its indirect influence on human life through the medium of these sciences. But there was no direct influence of mathematics upon the general thought of the age.

In reviewing this rapid sketch of the influence of mathematics throughout European history, we see that it had two great periods of direct influence upon general thought, both periods lasting for about two hundred years. The first period was that stretching from Pythagoras to Plato, when the possibility of the science, and its general character, first dawned upon the Grecian thinkers. The second period comprised the seventeenth and eighteenth centuries of our modern epoch. Both periods had certain common characteristics. In the earlier, as in the later period, the general categories of thought in many spheres of human interest, were in a state of disintegration. In the age of Pythagoras, the unconscious Paganism, with its traditional clothing of beautiful ritual and of magical rites, was passing into a new phase under two influences. There were waves of religious enthusiasm, seeking direct enlightenment into the secret depths of being; and at the opposite pole, there was the awakening of critical analytical thought, probing with cool dispassionateness into ultimate meanings. In both influences, so diverse in their outcome, there was one common element—an awakened curiosity, and a movement towards the reconstruction of traditional ways. The pagan mysteries may be compared to the Puritan reaction and to the Catholic reaction; critical scientific interest was alike in both epochs, though with minor differences of substantial importance.

In each stage, the earlier stages were placed in periods of rising prosperity, and of new opportunities. In this respect, they differed from the period of gradual declension in the second and third centuries when Christianity was advancing to the conquest of the Roman world. It is only in a period, fortunate both in its opportunities for disengagement from the immediate pressure of circumstances, and in its eager curiosity, that the Age-Spirit can undertake any direct revision of those final abstractions which lie hidden in the more concrete concepts from which the serious thought of an age takes its start. In the rare periods when this task can be undertaken, mathematics becomes relevant to philosophy. For mathematics is the science of the most complete abstractions to which the human mind can attain.

The parallel between the two epochs must not be pressed too far. The modern world is larger and more complex than the ancient civilisation round the shores of the Mediterranean, or even than that of the Europe which sent Columbus and the Pilgrim Fathers across the ocean. We cannot now explain our age by some simple formula which becomes dominant and will then be laid to rest for a thousand years. Thus the temporary submergence of the mathematical mentality from the time of Rousseau onwards appears already to be at an end. We are entering upon an age of reconstruction, in religion, in science, and in political thought. Such ages, if they are to avoid mere ignorant oscillation between extremes, must seek truth in its ultimate depths. There can be no vision of this depth of truth apart from a philosophy which takes full account of those ultimate abstractions, whose interconnections it is the business of mathematics to explore.

In order to explain exactly how mathematics is gaining in general importance at the present time, let us start from a particular scientific perplexity and consider the notions to which we are naturally led by some attempt to unravel its difficulties. At present physics is troubled by the quantum theory. I need not now explain what this theory is, to those who are not already familiar with it. But the point is that one of the most hopeful lines of explanation is to assume that an electron does not continuously traverse its path in space. The alternative notion as to its mode of existence is that it appears at a series of discrete positions in space which it occupies for successive durations of time. It is as though an automobile, moving at the average rate of thirty miles an hour along a road, did not traverse the road continuously; but appeared successively at the successive milestones, remaining for two minutes at each milestone.

In the first place there is required the purely technical use of mathematics to determine whether this conception does in fact explain the many perplexing characteristics of the quantum theory. If the notion survives this test, undoubtedly physics will adopt it. So far the question is purely one for mathematics and physical science to settle between them, on the basis of mathematical calculations and physical observations.

But now a problem is handed over to the philosophers. This discontinuous existence in space, thus assigned to electrons, is very unlike the continuous existence of material entities which we habitually assume as obvious. The electron seems to be borrowing the character which some people have assigned to the Mahatmas of Tibet. These electrons, with the correlative protons, are now conceived as being the fundamental entities out of which the material bodies of ordinary experience are composed. Accordingly if this explanation is allowed, we have to revise all our notions of the ultimate character of material existence. For when we penetrate to these final entities, this startling discontinuity of spatial existence discloses itself.

There is no difficulty in explaining the paradox, if we consent to apply to the apparently steady undifferentiated endurance of matter the same principles as those now accepted for sound and light. A steadily sounding note is explained as the outcome of vibrations in the air: a steady colour is explained as the outcome of vibrations in ether. If we explain the steady endurance of matter on the same principle, we shall conceive each primordial element as a vibratory ebb and flow of an underlying energy, or activity. Suppose we keep to the physical idea of energy: then each primordial element will be an organised system of vibratory streaming of energy. Accordingly there will be a definite period associated with each element; and within that period the stream-system will sway from one stationary maximum to another stationary maximum—or, taking a metaphor from the ocean tides, the system will sway from one high tide to another high tide. This system, forming the primordial element, is nothing at any instant. It requires its whole period in which to manifest itself. In an analogous way, a note of music is nothing at an instant, but it also requires its whole period in which to manifest itself.

Accordingly, in asking where the primordial element is, we must settle on its average position at the centre of each period. If we divide time into smaller elements, the vibratory system as one electronic entity has no existence. The path in space of such a vibratory entity—where the entity is *constituted by* the vibrations—must be represented by a series of detached positions in space, analogously to the automobile which is found at successive milestones and at nowhere between.

We first must ask whether there is any evidence to associate the quantum theory with vibration. This question is immediately answered in the affirmative. The whole theory centres round the radiant energy from an atom, and is intimately associated with the periods of the radiant wave-systems. It seems, therefore, that the hypothesis of essentially vibratory existence is the most hopeful way of explaining the paradox of the discontinuous orbit.

In the second place, a new problem is now placed before philosophers and physicists, if we entertain the hypothesis that the ultimate elements of matter are in their essence vibratory. By this I mean that apart from being a periodic system, such an element would have no existence. With this hypothesis we have to ask, what are the ingredients which form the vibratory organism. We have already got rid of the matter with its appearance of undifferentiated endurance. Apart from some metaphysical compulsion, there is no reason to provide another more subtle stuff to take the place of the matter which has just been explained away. The field is now open for the introduction of some new doctrine of organism which may take the place of the materialism with which, since the seventeenth century, science has saddled philosophy. It must be remembered that the physicists' energy is obviously an abstraction. The concrete fact, which is the organism, must be a complete expression of the character of a real occurrence. Such a displacement of scientific materialism, if it ever takes place, cannot fail to have important consequences in every field of thought.

Finally, our last reflection must be, that we have in the end come back to a version of the doctrine of old Pythagoras, from whom mathematics, and mathematical physics, took their rise. He discovered the importance of dealing with abstractions; and in particular directed attention to number as characterising the periodicities of notes of music. The importance of the abstract idea of periodicity was thus present at the very beginning both of mathematics and of European philosophy.

In the seventeenth century, the birth of modern science required a new mathematics, more fully equipped for the purpose of analysing the characteristics of vibratory existence. And now in the twentieth century we find physicists largely engaged in analysing the periodicities of atoms. Truly, Pythagoras in founding European philosophy and European mathematics, endowed them with the luckiest of lucky guesses—or, was it a flash of divine genius, penetrating to the inmost nature of things?

PART III

Arithmetic, Numbers and the Art of Counting

Poppy Seeds and Large Numbers

A MONG Archimedes' many contributions to mathematics, *The Sand Reckoner* stands high as a book of wide interest and historical importance. It describes a new system for generating and expressing very large numbers.

The Greeks used the twenty-seven letters of their alphabet for numerals, a perfectly satisfactory system whose only real drawback was the absence of a sign for zero. For the number 10,000 which the Greeks called a *myriad*, the letter M was used; this could be combined with other letters and signs to make bigger numbers, for example:

$$\text{T} \atop \text{M}\rho\nu.,\zeta_{\text{Ꙍ}}\pi\delta = 1507984$$

Very large numbers are of course clumsy to express in any ordinary notation, ours as well as the Greeks'. Archimedes contrived a procedure involving indices to reduce the problem to more wieldy dimensions. To the point he developed it Archimedes's system of *octads* yielded a number which in our notation would require 80,000 million million digits. This arithmetic could be elaborated (beyond the octads) to grind out numbers as large as desired, but the monster Archimedes had produced amply served him in his demonstration of the grains of sand it would take to fill the entire universe. Starting with the number of grains of sand that, placed side by side, would measure the width of a poppy seed, Archimedes presses forward with his exercise until he has the entire universe as stuffed as a child's beach pail. Big numbers apparently had a special fascination for Archimedes as shown not alone by this *tour de force*, but by his famous "cattle problem" the solution of which consisted of eight numbers which, it has been estimated, when written out would require almost 700 pages of the kind of closely printed book used for logarithmic tables. One may be permitted to doubt that Archimedes worked out the problem to the last digit.

A word should be said about the astronomical measurements used in the proof in *The Sand Reckoner*. Several astronomers had by the middle of the third century B.C. estimated the diameters of the sun, moon and earth. Archimedes refers to the work of Eudoxus, of Phidias (his own father) and of Aristarchus of Samos, sometimes called the Copernicus of antiquity. Here arises the historical importance of *The Sand Reckoner*. For it is from this work, as Sir Thomas Heath points out, that one learns that Aristarchus conceived the universe with the sun in the center and the planets, including the earth, revolving around it, and that he had dis-covered the angular diameter of the sun to be $\frac{1}{720}$th of the circle of the Zodiac or half a degree—an admirably close estimate. Almost 1800 years then elapsed before Aristarchus' profound conjectures of cosmology, abandoned soon after they had been presented, were revived by Copernicus himself.

1   The Sand Reckoner

By ARCHIMEDES

POPPY SEEDS AND THE UNIVERSE

"THERE are some, king Gelon, who think that the number of the sand is infinite in multitude; and I mean by the sand not only that which exists about Syracuse and the rest of Sicily but also that which is found in every region whether inhabited or uninhabited. Again there are some who, without regarding it as infinite, yet think that no number has been named which is great enough to exceed its multitude. And it is clear that they who hold this view, if they imagined a mass made up of sand in other respects as large as the mass of the earth, including in it all the seas and the hollows of the earth filled up to a height equal to that of the highest of the mountains, would be many times further still from recognising that any number could be expressed which exceeded the multitude of the sand so taken. But I will try to show you by means of geometrical proofs, which you will be able to follow, that, of the numbers named by me and given in the work which I sent to Zeuxippus, some exceed not only the number of the mass of sand equal in magnitude to the earth filled up in the way described, but also that of a mass equal in magnitude to the universe. Now you are aware that 'universe' is the name given by most astronomers to the sphere whose centre is the centre of the earth and whose radius is equal to the straight line between the centre of the sun and the centre of the earth. This is the common account ($\tau\grave{\alpha}$ $\gamma\rho\alpha\phi\acute{o}\mu\epsilon\nu\alpha$), as you have heard from astronomers. But Aristarchus of Samos brought out a book consisting of some hypotheses, in which the premises lead to the result that the universe is many times greater than that now so called. His hypotheses are that the fixed stars and the sun remain unmoved, that the earth revolves about the sun in the circumference of a circle, the sun lying in the middle of the orbit, and that the sphere of the fixed stars, situated about the same centre as the sun, is so great that the circle in which he supposes the earth to revolve bears such a proportion to the distance of the fixed stars as the centre of the sphere bears to its surface. Now it is easy to see that this is impossible; for, since the centre of the sphere has no magnitude, we cannot conceive it to bear any ratio whatever to the surface of the sphere. We must however take Aristarchus to mean this: since we conceive the earth to be, as it were, the centre of the universe, the ratio which the earth bears to what we describe as the 'universe' is the same as the ratio which the sphere containing the circle in which he supposes the earth to revolve bears to the sphere of the fixed stars. For he adapts the proofs of his results to a hypothesis of this kind, and in particular he appears to suppose the magnitude of the sphere in which he represents the earth as moving to be equal to what we call the 'universe.'

I say then that, even if a sphere were made up of the sand, as great as Aristarchus supposes the sphere of the fixed stars to be, I shall still prove that, of the numbers named in the *Principles*, some exceed in multitude the number of the sand which is equal in magnitude to the sphere referred to, provided that the following assumptions be made.

1. *The perimeter of the earth is about 3,000,000 stadia and not greater.*

It is true that some have tried, as you are of course aware, to prove that the said perimeter is about 300,000 stadia. But I go further and, putting the magnitude of the earth at ten times the size that my predecessors thought it, I suppose its perimeter to be about 3,000,000 stadia and not greater.

2. *The diameter of the earth is greater than the diameter of the moon, and the diameter of the sun is greater than the diameter of the earth.*

In this assumption I follow most of the earlier astronomers.

3. *The diameter of the sun is about 30 times the diameter of the moon and not greater.*

It is true that, of the earlier astronomers, Eudoxus declared it to be about nine times as great, and Pheidias my father twelve times, while Aristarchus tried to prove that the diameter of the sun is greater than 18 times but less than 20 times the diameter of the moon. But I go even further than Aristarchus, in order that the truth of my proposition may be established beyond dispute, and I suppose the diameter of the sun to be about 30 times that of the moon and not greater.

4. *The diameter of the sun is greater than the side of the chiliagon inscribed in the greatest circle in the (sphere of the) universe.*

I make this assumption [1] because Aristarchus discovered that the sun appeared to be about $\frac{1}{720}$th part of the circle of the zodiac, and I myself

---

[1] This is not, strictly speaking, an assumption; it is a proposition proved later by means of the result of an experiment about to be described.

tried, by a method which I will now describe, to find experimentallv (ὀργανικῶς) the angle subtended by the sun and having its vertex at the eye (τὰν γωνίαν, εἰς ἂν ὁ ἅλιος ἐναρμόζει τὰν κορυφὰν ἔχουσαν ποτὶ τᾷ ὄψει)."

[Up to this point the treatise has been literally translated because of the historical interest attaching to the *ipsissima verba* of Archimedes on such a subject. The rest of the work can now be more freely reproduced, and, before proceeding to the mathematical contents of it, it is only necessary to remark that Archimedes next describes how he arrived at a higher and a lower limit for the angle subtended by the sun. This he did by taking a long rod or ruler (κανών), fastening on the end of it a small cylinder or disc, pointing the rod in the direction of the sun just after its rising (so that it was possible to look directly at it), then putting the cylinder at such a distance that it just concealed, and just failed to conceal, the sun, and lastly measuring the angles subtended by the cylinder. He explains also the correction which he thought it necessary to make because "the eye does not see from one point but from a certain area" (ἐπεὶ αἱ ὄψιες αὐκ ἀφ᾽ ἑνὸς σαμείου βλέποντι, ἀλλὰ ἀπό τινος μεγέθεος).]

The result of the experiment was to show that the angle subtended by the diameter of the sun was less than $\frac{1}{164}$th part, and greater than $\frac{1}{200}$th part, of a right angle.

To prove that (on this assumption) the diameter of the sun is greater than the side of a chiliagon, or figure with 1000 *equal sides, inscribed in a great circle of the 'universe.'*

Suppose the plane of the paper to be the plane passing through the centre of the sun, the centre of the earth and the eye, at the time when the sun has just risen above the horizon. Let the plane cut the earth in the circle *EHL* and the sun in the circle *FKG*, the centres of the earth and sun being *C*, *O* respectively, and *E* being the position of the eye.

Further, let the plane cut the sphere of the 'universe' (i.e. the sphere whose centre is *C* and radius *CO*) in the great circle *AOB*.

Draw from *E* two tangents to the circle *FKG* touching it at *P*, *Q*, and from *C* draw two other tangents to the same circle touching it in *F*, *G* respectively.

Let *CO* meet the sections of the earth and sun in *H*, *K* respectively; and let *CF*, *CG* produced meet the great circle *AOB* in *A*, *B*.

Join *EO*, *OF*, *OG*, *OP*, *OQ*, *AB*, and let *AB* meet *CO* in *M*.

Now *CO* > *EO*, since the sun is just above the horizon. Therefore

$$\angle PEQ > \angle FCG.$$

And $\left.\begin{array}{l} \angle PEQ > \frac{1}{200}R \\ \qquad\quad < \frac{1}{164}R \end{array}\right\}$ where *R* represents a right angle.
but

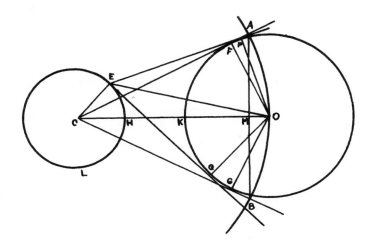

Thus $\angle FCG < \frac{1}{164}R$, *a fortiori,*

and the chord AB subtends an arc of the great circle which is less than $\frac{1}{656}$th of the circumference of that circle, i.e.

AB < (side of 656-sided polygon inscribed in the circle).

Now the perimeter of any polygon inscribed in the great circle is less than $4\frac{1}{7}CO$. [Cf. *Measurement of a circle*, Prop. 3.]

Therefore $AB : CO < 11 : 1148,$

and, *a fortiori,* $AB < \frac{1}{100}CO$ .... (a).

Again, since $CA = CO$, and AM is perpendicular to CO, while OF is perpendicular to CA,

$$AM = OF.$$

Therefore $AB = 2AM = $ (diameter of sun).

Thus   (diameter of sun) $< \frac{1}{100}CO$, by (a),

and, *a fortiori,*

(diameter of earth) $< \frac{1}{100}CO$.   [Assumption 2]

Hence $CH + OK < \frac{1}{100}CO,$

so that $HK > \frac{99}{100}CO,$

or $CO : HK < 100 : 99.$

And $CO > CF,$

while $HK < EQ.$

Therefore $CF : EQ < 100 : 99$ .... (β).

Now in the right-angled triangles *CFO, EQO*, of the sides about the right angles,

$$OF = OQ, \text{ but } EQ < CF \text{ (since } EO < CO).$$

Therefore $\angle OEQ : \angle OCF > CO :_, EO,$

but $< CF : EQ.$[2]

Doubling the angles,

$$\angle PEQ : \angle ACB < CF : EQ$$

$$< 100 : 99, \text{ by } (\beta) \text{ above.}$$

But $\angle PEQ > \tfrac{1}{200}R, \text{ by hypothesis.}$

Therefore $\angle ACB > \tfrac{99}{20000}R$

$$> \tfrac{1}{203}R.$$

It follows that the arc *AB* is greater than $\tfrac{1}{812}$th of the circumference of the great circle *AOB*.

Hence, *a fortiori*,

$AB >$ (side of chiliagon inscribed in great circle),

and *AB* is equal to the diameter of the sun, as proved above.

The following results can now be proved:

    (*diameter of 'universe'*) < 10,000 (*diameter of earth*);

and (*diameter of 'universe'*) < 10,000,000,000 *stadia.*

(1) Suppose, for brevity, that d_u represents the diameter of the 'universe,' d_s that of the sun, d_e that of the earth, and d_m that of the moon.

By hypothesis, $d_s \not> 30d_m,$ [Assumption 3]

and $d_e > d_m;$ [Assumption 2]

therefore $d_s < 30d_e.$

Now, by the last proposition,

$$d_s > \text{(side of chiliagon inscribed in great circle),}$$

so that   (perimeter of chiliagon) $< 1000d_s$

$$< 30,000d_e.$$

But the perimeter of any regular polygon with more sides than 6 inscribed in a circle is greater than that of the inscribed regular hexagon, and therefore greater than three times the diameter. Hence

$$\text{(perimeter of chiliagon)} > 3d_u.$$

It follows that $d_u < 10,000d_e.$

(2)   (Perimeter of earth) $\not> 3,000,000$ stadia.

[Asssumption 1]

---

[2] The proposition here assumed is of course equivalent to the trigonometrical formula which states that, if α, β are the circular measures of two angles, each less than a right angle, of which α is the greater, then

$$\frac{\tan \alpha}{\tan \beta} > \frac{\alpha}{\beta} > \frac{\sin \alpha}{\sin \beta}.$$

and $\qquad$ (perimeter of earth) $> 3d_e$.

Therefore $\qquad\qquad\qquad d_e < 1,000,000$ stadia,

whence $\qquad\qquad\qquad d_u < 10,000,000,000$ stadia.

Assumption 5.

Suppose a quantity of sand taken not greater than a poppy seed, and suppose that it contains not more than 10,000 grains.

Next suppose the diameter of the poppy seed to be not less than 1/40th of a finger-breadth.

ORDERS AND PERIODS OF NUMBERS

I. We have traditional names for numbers up to a myriad (10,000); we can therefore express numbers up to a myriad myriads (100,000,000). Let these numbers be called numbers of the *first order.*

Suppose the 100,000,000 to be the unit of the *second order*, and let the *second order* consist of the numbers from that unit up to $(100,000,000)^2$.

Let this again be the unit of the *third order* of numbers ending with $(100,000,000)^3$; and so on, until we reach the 100,000,000*th order* of numbers ending with $(100,000,000)^{100,000,000}$, which we will call P.

II. Suppose the numbers from 1 to P just described to form the *first period.*

Let P be the unit of the *first order of the second period*, and let this consist of the numbers from P up to $100,000,000\,P$.

Let the last number be the unit of the *second order of the second period*, and let this end with $(100,000,000)^2\,P$.

We can go on in this way till we reach the 100,000,000*th order of the second period* ending with $(100,000,000)^{100,000,000}\,P$, or P^2.

III. Taking P^2 as the unit of the *first order of the third period*, we proceed in the same way till we reach the 100,000,000*th order of the third period* ending with P^3.

IV. Taking P^3 as the unit of the *first order of the fourth period*, we continue the same process until we arrive at the 100,000,000*th order of the* 100,000,000*th period* ending with $P^{100,000,000}$. This last number is expressed by Archimedes as "a myriad-myriad units of the myriad-myriad-th order of the myriad-myriad-th period ($α\dot{i}$ $μυριακισμυριοστᾶς$ $περιόδου$ $μυρια$-$κισμυριοστῶν$ $ἀριθμῶν$ $μυρίαι$ $μυριάδες$)," which is easily seen to be 100,000,000 times the product of $(100,000,000)^{99,999,999}$ and $P^{99,999,999}$, i.e., $P^{100,000,000}$.

[The scheme of numbers thus described can be exhibited more clearly by means of *indices* as follows.

FIRST PERIOD

First order.	Numbers from 1 to 10^8.	
Second order.	" "	10^8 to 10^{16}.
Third order.	" "	10^{16} to 10^{24}.

$\vdots$

$(10^8)th$ *order.*   "   " $10^{8.(10^8-1)}$ to $10^{8.10^8}$ (P, say).

SECOND PERIOD

First order.	" "	$P \cdot 1$ to $P \cdot 10^8$.
Second order	" "	$P \cdot 10^8$ to $P \cdot 10^{16}$.

$\vdots$

$(10^8)th$ *order.*   "   " $P \cdot 10^{8.(10^8-1)}$ to
$P \cdot 10^{8.10^8}$ (or P^2).

$\vdots$

(10^8)TH PERIOD

First order.	" "	$P^{10^8-1} \cdot 1$ to $P^{10^8-1} \cdot 10^8$.
Second order.	" "	$P^{10^8-1} \cdot 10^8$ to $P^{10^8-1} \cdot 10^{16}$.

$\vdots$

$(10^8)th$ *order.*   "   " $P^{10^8-1} \cdot 10^{8.(10^8-1)}$ to
$P^{10^8-1} \cdot 10^{8.10^8}$ (i.e. P^{10^8}).

The prodigious extent of this scheme will be appreciated when it is considered that the last number in the *first period* would be represented now by 1 followed by 800,000,000 ciphers, while the last number of the $(10^8)th$ period would require 100,000,000 times as many ciphers, i.e. 80,000 million millions of ciphers.]

OCTADS

Consider the series of terms in continued proportion of which the first is 1 and the second 10 [i.e. the geometrical progression 1, 10^1, 10^2, 10^3, . . .]. The *first octad* of these terms [i.e. 1, 10^1, 10^2, . . . 10^7] fall accordingly under the *first order of the first period* above described, the *second* octad [i.e. 10^8, 10^9, . . . 10^{15}] under the *second order of the first period*, the first term of the octad being the unit of the corresponding order in each case. Similarly for the *third octad*, and so on. We can, in the same way, place any number of octads.

THEOREM

If there be any number of terms of a series in continued proportion, say A_1, A_2, A_3, ... A_m, ... A_n, ... A_{m+n-1}, ... *of which* $A_1 = 1$, $A_2 = 10$ [so that the series forms the geometrical progression 1, 10^1, 10^2, ... 10^{m-1}, ... 10^{n-1}, ... 10^{m+n-2}, ...], *and if any two terms as* A_m, A_n *be taken and multiplied, the product* $A_m \cdot A_n$ *will be a term in the same series and will be as many terms distant from* A_n *as* A_m *is distant from* A_1; *also it will be distant from* A_1 *by a number of terms less by one than the sum of the numbers of terms by which* A_m *and* A_n *respectively are distant from* A_1.

Take the term which is distant from A_n by the same number of terms as A_m is distant from A_1. This number of terms is m (the first and last being both counted). Thus the term to be taken is m terms distant from A_n, and is therefore the term A_{m+n-1}.

We have therefore to prove that

$$A_m \cdot A_n = A_{m+n-1}.$$

Now terms equally distant from other terms in the continued proportion are proportional.

Thus
$$\frac{A_m}{A_1} = \frac{A_{m+n-1}}{A_n}.$$

But $\quad A_m = A_m \cdot A_1$, since $A_1 = 1$.

Therefore $\quad A_{m+n-1} = A_m \cdot A_n \ldots\ldots\ldots\ldots\ldots (1)$.

The second result is now obvious, since A_m is m terms distant from A_1, A_n is n terms distant from A_1 and A_{m+n-1} is $(m + n - 1)$ terms distant from A_1.

APPLICATION TO THE NUMBER OF THE SAND

By Assumption 5 [p. 425],

(diam. of poppy seed) $\nless$ ¼₀ (finger-breadth);

and, since spheres are to one another in the triplicate ratio of their diameters, it follows that

(sphere of diam. 1 finger-breadth)

$\leqslant$ 64,000 poppy seeds

$\leqslant$ 64,000 × 10,000

$\leqslant$ 640,000,000

$\leqslant$ 6 units of *second order* + 40,000,000 units of *first order*

(*a fortiori*) < 10 units of *second order* of numbers.

} grains of sand.

We now gradually increase the diameter of the supposed sphere, multiplying it by 100 each time. Thus, remembering that the sphere is thereby multiplied by 100^3 or 1,000,000, the number of grains of sand which would be contained in a sphere with each successive diameter may be arrived at as follows.

Diameter of Sphere	*Corresponding Number of Grains of Sand*
(1) 100 finger-breadths	< 1,000,000 × 10 units of *second order* < (7th term of series) × (10th term of series) < 16th term of series　　　　[i.e. 10^{15}] < [10^7 or] 10,000,000 units of the *second order.*
(2) 10,000 finger-breadths	< 1,000,000 × (last number) < (7th term of series) × (16th term) < 22nd term of series　　　　[i.e. 10^{21}] < [10^5 or] 100,000 units of *third order.*
(3) 1 stadium (< 10,000 finger-breadths)	< 100,000 units of *third order.*
(4) 100 stadia	< 1,000,000 × (last number) < (7th term of series) × (22nd term) < 28th term of series　　　　　[10^{27}] < [10^3 or] 1,000 units of *fourth order.*
(5) 10,000 stadia	< 1,000,000 × (last number) < (7th term of series) × (28th term) < 34th term of series　　　　　[10^{33}] < 10 units of *fifth order.*
(6) 1,000,000 stadia	< (7th term of series) × (34th term) < 40th term　　　　　　　　　[10^{39}] < [10^7 or] 10,000,000 units of *fifth order.*
(7) 100,000,000 stadia	< (7th term of series) × (40th term) < 46th term　　　　　　　　　[10^{45}] < [10^5 or] 100,000 units of *sixth order.*
(8) 10,000,000,000 stadia	< (7th term of series) × (46th term) < 52nd term of series　　　　　[10^{51}] < [10^3 or] 1,000 units of *seventh order.*

But, by the proposition above [p. 425],

(diameter of 'universe') < 10,000,000,000 stadia.

Hence *the number of grains of sand which could be contained in a sphere of the size of our 'universe' is less than 1,000 units of the seventh order of numbers* [or 10^{51}].

From this we can prove further that *a sphere of the size attributed by*

Aristarchus to the sphere of the fixed stars would contain a number of grains of sand less than 10,000,000 *units of the eighth order of numbers* [or $10^{56+7} = 10^{63}$].

For, by hypothesis,

(earth) : ('universe') = ('universe') : (sphere of fixed stars).

And [p. 425]

(diameter of 'universe') < 10,000 (diam. of earth);

whence

(diam. of sphere of fixed stars) < 10,000 (diam. of 'universe').

Therefore

(sphere of fixed stars) < $(10,000)^3$. ('universe').

It follows that the number of grains of sand which would be contained in a sphere equal to the sphere of the fixed stars

< $(10,000)^3 \times 1,000$ units of *seventh order*

< (13th term of series) × (52nd term of series)

< 64th term of series   [i.e. 10^{63}]

< [10^7 or] 10,000,000 units of *eighth order* of numbers.

CONCLUSION

"I conceive that these things, king Gelon, will appear incredible to the great majority of people who have not studied mathematics, but that to those who are conversant therewith and have given thought to the question of the distances and sizes of the earth the sun and moon and the whole universe the proof will carry conviction. And it was for this reason that I thought the subject would not be inappropriate for your consideration."

The Art of Counting

PEOPLE knew how to count long before they wrote numerals; simple arithmetic evolved slowly from the operation of counting. Historians, archaeologists and others who study the origins of human culture disagree as to when numbers were first named, when numerals were first written and when arithmetic began. Like writing, these activities denote a high degree of civilization and it is reasonable to assume that the need for word and number symbols arose during a period when cities grew, trade and commerce prospered and governments began to keep records of taxes and other administrative data.

The Sumerians of Mesopotamia have obliged the historian by leaving behind evidence sufficient for him at least to get his bearings in this obscure and complicated business. They could read and write and had a respectable method of handling numbers and numerals; that was about 5,000 years ago and histories of arithmetic usually take off from this epoch. The ability of human beings to count goes back so much further that it is altogether futile to speculate on its origins. Not that specialists in prehistory have therefore refrained from trying.

The two selections offered herewith on the origins of counting require little comment. The first deals with counting as practiced among primitive and savage peoples. These groups have interested anthropologists because their cultures are thought to provide clues to the early culture of our own society. This is still only an assumption, but a good deal of the research is valuable whether or not the theory can be confirmed. Conant's essay on the number concept among primitive tribes was written fifty years ago but I am told that it has lost none of its standing, though a vast amount of work in the field has been done since. Conant was for many years professor of mathematics at the Worcester Polytechnic Institute, and from 1911–1913 its acting president. He was the author of numerous mathematics texts and active as an educationist. He died in 1916. The excerpt by Smith and Ginsburg is taken from a delightful monograph, *Numbers and Numerals*, which can be read by young people as well as adults. I recommend it warmly. The late David Eugene Smith, for many years a professor at Columbia University, was an authority on the teaching of mathematics, a pioneer in the study of early arithmetical works, and a foremost historian of arithmetic and other branches of the science. He is known to thousands as the author of many standard texts of elementary mathematics; I shall not say that he was loved for these primers. Jekuthiel Ginsburg is a genial, enormously erudite scholar who founded and edits the journal *Scripta Mathematica*. He is also head of the mathematics department at Yeshiva College. Professor Ginsburg has a taste for the uncommon, little-traveled byways of mathematics. He has contributed significantly to the diffusion among nonmathematicians of a knowledge of the culture, history and philosophy of mathematics.

"What's one and one and one and one and one and one and one and one and one and one?"

"I don't know," said Alice. "I lost count."

"She can't do addition," said the Red Queen.   —LEWIS CARROLL

2   Counting

By LEVI LEONARD CONANT

AMONG the speculative questions which arise in connection with the study of arithmetic from a historical standpoint, the origin of number is one that has provoked much lively discussion, and has led to a great amount of learned research among the primitive and savage languages of the human race. A few simple considerations will, however, show that such research must necessarily leave this question entirely unsettled, and will indicate clearly that it is, from the very nature of things, a question to which no definite and final answer can be given.

Among the barbarous tribes whose languages have been studied, even in a most cursory manner, none have ever been discovered which did not show some familiarity with the number concept. The knowledge thus indicated has often proved to be most limited; not extending beyond the numbers 1 and 2, or 1, 2, and 3. Examples of this poverty of number knowledge are found among the forest tribes of Brazil, the native races of Australia and elsewhere, and they are considered in some detail in the next chapter. At first thought it seems quite inconceivable that any human being should be destitute of the power of counting beyond 2. But such is the case; and in a few instances languages have been found to be absolutely destitute of pure numeral words. The Chiquitos of Bolivia had no real numerals whatever,[1] but expressed their idea for "one" by the word *etama,* meaning alone. The Tacanas of the same country have no numerals except those borrowed from Spanish, or from Aymara or Peno, languages with which they have long been in contact.[2] A few other South American languages are almost equally destitute of numeral words. But even here, rudimentary as the number sense undoubtedly is, it is not wholly lacking; and some indirect expression, or some form of circumlocution, shows a conception of the difference between *one* and *two,* or at least, between *one* and *many.*

These facts must of necessity deter the mathematician from seeking to push his investigation too far back toward the very origin of number. Philosophers have endeavoured to establish certain propositions concern-

---

[1] Brinton, D. G., *Essays of an Americanist*, p. 406; and *American Race*, p. 359.
[2] This information I received from Dr. Brinton by letter.

ing this subject, but, as might have been expected, have failed to reach any common ground of agreement. Whewell has maintained that "such propositions as that two and three make five are necessary truths, containing in them an element of certainty beyond that which mere experience can give." Mill, on the other hand, argues that any such statement merely expresses a truth derived from early and constant experience; and in this view he is heartily supported by Tylor.[3] But why this question should provoke controversy, it is difficult for the mathematician to understand. Either view would seem to be correct, according to the standpoint from which the question is approached. We know of no language in which the suggestion of number does not appear, and we must admit that the words which give expression to the number sense would be among the early words to be formed in any language. They express ideas which are, at first, wholly concrete, which are of the greatest possible simplicity, and which seem in many ways to be clearly understood, even by the higher orders of the brute creation. The origin of number would in itself, then, appear to lie beyond the proper limits of inquiry; and the primitive conception of number to be fundamental with human thought.

In connection with the assertion that the idea of number seems to be understood by the higher orders of animals, the following brief quotation from a paper by Sir John Lubbock may not be out of place: "Leroy . . . mentions a case in which a man was anxious to shoot a crow. 'To deceive this suspicious bird, the plan was hit upon of sending two men to the watch house, one of whom passed on, while the other remained; but the crow counted and kept her distance. The next day three went, and again she perceived that only two retired. In fine, it was found necessary to send five or six men to the watch house to put her out in her calculation. The crow, thinking that this number of men had passed by, lost no time in returning.' From this he inferred that crows could count up to four. Lichtenberg mentions a nightingale which was said to count up to three. Every day he gave it three mealworms, one at a time. When it had finished one it returned for another, but after the third it knew that the feast was over. There is an amusing and suggestive remark in Mr. Galton's interesting *Narrative of an Explorer in Tropical South Africa*. After describing the Demara's weakness in calculations, he says: 'Once while I watched a Demara floundering hopelessly in a calculation on one side of me, I observed, "Dinah," my spaniel, equally embarrassed on the other; she was overlooking half a dozen of her new-born puppies, which had been removed two or three times from her, and her anxiety was excessive, as she tried to find out if they were all present, or if any were still missing. She kept puzzling and running her eyes over them backwards and forwards, but could not satisfy herself. She evidently had a vague notion of counting, but

[3] Tylor, *Primitive Culture*, Vol. I, p. 240.

the figure was too large for her brain. Taking the two as they stood, dog and Demara, the comparison reflected no great honour on the man. . . .' According to my bird-nesting recollections, which I have refreshed by more recent experience, if a nest contains four eggs, one may safely be taken; but if two are removed, the bird generally deserts. Here, then, it would seem as if we had some reason for supposing that there is sufficient intelligence to distinguish two from four. An interesting consideration arises with reference to the number of the victims allotted to each cell by the solitary wasps. One species of Ammophila considers one large cater-pillar of *Noctua segetum* enough; one species of Eumenes supplies its young with five victims; another 10, 15, and even up to 24. The number appears to be constant in each species. How does the insect know when her task is fulfilled? Not by the cell being filled, for if some be removed, she does not replace them. When she has brought her complement she considers her task accomplished, whether the victims are still there or not. How, then, does she know when she has made up the number 24? Perhaps it will be said that each species feels some mysterious and innate tendency to provide a certain number of victims. This would, under no circum-stances, be any explanation; but it is not in accordance with the facts. In the genus Eumenes the males are much smaller than the females. . . . If the egg is male, she supplies five; if female, 10 victims. Does she count? Certainly this seems very like a commencement of arithmetic." [4]

Many writers do not agree with the conclusions which Lubbock reaches; maintaining that there is, in all such instances, a perception of greater or less quantity rather than any idea of number. But a careful consideration of the objections offered fails entirely to weaken the argument. Example after example of a nature similar to those just quoted might be given, in-dicating on the part of animals a perception of the difference between 1 and 2, or between 2 and 3 and 4; and any reasoning which tends to show that it is quantity rather than number which the animal perceives, will apply with equal force to the Demara, the Chiquito, and the Australian. Hence the actual origin of number may safely be excluded from the limits of investigation, and, for the present, be left in the field of pure specu-lation.

A most inviting field for research is, however, furnished by the primi-tive methods of counting and of giving visible expression to the idea of number. Our starting-point must, of course, be the sign language, which always precedes intelligible speech; and which is so convenient and so ex-pressive a method of communication that the human family, even in its most highly developed branches, never wholly lays it aside. It may, indeed, be stated as a universal law, that some practical method of numeration

[4] *Nature*, Vol. XXXIII, p. 45.

has, in the childhood of every nation or tribe, preceded the formation of numeral words.

Practical methods of numeration are many in number and diverse in kind. But the one primitive method of counting which seems to have been almost universal throughout all time is the finger method. It is a matter of common experience and observation that every child, when he begins to count, turns instinctively to his fingers; and, with these convenient aids as counters, tallies off the little number he has in mind. This method is at once so natural and obvious that there can be no doubt that it has always been employed by savage tribes, since the first appearance of the human race in remote antiquity. All research among uncivilized peoples has tended to confirm this view, were confirmation needed of anything so patent. Occasionally some exception to this rule is found; or some variation, such as is presented by the forest tribes of Brazil, who, instead of counting on the fingers themselves, count on the joints of their fingers.[5] As the entire number system of these tribes appears to be limited to *three*, this variation is no cause for surprise.

The variety in practical methods of numeration observed among savage races, and among civilized peoples as well, is so great that any detailed account of them would be almost impossible. In one region we find sticks or splints used; in another, pebbles or shells; in another, simple scratches, or notches cut in a stick, Robinson Crusoe fashion; in another, kernels or little heaps of grain; in another, knots on a string; and so on, in diversity of method almost endless. Such are the devices which have been, and still are, to be found in the daily habit of great numbers of Indian, negro, Mongolian, and Malay tribes; while, to pass at a single step to the other extremity of intellectual development, the German student keeps his beer score by chalk marks on the table or on the wall. But back of all these devices, and forming a common origin to which all may be referred, is the universal finger method; the method with which all begin, and which all find too convenient ever to relinquish entirely, even though their civilization be of the highest type. Any such mode of counting, whether involving the use of the fingers or not, is to be regarded simply as an extraneous aid in the expression or comprehension of an idea which the mind cannot grasp, or cannot retain, without assistance. The German student scores his reckoning with chalk marks because he might otherwise forget; while the Andaman Islander counts on his fingers because he has no other method of counting,—or, in other words, of grasping the idea of number. A single illustration may be given which typifies all practical methods of numeration. More than a century ago travellers in Mada-

[5] Spix and Martius, *Travels in Brazil*, Tr. from German by H. E. Lloyd, Vol. II, p. 255.

gascar observed a curious but simple mode of ascertaining the number of soldiers in an army.[6] Each soldier was made to go through a passage in the presence of the principal chiefs; and as he went through, a pebble was dropped on the ground. This continued until a heap of 10 was obtained, when one was set aside and a new heap begun. Upon the completion of 10 heaps, a pebble was set aside to indicate 100; and so on until the entire army had been numbered. Another illustration, taken from the very antipodes of Madagascar, recently found its way into print in an incidental manner,[7] and is so good that it deserves a place beside de Flacourt's time-honoured example. Mom Cely, a Southern negro of unknown age, finds herself in debt to the storekeeper; and, unwilling to believe that the amount is as great as he represents, she proceeds to investigate the matter in her own peculiar way. She had "kept a tally of these purchases by means of a string, in which she tied commemorative knots." When her creditor "undertook to make the matter clear to Cely's comprehension, he had to proceed upon a system of her own devising. A small notch was cut in a smooth white stick for every dime she owed, and a large notch when the dimes amounted to a dollar; for every five dollars a string was tied in the fifth big notch, Cely keeping tally by the knots in her bit of twine; thus, when two strings were tied about the stick, the ten dollars were seen to be an indisputable fact." This interesting method of computing 'the amount of her debt, whether an invention of her own or a survival of the African life of her parents, served the old negro woman's purpose perfectly; and it illustrates, as well as a score of examples could, the methods of numeration to which the children of barbarism resort when any number is to be expressed which exceeds the number of counters with which nature has provided them. The fingers are, however, often employed in counting numbers far above the first decade. After giving the Il-Oigob numerals up to 60, Müller adds:[8] "Above 60 all numbers indicated by the proper figure pantomime, are expressed by means of the word *ipi*." We know, moreover, that many of the American Indian tribes count one ten after another on their fingers; so that, whatever number they are endeavouring to indicate, we need feel no surprise if the savage continues to use his fingers throughout the entire extent of his counts. In rare instances we find tribes which, like the Mairassis of the interior of New Guinea, appear to use nothing but finger pantomime.[9] This tribe, though by no means destitute of the number sense, is said to have no numerals whatever, but to use the single word *awari* with each show of fingers, no matter how few or how many are displayed.

[6] De Flacourt, *Histoire de le grande Isle de Madagascar*, ch. xxviii. Quoted by Peacock, *Encyc. Met.*, Vol. I, p. 393.
[7] Bellamy, Elizabeth W., *Atlantic Monthly*, March, 1893, p. 317.
[8] *Grundriss der Sprachwissenschaft*, Bd. III. Abt. i., p. 94.
[9] Pruner-Bey, *Bulletin de la Société d'Anthr. de Paris*, 1861, p. 462.

In the methods of finger counting employed by savages a considerable degree of uniformity has been observed. Not only does he use his fingers to assist him in his tally, but he almost always begins with the little finger of his left hand, thence proceeding towards the thumb, which is 5. From this point onward the method varies. Sometimes the second 5 also is told off on the left hand, the same order being observed as in the first 5; but oftener the fingers of the right hand are used, with a reversal of the order previously employed; *i.e.* the thumb denotes 6, the index finger 7, and so on to the little finger, which completes the count to 10.

At first thought there would seem to be no good reason for any marked uniformity of method in finger counting. Observation among children fails to detect any such thing; the child beginning, with almost entire indifference, on the thumb or on the little finger of the left hand. My own observation leads to the conclusion that very young children have a slight, though not decided preference for beginning with the thumb. Experiments in five different primary rooms in the public schools of Worcester, Mass., showed that out of a total of 206 children, 57 began with the little finger and 149 with the thumb. But the fact that nearly three-fourths of the children began with the thumb, and but one-fourth with the little finger, is really far less significant than would appear at first thought. Children of this age, four to eight years, will count in either way, and sometimes seem at a loss themselves to know where to begin. In one school room where this experiment was tried the teacher incautiously asked one child to count on his fingers, while all the other children in the room watched eagerly to see what he would do. He began with the little finger—and so did every child in the room after him. In another case the same error was made by the teacher, and the child first asked began with the thumb. Every other child in the room did the same, each following, consciously or unconsciously, the example of the leader. The results from these two schools were of course rejected from the totals which are given above; but they serve an excellent purpose in showing how slight is the preference which very young children have in this particular. So slight is it that no definite law can be postulated of this age; but the tendency seems to be to hold the palm of the hand downward, and then begin with the thumb. The writer once saw a boy about seven years old trying to multiply 3 by 6; and his method of procedure was as follows: holding his left hand with its palm down, he touched with the forefinger of his right hand the thumb, forefinger, and middle finger successively of his left hand. Then returning to his starting-point, he told off a second three in the same manner. This process he continued until he had obtained 6 threes, and then he announced his result correctly. If he had been a few years older, he might not have turned so readily to his thumb as a starting-point for any digital count. The indifference manifested by very young children gradually disappears, and at the age of twelve or thirteen the tendency is decidedly in the direction of beginning with the little finger. Fully three-fourths of all persons above that age will be found to count from the little finger toward the thumb, thus reversing the proportion that was found to obtain in the primary school rooms examined.

With respect to finger counting among civilized peoples, we fail, then, to find any universal law; the most that can be said is that more begin with the little finger than with the thumb. But when we proceed to the study of this slight but important particular among savages, we find them employing a certain order of succession with such substantial uniformity that the conclusion is inevitable that there must lie back of this some well-defined reason, or perhaps instinct, which guides them in their choice. This instinct is undoubtedly the outgrowth of the almost universal right-handedness of the human race. In finger counting, whether among children or adults, the beginning is made on the left hand, except in the case of left-handed individuals; and even then the start is almost as likely to be on the left hand as on the right. Savage tribes, as might be expected, begin with the left hand. Not only is this custom almost invariable, when tribes as a whole are considered, but the little finger is nearly always called into requisition first. To account for this uniformity, Lieutenant Cushing gives the following theory,[10] which is well considered, and is based on the results of careful study and observation among the Zuñi Indians of the Southwest: "Primitive man when abroad never lightly quit hold of his weapons. If he wanted to count, he did as the Zuñi afield does to-day; he tucked his instrument under his left arm, thus constraining the latter, but leaving the right hand free, that he might check off with it the fingers of the rigidly elevated left hand. From the nature of this position, however, the palm of the left hand was presented to the face of the counter, so that he had to begin his score on the little finger of it, and continue his counting from the right leftward. An inheritance of this may be detected to-day in the confirmed habit the Zuñi has of gesticulating from the right leftward, with the fingers of the right hand over those of the left, whether he be counting and summing up, or relating in any orderly manner." Here, then, is the reason for this otherwise unaccountable phenomenon. If savage man is universally right-handed, he will almost inevitably use the index finger of his right hand to mark the fingers counted, and he will begin his count just where it is most convenient. In his case it is with the little finger of the left hand. In the case of the child trying to multiply 3 by 6, it was with the thumb of the same hand. He had nothing to tuck under his arm; so, in raising his left hand to a position where both eye and counting finger could readily run over its fingers, he held the palm turned away from his face. The same choice of starting-point then fol-

[10] "Manual Concepts," *Am. Anthropologist*, 1892, p. 292.

lowed as with the savage—the finger nearest his right hand; only in this case the finger was a thumb. The deaf mute is sometimes taught in this manner, which is for him an entirely natural manner. A left-handed child might be expected to count in a left-to-right manner, beginning, probably, with the thumb of his right hand.

To the law just given, that savages begin to count on the little finger of the left hand, there have been a few exceptions noted; and it has been observed that the method of progression on the second hand is by no means as invariable as on the first. The Otomacs [11] of South America began their count with the thumb, and to express the number 3 would use the thumb, forefinger, and middle finger. The Maipures,[12] oddly enough, seem to have begun, in some cases at least, with the forefinger; for they are reported as expressing 3 by means of the fore, middle, and ring fingers. The Andamans [13] begin with the little finger of either hand, tapping the nose with each finger in succession. If they have but one to express, they use the forefinger of either hand, pronouncing at the same time the proper word. The Bahnars,[14] one of the native tribes of the interior of Cochin China, exhibit no particular order in the sequence of fingers used, though they employ their digits freely to assist them in counting. Among certain of the negro tribes of South Africa [15] the little finger of the right hand is used for 1, and their count proceeds from right to left. With them, 6 is the thumb of the left hand, 7 the forefinger, and so on. They hold the palm downward instead of upward, and thus form a complete and striking exception to the law which has been found to obtain with such substantial uniformity in other parts of the uncivilized world. In Melanesia a few examples of preference for beginning with the thumb may also be noticed. In the Banks Islands the natives begin by turning down the thumb of the right hand, and then the fingers in succession to the little finger, which is 5. This is followed by the fingers of the left hand, both hands with closed fists being held up to show the completed 10. In Lepers' Island, they begin with the thumb, but having reached 5 with the little finger, they do not pass to the other hand, but throw up the fingers they have turned down, beginning with the forefinger and keeping the thumb for 10.[16] In the use of the single hand this people is quite peculiar. The second 5 is almost invariably told off by savage tribes on the second hand, though in passing from the one to the other primitive man does not follow any invariable law. He marks 6 with either the thumb or the little finger. Probably

[11] Tylor, *Primitive Culture*, Vol. I, p. 245.
[12] *Op. cit., loc. cit.*
[13] "Aboriginal Inhabitants of Andaman Islands," *Journ. Anth. Inst.*, 1882, p. 100.
[14] Morice, A., *Revue d'Anthropologie*, 1878, p. 634.
[15] Macdonald, J., "Manners, Customs, etc., of South African Tribes," *Journ. Anthr. Inst.*, 1889, p. 290. About a dozen tribes are enumerated by Mr. Macdonald: Pondos, Tembucs, Bacas, Tolas, etc.
[16] Codrington, R. H., *Melanesians, their Anthropology and Folk-Lore*, p. 353.

the former is the more common practice, but the statement cannot be made with any degree of certainty. Among the Zulus the sequence is from thumb to thumb, as is the case among the other South African tribes just mentioned; while the Veis and numerous other African tribes pass from thumb to little finger. The Eskimo, and nearly all the American Indian tribes, use the correspondence between 6 and the thumb; but this habit is by no means universal. Respecting progression from right to left or left to right on the toes, there is no general law with which the author is familiar. Many tribes never use the toes in counting, but signify the close of the first 10 by clapping the hands together, by a wave of the right hand, or by designating some object; after which the fingers are again used as before.

One other detail in finger counting is worthy of a moment's notice. It seems to have been the opinion of earlier investigators that in his passage from one finger to the next, the savage would invariably bend down, or close, the last finger used; that is, that the count began with the fingers open and outspread. This opinion is, however, erroneous. Several of the Indian tribes of the West [17] begin with the hand clenched, and open the fingers one by one as they proceed. This method is much less common than the other, but that it exists is beyond question.

In the Muralug Island, in the western part of Torres Strait, a somewhat remarkable method of counting formerly existed, which grew out of, and is to be regarded as an extension of, the digital method. Beginning with the little finger of the left hand, the natives counted up to 5 in the usual manner, and then, instead of passing to the other hand, or repeating the count on the same fingers, they expressed the numbers from 6 to 10 by touching and naming successively the left wrist, left elbow, left shoulder, left breast, and sternum. Then the numbers from 11 to 19 were indicated by the use, in inverse order, of the corresponding portions of the right side, arm, and hand, the little finger of the right hand signifying 19. The words used were in each case the actual names of the parts touched; the same word, for example, standing for 6 and 14; but they were never used in the numerical sense unless accompanied by the proper gesture, and bear no resemblance to the common numerals, which are but few in number. This method of counting is rapidly dying out among the natives of the island, and is at the present time used only by old people.[18] Variations on this most unusual custom have been found to exist in others of the neighbouring islands, but none were exactly similar to it. One is also reminded by it of a custom [19] which has for centuries prevailed among bargainers in the East, of signifying numbers by touching the joints of

---

[17] *e.g.*, the Zuñis. See Cushing's paper quoted above.
[18] Haddon, A. C., "Ethnography Western Tribes Torres Strait," *Journ. Anth. Inst.*, 1889, p. 305. For a similar method, see *Life in the Southern Isles*, by W. W. Gill.
[19] Tylor, *Primitive Culture*, Vol. I, p. 246.

each other's fingers under a cloth. Every joint has a special signification; and the entire system is undoubtedly a development from finger counting. The buyer or seller will by this method express 6 or 60 by stretching out the thumb and little finger and closing the rest of the fingers. The addition of the fourth finger to the two thus used signifies 7 or 70; and so on. It is said that between two brokers settling a price by thus snipping with the fingers, cleverness in bargaining, offering a little more, hesitating, expressing an obstinate refusal to go further, etc., are as clearly indicated as though the bargaining were being carried on in words.

The place occupied, in the intellectual development of man, by finger counting and by the many other artificial methods of reckoning,—pebbles, shells, knots, the abacus, etc.,—seems to be this: The abstract processes of addition, subtraction, multiplication, division, and even counting itself, present to the mind a certain degree of difficulty. To assist in overcoming that difficulty, these artificial aids are called in; and, among savages of a low degree of uevelopment, like the Australians, they make counting possible. A little higher in the intellectual scale, among the American Indians, for example, they are employed merely as an artificial aid to what could be done by mental effort alone. Finally, among semi-civilized and civilized peoples, the same processes are retained, and form a part of the daily life of almost every person who has to do with counting, reckoning, or keeping tally in any manner whatever. They are no longer necessary, but they are so convenient and so useful that civilization can never dispense with them. The use of the abacus, in the form of the ordinary numeral frame, has increased greatly within the past few years; and the time may come when the abacus in its proper form will again find in civilized countries a use as common as that of five centuries ago.

In the elaborate calculating machines of the present, such as are used by life insurance actuaries and others having difficult computations to make, we have the extreme of development in the direction of artificial aid to reckoning. But instead of appearing merely as an extraneous aid to a defective intelligence, it now presents itself as a machine so complex that a high degree of intellectual power is required for the mere grasp of its construction and method of working.

It is a profoundly erroneous truism, repeated by all copy books and by eminent people when they are making speeches, that we should cultivate the habit of thinking of what we are doing. The precise opposite is the case. Civilization advances by extending the number of important operations which we can perform without thinking about them.

—ALFRED NORTH WHITEHEAD

The point about zero is that we do not need to use it in the operations of daily life. No one goes out to buy zero fish. It is in a way the most civilized of all the cardinals, and its use is only forced on us by the needs of culti-vated modes of thought. —ALFRED NORTH WHITEHEAD

3   From Numbers to Numerals and From Numerals to Computation

By DAVID EUGENE SMITH and JEKUTHIEL GINSBURG

FROM NUMBERS TO NUMERALS

WE do not know how long ago it was that human beings first began to make their thoughts known to one another by means of speech; but it seems probable that people learned to use words in talking many thousands of years before they learned to set down these words in writing. In the same way, after people learned to name numbers it took a long time for them to learn to use signs for the numbers; for example, to use the numeral "2" instead of the word "two."

Where and when did the use of numerals begin? This question takes us back to the very beginnings of history. If you look at the map you will see Egypt, lying along the valley of the Nile. As long ago as 3000 B.C.—per-haps even earlier—there were prosperous cities in Egypt with markets and business houses, and with an established government over all the land. The keeping of the commercial and government records necessitated the use of large numbers. So the Egyptians made up a set of numerals by which they could express numbers of different values from units up to hundreds of thousands.

In another valley, between the Tigris and Euphrates Rivers in Asia, there lies a part of the territory now known as Iraq. This country was for-merly called Mesopotamia, or "the land between the rivers." Five thousand years ago the Sumerians, a strange people who seem to have come from the mountainous regions of Iran (Persia), had developed a high degree of civilization in this land. They could read and write and had a usable system of numbers and numerals. It will help us to realize what a long, long time ago this was if we recall that Abraham, the Hebrew patriarch, lived as a boy in this same country "between the rivers," a thousand years later than this; that is, about 2000 B.C. By the time that Abraham was born another race of people, the Babylonians, were taking the power away from the Sumerians. Meanwhile, however, the Babylonians were learning from the Sumerians how they carried on trade with other nations, and how they built their houses with bricks of baked clay, and wrote their letters and their historical records on the same sort of bricks. They also learned how to use the number symbols which the Sumerians seem to have invented. Inscriptions showing how both Sumerians and Babylonians kept their accounts are still in existence. In India, also, numerals were used in ancient days; and the Greeks and Romans, too, had their own ways of making number symbols.

To express the number "one" all these ancient peoples at times made use of a numeral like our "1." The Egyptians painted it on pottery and cut it on stone at least five thousand years ago; and somewhat later the Sumerians taught the Babylonians to stamp it on their clay tablets. More than two thousand years ago the Greeks and the Romans used this symbol, and in most of India the natives did the same. This numeral probably came from the lifted finger, which seems to be the most easy and natural way of showing that we mean "one."

But you will remember that we found several different ways of showing the savage poultrymen that we wanted to buy one chicken. We could lift one finger, or we could lay down one pebble or one stick on the ground. In the same way, the "one" was sometimes expressed by a symbol representing a pebble or a bead, and sometimes by a line like this —, representing a stick laid down.

The "two" was commonly expressed by two fingers $||$ or two lines $=$. If you write $||$ rapidly you will have N, from which has come the figure $\digamma$ used by the Arabs and Persians. The $=$ written rapidly becomes Z, which developed into our 2. The symbol $=$ is used in China and Japan today. The people of India have many languages and various ways of writing numbers, but generally in early times they preferred the $=$ to the $||$, so that our 2 was the favorite in many eastern countries, although the Arabic and Iranian (Persian) figure is even today almost universally used in the Mohammedan lands.

Cuneiform, or wedge-shaped, writing was much in use in ancient times. This developed in the Mesopotamian valley, where the lack of other writing materials led the people to stamp inscriptions on clay bricks with sticks which usually were triangular with sharp edges. The cuneiform numerals for 1, 2, and 3 are ▼ ▼▼ ▼▼▼. These numerals first appear on the clay bricks of the Sumerians and the Chaldeans, but they were used afterwards by the Babylonians, the Hittites, the Assyrians, and other ancient races. They have been found as far west as Egypt, as far north as Asia Minor (Anatolia), and as far east as Iran (Persia). They are known to have been used about five thousand years ago, and to have continued in use for about three thousand years.

Sumerian Tablet in the State Museum, Berlin. In the second column, line 2, is the number 60 + 10 + 10 + 10 + 10, or 100. (From Menninger, *Zahlwort und Ziffer*.)

In writing numerals the Babylonians sometimes used a stick with a circular cross section, which gave the "one" the shape of a pebble or a bead. They thus had two types of numerals, as may be understood from the following table:

Triangle	Υ	ᐊ	Υ	⊨
Circle	D	O	D	DD
Value	1	10	60	60 × 10

The "three" was commonly represented by three lines, $|\,|\,|$, or more commonly in the Far East, ☰. In China and Japan, even today, the most common form is ☰. When people began to write with some kind of pen they joined these lines together in a form like this: Ꝫ ; and from this came our 3. In the same way, when they came to write $|\,|\,|$ rapidly, they used forms like ᴌ , which finally became the Iranian (Persian) and Arabic ᴌ , used by millions of people today. It is interesting to see that when this is turned over on its side we have ᴈ , and this accounts for the shape of the three in many East Indian regions where we find the Sanskrit ᴈ and many similar forms.

Since in most countries the form of the numerals meaning one, two, and three developed from the arrangement of sticks $|$ $||$ $|||$ or $-$ $=$ $≡$, we might naturally expect four to be written in the same way. It is true that this was done in Egypt and Rome (||||), in Babylonia (▼▼▼▼), and in a few other countries; but the lines were generally not joined together. Of all the ancient numerals signifying four, only the Arabic Σ was made by joining four lines. This Arabic numeral appears today as ٤ .

Combining numerals and writing large numbers presented a different sort of problem. Five thousand or more years ago the Egyptians used numerical symbols like these:

The number 27,529 as written by the early Egyptians.

In the above a bent line is used to represent tens of thousands. But in the next inscription, cut in stone, the line is straight and the symbol for hundreds (in the bottom row of numerals) is curved somewhat differently from the one used in the number given above.

This inscription is from a monument now in Cairo, and is part of an inventory of the king's herds of cattle. The number at the top is 223,000, and the one at the bottom is 232,413. (From Menninger, *Zahlwort und Ziffer*.)

Among the oldest systems of numerals are those used by the Chinese, and later adopted by the Japanese. These have naturally changed somewhat in the course of the centuries, but it is necessary to mention only two types. The first is based on the use of sticks laid upon a table and used for the purposes of calculating, but it was also used in written documents. It is as follows:

Ι	ΙΙ	ΙΙΙ	ΙΙΙΙ	ΙΙΙΙΙ	T	TT	TTT	TTTT
1	2	3	4	5	6	7	8	9

—	=	≡	≣	≣	⊥	⊥	≟	≟
10	20	30	40	50	60	70	80	90

The system commonly used to represent the first ten numbers is as follows:

	1		2		3		4		5

	6		7		8		9		10

The Greeks had several ways of writing their numbers, but we shall consider only two of them. In one method they simply used the initial letters of the number names, but their letters were quite different from ours. For example, the following early letters and names were used:

Number	Name	Letter
1000	kilo or chilo	X, our *ch*
100	hekto	H, early form
10	deka	Δ, our *d*
5	penta	Π, or Γ, our *p*

These were often combined as shown below, the number being 2977.

XXΓᴴHHHHΓᴾΔΔΓΙΙ

2000	500	400	50	27

The later Greeks, about two thousand years ago, generally used the first ten letters of their alphabet to represent the first ten numbers. For larger numbers they used other letters: K′ for 20, Λ′ for 30, and so on. They often placed a mark (/ or ′) by each letter to show that it stood for a number. The letters, representing numerals from 1 to 9, were then as follows:

A′	B′	Γ′	Δ′	E′	F′	Z′	H′	Θ′
1	2	3	4	5	6	7	8	9

The tens were the next nine letters:

I′	K′	Λ′	M′	N′	Ξ′	O′	Π′	Q′
10	20	30	40	50	60	70	80	90

The hundreds were the next nine letters:

P′	Σ′	T′	Υ′	Φ′	X′	Ψ′	Ω′	Z′
100	200	300	400	500	600	700	800	900

(Q and Z are here used in place of two ancient Greek letters not in our alphabet.)

One of the most interesting evidences of the early use of the initial Greek numerals is seen in a vase now in the Museum at Naples. The picture refers to the Persian wars of the time of Darius, about 500 B.C. In the lower row of figures there is, at the left, a man seated at a table and holding a wax diptych (two-winged) tablet on which are letters which represent 100 *tálenta* (talents of money). On the table are the letters MXHΔΠO<T. The first five of these letters represent the number of talents: 10,000, 1000, 100, 10, and 5. The last three letters stand for 1 *obol* (a fraction of a talent), ½ *obol*, and ¼ *obol*.

The drawing below, which has been enlarged from the figure appearing on the vase, shows the numbers more distinctly. The man seems to be a receiver of taxes or a money changer. There is a possibility that the table was used for computing by counters.

We are all familiar with such numerals as I, II, III, and IIII or IV as

seen on the faces of clocks. These have come to us through many centuries, having been used by the ancient Romans. We speak of them as "Roman numerals." There are seven such characters which are used at present in numbering chapters of books, volumes, the main divisions of outlines, and the like. They are our common letters with values as follows:

I	V	X	L	C	D	M
1	5	10	50	100	500	1000

They have changed their shapes from time to time, the following being one of the early forms:

I	V	X	L	C	D	⊂⊃∞
1	5	10	50	100	500	1000

This shows how the number 2,752,899 might have been written in the late Roman form:

$$\overline{\text{CIƆCIƆDCCL}}\text{CIƆCIƆDCCCLXXXXVIIII}$$

In the earliest inscriptions on stone monuments the "one" was a vertical stroke I, as in the other systems in western Europe. In the Middle Ages, after the small letters became common, i and j were used, the j being usually placed at the end of the number, as in vij for seven.

The earliest and most interesting use of the large Roman numerals is found in a monument set up in Rome to commemorate the victory over the Carthaginians. In the second and third lines from the bottom of the picture on page 449 is the numeral for 100,000, repeated twenty-three times, making 2,300,000. It shows the awkwardness of the Roman numerals as written about twenty-two hundred years ago.

The "five" was generally V, perhaps as representing a hand. This naturally suggests X (two V's) for ten. It is quite as probable, however, that the X came from the crossing off of ten ones. There are various other suggestions in regard to the origin of such Roman numerals. In the Middle Ages the U, which was only another form for V, was used for five, as in uiij for eight.

Fifty was represented in several ways, but L was most commonly used. In the Middle Ages we find many such forms as Mlxj for 1061.

The Roman word for 100 was *centum*, and that for 1000 was *mille*. This probably accounts for the use of C for 100, and M for 1000, although other forms, from which these seem to have been derived, were used in early times.

The Romans often wrote four as IIII, and less often as IV, that is,

Numerals on the Columna Rostrata, 260 B.C. In the Palazzo dei Conservatori, in Rome.

I from V. On clock faces we find both of these forms even today. It was easier for the ancients to think of "five (fingers) less one" than of "four," and of "ten less one" than of "nine." We show this when we say "15 minutes to 10" instead of "45 minutes after 9," or "a quarter to 3" instead of "three quarters after 2." In writing IV, in which I is taken from V, we use what is called the "subtractive principle." This is found in various systems besides the Roman. The first trace that we find is in the Babylonian clay tablets of 2000 B.C. and earlier. In these there frequently occurs the word lal (𒑊𒁹) to indicate subtraction. Thus we have [1]

$$𒌋𒌋𒁹𒑊 \quad \text{for } 10 + 10 - 1, \text{ or } 19$$
$$𒑊𒁹𒌋𒌋 \quad \text{for } 10 + 10 + 10 + 10 - 3, \text{ or } 37$$

The following notes relating to the Roman numerals may be interesting:

9 was written IX (that is, I from X) but until the beginning of printing it appeared quite as often in the form VIIII.

19 was written XIX (that is, X + IX), but it also appeared as IXX (that is, 1 from 20).

18 commonly appears as XVIII, but IIXX was also used.

C|Ɔ was a favorite way of writing 1000, but was later changed to

[1] See O. Neugebauer, *Vorlesungen über Geschichte der antiken mathematischen Wissenschaften*, Bd. I, *Vorgriechische Mathematik*, Berlin, 1934, p. 17.

M, the initial of *mille,* 1000. Half of this symbol, either C or D ,
led to the use of D for 500.

In writing larger numbers the Romans made use of the following forms:

$$\mathsf{C I D} \quad \text{for } 1000$$
$$\mathsf{C C I D D} \quad \text{for } 10,000$$
$$\mathsf{C C C I D D D} \quad \text{for } 100,000$$

One of the most prominent arithmetics published in England in the
sixteenth century (Baker's 1568) gives this curious way of writing
451,234,678,567:

four CliM, two Cxxxiiii, millions, sixe ClxxviiiM, five Clxvii.

It would seem that this is enough to puzzle both teachers and pupils, but
when we remember that our word "billion" ("thousand million" in most
European countries) was not in common use in the sixteenth century, we
can understand that it means

[(4C + 51)1000's + 2C + 34]1000000's + (6C + 78)1000's + 5C + 67

and there is no difficulty in seeing that the latter expression means
451,234 × 1000000 + 678 × 1000 + 567.

Quæpam fuerunt notæ Romæ
norúm?

I.   1.
V.   5.
X.   10.
L.   50.
C.   100.

D. ID. 500.   *Quingenta.*

CXD. ∞ . CID. 1000.   Χίλια. *Mille.*

.   IDD. 5000.   *Quinque millia.*

CMD. . CCIDD. 10000.   Μύεια. *Decem millia.*

. IDDD.50000.   *Quinquaginta millia.*

. CCCIDDD. 100000. *Centum millia.*

IDDDD.   500000. *Quingenta millia.*

CCCCIDDDD. . CCCCIDDDD.1000000. *Decies*
centena millia.

Romani numeri non progrediuntur ultra decies centena
millia illa et cû plura significare uolunt, duplicant notas: ut,

∞ .   ∞., 2000.

CID. CID. CID. 3000.

CID, ID, 1500, ∞ . D.

The preceding from the work of a Swiss scholar, Freigius, published in 1582, shows the forms of the Roman numerals recognized in his time.

The Roman numerals were commonly used in bookkeeping in European countries until the eighteenth century, although our modern numerals were generally known in Europe at least as early as the year 1000. In 1300 the use of our numerals was forbidden in the banks of certain European cities, and in commercial documents. The argument was that they were more easily forged or falsified than Roman numerals; since, for example, the 0 could be changed into a 6 or a 9 by a single stroke. When books began to be printed, however, they made rapid progress, although the Roman numerals continued in use in some schools until about 1600, and in commercial bookkeeping for another century.

One reason why the Roman numerals were preferred in bookkeeping was that it is easier to add and subtract with them than with our modern numerals. This may be seen in these two cases:

Addition		Subtraction	
DCCLXXVII	(777)	DCCLXXVII	(777)
CC X VI	(216)	CC X VI	(216)
DCCCCLXXXXIII	(993)	D L X I	(561)

In such work as this it is unnecessary to learn any addition or subtraction facts; simply V and V make X, CC + CC = CCCC, and so on. The only advantage of our numerals in addition and subtraction is that ours are easier to write. As to multiplication and division, however, our numerals are far superior. The ancient Romans used to perform these operations by the use of counters.

The Hebrews used their alphabet in writing numerals in the same way as the Greeks; that is, the first ten letters represented the first ten numbers, as shown below.

א ב ג ד ה ו ז ח ט י

1 2 3 4 5 6 7 8 9 10

The letters and numerals here shown are arranged from right to left, this being the way of writing used by the Hebrews.

Another interesting set of alphabetic numerals was used by the Goths, a people first known in Poland and Germany, who later conquered a considerable part of Europe. These numerals are for the most part of Greek origin. They are shown in the illustration at the top of the next page. Just below to the left is a page from a Bible translated into Gothic by Bishop

Ulfilas in the fourth century. On its left margin is a row of numerals. These numerals, enlarged, are shown in the small picture to the right. If you compare these with the alphabetic numerals in the picture above you can easily read them. The first is $300 + 40 + 3$, or 343. The second is 344, and the third is 345.

We now come to the numerals that are used in Europe and the Amer-

1	**A**	10	**I ï**	100	**Ҟ**
2	**B**	20	**K**	200	**S**
3	**Γ**	30	**Λ**	300	**T**
4	**d**	40	**M**	400	**Y**
5	**Є**	50	**N**	500	**F**
6	**u**	60	**G**	600	**x**
7	**z**	70	**n**	700	**Θ**
8	**h**	80	**п**	800	**Ω**
9	**ф**	90	**ч**	900	**↑**

Alphabetic numerals used by the Goths. (In part from Menninger, *Zahlwort und Ziffer*.)

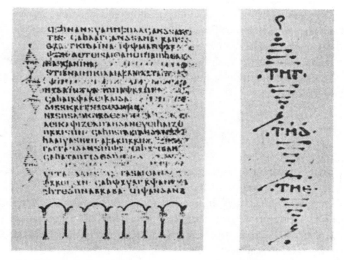

Page from the *Codex Argenteus* (Silver Manuscript), a Bible translated into Gothic by Bishop Ulfilas in the fourth century. (From Menninger, *Zahlwort und Ziffer*.)

icas today, as well as in certain parts of Asia and Africa and regions such as Australasia which were settled by Europeans. First of all, it is necessary to understand that although our European and American numerals are often spoken of as Arabic, they have never been used by the Arabs. They came to us by means of a book on arithmetic which apparently was written in India about twelve hundred years ago, and was translated into Arabic soon afterward. By chance this book was carried by merchants to Europe and there was translated from the Arabic into Latin. This was hundreds of years before books were first printed in Europe, and this arithmetic book was known only in manuscript form. Since it had been translated from Arabic, the numerals were supposed to be those used by the Arabs, but this was not the case. They might be called Hindu-Arabic, but since they took their present shapes in Europe they may better be called European or Modern numerals.

We have seen that our modern numerals 1, 2, and 3 have a long history. The four as we make it, 4, is not as old as these. It is first found in fairly common use in Europe about seven hundred years ago. Among the forms for four in common use in India two thousand years ago were × and +, but there were and still are many other forms used in Asia. The origin of the rest of the numerals is generally unknown. Since in most countries in early days the priests were practically the only educated persons, and since travel was so difficult that different tribes developed different languages, the priests simply invented their own letters and numerals. As travel became easier and as merchants and rulers felt the need for writing, the numerals of the various tribes tended to become more alike. Today, international trade has made the European numerals quite generally known all over the world, although the Chinese or the Arabic forms are still the ones most commonly used by many millions of people.

Our present numerals have changed a great deal from their original forms. Following are some early Hindu characters, found in a cave in India and dating from the second or third century B.C. Some of these are certainly numerals, and the others probably are.

1	2	4	6	7	9	10	10	10

20	60	80	100	100	100	200	400

700	1000	4000	6000	10,000	20,000

The Hindu numerals which are in some cases closely related to our own, and which were taken to Baghdad, in Iraq, about a thousand years ago, are as follows:

The Arabic numerals used then and now are as follows:

You will observe that the zero is simply a dot and that the five is quite like our zero. The Iranian (Persian) numerals are substantially the same except for the four and the five.

Coming now more closely to our modern numerals we have here the oldest example of all these forms (lacking zero) known in any European manuscript. This was written in Spain in the year 976.

The following table shows the changes in our numerals from the time of their first use in Europe to the beginning of printing.

1	2	3	4	5	6	7	8	9	0	
										Twelfth century
										1197 A.D.
										1275 A.D.
										c. 1294 A.D.
										c. 1303 A.D.
										c. 1360 A.D.
										c. 1442 A.D.

After they began to appear in books there were few changes of any significance, the chief ones being in the numerals for four and five.[2] Even at the present time the forms of our numerals frequently change in the attempt to find which kind is the most easily read. For example, consider the following specimens and decide which is the easiest for you to read:

1234567890   1234567890   1234567890

[2] See Smith and Karpinski, *The Hindu-Arabic Numerals* (Boston, 1911); Sir G. F. Hill, *The Development of Arabic Numerals in Europe* (Oxford, 1915); D. E. Smith, *History of Mathematics* (2 vol., Boston, 1923, 1925).

The names for large numbers have also changed from time to time. For example, the word *million* seems not to have been used before the thirteenth century. It means "a big thousand," *mille* being the Latin for thousand, and *-on* meaning (in Italian) *big*. The word started in Italy, was taken over by France in the fifteenth century or earlier, and was thereafter used in England. Until the seventeenth century, people generally spoke of "a thousand thousand" rather than of "a million," and they do so today in certain parts of Europe.

"Billion" is a relatively new word. It comes from the Italian, and is first found as *bimillion, bilioni,* and *byllion.* It originally meant a million million, and in England it is generally still so understood, athough the American use of the term to mean a thousand million has come to be generally understood of late years, largely because of the enormous numbers now in use in financial transactions which concern all countries. The larger numbers have names like trillions, quadrillions, quintillions, and so on, but these are seldom used.

Besides the written numerals which have been described, finger numerals were used during many centuries and by many peoples. The ancient Greeks and Romans used them as did the Europeans of the Middle Ages; and the Asiatics in later times. Even today they are not infrequently used for bargaining in the market places. Indeed, "counting on the fingers" and even multiplying and dividing by these means are known in certain countries, but not commonly in Western Europe and the American continents.

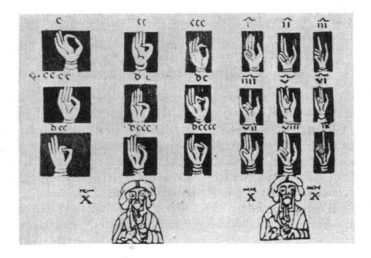

Finger numerals. Spanish, Thirteenth Century. The top row represents 100, 200, 300, 1000, 2000, 3000.

The preceding illustration shows a few of the finger symbols as they appear in a manuscript written in Spain in the thirteenth century.

The finger symbols below are from a book printed in the sixteenth century, three hundred years after the Spanish manuscript mentioned above. Each represents only part of the system. It may be observed, how-

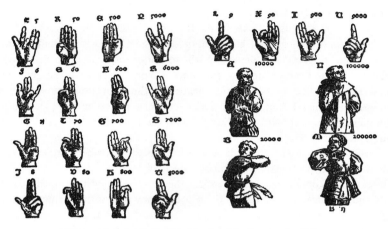

Finger symbols from an arithmetic printed in Germany in 1532.

ever, that the 500 (d) is the same in both illustrations, as are also the 600 (dc), 700 (dcc), 800 (dccc), and 900 (dcccc). There is a similar correspondence in the other numerals not given in these partial illustrations.

FROM NUMERALS TO COMPUTATION

When people first began to use numbers they knew only one way to work with them; that was, to count. Little by little they found out how to add, subtract, and multiply; but this was slow work and in some countries special devices were invented to make computation easier, especially in dealing with large numbers. The Romans used a counting table, or abacus, in which units, fives, tens, and so on were represented by beads which could be moved in grooves, as shown in this illustration.

They called these beads *calculi,* which is the plural of *calculus,* or pebble. We see here the origin of our word "calculate." Since the syllable *calc* means lime, and marble is a kind of limestone, we see that a calculus was a small piece of marble, probably much like those used in playing marbles today. Sometimes, as in the Chinese abacus described below, the calculi slid along on rods. This kind of abacus is called a *suan-pan,* and it

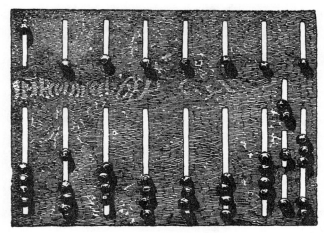

Ancient bronze abacus used by the Romans.

is used today in all parts of China. In the one shown in the following illustration the beads are arranged so as to make the number 27,091, each bead at the top representing five of the different orders (units, tens, and so on).

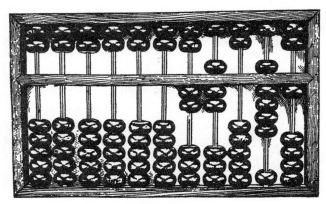

Chinese abacus.

The Japanese use a similar instrument known as the *soroban*. The Chinese and Japanese can add and subtract on the abacus, or counting board, much more rapidly than we can with pencil and paper, but they cannot multiply and divide as quickly as we can. In the *soroban* here shown, beginning in the twelfth column from the right, the number represented is 90,278. The other columns are not being used.

In Russia there is still used a type of abacus known as the *s'choty*, and

Japanese abacus.

not long ago a similar one was used in Turkey (the *coulba*) and also in Armenia (the *choreb*).

Now, to come nearer home, you have often bought things over a "counter" in a store, but did you know that the "counter" tells part of the story of addition and subtraction? Let us see what this story is.

We have already seen that Roman numerals, I, V, X, and so on, were in common use in Europe for nearly two thousand years. It was difficult, however, to write large numbers with these numerals. For example, 98,549 might be written in this way: lxxxviiiMDXLVIIII. There were other ways of writing this number, but they were equally clumsy. The merchants therefore invented an easier method of expressing large numbers. They drew lines on a board, with spaces between the lines, and used disks (small circular pieces like checkers) to count with. On the lowest line there might be from one to four disks, each disk having the value of 1.

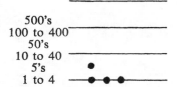

A disk in the space above had the value of 5, and this combined with the disks on the line below could give 6, 7, 8, or 9. In this illustration $5 + 3$ is represented. Larger numbers were handled in the same way on the upper lines and spaces. Sometimes the counters were slid along the rods.

```
1000's   ———●—●—●———
 500's
 100's   ———————●———————
  50's       ●
  10's   ———●—●—●———
   5's       ●
   1's   ———●—●—●———
```

Now look at this figure. There are four disks on the thousand's line, none in the five hundred's space, and so on; that is, you have $4000 + $ no 500's $+ 100 + 50 + 40 + 5 + 3$, or 4198.

You may care to see how such a counting board, with its counters, looked in one of the oldest English arithmetics about four hundred years ago. Maybe you would like to read the old English words as they were printed in the book. The problem is to add 2659 and 8342. The two numbers are not written but are expressed by counters. The star is put on the board so that the eye may easily see where the thousands come, just as we write a comma in a number like 2,659 to show that the left-hand figure is 2 thousand. (This is usually omitted in a date like 1937, and in many other cases, especially where there are numbers of only four figures.)

A D D I T I O N.
Maſter.

Ꞇ he eaſieſt way in this arte, is to adde
but two ſummes at ones togyther:
how be it, you maye adde more, as I wil tel
you anone. therefore whenne you wyll:
adde two ſummes, you ſhall ſpyſte ſet downe
one of them, it forceth not whiche, and then
by it draw a lyne croſſe the other lynes. And
afterwarde ſette doune the other ſumme, ſo
that that lyne maye
be betwene them; as
if you woulde adde
2659 to 8342, you
muſt ſet your ſumes
as you ſee here.

And then if you
lyſt, you maye adde
the one to the other in the ſame place, or els
you may adde them both togither in a new
place: which way, bycauſe it is moſt pypycſt

Page from Robert Record's *Ground of Artes*, printed nearly four hundred years ago.

Because these disks were used in counting they were called *counters*, and the board was sometimes called a *counter board*. When the European countries gave up using counters of this kind (quite generally four hundred years ago) they called the boards used in the shops and banks "counters," and this name has since been commonly used for the bench on which goods are shown in stores. The expression "counting house" is still used in some places to designate the room in which accounts are kept.

One reason for using the counters was that paper was not generally known in Europe until about the eleventh century. Boards covered with a thin coat of wax had been used from the time of the Greeks and Romans, more than a thousand years before. On these it was possible to scratch numbers and words, erasing them by smoothing the wax with a spoon-shaped eraser, but it was very slow work. Slates were used in some parts of Europe, but usable slate quarries were not common and therefore slates could not readily be used elsewhere.

When blackboards were first made, chalk was not always easily found, and so written addition was not so common as addition by counters. When slates, blackboards, and paper all came into use, people added about as we do now. Since all careful computers "check" their work by adding from the bottom up and then from the top down so as to find any mistakes, pupils today add both ways, and there is no reason for teaching addition in only one direction.

Subtraction was done on the counting boards in much the same way as addition. The numbers were represented by counters and were taken away as the problem required. The terms "carry" and "borrow" had more meaning than at present, because a counter was actually lifted up and carried to the next place. If one was borrowed from the next place, it was actually paid back.

Today we learn the multiplication facts just as we learn to read words. If we need to use 7×8, we simply think "56," just as we think "cat" when we see the word CAT. Formerly, however, the "multiplication table" was first written down and then learned as a whole. On the following page are two of these tables from one of the oldest printed arithmetics, a German book of 1489 by Johann Widman. You may wish to see how they were arranged and how to find, in each table, the product of 8×9.

You may also like to see how multiplication looked in 1478, and to find the meaning of the four cases on the right. It will be easier for you if

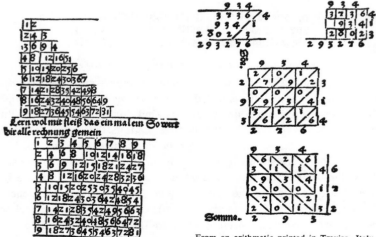

From Johann Widman's Arithmetic of 1489.

From an arithmetic printed in Treviso, Italy, in 1478.

you are told that in each one the problem is to multiply 934 by 314, the product being 293,276, the comma not being used at that time. Do you see how it was found?

Below left another example of multiplication reproduced from a manuscript of the fifteenth century showing how the product of 456,789 × 987,654 was found. The product of these figures is here shown to be 451,149,483,006. The picture looks something like a grating, and so this

$$\begin{array}{l} 2 \\ 38 \\ 1728 (12 \\ 1444 \\ 14 \end{array}$$

was sometimes called the "grating method." You may care to see if you can multiply 345 by 678 using this method. You may also care to try multiplying a 4-figure number by (say) a 3-figure number. If you understand the above illustration, it is not difficult to make such multiplication.

Division was rarely used in ancient times except where the divisor was very small. Indeed, at the present time it is not often needed in comparison with multiplication, and it is far more rarely employed than addition and subtraction. On the abacus it was often done by subtraction; that is, to find how many times 37 is contained in 74, we see that $74 - 37 = 37$, and $37 - 37 = 0$, so that 37 is contained twice in 74.

Above right another way of dividing that was the most common of any in the fifteenth century. It shows the division of 1728 by 144. As fast as the numbers had been used they were scratched out, and so this was often called the "scratch method." The number 1728 was first written; then 144 was written below it. Since the first figure in the quotient is 1, the numbers 144 and 172 are scratched out, and 144 was again written below. The remainders are written above. Divide 1728 by 144 as you would do it today, and compare your method with this.

Our present method, often called "long division," began to be used in the fifteenth century. It first appeared in print in Calandri's arithmetic, published in Florence, Italy, in 1491, a year before Columbus discovered America. The first example shown gives the division of 53,497 by 83, the result being $644\frac{45}{83}$. The second example is the division of $\frac{3}{8}$ by 60, the result being $\frac{1}{160}$. The other three are

$$137\frac{1}{2} \div 12 = 11\frac{11}{24} \qquad 60 \div \frac{3}{8} = 160 \qquad \frac{2}{3} \div \frac{7}{9} = \frac{18}{3} \div 7 = \frac{18}{35}$$

You may be interested in following each step in these examples from this very old arithmetic.

This interesting picture below is from a book that was well known about four hundred years ago. The book was first printed in 1503 and it shows two styles of computing at that time—the counters and the numerals. The number on the counting board at the right is 1241. The one at the left represents the attempted division of 1234 by 97, unsuccessful because decimal fractions such as we know were not yet invented.

A method of computing by counters on a board seems to have been invented by the Indians of South America before the advent of the European discoverers, although our first evidence of the fact appeared in a Spanish work of 1590. In this a Jesuit priest, Joseph de Acosta, tells us:

"In order to effect a very difficult computation for which an able calculator would require pen and ink . . . these Indians [of Peru] made use of their kernels [*sus granos*] of wheat. They place one here, three somewhere else and eight I know not where. They move one kernel here and three there and the fact is that they are able to complete their computation without making the smallest mistake. As a matter of fact, they are better

From the *Margarita Philosophica*, by Gregorius Reisch.

at calculating what each one is due to pay or give than we should be with pen and ink." [3]

In South America, apparently long before the European conquerors arrived, the natives of Peru and other countries used knotted cords for keeping accounts. These were called *quipus*, and were used to record the results found on the counting table. How old this use of the *quipu* may be, together with some kind of abacus, we do not know. A manuscript written in Spanish by a Peruvian Indian, Don Felipe Huaman Poma de Ayala, between 1583 and 1613 has recently been discovered, and is now in the Royal Library at Copenhagen. It contains a large number of pen-and-ink sketches. One of these is here reproduced from a booklet published in 1931.[3] This portrays the accountant and treasurer (*Cōtador maior i tezorero*) of the Inca, holding a *quipu*. In the lower left-hand corner is a counting board with counters and holes for pebbles or kernels.

The use of knots to designate numbers is found in Germany in connection with the number of measures of grain in a sack. Menninger shows the shapes of the knots and their numerical meaning as follows:

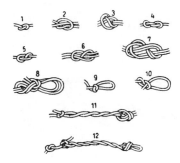

[3] Henry Wassén, *The Ancient Peruvian Abacus*, Göteborg, 1931; reprinted from *Comparative Ethnographical Studies*, vol. 9.

COMMENTARY ON
Idiot Savants

TO do sums in one's head is no great feat even when the results are correct. But to execute mentally, at high speed, long and complicated numerical calculations—extracting roots, raising to high powers, multiplying and dividing by numbers of ten or twenty digits—is a rare and strange talent. Indeed the performance of a calculating prodigy is more apt to amaze us than that of a young chess wizard, juvenile musician or otherwise gifted child.

The nature of this power is little understood. Studies of the early mental traits of geniuses indicate that very few in this class, even among the mathematicians, possess exceptional abilities as calculators. Ampère exhibited astonishing mathematical proficiency at the age of three; James Watt performed geometrical exercises when he was six—nine years before he started brooding fruitfully over the steam emerging from his aunt's teakettle. Pascal, before he was twelve and without knowing any mathematical terms (he called circles "rounds" and lines "bars"), made up his own axioms and wrote out perfect demonstrations of numerous important geometrical theorems; Gauss is said to have corrected his father's mistakes in payroll reckoning when he was three years old. Yet it is doubtful that any of these precocious children could have competed in lightning arithmetic with the self-taught English boy George Parker Bidder, or with Zerah Colburn, the son of a nineteenth-century Vermont farmer. "What is the compound interest on £4444 for 4444 days at 4½%?": Bidder, aged ten, performed the exercise (put to him at a séance of learned men) in two minutes. Colburn, when he was eight, was asked to raise the number 8 to the sixteenth power; he announced the answer (281,474,976,710,656) "promptly and with facility," causing the academic audience to weep.

Calculating prodigies exhibit wide differences as regards heredity, education, general intelligence, retention of their powers over many years, success in other fields of work, longevity and the like. The American Negro Tom Fuller, kidnaped from Africa by slave traders, never learned (he lived to be eighty) to read or write but was a marvel in arithmetic; Bidder, a stonemason's son, not only retained his calculating power throughout his life but became an eminent civil engineer; Colburn's powers fell off at an early age and he died obscurely at thirty-six; Jedediah Buxton and Johann Dase were typical *idiot savants*: brilliant in handling immense numbers, stupid in everything else, including mathematics. Several of the calculators have tried to explain their ability, but without conspicuous success. It is no more enlightening to hear from Bidder that multiplying

89 by 73 was an "instantaneous operation" ("I multiply 80 by 70, 80 by 3, 9 by 70 and 9 by 3 . . . and the answer comes immediately to mind") than to be told by a Dr. Ferrol that when asked "any question . . . the result immediately proceeded from my sensibility . . . I have often the sensation of somebody beside me whispering the right way to find the desired result." One point at least is clear: difficult problems are difficult even for prodigies. Take note of the Rev. H. W. Adams' description of the operating behavior of ten-year-old Truman Henry Safford (1836–1901). "Multiply in your head" (ordered the compassionate Dr. Adams) "365,365,365,365,365,365 by 365,365,365,365,365,365. He flew around the room like a top, pulled his pantaloons over the tops of his boots, bit his hands, rolled his eyes in their sockets, sometimes smiling and talking, and then seeming to be in an agony, until, in not more than one minute, said he, 133,491,850,208,566,925,016,658,299,941,583,225!" An electronic computer might do the job a little faster but it wouldn't be as much fun to watch.

*   *   *   *   *

The following selection is from W. W. Rouse Ball's well-loved classic, *Mathematical Recreations and Essays.* Ball was a fellow of Trinity College, Cambridge, for many years (1878–1905) a mathematical lecturer at the College and a tutor, a member of the bar (although he did not practice) and a mathematical historian. *A Short History of Mathematics* which he published in 1888 presents biographical material on many mathematicians and an interesting and agreeably written account of the subject to the end of the nineteenth century. *Mathematical Recreations and Essays* is a delightful grab bag of puzzles, paradoxes, problems, magic squares, arithmetical and geometrical exercises and the like. It appeared first in 1892 and is now in its thirteenth edition. Ball died in 1925, aged seventy-five.[1]

[1] For obituaries, see *The Cambridge Review,* April 24, 1925, pp. 341–342, and *Nature,* May 23, 1925, Vol. 115, pp. 808–809.

The power of dealing with numbers is a kind of "detached lever" arrangement, which may be put into a mighty poor watch. I suppose it is about as common as the power of moving ears voluntarily, which is a moderately rare endowment.   —OLIVER WENDELL HOLMES

4 Calculating Prodigies

By W. W. ROUSE BALL

AT rare intervals there have appeared lads who possess extraordinary powers of mental calculation.[1] In a few seconds they gave the answers to questions connected with the multiplication of numbers and the extraction of roots of numbers, which an expert mathematician could obtain only in a longer time and with the aid of pen and paper. Nor were their powers always limited to such simple problems. More difficult questions, dealing for instance with factors, compound interest, annuities, the civil and ecclesiastical calendars, and the solution of equations, were solved by some of them with facility as soon as the meaning of what was wanted had been grasped. In most cases these lads were illiterate, and usually their rules of working were of their own invention.

The performances were so remarkable that some observers held that these prodigies possessed powers differing in kind from those of their contemporaries. For such a view there is no foundation. Any lad with an excellent memory and a natural turn for arithmetic can, if he continuously gives his undivided attention to the consideration of numbers, and indulges in constant practice, attain great proficiency in mental arithmetic, and of course the performances of those that are specially gifted are exceptionally astonishing.

In this chapter I propose to describe briefly the doings of the more famous calculating prodigies. It will be seen that their performances were of much the same general character, though carried to different extents, hence in the later cases it will be enough to indicate briefly peculiarities of the particular calculators.

I confine myself to self-taught calculators, and thus exclude the consideration of a few public performers who by practice, arithmetical de-

[1] Most of the facts about calculating prodigies have been collected by E. W. Scripture, *American Journal of Psychology*, 1891, vol. IV, pp. 1–59; by F. D. Mitchell, *Ibid.*, 1907, vol. XVIII, pp. 61–143; and G. E. Müller, *Zur Analyse der Gedächtnistätigkeit und des Vorstellungsverlaufes*, Leipzig, 1911. I have used these papers freely, and in some cases where authorities are quoted of which I have no first-hand information have relied exclusively on them. These articles should be consulted for bibliographical notes on the numerous original authorities.
[For a more recent study in this field see Harvey C. Lehman, *Age and Achievement*, Princeton, 1953; also Max Wertheimer, *Productive Thinking*, Harper and Brothers, 1945. ED.]

vices, and the tricks of the showman have simulated like powers. I also concern myself only with those who showed the power in youth. As far as I know the only self-taught mathematician of advanced years whom I thus exclude is John Wallis, 1616–1703, the Savilian Professor at Oxford, who in middle-life developed, for his own amusement, his powers in mental arithmetic. As an illustration of his achievements, I note that on 22 December 1669 he, when in bed, occupied himself in finding (mentally) the integral part of the square root of 3×10^{40}; and several hours afterwards wrote down the result from memory. This fact having attracted notice, two months later he was challenged to extract the square root of a number of fifty-three digits; this he performed mentally, and a month later he dictated the answer which he had not meantime committed to writing. Such efforts of calculation and memory are typical of calculating prodigies.

One of the earliest of these prodigies, of whom we have records, was *Jedediah Buxton*, who was born in or about 1707 at Elmton, Derbyshire. Although a son of the village schoolmaster, his education was neglected, and he never learnt to write or cipher. With the exception of his power of dealing with large numbers, his mental faculties were of a low order: he had no ambition, and remained throughout his life a farm-labourer, nor did his exceptional skill with figures bring him any material advantage other than that of occasionally receiving small sums of money from those who induced him to exhibit his peculiar gift. He does not seem to have given public exhibitions. He died in 1772.

He had no recollection as to when or how he was first attracted by mental calculations, and of his performances in early life we have no reliable details. Mere numbers however seem always to have had a strange fascination for him. If the size of an object was stated, he began at once to compute how many inches or hair-breadths it contained; if a period of time was mentioned, he calculated the number of minutes in it; if he heard a sermon, he thought only of the number of words or syllables in it. No doubt his powers in these matters increased by incessant practice, but his ideas were childish, and do not seem to have gone beyond pride in being able to state accurately the results of such calculations. He was slow witted, and took far longer to answer arithmetical questions than most of these prodigies. The only practical accomplishment to which his powers led him was the ability to estimate by inspection the acreage of a field of irregular shape.

His fame gradually spread through Derbyshire. Among many questions put to him by local visitors were the following, which fairly indicate his powers when a young man:—How many acres are there in a rectangular field 351 yards long and 261 wide; answered in 11 minutes. How many cubic yards of earth must be removed in order to make a pond 426 feet long, 263 feet wide, and 2½ feet deep; answered in 15 minutes. If sound travels 1142 feet in one second, how long will it take to travel 5 miles; answered in 15 minutes. Such questions involve no difficulties of principle.

Here are a few of the harder problems solved by Buxton when his powers were fully developed. He calculated to what sum a farthing would amount if doubled 140 times: the answer is a number of pounds sterling which requires thirty-nine digits to represent it with 2*s*. 8*d*. over. He was then asked to multiply this number of thirty-nine digits by itself: to this he gave the answer two and a half months later, and he said he had carried on the calculation at intervals during that period. In 1751 he calculated how many cubic inches there are in a right-angled block of stone 23,145,789 yards long, 5,642,732 yards wide, and 54,965 yards thick; how many grains of corn would be required to fill a cube whose volume is 202,680,000,360 cubic miles; and how many hairs one inch long would be required to fill the same space—the dimensions of a grain and a hair being given. These problems involve high numbers, but are not intrinsically difficult, though they could not be solved mentally unless the calculator had a phenomenally good memory. In each case he gave the correct answer, though only after considerable effort. In 1753 he was asked to give the dimensions of a cubical cornbin, which holds exactly one quarter of malt. He recognized that to answer this required a process equivalent to the extraction of a cube root, which was a novel idea to him, but in an hour he said that the edge of the cube would be between 25½ and 26 inches, which is correct: it has been suggested that he got this answer by trying various numbers.

Accounts of his performances were published, and his reputation reached London, which he visited in 1754. During his stay there he was examined by various members of the Royal Society, who were satisfied as to the genuineness of his performances. Some of his acquaintances took him to Drury Lane Theatre to see Garrick, being curious to see how a play would impress his imagination. He was entirely unaffected by the scene, but on coming out informed his hosts of the exact number of words uttered by the various actors, and of the number of steps taken by others in their dances.

It was only in rare cases that he was able to explain his methods of work, but enough is known of them to enable us to say that they were clumsy. He described the process by which he arrived at the product of 456 and 378: shortly it was as follows:—If we denote the former of these numbers by a, he proceeded first to find $5a =$ (say) b; then to find $20b =$ (say) c; and then to find $3c =$ (say) d. He next formed $15b =$ (say) e, which he added to d. Lastly he formed $3a$ which, added to the sum last obtained, gave the result. This is equivalent to saying that he used the multiplier 378 in the form $(5 \times 20 \times 3) + (5 \times 15) + 3$. Mitchell suggests that this may mean that Buxton counted by multiples of 60 and of

15, and thus reduced the multiplication to addition. It may be so, for it is difficult to suppose that he did not realize that successive multiplications by 5 and 20 are equivalent to a multiplication by 100, of which the result can be at once obtained. Of billions, trillions, &c., he had never heard, and in order to represent the high numbers required in some of the questions proposed to him he invented a notation of his own, calling 10^{18} a tribe and 10^{36} a cramp.

As in the case of all these calculators, his memory was exceptionally good, and in time he got to know a large number of facts (such as the products of certain constantly recurring numbers, the number of minutes in a year, and the number of hair-breadths in a mile) which greatly facilitated his calculations. A curious and perhaps unique feature in his case was that he could stop in the middle of a piece of mental calculation, take up other subjects, and after an interval, sometimes of weeks, could resume the consideration of the problem. He could answer simple questions when two or more were proposed simultaneously.

Another eighteenth-century prodigy was *Thomas Fuller*, a negro, born in 1710 in Africa. He was captured there in 1724, and exported as a slave to Virginia, U.S.A., where he lived till his death in 1790. Like Buxton, Fuller never learnt to read or write, and his abilities were confined to mental arithmetic. He could multiply together two numbers, if each contained not more than nine digits, could state the number of seconds in a given period of time, the number of grains of corn in a given mass, and so on—in short, answer the stock problems commonly proposed to these prodigies, as long as they involved only multiplications and the solutions of problems by rule of three. Although more rapid than Buxton, he was a slow worker as compared with some of those whose doings are described below.

I mention next the case of two mathematicians of note who showed similar aptitude in early years. The first of these was *André Marie Ampère*, 1775–1836, who, when a child some four years old, was accustomed to perform long mental calculations, which he effected by means of rules learnt from playing with arrangements of pebbles. But though always expert at mental arithmetic, and endowed with a phenomenal memory for figures, he did not specially cultivate this arithmetical power. It is more difficult to say whether *Carl Friedrich Gauss*, 1777–1855, should be reckoned among these calculating prodigies. He had, when three years old, taught himself some arithmetical processes, and astonished his father by correcting him in his calculations of certain payments for overtime; perhaps, however, this is only evidence of the early age at which his consummate abilities began to develop. Another remarkable case is that of *Richard Whately*, 1787–1863, afterwards Archbishop of Dublin. When he was between five or six years old he showed considerable skill in mental arithmetic: it disappeared in about three years. I soon, said he, "got to do the most difficult sums, always in my head, for I knew nothing of figures beyond numeration, nor had I any names for the different processes I employed. But I believe my sums were chiefly in multiplication, division, and the rule of three . . . I did these sums much quicker than any one could upon paper, and I never remember committing the smallest error. I was engaged either in calculating or in castle-building . . . morning, noon, and night . . . When I went to school, at which time the passion was worn off, I was a perfect dunce at ciphering, and so have continued ever since." The archbishop's arithmetical powers were, however, greater in after-life than he here allows.

The performances of *Zerah Colburn* in London, in 1812, were more remarkable. Colburn,[2] born in 1804, at Cabut, Vermont, U.S.A., was the son of a small farmer. While still less than six years old he showed extraordinary powers of mental calculation, which were displayed in a tour in America. Two years later he was brought to England, where he was repeatedly examined by competent observers. He could instantly give the product of two numbers each of four digits, but hesitated if both numbers exceeded 10,000. Among questions asked him at this time were to raise 8 to the 16th power; in a few seconds he gave the answer 281,474,976,710,-656, which is correct. He was next asked to raise the numbers 2, 3, . . . 9 to the 10th power: and he gave the answers so rapidly that the gentleman who was taking them down was obliged to ask him to repeat them more slowly; but he worked less quickly when asked to raise numbers of two digits like 37 or 59 to high powers. He gave instantaneously the square roots and cube roots (when they were integers) of high numbers, *e.g.*, the square root of 106,929 and the cube root of 268,336,125, such integral roots can, however, be obtained easily by various methods. More remarkable are his answers to questions on the factors of numbers. Asked for the factors of 247,483 he replied 941 and 263; asked for the factors of 171,395 he gave 5, 7, 59, and 83; asked for the factors of 36,083 he said there were none. He, however, found it difficult to answer questions about the factors of numbers higher than 1,000,000. His power of factorizing high numbers was exceptional and depended largely on the method of two-digit terminals described below. Like all these public performers he had to face buffoons who tried to make fun of him, but he was generally equal to them. Asked on one such occasion how many black beans were required to make three white ones, he is said to have at once replied "three, if you skin them"—this, however, has much the appearance of a pre-arranged show.

It was clear to observers that the child operated by certain rules, and

---

[2] To the authorities mentioned by E. W. Scripture and F. D. Mitchell should be added *The Annual Register*, London, 1812, p. 507 *et seq.*

during his calculations his lips moved as if he was expressing the process in words. Of his honesty there seems to have been no doubt. In a few cases he was able to explain the method of operation. Asked for the square of 4,395 he hesitated, but on the question being repeated he gave the correct answer, namely 19,395,025. Questioned as to the cause of his hesitation, he said he did not like to multiply four figures by four figures, but said he, "I found out another way; I multiplied 293 by 293 and then multiplied this product twice by the number 15." On another occasion when asked for the product of 21,734 by 543 he immediately replied 11,801,562; and on being questioned explained that he had arrived at this by multiplying 65,202 by 181. These remarks suggest that whenever convenient he factorized the numbers with which he was dealing.

In 1814 he was taken to Paris, but amid the political turmoil of the time his exhibitions fell flat. His English and American friends however raised money for his education, and he was sent in succession to the Lycée Napoleon in Paris and Westminster School in London. With education his calculating powers fell off, and he lost the frankness which when a boy had charmed observers. His subsequent career was diversified and not altogether successful. He commenced with the stage, then tried school-mastering, then became an itinerant preacher in America, and finally a "professor" of languages. He wrote his own biography which contains an account of the methods he used. He died in 1840.

Contemporary with Colburn we find another instance of a self-taught boy, *George Parker Bidder*, who possessed quite exceptional powers of this kind. He is perhaps the most interesting of these prodigies because he subsequently received a liberal education, retained his calculating powers, and in later life analyzed and explained the methods he had invented and used.

Bidder was born in 1806 at Moreton Hampstead, Devonshire, where his father was a stone-mason. At the age of six he was taught to count up to 100, but though sent to the village school learnt little there, and at the beginning of his career was ignorant of the meaning of arithmetical terms and of numerical symbols. Equipped solely with this knowledge of counting he taught himself the results of addition, subtraction, and multiplication of numbers (less than 100) by arranging and rearranging marbles, buttons, and shot in patterns. In after-life he attached great importance to such concrete representations, and believed that his arithmetical powers were strengthened by the fact that at that time he knew nothing about the symbols for numbers. When seven years old he heard a dispute between two of his neighbours about the price of something which was being sold by the pound, and to their astonishment remarked that they were both wrong, mentioning the correct price. After this exhibition the villagers delighted in trying to pose him with arithmetical problems.

His reputation increased and, before he was nine years old, his father found it profitable to take him about the country to exhibit his powers. A couple of distinguished Cambridge graduates (Thomas Jephson, then tutor of St. John's, and John Herschel) saw him in 1817, and were so impressed by his general intelligence that they raised a fund for his education, and induced his father to give up the rôle of showman; but after a few months Bidder senior repented of his abandonment of money so easily earned, insisted on his son's return, and began again to make an exhibition of the boy's powers. In 1818, in the course of a tour young Bidder was pitted against Colburn and on the whole proved the abler calculator. Finally the father and son came to Edinburgh, where some members of that University intervened and persuaded his father to leave the lad in their care to be educated. Bidder remained with them, and in due course graduated at Edinburgh, shortly afterwards entering the profession of civil engineering in which he rose to high distinction. He died in 1878.

With practice Bidder's powers steadily developed. His earlier performances seem to have been of the same type as those of Buxton and Colburn which have been already described. In addition to answering questions on products of numbers and the number of specified units in given quantities, he was, after 1819, ready in finding square roots, cube roots, &c. of high numbers, it being assumed that the root is an integer, and later explained his method which is easy of application: this method is the same as that used by Colburn. By this time he was able to give immediate solutions of easy problems on compound interest and annuities which seemed to his contemporaries the most astonishing of all his feats. In factorizing numbers he was less successful than Colburn and was generally unable to deal at sight with numbers higher than 10,000. As in the case of Colburn, attempts to be witty at his expense were often made, but he could hold his own. Asked at one of his performances in London in 1818, how many bulls' tails were wanted to reach to the moon, he immediately answered one, if it is long enough.

Here are some typical questions put to and answered by him in his exhibitions during the years 1815 to 1819—they are taken from authenticated lists which comprise some hundreds of such problems: few, if any, are inherently difficult. His rapidity of work was remarkable, but the time limits given were taken by unskilled observers and can be regarded as only approximately correct. Of course all the calculations were mental without the aid of books, pencil, or paper. In 1815, being then nine years old, he was asked:—If the moon be distant from the earth 123,256 miles, and sound travels at the rate of 4 miles a minute,* how long would it be before the inhabitants of the moon could hear of the battle of Waterloo: answer, 21 days, 9 hours, 34 minutes, given in less than one minute. In 1816, being then ten years old, just learning to write, but unable to form figures,

* [The speed of sound in air is 1087 ft. per second; the moon's mean distance from the earth is 238,857 miles. ED.]

he answered questions such as the following:—What is the interest on £11,111 for 11,111 days at 5 per cent. a year: answer, £16,911. 11s., given in one minute. How many hogsheads of cider can be made from a million of apples, if 30 apples make one quart: answer, 132 hogsheads, 17 gallons, 1 quart, and 10 apples over, given in 35 seconds. If a coach-wheel is 5 feet 10 inches in circumference, how many times will it revolve in running 800,000,000 miles: answer, 724,114,285,704 times and 20 inches remaining, given in 50 seconds. What is the square root of 119,-550,669,121: answer 345,761, given in 30 seconds. In 1817, being then eleven years old, he was asked:—How long would it take to fill a reservoir whose volume is one cubic mile if there flowed into it from a river 120 gallons of water a minute: answered in 2 minutes. Assuming that light travels from the sun to the earth in 8 minutes, and that the sun is 98,000,-000 miies off, if light takes 6 years 4 months travelling from the nearest fixed star to the earth, what is the distance of that star, reckoning 365 days 6 hours to each year and 28 days to each month—asked by Sir William Herschel: answer, 40,633,740,000,000 miles. In 1818, at one of his performances, he was asked:—If the pendulum of a clock vibrates the distance of 9¾ inches in a second of time, how many inches will it vibrate in 7 years 14 days 2 hours 1 minute 56 seconds, each year containing 365 days 5 hours 48 minutes 55 seconds: answer, 2,165,625,744¾ inches, given in less than a minute. If I have 42 watches for sale and I sell the first for a farthing, and double the price for every succeeding watch I sell, what will be the price of the last watch: answer, £2,290,649,224. 10s. 8d. If the diameter of a penny piece is 1⅝ inches, and if the world is girdled with a ring of pence put side by side, what is their value sterling, supposing the distance to be 360 degrees, and a degree to contain 69.5 miles: answer, £4,803,340, given in one minute. Find two numbers, whose difference is 12, and whose product, multiplied by their sum, is equal to 14,560: answer, 14 and 26. In 1819, when fourteen years old, he was asked:—Find a number whose cube less 19 multiplied by its cube shall be equal to the cube of 6: answer, 3, given instantly. What will it cost to make a road for 21 miles 5 furlongs 37 poles 4 yards, at the rate of £123. 14s. 6d. a mile: answer, £2688. 13s. 9¾d., given in 2 minutes. If you are now 14 years old and you live 50 years longer and spend half-a-crown a day, how many farthings will you spend in your life: answer, 2,805,120, given in 15 seconds. Mr. Moor contracted to illuminate the city of London with 22,965,321 lamps, the expense of trimming and lighting was 7 farthings a lamp, the oil consumed was ⅔ths of a pint for every three lamps, and the oil cost 3s. 7½d. a gallon; he gained 16½ per cent. on his outlay: how many gallons of oil were consumed, what was the cost to him, and what was the amount of the contract: answer, he used 212,641 gallons of oil, the cost was £205,996. 16s. 1¾d., and the amount of the contract was £239,986. 13*s*. 2*d*. If the distance of the earth from the moon be 29,531,531¼ yards, what is the weight of a thread which will extend that distance, supposing 7^{11}⁄$_{16}$ yards of it weigh ⅛th part of a drachm: answer, 8 cwt. 1 qr. 13 lbs. 9 oz. 1 dr. and ^{13}⁄$_{16}$ths of a drachm.

It should be noted that Bidder did not visualize a number like 984 in symbols, but thought of it in a concrete way as so many units which could be arranged in 24 groups of 41 each. It should also be observed that he, like Inaudi whom I mention later, relied largely on the auditory sense to enable him to recollect numbers. "For my own part," he wrote, in later life, "though much accustomed to see sums and quantities expressed by the usual symbols, yet if I endeavour to get any number of figures that are represented on paper fixed in my memory, it takes me a much longer time and a very great deal more exertion than when they are expressed or enumerated verbally." For instance suppose a question put to find the product of two numbers each of nine digits, if they were "read to me, I should not require this to be done more than once; but if they were represented in the usual way, and put into my hands, it would probably take me four times to peruse them before it would be in my power to repeat them, and after all they would not be impressed so vividly on my imagination."

Bidder retained his power of rapid mental calculation to the end of his life, and as a constant parliamentary witness in matters connected with engineering it proved a valuable accomplishment. Just before his death an illustration of his powers was given to a friend who talking of then recent discoveries remarked that if 36,918 waves of red light which only occupy one inch are required to give the impression of red, and if light travels at 190,000 miles a second, how immense must be the number of waves which must strike the eye in one second to give the impression of red. "You need not work it out," said Bidder, "the number will be 444,433,-651,200,000."

Other members of the Bidder family have also shown exceptional powers of a similar kind as well as extraordinary memories. Of Bidder's elder brothers, one became an actuary, and on his books being burnt in a fire he rewrote them in six months from memory but, it is said, died of consequent brain fever; another was a Plymouth Brother and knew the whole Bible by heart, being able to give chapter and verse for any text quoted. Bidder's eldest son, a lawyer of eminence, was able to multiply together two numbers each of fifteen digits. Neither in accuracy nor rapidity was he equal to his father, but then he never steadily and continuously devoted himself to developing his abilities in this direction. He remarked that in his mental arithmetic he worked with pictures of the figures, and said "If I perform a sum mentally it always proceeds in a visible form in my mind; indeed I can conceive no other way possible of doing mental arithmetic": this it will be noticed is opposed to his father's method. Two of his children, one son and one daughter representing a third generation, have inherited analogous powers.

I mention next the names of *Henri Mondeux*, and *Vito Mangiamele.* Both were born in 1826 in humble circumstances, were sheep-herds, and became when children, noticeable for feats in calculation which deservedly procured for them local fame. In 1839 and 1840 respectively they were brought to Paris where their powers were displayed in public, and tested by Arago, Cauchy, and others. Mondeux's performances were the more striking. One question put to him was to solve the equation $x^3 + 84 = 37x$: to this he at once gave the answer 3 and 4, but did not detect the third root, namely, -7. Another question asked was to find solutions of the indeterminate equation $x^2 - y^2 = 133$: to this he replied immediately 66 and 67; asked for a simpler solution he said after an instant 6 and 13. I do not however propose to discuss their feats in detail, for there was at least a suspicion that these lads were not frank, and that those who were exploiting them had taught them rules which enabled them to simulate powers they did not really possess. Finally both returned to farm work, and ceased to interest the scientific world. If Mondeux was self-taught we must credit him with a discovery of some algebraic theorems which would entitle him to rank as a mathematical genius, but in that case it is inconceivable that he never did anything more, and that his powers appeared to be limited to the particular problems solved by him.

Johann Martin Zacharias Dase, whom I next mention, is a far more interesting example of these calculating prodigies. Dase was born in 1824 at Hamburg. He had a fair education, and was afforded every opportunity to develop his powers, but save in matters connected with reckoning and numbers he made little progress and struck all observers as dull. Of geometry and any language but German he remained ignorant to the end of his days. He was trustworthy and filled various small official posts in Germany. He gave exhibitions of his calculating powers in Germany, Austria, and England. He died in 1861.

When exhibiting in Vienna in 1840, he made the acquaintance of Strasznicky who urged him to apply his powers to scientific purposes. This Dase gladly agreed to do, and so became acquainted with Gauss, Schumacher, Petersen, and Encke. To his contributions to science I allude later. In mental arithmetic the only problems to which I find allusions are straightforward examples like the following:—Multiply 79,532,853 by 93,758,479: asked by Schumacher, answered in 54 seconds. In answer to a similar request to find the product of two numbers each of twenty digits he took 6 minutes; to find the product of two numbers each of forty digits he took 40 minutes; to find the product of two numbers each of a hundred digits he took 8 hours 45 minutes. Gauss thought that perhaps on paper the last of these problems could be solved in half this time by a skilled computator. Dase once extracted the square root of a number of a hundred digits in 52 minutes. These feats far surpass all other records of the kind, the only calculations comparable to them being Buxton's squaring of a number of thirty-nine digits, and Wallis' extraction of the square root of a number of fifty-three digits. Dase's mental work however was not always accurate, and once (in 1845) he gave incorrect answers to every question put to him, but on that occasion he had a headache, and there is nothing astonishing in his failure.

Like all these calculating prodigies he had a wonderful memory, and an hour or two after a performance could repeat all the numbers mentioned in it. He had also the peculiar gift of being able after a single glance to state the number (up to about 30) of sheep in a flock, of books in a case, and so on; and of visualizing and recollecting a large number of objects. For instance, after a second's look at some dominoes he gave the sum (117) of their points; asked how many letters were in a certain line of print chosen at random in a quarto page he instantly gave the correct number (63); shown twelve digits he had in half a second memorized them and their positions so as to be able to name instantly the particular digit occupying any assigned place. It is to be regretted that we do not know more of these performances. Those who are acquainted with the delightful autobiography of Robert-Houdin will recollect how he cultivated a similar power, and how valuable he found it in the exercise of his art.

Dase's calculations, when also allowed the use of paper and pencil, were almost incredibly rapid, and invariably accurate. When he was sixteen years old Strasznicky taught him the use of the familiar formula $\pi/4 = \tan^{-1}(\frac{1}{2}) + \tan^{-1}(\frac{1}{5}) + \tan^{-1}(\frac{1}{8})$, and asked him thence to calculate π. In two months he carried the approximation to 205 places of decimals, of which 200 are correct.[3] Dase's next achievement was to calculate the natural logarithms of the first 1,005,000 numbers to 7 places of decimals; he did this in his off-time from 1844 to 1847, when occupied by the Prussian survey. During the next two years he compiled in his spare time a hyperbolic table which was published by the Austrian Government in 1857. Later he offered to make tables of the factors of all numbers from 7,000,000 to 10,000,000 and, on the recommendation of Gauss, the Hamburg Academy of Sciences agreed to assist him so that he might have leisure for the purpose, but he lived only long enough to finish about half the work.

Truman Henry Safford, born in 1836 at Royalton, Vermont, U.S.A., was another calculating prodigy. He was of a somewhat different type for he received a good education, graduated in due course at Harvard,

[3] The result was published in *Crelle's Journal*, 1844, vol. xxvii, p. 198.

and ultimately took up astronomy in which subject he held a professional post. I gather that though always a rapid calculator, he gradually lost the exceptional powers shown in his youth. He died in 1901.

Safford never exhibited his calculating powers in public, and I know of them only through the accounts quoted by Scripture and Mitchell, but they seem to have been typical of these calculators. In 1842, he amused and astonished his family by mental calculations. In 1846, when ten years old, he was examined, and here are some of the questions then put to him:—Extract the cube root of a certain number of seven digits; answered instantly. What number is that which being divided by the product of its digits, the quotient is three, and if 18 be added the digits will be inverted: answer 24, given in about a minute. What is the surface of a regular pyramid whose slant height is 17 feet, and the base a pentagon of which each side is 33·5 feet: answer 3354·5558 square feet, given in two minutes. Asked to square a number of eighteen digits he gave the answer in a minute or less, but the question was made the more easy as the number consisted of the digits 365 repeated six times. Like Colburn he factorized high numbers with ease. In such examples his processes were empirical, he selected (he could not tell how) likely factors and tested the matter in a few seconds by actual division.

There are to-day four calculators of some note. These are *Ugo Zamebone*, an Italian, born in 1867; *Pericles Diamandi*, a Greek, born in 1868; *Carl Rückle*, a German; and *Jacques Inaudi*, born in 1867. The three first mentioned are of the normal type and I do not propose to describe their performances, but Inaudi's performances merit a fuller treatment.

Jacques Inaudi [4] was born in 1867 at Onorato in Italy. He was employed in early years as a sheep-herd, and spent the long idle hours in which he had no active duties in pondering on numbers, but used for them no concrete representations such as pebbles. His calculating powers first attracted notice about 1873. Shortly afterwards his elder brother sought his fortune as an organ grinder in Provence, and young Inaudi, accompanying him, came into a wider world, and earned a few coppers for himself by street exhibitions of his powers. His ability was exploited by showmen, and thus in 1880 he visited Paris where he gave exhibitions: in these he impressed all observers as being modest, frank, and straightforward. He was then ignorant of reading and writing: these arts he subsequently acquired.

His earlier performances were not specially remarkable as compared with those of similar calculating prodigies, but with continual practice he improved. Thus at Lyons in 1873 he could multiply together almost

[4] See Charcot and Darboux, *Mémoires de l'Institut, Comptes Rendus*, 1892, vol. CXIV, pp. 275, 528, 578; and Binet, *Révue des deux Mondes*, 1892, vol. CXI, pp. 905–924. ·

instantaneously two numbers of three digits. In 1874 he was able to multiply a number of six digits by another number of six digits. Nine years later he could work rapidly with numbers of nine or ten digits. Still later, in Paris, asked by Darboux to cube 27, he gave the answer in 10 seconds. In 13 seconds he calculated how many seconds are contained in 18 years 7 months 21 days 3 hours: and he gave immediately the square root of one-sixth of the difference between the square of 4801 and unity. He also calculated with ease the amount of wheat due according to the traditional story to Sessa who, for inventing chess, was to receive 1 grain on the first cell of a chess-board, 2 on the second, 4 on the third, and so on in geometrical progression.

He can find the integral roots of equations and integral solutions of problems, but proceeds only by trial and error. His most remarkable feat is the expression of numbers less than 10^5 in the form of a sum of four squares, which he can usually do in a minute or two; this power is peculiar to him. Such problems have been repeatedly solved at private performances, but the mental strain caused by them is considerable.

A performance before the general public rarely lasts more than 12 minutes, and is a much simpler affair. A normal programme includes the subtraction of one number of twenty-one digits from another number of twenty-one digits: the addition of five numbers each of six digits: the multiplying of a number of four digits by another number of four digits: the extraction of the cube root of a number of nine digits, and of the fifth root of a number of twelve digits: the determination of the number of seconds in a period of time, and the day of the week on which a given date falls. Of course the questions are put by members of the audience. To a professional calculator these problems are not particularly difficult. As each number is announced, Inaudi repeats it slowly to his assistant, who writes it on a blackboard, and then slowly reads it aloud to make sure that it is right. Inaudi then repeats the number once more. By this time he has generally solved the problem, but if he wants longer time he makes a few remarks of a general character, which he is able to do without interfering with his mental calculations. Throughout the exhibition he faces the audience: the fact that he never even glances at the blackboard adds to the effect.

It is probable that the majority of calculating prodigies rely on the speech muscles as well as on the eye and the ear to help them to recollect the figures with which they are dealing. It was formerly believed that they all visualized the numbers proposed to them, and certainly some have done so. Inaudi however trusts mainly to the ear and to articulation. Bidder also relied partly on the ear, and when he visualized a number it was not as a collection of digits but as a concrete collection of units divisible, if the number was composite, into definite groups. Rückle relies mainly on visualizing the numbers. So it would seem that there are different types of the memories of calculators. Inaudi can reproduce mentally the sound of the repetition of the digits of the number in his own voice, and is confused, rather than helped, if the numbers are shown to him in writing. The articulation of the digits of the number also seems necessary to enable him fully to exhibit his powers, and he is accustomed to repeat the numbers aloud before beginning to work on them—the sequence of sounds being important. A number of twenty-four digits being read to him, in 59 seconds he memorized the sound of it, so that he could give the sequence of digits forwards or backwards from selected points—a feat which Mondeux had taken 5 minutes to perform. Numbers of about a hundred digits were similarly memorized by Inaudi in 12 minutes, by Diamandi in 25 minutes, and by Rückle in under 5 minutes. This power is confined to numbers, and calculators cannot usually recollect a long sequence of letters. Numbers are ever before Inaudi: he thinks of little else, he dreams of them, and sometimes even solves problems in his sleep. His memory is excellent for numbers, but normal or subnormal for other things. At the end of a séance he can repeat the questions which have been put to him and his answers, involving hundreds of digits in all. Nor is his memory in such matters limited to a few hours. Once eight days after he had been given a question on a number of twenty-two digits, he was unexpectedly asked about it, and at once repeated the number. He has been repeatedly examined, and we know more of his work than of any of his predecessors, with the possible exception of Bidder.

Most of these calculating prodigies find it difficult or impossible to explain their methods. But we have a few analyses by competent observers of the processes used: notably one by Bidder on his own work; another by Colburn on his work; and others by Müller and Darboux on the work of Rückle and Inaudi respectively. That by Bidder is the most complete, and the others are on much the same general lines.

Bidder's account of the processes he had discovered and used is contained in a lecture [5] given by him in 1856 to the Institution of Civil Engineers. Before describing these processes there are two remarks of a general character which should, I think, be borne in mind when reading his statement. In the first place he gives his methods in their perfected form, and not necessarily in that which he used in boyhood: moreover it is probable that in practice he employed devices to shorten the work which he did not set out in his lecture. In the second place it is certain, in spite of his belief to the contrary, that he, like most of these prodigies, had an exceptionally good memory, which was strengthened by incessant

[5] *Institution of Civil Engineers, Proceedings*, London, 1856, vol. xv, pp. 251–280. An early draft of the lecture is extant in MS.; the variations made in it are interesting, as showing the history of his mental development, but are not sufficiently important to need detailed notice here.

practice. One example will suffice. In 1816, at a performance, a number was read to him backwards: he at once gave it in its normal form. An hour later he was asked if he remembered it: he immediately repeated it correctly. The number was:—2,563,721,987,653,461,598,746,231,905,-607,541,128,975,231.

Of the four fundamental processes, addition and subtraction present no difficulty and are of little interest. The only point to which it seems worth calling attention is that Bidder, in adding three or more numbers together, always added them one at a time, as is illustrated in the examples given below. Rapid mental arithmetic depended, in his opinion, on the arrangement of the work whenever possible, in such a way that only one fact had to be dealt with at a time. This is also noticeable in Inaudi's work.

The multiplication of one number by another was, naturally enough, the earliest problem Bidder came across, and by the time he was six years old he had taught himself the multiplication table up to 10 times 10. He soon had practice in harder sums, for, being a favourite of the village blacksmith, and constantly in the smithy, it became customary for the men sitting round the forge-fire to ask him multiplication sums. From products of numbers of two digits, which he would give without any appreciable pause for thought, he rose to numbers of three and then of four digits. Halfpence rewarded his efforts, and by the time he was eight years old, he could multiply together two numbers each of six digits. In one case he even multiplied together two numbers each of twelve digits, but, he says, "it required much time," and "was a great and distressing effort."

The method that he used is, in principle, the same as that explained in the usual text-books, except that he added his results as he went on. Thus to multiply 397 by 173 he proceeded as follows:—

We have	$100 \times 397 = 39700,$		
to this must be added	$70 \times 300 = 21000$	making	60,700,
,, ,, ,, ,, ,,	$70 \times 90 = 6300$	,,	67,000,
,, ,, ,, ,, ,,	$70 \times 7 = 490$	,,	67,490,
,, ,, ,, ,, ,,	$3 \times 300 = 900$	,,	68,390,
,, ,, ,, ,, ,,	$3 \times 90 = 270$	,,	68,660,
,, ,, ,, ,, ,,	$3 \times 7 = 21$	,,	68,681.

We shall underrate his rapidity if we allow as much as a second for each of these steps, but even if we take this low standard of his speed of working, he would have given the answer in 7 seconds. By this method he never had at one time more than two numbers to add together, and the factors are arranged so that each of them has only one significant digit: this is the common practice of mental calculators. It will also be observed that here, as always, Bidder worked from left to right: this, though not usually taught in our schools, is the natural and most convenient way. In effect he formed the product of $(100 + 70 + 3)$ and $(300 + 90 + 7)$, or $(a + b + c)$ and $(d + e + f)$ in the form $ad + ae \ldots + ef$.

The result of a multiplication like that given above was attained so rapidly as to seem instantaneous, and practically gave him the use of a multiplication table up to 1000 by 1000. On this basis, when dealing with much larger numbers, for instance, when multiplying 965,446,371 by 843,409,133, he worked by numbers forming groups of 3 digits, proceeding as if 965,446, &c., were digits in a scale whose radix was 1000: in middle life he would solve a problem like this in about 6 minutes. Such difficulty as he experienced in these multiplications seems to have been rather in recalling the result of the previous step than in making the actual multiplications.

Inaudi also multiplies in this way, but he is content if one of the factors has only one significant digit: he also sometimes makes use of negative quantities: for instance he thinks of 27×729 as $27(730 - 1)$; so, too, he thinks of 25×841 in the form $84100/4$: and in squaring numbers he is accustomed to think of the number in the form $a + b$, choosing a and b of convenient forms, and then to calculate the result in the form $a^2 + 2ab + b^2$.

In multiplying concrete data by a number Bidder worked on similar lines to those explained above in the multiplication of two numbers. Thus to multiply £14. 15s. 6¾d. by 787 he proceeded thus:

We have   £(787)(14) = £11018. 0s. 0d.
to which we add (787)(15) shillings = £590. 5s. 0d. making £11608. 5s. 0d.
to which we add (787)(27) farthings = £22. 2s. 8¼d. making £11630. 7s. 8¼d.

Division was performed by Bidder much as taught in schoolbooks, except that his power of multiplying large numbers at sight enabled him to guess intelligently and so save unnecessary work. This also is Inaudi's method. A division sum with a remainder presents more difficulty. Bidder was better skilled in dealing with such questions than most of these prodigies, but even in his prime he never solved such problems with the same rapidity as those with no remainder. In public performances difficult questions on division are generally precluded by the rules of the game.

If, in a division sum, Bidder knew that there was no remainder he often proceeded by a system of two-digit terminals. Thus, for example, in dividing (say) 25,696 by 176, he first argued that the answer must be a number of three digits, and obviously the left-hand digit must be 1. Next he had noticed that there are only 4 numbers of two digits (namely, 21, 46, 71, 96) which when multiplied by 76 give a number which ends in 96. Hence the answer must be 121, or 146, or 171, or 196; and experience enabled him to say without calculation that 121 was too small and 171 too large. Hence the answer must be 146. If he felt any hesitation he mentally multiplied 146 by 176 (which he said he could do "instantaneously") and thus checked the result. It is noticeable that when Bidder, Colburn, and some other calculating prodigies knew the last two digits of a product of two numbers they also knew, perhaps subconsciously, that the last two digits of the separate numbers were necessarily of certain forms. The theory of these two-digit arrangements has been discussed by Mitchell.

Frequently also in division, Bidder used what I will call a digital process, which *a priori* would seem far more laborious than the normal method, though in his hands the method was extraordinarily rapid: this method was, I think, peculiar to him. I define the digital of a number as the digit obtained by finding the sum of the digits of the original number, the sum of the digits of this number, and so on, until the sum is less than 10. The digital of a number is the same as the digital of the product of the digitals of its factors. Let us apply this in Bidder's way to see if 71 is an exact divisor of 23,141. The digital of 23,141 is 2. The digital of 71 is 8. Hence if 71 is a factor the digital of the other factor must be 7, since 7 times 8 is the only multiple of 8 whose digital is 2. Now the only number which multiplied by 71 will give 41 as terminal digits is 71. And since the other factor must be one of three digits and its digital must be 7, this factor (if any) must be 871. But a cursory glance shows that 871 is too large. Hence 71 is not a factor of 23,141. Bidder found this process far more rapid than testing the matter by dividing by 71. As another example let us see if 73 is a factor of 23,141. The digital of 23,141 is 2; the digital of 73 is 1; hence the digital of the other factor (if any) must be 2. But since the last two digits of the number are 41, the last two digits of this factor (if any) must be 17. And since this factor is a number of three digits and its digital is 2, such a factor, if it exists, must be 317. This on testing (by multiplying it by 73) is found to be a factor.

When he began to exhibit his powers in public, questions concerning weights and measures were, of course, constantly proposed to him. In solving these he knew by heart many facts which frequently entered into such problems, such as the number of seconds in a year, the number of ounces in a ton, the number of square inches in an acre, the number of pence in a hundred pounds, the elementary rules about the civil and ecclesiastical calendars, and so on. A collection of such data is part of the equipment of all calculating prodigies.

In his exhibitions Bidder was often asked questions concerning square roots and cube roots, and at a later period higher roots. That he could at once give the answer excited unqualified astonishment in an uncritical audience; if, however, the answer is integral, this is a mere sleight of art which anyone can acquire. Without setting out the rules at length, a few examples will illustrate his method.

He was asked to find the square root of 337,561. It is obvious that the root is a number of three digits. Since the given number lies between 500^2 or 250,000 and 600^2 or 360,000, the left-hand digit of the root must be a 5. Reflection had shown him that the only numbers of two digits, whose squares end in 61 are 19, 31, 69, 81, and he was familiar with this fact. Hence the answer was 519, or 531, or 569, or 581. But he argued that as 581 was nearly in the same ratio to 500 and 600 as 337,561 was to 250,000 and 360,000, the answer must be 581, a result which he verified by direct multiplication in a couple of seconds. Similarly in extracting the square root of 442,225, he saw at once that the left-hand digit of the answer was 6, and since the number ended in 225 the last two digits of the answer were 15 or 35, or 65 or 85. The position of 442,225 between $(600)^2$ and $(700)^2$ indicates that 65 should be taken. Thus the answer is 665, which he verified, before announcing it. Other calculators have worked out similar rules for the extraction of roots.

For exact cube roots the process is more rapid. For example, asked to extract the cube root of 188,132,517, Bidder saw at once that the answer was a number of three digits, and since $5^3 = 125$ and $6^3 = 216$, the left-hand digit was 5. The only number of two digits whose cube ends in 17 is 73. Hence the answer is 573. Similarly the cube root of 180,362,125 must be a number of three digits, of which the left-hand digit is a 5, and the two right-hand digits were either 65 or 85. To see which of these was required he mentally cubed 560, and seeing it was near the given number, assumed that 565 was the required answer, which he verified by cubing it. In general a cube root that ends in a 5 is a trifle more difficult to detect at sight by this method than one that ends in some other digit, but since 5^3 must be a factor of such numbers we can divide by that and apply the process to the resulting number. Thus the above number 180,362,125 is equal to $5^3 \times 1,442,897$ of which the cube root is at once found to be 5 (113), that is, 565.

For still higher exact roots the process is even simpler, and for fifth roots it is almost absurdly easy, since the last digit of the number is always the same as the last digit of the root. Thus if the number proposed is less than 10^{10} the answer consists of a number of two digits. Knowing the fifth powers to 10, 20, . . . 90 we have, in order to know the first digit of the answer, only to see between which of these powers the number proposed lies, and the last digit being obvious we can give the answer instantly. If the number is higher, but less than 10^{15}, the answer is a number of three digits, of which the middle digit can be found almost as rapidly as the others. This is rather a trick than a matter of mental calculation.

In his later exhibitions, Bidder was sometimes asked to extract roots, correct to the nearest integer, the exact root involving a fraction. If he suspected this he tested it by "casting out the nines," and if satisfied that the answer was not an integer proceeded tentatively as best he could.

Such a question, if the answer is a number of three or more digits, is a severe tax on the powers of a mental calculator, and is usually disallowed in public exhibitions.

Colburn's remarkable feats in factorizing numbers led to similar questions being put to Bidder, and gradually he evolved some rules, but in this branch of mental arithmetic I do not think he ever became proficient. Of course a factor which is a power of 2 or of 5 can be obtained at once, and powers of 3 can be obtained almost as rapidly. For factors near the square root of a number he always tried the usual method of expressing the number in the form $a^2 - b^2$, in which case the factors are obvious. For other factors he tried the digital method already described.

Bidder was successful in giving almost instantaneously the answers to questions about compound interest and annuities: this was peculiar to him, but his method is quite simple, and may be illustrated by his determination of the compound interest on £100 at 5 per cent. for 14 years. He argued that the simple interest amounted to £$(14)(5)$, *i.e.* to £70. At the end of the first year the capital was increased by £5, the annual interest on this was 5*s*. or one crown, and this ran for 13 years, at the end of the second year another £5 was due, and the 5*s*. interest on this ran for 12 years. Continuing this argument he had to add to the £70 a sum of $(13 + 12 + \ldots + 1)$ crowns, *i.e.* $(13/2)(14)(5)$ shillings, *i.e.* £22. 15*s*. 0*d*., which, added to the £70 before mentioned, made £92. 15*s*. 0*d*. Next the 5*s*. due at the end of the second year (as interest on the £5 due at the end of the first year) produced in the same way an annual interest of 3*d*. All these three-pences amount to $(12/3)(13/2)(14)(3)$ pence, *i.e.* £4. 11*s*. 0*d*. which, added to the previous sum of £92. 15*s*. 0*d*., made £97. 6*s*. 0*d*. To this we have similarly to add $(11/4)(12/3)(13/2)(14)(3/20)$ pence, *i.e.* 12*s*. 6*d*., which made a total of £97. 18*s*. 6*d*. To this again we have to add $(10/5)(11/4)(12/3)(13/2)(14)(3/400)$ pence, *i.e.* 1*s*. 3*d*., which made a total of £97. 19*s*. 9*d*. To this again we have to add $(9/6)(10/5)(11/4)(12/3)(13/2)(14)(3/8000)$ pence, *i.e.* 1*d*., which made a total of £97. 19*s*. 10*d*. The remaining sum to be added cannot amount to a farthing, so he at once gave the answer as £97. 19*s*. 10*d*. The work in this particular example did in fact occupy him less than one minute—a much shorter time than most mathematicians would take to work it by aid of a table of logarithms. It will be noticed that in the course of his analysis he summed various series.

In the ordinary notation, the sum at compound interest amounts to £$(1{\cdot}05)^{14} \times 100$. If we denote £100 by P and $\cdot 05$ by r, this is equal to $P(1 + r)^{14}$ or $P(1 + 14r + 91r^2 + \ldots)$, which, as r is small, is rapidly convergent. Bidder in effect arrived by reasoning at the successive terms of the series, and rejected the later terms as soon as they were sufficiently small.

In the course of this lecture Bidder remarked that if his ability to recollect results had been equal to his other intellectual powers he could easily have calculated logarithms. A few weeks later he attacked this problem, and devised a mental method of obtaining the values of logarithms to seven or eight places of decimals. He asked a friend to test his accuracy, and in answer to questions gave successively the logarithms of 71, 97, 659, 877, 1297, 8963, 9973, 115249, 175349, 290011, 350107, 229847, 369353, to eight places of decimals; taking from thirty seconds to four minutes to make the various calculations. All these numbers are primes. The greater part of the answers were correct, but in a few cases there was an error, though generally of only one digit: such mistakes were at once corrected on his being told that his result was wrong. This remarkable performance took place when Bidder was over 50.

His method of calculating logarithms is set out in a paper [6] by W. Pole. It was, of course, only necessary for him to deal with prime numbers, and Bidder began by memorizing the logarithms of all primes less than 100. For a prime higher than this he took a composite number as near it as he could, and calculated the approximate addition which would have to be added to the logarithm: his rules for effecting this addition are set out by Pole, and, ingenious though they are, need not detain us here. They rest on the theorems that, to the number of places of decimals quoted, if $\log n$ is p, then $\log (n + n/10^2)$ is $p + \log 1 \cdot 01$, *i.e.* is $p + 0 \cdot 0043214$, $\log (n + n/10^3)$ is $p + 0 \cdot 00043407$, $\log (n + n/10^4)$ is $p + 0 \cdot 0000434$, $\log (n + n/10^5)$ is $p + 0 \cdot 0000043$, and so on.

The last two methods, dealing with compound interest and logarithms, are peculiar to Bidder, and show real mathematical skill. For the other problems mentioned his methods are much the same in principle as those used by other calculators, though details vary. Bidder, however, has set them out so clearly that I need not discuss further the methods generally used.

A curious question has been raised as to whether a law for the rapidity of the mental work of these prodigies can be found. Personally I do not think we have sufficient data to enable us to draw any conclusion, but I mention briefly the opinions of others. We shall do well to confine ourselves to the simplest case, that of the multiplication of a number of n digits by another number of n digits. Bidder stated that in solving such a problem he believed that the strain on his mind (which he assumed to be proportional to the time taken in answering the question) varied as n^4, but in fact it seems in his case according to this time test to have varied approximately as n^5. In 1855 he worked at least half as quickly again as in 1819, but the law of rapidity for different values of n is said to have been about the same. In Dase's case, if the time occupied is proportional to n^x, we must have x less than 3. From this, some have inferred that probably Dase's methods were different in character to those used by Bidder, and it is suggested that the results tend to imply that Dase visualized recorded numerals, working in much the same way as with pencil and paper, while Bidder made no use of symbols, and recorded successive results verbally in a sort of cinematograph way; but it would seem that we shall need more detailed observations before we can frame a theory on this subject.

The cases of calculating prodigies here mentioned, and as far as I know the few others of which records exist, do not differ in kind. In most of them the calculators were uneducated and self-taught. Blessed with excellent memories for numbers, self-confident, stimulated by the astonishment their performances excited, the odd coppers thus put in their pockets and the praise of their neighbours, they pondered incessantly on numbers and their properties; discovered (or in a few cases were taught) the fundamental arithmetical processes, applied them to problems of ever increasing difficulty, and soon acquired a stock of information which shortened their work. Probably *constant practice* and *undivided devotion to mental calculation* are essential to the maintenance of the power, and this may explain why a general education has so often proved destructive to it. The performances of these calculators are remarkable, but, in the light of Bidder's analysis, are not more than might be expected occasionally from lads of exceptional abilities.

COMMENTARY ON
Gifted Birds

ARISTOTLE was of the opinion that man is a rational animal because he can count. This may not seem to us a very impressive argument. Arithmetic is easier than it was in ancient times; the number system has been improved and better methods of calculation have been invented. We use machines which are far more proficient at arithmetic than even the cleverest human computer. It is not surprising, therefore, that arithmetic has lost caste. Bertrand Russell points out that "though many philosophers continue to tell us what fine fellows we are, it is no longer on account of our arithmetical skill that they praise us." [1]

Until recently, however, it was thought that man alone could count; now even this small remnant of pre-Darwinian eminence has been snatched from him. Everyone has seen performances by mathematically gifted horses and dogs (some of them also tell fortunes). But there is little ground for believing that these depressing animals have the faintest idea what they are doing when they stamp or bark out the correct answers to problems of arithmetic. Possessing an initial aptitude for learning commands, they merely obey signals given skillfully (and perhaps even involuntarily) by their masters.

The scientific experiments reported in the following paper by the noted zoölogist, Professor O. Koehler of the University of Freiburg, contain evidence of a different order. They seem to show that animals can be taught to count in the real sense of the word—to recognize differences between unequal groups of dots, or rows of marks, and to take action, to reason, on the basis of quantitative differences. Koehler's pupils have been birds; it is possible that other experimenters have achieved similar results working with dogs, seals, pigs or rats. Konrad Lorenz, the famous Austrian naturalist, has persuaded baby mallards to regard him as their mother by quacking like a duck; Karl von Frisch has demonstrated that bees communicate by a special kind of dance. The skill, imagination and extraordinary patience of these men is exhibited also by Koehler. His studies are important not alone for what they teach about animal behavior, but because they enlarge our understanding of the nature of thinking, of the number concept and of the psychological basis of mathematical reasoning. That is a good deal to learn from the behavior of jackdaws and budgereegahs. St. Francis would be pleased.

[1] Bertrand Russell, *Unpopular Essays*, New York, 1951, p. 72.

A rare bird.—JUVENAL

I heard the little bird say so.—JONATHAN SWIFT (Letter to Stella)

For there is an upstart crow . . .—ROBERT GREENE (1560–1592)

5 The Ability of Birds to "Count"

By O. KOEHLER

SO-CALLED "clever" horses and dogs, like the Loch Ness monster, appear from time to time. The masters of these animals are no doubt often sincere in their belief that their pets really display intelligence of the highest order; for they are quite unable to imagine that the animals can really perceive small involuntary signs and signals which tell them when the right answer has been reached, and so when they must stop barking or stamping, as the case may be. Yet it is now quite certain that that is precisely what these "clever" animals are doing. It was not the stallion 'Zarif' of Elberfeld who reproduced Descartes's "cogito ergo sum" in ill-spelt German, nor was it the dachshund 'Kuno' who answered the professor's question as to what creed he had with the words "Mine is yours!" No, these "clever" animals never solve any problems themselves; they only obey their masters' signs which, given at the proper moment, stop the barking or stamping by which they spell out their answers in code.

In the work here described all possible precautions have been taken to avoid the danger of self-deceit to which the owners of "clever" animals have succumbed in the past, and I hope we have been successful. At any rate, it will be evident from our account that the achievements even of the cleverest of our birds are insignificant compared with those alleged for the "clever" animals. But whereas the performances of the latter are spurious we hope we have proved conclusively that the achievements of our birds are examples of true learning and in no way dependent on clues, voluntary or involuntary, given by the experimenter.

During training and "spontaneous" experiments (see below) the observer and the bird were always separated by a partition so that the experimenter was never seen by it; while at work he was able to watch it through the viewfinder of a cine-camera fitted into the wall. Moreover, when a deterrent was required in conditioning, it was only one degree of punishment and this was always the same. It was thus impossible for the experimenter inadvertently to give signs by means of the punishing apparatus since the latter functioned along all-or-none lines.

After thorough conditioning we made so-called "spontaneous" experiments in which no punishing apparatus was used at all. In these experiments the animal was left entirely to itself and the results recorded automatically, so that all possibility of clues from the investigator was ruled out. Such procedure is, of course, fundamentally different from the process of circus training, where learning continues all the time in front of the spectators. We never see the animal perform alone but are always shown the trainer and animal in co-operation. Our birds learn many tasks without any punishment, merely by self-conditioning. That is by omitting unsuccessful movements and in such cases there is therefore no difference between training and spontaneous experiments; in fact every experiment is spontaneous. Many spontaneous experiments were made by the birds *in complete absence of any human being.* They were left alone with their task in the big garden while the experimenter was occupied far away in the house. He could not see the birds, neither could the birds see him. Then, perhaps after an hour, he would come back and find the problems solved. Thus even telepathic communication between man and animal, if such be a possibility, is excluded. Nevertheless, the results obtained under such conditions with the observer absent were often amongst the best. Nearly all the problems solved by the birds in spontaneous experiments were recorded automatically by cinema. The total length of film required to record such successful solutions has amounted to about 3 Km. Unfortunately the vast majority of this documentary evidence was destroyed by fire during the war, and only four films made for teaching purposes were rescued. Parts of these are what you see to-day. These four films refer only to experiments already published with full statistical details. There remain from the old series, records of experiments of about ten more ravens and one gray parrot, Jake, still to be published. Now we have commenced work again with parrots, magpies and some other species.

It may be asked why we take all this trouble with a problem which is apparently without biological significance since we know no example of any wild bird counting in its normal state? Though this is so, I believe nevertheless the problem is of interest in regard to the old question of the origin of the human *language.* In spite of Revesz' (1946) most valuable book, I believe it is true to say that apart from our experiments, comparative psychology has no evidence to offer towards the solution of this problem. This present congress has given example after example of striking similarities between human and animal action at the lower levels of sensation, orientation, locomotory mechanisms, instinctive behaviour, releasing patterns, social behaviour, moods and learning. Only language appears to be a technique peculiar to man. It therefore seems worth while to suggest as a working hypothesis that all the characteristics which make for the tremendous difference between human and animal behaviour are the consequences of this one human prerogative, viz. language. It is only man that has given *names* to things, ideas, relations, etc. The human child *learns* words just like a parrot; but after having learnt the first few words it may at once form new sentences with them, sentences which express some true relation, which express wishes and which later ask questions. This ability to form new sensible sentences is not learnt in the ordinary sense; on the contrary, this capacity to make reasonable use of learned words is *innate* in the child. This innate capacity is the greatest, indeed the decisive prerogative of man over all animals. As far as I know, no animal has ever learnt to make a similar use of words.

The question arises, however, how it is that our words are fit to be used for forming true sentences, for speaking, ordering and asking; in short, why our language fits our surroundings and why it is of more use than the warblings of a bird. I will try to answer this question by giving an example. *Our birds did not count*, for they lack words. They could not name the numbers that they are able to perceive and to act upon, but in actual fact they learn to *"think unnamed numbers."* The difference between birds and ourselves is perhaps best shown by an instance personally known to me, of how a child learned to count. When nearly four years old, she had learnt the words, one, two three up to seven in the right order and she had learnt to read the figures. She was taught this during a very short time by a reliable friend of the parents who assured them that the lesson had been restricted to exactly these two things (seven words and seven figures, both in the right sequence and co-ordination); nothing more than this was taught. Some days later the child, in the presence of her mother, saw a tram coming, and said, "Look, there comes the three and it has three cars; that fits." It was route No. 3 and the tram had three cars; three was the figure she had read; three cars was the sum of three similar objects, which she had seen and counted. When she further remarked that the figure fits the sum, then that fits meant *"number,"* in the sense that the figure seen and the sum of the objects seen had in common the number *named* three. As a philosopher might say, she had grasped practically the idea of *named numbers*. There is of course nothing abnormal in this. It is merely one example of the innate capacity of every normal human child for using learnt words and at once, without any further learning process, applying them in the right sense; in this instance for counting everything and for calculating without the least instruction. The child had no further contact with her former teacher and the parents themselves gave no further instruction. Indeed they deliberately avoided doing so, but nevertheless she could produce sentences like the following: "I see six is five with one more," "Two and five, I can do that: three (one), four (two) . . ., . . ., seven (five), seven, is right." The parents never encouraged the child and she soon dropped this game. Two years later in school she did not remember her early achievement in mental arithmetic and began to learn counting and, like her schoolmates, with the help of her fingers.

Not even the cleverest of our birds showed inborn capacities like that. But they did learn *unnamed numbers* which were presented to them by two principal methods: (i) simultaneous presentation and (ii) successive presentation. The hypothesis here put forward is that man would never have started counting, i.e., to name numbers, without the following two prelinguistic abilities which he has in common with birds. The first of these abilities is that of being able to compare groups of units presented *simultaneously* side by side by seeing numbers of those units only, excluding all other clues. This problem was presented to birds in quite different ways, starting with only two groups of edible units (grains of corn, fruit, pieces of meat, etc.), the two groups different by one unit only, and at the other extreme rising to choice according to pattern (N. Kohts). A raven, named Jacob, and a gray parrot named Geier were presented with five small boxes covered with lids bearing 2, 3, 4, 5 and 6 spots each respectively—the key being a lid with one of these numbers of spots lying on the ground in front of the boxes. Both birds learned to open only that one of the five lids which had the same number of spots as the key pattern. As a control of the experiment everything was changed in a random manner from one experiment to the next. There were 15 positions of the five boxes and very many different positions of the key pattern; the number of units in the key pattern changed with each experiment and there were five places for the "positive" number of spots on a lid corresponding to that of the key. Moreover there were 24 permutations of the four negative numbers. Also the relative situation of the spots of one group and with Jacob only, the size and form of those spots was changed with absolute irregularity. In the final series of experiments, for each trial I broke afresh a flat plasticine cake into pieces of highly irregular and broken outlines, the size of a piece varying from 1 to 50 units of surface area, and I always took care to make the general patterns of the positive number on the lid as unlike as possible to that of the key pattern. In an experiment such as this I hope that any clue foreign to the problem was excluded, and yet Jacob solved the problem by choosing the positive lid according to the only item which was not changing through all the experiments, i.e. the number of spots equal to that of the particular key pattern presented.

The second ability with which we are concerned is to estimate, i.e. to remember numbers of incidents following each other and thus to keep in mind *numbers presented successively in time,* independent of rhythm or any other clue which might be helpful. Here again many different methods were employed. At first birds were trained to eat only "x" grains out of many offered without any help being given from a figure. For instance, if a hundred or more grains are given in a big heap, one grain touching many others, there are no figures at all for no gaps are to be seen between them (in the budgerigar film you will see birds trained to eat only two or three grains out of a big heap). Another task was to eat only "x" peas which were rolled into a cup, one after another, at intervals ranging from 1-60 seconds. Here again there are no figures at all; the peas are delivered as if by a slot machine, one by one, so the pigeon never sees more than one pea in the cup. The bird always has the same view and there is absolutely no visible clue for distinguishing the last permitted pea from the first forbidden one. The same applies to opening lids of boxes standing in a long row until "x" baits have been secured; since the baits were arranged in the boxes in 20 or more distributions from one experiment to the next, the number of the lids to be opened is constantly changing. For instance, if the bird is trained to take five baits it may have to open any number of lids between one and seven. The bird thus opens lids up to 5 baits, i.e., it "acts upon five," but the number of lids it has to open to get five baits changes with each experiment. Still more remarkable is the fact that birds learned to master up to four problems of this kind at the same time. For instance, a jackdaw learned to open black lids until it had secured two baits, green lids up to three, red lids up to four, and white lids until it had secured five baits. Similarly budgerigars learned to "act upon two" when the experimentor said "dyo dyo dyo" and upon three (take only three grains from the heap) when he uttered the words "treis treis treis . . .". A similar result could be obtained with a bell indicating two and a buzzer indicating three.

Since, as far as we could see, all external clues were carefully excluded in this experiment, only an inner token can have been responsible for the birds ceasing action when the number was reached. Here we are forced to use psychological terms. It seems as if the bird does some inward marking of the units he is acting upon, sometimes those supposed "inward marks" show themselves in external behaviour in the form of intention movements. Thus a jackdaw, given the task to open lids until five baits had been secured, which in this case were distributed in the first five boxes in the order 1, 2, 1, 0, 1, went home to its cage after having opened only the first three lids and having consequently eaten only four baits. The experimenter was just about to record "one too few, incorrect solution," when the jackdaw came back to the line of boxes. Then the bird went through a most remarkable performance: it bowed its head once before the first box it had emptied, made two bows in front of the second box, one before the third, then went further along the line, opened the fourth lid (no bait) and the fifth, and took out the last (fifth bait). Having done this it left the rest of the line of boxes untouched and went home with an air of finality. This "intention" bowing, repeated the same number of times before each open box as on the first occasion when it had found baits in them, seems to prove that the bird remembered its previous actions. It looks as if after its first departure it became aware it had not finished the task, so it came back and started again, "picking up", with intention movements, baits it had already picked up in actuality; when, however, it came to the last two boxes, which by mistake it had omitted to open on its first trip, it performed the full movements and thus completed its task.

The simplest explanation that I have to offer for this inner marking is that it may consist in *equal* marks as if we were to think or give one nod of the head for one, two for two, and so on. This I propose to call *"thinking un-named numbers."* On the contrary, the most prodigal assumption which one could make and one which, I believe, is unnecessary, would be that the bird did the marking by *inequal* or *qualitatively different* marks in fixed order, as if we think 1, 2, 3, 4, 5, or alternatively a phrase of five different syllables. That of course would be *named numbers*, in other words, real counting. If my simplest assumption, that of equal marks, be true, one might expect that an *anticipated* punishment during the process of acting up to "x" for instance, would be without effect. Thus one might expect that if in the course of being trained for five the bird was punished after having taken two grains, it would continue to act upon five; and that is indeed what the pigeons did. But budgerigars and still better, jackdaws, were able to adhere to a new order inculcated by only one anticipated punishment, and one jackdaw adhered to this new order for a long series of spontaneous experiments before finally returning to the old higher value of "x." Notwithstanding this result, I hesitate to accept this more prodigal assumption, because I have only this one experiment pointing in that direction. All other tests indicate the contrary. Thus there is no sign of "number intelligence" at all, no obvious progress of learning ability as in the case of children, but if the marks were quite equal, according to our first assumption, it is hard to understand how the jackdaw could grasp the new lower number at once, indicated by only one anticipated deterrent which occurs quite without warning. Perhaps the assumption of purely *quantitative* differences between the marks which are qualitatively identical, would serve the explanation; for instance, a crescendo of expectancy of punishment or in the case of self-conditioning without punishment, a decreasing avidity for the baits as the bird passed along the line of marks.

Pigeons and budgerigars were hopelessly upset by changes in experimental conditions during training for a problem, whereas jackdaws grasped the transposition at once. One bird, after having learnt to open the lid with three spots and not touch the lid with four spots, was presented with three and four mealworms moving on the lids instead of spots. The film shows the first presentation of this new kind of clue in a spontaneous experiment. The bird is anxious and hesitates a long time, but finally it goes straight to the three mealworms and leaves the four alone, and instead of opening the lid it eats the mealworms from it without opening it at all. To summarise: a given species of bird would show the same ability of grasping unnamed numbers where they are presented simultaneously or successively, but the ability differs with the species. Thus with pigeons it may be five or six according to experimental conditions, with jackdaws it is six and with ravens and parrots, seven.

Similar tests of the same ability have been carried out with man. In this case the groups of figures are presented tachistoscopically by lantern slides, shown simultaneously and carefully timed so that while they may be fully seen the time is too short for counting to take place. It is remarkable that, when this is done with human beings, the limit of achievement is of the same order as that shown by birds. Thus few persons reach eight and many, like pigeons, get no further than five. As to the second ability, experiments are still being carried out, but as far as we can see there is no expectation that man can achieve better results than the birds provided named counting is excluded. This similarity of the limit in both abilities, as in the six species including man, seems to be a most important fact, and we can hardly assume such a correlation to be due to chance. For ourselves, the two abilities, to grasp unnamed numbers presented simultaneously (differentiating between groups seen side by side) or successively (acting upon "x"), have in common only the absolute idea of unnamed numbers. Since birds do not speak and have names neither for things nor numbers, we must ask whether a bird may not learn to associate the two abilities. This can be decided in two ways, first by testing the power of action upon seen numbers and second by choosing a certain seen number equalling a number just acted upon. We are working along both these lines. As to the first, there has been one positive result. After many failures a jackdaw learned to take two mealworms out of a circle containing many if there were two points in the centre of the circle, and four mealworms if there were four points.

Two pre-linguistic faculties common to some species of birds and to man have been demonstrated. "Unnamed thinking," which I would like to suggest may be common with us. Our "brainwaves" are very often wordless. We often quite suddenly perceive correct solutions of problems with which we have been striving for a long time before we have decided we will name them in one language or another by a formula or a sketch. Very many ideas come to us in wordless form, just as most of the ideas which an animal may grasp; the difference merely is that we can recount it afterwards in words and animals cannot. Fighting, playing, courting, etc., in both animals and man may have many items in common, items which probably are wordless, that is, in both cases.

This attempt may encourage more research towards the detection of "unnamed thinking" in other directions. The question "what is it enables our language to fit nature as it does?" may perhaps be partly answered by pointing to forms of pre-linguistic thinking which higher animals and man have in common. So far we seem to have found two of them.

REFERENCES

Koehler, O. (1941) Vom Erlernen unbenannter Anzahlen bei Vögeln. Die Naturwissenschaften 29, 201–218. Photos taken from the films, former literature cited.
Koehler, O. 4 films to be had from Institut für den Unterrichtsfilm in München 23, Leopoldstrasse 175,
    C 281, 1939, Können Tauben zählen?
    B 440, 1940, Vom Erlernen unbenannter Anzahlen bei Tauben.
    B 442, 1940, Wellensittiche erlernen unbenannte Anzahlen.
    B 467, 1940, Dohlen erlernen unbenannte Anzahlen.
Koehler, O. "Zähl"—versuche an einem Kohlkraben und Vergleichsversuche an Menschen. Zeitschift für Tierpsychologie 5, 1943, 575–712.
Kohts, N. N. Ladygina, (1921) *Berichte des Zoologischen Laboratoriums beim Darwin-Museum fur die Zeit* 1914–1920, Staatsverlag. 15 pages, 5 tables, Russian literature cited.
Revesz, G. (1946), *Ursprung und Vorgeschichte der Sprache.* Bern, A. Francke AG. Verlag.

The Mysteries of Arithmetic

A RITHMETIC is commonly supposed to be the simplest branch of mathematics. Nothing could be further from the truth. The subject is difficult from the ground up, though the practice of elementary arithmetic is admittedly easy enough. The same can be said of most sciences: it is unnecessary to understand electromagnetic theory before wiring a lamp or to study physics in order to repair a pump. We count on our fingers and give no heed to the proliferating implications of the act.

The fundamental rules and operations of arithmetic are extraordinarily hard to define. The concepts of counting and number have taxed the powers of the subtlest thinkers and problems of number theory which can be stated so that a child can grasp their meaning have for centuries withstood attempts at solution. It is strange that we know so little about the properties of numbers. They are our handiwork, yet they baffle us; we can fathom only a few of their intricacies. Having defined their attributes and prescribed their behavior, we are hard pressed to perceive the implications of our formulas.

E. T. Bell offers a pretty survey of a vast, rambling subject. The excerpt comes from *Mathematics—Queen and Servant of Science.* He discusses, among other things, the curious ways of prime numbers, fermat and mersenne numbers, the work of Diophantus of Alexandria, algebraic and transcendental numbers, the "notorious" guess of Goldbach, the ingenious conjecture of Waring. The theory of numbers, Bell says, "is the last great uncivilized continent of mathematics." Out of this theory, as out of Africa, there is always something new. For 2,500 years amateurs as well as professionals have explored it, yet there is every reason to expect that future discoveries, with and without the help of machines, will "far surpass" those of the past. You will agree, after reading Bell, that arithmetic is not simple.

6   The Queen of Mathematics

By ERIC TEMPLE BELL

AN UNRULY DOMAIN

GAUSS, I recall, crowned arithmetic the Queen of Mathematics. Gauss
lived from 1777 to 1855, and to his profound inventiveness is due more
than one deep and broad river of mathematical progress during the nineteenth and twentieth centuries. In the higher arithmetic, or the theory of
numbers, his work is as vital as it was in 1801, when he published his
Disquisitiones arithmeticae. Some of his innovations in that masterpiece,
for example the concept of congruence, have had a far wider significance
than even he may have foreseen. The like holds for his contributions to
geometry, which Riemann was to generalize in preparation for the physics
of the twentieth century. He also made outstanding contributions to the
science of his time, notably to electromagnetism and astronomy. His
opinions therefore are respected by all mathematicians and by some
scientists.

Arithmetic to Gauss, as to the Greeks, was primarily the study of the
properties of the whole numbers. The Greeks, it may be remembered,
used a different word for calculation and its applications to trade. For this
practical kind of arithmetic the aristocratic, slave-owning Greeks seem to
have had a sort of contempt. They called it *logistica*, a name which survives in the logistics of one modern school in the logic and foundations of
mathematics. However, without this despised drudge of the Queen even
the most aloof of the great Greek astronomers could not have gone very
far with his "System of the World"—in Laplace's phrase.

In arithmetic, as in all fields of mathematics since 1920, discovery went
wide and far. But there was one most significant difference between this
advance and the others. Geometry, analysis, and algebra each acquired
one or more vantage points from which to survey its whole domain; arithmetic did not.

The Greeks left no problem in geometry which the moderns have failed
to dispose of. Faced by some of the trifles, like that of 'perfect numbers,'
which the Greeks left in arithmetic we are still baffled. For instance, give
a rule for finding all those numbers which, like 6, are the sums of all their
divisors less than themselves, $6 = 1 + 2 + 3$, and prove or disprove that
no odd number has this property. Such a number is called *perfect*; the

next perfect number after 6 is 28. We shall meet these numbers again. To say that arithmetic is mistress of its own domain when it cannot subdue a childish thing like this is undeserved flattery.

The theory of numbers is the last great uncivilized continent of mathematics. It is split up into innumerable countries, fertile enough in themselves, but all more or less indifferent to one another's welfare and without a vestige of a central, intelligent government. If any young Alexander is weeping for a new world to conquer, it lies before him. Arithmetic has not yet had its Descartes, to say nothing of its Newton.

Lest this estimate seem unduly pessimistic, let us not forget that in each of the several countries of arithmetic there was remarkable progress since the time of Gauss, and especially since 1914, when modern analysis was applied to problems in the theory of numbers that had withstood the strongest efforts of Gauss's successors for over a century. Indeed, two or three of the splendid things done are comparable to anything in geometry, with this qualification, however: no one advance affected the whole course of development. This possibly is due to the very nature of the subject.

Among the notable advances is that which revealed one source of some of those mysterious harmonies which Gauss admired in the properties of whole numbers. This was the creation by E. E. Kummer (1810–1893), Dedekind, and Kronecker of the theory of algebraic numbers. In this particular field Kummer's invention of ideal numbers is comparable to that of non-Euclidean geometry, and likewise for Dedekind's creation of the theory of algebraic numbers which was the source of several ideas in modern abstract algebra. Another striking advance was the brilliant development of the analytic theory of numbers since the early 1900s. Of isolated problems inherited from the past that have been successfully grappled with, we may mention in particular E. Waring's (1734–1798) of the eighteenth century, to which we shall recur, and so for C. Goldbach's (1690–1764) conjecture. Another result of singular interest was the proof that certain numbers are transcendental, and the construction of many such numbers. We shall briefly indicate the nature of all these things presently. These preliminaries may well be closed with the following quotations and a note on amateurs of arithmetic.

The higher arithmetic [Gauss wrote in 1849] presents us with an inexhaustible storehouse of interesting truths—of truths, too, which are not isolated, but stand in the closest relation to one another, and between which, with each successive advance of the science, we continually discover new and wholly unexpected points of contact. A great part of the theories of arithmetic derive an additional charm from the peculiarity that we easily arrive by induction at important propositions, which have the stamp of simplicity upon them, but the demonstration of which lies so deep as not to be discovered until after many fruitless efforts; and even then it is obtained by some tedious and artificial process, while the simpler methods of proof long remain hidden from us.

In the preface to his *Introduction to the theory of numbers* (1929), L. E. Dickson (1874–) says,

During twenty centuries the theory of numbers has been a favorite subject of research by leading mathematicians and thousands of amateurs. Recent investigations compare favorably with the older ones. Future discoveries will far surpass those of the past.

Among other things I have included from the vast mass of results accumulated during the past 2,500 years are several items that still intrigue amateurs. The simplicity of their statements is no index of their difficulty. In fact it seems almost as if amateurs have been attracted by some of the hardest problems that continue to rebuff professionals. Occasionally some observant amateur notices an interesting property of numbers that the professionals had overlooked. The professionals then have new problems to worry them, sometimes for a century or more.

FERMAT AND MERSENNE NUMBERS

The theory of numbers as an independent discipline of mathematics originated with P. S. (de) Fermat (1601–1665). By profession Fermat was a jurist and parliamentarian. His diversion was mathematics, which he approached as an amateur but developed as a master of masters. Although he was one of the founders of analytic geometry and the calculus he is remembered today principally for his work in the theory of numbers. He was the born arithmetician, and it is doubtful whether his penetrating insight into the intrinsic properties of the natural numbers 1, 2, 3, . . . has yet been equalled; certainly it has not been surpassed. It should be remembered that when Fermat did his greatest work he had only a clumsy substitute for the concise and creative algebraic symbolism which today is learned in a few weeks by beginners in school. Some of his deepest discoveries were reasoned out verbally with very few if any symbols, and those for the most part mere abbreviations of words. Any impatient student of mathematics or science or engineering who is irked by having algebraic symbolism thrust on him should try to get on without it for a week.

Fermat, Pascal, and other French mathematicians of the early seventeenth century had a mutual friend in Father M. Mersenne (1588–1648), who acted as a sort of post office, transmitting the mathematical and scientific letters of his friends from one to another. Mersenne also was an amateur. He survives in arithmetic through the numbers named after him. Some historians have conjectured that it was Fermat who really was responsible for these famous and mysterious numbers. But this seems unlikely, if for no other reason than that Mersenne was not given to stealing his friends' ideas.

Fermat's numbers, denoted by F_n, are defined by

$$F_n = 2^{2^n} + 1, \; n = 1, 2, 3, \ldots,$$

so that F_4, for instance, is $2^{16} + 1$, F_6 is $2^{64} + 1$; thus

$$F_1 = 5, \; F_2 = 17, \; F_3 = 257, \; F_4 = 65{,}537,$$

all four of which are readily verified to be primes. The next is the sizable number

$$F_5 = 4{,}294{,}967{,}297.$$

Fermat's reason for investigating these numbers was his hope of finding a formula involving the integer n which for $n = 1, 2, 3, \ldots$ would yield *only* prime numbers. He mistakenly thought that F_n was such a formula, but said explicitly that he could not prove that F_n is prime for all n. This is of crucial historical importance, because Fermat stated many of his results without saying how he had derived them and without indicating proofs, although he either said or implied that he had proofs. With one very famous exception, to be noted later, all of his asserted theorems have been proved. He was mistaken in his conjecture about F_n, and this one definite misjudgment is presumptive evidence that he knew when he had a proof and when he had not.

Euler (1732) factored F_5,

$$F_5 = 641 \times 6{,}700{,}417.$$

Next (1880), F_6, which is too long to write out here in unfactored form, was shown to be not prime,

$$F_6 = 274{,}177 \times 67{,}280{,}421{,}310{,}721.$$

Later it was proved that F_n is not prime for

$$n = 7, 8, 9, 11, 12, 18, 23, 36, 38, 73.$$

With a few reasonable assumptions regarding typography and a little common arithmetic, supplemented by logarithms if convenient, the reader may convince himself that if F_{73} were written out in full not all the libraries in the world could hold it. The last prime F_n so far found is F_4, and some hardy guessers have conjectured that there are no more primes F_n. The problem of proving or disproving that there are only a *finite* number of primes F_n is open. If there are infinitely many primes F_n this, conceivably, might be proved by some simple device such as Euclid's in his proof that there is no end to the primes 2, 3, 5, 7, 11, 13, ... In this same direction F. M. G. Eisenstein (1823–1852), a first-rate arithmetician, stated (1844) as a problem that there are an infinity of primes in the sequence

$$2^2 + 1, \; 2^{2^2} + 1, \; 2^{2^{2^2}} + 1, \; \ldots .$$

Doubtless he had a proof. This looks like the sort of thing an ingenious amateur might settle. If anyone asks why I have not done it myself—I am neither an amateur nor ingenious. In passing, I recall the extraordinary statement attributed to Gauss, "There have been only three epoch-making mathematicians in history, Archimedes, Newton, and Eisenstein."

The Fermat numbers offer a beautiful example of "the new and wholly unexpected points of contact" (between the theory of numbers and other departments of mathematics) remarked by Gauss in his tribute to arithmetic. He may have been thinking of the spectacular discovery which he made at the age of eighteen and which decided him to devote his genius to mathematics rather than to languages and philology—for which he had a rare gift. A geometrical construction using only a straightedge and a pair of compasses is called *Euclidean*. A polygon of which all the sides are equal and all the angles are equal is said to be *regular*. By 350 B.C. the Greeks knew Euclidean constructions for the regular polygons of 4, 8, 16, . . . sides and for those of 3 and 5 sides—the equilateral triangle and the regular pentagon. From these it was easy to construct regular polygons of, $2^c \times 3$, $2^c \times 5$, $2^c \times 3 \times 5$ sides, where c is any positive integer, and the Greeks in effect showed how. They got no farther. Young Gauss proved that if N is of the form 2^c or 2^c times a product of *different* Fermat primes F_n, then there is a Euclidean construction for a regular polygon of N sides. This form of N is both necessary and sufficient for the possibility of a Euclidean construction. It would be interesting if F_4 really is the last Fermat prime—it is always satisfying to know that some problem is definitely done with. Simple Euclidean constructions for the regular polygons of 17 and 257 sides are available, and an industrious algebraist expended the better part of his years and a mass of paper in attempting to construct the F_4 regular polygon of 65,537 sides. The unfinished outcome of all this grueling labor was piously deposited in the library of a German university. Could misguided zeal go farther? It often has.

There is a baseless legend that Gauss's tombstone is inscribed with the diagram of his construction for the regular polygon of 17 sides. It may possibly be true that at one time Gauss, remembering the tomb of Archimedes with its diagram of the quadrature of the sphere, described by Cicero, wished such a memorial. If he had requested the like for the regular polygon of 65,537 sides, his executors might have had to duplicate or surpass the Great Pyramid. Of course this is an exaggeration. But nobody yet has been so pertinaciously stupid as actually to carry out the straightedge-and-compass construction for 65,537. Happily a Euclidean construction for 4,294,967,297 is impossible.

Mersenne's numbers M_p are as famous as Fermat's. They are defined by

$$M_p = 2^p - 1, \quad p = 2, 3, 5, 7, 11, 13, \ . \ . \ . \ , \ 257, \ . \ . \ . \ ,$$

where p is prime. Mersenne asserted (1644) that the only p's for which M_p is prime are

$$p = 2, 3, 5, 7, 13, 17, 19, 31, 67, 127, 257.$$

(If p were composite, $2^p - 1$ would be immediately factorable. For example, $2^6 - 1 = (2^3 + 1)(2^3 - 1)$, and so on.) There are 44 other primes p less than 257. So, according to Mersenne, M_p is not prime for any of these. It would be of great interest to know what grounds Mersenne thought he had for his assertions. It seems unlikely that he was just crudely guessing. He was an honest man and he was not a fool. Nevertheless he was far wrong. His first error was detected in the 1880s, when M_{61} was proved to be prime. This was shrugged off by those who still believed that the mysterious Father Mersenne really knew what he was talking about; 61 was just some careless copyist's mistake for 67. But in 1903 F. N. Cole (1861–1927) proved that M_{67} is not prime.

I should like here to preserve a small bit of history before all the American mathematicians of the first half of the twentieth century are gone. When I asked Cole in 1911 how long it had taken him to crack M_{67} he said "three years of Sundays." But this, though interesting, is not the history. At the October, 1903, meeting in New York of the American Mathematical Society, Cole had a paper on the program with the modest title *On the factorization of large numbers.* When the chairman called on him for his paper, Cole—who was always a man of very few words—walked to the board and, saying nothing, proceeded to chalk up the arithmetic for raising 2 to the sixty-seventh power. Then he carefully subtracted 1. Without a word he moved over to a clear space on the board and multiplied out, by longhand,

$$193,707,721 \times 761,838,257,287.$$

The two calculations agreed. Mersenne's conjecture—if such it was—vanished into the limbo of mathematical mythology. For the first and only time on record, an audience of the American Mathematical Society vigorously applauded the author of a paper delivered before it. Cole took his seat without having uttered a word. Nobody asked him a question.

Another critical prime p was 257, the largest claimed by Mersenne for his prime M_p's. D. H. Lehmer (son of D. N. Lehmer, whose factor tables for the first ten million numbers set an unsurpassed record for completeness and accuracy) proved (1931) that M_{257} is not prime, although the method used did not produce a pair of factors. In the fall of 1932 I was one of some friends of D. H. Lehmer who witnessed (in Pasadena) a test of his electronic machine for seeking the factors of large numbers. It was a dramatic demonstration of brains plus machinery, with the emphasis on brains. For a graphic account of that seance, the reader may consult D. N. Lehmer's Hunting Big Game in the Theory of Numbers (*Scripta Mathematica*, Vol. 1, pp. 229-235, 1932). As this is written, there are rumors that an electronic machine in Manchester (England) is being used to test Mersenne numbers.[1]

As someone may wish to push out farther, I state E. Lucas's (1842–1891) criterion, as much sharpened by D. H. Lehmer, for the primality of M_p. In the sequence

$$S_1 = 4,\ S_2 = 14,\ S_3 = 194,\ \ldots,\ S_{t+1}^2 = S_t^2 - 2,\ \ldots,$$

each term after the first is equal to the square of the preceding term minus 2. Thus

$$S_2 = 4^2 - 2 = 14,\ S_3 = 14^2 - 2 = 194,\ S^4 = 194^2 - 2 = 27634,\ \ldots,$$

and so on. The terms of the sequence increase with terrific rapidity, as the reader may see by calculating a few more. The criterion is: If p is a prime greater than 2, the Mersenne number $M_p = 2^p - 1$ is prime if, and only if, M_p divides the term of rank $(p - 1)$ in the sequence. Anyone who attempts to apply this to M_{257} will soon admit that some mechanical aid is desirable. Of course the terms of the sequence are not actually calculated out—even the hugest calculating machine in existence could not produce the sequence required for M_{257}. Short cuts provided by the theory of numbers make the criterion usable.

Having mentioned Lucas, I may recall that he was another of the great amateurs, in the sense that, although he was conversant with much of the higher mathematics of his day, he refrained from working in the fashionable things of his time in order to give his instinct for arithmetic free play. His *Théorie des nombres*, première partie, 1891 (all issued, unfortunately), is a fascinating book for amateurs and the less academic professionals in the theory of numbers. His widely scattered writings should be collected, and his unpublished manuscripts sifted and edited.[2]

When we come presently to Fermat's baffling 'Last Theorem' we shall note a curious connection between it and the Fermat and Mersenne numbers. For the moment I close Mersenne's account with immortality by quoting from D. H. Lehmer's paper in *The Bulletin of the American Mathematical Society,* Vol. 53, p. 167, 1947. M_p as before is $2^p - 1$, where p is prime.

(1) For the following primes p, M_p is prime,

$$2,\ 3,\ 5,\ 7,\ 13,\ 17,\ 19,\ 31,\ 61,\ 89,\ 107,\ 127.$$

---

[1] (Added in proof) Since 1947, when Mersenne was settled, the search for M_p's has been pushed to about 10,000.

[2] Some years ago the fantastic price of thirty thousand dollars was being asked for Lucas's manuscripts. In all his life Lucas never had that much money.

(2) For the following primes p, M_p is not prime and has been completely factored,

11, 23, 29, 37, 41, 43, 47, 53, 59, 67, 71, 73, 79, 113.

(3) For the following primes p, M_p is not prime and two or more factors are known,

151, 163, 173, 179, 181, 223, 233, 239, 251.

(4) For the following primes p, M_p is not prime and only one prime factor is known,

83, 97, 131, 167, 191, 211, 229.

(5) For the following primes p, M_p is not prime but no factor is known,

101, 103, 109, 137, 139, 149, 157, 199, 241, 257.

In Lehmer's summary (6) the primes $p = 193$, $p = 227$ were undecided. H. S. Uhler (1872–), in 1948 settled these two.* Neither is prime. Mersenne's final score, then, is five mistakes—the *inclusion* of 67, 257 and the *exclusion* of 61, 89, 107. What can the man have thought he was thinking about? It took 304 years to set him right.

From (3), (4), (5), it appears that something still remains to be done about Mersenne's mishaps. Modern calculating machines may do it—with the humble assistance of human brains.

A LITTLE ABOUT PRIMES

The (rational) primes 2, 3, 5, 7, 11, 13, 17, 19, 23, . . . are the building blocks for the *multiplicative* division of the (elementary) theory of numbers. This is concerned with the consequences of *the fundamental theorem of arithmetic: A number* (positive integer) *is a product of primes in essentially one way only.* 'Essentially' means that two products of primes differing only in the arrangement of their factors are not counted as distinct. For example,

$$105 = 3 \times 5 \times 7 = 5 \times 7 \times 3,$$

and so on. The theorem can be proved by mathematical induction plus some simple but skilled ingenuity. E. Zermelo (1871–) gave such a proof in 1912 and published it in 1934. Others imitated him.

Of older theorems on primes, one of Fermat's may be specially mentioned, first because in itself it is prized as a gem by arithmeticians, and second because Fermat proved it by his method of *'infinite descent.'* [3] Every prime of the form $4n + 1$ is a sum of two squares in one way only.

---

* [In 1935 six Mersenne numbers remained untested. All of these were first shown by Uhler to be composite. ED.]

[3] Actually first used by Euclid.

For example, $5 = 1 + 4$, $13 = 4 + 9$, $17 = 1 + 16$, $29 = 4 + 25$, $101 =$ $1 + 100$. Moreover, as is seen immediately, no number of the form $4n + 3$ is a sum of two squares. The proof by descent assumes that there is a smaller prime of the same form for which the theorem is false. Descending thus we reach the conclusion that the theorem is false for 5. But $5 = 1 + 4 = 1^2 + 2^2$. This contradiction establishes the theorem. The method of descent works infallibly when it can be applied, but usually it is difficult to find the essential step down.

A cornerstone of the theory of numbers is Fermat's ('little,' or 'lesser') theorem of 1640, which states that if n is any integer not divisible by the prime p, then $n^{p-1} - 1$ is divisible by p. This (like nearly everything else inherited from the past in mathematics) has been extended and generalized in several ways, by no means all trivial. As simple proofs of the theorem are readily accessible in college algebras and elsewhere, I shall not describe them. Any student of modern algebra will know the far-reaching consequences of this theorem.

We remarked Fermat's attempt to find a formula yielding only primes. A closely similar problem is to find a criterion for primality. This was achieved before 1770 by J. Wilson (1741–1793), who stated that n is prime if, and only if,

$$1 + 2 \times 3 \times 4 \times \ldots \times (n - 1)$$

is divisible by n. For $n = 6$ this gives $1 + 120 = 121$, not divisible by 6, so 6 is not prime; for $n = 7$, $1 + 720 = 721$, divisible by 7, so 7 is prime. Unfortunately Wilson's absolute criterion is unusable for even very moderately large numbers—try it for $n = 101$.

There is an interesting connection between Mersenne primes and the so-called perfect numbers already mentioned. Perfect numbers entered arithmetic with the Pythagoreans, who attributed mystical and slightly nonsensical virtues to them. If $S(n)$ denotes the sum of *all* the divisors of n, including 1 and n itself, n is called *perfect* if $S(n) = 2n$. This evidently agrees with the former definition. For example, 6 is perfect since 1, 2, 3, 6 are all the divisors of 6, and the sum of these is 12, or 2×6. So for 28: all the divisors of 28 are 1, 2, 4, 7, 14, 28, whose sum is 56 or 2×28. Euclid and Euler between them proved that an *even* number is perfect if, and only if, it is of the form $2^c(2^{c+1} - 1)$, where $2^{c+1} - 1$ *is prime*. So to every Mersenne prime there corresponds an *even* perfect number. But what about *odd* perfect numbers? Are there any? The question was still unanswered in 1950 after about 2,300 years. Some progress had been made, however, in proving such things as that if an odd perfect number exists it must have at least six different prime factors as shown by Sylvester. Other negative results were proved in 1947 and later, but as they did not come near to settling the question I shall not describe them.

Almost the first question anyone, amateur or professional, might ask about primes is, "How many primes are there not exceeding any prescribed limit, say a billion, or x?" A more accessible question is, how many primes, *in the long run*, are there not exceeding x? More precisely, if $P(x)$ denotes the number of primes not exceeding x, is there a function of x, say $L(x)$, such that, as x tends to infinity—becomes indefinitely large—$P(x)/L(x)$ gets closer and closer to 1? If so, we say that $P(x)$ is *asymptotic* to $L(x)$. It was proved independently and almost simultaneously in 1896 by J. Hadamard (1865–) and C. J. de la Vallée Poussin (1866–) that $P(x)$ is asymptotic to $x/\log x$. (The log here is the natural logarithm.) This is the 'prime number theorem.' Its proof was a triumph of delicate mathematical analysis. Numerous modifications of the original proof engendered a vast literature of interest chiefly to specialists. Then, quite unexpectedly, A. Selberg in 1949 published an elementary proof. 'Elementary' does not mean 'easy.' The proof is elementary in a somewhat technical sense. It probably will be simplified.

Another famous theorem about primes is P. G. L. Dirichlet's (1805–1859) which states that in any arithmetic progression

$$an + b, n = 0, 1, 2, 3, \ldots ,$$

where a, b are positive integers having no common factor greater than 1, there are an infinity of primes. For example, there are an infinity of primes of the form $6n + 1$. Dirichlet proved this (1837) by difficult analysis. His proof was the real beginning of the modern analytic theory of numbers, in which *analysis*, the mathematics of *continuity*, is applied to problems concerning the *discrete* domain of the integers. Again most unexpectedly, Selberg published (1949) an 'elementary' proof of this theorem of Dirichlet's.

On possibly a higher level of difficulty is the simple-looking question whether or not there are an infinity of primes of the form $n^2 + 1$. Almost anyone can guess about such problems that nobody yet has dented, much less split. Few reputable mathematicians today publish their unsubstantiated conjectures. It used to be imagined by some of the more romantic arithmeticians that the reckless guessers of the past possessed mysterious 'lost' methods of extraordinary power which enabled them to discover truths far beyond the reach of modern mathematics. The deflation of Mersenne has made guessing unpopular in some quarters, but by no means in all.

A notorious guess was that of Goldbach in 1742. On only the scantiest numerical evidence he asserted that every even number greater than 2 is a sum of two primes, for example $30 = 13 + 17$. There was no significant progress toward a decision of this conjecture until 1937, when I. M. Vinogradov proved that every 'sufficiently large' odd number is a sum of three odd primes. Theoretically, the finite gap implied by 'large' could be closed in a reasonable time by modern calculating machines, and doubtless will be unless the exact Goldbach guess is disposed of before Vinogradov's theorem is outdated.

The prime-number theorem belongs to what is called the analytic theory of numbers. This vast and intricate structure, mostly a creation of the twentieth century, is largely concerned with determining the *order* (relative size) of the errors made if we take an *approximate* enumeration in a particular problem concerning a class of numbers instead of the exact enumeration. In this the leaders were Hardy, E. Landau (1877–1938), and J. E. Littlewood (1885–). As a significant by-product of this analytic theory of (discrete) numbers, we may recall the extremely useful and simple concept of *order functions*, now a commonplace in treatises on analysis. This is indispensable in sufficiently accurate approximative calculation in many departments of applied mathematics, such as statistical mechanics, where *exact* results are sometimes humanly unattainable, even with calculating machines. The theory of *order* (magnitude of error, or of approximation) in this sense originated in 1892 in problems of the theory of numbers having no discernible connection with physics. Specifically, the 'O' function was introduced by P. Bachmann in his *Analytische Zahlentheorie* (1892). The physicist R. H. Fowler (1889–1943) told me that he acquired the great skill in approximative calculation exhibited in his *Statistical mechanics* of 1929 in a course at Cambridge under Hardy in the analytic theory of numbers.

The broader significance of all this work is its fusion of modern analysis and arithmetic into a powerful method of research in the theory of numbers.

DIOPHANTINE ANALYSIS

As the name implies, this vast domain of the theory of numbers goes back to Diophantus of Alexandria, the earliest known master of the subject and, with Euler, Lagrange, and Gauss, one of the greatest. Diophantine analysis deals with the solution *in integers*, or *in rational numbers* (common fractions), of single equations or systems of equations in two or more unknowns.

The stated restrictions on the solutions are the sources of difficulty. For instance, in school algebra $2x + 3y = 5$ is solved for y thus, $y = (5 - 2x) \div 3$, where x may be *any* number. But *if only integers x, y are permitted*, the problem is not so easy. This particular equation can be solved 'by inspection': $x = 1$, $y = 1$ is a particular solution; if t is any integer, $x = 1$

$+ 3t$, $y = 1 - 2t$ is the complete solution. If *positive* solutions are required, t must be zero. But, as the reader may satisfy himself, such an equation as

$$173x + 201y + 257z = 11,001$$

can hardly be solved by inspection.

There is a complete theory for systems of diophantine equations of the first degree. It is the final outcome of some fifteen centuries of effort by numerous arithmeticians beginning with the remote Hindus but essentially finished only in the 1860s by H. J. S. Smith (1826–1883). For equations of degree higher than the first, very little of any generality is known, and one of the outstanding unsolved problems of the theory of numbers—indeed of all pure mathematics, as emphasized in 1902 by Hilbert—is to devise usable criteria for distinguishing solvable from unsolvable diophantine equations. The corresponding problem for algebraic equations is at least approachable, if not yet solved in detail, by the Galois theory and its modern extensions.

To keep this chapter within reasonable bounds I must omit much of interest and note only the most celebrated of all diophantine problems, Fermat's of about 1637. Fermat was accustomed to record some of his discoveries on the margin of his copy of C. G. Bachet's (1581–1638) edition of Diophantus' *Arithmetica*. What follows is Fermat's ever-memorable enunciation of his 'Last Theorem.' I quote Vera Sanford's translation of Fermat's Latin from *A Source Book in Mathematics* (McGraw-Hill, 1929).

To divide a cube into two other cubes, a fourth power, or in general any power whatever into two powers of the same denomination above the second is impossible, and I have assuredly found an admirable proof of this, but the margin is too narrow to contain it.

Over three centuries of sustained efforts by some of the greatest mathematicians in history have failed to amplify that narrow margin to a complete proof that

$$x^n + y^n = z^n$$

is impossible in integers x, y, z all different from zero if n is an integer greater than 2. The exception $n = 2$ is necessary, as we noticed in connection with the ancient Babylonians. Many a would-be disposer of the famous Last Theorem has wrecked himself by 'proving' the impossibility of the equation for all n's greater than 1.

In passing, may I request any reader of this section who imagines he has a proof not to send it to me? I have examined well over a hundred fallacious attempts, and I feel that I have done my share. One such, many years ago, stuck me for three weeks. I felt that there was a mistake, but couldn't find it. In desperation I turned the author's manuscript over to a very bright girl in my trigonometry class, who detected the blunder in half an hour. This was not as humiliating as it might have been. C. L. F. Lindemann (1852–1939), who in 1882 immortalized himself by proving the transcendence of π (to be noted presently), toward the end of his life published at his own expense a long alleged proof. The fatal mistake was almost at the beginning of the argument. Anyone contemplating a proof may be interested in what Hilbert said in 1920 when asked why he did not try: "Before beginning I should put in three years of intensive study, and I haven't that much time to squander on a probable failure."

Fermat left a proof by descent for the case $n = 4$, and Euler (1770) gave an incomplete proof for $n = 3$, subsequently completed by others. Since $x^{mn} = (x^m)^n$, and likewise for y^{mn}, z^{mn}, it would be sufficient now to prove the theorem for *odd prime* exponents n greater than 3. So far all attempts at a general proof have had to distinguish two cases: n does not divide any of x, y, z; n does divide one of them. The second appears to be much the more difficult. For the first case, J. B. Rosser (1907–) proved (1940) that the theorem is true for all odd primes not exceeding 41,000,-000. This was bettered (1941) by D. H. Lehmer and E. Lehmer (1905–) to 253,747,889. For the second possibility, the limit up to 1950 was the 607 of H. S. Vandiver (1882–). For a summary of what was known up to 1946, the reader may consult the paper in the *American Mathematical Monthly*, vol. 53, 1946, pp. 555–578, by Vandiver, the arithmetician who since Kummer went more deeply than anyone else into Fermat's Last Theorem.

Three special results may be mentioned for their suggestive connections with problems that interested Fermat. A. Wieferich proved (1909) that, if in the first case there are solutions x, y, z, the prime n must be such that $2^{n-1} - 1$ is divisible by n^2. The only n less than 2,000 that fits is 1,093. Numerous similar criteria were discovered after 1909. From Wieferich's criterion, E. Gottschalk deduced (1938) that there are no solutions in the first case if n is a Fermat prime $2^{2^a} + 1$ or a Mersenne prime $2^b - 1$. Is this historically suggestive? Or isn't it?

Finally, H. Kapferer (1888–) proved (1933) the astonishing result that the existence of a solution of the equation

$$z^3 - y^2 = 3^3 \cdot 2^{2n-2} \cdot x^{2n}$$

in rational integers x, y, z, any two of which have no common factor greater than 1, is equivalent to the existence of a solution of Fermat's equation

$$u^n - v^n = w^n.$$

ALGEBRAIC NUMBERS

The greatest service Fermat's Last Theorem so far has rendered mathematics was its instigation of the theory of algebraic numbers. This theory was responsible for some of the guiding concepts—for instance, ideals—in modern algebra, and these in turn have reacted on modern mathematical physics.

The positive, zero, and negative whole numbers of common arithmetic are called *rational* integers, to distinguish them from *algebraic* integers, which are defined as follows.

Let $a_0, a_1, a_2, \ldots, a_{n-1}, a_n$ be $n+1$ given rational integers, of which a_0 is not zero, and not all of which have a common divisor greater than 1. It is known from the fundamental theorem of algebra (first proved in 1799 by Gauss) that the equation

$$a_0 x^n + a_1 x^{n-1} + \ldots + a_{n-1} x + a_n = 0$$

has exactly n roots. That is, there are exactly n real or complex numbers, say $x_1, x_2, \ldots, x_n$, such that if any one of these be put for x in the equation, the left-hand side becomes zero. Notice that no kind of number beyond the complex has to be created to solve the equation. If $n = 2$, we have the familiar fact that a quadratic equation has precisely two roots. For clarity I repeat that $a_0, a_1, a_2, \ldots, a_n$ in the present discussion are rational integers, and that a_0 is not zero. The n roots $x_1, x_2, \ldots, x_n$ are called *algebraic numbers*. If a_0 *is* 1, these algebraic numbers are called *algebraic integers*, which are a *generalization* of the rational integers, as seen in a moment. For instance, the two roots of $3x^2 + 5x + 7 = 0$ are algebraic numbers; the two roots of $x^2 + 5x + 7 = 0$ are algebraic integers.

A rational integer, say n, is also an algebraic integer, for it is the root of $x - n = 0$, and so satisfies the general definition. But an algebraic integer is not necessarily rational. For instance, neither of the roots of $x^2 + x + 5 = 0$ is a rational number although both, according to the definition, are algebraic integers. In the study of algebraic numbers and integers we have another instance of the tendency to generalization which distinguishes modern mathematics.

Omitting technical details and refinements, I shall give some idea of a radical distinction between rational integers and those algebraic integers which are not rational. First we must see what a field of algebraic numbers is. I shall state the basic definitions.

If the left-hand side of the given equation

$$a_0 x^n + a_1 x^{n-1} \ldots + a_n = 0,$$

in which $a_0, a_1, \ldots, a_n$ are rational numbers, can *not* be split into two factors each of which has again rational numbers as coefficients, the equation is called *irreducible* of *degree n*.

Now consider all the expressions which can be made by starting with a particular root of an irreducible equation of degree n (as above) and operating on that root by addition, multiplication, subtraction, and division (division by zero excluded). Say the root chosen is r; as specimens of the result we get $r + r$, or $2r$, r/r or 1, $r \times r$ or r^2, then $2r^2$, and so on indefinitely. The set of all such expressions is evidently a field, according to our previous definition; it is called the *algebraic number field* of degree n generated by r. It can be proved that any element of the field generated by r is expressible as a *polynomial* in r of degree not exceeding $n - 1$, where n is the degree of the irreducible equation having r as one of its roots. The field will contain algebraic numbers and algebraic integers. It is these integers at which we must look, after a slight digression on rational integers.

The *rational* primes are 2, 3, 5, 7, 11, 13, 17, 19, 23, 29, . . . , namely, the numbers greater than 1 which have only 1 and themselves as divisors. The *fundamental theorem of (rational) arithmetic*, I recall, states that a rational integer greater than 1 either is a prime or can be built up by multiplying primes in essentially *one way only*. This is so well known that some writers of schoolbooks assert it to be 'self-evident,' which is a signal instance of the danger of the 'obvious' in mathematics. Whoever can prove it without cribbing from a book may have the stuff of a real mathematician in him. I have mentioned Zermelo's proof.

Primes in algebraic numbers are defined exactly as in common arithmetic. But the 'self-evident' theorem that *every* integer in *every* algebraic number field can be built up in essentially one way only by multiplying primes is, unfortunately, *false*. The foundation has vanished and the whole superstructure has gone to smash.

One should not feel unduly humiliated at having jumped to this particular 'obvious' but wrong conclusion. More than one first-rank mathematician of the nineteenth century did the same. One of them was Cauchy, but he soon pulled himself up short. In *some* algebraic number fields an algebraic integer can be built up in more than one way by multiplying primes together. This is chaos, and the way back to order demanded high genius for its discovery.

The theory of algebraic numbers originated with Kummer's attempt to prove Fermat's Last Theorem. About 1845 he thought he had succeeded. His friend Dirichlet pointed out the mistake. Kummer had assumed the truth of that 'obvious' but not always true theorem about the prime factors of algebraic integers. He set to work to restore order to the chaos in which arithmetic found itself, and in 1847 published his *restoration* of the fundamental law of arithmetic for the particular fields connected with Fermat's assertion. This achievement is usually rated as of greater mathematical importance than would be a proof of Fermat's theorem. To restore unique factorization into primes in his fields, Kummer created a totally new species of number, which he called *ideal*.

In 1871 Dedekind did the like by a simpler method which is applicable to the integers of *any* algebraic number field. Rational arithmetic was thereby truly generalized, for the *rational* integers are the algebraic integers in the field generated by 1 (according to our previous definitions).

Dedekind's 'ideals,' which replace numbers, stand out as one of the memorable landmarks of the nineteenth century. I can recall no instance in mathematics where such intense penetration was necessary to see the underlying regular pattern beneath the apparent complexity and chaos of the facts, and where the thing seen was of such shining simplicity.

The first algebraic number field beyond the rational to be investigated was that generated by a root of $i^2 + 1 = 0$. The integers in this field are of the form $a + bi$, where a, b are rational integers. They are called *Gaussian integers*, as Gauss introduced them in 1828–32. The fundamental theorem of arithmetic holds for these integers, as also does the so-called *Euclidean algorithm* for finding the G.C.D. of two numbers, but of course in a modified and generalized form. I state the kernel of the matter to prepare for a striking theorem proved in 1947. If we divide one positive rational integer by another, say 12 by 5, the least positive remainder, here 2, is less than the divisor 5. How could this be generalized to Gaussian integers? It is meaningless to speak of one complex number being less than another, because the complex numbers cannot be arranged in linear order. But if a unique *real* number could be 'associated with' the complex number $a + bi$, we might hope to continue.

As so often in the theory of numbers, it took unusual insight, not to say genius, to see the right thing to do. The *conjugate* of $a + bi$ is by definition $a - bi$; the product, $a^2 + b^2$, of these is called the *norm* of either of them. Gauss proved that in dividing one Gaussian integer by another the process can be so arranged that the norm of the remainder is always less than the norm of the divisor, exactly as in rational arithmetic. From this it can be shown that the fundamental theorem of arithmetic holds for Gaussian integers. The field defined by $i^2 + 1 = 0$ is *quadratic*, since the defining equation is of the *second* degree, and *imaginary*, since the roots of the equation are not real numbers, and finally *Euclidean*, since, by the result just described, Euclid's algorithm for the G.C.D. holds for the integers of the field. How many quadratic fields are there in which there is a Euclidean algorithm? The question is really difficult. It was finally answered in 1947 by K. Inkeri, who proved that there are exactly twenty-two such fields. This is one of those completely satisfying results in the theory of numbers where a problem outstanding for many years is solved once and for all.

TRANSCENDENTAL NUMBERS

A mere glance at these must suffice. A number which is *not algebraic* is called *transcendental*. Otherwise stated, a transcendental number satisfies no algebraic equation whose coefficients are rational numbers. It was only in 1844 that the existence of transcendentals was proved, by Liouville. The transcendental numbers, hard as they are to find individually, are *infinitely more numerous* than algebraic numbers. This was first proved by G. Cantor, much to the astonishment of mathematicians.

A very famous transcendental is π (pi), the ratio of the circumference of a circle to its diameter. To 7 decimals $\pi = 3.1415926 \ldots$, and it was somewhat uselessly computed in 1874 to 707.[4] In 1882 Lindemann, using a method devised in 1873 by Hermite, proved that π is transcendental, thus destroying for ever the last slim hope of those who would square the circle—although many of them don't know even yet that the ancient Hebrew value 3 of π was knocked from under them centuries ago.

In 1900 Hilbert emphasized what was then an outstanding problem, to prove or disprove that $2^{\sqrt{2}}$ is transcendental. The rapidity of modern progress can be judged from the fact that R. Kusmin in 1930 proved a whole infinity of numbers, one of which is Hilbert's, to be transcendental. The proof is relatively simple. Then, in 1934, A. Gelfond proved the vastly more inclusive theorem that a^b, where a is any algebraic number other than 0 or 1, and b is any irrational algebraic number, is transcendental. The proof of this is not easy.

WARING'S CONJECTURE

Fermat proved that every rational integer is a sum of *four* rational integer squares (zero is included as a possibility). Thus $10 = 0^2 + 0^2 + 1^2 + 3^2$, $293 = 2^2 + 8^2 + 9^2 + 12^2$, and so on. In 1770 Waring guessed that every rational integer is the sum of a *fixed* number N of nth powers of rational integers, where n is any given positive integer and N depends only upon n. For $n = 3$, the required N is 9; for $n = 4$, it is known that N is not greater than 21.

Hilbert in 1909, by most ingenious reasoning, proved Waring's conjecture to be correct. But his proof did not indicate the number of nth powers required. Like much of Hilbert's work, the proof was one of 'existence.' This is slightly ironical, as the logic of existence theorems, among other things, was to force Hilbert in the 1930s to abandon his apparently promising attempt to prove the consistency of mathematical analysis.

---

[4] (Added in proof.) Since this was written, one of the new (ENIAC) calculating machines, in about 70 hours, computed π to 2035 decimals. Such a computation by hand might take all of a normal lifetime of hard labor.

In 1919, Hardy, applying the powerful machinery of modern analysis, gave a deeper proof, the spirit of which is applicable to many other extremely difficult questions in arithmetic. This advance was highly significant for its joining of two widely separated fields of mathematics, *analysis*, which deals with the *uncountable*, or *continuous*, and *arithmetic*, which deals with the *countable*, or *discrete*.

Finally, beginning in 1923, Vinogradov brought some of these extremely difficult matters within the scope of comparatively elementary methods. Proceeding partly from Vinogradov's results, Dickson and S. S. Pillai almost simultaneously disposed of the entire Waring problem except for certain obstinate cases which were settled by I. M. Niven (1915–) in 1943. Conquests such as this would have seemed to the mathematicians of the 1840s to be centuries beyond them. I have dwelt on this episode—which after all is only an episode, though a brilliant one, in the general advance of mathematics—because it is a typical example of the internationalism of mathematics. A conjecture propounded by an Englishman was completely settled by the combined efforts of an Englishman, a German, a Russian, an Indian, a Texan, and a Canadian now a United States citizen. There is no national prejudice in mathematics.

Waring's problem belongs to the *additive* theory of numbers, in which it is required to express numbers of one specified class as *sums* of numbers in another specified class. For example, if each of the classes is that of all non-negative integers, we ask in how many ways is any given integer a sum of integers. This apparently idle question is less impractical than it may seem. It and its simpler variants are of use in statistical mechanics and the kinetic theory of gases. Euler initiated the theory of partitions in 1741. The modern theory falls into two main divisions, the algebraic and the analytic. The latter was elaborated in the 1920s and 1930s to facilitate actual numerical calculation. Here again asymptotic formulas were sought and found by advanced and difficult mathematical analysis.

THE QUEEN OF QUEENS' SLAVES

Some of my unmathematical friends have incautiously urged me to include a note about the origin of modern calculating machines. This is the proper place to do so, as the Queen of queens has enslaved a few of these infernal things to do some of her more repulsive drudgery. What I shall say about these marvelous aids to the feeble human intelligence will be little indeed, for two reasons: I have always hated machinery, and the only machine I ever understood was a wheelbarrow, and that but imperfectly.

The first definitely recorded calculating machine (if we ignore our ten fingers and the abacus, which hardly rate) was Pascal's, completed in 1642—the year of Newton's birth, which is suggestive in view of what

calculating machines may do for Newton's outstanding problem of three
bodies and its generalizations. The merit of Pascal's wheels and ratchets
was that they performed the carrying of tens automatically. Further
description should be superfluous for any normal youngster in our gadget-
ridden age. Pascal, not living in a perpetual whirr of wheels within wheels,
thought it necessary to guide his "dear reader" by the hand in leading
him or her though the elementary mysteries of a train of gears. What cost
the mathematical genius Pascal a tremendous effort is now an easily understood toy for any twelve-year-old with half a dollar to spend. Anyone
with a quarter of a million to put out may buy an improved model and
play with it at his leisure.

Pascal's gears were adapted to addition. Leibniz about 1671 invented
a more versatile machine, capable, as he said, of "counting, addition,
subtraction, multiplication, and division." But this was not Leibniz' main
contribution to modern calculating machinery, which he made more or
less in spite of himself and his bizarre theology. The main contribution
was his recognition of the advantages of the binary scale, or notation,
over the denary.

Probably it was not Leibniz, but may have been the ancient Chinese,
who first observed that the simplest and inherently most natural way of
representing integers is as sums of powers of 2, each power with a coefficient zero or one, instead of, as in our prevalent denary system, as a sum
of powers of 10, each power with a zero or positive coefficient less than
10. In fact the Chinese either guessed or proved the special case of
Fermat's 'lesser' theorem that $2^{p-1} - 1$ is divisible by p when p is prime.
This can be inferred from the representation of integers in the binary
scale. But they slipped in supposing that this is a sufficient condition that
p be prime. One of the first purely mathematical tasks of a certain modern
machine devised for military purposes was an extensive correction of the
ancient Chinese mistake. This of course was done while the machine was
off duty. The machine did in a few hours what calculation 'by hand'
would have taken several years.

To give an example of the two scales, 2,456 in the denary scale means
$(2 \times 10^3) + (4 \times 10^2) + (5 \times 10) + 6$. In the binary scale no such sequence of digits as 2,456 can occur, since each digit in the number represented must be either 0 or 1. To express 2,456 in the binary scale we note
the highest power of 2 that does not exceed 2,456. It is 2^{11}, and $2,456 =
2^{11} + 408$. We then proceed in the same way with the remainder 408,
and so on: $408 = 2^8 + 152$; $152 = 2^7 + 24$; $24 = 2^4 + 2^3$; thus

$$2,456 = 2^{11} + 2^8 + 2^7 + 2^4 + 2^3 = 100,110,011,000$$

in binary notation.

There are easy routines for converting a number from any one scale to
any other. In the 1890s some school algebras included a short chapter on

scales of notation. It disappeared from the texts because it seemed to be useless. Today binary arithmetic is the mathematics behind one efficient type of 'digital' computing machines. These machines rely on simple counting. The binary scale may also be the ultimate secret of the creation of the universe. For the fact that zeros and ones suffice in binary arithmetic for the expression of any integer convinced Leibniz that God had created the universe (1) out of nothing (0). Leibniz was not only a great mathematician but a great philosopher as well. However, he confused 'nothing' with 'zero'—a remarkable feat for a mathematician and logician.

Between Pascal and Leibniz and our own times there was another pioneer who deserves mention, C. Babbage (1792–1871), who invented the first really modern calculating machine. Babbage was another of those British nonconformists, like De Morgan, who cared little for the fashions of his day. In his youth he was one of the brash young founders (1815) of the Analytical Society at Cambridge, where patriotic idolatry of Newton had put the British mathematicians a century behind their Continental rivals. The Society raised British mathematics from the dead and gave it at least a semblance of life until it almost succeeded in committing suicide in the stupidest examination system—except possibly the mandarin Chinese—in history. Babbage's 'analytical engine' was capable of tabulating the values of any function and printing the results. Only a part of the machine was ever actually constructed. The project was financed by the British government because of possible use by the Admiralty. The expense was far greater than had been anticipated and the government refused to continue, although a scientific commission reported favorably. I have seen it stated somewhere that the government put £100,000 into the 'engine' before consigning it to the Kensington Museum, but this is hard to believe. In spite of the commission's favorable report, it is doubtful whether the machine tools of the time were adequate for finishing the job with sufficient accuracy.

I shall not attempt to describe a digital computer, partly for the reasons already given and partly because as this is written new and improved types of machines are being invented and manufactured in rapid succession. Commercial competition has stimulated invention, as usual. The underlying mathematics is simple enough compared to the physical and engineering problems that must be solved for actual production. Electronics is (was?) one of the most effective sciences applied in design. I have already noted the connection with a two-valued logic, an open or closed circuit, a 'yes-no,' and hence a *binary* mechanism.[5]

There is one feature in which some of the modern machines differ from all their immediate predecessors. They exhibit a fair but far from perfect

[5] [Pp. 56–66 of Bell's *Mathematics, Queen and Servant of Science*, from which this extract was taken. ED.]

imitation of human memory. Numbers are stored in the machine by one ingenious device or another for future use; the machine 'remembers' them and automatically puts them into the calculation. The electronic tubes accept and retain numbers for future use. But the electronic memory, like the human, is finite, and a not completely solved problem (at this writing) is to improve the memory of a machine without diminishing its speed. Similar problems arise in the transmission of messages of any kind. There is one limitation of speed no machine is likely to overcome unless Einstein is wrong; the speed of light is an upper limit to the speed of whatever can move.

I shall only mention another type of machine, the 'analogue computers.' If there is a basic mathematics for these, it may be kinematics or topology. Older examples of analogue computers are slide rules, planimeters, differential analysers, and tide predictors. The analogue machines are not counters like the digital.

For those who wish further information on this topic there are two articles in the *Scientific American*, April and July, 1949—Mathematical Machines, by H. M. Davis, and The Mathematics of Communication, by W. Weaver (1894–   ). Interesting and impressive as these accounts are, I cannot see that the machines have dethroned the Queen. Mathematicians who would dispense entirely with brains possibly have no need of any.

COMMENTARY ON

$(P+PQ)^{m/n}$

NEWTON is well represented in this volume by articles about him and his accomplishments. I wanted, in addition, to furnish at least one example of his work, and have had some trouble choosing an appropriate one. His mathematical writings are apt to be both difficult in content and unfamiliar in expression. However, the material finally selected should not baffle anyone who has studied algebra and remembers a reasonable amount of what he learned. The subject is the binomial theorem.

Newton's first important invention in mathematics was an expression of this theorem for negative and fractional exponents. The expansion of a binomial, $1 - X^2$ say, by a fractional exponent, ½, for example, leads, as Newton discovered, to an infinite series. (The expansion, in this case, is accomplished by taking the square root of $(1 - X^2)$.) He tested the formula,[1] or Rule, as he called it, which he divined soon after attacking the problem, by a set of meticulous operations. These included, first, multiplying the series for $(1 - X^2)^{1/2}$ by itself, term by term, whence he found that it yielded $1 - X^2$ "as he hoped it would, and never a further term however far he took it." Then, "with the characteristic sense of balance that marked the progress of all his ideas he immediately swung over to the converse operation and took the square root of $1 - X^2$ in the ordinary arithmetical way—a far from obvious step to take—another flash of genius—and once more the series steadily reappeared term by term. He thereupon set about using this homely weapon to the utmost, and sought to arithmetize algebra by developing this process of root extraction, followed by the simpler process of long division, and a more intricate general process applicable to the solution of higher equations." [2]

Newton's Rule was first communicated, June 13, 1676, in a letter to Oldenburg, the Secretary of the Royal Society, for transmission to Leibniz who had made inquiries as to Newton's work in infinite series. In response

---

[1] The theorem was stated in the form:

$$(P + PQ)^{m/n} = P^{m/n} + \frac{m}{n}AQ + \frac{m-n}{2n}BQ + \frac{m-2n}{3n}CQ, \text{ etc.}$$

where each of A, B, C denotes the term immediately preceding.

[2] The quotations are from an excellent short survey of Newton's mathematical work, H. W. Turnbull's "Newton: The Algebraist and Geometer," in *The Royal Society: Newton Tercentenary Celebrations*; Cambridge; 1947, p. 68. I call the reader's attention to a much less admiring appraisal of Newton's advances in this field, namely, J. L. Coolidge, "The Story of the Binomial Theorem," *The American Mathematical Monthly*, Vol. 56, No. 3, March, 1949, pp. 147–157. Prof. Coolidge, after tracing the filiation of Newton's work and describing it, concludes, "I cannot see . . . brilliant as was his genius in other matters . . . that he [Newton] deserves extraordinary credit for his contribution to the binomial theorem." For my part, I disagree.

to Leibniz's request for further information, Newton wrote a second letter, October 24, 1676, in which he explains in some detail how he made his early discoveries, and discloses that his binomial rule was formulated twelve years earlier, in 1664, while he was an undergraduate at Cambridge.[3] The two letters, portions of which are reproduced in the selection following, show the influence of John Wallis, Savilian Professor of Geometry in Oxford. Newton had encountered at least one of his works, the *Arithmetica Infinitorum* (1656). Turnbull suggests that Isaac Barrow, Lucasian Professor of Mathematics in Cambridge, also influenced Newton to turn toward this path of inquiry.[4] The theorem itself, though overshadowed by such formidable Newtonian inventions as the calculus, is a great building block, a major contribution "to mathematical technique and procedure."

The exposition below is admirably lucid and there is excitement and satisfaction to be gained from following it along step by step. The symbol $\rceil$ is the same as the present ( ); XX is X^2. I suggest, if you are alarmed by the symbols, or bog down after a brave start, that you turn to any good algebra book to help you through. It is worth making the effort.

[3] See David Eugene Smith, *A Source Book in Mathematics*, New York, 1929, p. 224; also H. W. Turnbull, *The Mathematical Discoveries of Newton*; London and Glasgow, 1945, pp. 12–18.

[4] "In 1663 he (Newton) had read Descartes' *Geometry* and his marginal notes such as 'error, error, non est geometria' show perhaps the influence of Barrow with his dislike of any but the purest geometrical methods. Newton thereupon turned to the far more arithmetical Wallis. . . ." Turnbull, *op. cit.*, p. 13. For a fuller account of the circumstances of the letters and their content, I recommend the entire discussion in Turnbull's little book.

> *One cannot escape the feeling that these mathematical formulae have an independent existence and an intelligence of their own, that they are wiser than we are, wiser even than their discoverers, that we get more out of them than was originally put into them.* —HEINRICH HERTZ

7 On the Binomial Theorem for Fractional and Negative Exponents

By ISAAC NEWTON

LETTER OF JUNE 13, 1676 [1]

ALTHOUGH the modesty of Dr. Leibniz in the Excerpts which you recently sent me from his Letter, attributes much to my work in certain Speculations regarding *Infinite Series*,[2] rumor of which is already beginning to spread, I have no doubt that he has found not only a method of reducing any Quantities whatsoever into Series of this type, *as he himself asserts*, but also that he has found various Compendia, similar to ours if not even better.

Since, however, he may wish to know the discoveries that have been made in this direction by the English (I myself fell into this Speculation some years ago) and in order to satisfy his wishes to some degree at least, I have sent you certain of the points which have occurred to me.

Fractions may be reduced to Infinite Series by Division, and Radical Quantities may be so reduced by the Extraction of Roots. These Operations may be extended to Species [3] in the same way as that in which they apply to Decimal Numbers. These are the Foundations of the Reductions.

The Extractions of Roots are much shortened by the Theorem

$$\overline{P + PQ}|\frac{m}{n} = P\frac{m}{n} + \frac{m}{n}AQ + \frac{m-n}{2n}BQ + \frac{m-2n}{4n}{}^{*}CQ + \frac{m-3n}{4n}DQ + \&c.$$

---

* Evidently a misprint for $3n$.

[1] *Commercium Epistolicum* (1712; 1725 edition, pp. 131–132).

[2] Probably as early as 1666, Newton had told Barrow and others of his work in infinite series in connection with the problem of finding the area under a curve, but this work was not published until 1704 when it appeared as an appendix to Newton's *Opticks*.

[3] That is "to algebraic numbers." In his *Arithmetica Universalis* (1707; 1728 edition) Newton says, "Computation is either perform'd by *Numbers*, as in Vulgar Arithmetick, or by *Species*, as usual among Algebraists . . ."

where $P + PQ$ stands for a Quantity whose Root or Power or whose Root of a Power is to be found, P being the first Term of that quantity, Q being the remaining terms divided by the first term, and $\dfrac{m}{n}$ the numerical Index of the powers of $P + PQ$. This may be a Whole Number or (so to speak) a Broken Number; a positive number or a negative one. For, as the Analysts write a^2 and a^3 &c. for aa and aaa, so for $\sqrt{a}$, $\sqrt{a^3}$, $\sqrt{c \cdot a^5}$, &c.

I write $a^{1/2}$, $a^{3/2}$, $a^{5/3}$, &c.; for $\dfrac{1}{a}$, $\dfrac{1}{aa}$, $\dfrac{1}{aaa}$, a^{-1}, a^{-2}, a^{-3}; for $\dfrac{aa}{\sqrt{c \cdot a^3 + bbx}}$,

$aa \times \overline{a^3 + bbx} \rvert^{-1/3}$; and for $\dfrac{aab}{\sqrt{c : a^3 + bbx \times a^3 + bbx}}$, I write $aab \times$

$\overline{a^3 + bbx}\rvert^{-2/3}$. In this last case, if $\overline{a^3 + bbx}\rvert^{-2/3}$ be taken to mean $P + PQ$ in the Formula, then will $P = a^3$, $Q = bbx/a^3$, $m = -2$, $n = 3$. Finally, in place of the terms that occur in the course of the work in the Quotient, I shall use A, B, C, D, &c. Thus A stands for the first term $P^{m/n}$; B for the second term $\dfrac{m}{n} AQ$; and so on. The use of this Formula will become clear through Examples.[4]

*   *   *   *   *

LETTER OF OCTOBER 24, 1676 [5]

One of my own [methods of deriving infinite series] I described before; and now I shall add another, namely, the way in which I discovered these Series, for I found them before I knew the Divisions and Extractions of Roots which I now use. The explanation of this method will give the basis of the Theorem given at the beginning of my former Letter which Dr. Leibniz desires of me.

Towards the beginning of my study of Mathematics, I happened on the works of our most Celebrated Wallis [6] and his considerations of the Series by whose intercalation he himself shows the values of the Area of a Circle and Hyperbola, and of that series of curves that have a common Base or Axis x and whose Ordinates are in the Form $\overline{1 - xx}\rvert^0 \cdot \overline{1 - xx}\rvert^{1/2}$ $\cdot \overline{1 - xx}\rvert^{2/2} \cdot \overline{1 - xx}\rvert^{3/2} \cdot \overline{1 - xx}\rvert^{4/2} \cdot \overline{1 - xx}\rvert^{5/2}$. &c. Then if the Areas of the

[4] The examples show the application of the formula in cases in which the exponents are ½, ⅓, −⅓, ⅔, −⅔.

[5] *Commercium Epistolicum* (1712; 1725 edition, pp. 142–145). This letter begins with a note of appreciation of the work in series done by Leibniz.

[6] John Wallis (1616–1703) Savilian professor at Oxford, whose *Arithmetica Infinitorum* appeared in 1655. At a later date, Newton wrote an appendix to Wallis's *Algebra*.

alternate ones which are x, $x - \frac{1}{3}x^3$, $x - \frac{2}{3}x^3 + \frac{1}{5}x^5$, $x - \frac{3}{3}x^3 + \frac{3}{5}x^5 - \frac{1}{7}x^7$, &c. could have values interpolated between these terms, we should have the Areas of the intermediates, the first of which $\overline{1 - xx}|^{\frac{1}{2}}$ is the Circle. For these interpolations, I noticed that the first term in each is x and that the second term $\frac{0}{3}x^3$, $\frac{1}{3}x^3$, $\frac{2}{3}x^3$, $\frac{3}{3}x^3$, &c., are in Arithmetic progression. Thus the two first terms of the Series to be intercalated should be $x - \dfrac{\frac{1}{2}x^3}{3}$, $x - \dfrac{\frac{3}{2}x^3}{3}$, $x - \dfrac{\frac{5}{2}x^3}{5}$, &c.

For intercalating the rest, I considered that the Denominators 1, 3, 5, 7, &c. were in Arithmetic progression and so only the Numerical Coefficients of the Numerators would require investigation. Moreover, in the alternate Areas given, these were the figures of the powers of the eleventh number, namely 11^0, 11^1, 11^2, 11^3, 11^4. That is, first 1, then 1, 1, thirdly 1, 2, 1, fourth, 1, 3, 3, 1, fifth 1, 4, 6, 4, 1, &c. Therefore, I sought a method of deriving the remaining elements in these Series, having given the two first figures. I found that when the second figure m was supplied, the rest would be produced by continuous multiplication of the terms of this Series:

$$\frac{m - 0}{1} \times \frac{m - 1}{2} \times \frac{m - 2}{3} \times \frac{m - 3}{4} \times \frac{m - 4}{5} \text{ &c.}$$

For Example: Let (second term) $m = 4$, then the third term will be $4 \times \dfrac{m - 1}{2}$, that is 6; and $6 \times \dfrac{m - 2}{3}$, that is 4, the fourth; and $4 \times \dfrac{m - 3}{4}$ that is 1, the fifth; and $1 \times \dfrac{m - 4}{5}$, that is 0, the sixth at which the series ended in this case.

I therefore, applied this Rule to the Series to be inserted. Thus for a Circle, the second term would be $\dfrac{\frac{1}{2}x^3}{3}$, I then let $m = \frac{1}{2}$, and the terms which resulted were $\frac{1}{2} \times \dfrac{\frac{1}{2} - 1}{2}$ or $-\frac{1}{8}$, $-\frac{1}{8} \times \dfrac{\frac{1}{2} - 2}{3}$ or $+\frac{1}{16}$, $+\frac{1}{16} \times \dfrac{\frac{1}{2} - 3}{4}$ or $-\frac{5}{128}$, and so infinity. From this I learned that the desired Area of a segment of a Circle is

$$x - \frac{\frac{1}{2}x^3}{3} - \frac{\frac{1}{8}x^5}{5} - \frac{\frac{1}{16}x^7}{7} - \frac{\frac{5}{128}x^9}{9} \text{ &c.}$$

By the same process the areas of the remaining Curves to be inserted were found, as the area of a Hyperbola, and of the other alternates in this Series $\overline{1+xx}|^{9/2}$, $\overline{1+xx}|^{1/2}$, $\overline{1+xx}|^{3/2}$, $\overline{1+xx}|^{3/2}$, &c.

The same method may be used for intercalating other Series, even with intervals of two or more terms lacking at once.

This was my first entry into these studies; which would surely have slipped from my memory had I not referred to certain notes a few weeks ago.

But when I had learned this, I soon considered that the terms $\overline{1-xx}|^{9/2}$, $\overline{1-xx}|^{3/2}$, $\overline{1-xx}|^{1/2}$, $\overline{1-xx}|^{9/2}$, &c. that is 1, $1-xx$, $1-2xx+x^4$, $1-3xx+3x^4-x^6$, &c. could be interpolated in the same way and areas could be derived from them; and that for this nothing more is required than the omission of the denominators 1, 3, 5, 7, &c. in the terms expressing the areas, that is, the coefficients of the terms of the quantity to be intercalated $\overline{1-xx}|^{1/2}$, or $\overline{1-xx}|^{3/2}$, or more generally $\overline{1-xx}|^{m}$ could be produced by continuous multiplication of the terms of this Series $m \times$

$$\frac{m-1}{2} \times \frac{m-2}{3} \times \frac{m-3}{4} \text{ &c.}$$

Thus, (for example), $\overline{1-xx}|^{1/2}$ would amount to $1 - \frac{1}{2}x^2 - \frac{1}{8}x^4 - \frac{1}{16}x^6$ &c. And $\overline{1-xx}|^{3/2}$ would come to $1 - \frac{1}{2}x^2 + \frac{1}{8}x^4 + \frac{1}{16}x^6$ &c. And $\overline{1-xx}|^{1/3}$ would be $1 - \frac{1}{3}xx - \frac{1}{9}x^4 - \frac{1}{81}x^6$ &c.

Thus the general Reduction of Radicals into infinite Series became known to me through the Rule which I set at the beginning of the former Letter, before I knew the Extractions of Roots.

But, having learned this, the other could not long remain hidden from me. To prove these operations, I multiplied $1 - \frac{1}{2}x^2 - \frac{1}{8}x^4 - \frac{1}{16}x^6$ &c. by itself, and $1 - xx$ resulted, the remaining terms vanishing into infinity by the continuance of the series. Similarly $1 - \frac{1}{3}xx - \frac{1}{9}x^4 - \frac{5}{81}x^6$ &c. twice multiplied by itself produced $1 - xx$. Which, that these might be a Demonstration of these conclusions, led me naturally to try the converse, to see whether these Series which it was certain were Roots of the quantity $1 - xx$ could not be extracted by Arithmetical means. The attempt succeeded well . . .

Having discovered this, I gave up entirely the interpolation of Series, and used these operations alone as a more genuine basis, nor did I fail to discover Reduction by Division, a method certainly easier.

The Number Concept

THE discovery that the square root of 2 is not a rational number is said to have been celebrated by the Greek philosophers with the sacrifice of 100 oxen. How their colleagues in mathematics responded to the discovery is not recorded but the probability is that their joy was less exuberant. Indeed one may conjecture that the advent of incommensurables produced what the French mathematician Paul Tannery has called "un véritable scandale logique." The earlier geometry depended in its proofs on the Pythagorean theory of proportion and that is a numerical theory which applies only to commensurables. Thus all the geometric proofs suddenly became "inconclusive." (I quote from Sir Thomas Heath.[1]) The situation was retrieved by Eudoxus, whose famous definition of ratios, given by Euclid, enabled mathematicians to go forward with their geometry—after its temporary "paralysis"—and to use irrational numbers as comfortably and as rigorously as they had used rational numbers.[2]

For more than 2,000 years Eudoxus' definition provided the only basis for handling irrational numbers. Nevertheless, mathematicians were never completely satisfied with the tidiness of the concept. They grew especially uneasy on this score in the eighteenth century, a period when they were increasingly interested in re-examining the foundations of mathematics to make certain that all was secure and that the superstructure could be safely extended. In 1872, a German mathematician, Richard Dedekind, published a little book, *Continuity and Irrational Numbers*, in which he attempted to remove all ambiguities and doubts as to how irrational numbers fitted into the domain of arithmetic and what place they merited in a rigorous and logical formulation of mathematical continuity. Some of the considerations bearing on the task were as follows:

1. A rational number can be expressed in the form of a fraction a/b, where a and b are integers.

2. A number which *cannot* be expressed as a rational fraction is an irrational number. For example, $\sqrt{2}$, $\sqrt{3}$, $\sqrt{6}$, e, π. The class of real numbers is made up of rationals such as $1(= 1/1)$, $2(=2/1)$, $3(= 3/1)$, $1/4$, $15/73$, and irrationals as above.

---

[1] *A History of Greek Mathematics*, Oxford, 1921, vol. 1, p. 326.
[2] Here is Eudoxus' definition as given in Euclid V, Def. 5; I give it but cannot tarry over it: "Magnitudes are said to be in the same ratio, the first to the second and the third to the fourth, when if any equimultiples (the same multiples) whatever be taken of the first and third, and any equimultiples whatever of the second and fourth, the former equimultiples alike exceed, are alike equal to, or alike fall short of, the latter equimultiples taken in corresponding order."

3. A rational number can be expressed in decimal notation and where the decimal does not "terminate" (i.e., end in zeros), it is recurrent, that is, it repeats itself periodically; for example $10/13 = .769230.769230.769230$. $14/11 = 1.272727$. An irrational number when expressed as a decimal neither terminates nor exhibits such periods. It is impossible in other words *exactly* to express numbers such as $\sqrt{2}$ or $\sqrt{3}$ as decimals; one can approximate their value as closely as desired but the decimal can never express the root exactly or periodically.

4. It is difficult to perform arithmetic operations with magnitudes incapable of exact expression. The point is well summarized by E. T. Bell: "If two rational numbers are equal, it is no doubt obvious that their square roots are equal. Thus 2×3 and 6 are equal; so also then are $\sqrt{2 \times 3}$ and $\sqrt{6}$. But it is *not* obvious that $\sqrt{2} \times \sqrt{3} = \sqrt{6}$. The unobviousness of this simple assumed equality, $\sqrt{2} \times \sqrt{3} = \sqrt{6}$, taken for granted in school arithmetic, is evident if we visualize what the equality implies: the 'lawless' square roots of 2, 3, 6 are to be extracted, the first two of these are then to be multiplied together, and the result is to come out equal to the third. As not one of these three roots can be extracted exactly, no matter to how many decimal places the computation is carried, it is clear that the verification by multiplication as just described will never be complete. The whole human race toiling incessantly through all its existence could never *prove* in this way that $\sqrt{2} \times \sqrt{3} = \sqrt{6}$."[3]

Dedekind's accomplishment was to define irrational numbers in terms of rationals. The core of his method is his concept of the "cut." An irrational number is a cut separating *all* rational numbers into two classes, an upper and lower class, all the numbers of the lower class being smaller than the numbers of the upper class. The square root of 2, for example, is an irrational number *defined by a Dedekind cut* dividing the set of rationals into an upper class whose members are greater than $\sqrt{2}$ (to express it differently, the square of every number in the upper class is greater than 2) and a lower class containing all other rational numbers. Thus $\sqrt{2}$ stands alone yet is flanked by the other members of the number system and, in effect, fills a gap in their train.[4] In a co-ordinate system

[3] E. T. Bell, *Men of Mathematics*, New York, 1937, p. 518.

[4] Dedekind's idea may be further amplified as follows. If the rational numbers are divided by a "cut" into two classes, A and B, there are three possibilities:

(1) Class A has a largest element a, e.g., if A contains all rational numbers less than or equal to 1, and B contains all other rational numbers, then $a = 1$.

(2) Class B has a smallest element b, e.g., if A contains all rational numbers less than 1, and B contains all rational numbers greater than or equal to 1, then $b = 1$.

(3) Class A has no largest element and class B no smallest element, e.g., if A contains all rationals less than $\sqrt{2}$, and B all rationals greater than $\sqrt{2}$. Then, the cut $\sqrt{2}$ defines or *is* an *irrational* number, a demarcation separating classes A and B but belonging to neither.

This presentation follows Courant and Robbins, *What Is Mathematics?*, New York, 1941, pp. 71-72.

where to each point of a straight line corresponds a number, the measure of distance length from a fixed point, the irrational number $\sqrt{2}$ corresponds to the point whose distance is exactly $\sqrt{2}$ units from an agreed-upon origin. If the point were missing the line would be discontinuous; the irrationals therefore support the intuitive requirement that the line be smooth, unbroken—as if "traced out by the continuous motion of a point." Dedekind's theory is subtle and it presents a beautiful example of mathematical reasoning and mathematical creation. It is not easy but it will yield to close attention. Observe particularly the extent to which he makes use of Eudoxus' magnificent formula for determining equality.

Before giving the excerpt itself let me say a few words about the man. Dedekind, the son of a professor of law, was born at Brunswick in 1831. He studied at Göttingen, taught there and at the Zürich polytechnic for a few years, and then bedded down for half a century as professor of mathematics in the technical high school of his native city. Why he occupied this relatively obscure place while the choice university chairs were held by men much inferior to him is not explained. At any rate, the post agreed with him. He made original and important contributions to the theory of algebraic numbers, enjoyed the care traditionally lavished upon a bachelor by an unmarried sister (Julie, a novelist, with whom he lived until her death in 1914) and kept his health and his sense of humor to his eighty-fifth year when he died (1916). It is unlikely that he could have done much better at an important university.

8   Irrational Numbers

By RICHARD DEDEKIND

THE analogy between rational numbers and the points of a straight line, as is well known, becomes a real correspondence when we select upon the straight line L a definite origin or zero-point o and a definite unit of length for the measurement of segments. With the aid of the latter to every rational number a a corresponding length can be constructed and if we lay this off upon the straight line to the right or left of o according as a is positive or negative, we obtain a definite end-point p, which may be regarded as the point corresponding to the number a; to the rational number zero corresponds the point o. In this way to every rational number a, in the domain R of rational numbers, corresponds one and only one point p, i. e., an individual in L. To the two numbers a, b respectively correspond the two points p, q, and if $a > b$, then p lies to the right of q.

CONTINUITY OF THE STRAIGHT LINE

Of the greatest importance, however, is the fact that in the straight line L there are infinitely many points which correspond to no rational number. If the point p corresponds to the rational number a, then, as is well known, the length op is commensurable with the invariable unit of measure used in the construction, i. e., there exists a third length, a so-called common measure, of which these two lengths are integral multiples. But the ancient Greeks already knew and had demonstrated that there are lengths incommensurable with a given unit of length, e. g., the diagonal of the square whose side is the unit of length. If we lay off such a length from the point o upon the line we obtain an end-point which corresponds to no rational number. Since further it can be easily shown that there are infinitely many lengths which are incommensurable with the unit of length, we may affirm: The straight line L is infinitely richer in point-individuals than the domain R of rational numbers in number-individuals.

If now, as is our desire, we try to follow up arithmetically all phenomena in the straight line, the domain of rational numbers is insufficient and it becomes absolutely necessary that the instrument R constructed by the creation of the rational numbers be essentially improved by the creation of new numbers such that the domain of numbers shall gain the same completeness, or as we may say at once, the same *continuity*, as the straight line.

The previous considerations are so familiar and well known to all that many will regard their repetition quite superfluous. Still I regarded this recapitulation as necessary to prepare properly for the main question. For, the way in which the irrational numbers are usually introduced is based directly upon the conception of extensive magnitudes—which itself is nowhere carefully defined—and explains number as the result of measuring such a magnitude by another of the same kind.[1] Instead of this I demand that arithmetic shall be developed out of itself.

That such comparisons with non-arithmetic notions have furnished the immediate occasion for the extension of the number-concept may, in a general way, be granted (though this was certainly not the case in the introduction of complex numbers); but this surely is no sufficient ground for introducing these foreign notions into arithmetic, the science of numbers. Just as negative and fractional rational numbers are formed by a new creation, and as the laws of operating with these numbers must and can be reduced to the laws of operating with positive integers, so we must endeavor completely to define irrational numbers by means of the rational numbers alone. The question only remains how to do this.

The comparison of the domain R of rational numbers with a straight line has led to the recognition of the existence of gaps, of a certain incompleteness or discontinuity of the former, while we ascribe to the straight line completeness, absence of gaps, or continuity. In what then does this continuity consist? Everything must depend on the answer to this question, and only through it shall we obtain a scientific basis for the investigation of *all* continuous domains. By vague remarks upon the unbroken connection in the smallest parts obviously nothing is gained; the problem is to indicate a precise characteristic of continuity that can serve as the basis for valid deductions. For a long time I pondered over this in vain, but finally I found what I was seeking. This discovery will, perhaps, be differently estimated by different people; the majority may find its substance very commonplace. It consists of the following. In the preceding section attention was called to the fact that every point p of the straight line produces a separation of the same into two portions such that every point of one portion lies to the left of every point of the other. I find the essence of continuity in the converse, i. e., in the following principle:

"If all points of the straight line fall into two classes such that every point of the first class lies to the left of every point of the second class, then there exists one and only one point which produces this division of all points into two classes, this severing of the straight line into two portions."

[1] The apparent advantage of the generality of this definition of number disappears as soon as we consider complex numbers. According to my view, on the other hand, the notion of the ratio between two numbers of the same kind can be clearly developed only after the introduction of irrational numbers.

As already said I think I shall not err in assuming that every one will at once grant the truth of this statement; the majority of my readers will be very much disappointed in learning that by this commonplace remark the secret of continuity is to be revealed. To this I may say that I am glad if every one finds the above principle so obvious and so in harmony with his own ideas of a line; for I am utterly unable to adduce any proof of its correctness, nor has any one the power. The assumption of this property of the line is nothing else than an axiom by which we attribute to the line its continuity, by which we find continuity in the line. If space has at all a real existence it is *not* necessary for it to be continuous; many of its properties would remain the same even were it discontinuous. And if we knew for certain that space was discontinuous there would be nothing to prevent us, in case we so desired, from filling up its gaps, in thought, and thus making it continuous; this filling up would consist in a creation of new point-individuals and would have to be effected in accordance with the above principle.

CREATION OF IRRATIONAL NUMBERS

From the last remarks it is sufficiently obvious how the discontinuous domain R of rational numbers may be rendered complete so as to form a continuous domain. It has been pointed out that every rational number a effects a separation of the system R into two classes such that every number a_1 of the first class A_1 is less than every number a_2 of the second class A_2; the number a is either the greatest number of the class A_1 or the least number of the class A_2. If now any separation of the system R into two classes A_1, A_2, is given which possesses only *this* characteristic property that every number a_1 in A_1 is less than every number a_2 in A_2, then for brevity we shall call such a separation a *cut* [Schnitt] and designate it by (A_1, A_2). We can then say that every rational number a produces one cut or, strictly speaking, two cuts, which, however, we shall not look upon as essentially different; this cut possesses, *besides*, the property that either among the numbers of the first class there exists a greatest or among the numbers of the second class a least number. And conversely, if a cut possesses this property, then it is produced by this greatest or least rational number.

But it is easy to show that there exist infinitely many cuts not produced by rational numbers. The following example suggests itself most readily.

Let D be a positive integer but not the square of an integer, then there exists a positive integer λ such that

$$\lambda^2 < D < (\lambda + 1)^2.$$

If we assign to the second class A_2, every positive rational number a_2 whose square is $> D$, to the first class A_1 all other rational numbers a_1, this separation forms a cut (A_1, A_2), i. e., every number a_1 is less than every number a_2. For if $a_1 = 0$, or is negative, then on that ground a_1 is less than any number a_2, because, by definition, this last is positive; if a_1 is positive, then is its square $\leqq D$, and hence a_1 is less than any positive number a_2 whose square is $> D$.

But this cut is produced by no rational number. To demonstrate this it must be shown first of all that there exists no rational number whose square $= D$. Although this is known from the first elements of the theory of numbers, still the following indirect proof may find place here. If there exist a rational number whose square $= D$, then there exist two positive integers t, u, that satisfy the equation

$$t^2 - Du^2 = 0,$$

and we may assume that u is the *least* positive integer possessing the property that its square, by multiplication by D, may be converted into the square of an integer t. Since evidently

$$\lambda u < t < (\lambda + 1)\, u,$$

the number $u' = t - \lambda u$ is a positive integer certainly *less* than u. If further we put

$$t' = Du - \lambda t,$$

t' is likewise a positive integer, and we have

$$t'^2 - Du'^2 = (\lambda^2 - D)\,(t^2 - Du^2) = 0,$$

which is contrary to the assumption respecting u.

Hence the square of every rational number x is either $< D$ or $> D$. From this it easily follows that there is neither in the class A_1 a greatest, nor in the class A_2 a least number. For if we put

$$y = \frac{x(x^2 + 3D)}{3x^2 + D},$$

we have

$$y - x = \frac{2x(D - x^2)}{3x^2 + D}$$

and

$$y^2 - D = \frac{(x^2 - D)^3}{(3x^2 + D)^2}.$$

If in this we assume x to be a positive number from the class A_1, then $x^2 < D$, and hence $y > x$ and $y^2 < D$. Therefore y likewise belongs to the class A_1. But if we assume x to be a number from the class A_2, then

$x^2 > D$, and hence $y < x$, $y > 0$, and $y^2 > D$. Therefore y likewise belongs to the class A_2. This cut is therefore produced by no rational number.

In this property that not all cuts are produced by rational numbers consists the incompleteness or discontinuity of the domain R of all rational numbers.

Whenever, then, we have to do with a cut (A_1, A_2) produced by no rational number, we create a new, an *irrational* number α, which we regard as completely defined by this cut (A_1, A_2); we shall say that the number α corresponds to this cut, or that it produces this cut. From now on, therefore, to every definite cut there corresponds a definite rational or irrational number, and we regard two numbers as *different* or *unequal* always and only when they correspond to essentially different cuts.

In order to obtain a basis for the orderly arrangement of all *real*, i. e., of all rational and irrational numbers we must investigate the relation between any two cuts (A_1, A_2) and (B_1, B_2) produced by any two numbers α and β. Obviously a cut (A_1, A_2) is given completely when one of the two classes, e. g., the first A_1 is known, because the second A_2 consists of all rational numbers not contained in A_1, and the characteristic property of such a first class lies in this that if the number a_1 is contained in it, it also contains all numbers less than a_1. If now we compare two such first classes A_1, B_1 with each other, it may happen

1. That they are perfectly identical, i. e., that every number contained in A_1 is also contained in B_1, and that every number contained in B_1 is also contained in A_1. In this case A_2 is necessarily identical with B_2, and the two cuts are perfectly identical, which we denote in symbols by $\alpha = \beta$ or $\beta = \alpha$.

But if the two classes A_1, B_1 are not identical, then there exists in the one, e. g., in A_1, a number $a'_1 = b'_2$ not contained in the other B_1 and consequently found in B_2; hence all numbers b_1 contained in B_1 are certainly less than this number $a'_1 = b'_2$ and therefore all numbers b_1 are contained in A_1.

2. If now this number a'_1 is the only one in A_1 that is not contained in B_1, then is every other number a_1 contained in A_1 also contained in B_1 and is consequently $< a'_1$, i. e., a'_1 is the greatest among all the numbers a_1, hence the cut (A_1, A_2) is produced by the rational number $\alpha = a'_1 = b'_2$. Concerning the other cut (B_1, B_2) we know already that all numbers b_1 in B_1 are also contained in A_1 and are less than the number $a'_1 = b'_2$ which is contained in B_2; every other number b_2 contained in B_2 must, however, be greater than b'_2, for otherwise it would be less than a'_1, therefore contained in A_1 and hence in B_1; hence b'_2 is the least among all numbers contained in B_2, and consequently the cut (B_1, B_2) is

produced by the same rational number $\beta = b'_2 = a'_1 = a$. The two cuts are then only unessentially different.

3. If, however, there exist in A_1 at least two different numbers $a'_1 = b'_2$ and $a''_1 = b''_2$, which are not contained in B_1, then there exist infinitely many of them, because all the infinitely many numbers lying between a'_1 and a''_1 are obviously contained in A_1 but not in B_1. In this case we say that the numbers a and β corresponding to these two essentially different cuts (A_1, A_2) and (B_1, B_2) are *different*, and further that a is *greater* than β, that β is *less* than a, which we express in symbols by $a > \beta$ as well as $\beta < a$. It is to be noticed that this definition coincides completely with the one given earlier, when a, β are rational.

The remaining possible cases are these:

4. If there exists B_1 one and only one number $b'_1 = a'_2$, that is not contained in A_1 then the two cuts (A_1, A_2) and (B_1, B_2) are only unessentially different and they are produced by one and the same rational number $a = a'_2 = b'_1 = \beta$.

5. But if there are in B_1 at least two numbers which are not contained in A_1 then $\beta > a$, $a < \beta$.

As this exhausts the possible cases, it follows that of two different numbers one is necessarily the greater, the other the less, which gives two possibilities. A third case is impossible. This was indeed involved in the use of the *comparative* (greater, less) to designate the relation between a, β; but this use has only now been justified. In just such investigations one needs to exercise the greatest care so that even with the best intention to be honest he shall not, through a hasty choice of expressions borrowed from other notions already developed, allow himself to be led into the use of inadmissible transfers from one domain to the other.

If now we consider again somewhat carefully the case $a > \beta$ it is obvious that the less number β, if rational, certainly belongs to the class A_1; for since there is in A_1 a number $a'_1 = b'_2$ which belongs to the class B_2, it follows that the number β, whether the greatest number in B_1 or the least in B_2 is certainly $\leq a'_1$ and hence contained in A_1. Likewise it is obvious from $a > \beta$ that the greater number a, if rational, certainly belongs to the class B_2, *because* $a \geqq a'_1$. Combining these two considerations we get the following result: If a cut is produced by the number a then any rational number belongs to the class A_1 or to the class A_2 according as it is less or greater than a; if the number a is itself rational it may belong to either class.

From this we obtain finally the following: If $a > \beta$, i. e., if there are infinitely many numbers in A_1 not contained in B_1 then there are infinitely many such numbers that at the same time are different from a and from β; every such rational number c is $< a$, because it is contained in A_1 and at the same time it is $> \beta$ because contained in B_2.

CONTINUITY OF THE DOMAIN OF REAL NUMBERS

In consequence of the distinctions just established the system $\mathfrak{R}$ of all real numbers forms a well-arranged domain of one dimension; this is to mean merely that the following laws prevail:

I. If $\alpha > \beta$, and $\beta > \gamma$, then is also $\alpha > \gamma$. We shall say that the number β lies between α and γ.

II. If α, γ are any two different numbers, then there exist infinitely many different numbers β lying between α, γ.

III. If α is any definite number then all numbers of the system $\mathfrak{R}$ fall into two classes $\mathfrak{U}_1$ and $\mathfrak{U}_2$ each of which contains infinitely many individuals; the first class $\mathfrak{U}_1$ comprises all the numbers α_1 that are less than α, the second $\mathfrak{U}_2$ comprises all the numbers α_2 that are greater than α; the number α itself may be assigned at pleasure to the first class or to the second, and it is respectively the greatest of the first or the least of the second class. In each case the separation of the system $\mathfrak{R}$ into the two classes $\mathfrak{U}_1$, $\mathfrak{U}_2$ is such that every number of the first class $\mathfrak{U}_1$ is smaller than every number of the second class $\mathfrak{U}_2$ and we say that this separation is produced by the number α.

For brevity and in order not to weary the reader I suppress the proofs of these theorems which follow immediately from the definitions of the previous section.

Beside these properties, however, the domain $\mathfrak{R}$ possesses also *continuity*; i. e., the following theorem is true:

IV. If the system $\mathfrak{R}$ of all real numbers breaks up into two classes $\mathfrak{U}_1$, $\mathfrak{U}_2$ such that every number α_1 of the class $\mathfrak{U}_1$ is less than every number α_2 of the class $\mathfrak{U}_2$ then there exists one and only one number α by which this separation is produced.

Proof. By the separation or the cut of $\mathfrak{R}$ into $\mathfrak{U}_1$ and $\mathfrak{U}_2$ we obtain at the same time a cut (A_1, A_2) of the system R of all rational numbers which is defined by this that A_1 contains all rational numbers of the class $\mathfrak{U}_1$ and A_2 all other rational numbers, i. e., all rational numbers of the class $\mathfrak{U}_2$. Let α be the perfectly definite number which produces this cut (A_1, A_2). If β is any number different from α, there are always infinitely many rational numbers c lying between α and β. If $\beta < \alpha$, then $c < \alpha$; hence c belongs to the class A_1 and consequently also to the class $\mathfrak{U}_1$, and since at the same time $\beta < c$ then β also belongs to the same class $\mathfrak{U}_1$, because every number in $\mathfrak{U}_2$ is greater than every number c in $\mathfrak{U}_1$. But if $\beta > \alpha$, then is $c > \alpha$; hence c belongs to the class A_2 and consequently also to the class $\mathfrak{U}_2$, and since at the same time $\beta > c$, then β also belongs to the same class $\mathfrak{U}_2$, because every number in $\mathfrak{U}_1$ is less than every number c in $\mathfrak{U}_2$. Hence every number β different from α belongs to the class $\mathfrak{U}_1$ or to the class $\mathfrak{U}_2$ according as $\beta < \alpha$ or $\beta > \alpha$;

consequently α itself is either the greatest number in $\mathfrak{U}_1$ or the least number in $\mathfrak{U}_2$, i. e., α is one and obviously the only number by which the separation of R into the classes $\mathfrak{U}_1$, $\mathfrak{U}_2$ is produced. Which was to be proved.

OPERATIONS WITH REAL NUMBERS

To reduce any operation with two real numbers α, β to operations with rational numbers, it is only necessary from the cuts (A_1, A_2), (B_1, B_2) produced by the numbers α and β in the system R to define the cut (C_1, C_2) which is to correspond to the result of the operation, γ. I confine myself here to the discussion of the simplest case, that of addition.

If c is any rational number, we put it into the class C_1, provided there are two numbers one a_1 in A_1 and one b_1 in B_1 such that their sum $a_1 + b_1 \geqq c$; all other rational numbers shall be put into the class C_2. This separation of all rational numbers into the two classes C_1, C_2 evidently forms a cut, since every number c_1 in C_1 is less than every number c_2 in C_2. If both α and β are rational, then every number c_1 contained in C_1 is $\leqq \alpha + \beta$, because $a_1 \leqq \alpha$, $b_1 \leqq \beta$, and therefore $a_1 + b_1 \leqq \alpha + \beta$; further, if there were contained in C_2 a number $c_2 < \alpha + \beta$, hence $\alpha + \beta = c_2 + p$, where p is a positive rational number, then we should have

$$c_2 = (\alpha - \tfrac{1}{2}p) + (\beta - \tfrac{1}{2}p),$$

which contradicts the definition of the number c_2, because $\alpha - \tfrac{1}{2}p$ is a number in A_1, and $\beta - \tfrac{1}{2}p$ a number in B_1; consequently every number c_2 contained in C_2 is $\geqq \alpha + \beta$. Therefore in this case the cut (C_1, C_2) is produced by the sum $\alpha + \beta$. Thus we shall not violate the definition which holds in the arithmetic of rational numbers if in all cases we understand by the sum $\alpha + \beta$ of any two real numbers α, β that number γ by which the cut (C_1, C_2) is produced. Further, if only one of the two numbers α, β is rational, e. g., α, it is easy to see that it makes no difference with the sum $\gamma = \alpha + \beta$ whether the number α is put into the class A_1 or into the class A_2.

Just as addition is defined, so can the other operations of the so-called elementary arithmetic be defined, viz., the formation of differences, products, quotients, powers, roots, logarithms, and in this way we arrive at real proofs of theorems (as, e. g., $\sqrt{2} \cdot \sqrt{3} = \sqrt{6}$), which to the best of my knowledge have never been established before. The excessive length that is to be feared in the definitions of the more complicated operations is partly inherent in the nature of the subject but can for the most part be avoided. Very useful in this connection is the notion of an *interval*, i. e., a system A of rational numbers possessing the following characteristic property: if a and a' are numbers of the system A, then are all rational numbers lying between a and a' contained in A. The system R of all rational numbers, and also the two classes of any cut are intervals. If there exist a rational number a_1 which is less and a rational number a_2 which is greater than every number of the interval A, then A is called a finite interval; there then exist infinitely many numbers in the same condition as a_1 and infinitely many in the same condition as a_2; the whole domain R breaks up into three parts A_1, A, A_2 and there enter two perfectly definite rational or irrational numbers a_1, a_2 which may be called respectively the lower and upper (or the less and greater) *limits* of the interval; the lower limit a_1 is determined by the cut for which the system A_1 forms the first class and the upper a_2 by the cut for which the system A_2 forms the second class. Of every rational or irrational number a lying between a_1 and a_2 it may be said that it lies *within* the interval A. If all numbers of an interval A are also numbers of an interval B, then A is called a portion of B.

Still lengthier considerations seem to loom up when we attempt to adapt the numerous theorems of the arithmetic of rational numbers [as, e. g., the theorem $(a + b)c = ac + bc$] to any real numbers. This, however, is not the case. It is easy to see that it all reduces to showing that the arithmetic operations possess a certain continuity. What I mean by this statement may be expressed in the form of a general theorem:

"If the number λ is the result of an operation performed on the numbers $a, \beta, \gamma, \ldots$ and λ lies within the interval L, then intervals $A, B, C, \ldots$ can be taken within which lie the numbers $a, \beta, \gamma, \ldots$ such that the result of the same operation in which the numbers $a, \beta, \gamma, \ldots$ are replaced by arbitrary numbers of the intervals $A, B, C, \ldots$ is always a number lying within the interval L." The forbidding clumsiness, however, which marks the statement of such a theorem convinces us that something must be brought in as an aid to expression; this is, in fact, attained in the most satisfactory way by introducing the ideas of *variable magnitudes, functions, limiting values,* and it would be best to base the definitions of even the simplest arithmetic operations upon these ideas, a matter which, however, cannot be carried further here.

Science, being human enquiry, can hear no answer except an answer couched somehow in human tones. Primitive man stood in the mountains and shouted against a cliff; the echo brought back his own voice, and he believed in a disembodied spirit. The scientist of to-day stands counting out loud in the face of the unknown. Numbers come back to him—and he believes in the Great Mathematician. —RICHARD HUGHES

9   Definition of Number

By BERTRAND RUSSELL

[*For a commentary on, and an autobiographical article by, Bertrand Russell, see pages 377–391.*]

THE question "What is a number?" is one which has been often asked, but has only been correctly answered in our own time. The answer was given by Frege in 1884, in his *Grundlagen der Arithmetik*.[1] Although this book is quite short, not difficult, and of the very highest importance, it attracted almost no attention, and the definition of number which it contains remained practically unknown until it was rediscovered by the present author in 1901.

In seeking a definition of number, the first thing to be clear about is what we may call the grammar of our inquiry. Many philosophers, when attempting to define number, are really setting to work to define plurality, which is quite a different thing. *Number* is what is characteristic of numbers, as *man* is what is characteristic of men. A plurality is not an instance of number, but of some particular number. A trio of men, for example, is an instance of the number 3, and the number 3 is an instance of number; but the trio is not an instance of number. This point may seem elementary and scarcely worth mentioning; yet it has proved too subtle for the philosophers, with few exceptions.

A particular number is not identical with any collection of terms having that number: the number 3 is not identical with the trio consisting of Brown, Jones, and Robinson. The number 3 is something which all trios have in common, and which distinguishes them from other collections. A number is something that characterises certain collections, namely, those that have that number.

Instead of speaking of a "collection," we shall as a rule speak of a "class," or sometimes a "set." Other words used in mathematics for the same thing are "aggregate" and "manifold." We shall have much to say later on about classes. For the present, we will say as little as possible. But there are some remarks that must be made immediately.

A class or collection may be defined in two ways that at first sight

[1] The same answer is given more fully and with more development in his *Grundgesetze der Arithmetik*, vol. i., 1893.

seem quite distinct. We may enumerate its members, as when we say, "The collection I mean is Brown, Jones, and Robinson." Or we may mention a defining property, as when we speak of "mankind" or "the inhabitants of London." The definition which enumerates is called a definition by "extension," and the one which mentions a defining property is called a definition by "intension." Of these two kinds of definition, the one by intension is logically more fundamental. This is shown by two considerations: (1) that the extensional definition can always be reduced to an intensional one; (2) that the intensional one often cannot even theoretically be reduced to the extensional one. Each of these points needs a word of explanation.

(1) Brown, Jones, and Robinson all of them possess a certain property which is possessed by nothing else in the whole universe, namely, the property of being either Brown or Jones or Robinson. This property can be used to give a definition by intension of the class consisting of Brown and Jones and Robinson. Consider such a formula as "x is Brown or x is Jones or x is Robinson." This formula will be true for just three x's, namely, Brown and Jones and Robinson. In this respect it resembles a cubic equation with its three roots. It may be taken as assigning a property common to the members of the class consisting of these three men, and peculiar to them. A similar treatment can obviously be applied to any other class given in extension.

(2) It is obvious that in practice we can often know a great deal about a class without being able to enumerate its members. No one man could actually enumerate all men, or even all the inhabitants of London, yet a great deal is known about each of these classes. This is enough to show that definition by extension is not *necessary* to knowledge about a class. But when we come to consider infinite classes, we find that enumeration is not even theoretically possible for beings who only live for a finite time. We cannot enumerate all the natural numbers: they are 0, 1, 2, 3, *and so on*. At some point we must content ourselves with "and so on." We cannot enumerate all fractions or all irrational numbers, or all of any other infinite collection. Thus our knowledge in regard to all such collections can only be derived from a definition by intension.

These remarks are relevant, when we are seeking the definition of number, in three different ways. In the first place, numbers themselves form an infinite collection, and cannot therefore be defined by enumeration. In the second place, the collections having a given number of terms themselves presumably form an infinite collection: it is to be presumed, for example, that there are an infinite collection of trios in the world, for if this were not the case the total number of things in the world would be finite, which, though possible, seems unlikely. In the third place, we wish to define "number" in such a way that infinite numbers may be possible; thus we must be able to speak of the number of terms in an infinite collection, and such a collection must be defined by intension, *i.e.* by a property common to all its members and peculiar to them.

For many purposes, a class and a defining characteristic of it are practically interchangeable. The vital difference between the two consists in the fact that there is only one class having a given set of members, whereas there are always many different characteristics by which a given class may be defined. Men may be defined as featherless bipeds, or as rational animals, or (more correctly) by the traits by which Swift delineates the Yahoos. It is this fact that a defining characteristic is never unique which makes classes useful; otherwise we could be content with the properties common and peculiar to their members. Any one of these properties can be used in place of the class whenever uniqueness is not important.

Returning now to the definition of number, it is clear that number is a way of bringing together certain collections, namely, those that have a given number of terms. We can suppose all couples in one bundle, all trios in another, and so on. In this way we obtain various bundles of collections, each bundle consisting of all the collections that have a certain number of terms. Each bundle is a class whose members are collections, *i.e.* classes; thus each is a class of classes. The bundle consisting of all couples, for example, is a class of classes: each couple is a class with two members, and the whole bundle of couples is a class with an infinite number of members, each of which is a class of two members.

How shall we decide whether two collections are to belong to the same bundle? The answer that suggests itself is: "Find out how many members each has, and put them in the same bundle if they have the same number of members." But this presupposes that we have defined numbers, and that we know how to discover how many terms a collection has. We are so used to the operation of counting that such a presupposition might easily pass unnoticed. In fact, however, counting, though familiar, is logically a very complex operation; moreover it is only available, as a means of discovering how many terms a collection has, when the collection is finite. Our definition of number must not assume in advance that all numbers are finite; and we cannot in any case, without a vicious circle, use counting to define numbers, because numbers are used in counting. We need, therefore, some other method of deciding when two collections have the same number of terms.

In actual fact, it is simpler logically to find out whether two collections have the same number of terms than it is to define what that number is. An illustration will make this clear. If there were no polygamy or polyandry anywhere in the world, it is clear that the number of husbands living at any moment would be exactly the same as the number of wives. We do not need a census to assure us of this, nor do we need to know what is the actual number of husbands and of wives. We know the number must be the same in both collections, because each husband has one wife and each wife has one husband. The relation of husband and wife is what is called "one-one."

A relation is said to be "one-one" when, if x has the relation in question to y, no other term x' has the same relation to y, and x does not have the same relation to any term y' other than y. When only the first of these two conditions is fulfilled, the relation is called "one-many"; when only the second is fulfilled, it is called "many-one." It should be observed that the number 1 is not used in these definitions.

In Christian countries, the relation of husband to wife is one-one; in Mahometan countries it is one-many; in Tibet it is many-one. The relation of father to son is one-many; that of son to father is many-one, but that of eldest son to father is one-one. If n is any number, the relation of n to $n + 1$ is one-one; so is the relation of n to $2n$ or to $3n$. When we are considering only positive numbers, the relation of n to n^2 is one-one; but when negative numbers are admitted, it becomes two-one, since n and $-n$ have the same square. These instances should suffice to make clear the notions of one-one, one-many, and many-one relations, which play a great part in the principles of mathematics, not only in relation to the definition of numbers, but in many other connections.

Two classes are said to be "similar" when there is a one-one relation which correlates the terms of the one class each with one term of the other class, in the same manner in which the relation of marriage correlates husbands with wives. A few preliminary definitions will help us to state this definition more precisely. The class of those terms that have a given relation to something or other is called the *domain* of that relation: thus fathers are the domain of the relation of father to child, husbands are the domain of the relation of husband to wife, wives are the domain of the relation of wife to husband, and husbands and wives together are the domain of the relation of marriage. The relation of wife to husband is called the *converse* of the relation of husband to wife. Similarly *less* is the converse of *greater*, *later* is the converse of *earlier*, and so on. Generally, the converse of a given relation is that relation which holds between y and x whenever the given relation holds between x and y. The *converse domain* of a relation is the domain of its converse: thus the class of wives is the converse domain of the relation of husband to wife. We may now state our definition of similarity as follows:—

One class is said to be "similar" to another when there is a one-one relation of which the one class is the domain, while the other is the converse domain.

It is easy to prove (1) that every class is similar to itself, (2) that if a class a is similar to a class β, then β is similar to a, (3) that if a is similar to β and β to γ, then a is similar to γ. A relation is said to be *reflexive* when it possesses the first of these properties, *symmetrical* when it possesses the second, and *transitive* when it possesses the third. It is obvious that a relation which is symmetrical and transitive must be reflexive throughout its domain. Relations which possess these properties are an important kind, and it is worth while to note that similarity is one of this kind of relations.

It is obvious to common sense that two finite classes have the same number of terms if they are similar, but not otherwise. The act of counting consists in establishing a one-one correlation between the set of objects counted and the natural numbers (excluding 0) that are used up in the process. Accordingly common sense concludes that there are as many objects in the set to be counted as there are numbers up to the last number used in the counting. And we also know that, so long as we confine ourselves to finite numbers, there are just n numbers from 1 up to n. Hence it follows that the last number used in counting a collection is the number of terms in the collection, provided the collection is finite. But this result, besides being only applicable to finite collections, depends upon and assumes the fact that two classes which are similar have the same number of terms; for what we do when we count (say) 10 objects is to show that the set of these objects is similar to the set of numbers 1 to 10. The notion of similarity is logically presupposed in the operation of counting, and is logically simpler though less familiar. In counting, it is necessary to take the objects counted in a certain order, as first, second, third, etc., but order is not of the essence of number: it is an irrelevant addition, an unnecessary complication from the logical point of view. The notion of similarity does not demand an order: for example, we saw that the number of husbands is the same as the number of wives, without having to establish an order of precedence among them. The notion of similarity also does not require that the classes which are similar should be finite. Take, for example, the natural numbers (excluding 0) on the one hand, and the fractions which have 1 for their numerator on the other hand: it is obvious that we can correlate 2 with ½, 3 with ⅓, and so on, thus proving that the two classes are similar.

We may thus use the notion of "similarity" to decide when two collections are to belong to the same bundle, in the sense in which we were asking this question earlier in this chapter. We want to make one bundle containing the class that has no members: this will be for the number 0.

Then we want a bundle of all the classes that have one member: this will
be for the number 1. Then, for the number 2, we want a bundle consisting
of all couples; then one of all trios; and so on. Given any collection, we
can define the bundle it is to belong to as being the class of all those
collections that are "similar" to it. It is very easy to see that if (for
example) a collection has three members, the class of all those collections
that are similar to it will be the class of trios. And whatever number of
terms a collection may have, those collections that are "similar" to it
will have the same number of terms. We may take this as a *definition* of
"having the same number of terms." It is obvious that it gives results
conformable to usage so long as we confine ourselves to finite collections.

So far we have not suggested anything in the slightest degree paradox-
ical. But when we come to the actual definition of numbers we cannot
avoid what must at first sight seem a paradox, though this impression will
soon wear off. We naturally think that the class of couples (for example)
is something different from the number 2. But there is no doubt about
the class of couples: it is indubitable and not difficult to define, whereas
the number 2, in any other sense, is a metaphysical entity about which
we can never feel sure that it exists or that we have tracked it down. It is
therefore more prudent to content ourselves with the class of couples,
which we are sure of, than to hunt for a problematical number 2 which
must always remain elusive. Accordingly we set up the following
definition:—

*The number of a class is the class of all those classes that are similar
to it.*

Thus the number of a couple will be the class of all couples. In fact,
the class of all couples will *be* the number 2, according to our definition.
At the expense of a little oddity, this definition secures definiteness and
indubitableness; and it is not difficult to prove that numbers so defined
have all the properties that we expect numbers to have.

We may now go on to define numbers in general as any one of the
bundles into which similarity collects classes. A number will be a set of
classes such as that any two are similar to each other, and none outside
the set are similar to any inside the set. In other words, a number (in
general) is any collection which is the number of one of its members; or,
more simply still:

A number is anything which is the number of some class.

Such a definition has a verbal appearance of being circular, but in fact
it is not. We define "the number of a given class" without using the notion
of number in general; therefore we may define number in general in
terms of "the number of a given class" without committing any logical
error.

Definitions of this sort are in fact very common. The class of fathers, for example, would have to be defined by first defining what it is to be the father of somebody; then the class of fathers will be all those who are somebody's father. Similarly if we want to define square numbers (say), we must first define what we mean by saying that one number is the square of another, and then define square numbers as those that are the squares of other numbers. This kind of procedure is very common, and it is important to realise that it is legitimate and even often necessary.

PART IV

Mathematics of Space and Motion

COMMENTARY ON
WILLIAM KINGDON CLIFFORD

WILLIAM KINGDON CLIFFORD was born at Exeter (England) in 1845 and died of tuberculosis in Madeira at the age of thirty-four. In a tragically short life, with no more than fifteen working years, he enlarged scientific thought by a series of contributions as original as they were fertile, as far-reaching as they were lucid.

Clifford was one of the distinguished mathematicians of his century and a philosopher of considerable power. His mathematical work had a prophetic quality; his philosophical expression was rational and humane; he had a gift of clarity, as Bertrand Russell has said, "that comes of profound and orderly understanding by virtue of which principles become luminous and deductions look easy." [1] Clifford's mathematical interests lay principally in geometry. It was there that he did important work, that his "mathematical intuition appeared at its best."

The most exciting mathematical advance of the nineteenth century was in non-Euclidean geometry. For more than two thousand years the system perfected by Euclid had occupied a position of absolute authority. The rules he had laid down for geometric relations in space were assumed to be as inviolate as the multiplication table. Space obeyed Euclid, and Euclid obeyed space. This sovereignty was undermined by the investigations of several outstanding mathematicians, among them Gauss, Lobachevsky, Bolyai, Helmholtz, Riemann and Clifford. They decided, after analyzing its foundations, that while Euclid's geometry was unimpeachable as a system of ideal space and as an exercise in logic comparable to a game played with formal rules, its validity as regards actual space should be tested, not by mathematics, but by observation. Other geometries deduced from postulates differing from those framed by Euclid—especially his parallel postulate—were not only logically possible, but might turn out better suited to describe regions of space not normally accessible to our senses. The conclusion was disturbing but fruitful, for it extended enormously the horizons of mathematics.

Clifford allied himself firmly with the view that applied geometry is an experimental science, a proper part of physics. It is from the success of this challenge to established beliefs that the new concepts of space, time, energy and matter underlying modern physics have evolved. Besides his original researches, Clifford translated Riemann's epoch-making inaugural dissertation, *On the Hypotheses That Lie at the Bases of Geom-*

---

[1] Bertrand Russell, in a preface to Clifford's *The Common Sense of the Exact Sciences*, Newly Edited, with an Introduction, by James R. Newman; New York, 1946. p. V.

etry, and delivered a number of lectures to general audiences in which various scientific and philosophical ideas are beautifully explained. The lectures illustrate Clifford's singular power for making hard concepts understandable. He spoke extemporaneously and with ease; he liked his audience and appreciated both their appetite for knowledge and their limitations; he mastered the subject and refused to conceal or gloss over its complexities; he approached it from an angle that made its features distinct. The selections which follow have a common theme. The first, taken from "On the Aims and Instruments of Scientific Thought," [2] considers the exactness of mathematical laws, the uniformity of nature, the assumption of exactness in geometric relations, the revolution in thought brought about by the non-Euclidean heretics. The second excerpt is from a group of lectures, "The Philosophy of the Pure Sciences." [3] It deals in some detail with the postulates of Euclid's geometry which determine its peculiar characteristics, and how changes in these postulates lead to other geometries, primarily Lobachevsky's and Riemann's. The third excerpt is an abstract (I have given all of it) of a remarkable contribution to the Cambridge Philosophical Society "On the Space Theory of Matter." This paper, written three years before Clifford's death when he was already gravely ill, is in a sense the crown of his reflections on geometry. It is hard to realize that it was published forty years before Einstein announced his theory of gravitation.

[2] Given before the British Association, 1872.
[3] Given before the Royal Institution, 1873.

The cowboys have a way of trussing up a steer or a pugnacious bronco which fixes the brute so that it can neither move nor think. This is the hog-tie, and it is what Euclid did to geometry. —ERIC TEMPLE BELL

And for mathematical science, he that doubts their certainty hath need of a dose of hellebore. —JOSEPH GLANVILL

1 The Exactness of Mathematical Laws

By WILLIAM KINGDON CLIFFORD

WHEN a student is first introduced to those sciences which have come under the dominion of mathematics, a new and wonderful aspect of Nature bursts upon his view. He has been accustomed to regard things as essentially more or less vague. All the facts that he has hitherto known have been expressed qualitatively, with a little allowance for error on either side. Things which are let go fall to the ground. A very observant man may know also that they fall faster as they go along. But our student is shown that, after falling for one second in a vacuum, a body is going at the rate of thirty-two feet per second, that after falling for two seconds it is going twice as fast, after going two and a half seconds two and a half times as fast. If he makes the experiment, and finds a single inch per second too much or too little in the rate, one of two things must have happened: either the law of falling bodies has been wrongly stated, or the experiment is not accurate—there is some mistake. He finds reason to think that the latter is always the case; the more carefully he goes to work, the more of the error turns out to belong to the experiment. Again, he may know that water consists of two gases, oxygen and hydrogen, combined; but he now learns that two pints of steam at a temperature of 150° Centigrade will always make two pints of hydrogen and one pint of oxygen at the same temperature, all of them being pressed as much as the atmosphere is pressed. If he makes the experiment and gets rather more or less than a pint of oxygen, is the law disproved? No; the steam was impure, or there was some mistake. Myriads of analyses attest the law of combining volumes; the more carefully they are made, the more nearly they coincide with it. The aspects of the faces of a crystal are connected together by a geometrical law, by which, four of them being given, the rest can be found. The place of a planet at a given time is calculated by the law of gravitation; if it is half a second wrong, the fault is in the instrument, the observer, the clock, or the law; now, the more observations are made, the more of this fault is brought home to the instrument, the observer, and the clock. It is no wonder, then, that our student, contemplating these and many like instances, should be led to say, 'I have been shortsighted; but I have now put on the spectacles of science which Nature had prepared for my eyes; I see that things have definite outlines, that the world is ruled by exact and rigid mathematical laws; καὶ σύ, θεός, γεωμετρεῖς.' It is our business to consider whether he is right in so concluding. Is the uniformity of Nature absolutely exact, or only more exact than our experiments?

At this point we have to make a very important distinction. There are two ways in which a law may be inaccurate. The first way is exemplified by the law of Galileo that a body falling *in vacuo* acquires equal increase in velocity in equal times. No matter how many feet per second it is going, after an interval of a second it will be going thirty-two *more* feet per second. We now know that this rate of increase is not exactly the same at different heights, that it depends upon the distance of the body from the centre of the earth; so that the law is only approximate; instead of the increase of velocity being exactly *equal* in equal times, it itself increases very slowly as the body falls. We know also that this variation of the law from the truth is *too small to be perceived* by direct observation on the change of velocity. But suppose we have invented means for observing this, and have verified that the increase of velocity is inversely as the squared distance from the earth's centre. Still the law is not accurate; for the earth does not attract accurately towards her centre, and the direction of attraction is continually varying with the motion of the sea; the body will not even fall in a straight line. The sun and the planets, too, especially the moon, will produce deviations; yet the sum of all these errors will escape our new process of observation, by being a great deal smaller than the necessary errors of that observation. But when these again have been allowed for, there is still the influence of the stars. In this case, however, we only give up one exact law for another. It may still be held that if the effect of every particle of matter in the universe on the falling body were calculated according to the law of gravitation, the body would move exactly as this calculation required. And if it were objected that the body must be slightly magnetic or diamagnetic, while there are magnets not an infinite way off; that a very minute repulsion, even at sensible distances, accompanies the attraction; it might be replied that these phenomena are themselves subject to exact laws, and that when *all* the laws have been taken into account, the actual motion will exactly correspond with the calculated motion.

I suppose there is hardly a physical student (unless he has specially considered the matter) who would not at once assent to the statement I have just made; that if we knew all about it, Nature would be found universally subject to exact numerical laws. But let us just consider for another moment what this means.

The word 'exact' has a practical and a theoretical meaning. When a grocer weighs you out a certain quantity of sugar very carefully, and says it is exactly a pound, he means that the difference between the mass of the sugar and that of the pound weight he employs is too small to be detected by his scales. If a chemist had made a special investigation, wishing to be as accurate as he could, and told you this was exactly a pound of sugar, he would mean that the mass of the sugar differed from that of a certain standard piece of platinum by a quantity too small to be detected by *his* means of weighing, which are a thousandfold more accurate than the grocer's. But what would a mathematician mean, if he made the same statement? He would mean this. Suppose the mass of the standard pound to be represented by a length, say a foot, measured on a certain line; so that half a pound would be represented by six inches, and so on. And let the difference between the mass of the sugar and that of the standard pound be drawn upon the same line to the same scale. Then, if that difference were magnified an infinite number of times, it would still be invisible. This is the theoretical meaning of exactness; the practical meaning is only very close approximation; *how* close, depends upon the circumstances. The knowledge then of an exact law in the theoretical sense would be equivalent to an infinite observation. I do not say that such knowledge is impossible to man; but I do say that it would be absolutely different in kind from any knowledge that we now possess.

I shall be told, no doubt, that we do possess a great deal of knowledge of this kind, in the form of geometry and mechanics; and that it is just the example of these sciences that has led men to look for exactness in other quarters. If this had been said to me in the last century, I should not have known what to reply. But it happens that about the beginning of the present century the foundations of geometry were criticised independently by two mathematicians, Lobatschewsky [1] and the immortal Gauss; [2] whose results have been extended and generalized more recently by Riemann [3] and Helmholtz. [4] And the conclusion to which these investigations lead is that, although the assumptions which were very properly made by the ancient geometers are practically exact—that is to say, more exact than experiment can be—for such finite things as we have to deal with, and such portions of space as we can reach; yet the truth of them

[1] *Geometrische Untersuchungen zur Theorie der Parallellinien.* Berlin, 1840. Translated by Hoüel. Gauthier-Villars, 1866.
[2] Letter to Schumacher, Nov. 28, 1846 (refers to 1792).
[3] *Ueber die Hypothesen welche der Geometrie zu Grunde liegen.* Göttingen, Abhandl., 1866–7. Translated by Hoüel in *Annali di Matematica*, Milan, vol. iii.
[4] *The Axioms of Geometry*, Academy, vol. i. p. 128 (a popular exposition). [This lecture, under the title "Origin and Significance of Geometrical Axioms," appears on pp. 647–668.—ED.]

for very much larger things, or very much smaller things, or parts of space which are at present beyond our reach, is a matter to be decided by experiment, when its powers are considerably increased. I want to make as clear as possible the real state of this question at present, because it is often supposed to be a question of words or metaphysics, whereas it is a very distinct and simple question of fact. I am supposed to know then that the three angles of a rectilinear triangle are exactly equal to two right angles. Now suppose that three points are taken in space, distant from one another as far as the Sun is from α Centauri, and that the shortest distances between these points are drawn so as to form a triangle. And suppose the angles of this triangle to be very accurately measured and added together; this can at present be done so accurately that the error shall certainly be less than one minute, less therefore than the five-thousandth part of a right angle. Then I do not know that this sum would differ at all from two right angles; but also I do not know that the difference would be less than ten degrees, or the ninth part of a right angle.[5] And I have reasons for not knowing.

This example is exceedingly important as showing the connexion between exactness and universality. It is found that the deviation if it exists must be nearly proportional to the area of the triangle. So that the error in the case of a triangle whose sides are a mile long would be obtained by dividing that in the case I have just been considering by four hundred quadrillions; the result must be a quantity inconceivably small, which no experiment could detect. But between this inconceivably small error and no error at all, there is fixed an enormous gulf; the gulf between practical and theoretical exactness, and, what is even more important, the gulf between what is practically universal and what is theoretically universal. I say that a law is practically universal which is more exact than experiment for all cases that might be got at by such experiments as we can make. We assume this kind of universality, and we find that it pays us to assume it. But a law would be theoretically universal if it were true of all cases whatever; and this is what we do not know of any law at all.

[5] Assuming that parallax observations prove the deviation less than half a second for a triangle whose vertex is at the star and base a diameter of the earth's orbit.

2   The Postulates of the Science of Space

By WILLIAM KINGDON CLIFFORD

IN my first lecture I said that, out of the pictures which are all that we can really see, we imagine a world of solid things; and that this world is constructed so as to fulfil a certain code of rules, some called axioms, and some called definitions, and some called postulates, and some assumed in the course of demonstration, but all laid down in one form or another in Euclid's Elements of Geometry. It is this code of rules that we have to consider to-day. I do not, however, propose to take this book that I have mentioned, and to examine one after another the rules as Euclid has laid them down or unconsciously assumed them; notwithstanding that many things might be said in favour of such a course. This book has been for nearly twenty-two centuries the encouragement and guide of that scientific thought which is one thing with the progress of man from a worse to a better state. The encouragement; for it contained a body of knowledge that was really known and could be relied on, and that moreover was growing in extent and application. For even at the time this book was written—shortly after the foundation of the Alexandrian Museum—Mathematic was no longer the merely ideal science of the Platonic school, but had started on her career of conquest over the whole world of Phenomena. The guide; for the aim of every scientific student of every subject was to bring his knowledge of that subject into a form as perfect as that which geometry had attained. Far up on the great mountain of Truth, which all the sciences hope to scale, the foremost of that sacred sisterhood was seen, beckoning to the rest to follow her. And hence she was called, in the dialect of the Pythagoreans, 'the purifier of the reasonable soul.' Being thus in itself at once the inspiration and the aspiration of scientific thought, this Book of Euclid's has had a history as chequered as that of human progress itself. It embodied and systematized the truest results of the search after truth that was made by Greek, Egyptian, and Hindu. It presided for nearly eight centuries over that promise of light and right that was made by the civilized Aryan races on the Mediterranean shores; that promise, whose abeyance for nearly as long an interval is so full of warning and of sadness for ourselves. It went into exile along with the intellectual activity and the goodness of Europe. It was taught, and commented upon, and illustrated, and supplemented, by Arab and Nestorian, in the Universities of Bagdad and of Cordova. From these it was brought back into barbaric Europe by terrified students who dared tell hardly any other thing of what they had learned among the Saracens. Translated from Arabic into Latin, it passed into the schools of Europe, spun out with additional cases for every possible variation of the figure, and bristling with words which had sounded to Greek ears like the babbling of birds in a hedge. At length the Greek text appeared and was translated; and, like other Greek authors, Euclid became an authority. There had not yet arisen in Europe 'that fruitful faculty,' as Mr. Winwood Reade calls it, 'with which kindred spirits contemplate each other's works; which not only takes, but gives; which produces from whatever it receives; which embraces to wrestle, and wrestles to embrace.' Yet it was coming; and though that criticism of first principles which Aristotle and Ptolemy and Galen underwent waited longer in Euclid's case than in theirs, it came for him at last. What Vesalius was to Galen, what Copernicus was to Ptolemy, that was Lobatchewsky to Euclid. There is, indeed, a somewhat instructive parallel between the last two cases. Copernicus and Lobatchewsky were both of Slavic origin. Each of them has brought about a revolution in scientific ideas so great that it can only be compared with that wrought by the other. And the reason of the transcendent importance of these two changes is that they are changes in the conception of the Cosmos. Before the time of Copernicus, men knew all about the Universe. They could tell you in the schools, pat off by heart, all that it was, and what it had been, and what it would be. There was the flat earth, with the blue vault of heaven resting on it like the dome of a cathedral, and the bright cold stars stuck into it; while the sun and planets moved in crystal spheres between. Or, among the better informed, the earth was a globe in the centre of the universe, heaven a sphere concentric with it; intermediate machinery as before. At any rate, if there was anything beyond heaven, it was a void space that needed no further description. The history of all this could be traced back to a certain definite time, when it began; behind that was a changeless eternity that needed no further history. Its future could be predicted in general terms as far forward as a certain epoch, about the precise determination of which there were, indeed, differences among the learned. But after that would come again a changeless eternity, which was fully accounted for and described. But in any case the Universe was a known thing. Now the enormous effect of the Copernican system, and of the astronomical discoveries that have followed it, is that, in place of this knowledge of a little, which was called knowledge of the Universe, of Eternity and Immensity, we have now got knowledge of a great deal more; but we only call it the knowledge of Here and Now. We can tell a great deal about the solar system; but, after all, it is our house, and not the city. We can tell something about the star-system to which our sun belongs; but, after all, it is our star-system, and not the Universe. We are talking about Here with the consciousness of a There beyond it, which we may know some time, but do not at all know now. And though the nebular hypothesis tells us a great deal about the history of the solar system, and traces it back for a period compared with which the old measure of the duration of the Universe from beginning to end is not a second to a century, yet we do not call this the history of eternity. We may put it all together and call it Now, with the consciousness of a Then before it, in which things were happening that may have left records; but we have not yet read them. This, then, was the change effected by Copernicus in the idea of the Universe. But there was left another to be made. For the laws of space and motion, that we are presently going to examine, implied an infinite space and an infinite duration, about whose properties as space and time everything was accurately known. The very constitution of those parts of it which are at an infinite distance from us, 'geometry upon the plane at infinity,' is just as well known, if the Euclidean assumptions are true, as the geometry of any portion of this room. In this infinite and thoroughly well-known space the Universe is situated during at least some portion of an infinite and thoroughly well-known time. So that here we have real knowledge of something at least that concerns the Cosmos; something that is true throughout the Immensities and the Eternities. That something Lobatchewsky and his successors have taken away. The geometer of to-day knows nothing about the nature of actually existing space at an infinite distance; he knows nothing about the properties of this present space in a past or a future eternity. He knows, indeed, that the laws assumed by Euclid are true with an accuracy that no direct experiment can approach, not only in this place where we are, but in places at a distance from us that no astronomer has conceived; but he knows this as of Here and Now; beyond his range is a There and Then of which he knows nothing at present, but may ultimately come to know more. So, you see, there is a real parallel between the work of Copernicus and his successors on the one hand, and the work of Lobatchewsky and his successors on the other. In both of these the knowledge of Immensity and Eternity is replaced by knowledge of Here and Now. And in virtue of these two revolutions the idea of the Universe, the Macrocosm, the All, as subject of human knowledge, and therefore of human interest, has fallen to pieces.

It will now, I think, be clear to you why it will not do to take for our present consideration the postulates of geometry as Euclid has laid them down. While they were all certainly true, there might be substituted for them some other group of equivalent propositions; and the choice of the particular set of statements that should be used as the groundwork of the science was to a certain extent arbitrary, being only guided by convenience of exposition. But from the moment that the actual truth of these assumptions becomes doubtful, they fall of themselves into a necessary order and classification; for we then begin to see which of them may be true independently of the others. And for the purpose of criticizing the evidence for them, it is essential that this natural order should be taken; for I think you will see presently that any other order would bring hopeless confusion into the discussion.

Space is divided into parts in many ways. If we consider any material thing, space is at once divided into the part where that thing is and the part where it is not. The water in this glass, for example, makes a distinction between the space where it is and the space where it is not. Now, in order to get from one of these to the other you must cross the *surface* of the water; this surface is the boundary of the space where the water is which separates it from the space where it is not. Every *thing*, considered as occupying a portion of space, has a surface which separates the space where it is from the space where it is not. But, again, a surface may be divided into parts in various ways. Part of the surface of this water is against the air, and part is against the glass. If you travel over the surface from one of these parts to the other, you have to cross the *line* which divides them; it is this circular edge where water, air, and glass meet. Every part of a surface is separated from the other parts by a line which bounds it. But now suppose, further, that this glass had been so constructed that the part towards you was blue and the part towards me was white, as it is now. Then this line, dividing two parts of the surface of the water, would itself be divided into two parts; there would be a part where it was against the blue glass, and a part where it was against the white glass. If you travel in thought along that line, so as to get from one of these two parts to the other, you have to cross a *point* which separates them, and is the boundary between them. Every part of a line is separated from the other parts by points which bound it. So we may say altogether—

The boundary of a solid (i.e., of a part of space) is a surface.

The boundary of a part of a surface is a line.

The boundaries of a part of a line are points.

And we are only settling the meanings in which words are to be used. But here we may make an observation which is true of all space that we are acquainted with: it is that the process ends here. There are no parts of a point which are separated from one another by the next link in the series. This is also indicated by the reverse process.

For I shall now suppose this point—the last thing that we got to—to move round the tumbler so as to trace out the line, or edge, where air, water, and glass meet. In this way I get a series of points, one after another; a series of such a nature that, starting from any one of them, only two changes are possible that will keep it within the series: it must go forwards or it must go backwards, and each of these is perfectly definite. The line may then be regarded as an aggregate of points. Now let us imagine, further, a change to take place in this line, which is nearly a circle. Let us suppose it to contract towards the centre of the circle, until it becomes indefinitely small, and disappears. In so doing it will trace out the upper surface of the water, the part of the surface where it is in contact with the air. In this way we shall get a series of circles one after another—a series of such a nature that, starting from any one of them, only two changes are possible that will keep it within the series: it must expand or it must contract. This series, therefore, of circles, is just similar to the series of points that make one circle; and just as the line is regarded as an aggregate of points, so we may regard this surface as an aggregate of lines. But this surface is also in another sense an aggregate of points, in being an aggregate of aggregates of points. But, starting from a point in the surface, more than two changes are possible that will keep it within the surface, for it may move in any direction. The surface, then, is an aggregate of points of a different kind from the line. We speak of the line as a point-aggregate of one dimension, because, starting from one point, there are only two possible directions of change; so that the line can be traced out in one motion. In the same way, a surface is a line-aggregate of one dimension, because it can be traced out by one motion of the line; but it is a point-aggregate of two dimensions, because, in order to build it up of points, we have first to aggregate points into a line, and then lines into a surface. It requires two motions of a point to trace it out.

Lastly, let us suppose this upper surface of the water to move downwards, remaining always horizontal till it becomes the under surface. In so doing it will trace out the part of space occupied by the water. We shall thus get a series of surfaces one after another, precisely analogous to the series of points which make a line, and the series of lines which make a surface. The piece of solid space is an aggregate of surfaces, and an aggregate of the same kind as the line is of points; it is a surface-aggregate of one dimension. But at the same time it is a line-aggregate of two dimensions, and a point-aggregate of three dimensions. For if you consider a particular line which has gone to make this solid, a circle partly contracted and part of the way down, there are more than two opposite changes which it can undergo. For it can ascend or descend, or expand or contract, or do both together in any proportion. It has just as great a variety of changes as a point in a surface. And the piece of space is called a point-aggregate of three dimensions, because it takes three distinct motions to get it from a point. We must first aggregate points into a line, then lines into a surface, then surfaces into a solid.

At this step it is clear, again, that the process must stop in all the space we know of. For it is not possible to move that piece of space in such a way as to change every point in it. When we moved our line or our surface, the new line or surface contained no point whatever that was in the old one; we started with one aggregate of points, and by moving it we got an entirely new aggregate, all the points of which were new. But this cannot be done with the solid; so that the process is at an end. We arrive, then, at the result that *space is of three dimensions.*

Is this, then, one of the postulates of the science of space? No; it is not. The science of space, as we have it, deals with relations of distance existing in a certain space of three dimensions, but it does not at all require us to assume that no relations of distance are possible in aggregates of more than three dimensions. The fact that there are only three dimensions does regulate the number of books that we write, and the parts of the subject that we study: but it is not itself a postulate of the science. We investigate a certain space of three dimensions, on the hypothesis that it has certain elementary properties; and it is the assumptions of these elementary properties that are the real postulates of the science of space. To these I now proceed.

The first of them is concerned with *points*, and with the relation of space to them. We spoke of a line as an aggregate of points. Now there are two kinds of aggregates, which are called respectively continuous and discrete. If you consider this line, the boundary of part of the surface of the water, you will find yourself believing that between any two points of it you can put more points of division, and between any two of these more again, and so on; and you do not believe there can be any end to the process. We may express that by saying you believe that between any two points of the line there is an infinite number of other points. But now here is an aggregate of marbles, which, regarded as an aggregate, has many characters of resemblance with the aggregate of points. It is a series of marbles, one after another; and if we take into account the relations of nextness or contiguity which they possess, then there are only two changes possible from one of them as we travel along the series: we must go to the next in front, or to the next behind. But yet it is not true that between any two of them here is an infinite number of other marbles; between these two, for example, there are only three. There, then, is a distinction at once between the two kinds of aggregates. But there is another, which was pointed out by Aristotle in his Physics and made the basis of a definition of continuity. I have here a row of two different kinds of marbles, some white and some black. This aggregate is divided into two parts, as we formerly supposed the line to be. In the case of the line the boundary between the two parts is a point which is the element of which the line is an aggregate. In this case before us, a marble is the element; but here we cannot say that the boundary between the two parts is a marble. The boundary of the white parts is a white marble, and the boundary of the black parts is a black marble; these two adjacent parts have different boundaries. Similarly, if instead of arranging my marbles in a series, I spread them out on a surface, I may have this aggregate divided into two portions—a white portion and a black portion; but the boundary of the white portion is a row of white marbles, and the boundary of the black portion is a row of black marbles. And lastly, if I made a heap of white marbles, and put black marbles on the top of them, I should have a discrete aggregate of three dimensions divided into two parts: the boundary of the white part would be a layer of white marbles, and the boundary of the black part would be a layer of black marbles. In all these cases of discrete aggregates, when they are divided into two parts, the two adjacent parts have different boundaries. But if you come to consider an aggregate that you believe to be continuous, you will see that you think of two adjacent parts as having the *same* boundary. What is the boundary between water and air here? Is it water? No; for there would still have to be a boundary to divide that water from the air. For the same reason it cannot be air. I do not want you at present to think of the actual physical facts by the aid of any molecular theories; I want you only to think of what appears to be, in order to understand clearly a conception that we all have. Suppose the things actually in contact. If, however much we magnified them, they still appeared to be thoroughly homogeneous, the water filling up a certain space, the air an adjacent space; if this held good indefinitely through all degrees of conceivable magnifying, then we could not say that the surface of the water was a layer of water and the surface of air a layer of air; we should have to say that the same surface was the surface of both of them, and was itself neither one nor the other—that this surface occupied *no* space at all. Accordingly, Aristotle defined the continuous as that of which two adjacent parts have the same boundary; and the discontinuous or discrete as that of which two adjacent parts have direct boundaries.[1]

Now the first postulate of the science of space is that space is a continuous aggregate of points, and not a discrete aggregate. And this postulate—which I shall call the postulate of continuity—is really involved in

---

[1] Phys. Ausc. V. 3, p. 227, ed. Bekker. Τὸ δὲ συνεχὲς ἔστι μὲν ὅπερ ἐχόμενόν τι, λέγω δ' εἶναι συνεχὲς ὅταν ταὐτὸ γένηται καὶ ἕν τὸ ἑκατέρου πέρας οἷς ἅπτονται, καὶ ὥσπερ σημαίνει τοὔνομα συνέχηται. Τοῦτο δ' οὐχ οἷόν τε δυοῖν ὄντοιν εἶναι τοῖν ἐσχάτοιν.

A little further on he makes the important remark that on the hypothesis of continuity a line is not *made up* of points in the same way that a whole is made up of parts, VI. 1, p. 231. 'Αδύνατον ἐ ξ ἀδιαιρέτων εἶναι τι συνεχές, οἷον γραμμὴν ἐκ στιγμῶν, εἴπερ ἡ γραμμὴ μὲν συνεχές, ἡ στιγμὴ δὲ ἀδιαίρετον.

those three of the six [2] postulates of Euclid for which Robert Simson has retained the name of postulate. You will see, on a little reflection, that a discrete aggregate of points could not be so arranged that any two of them should be relatively situated to one another in exactly the same manner, so that any two points might be joined by a straight line which should always bear the same definite relation to them. And the same difficulty occurs in regard to the other two postulates. But perhaps the most conclusive way of showing that this postulate is really assumed by Euclid is to adduce the proposition he proves, that every finite straight line may be bisected. Now this could not be the case if it consisted of an odd number of separate points. As the first of the postulates of the science of space, then, we must reckon this postulate of Continuity; according to which two adjacent portions of space, or of a surface, or of a line, have the *same* boundary, viz.—a surface, a line, or a point; and between every two points on a line there is an infinite number of intermediate points.

The next postulate is that of Elementary Flatness. You know that if you get hold of a small piece of a very large circle, it seems to you nearly straight. So, if you were to take any curved line, and magnify it very much, confining your attention to a small piece of it, that piece would seem straighter to you than the curve did before it was magnified. At least, you can easily conceive a curve possessing this property, that the more you magnify it, the straighter it gets. Such a curve would possess the property of elementary flatness. In the same way, if you perceive a portion of the surface of a very large sphere, such as the earth, it appears to you to be flat. If, then, you take a sphere of say a foot diameter, and magnify it more and more, you will find that the more you magnify it the flatter it gets. And you may easily suppose that this process would go on indefinitely; that the curvature would become less and less the more the surface was magnified. Any curved surface which is such that the more you magnify it the flatter it gets, is said to possess the property of elementary flatness. But if every succeeding power of our imaginary microscope disclosed new wrinkles and inequalities without end, then we should say that the surface did not possess the property of elementary flatness.

But how am I to explain how solid space can have this property of elementary flatness? Shall I leave it as a mere analogy, and say that it is the same kind of property as this of the curve and surface, only in three dimensions instead of one or two? I think I can get a little nearer to it than that; at all events I will try.

If we start to go out from a point on a surface, there is a certain choice of directions in which we may go. These directions make certain angles with one another. We may suppose a certain direction to start with, ʌnd

[2] See De Morgan, in Smith's Dict. of Biography and Mythology, Art. *Euclid*; and in the English Cyclopædia, Art. *Axiom*.

then gradually alter that by turning it round the point: we find thus a single series of directions in which we may start from the point. According to our first postulate, it is a continuous series of directions. Now when I speak of a direction from the point, I mean a direction of starting; I say nothing about the subsequent path. Two different paths may have the same direction at starting; in this case they will touch at the point; and there is an obvious difference between two paths which touch and two paths which meet and form an angle. Here, then, is an aggregate of directions, and they can be changed into one another. Moreover, the changes by which they pass into one another have magnitude, they constitute distance-relations; and the amount of change necessary to turn one of them into another is called the angle between them. It is involved in this postulate that we are considering, that angles can be compared in respect of magnitude. But this is not all. If we go on changing a direction of start, it will, after a certain amount of turning, come round into itself again, and be the same direction. On every surface which has the property of elementary flatness, the amount of turning necessary to take a direction all round into its first position is the same for all points of the surface. I will now show you a surface which at one point of it has not this property. I take this circle of paper from which a sector has been cut out, and bend it round so as to join the edges; in this way I form a surface which is called a *cone*. Now on all points of this surface but one, the law of elementary flatness holds good. At the vertex of the cone, however, notwithstanding that there is an aggregate of directions in which you may start, such that by continuously changing one of them you may get it round into its original position, yet the whole amount of change necessary to effect this is not the same at the vertex as it is at any other point of the surface. And this you can see at once when I unroll it; for only part of the directions in the plane have been included in the cone. At this point of the cone, then, it does not possess the property of elementary flatness; and no amount of magnifying would ever make a cone seem flat at its vertex.

To apply this to solid space, we must notice that here also there is a choice of directions in which you may go out from any point; but it is a much greater choice than a surface gives you. Whereas in a surface the aggregate of directions is only of one dimension, in solid space it is of two dimensions. But here also there are distance-relations, and the aggregate of directions may be divided into parts which have quantity. For example, the directions which start from the vertex of this cone are divided into those which go inside the cone, and those which go outside the cone. The part of the aggregate which is inside the cone is called a solid angle. Now in those spaces of three dimensions which have the property of elementary flatness, the whole amount of solid angle round one point is equal to the whole amount round another point. Although the space need not be exactly similar to itself in all parts, yet the aggregate of directions round one point is exactly similar to the aggregate of directions round another point, if the space has the property of elementary flatness.

How does Euclid assume this postulate of Elementary Flatness? In his fourth postulate he has expressed it so simply and clearly that you will wonder how anybody could make all this fuss. He says, 'All right angles are equal.'

Why could I not have adopted this at once, and saved a great deal of trouble? Because it assumes the knowledge of a surface possessing the property of elementary flatness in all its points. Unless such a surface is first made out to exist, and the definition of a right angle is restricted to lines drawn upon it—for there is no necessity for the word *straight* in that definition—the postulate in Euclid's form is obviously not true. I can make two lines cross at the vertex of a cone so that the four adjacent angles shall be equal, and yet not one of them equal to a right angle.

I pass on to the third postulate of the science of space—the postulate of Superposition. According to this postulate a body can be moved about in space without altering its size or shape. This seems obvious enough, but it is worth while to examine a little closely into the meaning of it. We must define what we mean by size and by shape. When we say that a body can be moved about without altering its size, we mean that it can be so moved as to keep unaltered the length of all the lines in it. This postulate therefore involves that lines can be compared in respect of magnitude, or that they have a length independent of position; precisely as the former one involved the comparison of angular magnitudes. And when we say that a body can be moved about without altering its shape, we mean that it can be so moved as to keep unaltered all the angles in it. It is not necessary to make mention of the motion of a body, although that is the easiest way of expressing and of conceiving this postulate; but we may, if we like, express it entirely in terms which belong to space, and that we should do in this way. Suppose a figure to have been constructed in some portion of space; say that a triangle has been drawn whose sides are the shortest distances between its angular points. Then if in any other portion of space two points are taken whose shortest distance is equal to a side of the triangle, and at one of them an angle is made equal to one of the angles adjacent to that side, and a line of shortest distance drawn equal to the corresponding side of the original triangle, the distance from the extremity of this to the other of the two points will be equal to the third side of the original triangle, and the two will be equal in all respects; or generally, if a figure has been constructed anywhere, another figure, with all its lines and all its angles equal to the corresponding lines and angles of the first, can be constructed anywhere else. Now this is exactly what is meant by the principle of superposition employed by Euclid to prove the proposition that I have just mentioned. And we may state it again in this short form—All parts of space are exactly alike.

But this postulate carries with it a most important consequence. It enables us to make a pair of most fundamental definitions—those of the plane and of the straight line. In order to explain how these come out of it when it is granted, and how they cannot be made when it is not granted, I must here say something more about the nature of the postulate itself, which might otherwise have been left until we come to criticize it.

We have stated the postulate as referring to solid space. But a similar property may exist in surfaces. Here, for instance, is part of the surface of a sphere. If I draw any figure I like upon this, I can suppose it to be moved about in any way upon the sphere, without alteration of its size or shape. If a figure has been drawn on any part of the surface of a sphere, a figure equal to it in all respects may be drawn on any other part of the surface. Now I say that this property belongs to the surface itself, is a part of its own internal economy, and does not depend in any way upon its relation to space of three dimensions. For I can pull it about and bend it in all manner of ways, so as altogether to alter its relation to solid space; and yet, if I do not stretch it or tear it, I make no difference whatever in the length of any lines upon it, or in the size of any angles upon it.[3] I do not in any way alter the figures drawn upon it, or the possibility of drawing figures upon it, *so far as their relations with the surface itself are concerned.* This property of the surface, then, could be ascertained by people who lived entirely in it, and were absolutely ignorant of a third dimension. As a point-aggregate of two dimensions, it has in itself properties determining the distance-relations of the points upon it, which are absolutely independent of the existence of any points which are not upon it.

Now here is a surface which has not that property. You observe that it is not of the same shape all over, and that some parts of it are more curved than other parts. If you drew a figure upon this surface, and then tried to move it about, you would find that it was impossible to do so without altering the size and shape of the figure. Some parts of it would have to expand, some to contract, the lengths of the lines could not all be kept the same, the angles would not hit off together. And this property of the surface—that its parts are different from one another—is a property of the surface itself, a part of its internal economy, absolutely independent of any relations it may have with space outside of it. For, as with the

[3] This figure was made of linen, starched upon a spherical surface, and taken off when dry. That mentioned in the next paragraph was similarly stretched upon the irregular surface of the head of a bust. For durability these models should be made of two thicknesses of linen starched together in such a way that the fibres of one bisect the angles between the fibres of the other, and the edge should be bound by a thin slip of paper. They will then retain their curvature unaltered for a long time.

other one, I can pull it about in all sorts of ways, and, so long as I do not stretch it or tear it, I make no alteration in the length of lines drawn upon it or in the size of the angles.

Here, then, is an intrinsic difference between these two surfaces, as surfaces. They are both point-aggregates of two dimensions; but the points in them have certain relations of distance (distance measured always *on* the surface), and these relations of distance are not the same in one case as they are in the other.

The supposed people living in the surface and having no idea of a third dimension might, without suspecting that third dimension at all, make a very accurate determination of the nature of their *locus in quo*. If the people who lived on the surface of the sphere were to measure the angles of a triangle, they would find them to exceed two right angles by a quantity proportional to the area of the triangle. This excess of the angles above two right angles, being divided by the area of the triangle, would be found to give exactly the same quotient at all parts of the sphere. That quotient is called the curvature of the surface; and we say that a sphere is a surface of uniform curvature. But if the people living on this irregular surface were to do the same thing, they would not find quite the same result. The sum of the angles would, indeed, differ from two right angles, but sometimes in excess, and sometimes in defect, according to the part of the surface where they were. And though for small triangles in any one neighbourhood the excess or defect would be nearly proportional to the area of the triangle, yet the quotient obtained by dividing this excess or defect by the area of the triangle would vary from one part of the surface to another. In other words, the curvature of this surface varies from point to point; it is sometimes positive, sometimes negative, sometimes nothing at all.

But now comes the important difference. When I speak of a triangle, what do I suppose the sides of that triangle to be?

If I take two points near enough together upon a surface, and stretch a string between them, that string will take up a certain definite position upon the surface, marking the line of shortest distance from one point to the other. Such a line is called a geodesic line. It is a line determined by the intrinsic properties of the surface, and not by its relations with external space. The line would still be the shortest line, however the surface were pulled about without stretching or tearing. A geodesic line may be *produced*, when a piece of it is given; for we may take one of the points, and, keeping the string stretched, make it go round in a sort of circle until the other end has turned through two right angles. The new position will then be a prolongation of the same geodesic line.

In speaking of a triangle, then, I meant a triangle whose sides are geodesic lines. But in the case of a spherical surface—or, more generally, of a surface of constant curvature—these geodesic lines have another and most important property. They are *straight*, so far as the surface is concerned. On this surface a figure may be moved about without altering its size or shape. It is possible, therefore, to draw a line which shall be of the same shape all along and on both sides. That is to say, if you take a piece of the surface on one side of such a line, you may slide it all along the line and it will fit; and you may turn it round and apply it to the other side, and it will fit there also. This is Leibnitz's definition of a straight line, and, you see, it has no meaning except in the case of a surface of constant curvature, a surface all parts of which are alike.

Now let us consider the corresponding things in solid space. In this also we may have geodesic lines; namely, lines formed by stretching a string between two points. But we may also have geodesic surfaces; and they are produced in this manner. Suppose we have a point on a surface, and this surface possesses the property of elementary flatness. Then among all the directions of starting from the point, there are some which start *in the surface*, and do not make an angle with it. Let all these be prolonged into geodesics; then we may imagine one of these geodesics to travel round and coincide with all the others in turn. In so doing it will trace out a surface which is called a geodesic surface. Now in the particular case where a space of three dimensions has the property of superposition, or is all over alike, these geodesic surfaces are *planes*. That is to say, since the space is all over alike, these surfaces are also of the same shape all over and on both sides; which is Leibnitz's definition of a plane. If you take a piece of space on one side of such a plane, partly bounded by the plane, you may slide it all over the plane, and it will fit; and you may turn it round and apply it to the other side, and it will fit there also. Now it is clear that this definition will have no meaning unless the third postulate be granted. So we may say that when the postulate of Superposition is true, then there are planes and straight lines; and they are defined as being of the same shape throughout and on both sides.

It is found that the whole geometry of a space of three dimensions is known when we know the curvature of three geodesic surfaces at every point. The third postulate requires that the curvature of all geodesic surfaces should be everywhere equal to the same quantity.

I pass to the fourth postulate, which I call the postulate of Similarity. According to this postulate, any figure may be magnified or diminished in any degree without altering its shape. If any figure has been constructed in one part of space, it may be reconstructed to any scale whatever in any other part of space, so that no one of the angles shall be altered though all the lengths of lines will of course be altered. This seems to be a sufficiently obvious induction from experience; for we have all frequently seen different sizes of the same shape; and it has the advantage of embodying the fifth and sixth of Euclid's postulates in a single principle, which bears a great resemblance in form to that of Superposition, and may be used in the same manner. It is easy to show that it involves the two postulates of Euclid: 'Two straight lines cannot enclose a space,' and 'Lines in one plane which never meet make equal angles with every other line.'

This fourth postulate is equivalent to the assumption that the constant curvature of the geodesic surfaces is zero; or the third and fourth may be put together, and we shall then say that the three curvatures of space are all of them zero at every point.

The supposition made by Lobatchewsky was, that the three first postulates were true, but not the fourth. Of the two Euclidean postulates included in this, he admitted one, viz., that two straight lines cannot enclose a space, or that two lines which once diverge go on diverging for ever. But he left out the postulate about parallels, which may be stated in this form. If through a point outside of a straight line there be drawn another, indefinitely produced both ways; and if we turn this second one round so as to make the point of intersection travel along the first line, then at the very instant that this point of intersection disappears at one end it will reappear at the other, and there is only one position in which the lines do not intersect. Lobatchewsky supposed, instead, that there was a finite angle through which the second line must be turned after the point of intersection had disappeared at one end, before it reappeared at the other. For all positions of the second line within this angle there is then no intersection. In the two limiting positions, when the lines have just done meeting at one end, and when they are just going to meet at the other, they are called parallel; so that two lines can be drawn through a fixed point parallel to a given straight line. The angle between these two depends in a certain way upon the distance of the point from the line. The sum of the angles of a triangle is less than two right angles by a quantity proportional to the area of the triangle. The whole of this geometry is worked out in the style of Euclid, and the most interesting conclusions are arrived at; particularly in the theory of solid space, in which a surface turns up which is not plane relatively to that space, but which, for purposes of drawing figures upon it, is identical with the Euclidean plane.

It was Riemann, however, who first accomplished the task of analysing all the assumptions of geometry, and showing which of them were independent. This very disentangling and separation of them is sufficient to deprive them for the geometer of their exactness and necessity; for the process by which it is effected consists in showing the possibility of conceiving these suppositions one by one to be untrue; whereby it is clearly made out how much is supposed. But it may be worth while to state formally the case for and against them.

When it is maintained that we know these postulates to be universally true, in virtue of certain deliverances of our consciousness, it is implied that these deliverances could not exist, except upon the supposition that the postulates are true. If it can be shown, then, from experience that our consciousness would tell us exactly the same things if the postulates are not true, the ground of their validity will be taken away. But this is a very easy thing to show.

That same faculty which tells you that space is continuous tells you that this water is continuous, and that the motion perceived in a wheel of life is continuous. Now we happen to know that if we could magnify this water as much again as the best microscopes can magnify it, we should perceive its granular structure. And what happens in a wheel of life is discovered by stopping the machine. Even apart, then, from our knowledge of the way nerves act in carrying messages, it appears that we have no means of knowing anything more about an aggregate than that it is too fine-grained for us to perceive its discontinuity, if it has any.

Nor can we, in general, receive a conception as positive knowledge which is itself founded merely upon inaction. For the conception of a continuous thing is of that which looks just the same however much you magnify it. We may conceive the magnifying to go on to a certain extent without change, and then, as it were, leave it going on, without taking the trouble to doubt about the changes that may ensue.

In regard to the second postulate, we have merely to point to the example of polished surfaces. The smoothest surface that can be made is the one most completely covered with the minutest ruts and furrows. Yet geometrical constructions can be made with extreme accuracy upon such a surface, on the supposition that it is an exact plane. If, therefore, the sharp points, edges, and furrows of space are only small enough, there will be nothing to hinder our conviction of its elementary flatness. It has even been remarked by Riemann that we must not shrink from this supposition if it is found useful in explaining physical phenomena.

The first two postulates may therefore be doubted on the side of the very small. We may put the third and fourth together, and doubt them on the side of the very great. For if the property of elementary flatness exist on the average, the deviations from it being, as we have supposed, too small to be perceived, then, whatever were the true nature of space, we should have exactly the conceptions of it which we now have, if only the regions we can get at were small in comparison with the areas of curvature. If we suppose the curvature to vary in an irregular manner, the effect of it might be very considerable in a triangle formed by the nearest fixed stars; but if we suppose it approximately uniform to the limit of telescopic reach, it will be restricted to very much narrower limits. I cannot perhaps do better than conclude by describing to you as well as I can what is the nature of things on the supposition that the curvature of all space is nearly uniform and positive.

In this case the Universe, as known, becomes again a valid conception; for the extent of space is a finite number of cubic miles.[4] And this comes about in a curious way. If you were to start in any direction whatever, and move in that direction in a perfect straight line according to the definition of Leibnitz; after travelling a most prodigious distance, to which the parallactic unit—200,000 times the diameter of the earth's orbit— would be only a few steps, you would arrive at—this place. Only, if you had started upwards, you would appear from below. Now, one of two things would be true. Either, when you had got half-way on your journey, you came to a place that is opposite to this, and which you must have gone through, whatever direction you started in; or else all paths you could have taken diverge entirely from each other till they meet again at this place. In the former case, every two straight lines in a plane meet in two points, in the latter they meet only in one. Upon this supposition of a positive curvature, the whole of geometry is far more complete and interesting; the principle of duality, instead of half breaking down over metric relations, applies to all propositions without exception. In fact, I do not mind confessing that I personally have often found relief from the dreary infinities of homaloidal space in the consoling hope that, after all. this other may be the true state of things.

[4] The assumptions here made about the *Zusammenhang* of space are the simplest ones, but even the finite extent does not follow necessarily from uniform positive curvature; as Riemann seems to have supposed.

Matter exists only as attraction and repulsion—attraction and repulsion are
matter.   —EDGAR ALLAN POE (*Eureka*)

These our actors,
As I foretold you, were all spirits and
Are melted into air, into thin air.   —SHAKESPEARE (*The Tempest*)

Common sense starts with the notion that there is matter where we can get
sensations of touch, but not elsewhere. Then it gets puzzled by wind, breath,
clouds, etc., whence it is led to the conception of "spirit"—I speak etymo-
logically. After "spirit" has been replaced by "gas," there is a further stage,
that of the aether.   —BERTRAND RUSSELL

3   On the Space Theory of Matter

By WILLIAM KINGDON CLIFFORD

(ABSTRACT)

RIEMANN has shown that as there are different kinds of lines and surfaces, so there are different kinds of space of three dimensions; and that
we can only find out by experience to which of these kinds the space in
which we live belongs. In particular, the axioms of plane geometry are
true within the limits of experiment on the surface of a sheet of paper,
and yet we know that the sheet is really covered with a number of small
ridges and furrows, upon which (the total curvature not being zero) these
axioms are not true. Similarly, he says although the axioms of solid geometry are true within the limits of experiment for finite portions of our
space, yet we have no reason to conclude that they are true for very small
portions; and if any help can be got thereby for the explanation of physical phenomena, we may have reason to conclude that they are not true
for very small portions of space.

I wish here to indicate a manner in which these speculations may be
applied to the investigation of physical phenomena. I hold in fact

(1) That small portions of space *are* in fact of a nature analogous to
little hills on a surface which is on the average flat; namely, that the ordinary laws of geometry are not valid in them.

(2) That this property of being curved or distorted is continually being
passed on from one portion of space to another after the manner of a
wave.

(3) That this variation of the curvature of space is what really happens
in that phenomenon which we call the *motion of matter*, whether
ponderable or etherial.

(4) That in the physical world nothing else takes place but this variation, subject (possibly) to the law of continuity.

I am endeavouring in a general way to explain the laws of double refraction on this hypothesis, but have not yet arrived at any results
sufficiently decisive to be communicated.

COMMENTARY ON
A Famous Problem

TOPOLOGY is the geometry of distortion. It deals with fundamental geometric properties that are unaffected when we stretch, twist or otherwise change an object's size and shape. It studies linear figures, surfaces or solids; anything from pretzels and knots to networks and maps. Another name for topology is *analysis situs:* analysis of position. Unlike the geometries of Euclid, Lobachevsky, Riemann and others, which measure lengths and angles and are therefore called *metric,* topology is a nonmetric or nonquantitative geometry. Its propositions hold as well for objects made of rubber as for the rigid figures encountered in metric geometry.

Topology seems a queer subject; it delves into strange implausible shapes and its propositions are either childishly obvious (that is, until you try to prove them) or so difficult and abstract that not even a topologist can explain their intuitive meaning. But topology is no queerer than the physical world as we now interpret it. A world made up entirely of erratic electrical gyrations in curved space requires a bizarre mathematics to do it justice. Euclidian geometry, despite its familiar appearance, is a little too bizarre for this world; it is concerned with wholly fictitious objects— perfectly rigid figures and bodies which suffer no change when moved about. Topology starts from the sound premise that there are no rigid objects, that everything in the world is a little askew, and is further deformed when its position is altered. The aim is to find the elements of order in this disorder, the permanence in this impermanence.

Mathematicians use the word *transformation* to describe changes of position, size or shape, and the word *invariant* to describe the properties unaffected by these changes.[1] In ordinary geometry metric properties are said to be invariant under the transformation of motion. Motion is assumed to have no distorting effect; my pen retains its dimensions as it moves over the paper, this book neither shrinks nor expands as the reader turns its pages. In topology the problem is to find the geometric properties invariant under distorting transformations. If a triangle is stretched into a circle, which of its geometric properties are retained? Is the hole

---

[1] The concept of transformation has a family likeness to the concepts of *relation* and *function* and is of the greatest importance in almost all branches of mathematics and logic. It has its root "in the power we have, when given any two objects of thought, to *associate* either of them with the other" and its special meaning in each of the branches of formal reasoning where it is used—algebra, geometry, group theory, logic and so on—is derived from this basic idea. For a full discussion of *transformation, invariant* and related terms, see Cassius J. Keyser, *Mathematical Philosophy*, New York, 1922 (Keyser's book is for "educated laymen"); also, pp. 1535–1537 [Introduction to Keyser piece on groups].

"inside" or "outside" the doughnut? How can the hole be removed? What is a knot? Can a surface be constructed which has only one side? Can a cylinder with a hole through it be squeezed into a sphere? Is it possible to make a bottle with no edges, no inside and no outside? These are examples of topological questions.

Topology got under way as a full-fledged branch of geometry in the nineteenth century. *Vorstudien zur Topologie*, published in 1847 by the German mathematician Listing, was the first systematic treatise in the field. Its origins, however, go back to major discoveries made by Descartes and Euler. Both had observed (Descartes in 1640, Euler in 1752) a fundamental relationship between the vertices, edges and faces of a simple polyhedron. Euler expressed this important geometric fact in a famous formula

$$V - E + F = 2$$

where V stands for vertices, E for edges, F for faces. He also solved the celebrated problem of the Königsberg Bridges in a memoir which must be regarded as one of the foundation stones of topology.

The material appearing below consists of a translation of Euler's memoir on the seven bridges, and an admirable survey of representative topological problems taken from the book *What Is Mathematics?* by Richard Courant and Herbert Robbins. Dr. Courant, a leading contemporary mathematician, is known for his work in function theory and for his fascinating researches on the mathematics of soap-films. From 1920 to 1933 he was director of the Mathematical Institute of Göttingen, where the great David Hilbert long ruled. Courant is now head of the department of mathematics at New York University; his collaborator, Dr. Robbins, is on the mathematics faculty of Columbia University.

Leonhard Euler (1707–1783) is the most eminent of the scientists born in Switzerland. He enriched mathematics in almost every department, and his energy was at least as remarkable as his genius. "Euler calculated without apparent effort, as men breathe, or as eagles sustain themselves in the wind." [2] He worked with such facility that it is said he dashed off memoirs in the half-hour between the first and second calls to dinner.[3] It has been estimated that sixty to eighty large quarto volumes will be needed for his collected works. He also had thirteen children. The details of the life of this Defoe of mathematics are well set forth in Turnbull's little book (*see* pp. 148–151); here I need only mention the circumstances of the memoir on the Königsberg puzzle. The problem—to cross the seven bridges in a continuous walk without recrossing any of them—was re-

[2] The French astronomer and physicist Arago, as quoted in E. T. Bell, *Men of Mathematics*, New York, 1937, p. 139.
[3] E. T. Bell, *op. cit.*, p. 146.

garded as a small amusement of the Königsberg townsfolk. Euler, however, discovered an important scientific principle concealed in the puzzle.[4] He presented his simple and ingenious solution to the Russian Academy at St. Petersburg in 1735. His method was to replace the land areas by points and the bridges by lines connecting these points. The points are called vertices; a vertex is called odd or even according as the number of lines leading from it are odd or even. The entire configuration is a *graph*; the problem of crossing the bridges reduces to that of traversing the graph with one continuous sweep of the pencil without lifting it from the paper. Euler discovered that this can be done if the graph has only *even* vertices. If the graph contains no more than two *odd* vertices, it may be traversed in one journey but it is not possible to return to the starting point. The general principle is that if the graph contains $2n$ odd vertices, where n is any integer, it will require exactly n distinct journeys to traverse it.

Thus began a "vast and intricate theory," still young and growing, yet already one of the great forces of modern mathematics.[5]

[4] See Moritz Cantor, *Vorlesungen über Geschichte der Mathematik*; Leipzig, 1901, second edition; vol. III, p. 552.

[5] For an entertaining and instructive account of topology, supplementary to the discussion by Courant and Robbins, see the article "Topology," by Albert S. Tucker and Herbert W. Bailey, Jr., in *Scientific American*, January 1950.

It is a pleasant surprise to him [the pure mathematician] and an added problem if he finds that the arts can use his calculations, or that the senses can verify them, much as if a composer found that the sailors could heave better when singing his songs. —GEORGE SANTAYANA

4   The Seven Bridges of Königsberg

By LEONHARD EULER

1. THE branch of geometry that deals with magnitudes has been zealously studied throughout the past, but there is another branch that has been almost unknown up to now; Leibnitz spoke of it first, calling it the "geometry of position" (*geometria situs*). This branch of geometry deals with relations dependent on position alone, and investigates the properties of position; it does not take magnitudes into consideration, nor does it involve calculation with quantities. But as yet no satisfactory definition has been given of the problems that belong to this geometry of position or of the method to be used in solving them. Recently there was announced a problem that, while it certainly seemed to belong to geometry, was nevertheless so designed that it did not call for the determination of a magnitude, nor could it be solved by quantitative calculation; consequently I did not hesitate to assign it to the geometry of position, especially since the solution required only the consideration of position, calculation being of no use. In this paper I shall give an account of the method that I discovered for solving this type of problem, which may serve as an example of the geometry of position.

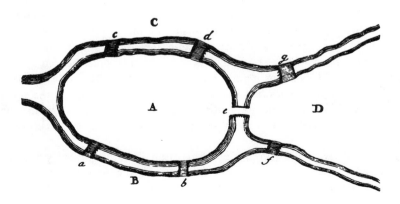

FIGURE 1

2. The problem, which I understand is quite well known, is stated as follows: In the town of Königsberg in Prussia there is an island A, called "Kneiphof," with the two branches of the river (Pregel) flowing around it, as shown in Figure 1. There are seven bridges, a, b, c, d, e, f and g, crossing the two branches. The question is whether a person can plan a walk in such a way that he will cross each of these bridges once but not more than once. I was told that while some denied the possibility of doing this and others were in doubt, there were none who maintained that it was actually possible. On the basis of the above I formulated the following very general problem for myself: Given any configuration of the river and the branches into which it may divide, as well as any number of bridges, to determine whether or not it is possible to cross each bridge exactly once.

3. The particular problem of the seven bridges of Königsberg could be solved by carefully tabulating all possible paths, thereby ascertaining by inspection which of them, if any, met the requirement. This method of solution, however, is too tedious and too difficult because of the large number of possible combinations, and in other problems where many more bridges are involved it could not be used at all. When the analysis is undertaken in the manner just described it yields a great many details that are irrelevant to the problem; undoubtedly this is the reason the method is so onerous. Hence I discarded it and searched for another more restricted in its scope; namely, a method which would show only whether a journey satisfying the prescribed condition could in the first instance be discovered; such an approach, I believed, would be much simpler.

4. My entire method rests on the appropriate and convenient way in which I denote the crossing of bridges, in that I use capital letters, A, B, C, D, to designate the various land areas that are separated from one another by the river. Thus when a person goes from area A to area B across bridge a or b, I denote this crossing by the letters AB, the first of which designates the area whence he came, the second the area where he arrives after crossing the bridge. If the traveller then crosses from B over bridge f into D, this crossing is denoted by the letters BD; the two crossings AB and BD performed in succession I denote simply by the three letters ABD, since the middle letter B designates the area into which the first crossing leads as well as the area out of which the second crossing leads.

5. Similarly, if the traveller proceeds from D across bridge g into C, I designate these three successive crossings by the four letters ABDC. These four letters signify that the traveller who was originally in A crossed over into B, then to D, and finally to C; and since these areas are separated from one another by the river the traveller must necessarily have

crossed three bridges. The crossing of four bridges will be represented by five letters, and if the traveller crosses an arbitrary number of bridges his journey will be described by a number of letters that is one greater than the number of bridges. For example, eight letters are needed to denote the crossing of seven bridges.

6. With this method I pay no attention to which bridges are used; that is to say, if the crossing from one area to another can be made by way of several bridges it makes no difference which one is used, so long as it leads to the desired area. Thus if a route could be laid out over the seven Königsberg bridges so that each bridge were crossed once and only once, we would be able to describe this route by using eight letters, and in this series of letters the combination AB (or BA) would have to occur twice, since there are two bridges a and b, connecting the regions A and B; similarly the combination AC would occur twice, while the combinations AD, BD, and CD would each occur once.

7. Our question is now reduced to whether from the four letters A, B, C, and D a series of eight letters can be formed in which all the combinations just mentioned occur the required number of times. Before making the effort, however, of trying to find such an arrangement we do well to consider whether its existence is even theoretically possible or not. For if it could be shown that such an arrangement is in fact impossible, then the effort expended on finding it would be wasted. Therefore I have sought for a rule that would determine without difficulty as regards this and all similar questions, whether the required arrangement of letters is feasible.

8. For the purpose of finding such a rule I take a single region A into which an arbitrary number of bridges, a, b, c, d, etc., leads (Figure 2).

FIGURE 2

Of these bridges I first consider only a. If the traveller crosses this bridge he must either have been in A before crossing or have reached A after crossing, so that according to the above method of denotation the letter A will appear exactly once. If there are three bridges, a, b, c, leading to A and the traveller crosses all three, then the letter A will occur twice in the expression for his route, whether it begins at A or not. And if there are five bridges leading to A the expression for a route that crosses them all will contain the letter A three times. If the number of bridges is odd, increase it by one, and take half the sum; the quotient represents the number of times the letter A appears.

9. Let us now return to the Königsberg problem (Figure 1). Since there are five bridges, a, b, c, d, e, leading to (and from) island A, the letter A must occur three times in the expression describing the route. The letter B must occur twice, since three bridges lead to B; similarly D and C must each occur twice. That is to say, the series of eight letters that represents the crossing of the seven bridges must contain A three times and B, C and D each twice; but this is quite impossible with a series of eight letters. Thus it is apparent that a crossing of the seven bridges of Königsberg in the manner required cannot be effected.

10. Using this method we are always able, whenever the number of bridges leading to a particular region is odd, to determine whether it is possible, in a journey, to cross each bridge exactly once. Such a route exists if the number of bridges plus one is equal to the sum of the numbers that indicate how often each individual letter must occur. On the other hand, if this sum is greater than the number of bridges plus one, as it is in our example, then the desired route cannot be constructed. The rule that I gave (section 8) for determining from the number of bridges that lead to A how often the letter A will occur in the route description is independent of whether these bridges all come from a single region B, as in Figure 2, or from several regions, because I am considering only the region A, and attempting to determine how often the letter A must occur.

11. When the number of bridges leading to A is even, we must take into account whether the route begins in A or not. For example, if there are two bridges that lead to A and the route starts from A, then the letter A will occur twice, once to indicate the departure from A by one of the bridges and a second time to indicate the return to A by the other bridge. However, if the traveller starts his journey in another region, the letter A will occur only once, since by my method of description the single occurrence of A indicates an entrance into as well as a departure from A.

12. Suppose, as in our case, there are four bridges leading into the region A, and the route is to begin at A. The letter A will then occur three times in the expression for the whole route, while if the journey had started in another region, A would occur only twice. With six bridges leading to A the letter A will occur four times if A is the starting point, otherwise only three times. In general, if the number of bridges is even, the number of occurrences of the letter A, when the starting region is not A, will be half the number of the bridges; one more than half, when the route starts from A.

13. Every route must, of course, start in some one region, thus from the number of bridges that lead to each region I determine the number

of times that the corresponding letter will occur in the expression for the entire route as follows: When the number of the bridges is odd I increase it by one and divide by two; when the number is even I simply divide it by two. Then if the sum of the resulting numbers is equal to the actual number of bridges plus one, the journey can be accomplished, though it must start in a region approached by an odd number of bridges. But if the sum is one less than the number of bridges plus one, the journey is feasible if its starting point is a region approached by an even number of bridges, for in that case the sum is again increased by one.

14. My procedure for determining whether in any given system of rivers and bridges it is possible to cross each bridge exactly once is as follows: 1. First I designate the individual regions separated from one another by the water as A, B, C, etc. 2. I take the total number of bridges, increase it by one, and write the resulting number uppermost. 3. Under this number I write the letters A, B, C, etc., and opposite each of these I note the number of bridges that lead to that particular region. 4. I place an asterisk next the letters that have even numbers opposite them. 5. Opposite each even number I write the half of that number and opposite each odd number I write half of the sum formed by that number plus one. 6. I add up the last column of numbers. If the sum is one less than, or equal to the number written at the top, I conclude that the required journey can be made. But it must be noted that when the sum is one less than the number at the top, the route must start from a region marked with an asterisk. And in the other case, when these two numbers are equal, it must start from a region that does not have an asterisk.

For the Königsberg problem I would set up the tabulation as follows:

Number of bridges 7, giving 8 (= 7 + 1) bridges

A,	5	3
B,	3	2
C,	3	2
D,	3	2

The last column now adds up to more than 8, and hence the required journey cannot be made.

15. Let us take an example of two islands, with four rivers forming the surrounding water, as shown in Figure 3. Fifteen bridges, marked a, b, c, d, etc., cross the water around the islands and the adjoining rivers; the question is whether a journey can be arranged that will pass over all the bridges, but not over any of them more than once. 1. I begin by marking all the regions that are separated from one another by the water with the letters A, B, C, D, E, F—there are six of them. 2. I take the number of bridges—15—add one and write this number—16—uppermost. 3. I write the letters A, B, C, etc. in a column and opposite each letter I

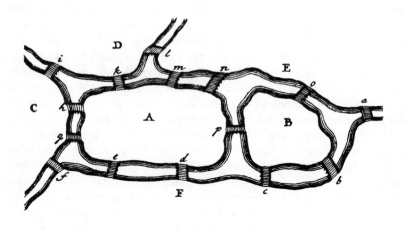

FIGURE 3

		16
A*,	8	4
B*,	4	2
C*,	4	2
D,	3	2
E,	5	3
F*,	6	3
		16

write the number of bridges connecting with that region, e.g., 8 bridges
for A, 4 for B, etc. 4. The letters that have even numbers opposite them
I mark with an asterisk. 5. In a third column I write the half of each
corresponding even number, or, if the number is odd, I add one to it,
and put down half the sum. 6. Finally I add the numbers in the third
column and get 16 as the sum. This is the same as the number 16 that
appears above, and hence it follows that the journey can be effected if it
begins in regions D or E, whose symbols have no asterisk. The following
expression represents such a route:

$$EaFbBcFdAeFfCgAhCiDkAmEnApBoElD.$$

Here I have also indicated, by small letters between the capitals, which
bridges are crossed.

16. By this method we can easily determine, even in cases of considerable complexity, whether a single crossing of each of the bridges in
sequence is actually possible. But I should now like to give another and
much simpler method, which follows quite easily from the preceding,

after a few preliminary remarks. In the first place, I note that the sum of all the numbers of bridges to each region, that are written down in the second column opposite the letters A, B, C, etc., is necessarily double the actual number of bridges. The reason is that in the tabulation of the bridges leading to the various regions each bridge is counted twice, once for each of the two regions that it connects.

17. From this observation it follows that the sum of the numbers in the second column must be an even number, since half of it represents the actual number of bridges. Hence it is impossible for exactly one of these numbers (indicating how many bridges connect with each region) to be odd, or, for that matter, three or five, etc. In other words, if *any* of the numbers opposite the letters A, B, C, etc., are odd, an even number of them must be odd. In the Königsberg problem, for instance, all four of the numbers opposite the letters A, B, C, D were odd, as explained in section 14, while in the example just given (section 15) only two of the numbers were odd, namely those opposite D and E.

18. Since the sum of the numbers opposite A, B, C, etc., is double the number of bridges, it is clear that if this sum is increased by two and then divided by 2 the result will be the number written at the top. When all the numbers in the second column are even, and the half of each is written down in the third column, the total of this column will be one less than the number at the top. In that case it will always be possible to cross all the bridges. For in whatever region the journey begins, there will be an even number of bridges leading to it, which is the requirement. In the Königsberg problem we could, for instance, arrange matters so that each bridge is crossed twice, which is equivalent to dividing each bridge into two, whence the number of bridges leading to each region would be even.

19. Further, when only two of the numbers opposite the letters are odd, and the others even, the required route is possible provided it begins in a region approached by an odd number of bridges. We take half of each even number, and likewise half of each odd number after adding one, as our procedure requires; the sum of these halves will then be one greater than the number of bridges, and hence equal to the number written at the top.

Similarly, where four, six, or eight, etc., of the numbers in the second column are odd it is evident that the sum of the numbers in the third column will be one, two, three, etc., greater than the top number, as the case may be, and hence the desired journey is impossible.

20. Thus for any configuration that may arise the easiest way of determining whether a single crossing of all the bridges is possible is to apply the following rules:

If there are more than two regions which are approached by an odd number of bridges, no route satisfying the required conditions can be found.

If, however, there are only two regions with an odd number of approach bridges the required journey can be completed provided it originates in one of the regions.

If, finally, there is no region with an odd number of approach bridges, the required journey can be effected, no matter where it begins. These rules solve completely the problem initially proposed.

21. After we have determined that a route actually exists we are left with the question how to find it. To this end the following rule will serve: Wherever possible we mentally eliminate any two bridges that connect the same two regions; this usually reduced the number of bridges considerably. Then—and this should not be difficult—we proceed to trace the required route across the remaining bridges. The pattern of this route, once we have found it, will not be substantially affected by the restoration of the bridges which were first eliminated from consideration—as a little thought will show; therefore I do not think I need say more about finding the routes themselves.

5   Topology

By RICHARD COURANT
and HERBERT ROBBINS

EULER'S FORMULA FOR POLYHEDRA

ALTHOUGH the study of polyhedra held a central place in Greek geometry, it remained for Descartes and Euler to discover the following fact: In a simple polyhedron let V denote the number of vertices, E the number of edges, and F the number of faces; then always

$$(1) \qquad\qquad V - E + F = 2.$$

By a *polyhedron* is meant a solid whose surface consists of a number of polygonal faces. In the case of the regular solids, all the polygons are congruent and all the angles at vertices are equal. A polyhedron is *simple* if there are no "holes" in it, so that its surface can be deformed continuously into the surface of a sphere. Figure 2 shows a simple polyhedron which is not regular, while Figure 3 shows a polyhedron which is not simple.

The reader should check the fact that Euler's formula holds for the simple polyhedra of Figures 1 and 2, but does not hold for the polyhedron of Figure 3.

To prove Euler's formula, let us imagine the given simple polyhedron to be hollow, with a surface made of thin rubber. Then if we cut out one of the faces of the hollow polyhedron, we can deform the remaining surface until it stretches out flat on a plane. Of course, the areas of the faces and the angles between the edges of the polyhedron will have changed in this process. But the network of vertices and edges in the plane will contain the same number of vertices and edges as did the original polyhedron, while the number of polygons will be one less than in the original polyhedron, since one face was removed. We shall now show that for the plane network, $V - E + F = 1$, so that, if the removed face is counted, the result is $V - E + F = 2$ for the original polyhedron.

First we "triangulate" the plane network in the following way: In some polygon of the network which is not already a triangle we draw a diagonal. The effect of this is to increase both E and F by 1, thus pre-

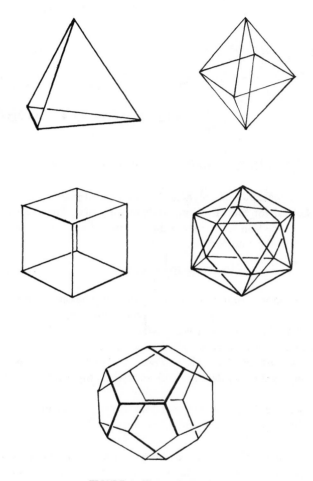

FIGURE 1—The regular polyhedra.

serving the value of $V - E + F$. We continue drawing diagonals joining pairs of points (Figure 4) until the figure consists entirely of triangles, as it must eventually. In the triangulated network, $V - E + F$ has the value that it had before the division into triangles, since the drawing of diagonals has not changed it. Some of the triangles have edges on the boundary of the plane network. Of these some, such as ABC, have only one edge on the boundary, while other triangles may have two edges on the boundary. We take any boundary triangle and remove that part of it which does not also belong to some other triangle. Thus, from ABC we remove the edge AC and the face, leaving the vertices A, B, C and the two edges AB and BC; while from DEF we remove the face, the two

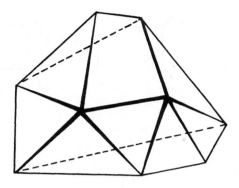

FIGURE 2—A simple polyhedron, $V - E + F = 9 - 18 + 11 = 2$.

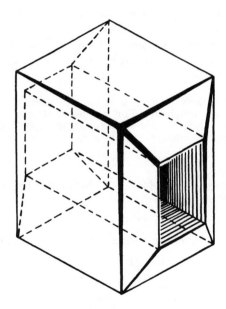

FIGURE 3—A non-simple polyhedron, $V - E + F = 16 - 32 + 16 = 0$.

edges DF and FE, and the vertex F. The removal of a triangle of type ABC decreases E and F by 1, while V is unaffected, so that $V - E + F$ remains the same. The removal of a triangle of type DEF decreases V by 1, E by 2, and F by 1, so that $V - E + F$ again remains the same. By a properly chosen sequence of these operations we can remove triangles with edges on the boundary (which changes with each removal), until finally only one triangle remains, with its three edges, three vertices, and one face. For this simple network, $V - E + F = 3 - 3 + 1 = 1$. But we

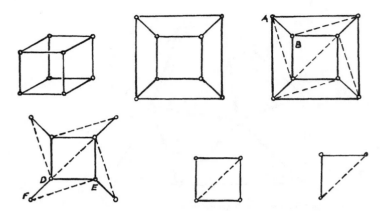

FIGURE 4—Proof of Euler's theorem.

have seen that by constantly erasing triangles $V - E + F$ was not altered. Therefore in the original plane network $V - E + F$ must equal 1 also, and thus equals 1 for the polyhedron with one face missing. We conclude that $V - E + F = 2$ for the complete polyhedron. This completes the proof of Euler's formula.

On the basis of Euler's formula it is easy to show that there are no more than five regular polyhedra. For suppose that a regular polyhedron has F faces, each of which is an n-sided regular polygon, and that r edges meet at each vertex. Counting edges by faces and vertices, we see that

$$(2) \qquad\qquad\qquad nF = 2E;$$

for each edge belongs to two faces, and hence is counted twice in the product nF; moreover,

$$(3) \qquad\qquad\qquad rV = 2E,$$

since each edge has two vertices. Hence from (1) we obtain the equation

$$\frac{2E}{n} + \frac{2E}{r} - E = 2$$

or

$$(4) \qquad\qquad\qquad \frac{1}{n} + \frac{1}{r} = \frac{1}{2} + \frac{1}{E}.$$

We know to begin with that $n \geq$ and $r \geq 3$, since a polygon must have at least three sides, and at least three sides must meet at each polyhedral angle. But n and r cannot both be *greater* than three, for then the left hand side of equation (4) could not exceed ½, which is impossible for any positive value of E. Therefore, let us see what values r may have when $n = 3$, and what values n may have when $r = 3$. The totality of polyhedra given by these two cases gives the number of possible regular polyhedra.

For $n = 3$, equation (4) becomes

$$\frac{1}{r} - \frac{1}{6} = \frac{1}{E};$$

r can thus equal 3, 4, or 5. (6, or any greater number, is obviously excluded, since $1/E$ is always positive.) For these values of n and r we get $E = 6$, 12, or 30, corresponding respectively to the tetrahedron, octahedron, and icosahedron. Likewise, for $r = 3$ we obtain the equation

$$\frac{1}{n} - \frac{1}{6} = \frac{1}{E},$$

from which it follows that $n = 3$, 4, or 5, and $E = 6$, 12, or 30, respectively. These values correspond respectively to the tetrahedron, cube, and dodecahedron. Substituting these values for n, r, and E in equations (2) and (3), we obtain the numbers of vertices and faces in the corresponding polyhedra.

TOPOLOGICAL PROPERTIES OF FIGURES
TOPOLOGICAL PROPERTIES

We have proved that the Euler formula holds for any simple polyhedron. But the range of validity of this formula goes far beyond the polyhedra of elementary geometry, with their flat faces and straight edges; the proof just given would apply equally well to a simple polyhedron with curved faces and edges, or to any subdivision of the surface of a sphere into regions bounded by curved arcs. Moreover, if we imagine the surface of the polyhedron or of the sphere to be made out of a thin sheet of rubber, the Euler formula will still hold if the surface is deformed by bending and stretching the rubber into any other shape, so long as the rubber is not torn in the process. For the formula is concerned only with the *numbers* of the vertices, edges, and faces, and not with lengths, areas, straightness, cross-ratios, or any of the usual concepts of elementary or projective geometry.

We recall that elementary geometry deals with the magnitudes (length, angle, and area) that are unchanged by the rigid motions, while projective geometry deals with the concepts (point, line, incidence, and cross-ratio) which are unchanged by the still larger group of projective transformations. But the rigid motions and the projections are both very special cases of what are called *topological transformations:* a topological transformation of one geometrical figure A into another figure A' is given by any correspondence

$$p \longleftrightarrow p'$$

between the points p of A and the points p' of A' which has the following two properties:

1. *The correspondence is biunique.* This means that to each point p of A corresponds just one point p' of A', and conversely.

2. *The correspondence is continuous in both directions.* This means that if we take any two points p, q of A and move p so that the distance between it and q approaches zero, then the distance between the corresponding points p', q' of A' will also approach zero, and conversely.

Any property of a geometrical figure A that holds as well for every figure into which A may be transformed by a topological transformation is called a *topological property* of A, and *topology* is the branch of geometry which deals only with the topological properties of figures. Imagine a figure to be copied "free-hand" by a conscientious but inexpert draftsman who makes straight lines curved and alters angles, distances and areas; then, although the metric and projective properties of the original figure would be lost, its topological properties would remain the same.

The most intuitive examples of general topological transformations are the *deformations*. Imagine a figure such as a sphere or a triangle to be made from or drawn upon a thin sheet of rubber, which is then stretched and twisted in any manner without tearing it and without bringing distinct points into actual coincidence. (Bringing distinct points into coincidence would violate condition 1. Tearing the sheet of rubber would violate condition 2, since two points of the original figure which tend toward coincidence from opposite sides of a line along which the sheet is torn would not tend towards coincidence in the torn figure.) The final position of the figure will then be a topological image of the original. A triangle can be deformed into any other triangle or into a circle or an ellipse, and hence these figures have exactly the same topological properties. But one

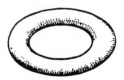 

FIGURE 5—Topologically equivalent surfaces.

FIGURE 6—Topologically non-equivalent surfaces.

cannot deform a circle into a line segment, nor the surface of a sphere into the surface of an inner tube.

The general concept of topological transformation is wider than the concept of deformation. For example, if a figure is cut during a deformation and the edges of the cut sewn together after the deformation in exactly the same way as before, the process still defines a topological transformation of the original figure, although it is not a deformation. Thus the two curves of Figure 12 [1] are topologically equivalent to each other or to a circle, since they may be cut, untwisted, and the cut sewn up. But it is impossible to deform one curve into the other or into a circle without first cutting the curve.

Topological properties of figures (such as are given by Euler's theorem and others to be discussed in this section) are of the greatest interest and importance in many mathematical investigations. They are in a sense the deepest and most fundamental of all geometrical properties, since they persist under the most drastic changes of shape.

CONNECTIVITY

As another example of two figures that are not topologically equivalent we may consider the plane domains of Figure 7. The first of these consists

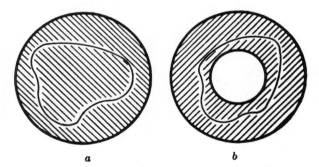

a   *b*

FIGURE 7—Simple and double connectivity.

of all points interior to a circle, while the second consists of all points contained between two concentric circles. Any closed curve lying in the domain *a* can be continuously deformed or "shrunk" down to a single point *within the domain*. A domain with this property is said to be *simply connected*. The domain *b* is not simply connected. For example, a circle concentric with the two boundary circles and midway between them cannot be shrunk to a single point within the domain, since during this process the curve would necessarily pass over the center of the circles, which is not a point of the domain. A domain which is not simply connected is said to be *multiply connected*. If the multiply connected domain

[1] [See p. 592, ED.]

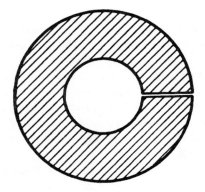

FIGURE 8—Cutting a doubly connected domain to make it simply connected.

b is cut along a radius, as in Figure 8, the resulting domain is simply connected.

More generally, we can construct domains with two, three, or more "holes," such as the domain of Figure 9. In order to convert this domain into a simply connected domain, two cuts are necessary. If $n - 1$ non-

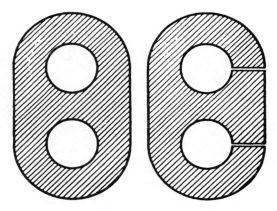

FIGURE 9—Reduction of a triply connected domain.

intersecting cuts from boundary to boundary are needed to convert a given multiply connected domain D into a simply connected domain, the domain D is said to be *n*-tuply connected. The degree of connectivity of a domain in the plane is an important topological invariant of the domain.

OTHER EXAMPLES OF TOPOLOGICAL THEOREMS
THE JORDAN CURVE THEOREM

A simple closed curve (one that does not intersect itself) is drawn in the plane. What property of this figure persists even if the plane is regarded as a sheet of rubber that can be deformed in any way? The length of the curve and the area that it encloses can be changed by a deformation. But there is a topological property of the configuration which is so simple that it may seem trivial: *A simple closed curve C in the plane divides the plane into exactly two domains, an inside and an outside.* By this is meant that the points of the plane fall into two classes— *A*, the outside of the curve, and *B*, the inside—such that any pair of points of the same class can be joined by a curve which does not cross *C*, while any curve joining a pair of points belonging to different classes must cross *C*. This statement is obviously true for a circle or an ellipse, but the self-evidence fades a little if one contemplates a complicated curve like the twisted polygon in Figure 10.

FIGURE 10—Which points of the plane are inside this polygon?

This theorem was first stated by Camille Jordan (1838–1922) in his famous *Cours d'Analyse*, from which a whole generation of mathematicians learned the modern concept of rigor in analysis. Strangely enough, the proof given by Jordan was neither short nor simple, and the surprise was even greater when it turned out that Jordan's proof was invalid and that considerable effort was necessary to fill the gaps in his reasoning. The first rigorous proofs of the theorem were quite complicated and hard to understand, even for many well-trained mathematicians. Only recently have comparatively simple proofs been found. One reason for the difficulty lies in the generality of the concept of "simple closed curve," which is not restricted to the class of polygons or "smooth" curves, but includes all curves which are topological images of a circle. On the other hand, many concepts such as "inside," "outside," etc., which are so clear to the intuition, must be made precise before a rigorous proof is possible. It is of the highest theoretical importance to analyze such concepts in their fullest generality, and much of modern topology is devoted to this task. But one should never forget that in the great majority of cases that arise from the study of concrete geometrical phenomena it is quite beside the point to work with concepts whose extreme generality creates unnecessary difficulties. As a matter of fact, the Jordan curve theorem is quite simple to prove for the reasonably well-behaved curves, such as polygons or curves with continuously turning tangents, which occur in most important problems.

THE FOUR COLOR PROBLEM

From the example of the Jordan curve theorem one might suppose that topology is concerned with providing rigorous proofs for the sort of obvious assertions that no sane person would doubt. On the contrary, there are many topological questions, some of them quite simple in form, to which the intuition gives no satisfactory answer. An example of this kind is the renowned "four color problem."

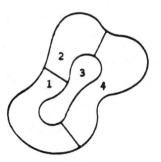

FIGURE 11—Coloring a map.

In coloring a geographical map it is customary to give different colors to any two countries that have a portion of their boundary in common. It has been found empirically that any map, no matter how many countries it contains nor how they are situated, can be so colored by using only *four* different colors. It is easy to see that no smaller number of colors will suffice for all cases. Figure 11 shows an island in the sea that certainly cannot be properly colored with less than four colors, since it contains four countries, each of which touches the other three.

The fact that no map has yet been found whose coloring requires more than four colors suggests the following mathematical theorem: *For any subdivision of the plane into non-overlapping regions, it is always possible to mark the regions with one of the numbers 1, 2, 3, 4 in such a way that no two adjacent regions receive the same number.* By "adjacent" regions we mean regions with a whole segment of boundary in common; two regions which meet at a single point only or at a finite number of points (such as the states of Colorado and Arizona) will not be called adjacent, since no confusion would arise if they were colored with the same color.

The problem of proving this theorem seems to have been first proposed by Moebius in 1840, later by DeMorgan in 1850, and again by Cayley in 1878. A "proof" was published by Kempe in 1879, but in 1890 Heawood found an error in Kempe's reasoning. By a revision of Kempe's proof, Heawood was able to show that *five* colors are always sufficient. Despite the efforts of many famous mathematicians, the matter essentially rests with this more modest result: It has been *proved* that five colors suffice for all maps and it is *conjectured* that four will likewise suffice. But, as in the case of the famous Fermat theorem neither a proof of this conjecture nor an example contradicting it has been produced, and it remains one of the great unsolved problems in mathematics. The four color theorem has indeed been proved for all maps containing less than thirty-eight regions. In view of this fact it appears that even if the general theorem is false it cannot be disproved by any very simple example.

In the four color problem the maps may be drawn either in the plane or on the surface of a sphere. The two cases are equivalent: any map on the sphere may be represented on the plane by boring a small hole through the interior of one of the regions *A* and deforming the resulting surface until it is flat, as in the proof of Euler's theorem. The resulting map in the plane will be that of an "island" consisting of the remaining regions, surrounded by a "sea" consisting of the region *A*. Conversely, by a reversal of this process, any map in the plane may be represented on the sphere. We may therefore confine ourselves to maps on the sphere. Furthermore, since deformations of the regions and their boundary lines do not affect the problem, we may suppose that the boundary of each region is a simple closed polygon composed of circular arcs. Even thus "regularized," the problem remains unsolved; the difficulties here, unlike those involved in the Jordan curve theorem, do not reside in the generality of the concepts of region and curve.

A remarkable fact connected with the four color problem is that for surfaces more complicated than the plane or the sphere the corresponding theorems have actually been proved, so that, paradoxically enough, the analysis of more complicated geometrical surfaces appears in this respect to be easier than that of the simplest cases. For example, on the surface of a torus (see Figure 5), whose shape is that of a doughnut or an inflated inner tube, it has been shown that any map may be colored by using seven colors, while maps may be constructed containing seven regions, each of which touches the other six.

KNOTS

As a final example it may be pointed out that the study of knots presents difficult mathematical problems of a topological character. A knot is formed by first looping and interlacing a piece of string and then joining the ends together. The resulting closed curve represents a geometrical figure that remains essentially the same even if it is deformed by pulling or twisting without breaking the string. But how is it possible to give an intrinsic characterization that will distinguish a knotted closed curve in space from an unknotted curve such as the circle? The answer is by no means simple, and still less so is the complete mathematical analysis of the various kinds of knots and the differences between them. Even for the simplest case this has proved to be a sizable task. Consider the two trefoil knots shown in Figure 12. These two knots are completely symmetrical "mirror images" of one another, and are topologically equivalent, but they are not congruent. The problem arises whether it is possible to deform one of these knots into the other in a continuous way. The answer is in the negative, but the proof of this fact requires considerably more knowledge of the technique of topology and group theory than can be presented here.

FIGURE 12—Topologically equivalent knots that are not deformable into one another.

THE TOPOLOGICAL CLASSIFICATION OF SURFACES
THE GENUS OF A SURFACE

Many simple but important topological facts arise in the study of two-dimensional surfaces. For example, let us compare the surface of a sphere with that of a torus. It is clear from Figure 13 that the two surfaces differ in a fundamental way: on the sphere, as in the plane, every simple closed curve such as C separates the surface into two parts. But on the torus there exist closed curves such as C' that do not separate the surface into two parts. To say that C separates the sphere into two parts means that if the sphere is cut along C it will fall into two distinct and unconnected pieces, or, what amounts to the same thing, that we can find two points on the sphere such that any curve on the sphere which joins them must intersect C. On the other hand, if the torus is cut along the closed curve

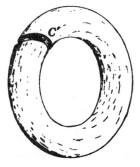

FIGURE 13—Cuts on sphere and torus.

C', the resulting surface still hangs together: any point of the surface can be joined to any other point by a curve that does not intersect C'. This difference between the sphere and the torus marks the two types of surfaces as topologically distinct, and shows that it is impossible to deform one into the other in a continuous way.

Next let us consider the surface with two holes shown in Figure 14. On this surface we can draw *two* non-intersecting closed curves A and B which do not separate the surface. The torus is always separated into two parts by any two such curves. On the other hand, *three* closed non-intersecting curves always separate the surface with two holes.

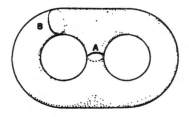

FIGURE 14—A surface of genus 2.

These facts suggest that we define the *genus* of a surface as the largest number of non-intersecting simple closed curves that can be drawn on the surface without separating it. The genus of the sphere is 0, that of the torus is 1, while that of the surface in Figure 14 is 2. A similar surface with p holes has the genus p. The genus is a topological property of a surface and remains the same if the surface is deformed. Conversely, it may be shown (we omit the proof) that if two closed surfaces have the same genus, then one may be deformed into the other, so that the genus $p = 0, 1, 2, \ldots$ of a closed surface characterizes it completely from the topological point of view. (We are assuming that the surfaces considered are ordinary "two-sided" closed surfaces. Later in this section we shall consider "one-sided" surfaces.) For example, the two-holed doughnut and the sphere with two "handles" of Figure 15 are both closed surfaces of genus 2, and it is clear that either of these surfaces may be

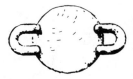

FIGURE 15—Surfaces of genus 2.

continuously deformed into the other. Since the doughnut with p holes, or its equivalent, the sphere with p handles, is of genus p, we may take either of these surfaces as the topological representative of all closed surfaces of genus p.

THE EULER CHARACTERISTIC OF A SURFACE

Suppose that a closed surface S of genus p is divided into a number of regions by marking a number of vertices on S and joining them by curved arcs. We shall show that

(1) $$V - E + F = 2 - 2p,$$

where V = number of vertices, E = number of arcs, and F = number of regions. The number $2 - 2p$ is called the *Euler characteristic* of the surface. We have already seen that for the sphere, $V - E + F = 2$, which agrees with (1), since $p = 0$ for the sphere.

To prove the general formula (1), we may assume that S is a sphere with p handles. For, as we have stated, any surface of genus p may be continuously deformed into such a surface, and during this deformation the numbers $V - E + F$ and $2 - 2p$ will not change. We shall choose the deformation so as to ensure that the closed curves A_1, A_2, B_1, B_2, . . . where the handles join the sphere consist of arcs of the given subdivision. (We refer to Figure 16, which illustrates the proof for the case $p = 2$.)

FIGURE 16

Now let us cut the surface S along the curves A_2, B_2, . . . and straighten the handles out. Each handle will have a free edge bounded by a new curve A^*, B^*, . . . with the same number of vertices and arcs as A_2, B_2, . . . respectively. Hence $V - E + F$ will not change, since the additional vertices exactly counterbalance the additional arcs, while no new regions are created. Next, we deform the surface by flattening out the projecting handles, until the resulting surface is simply a sphere from which $2p$ regions have been removed. Since $V - E + F$ is known to equal 2 for any subdivision of the whole sphere, we have

$$V - E + F = 2 - 2p$$

for the sphere with $2p$ regions removed, and hence for the original sphere with p handles, as was to be proved.

Figure 3 illustrates the application of formula (1) to a surface S consisting of flat polygons. This surface may be continuously deformed into a torus, so that the genus p is 1 and $2 - 2p = 2 - 2 = 0$. As predicted by formula (1),

$$V - E + F = 16 - 32 + 16 = 0.$$

ONE-SIDED SURFACES

An ordinary surface has two sides. This applies both to closed surfaces like the sphere or the torus and to surfaces with boundary curves, such as the disk or a torus from which a piece has been removed. The two sides of such a surface could be painted with different colors to distinguish them. If the surface is closed, the two colors never meet. If the surface has boundary curves, the two colors meet only along these curves. A bug crawling along such a surface and prevented from crossing boundary curves, if any exist, would always remain on the same side.

Moebius made the surprising discovery that there are surfaces with only *one* side. The simplest such surface is the so-called Moebius strip, formed by taking a long rectangular strip of paper and pasting its two ends together after giving one a half-twist, as in Figure 17. A bug crawling along this surface, keeping always to the middle of the strip, will return to its original position upside down (Figure 18). Anyone who contracts to paint one side of a Moebius strip could do it just as well by dipping the whole strip into a bucket of paint.

Another curious property of the Moebius strip is that it has only one edge, for its boundary consists of a single closed curve. The ordinary two-sided surface formed by pasting together the two ends of a rectangle without twisting has two distinct boundary curves. If the latter strip is cut along the center line it falls apart into two different strips of the same kind. But if the Moebius strip is cut along this line (shown in Figure 17) we find that it remains in one piece. It is rare for anyone not familiar with the Moebius strip to predict this behavior, so contrary to one's intuition of what "should" occur. If the surface that results from cutting the Moebius strip along the middle is again cut along its middle, two separate but intertwined strips are formed.

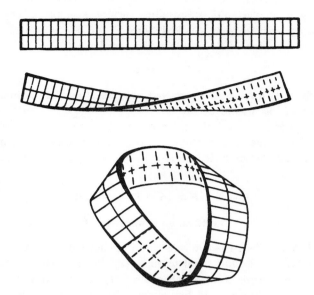

FIGURE 17—Forming a Moebius strip.

It is fascinating to play with such strips by cutting them along lines parallel to a boundary curve and ½, ⅓, etc. of the distance across. The Moebius strip certainly deserves a place in elementary geometrical instruction.

FIGURE 18—Reversal of up and down on traversing a Moebius strip.

The boundary of a Moebius strip is a simple and unknotted closed curve, and it is possible to deform it into a circle. During the deformation, however, the strip must be allowed to intersect itself. (Hence, such a

FIGURE 19—Cross-cap.

deformation of a "real" paper Moebius strip is only possible in the imagination.) The resulting self-intersecting and one-sided surface is known as a cross-cap (Figure 19). The line of intersection RS is regarded as two different lines, each belonging to one of the two portions of the surface which intersect there. The one-sidedness of the Moebius strip is preserved because this property is topological; a one-sided surface cannot be continuously deformed into a two-sided surface.

Another interesting one-sided surface is the "Klein bottle." This surface is closed, but it has no inside or outside. It is topologically equivalent to a pair of cross-caps with their boundaries coinciding.

FIGURE 20—Klein bottle.

It may be shown that any closed, *one-sided* surface of genus $p = 1, 2,$ . . . is topologically equivalent to a sphere from which p disks have been removed and replaced by cross-caps. From this it easily follows that the Euler characteristic $V - E + F$ of such a surface is related to p by the equation

$$V - E + F = 2 - p.$$

The proof is analogous to that for two-sided surfaces. First we show that the Euler characteristic of a cross-cap or Moebius strip is 0. To do this we observe that, by cutting across a Moebius strip which has been subdivided into a number of regions, we obtain a rectangle that contains two more vertices, one more edge, and the same number of regions as the Moebius strip. For the rectangle, $V - E + F = 1$, as we proved on pages 581–582. Hence for the Moebius strip $V - E + F = 0$. As an exercise, the reader may complete the proof.

It is considerably simpler to study the topological nature of surfaces such as these by means of plane polygons with certain pairs of edges conceptually identified. In the diagrams of Figure 21, parallel arrows are to be brought into coincidence—actual or conceptual—in position and direction.

This method of identification may also be used to define three-dimensional closed manifolds, analogous to the two-dimensional closed surfaces. For example, if we identify corresponding points of opposite faces of a

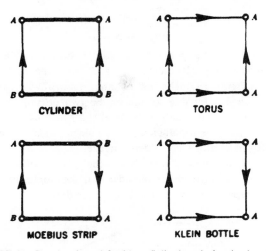

FIGURE 21—Closed surfaces defined by coördination of edges in plane figure.

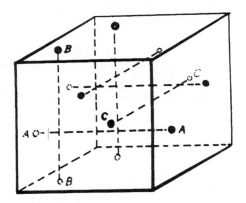

FIGURE 22—Three-dimensional torus defined by boundary identification.

cube (Figure 22), we obtain a closed, three-dimensional manifold called the three-dimensional torus. This manifold is topologically equivalent to the space between two concentric torus surfaces, one inside the other, in which corresponding points of the two torus surfaces are identified (Figure 23). For the latter manifold is obtained from the cube if two pairs of conceptually identified faces are brought together.

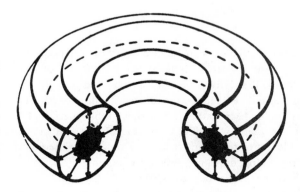

FIGURE 23—Another representation of three-dimensional torus. (Figure cut to show identification.)

DÜRER and the Mathematics of Painting

ALBRECHT DÜRER, as an English art critic observed some years ago, is one of the few good things to have come out of Nuremberg. He was born in 1471, the third of eighteen children and the favorite of his father, a goldsmith of Hungarian descent, who hoped Albrecht would follow this trade. "My father [he writes] took special pleasure in me because he saw that I was diligent to learn. So he sent me to school, and when I had learnt to read and write . . . taught me the goldsmith's craft. But . . . my liking drew me rather to painting . . . and in 1486 my father bound me apprentice to Michael Wohlgemut." It was a happy choice: Wohlgemut was the best painter in Nuremberg and the young Dürer learned much during the three years he served him. The next four years were spent wandering in Europe—the exact course of his travels is uncertain—working under various painters, engravers, wood block cutters and printers. He returned to Nuremberg in 1494, stayed long enough to acquire a wife plus a dowry of two hundred florins, and then set off for a few months' visit to Venice to improve his artistic training. For the next twenty years Dürer worked mainly in Nuremberg, with occasional visits to Italy and other European countries, gaining immense fame for his copper engravings, woodcuts, portraits and altar pieces. In 1520, on a journey to the Netherlands (vividly recorded in his sketch books and diary), he contracted the "strange disease"—probably a malarial fever—which afflicted him for the rest of his life. The last seven years (1521–1528) were spent in Nuremberg; despite ill health Dürer was immensely productive. He painted, among others, several superb portraits and two celebrated religious panels which he presented to the Nuremberg Council; made copper engravings of Erasmus and Melancthon—with whom, as with Luther, he was on close terms; published books on geometry and perspective, and on fortification, and wrote a remarkable work on human proportions which appeared the year after his death.[1] He died suddenly on the night of the sixth of April, 1528. His self-portraits and the descrip-

---

[1] The Nuremberg Council accepted the gift of the religious panels and voted Dürer a hundred florins, ten to his wife, and two to his apprentice. A century later the town sold the painting. This was typical of these worthy burghers who "never hesitated to convert Dürer's masterpieces into cash." At least seven or eight of his pictures were sold by the city to bidders from elsewhere. It has been said rightly that "few prophets have more sincerely loved their people than did Dürer, deserved more honor from them, or received less."

tion of him by contemporaries prove him to have been an unusually handsome man; there is ample evidence that the quality of his person matched his appearance. Camerarius of Nuremberg, who knew him well, wrote: "Nature bestowed on him a body remarkable in build and stature and not unworthy of the noble mind it contained. . . . His fingers, you would vow you had never seen anything more elegant. . . . An ardent zeal impelled him towards the attainment of all virtue in conduct and life, the display of which caused him to be deservedly held a most excellent man. Yet . . . whatever conduced to pleasantness and cheerfulness and was not inconsistent with honour and rectitude he cultivated all his life and approved even in his old age." [2]

The first of the selections which follow discusses the contribution to geometry and perspective made by Dürer in his treatise *Underweysung der Messung mit dem Zirckel uñ Richtscheyt, in Linien Ebnen und Ganzen Corporen* (Nuremberg, 1525). The excerpt is from a scholarly and comprehensive account of Dürer's life and works by the noted art historian Erwin Panofsky.[3] Dr. Panofsky has taught at the University of Hamburg and, since coming to the United States in 1934, at New York University, Harvard and Princeton. He is now professor at the Institute for Advanced Study.

Dürer was interested in geometry partly for intellectual, partly for practical reasons. The problem of representing three-dimensional figures in the plane has long interested mathematicians as well as painters. One of the methods of overcoming the difficulties of projection, so as to preserve a consistent scheme of metrical properties, is descriptive geometry. The invention of this science is generally ascribed to "the prince of teachers," the great French mathematician Gaspard Monge (1746–1818), but there is good reason for believing that the credit for its actual discovery belongs to Dürer.[4] The foundations of descriptive geometry are laid mainly in his posthumous treatise on human proportion.[5] The earlier geometry book made no pretense to being a complete survey but was intended primarily to help the draughtsman draw simple as well as complicated curves. It considers classical geometrical problems (e.g., the duplication of the cube), the inscription of polygons in a circle, conic sections, helical space

---

[2] I have taken this quotation and certain other details from a fine essay-review of Panofsky's biography of Dürer [see Note 3 below] appearing in the [London] *Times Literary Supplement*, February 9, 1946, pp. 61–62, under the title "The Good Nuremberger." The reviewer concludes with a quotation from Dürer himself which is strikingly apt: "God sometimes granteth unto a man to learn and know how to make things the like whereof in his day no other can contrive, and perhaps for a long time none hath been before him, and after him another cometh not soon."

[3] Erwin Panofsky, *Albrecht Dürer*, Princeton, second edition, revised, 1945.

[4] This point of view is supported by the noted historian of geometry, the late Julian Lowell Coolidge of Harvard, in his book, *The Mathematics of Great Amateurs*, Oxford, 1949, pp. 61–70.

[5] *De Symetria Partium in Rectis Formis Humanorum Corporum Libri, etc.* (Nuremberg, 1528).

curves; it also provides instructions for constructing regular solids by paper folding. According to Moritz Cantor, Dürer was probably the inventor of this instructive and amusing procedure which is still used in teaching geometry in schools as well as for parlor entertainment.[6] Dürer adorned his mathematical works with exquisite drawings, some of which are reproduced as Plates III, IV, V, VI and VII on pages 620–621.

The second selection deals with the origins of projective geometry, an important mathematical creation of the seventeenth century, which was "inspired by the art of painting." The essay is an expanded version of a chapter in *Mathematics in Western Culture* [7] by Morris Kline, professor of mathematics at New York University. Dr. Kline's book, one of the first of its kind, is an absorbing account of the influence of mathematical thought on Western philosophy, science, religion, literature and art. I commend this volume to you; as you will see from the essay which follows, Kline is a fine expositor of the concepts of mathematics and imparts a vivid picture of its role in the growth of ideas.

[6] Moritz Cantor, *Vorlesungen über die Geschichte der Mathematik*, second edition, Leipzig, 1913, Vol. II, p. 466; see also Coolidge, *op. cit.*, p. 68.
[7] Oxford Press, New York, 1953.

6   Dürer as a Mathematician

By ERWIN PANOFSKY

FOR perspective, in the sense of a scientific theory, Dürer was indeed much indebted to the Italians. During the early and high Middle Ages the reconstruction of three-dimensional space on a two-dimensional plane had not constituted a problem. The "picture" had been conceived as a material surface covered with lines and colors which could be interpreted as tokens or symbols of three-dimensional objects, and a wavy strip of green or brown had been sufficient to indicate the ground on which the figures, trees and houses were placed. In the course of the fourteenth century, however, the forms appearing *on* the surface came to be thought of as something existing *behind* the surface until Leone Battista Alberti could liken the picture to a "transparent window through which we look out into a section of the visible world." Objects of equal size began to diminish as they moved away from the beholder; the walls, floors and ceilings which delimited an interor, or the ground on which were disposed the elements of a landscape, began to "recede" toward the background; and such lines as were at right angles to the picture plane ("orthogonals") developed into perspective "vanishing lines" which tended to converge toward one center.

Around 1340 (in the Northern countries about thirty years later) this center had already assumed the character of a single "vanishing point" in which the "orthogonals" converged with mathematical precision, at least within one unobstructed plane; and as early as about 1420–25 it had been observed that the horizontal lines of a cubiform building set slantwise into space seemed to converge toward two points symmetrically located on one horizontal ("Perspectiva cornuta"). True, the "perspective" image thus developed—and, in part, already constructible by means of a ruler or a tight cord fastened to a pin—was not yet unified, and it was not yet possible to determine the correct sequence of equidistant transversals; Alberti expressly condemns a practice, apparently still in vogue about 1435, by which the intervals between one transversal and the next were mechanically, and of course mistakenly, diminished by one-third each. But even these problems could be solved, and were solved by the Flemish painters of the fifteenth century, on a purely empirical basis. In the 'fifties it was discovered that all orthogonals, and not only those located in one plane, had to converge in one "vanishing point" which thus established the "general horizon" of the picture; and the problem of determining the gradual diminution of equidistant transversals was solved by the simple device of running an oblique line across the converging orthogonals. Such an oblique line, it was reasoned, would cut the converging "orthogonals" so as to form the common diagonal of a continuous row of small, equal squares which would automatically determine the correct sequence of as many transversals (Figure 1). It was also discovered that this diagonal would intersect the "horizon" at the same point as the diagonals of the adjacent rows of small squares, thus constituting a "lateral vanishing point"; that the foreshortening of the whole system became the sharper the smaller the distance between this "lateral vanishing point" and the "central vanishing point" (that is to say, the point of convergence of the "orthogonals"); and that, if diagonals were drawn from left to right as well as from right to left, the two "lateral vanishing points" were equidistant from the "central vanishing point."

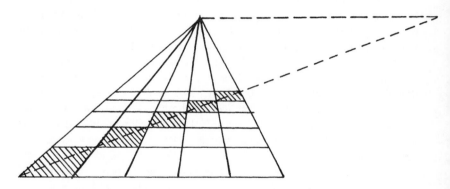

FIGURE 1—Empirical Perspective Construction of a "Checker-Board Floor."

These draftsmanlike methods were later codified by that Johannes Viator, or Jean Pélerin, whose *De artificiali perspectiva*, in spite of its late date (1505, second edition 1509), is still a representative of the pattern-and-recipe type of art-theoretical treatise. But they were known already by the last third of the fifteenth century, and they were quite sufficient to ensure perfectly "correct" perspective representations. Dirk Bouts's *Last Supper* of 1464–67 is already as impeccably constructed as any painting by an Italian Quattrocento master, and the same is true, as we remember, of everyone of Dürer's works executed after about 1500.

Yet Dürer undertook, in 1506, a special trip to Bologna—well over a hundred miles from Venice—"for the sake of 'art' in secret perspective which some one wants to teach me." After such engravings as *Weihnachten* and such woodcuts as the *Presentation* Dürer did not need any instruction in perspective as far as its practical application was concerned. What he was after was the theoretical foundation of a process which he had thus far learned only empirically, and this is exactly what he tried to express by saying that he wished to be instructed, not in perspective, but in *"Kunst* in heimlicher Perspectiva"—"Kunst" meaning, as we now know, theoretical knowledge or rational understanding as opposed to mere practice. This knowledge was indeed a "secret" insofar as it had not yet been divulged in any printed book; but it had been accessible to experts for almost exactly three-quarters of a century: it was the method traditionally, and perhaps justly, ascribed to Filippo Brunelleschi.

The procedure culminating in Dirk Bouts's *Last Supper* and Dürer's pre-Venetian prints had resulted from a more and more successful schematization of such patterns—foreshortened floors and ceilings, receding walls, etc.—as had been handed down by tradition or were perceived by direct observation. It had developed quite independently of that mathematical analysis of the process of vision which was known as 'Οπτική or "Optica" in classical Antiquity, and as "Prospectiva" or "Perspectiva" in the Latin Middle Ages. This discipline—formulated by Euclid, developed by Geminus, Damianus, Heliodorus of Larissa and others, transmitted to the western Middle Ages by the Arabs, and exhaustively treated by such scholastic writers as Robert Grosseteste, Roger Bacon, Vitellio and Peckham—attempted to express in geometrical theorems the exact relation between the real quantities found in objects and the apparent quantities which constitute our visual image. It was based on the assumption that objects are perceived by straight visual rays converging in the eye—so that the visual system can be described as a cone or pyramid having the object as its base and the eye as its apex—and that the apparent size of any real magnitude, and thereby the configuration of the entire visual image, depends on the width of the corresponding angle at the apex of said pyramid or cone ("visual angle").

Classical "Optica" and medieval "Prospectiva," then, were no more concerned with problems of artistic representation than the representational methods of Jan van Eyck, Petrus Cristus or Dirk Bouts were based on the doctrines of scholastic writers. A few of these, to wit Grosseteste and Roger Bacon, could already develop optical instruments—apparently not unlike our refracting telescopes—by means of which "that which is near and big can be made to appear very distant and small and vice versa, so that it is possible to read small letters from an incredible distance and to count grains of sand, seeds or other diminutive objects"; but no one thought of applying the Euclidian theory of vision to the problems of graphic representation. This was precisely what Brunelleschi and his followers proposed to do. They conceived the idea of intersecting the Euclidian pyramid by a plane inserted between the object and the eye, and thereby "projecting" the visual image on this surface just as a lens projects a picture on the screen or on a photographic film or plate. A pictorial representation thus came to be defined as "a cross-section through the visual pyramid or cone" ("l'intersegazione della piramide visiva," as Leone Battista Alberti puts it, or "a plane, transparent intersection of all those rays which travel from the eye to the object it sees," to translate the formula adopted by Dürer). To ensure perspective correctness one had only to evolve a method of constructing this cross-section by means of a compass and a ruler, and this was the essence of the new "Painter's Perspective" ("Prospectiva pingendi" or "Prospectiva artificialis") in contradistinction to which the old-time theory of vision came to be called "Prospectiva naturalis."

In its undiluted and comprehensive form—described, as far as the writings of the fifteenth century are concerned, only in the admirable *De prospectiva pingendi* by Piero della Francesca which was composed between 1470 and 1490 but was not printed until 1899—this construction requires two preparatory drawings, namely, the elevation and the groundplan of the whole visual system. In each of these, the visual pyramid or cone is represented by a triangle having its apex in a point standing for the eye while the projection (or picture) plane is represented by a vertical intersecting this triangle. In the elevation drawing the object has to be shown in a vertical diagram, and in the groundplan in a horizontal diagram. Either diagram is connected with the point representing the eye, and the points of intersection between the connecting lines and the vertical will determine the required set of values, namely, the vertical and transversal quantities of the perspective image. The latter can be constructed by simply combining these two sets in a third and final drawing (Figure 2).

Since this "costruzione legittima" requires two diagrams of the object, one horizontal and one vertical, it presupposes a familiarity with the method of parallel projection by which any required diagram can be developed from any two others, provided that they are located in planes at right angles to each other. This method had already been practiced by the medieval architects and had been dealt with in their treatises. But it had now to be applied to the human body in movement instead of being restricted to buildings and architectural details, and thus developed into a special branch of Renaissance art theory indispensable both for the study of human proportions and for the application of the "costruzione legittima." It was extensively treated in Piero della Francesca's *De prospectiva pingendi*; it was practiced by Leonardo and his pupils; and if we can believe Giovanni Paolo Lomazzo, it was almost a speciality with other Milanese theoreticians such as Vincenzo Foppa and Bartolommeo Suardi, called Bramantino.

Needless to say, this "costruzione legittima," requiring as it did so many

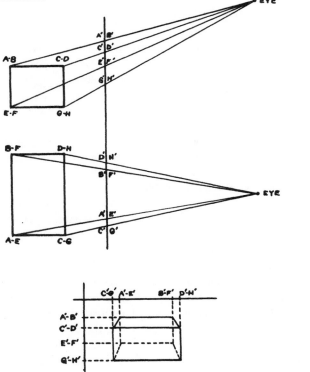

FIGURE 2—Systematic Perspective Construction of a Three-Dimensional Body ("Costruzione legittima").

tedious preliminaries, was too unwieldy for actual use. In practice, not even the most conscientious of painters would ever construct individual objects, let alone figures, by first developing two diagrams and then projecting these on the picture plane. It was deemed sufficient to build up a three-dimensional system of coordinates in foreshortening which enabled the artist to determine the relative magnitude, though not the shape, of any object he might wish to render. Such a system could easily be developed from a foreshortened square correctly divided into a number of smaller squares; and to obtain this basic square, or rather checkerboard—a problem solved by the Northern artists on purely empirical grounds, as we have seen—was the purpose of that "abbreviated construction" which was actually used by the Italian Quattrocento painters. It was described by Leone Battista Alberti, and, later on, by Piero della Francesca, Pomponius Gauricus and Leonardo da Vinci, and its practical application can be observed in drawings by Paolo Uccello and Leonardo himself. This abbreviated method begins with the procedure already followed in the "pre-Brunelleschian" period: the front line of the future basic square is divided into an arbitrary number of equal parts, and the dividing points are connected with the central "vanishing point" P. But now the sequence of the equidistant transversals is determined on a strictly Euclidian basis, that is to say, by constructing the now familiar profile elevation of the visual cone or pyramid: we erect a vertical—representing the picture plane —at one of the front corners of the future basic square and assume, on the horizontal determined by the vanishing point, a point representing the eye. When we connect this point with the terminals and dividing points of the front line of the future basic square, the points of intersection between the connecting lines and the vertical will indicate, on the latter, the correct sequence of equidistant transversals (Figure 3).

We do not know the name of Dürer's "teacher" in Bologna. But whoever he was, he must have been both well-informed and communicative. When Dürer returned from Venice he was acquainted with the "costruzione legittima" a written description of which could be found, as we have seen, only in the unpublished treatise by Piero della Francesca; with Piero's elegant method of transferring any given planimetrical figure from an unforeshortened square into a foreshortened one; and, more important, with his fundamental definition of "perspectiva artificialis" as such: "Perspective is a branch of painting which comprises five parts: the first is the organ of sight, *viz.*, the eye; the second is the form of the object seen; the third is the distance between the eye and the object; the fourth are the lines which start from the surface of the object and go to the eye; the fifth is the plane which is between the eye and the object wherever one intends to place [that is, on which one wishes to project] the objects." In addition, Dürer shows himself informed about such methods and devices as were common to all Italian theoreticians of perspective, but had re-

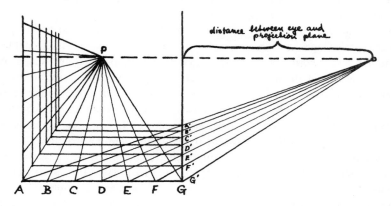

FIGURE 3—Systematic, but Abbreviated Perspective Construction of a "Checker-Board Floor" (called by Dürer "Der nähere Weg").

ceived, at least in part, the especial attention of the Milanese theoreticians. He knew the "abbreviated construction" first described by Alberti; parallel projection as applied to the human figure; apparatuses enabling the artist to draw directly from nature and yet to achieve "approximate" perspective accuracy; and the geometrical construction of cast-shadows, a speciality of Leonardo da Vinci. His "teacher," then, must have been a man familiar with both Piero della Francesca and the theorists of Milan. This applies to the two most plausible candidates thus far proposed, the mathematician Luca Pacioli and the great architect Donato Bramante; but it may also have been true of many a nameless painter or professor of Bologna University.

Characteristically, most of Dürer's drawings dealing with problems of perspective date from the years between 1510 and 1515—the period of the *Melencolia I* (1700–1706). The final presentation of the subject, however, is found at the end of the Fourth Book of his treatise on Geometry, the *Underweysung der Messung mit dem Zirckel uñ Richtscheyt* of 1525 (revised edition 1538). Here Dürer teaches: first, the "costruzione legittima," illustrated by a cube placed on a square and lighted so as to serve, at the same time, as an example for the construction of cast-shadows; second, the "abbreviated construction" which he happily calls "der nähere Weg" ("the shorter route"); third (in the revised edition only), Piero della Francesca's method of transferring planimetrical figures from the unforeshortened into the foreshortened square; fourth, two (in the revised edition, four) apparatuses to ensure an approximative correctness by mechanical instead of mathematical means.

Two of these apparatuses were already known to Alberti, Leonardo and Bramantino. The eye of the observer is fixed by a sight, and between it and the object is inserted either a glass plate (Figure 4) or a frame divided into small squares by a net of black thread ("graticola" or grill, as Alberti calls it). In the first case, an approximately correct picture can be obtained by simply copying the contours of the model as they appear on the glass plate and then transferring them to the panel or drawing sheet by means of tracing; in the second case the image perceived by the artist is divided into small units whose content can easily be entered upon a paper divided into a corresponding system of squares. The other two apparatuses—one invented by one Jacob Keser, the other apparently by Dürer himself—are nothing but improvements on the ones already described. Keser's device removes the difficulty that the distance between the eye and the glass plate can never exceed the length of the artist's arm, which entails an undesirably sharp foreshortening: the human eye is replaced by the eye of a big needle, driven into the wall, to which is fastened a piece of string with a sight at the other end; the operator can then "take aim" at the characteristic points of the object and mark them on the glass plate with the

FIGURE 4

perspective situation determined, not by the position of his eye but by that of the needle. The last apparatus eliminates the human eye altogether: it consists, again, of a needle driven into the wall and a piece of string, but the piece of string has a pin on one end and a weight on the other; between the eye of the needle and the object is placed a wooden frame within which every point can be determined by two movable threads crossing each other at right angles. When the pin is put on a certain point of the object the place where the string passes through the frame determines the location of that point within the future picture. This point is fixed by adjusting the two movable threads and is at once entered upon a piece of paper hinged to the frame; and by a repetition of this process the whole object may be transfered gradually to the drawing sheet (Figure 5).

Apart from this technical invention Dürer added nothing to the science of perspective as developed by the Italians. Yet the last section of his "Unterweisung der Messung" is memorable in two respects. First, it is the first literary document in which a strictly representational problem received a strictly scientific treatment at the hands of a Northerner; none of Dürer's forerunners and few of his contemporaries had any understanding of the fact that the rules for the construction of a perspective picture are based

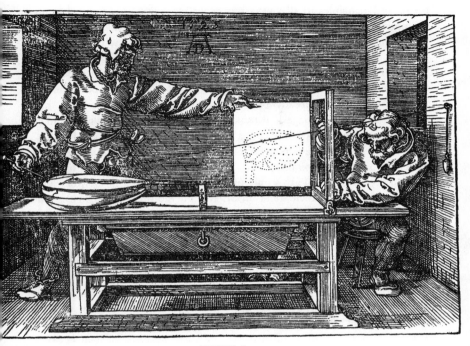

FIGURE 5

on the Euclidian concept of the visual pyramid or cone, and it is a remarkable fact that the methods taught in a treatise by one Hieronymus
Rodler, published as late as 1531 and even reprinted in 1546, are partly
purely empirical and partly downright wrong. Second, it emphasizes, by
its very place at the end of a "Course in the Art of Measurement with
Compass and Ruler," that perspective is not a technical discipline destined
to remain subsidiary to painting or architecture, but an important branch
of mathematics, capable of being developed into what is now known as
general projective geometry.

The Preface of this "Course in the Art of Measurement," which, like
the Treatise on Human Proportions, is dedicated to Pirckheimer, is the
first public statement of Dürer's lifelong conviction, to be repeated many
a time and alluded to in our discussions at various points: that the German painters were equal to all others in practical skill ("Brauch") and
power of imagination ("Gewalt"), but that they were inferior to the
Italians—who had "rediscovered, two hundred years ago, the art revered
by the Greeks and Romans and forgotten for a thousand years"—in a
rational knowledge ("Kunst") which would prevent them from "errors"
and "wrongness" in their work. "And since geometry is the right founda-

tion of all painting," Dürer continues, "I have decided to teach its rudiments and principles to all youngsters eager for art. . . . I hope that this my undertaking will not be criticized by any reasonable man, for . . . it may benefit not only the painters but also goldsmiths, sculptors, stonemasons, carpenters and all those who have to rely on measurement."

The "Unterweisung" is, therefore, still a book for practical use and not a treatise on pure mathematics. Dürer wanted to be understood by artists and artisans. It has already been mentioned that he appropriated their ancient technical language and took it as a model for his own. He liked to explain the practical implications of a given proposition even if he had to interrupt his systematic context, as when he teaches how to use the constructed spiral for capitals or for a foliated crozier. He refrains from learned divagations and gives only one example—the first in German literature—of a strict mathematical proof. But on the other hand his erudite friends kept him informed of those new ideas and problems which—to quote from the excellent Johannes Werner to whom Dürer appears to be indebted in more than one respect—"had wandered from Greece to the Latin geometricians of this age." Besides his first-hand knowledge of Euclid, he had established contact with the thought of Archimedes, Hero, Sporus, Ptolemy and Apollonius; and, more important, he was, himself, a natural-born geometrician. He had a clear idea of the infinite (for instance when he says that a straight line "can be prolonged unendingly or at least can be thought of in this way" or when he distinguished between parallelism and asymptotic convergence); he emphasized the basic difference between a geometrical figure in the abstract and its concrete realization in pen and ink (the mathematical point, he says, is not a "dot," however small, but can be "mentally located so high or so low that we cannot even reach there physically," and what applies to the point applies *a fortiori* to lines); he never confused exact with approximate constructions (the former ones being correct *"demonstrative,"* the latter ones merely *"mechanice"*); and he presented his material in perfect methodical order.

The First Book, beginning with the usual definitions, deals with the problems of linear geometry, from the straight line up to those algebraic curves which were to occupy the great mathematicians of the seventeenth century; Dürer even ventures upon the construction of helices, conchoids ("Muschellinie," "shell line") and epicycloids ("Spinnenlinie," "spider line"). One of the most interesting features of the First Book is the first discussion in German of conic sections, the theory of which had just been revived on the basis of classical sources. There can be little doubt that Dürer owes his familiarity with Apollonius's terms and definitions (parabola, hyperbola and ellipse) to the aforesaid Johannes Werner who lived in Nuremberg, and whose valuable *Libellus super viginti duobus elementis conicis* had appeared in 1522, three years before the "Unterweisung der

Messung." But Dürer approached the problem in a manner quite different from that of Werner, or, for that matter, of any professional mathematician. Instead of investigating the mathematical properties of the parabola, hyperbola and ellipse, he tried to construct them just as he had tried to construct his spirals and epicycloids; and this he achieved by the ingenious application of a method familiar to every architect and carpenter but never before applied to the solution of a purely mathematical problem, let alone the ultra-modern problem of the conic sections: the method of parallel projection. He represented the cone, cut as the case may be, in side elevation and groundplan and transferred a sufficient number of points from the former into the latter. Then the normal hyperbola—produced by a section parallel to the axis of the cone—can be directly read off when a front elevation is developed from the two other diagrams, while parabolas and ellipses, produced by oblique sections and therefore appearing in a reduction in any diagram except the side elevation, must be obtained by proportionately expanding their main axes. Crude though it is, this method, which may be called a genetic as opposed to a descriptive one, announces, in a way, the procedure of analytical geometry and did not fail to attract the attention of Kepler who corrected the only mistake committed by Dürer in his analysis. Like any schoolboy, Dürer found it hard to imagine that an ellipse is a perfectly symmetrical figure. He was unable to get away from the idea that it should widen in proportion with the widening of the cone, and he involuntarily twisted the construction until it resulted, not in an orthodox ellipse but in an "Eierlinie" ("egg line"), narrower at the top than at the bottom (Figure 6). Even with Dürer's primitive methods the error could have been easily avoided. That it was committed, not only illustrates a significant conflict between abstract geometrical thought and visual imagination, but also proves the independence of Dürer's researches. After the publication of his book he invented an ingenious compass which would have saved him from this error, but, needless to say, this instrument solves the problem of the ellipse only *"mechanice,"* not *"demonstrative."*

The Second Book proceeds from one-dimensional to two-dimensional figures, with special emphasis on the "quadratura circuli" and the construction of such regular polygons as cannot be developed from the square and the equilateral triangle, *viz.,* the pentagon, the enneagon, etc. Of these, only the pentagon (which also furnishes the decagon) and the pentecaidecagon, or fifteen-sided figure, had been treated in classical times; the pentagon because it is the basic element of one of the "Platonic" solids, namely, the dodecahedron; and the pentecaidecagon because it was necessary for the construction of an angle of 24 degrees, then generally considered as the correct measurement of the obliquity of the ecliptic. In the Middle Ages, however, the problem had gained a wider and more practical importance. Both Islamic and Gothic decoration—and, after the invention of firearms, fortification—required methods of constructing all kinds of regular polygons. Dürer, in fact, at once proceeds to develop these into tracery patterns and to combine them into "pavements" which anticipate Kepler's "Congruentia figurarum harmonicarum" in the Second Book of his *Har-*

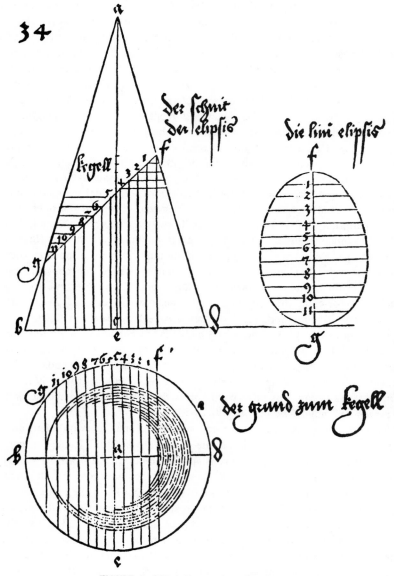

FIGURE 6—Dürer's Construction of the Ellipse.

monices mundi libri V. The medieval constructions of these polygons were, of course, approximate; but they were, and had to be, simple, preferably not even calling for a change in the opening of the compass; and it was Dürer rather than Leonardo—who also tried his hand at the construction of the more complicated regular polygons—who transmitted these constructions to the future.

The construction of the regular pentagon, for instance, is not described by Dürer according to Euclid. He gives, instead, the less well known construction of Ptolemy and, in addition, an approximate construction "with the opening of the compass unchanged" which, but for him, would have remained forever buried in the *Geometria deutsch*; and his construction of the enneagon, likewise an approximation, is not described in any written source but was taken over directly, as we happen to know, from the tradition of the "ordinary workmen" ("tägliche Arbeiter"). Thus the "Unterweisung der Messung," published in Latin in 1532, 1535 and 1605, served, so to speak, as a revolving door between the temple of mathematics and the market square. While it familiarized the coopers and cabinetmakers with Euclid and Ptolemy, it also familiarized the professional mathematicians with what may be called "workshop geometry." It is largely due to its influence that constructions "with the opening of the compass unchanged" became a kind of obsession with the Italian geometricians of the later sixteenth century, and Dürer's construction of the pentagon was to stimulate the imaginations of men like Cardano, Tartaglia, Benedetti, Galileo, Kepler and P. A. Cataldi, who wrote a whole monograph on the "Modo di formare un pentagono . . . descritto da Alberto Durero" (Bologna, 1570).

The Third Book of the "Unterweisung," on the other hand, is of purely practical character. It is intended to illustrate the application of geometry to the concrete tasks of architecture, engineering, decoration and typography. Dürer was an admiring student of Vitruvius—we still possess his German excerpts from several important chapters of the *De architectura* —and recommends him highly in the "Unterweisung," but he was far from being dogmatic about it. His praise of Vitruvius is directly followed by the statement that the German mind always demands "new patterns the like of which has never been seen before," and he proceeds to describe two columns not to be found in Vitruvius and submitted to the reader without obligations: "and let anyone cull therefrom what he likes, and do as he pleases." From a most interesting report unfortunately not included in the "Unterweisung" (1685) we learn that Dürer favored the classical roof, with a slope of little more than 20 degrees, against the steep Gothic one; and his ideas about city-planning—laid down in his Treatise on Fortification and possibly connected with one of the earliest "slum-clearing projects" in history, the Augsburg "Fuggerei" of 1519/20—reveal his familiarity with such modern theoreticians as Leone Battista Alberti and Francesco di Giorgio Martini. But his designs for capitals, bases, sun-dials and whole structures such as the tapering tower to be placed in the center of a market place are anything but classical. He also describes, among other things, triumphal monuments to be composed of actual guns, powder barrels, cannonballs and armor, yet accurately proportioned *more geometrico*. When celebrating a victory over rebellious peasants, these monuments are to be composed of rustic implements such as grain chests, milk cans, spades, pitchforks and crates; and a humorous extension of this principle leads—"von Abenteuer wegen" ("for the sake of curiosity")—to an epitaph in honor of a drunkard, consisting of a beer barrel, a checkerboard, a basket with food, etc. It should, however, be noted that these absurd contrivances met with the approval of François Blondel, and that at least one of them—a slender monument crowned by the figure of a wretched conquered peasant (see Plate I, page 620)—may have been inspired by certain fanciful designs of Leonardo's (see Plate II, page 620).

FIGURE 7—Construction of Roman Letters according to Sigismundus de Fantis, "Theorica et Pratica . . . de modo scribendi . . .," Venice (J. Rubeus), 1514.

At the end of the Third Book, Dürer familiarizes the Northern countries with another "secret" of the Renaissance, the geometrical construction of Roman letters, "litterae antiquae," as they were called by Lorenzo Ghiberti, thereby anticipating Geoffroy Tory's famous *Champ Fleury* by precisely four years. In Italy, this problem had been taken up by Felice Feliciano, the friend and archeological adviser of Andrea Mantegna, and had subsequently been treated by such authors as Damiano da Moile, or Damianus Moyllus (about 1480); Luca Pacioli (published 1509); Sigismundus de Fantis (published 1514); and—possibly—by Leonardo da Vinci. Dürer—introducing the subject by instructions on how to determine the suitable size of inscriptions high above eye level—was not in a position to improve on the methods of these Italian forerunners. As far as

FIGURE 8—Construction of Roman Letters according to Dürer's "Underweysung der Messung" of 1525.

the Roman letters are concerned he had to limit himself to the role of a middleman (Figures 7 and 8). The Gothic letters, however—or, to use his own expression, the "Textur" type—he constructed on a principle not to be found in any earlier source. Gismondo Fanti had dealt with them in the same fashion as with the "antiqua" type, that is to say he had inscribed each letter into a square, had established its proportions by dividing the sides of the square in a certain way, and had determined its contours by combining straight lines with circular arcs (Figure 9). Dürer, on the other hand, constructs his Gothic letters according to an entirely different principle. He dispenses with circular arcs altogether, and instead of inscribing the letter in a large square, he builds it up from a number of small geo-

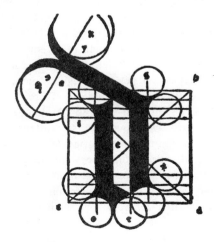

FIGURE 9—Construction of "Gothic" Letters according to Sigismundus de Fantis.

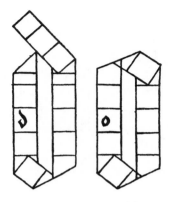

FIGURE 10—Construction of "Gothic" Letters according to Dürer.

metrical units such as squares, triangles or trapezoids (Figure 10). If of anything, this cumulative method is reminiscent of Arabic, and not of Italian, calligraphy.

After this excursion into the domain of the practical, the Fourth Book resumes the thread where the Second had left it: it deals with the geometry of three-dimensional bodies or stereometry, a field entirely disregarded during the Middle Ages. At the beginning, Dürer discusses the five regular or "Platonic" solids, a problem brought into the limelight by the revival of Platonic studies on the one hand and by the interest in perspective on the other. Whether or not Dürer was acquainted with the work of the two Italian specialists in this field, Luca Pacioli and Piero della Francesca, is an open question. Certain it is that, as in the case of the conic sections, he tackled the problem in an entirely independent way. Pacioli discusses, besides the five "Platonic" or regular bodies, only three of the thirteen "Archimedean" or semi-regular ones, and he illustrates them in perspective or stereographic images. Dürer treats seven—in the revised edition of 1538 even nine—of the "Archimedean" semi-regulars, plus several bodies of his own invention (for instance, one composed of eight dodecagons, twenty-four isosceles triangles and eight equilateral triangles), and instead of representing the solids in perspective or stereographic images, he devised the apparently original and, if one may say so, proto-topological method of developing them on the plane surface in such a way that the facets form a coherent "net" which, when cut out of paper and properly folded where two facets adjoin, will form an actual, three-dimensional model of the solid in question (Figure 11).

This section is followed by the discussion of another problem which had "wandered from Greece to the Latin geometricians of the time," namely, the problem of doubling the cube. This "Delian problem," as it was called, had been treated, almost simultaneously, by a mathematician named Heinrich Schreiber or Grammateus, who gives only one solution, and by the aforesaid Johannes Werner who, in a paraphrase of Eutocius's Commentary on Archimedes, gives no less than eleven. That Dürer, who gives three, made use of Werner's treatise is all the more probable as it is printed together with the *Libellus super viginti duobus elementis conicis.* Yet it can be shown by the very lettering of Dürer's figures that he also consulted Eutocius himself, with Pirckheimer dictating to him the text in a German translation.

As the "Delian problem" concerns the cube its discussion formed a wel-

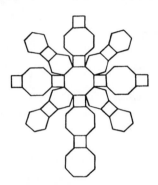

FIGURE 11—Dürer's "Net" of the "Cuboctahedron Truncum."

come transition to the last section of Dürer's *Underweysung der Messung,* the chapter on perspective which, we remember, exemplifies both the "costruzione legittima" and the "shorter route" by the construction of a lighted cube placed on a horizontal plane. To Dürer, as to many of his contemporaries, perspective meant the crown and keystone of the majestic edifice called Geometry.

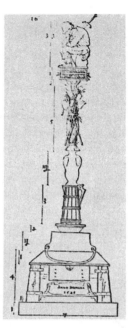

PLATE I—Dürer, Monument Celebrating a Victory over the Peasants. Woodcut from the "Underweysung der Messung . . . ," Nuremberg, 1525.

PLATE II—Leonardo da Vinci, Designs for Fountains. Windsor Castle, Royal Library.

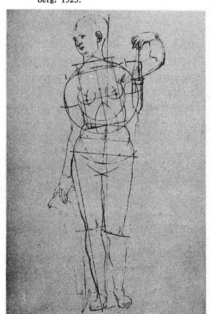

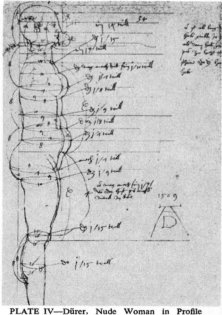

PLATE III—Dürer, Nude Woman (Construction), about 1500, Berlin, Kupferstichkabinett. Drawing L. 38 (1633), 307 by 208 mm.

PLATE IV—Dürer, Nude Woman in Profile (Study in Human Proportions), 1507 (corrected 1509), Dresden, Sächsische Landesbibliothek. Drawing (1641), 295 by 203 mm.

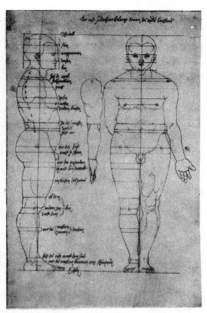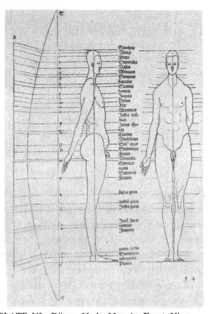

PLATE V—Dürer, Nude Man in Front View
and Profile (Auto-Tracing, Study for the
First Book of the "Vier Bücher von Menschlicher Proportion"), about 1523, Cambridge, Mass., Fogg Museum of Art. Drawing (1624 a), 286 by 178 mm.

PLATE VI—Dürer, Nude Man in Front View
and Profile, Distorted by Projection on a
Circular Curve. Woodcut from the Third
Book of the "Vier Bücher von Menschlicher
Proportion" (here reproduced from the Latin
Edition, 2nd volume, Nuremberg, 1534).

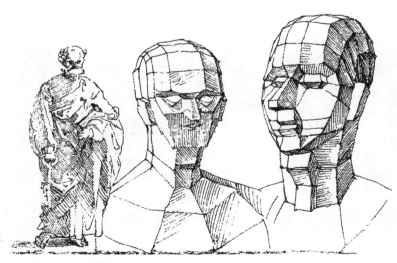

PLATE VII—Dürer, Two Heads Divided into Facets and St. Peter, 1519, Dresden, Sächsische
Landesbibliothek. Drawing T. 732 (839), 115 by 190 mm.

7 Projective Geometry

By MORRIS KLINE

IN the house of mathematics there are many mansions and of these the most elegant is projective geometry. The beauty of the concepts, the logical perfection of its structure, and its fundamental role in geometry recommend the subject to every student of mathematics. The fascination of projective geometry extends even to its very novel historical origins in the work of the Renaissance artists.

The medieval painters, who sought in the main to portray and embellish the central themes in the Christian drama, were content to express themselves in symbolic terms. They portrayed people and objects in a highly stylized manner, usually on a gold background, as if to emphasize that the subject of the painting had no connection with the real world. An excellent example of this style, indeed one that is regarded by critics as the flower of medieval painting, is Simone Martini's "The Annunciation."

However, the various forces and movements which are known collec-

"The Last Supper" by Leonardo da Vinci utilized projective geometry to create the illusion of three dimensions. Lines have been drawn on this reproduction to a point at infinity.

tively as the Renaissance drew the attention of the artists to the world of nature about them and induced a desire to paint realistically. At the same time the exhumation and study of Greek works revived the doctrine that the essence of nature is mathematical law. Because Renaissance painters became imbued with this belief, they struggled for over a hundred years to find a mathematical scheme which would enable them to depict the three-dimensional real world on a two-dimensional canvas. It was very fortunate that the painters were also architects and engineers and, in fact, the best mathematicians of the fifteenth century. Hence the task they set for themselves proved to be within their powers. The attempt to render space, distance, mass, volume, and visual effects succeeded. Though a full account of how well it succeeded would require a volume on Renaissance painting, for our purposes it may be sufficient merely to look at two of the best-known paintings of this era, Leonardo da Vinci's "The Last Supper" and Raphael's "School of Athens." The latter example displays also the Renaissance glorification of the Greek masters. Even a superficial comparison of these two paintings with Martini's "Annunciation" shows

"The School of Athens" by Raphael

that realism and a veridical description of three-dimensional scenes were achieved.

The key to three-dimensional representation was found in what is known as the principle of projection and section. What one sees in a particular scene depends of course upon the position of the viewer. The

Renaissance painter therefore supposed that he looked upon the scene from a fixed position. To simplify his problem he supposed further that he looked at the scene with one eye. This simplification is by no means trivial

"The Annunciation" by Simone Martini is an oustanding example of the flat, stylized painting of the medieval artists. The figures were symbolic and framed in a gold background.

and, to an extent, begs the problem because one eye sees only a flat or two-dimensional scene. We obtain the true perception of depth when our brains compare the sensations received by both eyes, each viewing the scene from a slightly different position. Nevertheless, a person looking with one eye at a reasonably familiar scene obtains a correct impression because his brain interprets the purely visual sensation.

Despite the simplification introduced by concentrating on the sensation received by one eye, the problem of reproducing a scene on canvas is still enormously difficult. The painter therefore imagined that a ray of light proceeded from each point in the scene to one eye. This collection of light rays—or, mathematically stated, this collection of converging lines—he called a projection. He then imagined that his canvas was a glass screen interposed between the scene and the eye. The collection of points where the lines of the projection intersected the glass screen was a "section." To achieve realism the painter had to reproduce on canvas the section that appeared on the glass screen.

Two woodcuts by the German painter Albrecht Dürer illustrate this principle of projection and section. In "The Designer of the Sitting Man," the artist is about to mark on a glass screen a point where one of the light rays from the scene to the artist's eye intersects the screen. The second woodcut, "The Designer of the Lute," shows the section marked out on the glass screen. The reader can visualize the entire process by looking at a scene through a window and noting where the lines from his eye to each point of the scene pass through the window. The proper section is then marked out on the window.

Please refer to Figures 4 and 5 on pages 610 and 611 in the preceding selection.

Of course the section depends not only upon where the artist stands but also where the glass screen is placed between the eye and scene. To be sure this means that there can be many different portrayals of the same scene. What matters is that when he has chosen his scene, his position, and the position of the glass screen the painter's task is to put on canvas precisely what the section contains. Since the artist's canvas is not transparent and since the scenes he paints sometimes exist only in his imagination, the principle of projection and section merely tells the artist what his canvas should contain, not how to put it there. For this practical task the Renaissance artists had to derive theorems which would specify how a scene would appear on the imaginary glass screen (the location, sizes and shapes of objects) so that it could be put on canvas. These theorems are, in a logical sense at least, part of Euclidean geometry, in that they are deduced from the axioms and theorems of Euclid and from the principle that the canvas must contain a section of the projection. We need not discuss these theorems here. They are incorporated in all modern textbooks on perspective prepared for students of painting.

The theorems raised questions which proved to be momentous for mathematics. Professional mathematicians took over the investigation of these questions and developed a geometry of great generality and power. Its name is projective geometry.

We have already observed that the section which appears on a glass screen depends upon the position of the viewer and the position of the screen. Suppose, for example, that the object under scrutiny is a square. If this figure is viewed from a point somewhat to the side of the square and if a glass screen is interposed between the eye and the object (Figure 1), the section on the screen will no longer be a square but an odd-shaped quadrilateral. For example, the floor tiles in Raphael's painting are not squares though the physical tiles he painted were squares. Moreover, if the position of the screen is changed but the position of the viewer kept fixed, the section on the screen changes but not *the impression created by the section on the eye*. Likewise, various sections of the projection of the

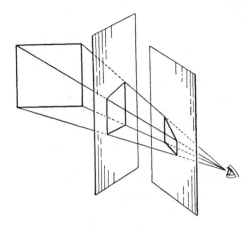

FIGURE 1

circle viewed from a fixed position differ considerably—they may be more or less flattened ellipses—but the impression created by all these sections on the eye will still be that created by the original circle at that fixed position.

To the intellectually curious mathematicians this phenomenon raised a question: should not the various sections presenting the same impression to the eye have certain geometrical properties in common? For that matter, should not sections of an object viewed from different positions also have some properties in common, since they all derive from the same object? In other words, the mathematicians were stimulated to seek geometrical properties common to all sections of the same projection and to sections of two different projections of a given scene. This problem is essentially the one that has been the chief concern of projective geometers in their development of the subject.

It is evident that just as the shape of a square or a circle varies in different sections of the same projection or in different projections of the figure, so also will the length of a line segment, the size of an angle or the size of an area. More than that, lines which are parallel in a physical scene are not parallel in a painting of it, but meet in one point; see, for example, the lines of the ceiling beams in Da Vinci's "The Last Supper," or the lines which follow the edges of the floor tiles from front to rear in Raphael's "School of Athens." Since neither length, angle, area, nor parallelism are transferred unaltered from section to section, we can see at once that two sections of the same or different projections of the same object cannot be congruent, similar, or equivalent—that is, have equal areas. In other words, the study of properties common to these various sections does not seem to lie within the province of the geometry concerned with these relationships—namely, Euclidean geometry.

Yet some rather simple properties that do carry over from section to section can at once be discerned. For example, a straight line will remain a line (that is, it will not become a curve) in all sections of all projections of it; a triangle will remain a triangle; a quadrilateral will remain a quadrilateral. This is not only intuitively evident but easily proved by Euclidean geometry. However, the discovery of these few fixed properties hardly elates the finder or adds appreciably to the structure and power of mathematics. Much deeper insight was required to obtain significant properties common to different sections.

The first man to supply such insight was Gérard Desargues, the self-educated architect and engineer, who worked during the first half of the seventeenth century. Desargues' motivation was to help the artists. He sought to combine the many theorems on perspective in compact form that would be useful to artists, engineers and stonecutters. He invented a special terminology which he thought would be more comprehensible than the language of mathematics, and he disseminated his findings through lectures and handbills. His interest in art extended to writing a book on how to teach children to sing well.

His chief result, still known as Desargues' theorem and still fundamental in the subject of projective geometry, states a significant property common to two sections of the same projection of a triangle. Desargues considered the following physical situation (Figure 2). Suppose that the eye at point O looks at triangle ABC. The lines of light from O to triangle

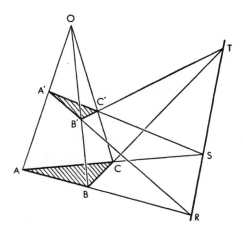

FIGURE 2

ABC constitute, as we know, a projection. If a section of this projection is made by a glass screen, the resulting figure cut out is triangle *A'B'C'*, where *A'* corresponds to *A*, *B'* to *B*, and *C'* to *C*. The relationship of the two triangles is described by saying that they are perspective from the point *O*. Desargues then asserted that each pair of corresponding sides of these two triangles will meet in a point, and, most important, these three points will lie on one straight line. With reference to the figure, the assertion is that *AB* and *A'B'* meet in a point, *R*, say, *AC* and *A'C'* meet in *S*, *BC* and *B'C'* meet in *T*, and *R*, *S*, and *T* lie on one straight line.

As stated, the theorem seems to describe a property which the original figure being viewed, triangle *ABC*, and a section, triangle *A'B'C'*, of a projection of this figure have in common. Alternatively we may regard triangles *ABC* and *A'B'C'* as two different sections of the projection of some third triangle (not shown in the figure) but such that *OAA'*, *OBB'*, and *OCC'* are lines of the projection of this third triangle from the point *O*. Moreover, as stated above, the two triangles *ABC* and *A'B'C'* are in different planes. However, as a theorem of pure mathematics Desargues' assertion holds even if triangles *ABC* and *A'B'C'* are in the same plane, e.g., the plane of this paper, though the proof of the theorem is different in the latter case.

The reader may be troubled about the assertion in Desargues' theorem that each pair of corresponding sides of the two triangles must meet in a point. He may ask: what about a case in which the sides happen to be parallel? Desargues disposed of such cases by invoking the mathematical convention that any set of parallel lines is to be regarded as having a point in common, which the student is often advised to think of as being at infinity—a bit of advice which essentially amounts to answering a question by not answering it. However, whether or not one can visualize this point at infinity is immaterial. It is logically possible to agree that parallel lines are to be regarded as having a point in common, which point is to be distinct from the usual finitely located points of the lines considered in Euclidean geometry. Hence we say of *any* two lines of projective geometry that they meet in one and only one point. In the case of parallel lines this common point is approached by traversing the parallel lines in either direction, so that the line of projective geometry has the structure of a circle, though it is not implied thereby that the length of the line is finite. We shall see later that length is an irrelevant concept in projective geometry.

One more agreement must be made about the new points which are introduced in projective geometry. The convention was adopted that any set of lines parallel to each other should have one point in common. Since there are many sets of parallel lines in a plane, there are many such points.

It is agreed that all the intersection points of the different sets of parallel lines in a given plane lie on one line, sometimes called the "line at infinity." Hence even if each of the three pairs of corresponding sides of the triangles involved in Desargues' theorem should consist of parallel lines, it would follow from our agreements that the three points of intersection lie on one line, the line at infinity.

These conventions or agreements not only are logically justifiable but also are recommended by the argument that projective geometry is concerned with problems which arise from the phenomenon of vision, and we never actually see parallel lines, as the familiar example of the apparently converging railroad tracks reminds us. Indeed, the property of parallelism plays no role in projective geometry.

Equally typical of theorems in projective geometry is one proved at the age of sixteen by the precocious French mathematician and philosopher Blaise Pascal. Pascal was a contemporary of Desargues and was urged by the latter to investigate the properties common to sections of projections. Pascal's major theorem has this to say: draw any six-sided polygon (a hexagon) with vertices on a circle and letter the vertices A, B, C, D, E, and F (Figure 3). Prolong a pair of opposite sides, AB and DE

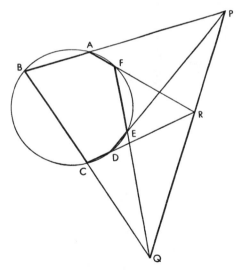

FIGURE 3

for example, until they meet in a point P. Prolong another pair of opposite sides, BC and EF, say, until they meet in a point Q. Finally, prolong the third pair of opposite sdes, CD and FA, until they meet in R. Then, Pascal asserts, P, Q, and R will lie on a straight line. (In words the theorem asserts that if the opposite sides of any hexagon inscribed in a circle are prolonged, the three points at which the extended pairs of lines meet will lie on a straight line.)

As stated, Pascal's theorem seems to have no bearing on the subject of projection and section. However let us visualize a projection of the figure involved in Pascal's theorem and then visualize a section of this projection (Figure 4). The projection of the circle is a cone, and in general a section of this cone will not be a circle but an ellipse, a hyperbola, or a parabola —that is, one of the curves usually called a conic section. Moreover, the

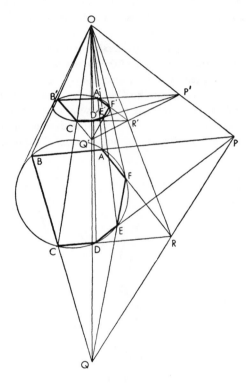

FIGURE 4

hexagon in the original circle will give rise to a hexagon inscribed in the conic section, and the remaining lines of the original figure will go over into lines in the new figures. Most important, each pair of opposite sides of the new hexagon will meet in a point and the three points of intersection will lie on one straight line—namely, the line which corresponds in the section to the line *PQR* of the original figure. Thus the theorem states a property of a circle which continues to hold in any section of any projection of that circle; it is clearly a theorem of projective geometry.

The theorems of Desargues and Pascal show that there are significant properties common to sections of any projection of a given figure. Thus these men gave some answers to the questions raised by the work of the painters. It would be pleasant to relate that the innovations of Desargues and Pascal were immediately appreciated by their fellow mathematicians and that the potentialities in their methods and ideas were eagerly seized upon and further developed. Actually this pleasure is denied us. Perhaps Desargues' novel terminology baffled mathematicians of his day, just as many people today are baffled and repelled by the language of mathematics. At any rate, all of Desargues' colleagues except René Descartes exhibited the usual reaction to radical ideas: they called Desargues crazy and dismissed projective geometry. Desargues himself became discouraged and returned to the practice of architecture and engineering. Every printed copy of Desargues' book, originally published in 1639, was lost.

Abraham Bosse, a pupil and friend of Desargues, published a book in 1648, *The Universal Method of Desargues for the Practice of Perspective,* and in an appendix to this book he reproduced Desargues' theorem and other of his results. Even this appendix was lost and was not rediscovered until 1804. Pascal's work on conics and his other work on projective geometry, published in 1640, likewise remained unknown until about 1800. Fortunately a pupil of Desargues, Phillippe de la Hire, made a manuscript copy of Desargues' book. In the nineteenth century this copy was picked up by accident in a bookshop by the geometer Michel Chasles, and thereby the world learned the full extent of Desargues' major work. In the meantime most of Desargues' and Pascal's discoveries had had to be remade independently by nineteenth-century geometers.

Another reason for the neglect of projective geometry during the seventeenth and eighteenth centuries was the fact that analytic geometry, created by Desargues' contemporaries, Descartes and Fermat, and the calculus, created chiefly by Newton and Leibniz, proved to be so useful in the rapidly expanding branches of physical science that mathematicians concentrated on these subjects. At any rate the study of projective geometry remained dormant for almost two hundred years. And then it was revived through a series of accidents and events almost as striking as those which gave rise to the subject.

The problem of designing fortifications attracted the geometrical talents of Gaspard Monge, the inventor of descriptive geometry. It is important to note that this subject, though distinct from projective geometry, uses projection and section. Monge was an inspiring teacher and gathered about him at the Ecole Polytechnique a host of bright pupils, among them Servois, Brianchon, Carnot and Poncelet. These men were greatly impressed by Monge's geometry. They sought to show that purely geometric methods could accomplish as much and more than the algebraic or analytical methods introduced by Descartes. Carnot in particular wished "to free geometry from the hieroglyphics of analysis." As if to take revenge on Descartes for the fact that his creation had caused the abandonment of pure geometry, the early nineteenth-century geometers made it their objective to beat Descartes.

It was Poncelet who revived projective geometry. As an officer in Napoleon's army during the invasion of Russia, he was captured and spent the year 1813–1814 in a Russian prison. There he reconstructed without the aid of any books all that he had learned from Monge and Carnot, and he then proceeded to create new results in projective geometry. He was perhaps the first mathematician to appreciate fully that this subject was indeed a totally new branch of mathematics. After he had reopened the subject, a whole group of French and, later, German mathematicians went on to develop it intensively.

One of the foundations on which they built was a concept already glimpsed by the Greeks but whose importance was not fully appreciated. Consider a section of the projection of a line divided by four points (Figure 5). Obviously the segments of the line in the section are not equal in length to those of the original line. One might venture that perhaps the

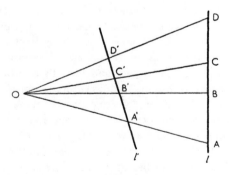

FIGURE 5

ratio of two segments, say, $A'C'/B'C'$, would equal the corresponding ratio AC/BC. This conjecture is incorrect. But the surprising fact is that the ratio of the ratios—namely $(A'C'/C'B')/(A'D'/D'B')$—will equal $(AC/CB)/(AD/DB)$. Thus this ratio of ratios, or cross-ratio as it is called, is a projective invariant—that is, it is the same for any section of the projection of l from any point O. It is necessary to note only that the lengths involved must be directed lengths; that is, if the direction from A to D is positive, then the length AD is positive but the length DB must be taken as negative.

The fact that any line intersecting the four lines OA, OB, OC and OD contains segments possessing the same cross-ratio as the original segments suggests that we assign to the four projection lines meeting in the point O a particular cross-ratio—namely the cross-ratio of the segments on any section. Moreover, the cross-ratio of the four lines is a projective invariant; that is, if a projection of these four lines is formed and a section made of this projection, the section will contain four concurrent lines whose cross-ratio is the same as that of the original four (Figure 6). Here in the section $O'A'B'C'D'$, formed in the projection of the figure $OABCD$ from the point O'', the four lines $O'A'$, $O'B'$, $O'C'$ and $O'D'$ have the same cross-ratio as OA, OB, OC and OD.

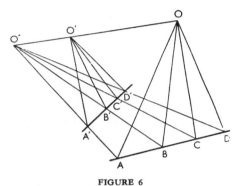

FIGURE 6

The projective invariance of cross-ratio was put to extensive use by the nineteenth-century geometers. We noted earlier in connection with Pascal's theorem that under projection and section a circle may become an ellipse, a hyperbola or a parabola—that is, any one of the conic sections. Now the fact that the figures are transformed is not surprising since projection and section distort the distance relations within a given figure. But it was astonishing that despite the distortion, conics under projection and section always give rise to conics. The geometers sought a common property which would account for this fact. They found the answer in terms of cross-ratio. Given a point O and four points A, B, C, D, the lines OA, OB, OC, and OD have a definite cross-ratio as noted on next page (Figure 7). If P is any other point on a conic section containing O, A, B, C, and D, then a remarkable theorem of projective geometry states that the lines PA, PB, PC, and PD have the same cross-ratio as OA, OB, OC, and OD. Conversely, if P is any point such that PA, PB, PC, and PD have the same cross-ratio as OA, OB, OC, and OD, then P must lie on the conic through O, A, B, C, and D. The essential point of this theorem and its converse is that a conic section is determined by the property of cross-ratio. This new characterization of a conic was most welcome not only because it utilized a projective property but also because it opened up a whole new line of investigation on the theory of conics.

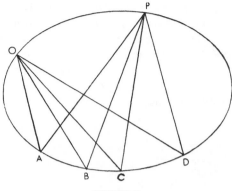

FIGURE 7

The satisfying accomplishments of projective geometry were capped
by the discovery of one of the most beautiful principles of all mathematics
—the principle of duality. It is true in projective geometry, as in Euclidean geometry, that any two points determine one line or, as we prefer to
put it, *any two points lie on one line.* But it is also true in projective
geometry that *any two lines* determine, or *lie on, one point.* (The reader
who has refused to accept the convention that parallel lines in Euclid's
sense are also to be regarded as having a point in common will have to
forego the next few paragraphs and pay for his stubbornness.) It will be
noted that the second italicized statement can be obtained from the first
merely by interchanging the words point and line. We say in projective
geometry that we have dualized the original statement. Thus we can speak
not only of a set of points on a line but also of a set of lines on a point

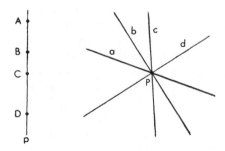

FIGURE 8

(Figure 8). Likewise the dual of the figure consisting of four points no
three of which lie on the same line is a figure of four lines no three of
which lie on the same point (Figure 9).

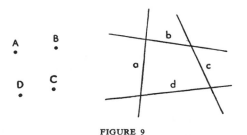

FIGURE 9

Let us attempt this rephrasing for a slightly more complicated figure. A triangle consists of three points not all on the same line and the lines joining these points. The dual statement would read: three lines not all on the same point and the points joining them—that is, the points in which the lines intersect. The figure we get by rephrasing the definition of a triangle is again a triangle, and so the triangle is called self-dual.

Now let us rephrase Desargues' theorem in dual terms, using the fact that the dual of a triangle is a triangle and assuming in this case that the two triangles and the point *O* lie in one plane.

Desargues' Theorem	*Dual of Desargues' Theorem*
If we have two triangles such that lines joining corresponding vertices pass through one point *O*, then the pairs of corresponding sides of the two triangles join in three points lying on one straight line.	If we have two triangles such that points which are the joins of corresponding sides lie on one line *O*, then the pairs of corresponding vertices of the two triangles are joined by three lines lying on one point.

We see that the dual statement is really the converse of Desargues' theorem; that is, it is the result of interchanging his hypothesis and his conclusion. Hence by interchanging point and line we have discovered the statement of a new theorem. It would be too much to ask that the proof of the new theorem should be obtainable from the proof of the old one by interchanging point and line. But if it is too much to ask, the gods have been generous beyond our merits, for the new proof can be obtained in precisely this way.

The principle of duality, as thus far described, tells us how to obtain a new statement or theorem from a given one involving points and lines. But projective geometry also deals with curves. How should one dualize a statement involving curves? The clue lies in the fact that a curve is, after all, but a collection of *points* satisfying a condition, as, for example, that involving cross-ratio. The principle of duality suggests then that the figure dual to a given curve might be a collection of *lines* satisfying the condition dual to the one defining the given curve. We arrive therefore at the notion of a curve consisting of a collection of lines, and the fact is such a collection suggests a curve as well as does a collection of points.

In the case of the conic sections, the dual figure to a point conic—that is, a conic regarded as a collection of points—turns out to be the collection of tangents to that point conic (Figure 10). If the conic section is a

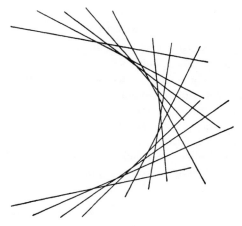

FIGURE 10

circle, the dual figure is the collection of tangents to that circle (Figure 11). This collection of tangents is called the line circle.

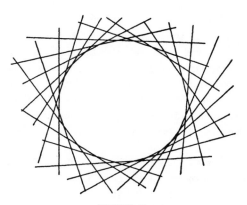

FIGURE 11

Now we can see whether the application of the principle of duality to curves is as suggestive as it was in the case of simpler figures. As a test we can dualize Pascal's theorem.

Pascal's Theorem	*Dual of Pascal's Theorem*
If we take six points, *A, B, C, D, E,* and *F* on the point circle, then the lines which join *A* and *B* and *D* and *E* join in a point *P*; the lines which join *B* and *C* and *E* and *F* join in a point *Q*; the lines which join *C* and *D* and *F* and *A* join in a point *R*. The three points *P, Q,* and *R* lie on one line *l*.	If we take six lines, *a, b, c, d, e,* and *f* on the line circle, then the points which join *a* and *b* and *d* and *e* are joined by the line *p*; the points which join *b* and *c* and *e* and *f* are joined by the line *q*; the points which join *c* and *d* and *f* and *a* are joined by the line *r*. The three lines *p, q,* and *r* lie on one point *L*.

The geometric meaning of the dual statement is as follows: since the line circle is the collection of tangents to the point circle, the six lines on the line circle are any six tangents to the point circle, and these six tangents form a hexagon circumscribed about the point circle. Hence the dual statement tells us that if we circumscribe a hexagon about a point circle, the lines joining opposite vertices of the hexagon, lines *p, q,* and *r* in the dual statement, meet in one point (Figure 12). This dual statement is indeed a theorem of projective geometry. It is called Brianchon's

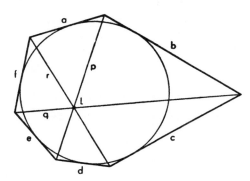

FIGURE 12

theorem after Monge's student, Charles Brianchon, who discovered it by applying the principle of duality to Pascal's theorem pretty much as we have done.

The principle of duality in projective geometry states that we can interchange point and line in a theorem about figures lying in one plane and obtain a meaningful statement. Moreover, the new or dual statement will itself be a theorem—that is, it can be proven. On the basis of what has been presented here we cannot see why this must always be the case for the dual statement. However, it is possible to show by one proof that every rephrasing of a theorem of projective geometry in accordance with the principle of duality must lead to a theorem. This principle is a remarkable characteristic of projective geometry. It reveals the symmetry in the roles that point and line play in the structure of that geometry. The principle of duality also gives us insight into the process of creating mathematics. Whereas the discovery of this principle, as well as of theorems such as Desargues' and Pascal's, calls for imagination and genius, the discovery of new theorems by means of the principle is an almost mechanical procedure.

The subject of projective geometry has many more exciting concepts to offer than we can consider here. I should mention that there is an analytical projective geometry just as there is the analytical Euclidean geometry due to Descartes and Fermat. The proofs which can be made by algebraic methods in projective geometry are among the most elegant in mathematics. However, the subject of projective geometry possesses significance which extends far beyond the concepts, methods, and theorems of the subject proper.

The properties which are invariant under projection and section deal with the collinearity of points (that is, when points lie on the same line), with the concurrence of lines (that is, when a set of lines meet in one point), with cross-ratio, and with the fundamental roles of point and line as exhibited by the principle of duality. On the other hand, Euclidean geometry deals with the equality of lengths, angles, areas. A comparison of these two classes of properties suggests that projective properties are simpler than those treated in Euclidean geometry. One might say that projective geometry deals with the very formation of the geometrical figures whose congruence, similarity and equivalence (equality of areas) are studied in Euclid. In other words, projective geometry is more fundamental than Euclidean geometry.

The clue to the relationship between the two geometries may be obtained by again considering projection and section. Consider the projection of a rectangle and a section in a plane parallel to the rectangle (Figure 13). The section is a rectangle similar to the original one. If now the

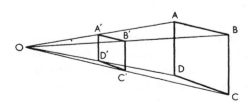

FIGURE 13

point *O* moves off indefinitely far to the left, the lines of the projection come closer and closer to parallelism with each other. When these lines become parallel and the center of the projection is the "point at infinity," the rectangles become not merely similar but congruent (Figure 14). In other words, from the standpoint of projective geometry the relationships

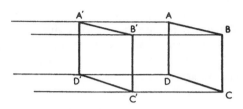

FIGURE 14

of congruence and similarity, which are so intensively studied in Euclidean geometry, can be studied through projection and section for special projections. We see, therefore, that Euclidean geometry is not only a logical subdivision of projective geometry, but one that can be looked upon in a new light—namely, as a study of geometric properties which are invariant under special projections.

If projective geometry is indeed logically fundamental to Euclidean geometry, then all the concepts of the latter geometry should be defined in terms of projective concepts. However, in projective geometry as described so far there is a logical blemish: our definition of cross-ratio, and hence concepts based on cross-ratio, rely on the notion of length, which should play no role in projective geometry proper because length is not an invariant under arbitrary projection and section. In turn the projective concepts such as the definition of a conic are faulty because they depend upon cross-ratio. This blemish was removed by the nineteenth-century geometer Karl G. C. von Staudt, who showed how to define cross-ratio in terms of projective concepts. Felix Klein then showed how to define length as well as the size of angles entirely in terms of cross-ratio. Hence it became possible to affirm that projective geometry is indeed logically prior to Euclidean geometry and that the latter can be built up as a special case. Both Klein and Arthur Cayley showed that the basic non-Euclidean geometries developed by Lobachevsky and Bolyai and the elliptic non-Euclidean geometry created by Riemann can also be derived as special cases of projective geometry. No wonder that Cayley exclaimed, "Projective geometry is all geometry."

It remained only to deduce the theorems of Euclidean and non-Euclidean geometry from axioms of projective geometry, and this geometers succeeded in doing in the late nineteenth and early twentieth centuries. What Euclid did to organize the work of three hundred years preceding his time, the projective geometers did recently for the investigations which Desargues and Pascal initiated.

Research in projective geometry is continuing today to a limited extent. Geometers are seeking to find simpler axioms and more elegant proofs. Some research is concerned with projective geometry in n-dimensional space. A vast new allied field is projective differential geometry, concerned with local or infinitesimal properties of curves and surfaces.

Projective geometry has had an important bearing on current mathematical research in several other fields. The relationship of projective geometry to Euclidean geometry and the non-Euclidean geometries, which was first divined and worked out during the latter half of the nineteenth century, suggested a totally new approach to geometry. Projection and section amount to what is called a transformation; that is, one starts with a given figure, forms a projection from some central point, and obtains a section of this projection. The entire process which takes one from the original figure to the section is the transformation, and projective geometry seeks invariants under this transformation. The question then suggested itself: are there other transformations more general than projection and section whose invariants might be studied? In recent times one new geometry has been developing by pursuing this line of thought—namely, topology. It would take us too far afield to consider topological transformations and to learn of the invariants which have been discovered in this branch of geometry. It must suffice here to state that topology considers transformations more general than projection and section and that it is now clear that topology is logically prior to projective geometry. Cayley was too hasty in affirming that projective geometry is all geometry.

The work of the projective geometers has had an important influence on modern physical science. When mathematical physicists working in the theory of relativity recognized that the laws of the universe may vary from one observer to another in a much more startling manner than had been suspected before the Michelson-Morley experiment, they were led to emphasize the concept of invariance of scientific laws under transformation from the co-ordinate system of one observer to that of another. This emphasis on invariance of scientific laws came naturally to the mathematical physicists because the concept of invariance had already played a prominent role in the work of the projective geometers.

Further, when physicists sought a mathematical way of expressing scientific laws which would exhibit their independence of co-ordinate systems, they found that the projective geometers had prepared the way for them in this respect too. During the latter half of the nineteenth century the projective geometers employed algebraic methods to expedite the search for invariants just as Descartes used algebra to advance the study of curves in Euclidean geometry. In algebraic terms the properties of geometrical figures are algebraic expressions, and the projective transformation from section to section is a change from one co-ordinate system to another. Under such co-ordinate transformations the projective invariants retain their algebraic form. Hence the projective geometers and other mathematicians studied the theory of invariance of algebraic forms under change of co-ordinate systems. The significant development in this direction which proved so useful to the physicists is the theory of tensors, or, as it is sometimes called, the calculus of tensors. This calculus proved to be the most convenient means for expressing scientific laws in a manner satisfying the requirement that they be invariant under change of co-ordinate system.

Thus, the projective geometers initiated the study of concepts and techniques employed in the modern theory of relativity, a contribution totally unforeseen by even the nineteenth-century projective geometers, to say nothing about Desargues and Pascal. In just this manner, however, does the purest of mathematical research advance science.

It is of course true that other branches of mathematics, notably differential equations, have contributed more to the advancement of science than has projective geometry. But no branch of mathematics competes with projective geometry in originality of ideas, co-ordination of intuition in discovery and rigor in proof, purity of thought, logical finish, elegance of proofs and comprehensiveness of concepts. The science born of art proved to be an art.

COMMENTARY ON
HERMANN VON HELMHOLTZ

HERMANN VON HELMHOLTZ was a scientist of such prodigious power and range that the mere listing of the subjects to which he made major contributions provokes disbelief. He was trained as a physician but practical medicine was the field that least interested him.[1] Instead, he undertook investigations in, among others, physiology, physiological optics and acoustics, electricity and magnetism, thermodynamics, theoretical mechanics, hydrodynamics, physical optics, the theory of heat, chemistry, mathematics, meteorology, biology and psychology. He was creative and courageous in philosophical outlook, a popularizer of first rank, a tolerant, elevated and rational man. All that was admirable in nineteenth-century German culture found expression in his life and work.

Helmholtz was born at Potsdam in 1821, the son of a gymnasium teacher of philology and philosophy, and of a "profoundly emotional" and intelligent mother descended from William Penn. He was an ailing child, thoughtful, precocious, and, under his father's influence, much given to study. The salary of a high-school teacher, even one who had the formidable title of Professor by Royal Patent, was small, and it was necessary therefore to enroll Hermann, who wanted to study physics, as a scholarship student in a school which turned out surgeons for the Prussian army. In return for the scholarship he was bound after graduation to give several years of military service. It was a stiff course of study, but Helmholtz thrived on it. There were forty-eight lectures a week ranging from splanchnology and physics to Latin and logic; the first class met at six in the morning; the afternoon was occupied by review classes and experimental work. Helmholtz managed, nevertheless, to play sonatas of Mozart and Beethoven, to attend the theater, to read Goethe and Byron "and sometimes for a change the integral calculus." [2]

In 1842 he received his M.D. degree and from 1843 to 1848 did his stint as a "Hussar Surgeon." The work was not exhausting and left Helmholtz with sufficient free time to pursue his mathematical reading, and, with the help of his good friends Ernest Brücke and Emil Du Bois-Reymond, both of whom later gained fame as physiologists, to set up in his barracks a small laboratory for physics and physiology. One of the

---

[1] A brief note on hay fever, from which he was a chronic sufferer, is the only medical paper by Helmholtz I have found.

[2] From a letter to his parents quoted in Leo Koenigsberger, *Hermann von Helmholtz*, Braunschweig, 1902; translated into English by Francis A. Welby, Oxford, 1906, p. 17. This is the standard biography on which a good deal of the above account is based. Page references below are to Welby's translation.

fruits of this establishment, a report on the *Theory of Animal Heat* (1845), foreshadowed his epochal paper on the conservation of energy read to the Physical Society of Berlin in 1847. To the elaboration of the principle that the matter and energy of the universe can neither be created nor destroyed—a principle which twentieth-century physics has severely mauled—Helmholtz devoted many years of his life.[3]

Having obtained a release from further military duty, Helmholtz was in 1849 appointed professor of physiology in Königsberg. There he remained until, in 1855, he took the chair of physiology at Bonn; from Bonn he removed in 1858 to Heidelberg and thirteen years later was called to the physics professorship at Berlin. Without relinquishing the Berlin chair, he assumed in 1887 the presidency of the newly formed physico-technical institute at Charlottenburg. He held these two positions until he died of a cerebral hemorrhage on September 8, 1894.

For more than fifty years, beginning with his inaugural dissertation announcing the discovery that the nerve fibers originate in the ganglion cells, a stream of scientific papers, many of the highest quality, flowed without interruption from Helmholtz's laboratory and study. During his tenure at Königsberg he experimented and wrote on the contraction of animal muscles, the rate of transmission of nerve impulses, the measurement of time intervals. He brought to bear on every problem the sharpest tools of applied and theoretical science, and if these were inadequate he often invented new ones. In 1851 he revolutionized the study of physiological optics by inventing the ophthalmoscope, a brilliantly ingenious instrument—he called it "this very precious egg of Columbus"—with which "it is possible to illuminate the dark background of the eye, through the pupil, without employing any dazzling light, and to obtain a view of all the elements of the retina at once." [4] The ophthalmoscope, like so many of Helmholtz's scientific advances, represented a synthesis of his experimental skill, his knowledge of physiology, his command of mathematics and physics. He seemed to possess a special mechanical sense "by means of which I almost feel how the stress and strain are distributed in a mechanical contrivance."

The monumental treatise on *Physiological Optics* appeared in installments from 1856 to 1866. Sandwiched in between these dates were his wonderful book, *Sensations of Tone* (1862), which has been called the

---

[3] The German physician Dr. Julius Robert Mayer had, in more restricted form, anticipated Helmholtz as regards the conservation principle—a fact which the latter always insisted, generously, on bringing to public notice. Others who contributed to its development were James Prescott Joule and Lord Kelvin—for many years Helmholtz's close friend.

[4] Koenigsberger, *op. cit.*, p. 74, quoting Helmholtz's description, "When the great ophthalmologist A. von Gräfe first saw the fundus of the living human eye, with its optic disc and blood vessels, his face flushed with excitement, and he cried, 'Helmholtz has unfolded to us a new world.'" *Encyclopaedia Britannica,* Eleventh Edition, article on Helmholtz.

Principia of physiological acoustics; researches and papers on the mechanism of the ear, pitch, the structure and perception of musical tone, perceptions of space, the sensations of sight and color vision, the duration and nature of induced electrical currents, animal electricity, the distribution of electrical currents in material conductors, "aerial vibrations in tubes with open ends," friction in fluids, the Persian and Arabian musical scales; and various philosophical discourses and popular lectures delivered on the Continent and in England. "My brain reels," Du Bois-Reymond wrote him on receiving several of Helmholtz's papers in a single post, "at your appalling industry and encyclopaedic knowledge."

At times even Helmholtz's brain reeled. He suffered frequently from severe migraine headaches, and periodically from complete exhaustion. Short vacations in Switzerland, Italy and at various watering places invariably restored his health. On trips to England he combined business with pleasure, seeking out the leaders of science, Tyndall, Huxley, Kelvin, Airy, Stokes, among them, and exchanging ideas on problems of common concern. A letter to his wife describes the "splendid moments" of a visit to Faraday. "He is as simple, charming and unaffected as a child; I have never seen a man with such winning ways. He was, moreover, extremely kind, and showed me all there was to see. That, indeed, was little enough, for a few wires and some old bits of wood and iron seem to serve him for the greatest discoveries." [5]

After completion of the *Physiological Optics*, Helmholtz inclined more and more to the investigation of problems in mathematical physics, and pure mathematics. He studied the phenomena of electrical oscillations, the principles of chemical and physical equilibrium, the kinetic-molecular theory of gases and the distribution of energy in mechanical systems, the origins of the solar system and the age of the sun (of which he made a famous estimate). One difficult experiment, measuring the length of electric waves, which he turned over to his favorite pupil Heinrich Hertz, led to a discovery which changed the world. Using an oscillating current given by the spark of an induction coil, Hertz succeeded in producing what Clerk Maxwell had theoretically predicted, electromagnetic waves (i.e., "wireless waves") possessing all the properties of light except visibility. None was happier over Hertz's achievement than Helmholtz.

In 1870, while at Heidelberg, Helmholtz gave a popular lecture, "On the Origin and Significance of Geometrical Axioms." This lecture, which appears as the next selection, offers to nonmathematicians the results of Helmholtz's reflections on non-Euclidean geometry, embodied earlier in an essay published by the Göttingen Scientific Society. Helmholtz's interest in this fundamental aspect of mathematical thought was stimulated by the work of Lobachevsky, Gauss and Bolyai; but chiefly by Riemann's

[5] Koenigsberger, *op. cit.*, pp. 111–112.

imperishable doctoral dissertation "On the Hypotheses That Underlie Geometry." The entire tone of the lecture, the attitude it expresses towards geometry is modern, and apposite, as E. A. Milne observed, to the science of today.[6] Helmholtz's conclusions were in line with his general empiricist, antimetaphysical philosophy.[7] He rejected Kant's view that the properties of space are integral parts of our understanding, determined by the "given form of our capacity of intuition." All we know about space, he said, is what we have learned from experience. If we lived in a spherical or pseudo-spherical space our sensible impressions of the world would dictate the adoption of the non-Euclidean geometries of Riemann or Lobachevsky; nothing in our intuition would require us to adopt a "flat-space" Euclidean system. The axioms of our geometry are, to be sure, partially suggested by experience; but definitions have crept in which have to do with ideal figures "to which the material figures of the actual world can only approximate." Thus the only test of the validity of the axioms in use is observation and measurement. If these procedures should prove the axioms incorrect, we should have to discard them; that is to say, discard them as accurate models of the physical world, though they might still serve as interesting objects of study for pure mathematicians, who have no interest in the "real" world.

One is struck by the resemblance between the opinions of Helmholtz, Clifford (see p. 548 *et seq.*) and, later, Poincaré.[8] Each of these men emphasized with admirable clarity the folly of attempting to deduce the "true" system of geometry from metaphysical postulates. Each regarded the choice

[6] E. A. Milne, *Sir James Jeans*, Cambridge, 1952, p. 82.

[7] Helmholtz distinguished sharply between philosophy and metaphysics. "In my opinion nothing has been so pernicious to philosophy as its repeated confusion with metaphysics. The latter has played much the same part in relation to the former as that which astrology has borne to astronomy. It has been metaphysics that turned the attention of the great majority of scientific amateurs to philosophy, and attracted troops of proselytes and disciples, who no doubt in many cases have wrought more harm than the bitterest opponents could have effected. They were led on by the delusive hope of obtaining insight, with little expenditure of time or trouble, into the deepest order of things and the nature of the human spirit, into the past and future of the world—in which lay the main interest that incited so many to take up the study of philosophy, just as the hope of prognostications for the future formerly led to the fostering of astronomy. What philosophy has so far been able to teach us, or with continued study of the facts involved, may one day be able to teach us, is of the utmost importance to the scientific thinker, who must know the exact capabilities of the instrument with which he is to work, that is, the human intellect. But as regards the satisfaction of this dilettante curiosity, or the still more frequent egoism of the individual, these severe and abstract studies will continue to yield only a small and reluctant response: just as the mathematical mechanics of the planetary system and the calculations of perturbation, are far less popular, despite their admirable systematic completeness, than was the astrological superstition of old days." Quoted in Koenigsberger, *op. cit.*, p. 427.

[8] "His [Helmholtz's] conclusions were in line with later views of Poincaré, namely that the axioms of geometry, including the axioms of non-Euclidean geometries, are compatible with any geometrical content whatever, but that, as soon as a framework has to be found for the principles of mechanics, we obtain a system of propositions which has real import, capable of verification or disproof by empirical observation." Milne, *op. cit.*, p. 82.

of the axioms of geometry as a matter, solely, of convenience and usefulness. Each showed that "it is possible to discuss mathematically the properties of non-Euclidean space, irrespective of the answer to the question whether such space is known to the senses." Each, in sum, labored to establish the principle of unlimited mathematical freedom. It is one of the agreeable paradoxes of the history of science, that the emancipation of geometry from purely physical considerations led to researches of the highest physical importance when Einstein formulated the theory of relativity.[9]

[9] See Sir William Cecil Dampier, *A History of Science*, Cambridge, 1949, p. 203.

Helmholtz—the physiologist who learned physics for the sake of his physiology, and mathematics for the sake of his physics, and is now in the first rank of all three.
                                              —WILLIAM KINGDON CLIFFORD

8   On the Origin and Significance of Geometrical Axioms

By HERMANN VON HELMHOLTZ

THE fact that a science can exist and can be developed as has been the case with geometry, has always attracted the closest attention among those who are interested in questions relating to the bases of the theory of cognition. Of all branches of human knowledge, there is none which, like it, has sprung as a completely armed Minerva from the head of Jupiter; none before whose death-dealing Aegis doubt and inconsistency have so little dared to raise their eyes. It escapes the tedious and troublesome task of collecting experimental facts, which is the province of the natural sciences in the strict sense of the word; the sole form of its scientific method is deduction. Conclusion is deduced from conclusion, and yet no one of common sense doubts but that these geometrical principles must find their practical application in the real world about us. Land surveying, as well as architecture, the construction of machinery no less than mathematical physics, are continually calculating relations of space of the most varied kind by geometrical principles; they expect that the success of their constructions and experiments shall agree with these calculations; and no case is known in which this expectation has been falsified, provided the calculations were made correctly and with sufficient data.

Indeed, the fact that geometry exists, and is capable of all this, has always been used as a prominent example in the discussion on that question, which forms, as it were, the centre of all antithesis of philosophical systems, that there can be a cognition of principles destitute of any bases drawn from experience. In the answer to Kant's celebrated question, 'How are synthetical principles *a priori* possible?' geometrical axioms are certainly those examples which appear to show most decisively that synthetical principles are *a priori* possible at all. The circumstance that such principles exist, and force themselves on our conviction, is regarded as a proof that space is an *a priori* mode of all external perception. It appears thereby to postulate, for this *a priori* form, not only the character of a purely formal scheme of itself quite unsubstantial, in which any given result experience would fit; but also to include certain peculiarities of the scheme, which bring it about that only a certain content, and one which, as it were, is strictly defined, could occupy it and be apprehended by us.

It is precisely this relation of geometry to the theory of cognition which emboldens me to speak to you on geometrical subjects in an assembly of those who for the most part have limited their mathematical studies to the ordinary instruction in schools. Fortunately, the amount of geometry taught in our gymnasia will enable you to follow, at any rate the tendency, of the principles I am about to discuss.

I intend to give you an account of a series of recent and closely connected mathematical researches which are concerned with the geometrical axioms, their relations to experience, with the question whether it is logically possible to replace them by others.

Seeing that the researches in question are more immediately designed to furnish proofs for experts in a region which, more than almost any other, requires a higher power of abstraction, and that they are virtually inaccessible to the non-mathematician, I will endeavour to explain to such a one the question at issue. I need scarcely remark that my explanation will give no proof of the correctness of the new views. He who seeks this proof must take the trouble to study the original researches.

Anyone who has entered the gates of the first elementary axioms of geometry, that is, the mathematical doctrine of space, finds on his path that unbroken chain of conclusions of which I just spoke, by which the ever more varied and more complicated figures are brought within the domain of law. But even in their first elements certain principles are laid down, with respect to which geometry confesses that she cannot prove them, and can only assume that anyone who understands the essence of these principles will at once admit their correctness. These are the so-called axioms.

For example, the proposition that if the shortest line drawn between two points is called a *straight* line, there can be only one such straight line. Again, it is an axiom that through any three points in space, not lying in a straight line, a plane may be drawn, i.e. a surface which will wholly include every straight line joining any two of its points. Another axiom, about which there has been much discussion, affirms that through a point lying without a straight line only one straight line can be drawn parallel to the first; two straight lines that lie in the same plane and never meet, however far they may be produced, being called parallel. There are also axioms that determine the number of dimensions of space and its surfaces, lines and points, showing how they are continuous; as in the propositions, that a solid is bounded by a surface, a surface by a line and a line by a point, that the point is indivisible, that by the movement of a point a line is described, by that of a line a line or a surface, by that of a surface a surface or a solid, but by the movement of a solid a solid and nothing else is described.

Now what is the origin of such propositions, unquestionably true yet incapable of proof in a science where everything else is reasoned conclusion? Are they inherited from the divine source of our reason as the idealistic philosophers think, or is it only that the ingenuity of mathematicians has hitherto not been penetrating enough to find the proof? Every new votary, coming with fresh zeal to geometry, naturally strives to succeed where all before him have failed. And it is quite right that each should make the trial afresh; for, as the question has hitherto stood, it is only by the fruitlessness of one's own efforts that one can be convinced of the impossibility of finding a proof. Meanwhile solitary inquirers are always from time to time appearing who become so deeply entangled in complicated trains of reasoning that they can no longer discover their mistakes and believe they have solved the problem. The axiom of parallels especially has called forth a great number of seeming demonstrations.

The main difficulty in these inquiries is, and always has been, the readiness with which results of everyday experience become mixed up as apparent necessities of thought with the logical processes, so long as Euclid's method of constructive intuition is exclusively followed in geometry. It is in particular extremely difficult, on this method, to be quite sure that in the steps prescribed for the demonstration we have not involuntarily and unconsciously drawn in some most general results of experience, which the power of executing certain parts of the operation has already taught us practically. In drawing any subsidiary line for the sake of his demonstration, the well-trained geometer always asks if it is possible to draw such a line. It is well known that problems of construction play an essential part in the system of geometry. At first sight, these appear to be practical operations, introduced for the training of learners; but in reality they establish the existence of definite figures. They show that points, straight lines, or circles such as the problem requires to be constructed are possible under all conditions, or they determine any exceptions that there may be. The point on which the investigations turn, that we are about to consider, is essentially of this nature. The foundation of all proof by Euclid's method consists in establishing the congruence of lines, angles, plane figures, solids, etc. To make the congruence evident, the geometrical figures are supposed to be applied to one another, of course without changing their form and dimensions. That this is in fact possible we have all experienced from our earliest youth. But, if we proceed to build necessities of thought upon this assumption of the free translation of fixed figures, with unchanged form, to every part of space, we must see whether the assumption does not involve some presupposition of which no logical proof is given. We shall see later on that it does indeed contain one of the most serious import. But if so, every proof by congruence rests upon a fact which is obtained from experience only.

I offer these remarks, at first only to show what difficulties attend the complete analysis of the presuppositions we make, in employing the common constructive method. We evade them when we apply, to the investigation of principles, the analytical method of modern algebraical geometry. The whole process of algebraical calculation is a purely logical operation; it can yield no relation between the quantities submitted to it that is not already contained in the equations which give occasion for its being applied. The recent investigations in question have accordingly been conducted almost exclusively by means of the purely abstract methods of analytical geometry.

However, after discovering by the abstract method what are the points in question, we shall best get a distinct view of them by taking a region of narrower limits than our own world of space. Let us, as we logically may, suppose reasoning beings of only two dimensions to live and move on the surface of some solid body. We will assume that they have not the power of perceiving anything outside this surface, but that upon it they have perceptions similar to ours. If such beings worked out a geometry, they would of course assign only two dimensions to their space. They would ascertain that a point in moving describes a line, and that a line in moving describes a surface. But they could as little represent to themselves what further spatial construction would be generated by a surface moving out of itself, as we can represent what would be generated by a solid moving out of the space we know. By the much-abused expression 'to represent' or 'to be able to think how something happens' I understand —and I do not see how anything else can be understood by it without loss of all meaning—the power of imagining the whole series of sensible impressions that would be had in such a case. Now as no sensible impression is known relating to such an unheard-of event, as the movement to a fourth dimension would be to us, or as a movement to our third dimension would be to the inhabitants of a surface, such a 'representation' is as impossible as the 'representation' of colours would be to one born blind, if a description of them in general terms could be given to him.

Our surface-beings would also be able to draw shortest lines in their superficial space. These would not necessarily be straight lines in our sense, but what are technically called *geodetic lines* of the surface on which they live; lines such as are described by a *tense* thread laid along the surface, and which can slide upon it freely. I will henceforth speak of such lines as the *straightest* lines of any particular surface or given space, so as to bring out their analogy with the straight line in a plane. I hope by this expression to make the conception more easy for the apprehension of my non-mathematical hearers without giving rise to misconception.

Now if beings of this kind lived on an infinite plane, their geometry would be exactly the same as our planimetry. They would affirm that only one straight line is possible between two points; that through a third point lying without this line only one line can be drawn parallel to it; that the ends of a straight line never meet though it is produced to infinity, and so on. Their space might be infinitely extended, but even if there were limits to their movement and perception, they would be able to represent to themselves a continuation beyond these limits; and thus their space would appear to them infinitely extended, just as ours does to us, although our bodies cannot leave the earth, and our sight only reaches as far as the visible fixed stars.

But intelligent beings of the kind supposed might also live on the surface of a sphere. Their shortest or straightest line between two points would then be an arc of the great circle passing through them. Every great circle, passing through two points, is by these divided into two parts; and if they are unequal, the shorter is certainly the shortest line on the sphere between the two points, but also the other or larger arc of the same great circle is a geodetic or straightest line, i.e. every smaller part of it is the shortest line between its ends. Thus the notion of the geodetic or straightest line is not quite identical with that of the shortest line. If the two given points are the ends of a diameter of the sphere, every plane passing through this diameter cuts semicircles, on the surface of the sphere, all of which are shortest lines between the ends; in which case there is an equal number of equal shortest lines between the given points. Accordingly, the axiom of there being only one shortest line between two points would not hold without a certain exception for the dwellers on a sphere.

Of parallel lines the sphere-dwellers would know nothing. They would maintain that any two straightest lines, sufficiently produced, must finally cut not in one only but in two points. The sum of the angles of a triangle would be always greater than two right angles, increasing as the surface of the triangle grew greater. They could thus have no conception of geometrical similarity between greater and smaller figures of the same kind, for with them a greater triangle must have different angles from a smaller one. Their space would be unlimited, but would be found to be finite or at least represented as such.

It is clear, then, that such beings must set up a very different system of geometrical axioms from that of the inhabitants of a plane, or from ours with our space of three dimensions, though the logical powers of all were the same; nor are more examples necessary to show that geometrical axioms must vary according to the kind of space inhabited by beings whose powers of reason are quite in conformity with ours. But let us proceed still farther.

Let us think of reasoning beings existing on the surface of an egg-shaped body. Shortest lines could be drawn between three points of such a surface and a triangle constructed. But if the attempt were made to construct congruent triangles at different parts of the surface, it would be found that two triangles, with three pairs of equal sides, would not have their angles equal. The sum of the angles of a triangle drawn at the sharper pole of the body would depart farther from two right angles than if the triangle were drawn at the blunter pole or at the equator. Hence it appears that not even such a simple figure as a triangle can be moved on such a surface without change of form. It would also be found that if circles of equal radii were constructed at different parts of such a surface (the length of the radii being always measured by shortest lines along the surface) the periphery would be greater at the blunter than at the sharper end.

We see accordingly that, if a surface admits of the figures lying on it being freely moved without change of any of their lines and angles as measured along it, the property is a special one and does not belong to every kind of surface. The condition under which a surface possesses this important property was pointed out by Gauss in his celebrated treatise on the curvature of surfaces.[1] The 'measure of curvature,' as he called it, i.e. the reciprocal of the product of the greatest and least radii of curvature, must be everywhere equal over the whole extent of the surface.

Gauss showed at the same time that this measure of curvature is not changed if the surface is bent without distension or contraction of any part of it. Thus we can roll up a flat sheet of paper into the form of a cylinder, or of a cone, without any change in the dimensions of the figures taken along the surface of the sheet. Or the hemispherical fundus of a bladder may be rolled into a spindle-shape without altering the dimensions on the surface. Geometry on a plane will therefore be the same as on a cylindrical surface; only in the latter case we must imagine that any number of layers of this surface, like the layers of a rolled sheet of paper, lie one upon another, and that after each entire revolution round the cylinder a new layer is reached different from the previous ones.

These observations are necessary to give the reader a notion of a kind of surface the geometry of which is on the whole similar to that of the plane, but in which the axiom of parallels does not hold good. This is a kind of curved surface which is, as it were, geometrically the counterpart of a sphere, and which has therefore been called the *pseudospherical surface* by the distinguished Italian mathematician E. Beltrami, who has in-

[1] Gauss, *Werke*, Bd. IV. p. 215, first published in *Commentationes Soc. Reg. Scientt. Gottengensis recentiores*, vol. vi., 1828.

vestigated its properties.[2] It is a saddle-shaped surface of which only limited pieces or strips can be connectedly represented in our space, but which may yet be thought of as infinitely continued in all directions, since each piece lying at the limit of the part constructed can be conceived as drawn back to the middle of it and then continued. The piece displaced must in the process change its flexure but not its dimensions, just as happens with a sheet of paper moved about a cone formed out of a plane rolled up. Such a sheet fits the conical surface in every part, but must be more bent near the vertex and cannot be so moved over the vertex as to be at the same time adapted to the existing cone and to its imaginary continuation beyond.

Like the plane and the sphere, pseudospherical surfaces have their measure of curvature constant, so that every piece of them can be exactly applied to every other piece, and therefore all figures constructed at one place on the surface can be transferred to any other place with perfect congruity of form, and perfect equality of all dimensions lying in the surface itself. The measure of curvature as laid down by Gauss, which is positive for the sphere and zero for the plane, would have a constant negative value for pseudospherical surfaces, because the two principal curvatures of a saddle-shaped surface have their concavity turned opposite ways.

A strip of a pseudospherical surface may, for example, be represented by the inner surface (turned towards the axis) of a solid anchor-ring. If the plane figure *aabb* (Figure 1) is made to revolve on its axis of symmetry AB, the two arcs *ab* will describe a pseudospherical concave-convex surface like that of the ring. Above and below, towards *aa* and *bb*, the surface will turn outwards with ever-increasing flexure, till it becomes perpendicular to the axis, and ends at the edge with one curvature infinite. Or, again, half of a pseudospherical surface may be rolled up into the shape of a champagne-glass (Figure 2), with tapering stem infinitely prolonged. But the surface is always necessarily bounded by a sharp edge beyond which it cannot be directly continued. Only by supposing each single piece of the edge cut loose and drawn along the surface of the ring or glass, can it be brought to places of different flexure, at which farther continuation of the piece is possible.

In this way too the straightest lines of the pseudospherical surface may be infinitely produced. They do not, like those on a sphere, return upon themselves, but, as on a plane, only one shortest line is possible between the two given points. The axiom of parallels does not, however, hold good. If a straightest line is given on the surface and a point without it, a whole

[2] *Saggio di Interpretazione della Geometria Non-Euclidea*, Napoli, 1868.—*Teoria fondamentale degli Spazii di Curvatura costante, Annali di Matematica*, Ser. II. Tom. II. pp. 232–55. Both have been translated into French by J. Hoüel, *Annales Scientifiques de l'Ecole Normale*, Tom. V., 1869.

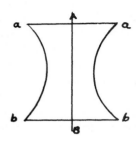

FIGURE 1   FIGURE 2

pencil of straightest lines may pass through the point, no one of which, though infinitely produced, cuts the first line; the pencil itself being limited by two straightest lines, one of which intersects one of the ends of the given line at an infinite distance, the other the other end.

Such a system of geometry, which excluded the axiom of parallels, was devised on Euclid's synthetic method, as far back as the year 1829, by N. J. Lobatchewsky, professor of mathematics at Kasan,[3] and it was proved that this system could be carried out as consistently as Euclid's. It agrees exactly with the geometry of the pseudospherical surfaces worked out recently by Beltrami.

Thus we see that in the geometry of two dimensions a surface is marked out as a plane, or a sphere, or a pseudospherical surface, by the assumption that any figure may be moved about in all directions without change of dimensions. The axiom, that there is only one shortest line between any two points, distinguishes the plane and the pseudospherical surface from the sphere, and the axiom of parallels marks off the plane from the pseudosphere. These three axioms are in fact necessary and sufficient, to define as a plane the surface to which Euclid's planimetry has reference, as distinguished from all other modes of space in two dimensions.

The difference between plane and spherical geometry has been long evident, but the meaning of the axiom of parallels could not be understood till Gauss had developed the notion of surfaces flexible without dilatation, and consequently that of the possibly infinite continuation of pseudospherical surfaces. Inhabiting, as we do, a space of three dimensions and endowed with organs of sense for their perception, we can represent to ourselves the various cases in which beings on a surface might have to develop their perception of space; for we have only to limit our own perceptions to a narrower field. It is easy to think away perceptions that we have; but it is very difficult to imagine perceptions to which there is nothing analogous in our experience. When, therefore, we pass to space

[3] *Principien der Geometrie,* Kasan, 1829–30.

of three dimensions, we are stopped in our power of representation, by the structure of our organs and the experiences got through them which correspond only to the space in which we live.

There is however another way of treating geometry scientifically. All known space-relations are measurable, that is, they may be brought to determination of magnitudes (lines, angles, surfaces, volumes). Problems in geometry can therefore be solved, by finding methods of calculation for arriving at unknown magnitudes from known ones. This is done in *analytical geometry*, where all forms of space are treated only as quantities and determined by means of other quantities. Even the axioms themselves make reference to magnitudes. The straight line is defined as the *shortest* between two points, which is a determination of quantity. The axiom of parallels declares that if two straight lines in a plane do not intersect (are parallel), the alternate angles, or the corresponding angles, made by a third line intersecting them, are equal; or it may be laid down instead that the sum of the angles of any triangle is equal to two right angles. These, also, are determinations of quantity.

Now we may start with this view of space, according to which the position of a point may be determined by measurements in relation to any given figure (system of co-ordinates), taken as fixed, and then inquire what are the special characteristics of our space as manifested in the measurements that have to be made, and how it differs from other extended quantities of like variety. This path was first entered by one too early lost to science, B. Riemann of Göttingen.[4] It has the peculiar advantage that all its operations consist in pure calculation of quantities, which quite obviates the danger of habitual perceptions being taken for necessities of thought.

The number of measurements necessary to give the position of a point, is equal to the number of dimensions of the space in question. In a line the distance from one fixed point is sufficient, that is to say, one quantity; in a surface the distances from two fixed points must be given; in space, the distances from three; or we require, as on the earth, longitude, latitude, and height above the sea, or, as is usual in analytical geometry, the distances from three co-ordinate planes. Riemann calls a system of differences in which one thing can be determined by n measurements an '$n$fold extended aggregate' or an 'aggregate of n dimensions.' Thus the space in which we live is a threefold, a surface is a twofold, and a line is a simple extended aggregate of points. Time also is an aggregate of one dimension. The system of colours is an aggregate of three dimensions, inasmuch as each colour, according to the investigations of Thomas Young and of

---

[4] Ueber die Hypothesen welche der Geometrie zu Grunde liegen, Habilitationsschrift vom 10 Juni 1854. (*Abhandl. der königl. Gesellsch. zu Göttingen*, Bd. XIII.)

Clerk Maxwell,[5] may be represented as a mixture of three primary colours, taken in definite quantities. The particular mixtures can be actually made with the colour-top.

In the same way we may consider the system of simple tones [6] as an aggregate of two dimensions, if we distinguish only pitch and intensity, and leave out of account differences of timbre. This generalisation of the idea is well suited to bring out the distinction between space of three dimensions and other aggregates. We can, as we know from daily experience, compare the vertical distance of two points with the horizontal distance of two others, because we can apply a measure first to the one pair and then to the other. But we cannot compare the difference between two tones of equal pitch and different intensity, with that between two tones of equal intensity and different pitch. Riemann showed, by considerations of this kind, that the essential foundation of any system of geometry, is the expression that it gives for the distance between two points lying in any direction towards one another, beginning with the infinitesimal interval. He took from analytical geometry the most general form for this expression, that, namely, which leaves altogether open the kind of measurements by which the position of any point is given.[7] Then he showed that the kind of free mobility without change of form which belongs to bodies in our space can only exist when certain quantities yielded by the calculation [8]—quantities that coincide wtih Gauss's measure of surface-curvature when they are expressed for surfaces—have everywhere an equal value. For this reason Riemann calls these quantities, when they have the same value in all directions for a particular spot, the measure of curvature of the space at this spot. To prevent misunderstanding, I will once more observe that this so-called measure of space-curvature is a quantity obtained by purely analytical calculation, and that its introduction involves no suggestion of relations that would have a meaning only for sense-perception. The name is merely taken, as a short expression for a complex relation, from the one case in which the quantity designated admits of sensible representation.

Now whenever the value of this measure of curvature in any space is everywhere zero, that space everywhere conforms to the axioms of Euclid; and it may be called a *flat* (*homaloid*) space in contradistinction to other spaces, analytically constructible, that may be called *curved*, because their measure of curvature has a value other than zero. Analytical geometry

---

[5] Helmholtz's *Popular Lectures*, Series I. p. 243.

[6] Ibid., p. 86.

[7] For the square of the distance of two infinitely near points the expression is a homogeneous quadric function of the differentials of their co-ordinates.

[8] They are algebraical expressions compounded from the coefficients of the various terms in the expression for the square of the distance of two contiguous points and from their differential quotients.

may be as completely and consistently worked out for such spaces as ordinary geometry can for our actually existing homaloid space.

If the measure of curvature is positive we have *spherical* space, in which straightest lines return upon themselves and there are no parallels. Such a space would, like the surface of a sphere, be unlimited but not infinitely great. A constant negative measure of curvature on the other hand gives *pseudospherical* space, in which straightest lines run out to infinity, and a pencil of straightest lines may be drawn, in any flattest surface, through any point which does not intersect another given straightest line in that surface.

Beltrami [9] has rendered these last relations imaginable by showing that the points, lines, and surfaces of a pseudospherical space of three dimensions, can be so portrayed in the interior of a sphere in Euclid's homaloid space, that every straightest line or flattest surface of the pseudospherical space is represented by a straight line or a plane, respectively, in the sphere. The surface itself of the sphere corresponds to the infinitely distant points of the pseudospherical space; and the different parts of this space, as represented in the sphere, become smaller, the nearer they lie to the spherical surface, diminishing more rapidly in the direction of the radii than in that perpendicular to them. Straight lines in the sphere, which only intersect beyond its surface, correspond to straightest lines of the pseudospherical space which never intersect.

Thus it appeared that space, considered as a region of measurable quantities, does not at all correspond with the most general conception of an aggregate of three dimensions, but involves also special conditions, depending on the perfectly free mobility of solid bodies without change of form to all parts of it and with all possible changes of direction; and, further, on the special value of the measure of curvature which for our actual space equals, or at least is not distinguishable from, zero. This latter definition is given in the axioms of straight lines and parallels.

Whilst Riemann entered upon this new field from the side of the most general and fundamental questions of analytical geometry, I myself arrived at similar conclusions,[10] partly from seeking to represent in space the system of colours, involving the comparison of one threefold extended aggregate with another, and partly from inquiries on the origin of our ocular measure for distances in the field of vision. Riemann starts by assuming the above-mentioned algebraical expression which represents in the most general form the distance between two infinitely near points, and deduces therefrom, the conditions of mobility of rigid figures. I, on the

---

[9] *Teoria fondamentale, &c., ut sup.*
[10] Ueber die Thatsachen die der Geometrie zum Grunde liegen (*Nachrichten von der königl. Ges. d. Wiss. zu Göttingen*, Juni 3, 1868).

other hand, starting from the observed fact that the movement of rigid figures is possible in our space, with the degree of freedom that we know, deduce the necessity of the algebraic expression taken by Riemann as an axiom. The assumptions that I had to make as the basis of the calculation were the following.

First, to make algebraical treatment at all possible, it must be assumed that the position of any point A can be determined, in relation to certain given figures taken as fixed bases, by measurement of some kind of magnitudes, as lines, angles between lines, angles between surfaces, and so forth. The measurements necessary for determining the position of A are known as its co-ordinates. In general, the number of co-ordinates necessary for the complete determination of the position of a point, marks the number of the dimensions of the space in question. It is further assumed that with the movement of the point A, the magnitudes used as co-ordinates vary continuously.

Secondly, the definition of a solid body, or rigid system of points, must be made in such a way as to admit of magnitudes being compared by congruence. As we must not, at this stage, assume any special methods for the measurement of magnitudes, our definition can, in the first instance, run only as follows: Between the co-ordinates of any two points belonging to a solid body, there must be an equation which, however the body is moved, expresses a constant spatial relation (proving at last to be the distance) between the two points, and which is the same for congruent pairs of points, that is to say, such pairs as can be made successively to coincide in space with the same fixed pair of points.

However indeterminate in appearance, this definition involves most important consequences, because with increase in the number of points, the number of equations increases much more quickly than the number of co-ordinates which they determine. Five points, A, B, C, D, E, give ten different pairs of points

$$AB, \ AC, \ AD, \ AE,$$
$$BC, \ BD, \ BE,$$
$$CD, \ CE,$$
$$DE,$$

and therefore ten equations, involving in space of three dimensions fifteen variable co-ordinates. But of these fifteen, six must remain arbitrary, if the system of five points is to admit of free movement and rotation, and thus the ten equations can determine only nine co-ordinates as functions of the six variables. With six points we obtain fifteen equations for twelve quantities, with seven points twenty-one equations for fifteen, and so on. Now from n independent equations we can determine n contained quantities, and if we have more than n equations, the superfluous ones must be deducible from the first n. Hence it follows that the equations which subsist between the co-ordinates of each pair of points of a solid body must have a special character, seeing that, when in space of three dimensions they are satisfied for nine pairs of points as formed out of any five points, the equation for the tenth pair follows by logical consequence. Thus our assumption for the definition of solidity, becomes quite sufficient to determine the kind of equations holding between the co-ordinates of two points rigidly connected.

Thirdly, the calculation must further be based on the fact of a peculiar circumstance in the movement of solid bodies, a fact so familiar to us that but for this inquiry it might never have been thought of as something that need not be. When in our space of three dimensions two points of a solid body are kept fixed, its movements are limited to rotations round the straight line connecting them. If we turn it completely round once, it again occupies exactly the position it had at first. This fact, that rotation in one direction always brings a solid body back into its original position, needs special mention. A system of geometry is possible without it. This is most easily seen in the geometry of a plane. Suppose that with every rotation of a plane figure its linear dimensions increased in proportion to the angle of rotation, the figure after one whole rotation through 360 degrees would no longer coincide with itself as it was originally. But any second figure that was congruent with the first in its original position might be made to coincide with it in its second position by being also turned through 360 degrees. A consistent system of geometry would be possible upon this supposition, which does not come under Riemann's formula.

On the other hand I have shown that the three assumptions taken together form a sufficient basis for the starting-point of Riemann's investigation, and thence for all his further results relating to the distinction of different spaces according to their measure of curvature.

It still remained to be seen whether the laws of motion, as dependent on moving forces, could also be consistently transferred to spherical or pseudospherical space. This investigation has been carried out by Professor Lipschitz of Bonn.[11] It is found that the comprehensive expression for all the laws of dynamics, Hamilton's principle, may be directly transferred to spaces of which the measure of curvature is other than zero. Accordingly, in this respect also, the disparate systems of geometry lead to no contradiction.

We have now to seek an explanation of the special characteristics of our own flat space, since it appears that they are not implied in the general motion of an extended quantity of three dimensions and of the free mo-

---

[11] 'Untersuchungen über die ganzen homogenen Functionen von n Differentialen' (Borchardt's *Journal für Mathematik*, Bd. lxx. 3, 71; lxxiii. 3, 1); 'Untersuchung eines Problems der Variationsrechnung' (*Ibid*. Bd. lxxiv.).

bility of bounded figures therein. *Necessities of thought*, such as are involved in the conception of such a variety, and its measurability, or from the most general of all ideas of a solid figure contained in it, and of its free mobility, they undoubtedly are not. Let us then examine the opposite assumption as to their origin being empirical, and see if they can be inferred from facts of experience and so established, or if, when tested by experience, they are perhaps to be rejected. If they are of empirical origin, we must be able to represent to ourselves connected series of facts, indicating a different value for the measure of curvature from that of Euclid's flat space. But if we can imagine such spaces of other sorts, it cannot be maintained that the axioms of geometry are necessary consequences of an *à priori* transcendental form of intuition, as Kant thought.

The distinction between spherical, pseudospherical, and Euclid's geometry depends, as was above observed, on the value of a certain constant called, by Riemann, the measure of curvature of the space in question. The value must be zero for Euclid's axioms to hold good. If it were not zero, the sum of the angles of a large triangle would differ from that of the angles of a small one, being larger in spherical, smaller in pseudospherical, space. Again, the geometrical similarity of large and small solids or figures is possible only in Euclid's space. All systems of practical mensuration that have been used for the angles of large rectilinear triangles, and especially all systems of astronomical measurement which make the parallax of the immeasurably distant fixed stars equal to zero (in pseudospherical space the parallax even of infinitely distant points would be positive), confirm empirically the axiom of parallels, and show the measure of curvature of our space thus far to be indistinguishable from zero. It remains, however, a question, as Riemann observed, whether the result might not be different if we could use other than our limited base-lines, the greatest of which is the major axis of the earth's orbit.

Meanwhile, we must not forget that all geometrical measurements rest ultimately upon the principle of congruence. We measure the distance between points by applying to them the compass, rule, or chain. We measure angles by bringing the divided circle or theodolite to the vertex of the angle. We also determine straight lines by the path of rays of light which in our experience is rectilinear; but that light travels in shortest lines as long as it continues in a medium of constant refraction would be equally true in space of a different measure of curvature. Thus all our geometrical measurements depend on our instruments being really, as we consider them, invariable in form, or at least on their undergoing no other than the small changes we know of, as arising from variation of temperature, or from gravity acting differently at different places.

In measuring, we only employ the best and surest means we know of to determine, what we otherwise are in the habit of making out by sight and touch or by pacing. Here our own body with its organs is the instrument we carry about in space. Now it is the hand, now the leg, that serves for a compass, or the eye turning in all directions is our theodolite for measuring arcs and angles in the visual field.

Every comparative estimate of magnitudes or measurement of their spatial relations proceeds therefore upon a supposition as to the behaviour of certain physical things, either the human body or other instruments employed. The supposition may be in the highest degree probable and in closest harmony with all other physical relations known to us, but yet it passes beyond the scope of pure space-intuition.

It is in fact possible to imagine conditions for bodies apparently solid such that the measurements in Euclid's space become what they would be in spherical or pseudospherical space. Let me first remind the reader that if all the linear dimensions of other bodies, and our own, at the same time were diminished or increased in like proportion, as for instance to half or double their size, we should with our means of space-perception be utterly unaware of the change. This would also be the case if the distension or contraction were different in different directions, provided that our own body changed in the same manner, and further that a body in rotating assumed at every moment, without suffering or exerting mechanical resistance, the amount of dilatation in its different dimensions corresponding to its position at the time. Think of the image of the world in a convex mirror. The common silvered globes set up in gardens give the essential features, only distorted by some optical irregularities. A well-made convex mirror of moderate aperture represents the objects in front of it as apparently solid and in fixed positions behind its surface. But the images of the distant horizon and of the sun in the sky lie behind the mirror at a limited distance, equal to its focal length. Between these and the surface of the mirror are found the images of all the other objects before it, but the images are diminished and flattened in proportion to the distance of their objects from the mirror. The flattening, or decrease in the third dimension, is relatively greater than the decrease of the surface-dimensions. Yet every straight line or every plane in the outer world is represented by a straight line or a plane in the image. The image of a man measuring with a rule a straight line from the mirror would contract more and more the farther he went, but with his shrunken rule the man in the image would count out exactly the same number of centimetres as the real man. And, in general, all geometrical measurements of lines or angles made with regularly varying images of real instruments would yield exactly the same results as in the outer world, all congruent bodies would coincide on being applied to one another in the mirror as in the outer world, all lines of sight in the outer world would be represented by straight lines of sight in the mirror. In short I do not see how men in the mirror are to discover that their bodies are not rigid solids and their experiences good examples of the correctness of Euclid's axioms. But if they could look out upon our world as we can look into theirs, without overstepping the boundary, they must declare it to be a picture in a spherical mirror, and would speak of us just as we speak of them; and if two inhabitants of the different worlds could communicate with one another, neither, so far as I can see, would be able to convince the other that he had the true, the other the distorted, relations. Indeed I cannot see that such a question would have any meaning at all, so long as mechanical considerations are not mixed up with it.

Now Beltrami's representation of pseudospherical space in a sphere of Euclid's space, is quite similar, except that the background is not a plane as in the convex mirror, but the surface of a sphere, and that the proportion in which the images as they approach the spherical surface contract, has a different mathematical expression.[12] If we imagine then, conversely, that in the sphere, for the interior of which Euclid's axioms hold good, moving bodies contract as they depart from the centre like the images in a convex mirror, and in such a way that their representatives in pseudospherical space retain their dimensions unchanged,—observers whose bodies were regularly subjected to the same change would obtain the same results from the geometrical measurements they could make as if they lived in pseudospherical space.

We can even go a step further, and infer how the objects in a pseudospherical world, were it possible to enter one, would appear to an observer, whose eye-measure and experiences of space had been gained like ours in Euclid's space. Such an observer would continue to look upon rays of light or the lines of vision as straight lines, such as are met with in flat space, and as they really are in the spherical representation of pseudospherical space. The visual image of the objects in pseudospherical space would thus make the same impression upon him as if he were at the centre of Beltrami's sphere. He would think he saw the most remote objects round about him at a finite distance,[13] let us suppose a hundred feet off. But as he approached these distant objects, they would dilate before him, though more in the third dimension than superficially, while behind him they would contract. He would know that his eye judged wrongly. If he saw two straight lines which in his estimate ran parallel for the hundred feet to his world's end, he would find on following them that the farther he advanced the more they diverged, because of the dilatation of all the objects to which he approached. On the other hand, behind him, their distance would seem to diminish, so that as he advanced they would appear always to diverge more and more. But two straight lines which from

---

[12] Compare the Appendix at the end of this Lecture.
[13] The reciprocal of the square of this distance, expressed in negative quantity, would be the measure of curvature of the pseudospherical space.

his first position seemed to converge to one and the same point of the background a hundred feet distant, would continue to do this however far he went, and he would never reach their point of intersection.

Now we can obtain exactly similar images of our real world, if we look through a large convex lens of corresponding negative focal length, or even through a pair of convex spectacles if ground somewhat prismatically to resemble pieces of one continuous larger lens. With these, like the convex mirror, we see remote objects as if near to us, the most remote appearing no farther distant than the focus of the lens. In going about with this lens before the eyes, we find that the objects we approach dilate exactly in the manner I have described for pseudospherical space. Now any one using a lens, were it even so strong as to have a focal length of only sixty inches, to say nothing of a hundred feet, would perhaps observe for the first moment that he saw objects brought nearer. But after going about a little the illusion would vanish, and in spite of the false images he would judge of the distances rightly. We have every reason to suppose that what happens in a few hours to any one beginning to wear spectacles would soon enough be experienced in pseudospherical space. In short, pseudospherical space would not seem to us very strange, comparatively speaking; we should only at first be subject to illusions in measuring by eye the size and distance of the more remote objects.

There would be illusions of an opposite description, if, with eyes practised to measure in Euclid's space, we entered a spherical space of three dimensions. We should suppose the more distant objects to be more remote and larger than they are, and should find on approaching them that we reached them more quickly than we expected from their appearance. But we should also see before us objects that we can fixate only with diverging lines of sight, namely, all those at a greater distance from us than the quadrant of a great circle. Such an aspect of things would hardly strike us as very extraordinary, for we can have it even as things are if we place before the eye a slightly prismatic glass with the thicker side towards the nose: the eyes must then become divergent to take in distant objects. This excites a certain feeling of unwonted strain in the eyes, but does not perceptibly change the appearance of the objects thus seen. The strangest sight, however, in the spherical world would be the back of our own head, in which all visual lines not stopped by other objects would meet again, and which must fill the extreme background of the whole perspective picture.

At the same time it must be noted that as a small elastic flat disk, say of india-rubber, can only be fitted to a slightly curved spherical surface with relative contraction of its border and distension of its centre, so our bodies, developed in Euclid's flat space, could not pass into curved space without undergoing similar distensions and contractions of their parts, their coherence being of course maintained only in as far as their elasticity permitted their bending without breaking. The kind of distension must be the same as in passing from a small body imagined at the center of Beltrami's sphere to its pseudospherical or spherical representation. For such passage to appear possible, it will always have to be assumed that the body is sufficiently elastic and small in comparison with the real or imaginary radius of curvature of the curved space into which it is to pass.

These remarks will suffice to show the way in which we can infer from the known laws of our sensible perceptions the series of sensible impressions which a spherical or pseudospherical world would give us, if it existed. In doing so, we nowhere meet with inconsistency or impossibility any more than in the calculation of its metrical proportions. We can represent to ourselves the look of a pseudospherical world in all directions just as we can develop the conception of it. Therefore it cannot be allowed that the axioms of our geometry depend on the native form of our perceptive faculty, or are in any way connected with it.

It is different with the three dimensions of space. As all our means of sense-perception extend only to space of three dimensions, and a fourth is not merely a modification of what we have, but something perfectly new, we find ourselves by reason of our bodily organisation quite unable to represent a fourth dimension.

In conclusion, I would again urge that the axioms of geometry are not propositions pertaining only to the pure doctrine of space. As I said before, they are concerned with quantity. We can speak of quantities only when we know of some way by which we can compare, divide, and measure them. All space-measurements, and therefore in general all ideas of quantities applied to space, assume the possibility of figures moving without change of form or size. It is true we are accustomed in geometry to call such figures purely geometrical solids, surfaces, angles, and lines, because we abstract from all the other distinctions, physical and chemical, of natural bodies; but yet one physical quality, rigidity, is retained. Now we have no other mark of rigidity of bodies or figures but congruence, whenever they are applied to one another at any time or place, and after any revolution. We cannot, however, decide by pure geometry, and without mechanical considerations, whether the coinciding bodies may not both have varied in the same sense.

If it were useful for any purpose, we might with perfect consistency look upon the space in which we live as the apparent space behind a convex mirror with its shortened and contracted background; or we might consider a bounded sphere of our space, beyond the limits of which we perceive nothing further, as infinite pseudospherical space. Only then we should have to ascribe to the bodies which appear to us to be solid, and to our own body at the same time, corresponding distensions and contractions, and we should have to change our system of mechanical principles entirely; for even the proposition that every point in motion, if acted upon by no force, continues to move with unchanged velocity in a straight line, is not adapted to the image of the world in the convex-mirror. The path would indeed be straight, but the velocity would depend upon the place.

Thus the axioms of geometry are not concerned with space-relations only but also at the same time with the mechanical deportment of solidest bodies in motion. The notion of rigid geometrical figure might indeed be conceived as transcendental in Kant's sense, namely, as formed independently of actual experience, which need not exactly correspond therewith, any more than natural bodies do ever in fact correspond exactly to the abstract notion we have obtained of them by induction. Taking the notion of rigidity thus as a mere ideal, a strict Kantian might certainly look upon the geometrical axioms as propositions given, *à priori*, by transcendental intuition, which no experience could either confirm or refute, because it must first be decided by them whether any natural bodies can be considered as rigid. But then we should have to maintain that the axioms of geometry are not synthetic propositions, as Kant held them; they would merely define what qualities and deportment a body must have to be recognised as rigid.

But if to the geometrical axioms we add propositions relating to the mechanical properties of natural bodies, were it only the axiom of inertia, or the single proposition, that the mechanical and physical properties of bodies and their mutual reactions are, other circumstances remaining the same, independent of place, such a system of propositions has a real import which can be confirmed or refuted by experience, but just for the same reason can also be gained by experience. The mechanical axiom, just cited, is in fact of the utmost importance for the whole system of our mechanical and physical conceptions. That rigid solids, as we call them, which are really nothing else than elastic solids of great resistance, retain the same form in every part of space if no external force affects them, is a single case falling under the general principle.

In conclusion, I do not, of course, maintain that mankind first arrived. at space-intuitions, in agreement with the axioms of Euclid, by any carefully executed systems of exact measurement. It was rather a succession of everyday experiences, especially the perception of the geometrical similarity of great and small bodies, only possible in flat space, that led to the rejection, as impossible, of every geometrical representation at variance with this fact. For this no knowledge of the necessary logical connection between the observed fact of geometrical similarity and the axioms was needed; but only an intuitive apprehension of the typical relations between lines, planes, angles, &c., obtained by numerous and attentive observations

—an intuition of the kind the artist possesses of the objects he is to represent, and by means of which he decides with certainty and accuracy whether a new combination, which he tries, will correspond or not with their nature. It is true that we have no word but *intuition* to mark this; but it is knowledge empirically gained by the aggregation and reinforcement of similar recurrent impressions in memory, and not a transcendental form given before experience. That other such empirical intuitions of fixed typical relations, when not clearly comprehended, have frequently enough been taken by metaphysicians for *à priori* principles, is a point on which I need not insist.

To sum up, the final outcome of the whole inquiry may be thus expressed:—

(1) The axioms of geometry, taken by themselves out of all connection with mechanical propositions, represent no relations of real things. When thus isolated, if we regard them with Kant as forms of intuition transcendentally given, they constitute a form into which any empirical content whatever will fit, and which therefore does not in any way limit or determine beforehand the nature of the content. This is true, however, not only of Euclid's axioms, but also of the axioms of spherical and pseudospherical geometry.

(2) As soon as certain principles of mechanics are conjoined with the axioms of geometry, we obtain a system of propositions which has real import, and which can be verified or overturned by empirical observations, just as it can be inferred from experience. If such a system were to be taken as a transcendental form of intuition and thought, there must be assumed a pre-established harmony between form and reality.

APPENDIX

The elements of the geometry of spherical space are most easily obtained by putting for space of four dimensions the equation for the sphere

$$x^2 + y^2 + z^2 + t^2 = R^2 \quad \ldots \ldots \quad (1)$$

and for the distance ds between the points (x, y, z, t) and $[(x + dx)(y + dy)(z + dz)(t + dt)]$ the value

$$ds^2 = dx^2 + dy^2 + dz^2 + dt^2 \quad \ldots \ldots \quad (2)$$

It is easily found by means of the methods used for three dimensions that the shortest lines are given by equations of the form

$$\left. \begin{aligned} ax + by + cz + ft &= 0 \\ \alpha x + \beta y + \gamma z + \phi t &= 0 \end{aligned} \right\} \quad \ldots \ldots \quad (3)$$

in which a, b, c, f, as well as α, β, γ, ϕ, are constants.

The length of the shortest arc, s, between the points (x, y, z, t), and (ξ, η, ζ, τ) follows, as in the sphere, from the equation

$$\cos \frac{s}{R} = \frac{x\xi + y\eta + z\zeta + t\tau}{R^2} \quad \ldots \ldots (4)$$

One of the co-ordinates may be eliminated from the values given in 2 to 4, by means of equation 1, and the expressions then apply to space of three dimensions.

If we take the distances from the points

$$\xi = \eta = \zeta = 0$$

from which equation 1 gives $\tau = R$, then,

$$\sin \left(\frac{s_0}{R} \right) = \frac{\sigma}{R}$$

in which

$$\sigma = \sqrt{x^2 + y^2 + z^2}$$

or, $\qquad s_0 = R \cdot \text{arc sin} \left(\frac{\sigma}{R} \right) = R \cdot \text{arc tang} \left(\frac{\sigma}{t} \right) \quad \ldots \ldots (5)$

In this, s_0 is the distance of the point, x, y, z, measured from the centre of the co-ordinates.

If now we suppose the point x, y, z, of spherical space, to be projected in a point of plane space whose co-ordinates are respectively

$$\mathfrak{x} = \frac{Rx}{t}, \, \mathfrak{y} = \frac{Ry}{t}, \, \mathfrak{z} = \frac{Rz}{t}$$

$$\mathfrak{x}^2 + \mathfrak{y}^2 + \mathfrak{z}^2 = \mathfrak{r}^2 = \frac{R^2 \sigma^2}{t^2}$$

then in the plane space the equations 3, which belong to the straightest lines of spherical space, are equations of the straight line. Hence the shortest lines of spherical space are represented in the system of $\mathfrak{x}$, $\mathfrak{y}$, $\mathfrak{z}$, by straight lines. For very small values of x, y, z, $t = R$, and

$$\mathfrak{x} = x, \, \mathfrak{y} = y, \, \mathfrak{z} = z$$

Immediately about the centre of the co-ordinates, the measurements of both spaces coincide. On the other hand, we have for the distances from the centre

$$s_0 = R \cdot \text{arc tang} \left(\pm \frac{\mathfrak{r}}{R} \right) \quad \ldots \ldots (6)$$

In this, $\mathfrak{r}$ may be infinite; but every point of plane space must be the projection of two points of the sphere, one for which $s_0 < \frac{1}{2} R\pi$, and one for which $s_0 > \frac{1}{2} R\pi$. The extension in the direction of $\mathfrak{r}$ is then

$$\frac{ds_0}{d\mathfrak{r}} = \frac{R^2}{R^2 + \mathfrak{r}^2}$$

In order to obtain corresponding expressions for pseudospherical space, let R and t be imaginary; that is, $R = \mathfrak{R}i$, and $t = \mathfrak{t}i$. Equation 6 gives then

$$\text{tang}\,\frac{s_0}{i\mathfrak{R}} = \pm\,\frac{\mathfrak{r}}{i\mathfrak{R}}$$

from which, eliminating the imaginary form, we get

$$s_0 = \frac{1}{2}\,\mathfrak{R}\,\text{log. nat.}\,\frac{\mathfrak{R} + \mathfrak{r}}{\mathfrak{R} - \mathfrak{r}}$$

Here s_0 has real values only as long as $\mathfrak{r} = R$; for $\mathfrak{r} = \mathfrak{R}$ the distance s_0 in pseudospherical space is infinite. The image in plane space is, on the contrary, contained in the sphere of radius R, and every point of this sphere forms only one point of the infinite pseudospherical space. The extension in the direction of $\mathfrak{r}$ is

$$\frac{ds_0}{d\mathfrak{r}} = \frac{\mathfrak{R}^2}{\mathfrak{R}^2 - \mathfrak{r}^2}$$

For linear elements, on the contrary, whose direction is at right angles to $\mathfrak{r}$, and for which t is unchanged, we have in both cases

$$\frac{\sqrt{dx^2 + dy^2 + dz^2}}{\sqrt{d\mathfrak{x}^2 + d\mathfrak{y}^2 + d\mathfrak{z}^2}} = \frac{t}{R} = \frac{\mathfrak{t}}{\mathfrak{R}} = \frac{\sigma}{\mathfrak{r}}$$

$$= \frac{\sqrt{x^2 + y^2 + z^2}}{\sqrt{\mathfrak{x}^2 + \mathfrak{y}^2 + \mathfrak{z}^2}}$$

COMMENTARY ON
Symmetry

IN the everyday sense symmetry carries the meaning of balance, proportion, harmony, regularity of form. Beauty is sometimes linked
with symmetry, but the relationship is not very illuminating since beauty
is an even vaguer quality than symmetry. The protean character of the
concept of symmetry is indicated by its widely different uses. A statue, a
musical composition, a gaseous nebula, an ethical standard of action: each
may be described as displaying symmetry. The ethical standard is that of
moderation—a mean between extremes of action.

The concept first acquires precision in geometry. Vague notions are
replaced by the idea of bilateral symmetry, the symmetry of left and
right, exhibited in the higher animals. "A body, a spatial configuration, is
symmetric with respect to a given place E if it is carried into itself by
reflection in E." The mathematical concept lends itself to remarkable
elaboration. We are apt to think of symmetry as a static property; the
analytic approach which deals with the genesis of symmetric forms is far
more fruitful. Our understanding of these forms is deepened by regarding
them as the product of various transformations or motions by means of
which one pattern is converted, element by element, into another. Simple
forms, when subjected to translatory and rotational motions, are transformed into marvelously intricate designs; these designs in two and three
dimensions are encountered in art, biology, chemistry, astronomy, physics
and crystallography. The ruling principle of symmetry is applied to forms
and processes of every conceivable kind; that is to say, the forms and
processes are defined in terms of the combination of motions which gives
birth to them.

This method of extending our insight is typical, as Weyl says, for all
theoretic knowledge. "We begin with some general but vague principle
(symmetry in the first sense), then find an important case where we can
give that notion a concrete precise meaning (bilateral symmetry), and
from that case we gradually rise again to generality, guided more by
mathematical construction and abstraction than by the mirages of philosophy; and if we are lucky we end up with an idea no less universal than
the one from which we started."

On the eve of his retirement from the Institute for Advanced Study in
Princeton, Hermann Weyl, a commanding figure among the mathematicians of this century, delivered a series of lectures on symmetry.[1] They
present a masterful survey of the applications of the principle of symmetry

---

[1] For a biographical note on Weyl, see p. 1830.

in sculpture, painting, architecture, ornament and design; its manifestations in organic and inorganic nature; its philosophical and mathematical significance. Symmetry establishes a ridiculous and wonderful cousinship between objects, phenomena and theories outwardly unrelated: terrestrial magnetism, women's veils, polarized light, natural selection, the theory of groups, invariants and transformations, the work habits of bees in the hive, the structure of space, vase designs, quantum physics, scarabs, flower petals, X-ray interference patterns, cell division in sea urchins, equilibrium positions of crystals, Romanesque cathedrals, snowflakes, music, the theory of relativity. The structure of these relationships is depicted by Weyl in a remarkable sweep. The style is not always easy; neither is the subject. Nevertheless the book affords an entry into a profound and fascinating subject which demonstrates, perhaps uniquely, the working of the mathematical intellect, the evolution of intuitive concepts into grand systems of abstract ideas. I have selected the first two of the lectures—on bilateral and related symmetries; I was tempted to give the entire series. You will discover within a few pages why it was so hard to resist the inclination.

> ... *What immortal hand or eye,*
> *Dare frame thy fearful symmetry?*
> —WILLIAM BLAKE

9   Symmetry

By HERMANN WEYL

BILATERAL SYMMETRY

IF I am not mistaken the word *symmetry* is used in our everyday language in two meanings. In the one sense symmetric means something like well-proportioned, well-balanced, and symmetry denotes that sort of concordance of several parts by which they integrate into a whole. *Beauty* is bound up with symmetry. Thus Polykleitos, who wrote a book on proportion and whom the ancients praised for the harmonious perfection of his sculptures, uses the word, and Dürer follows him in setting down a canon of proportions for the human figure.[1] In this sense the idea is by no means restricted to spatial objects; the synonym "harmony" points more toward its acoustical and musical than its geometric applications. *Ebenmass* is a good German equivalent for the Greek symmetry; for like this it carries also the connotation of "middle measure," the mean toward which the virtuous should strive in their actions according to Aristotle's Nicomachean Ethics, and which Galen in *De temperamentis* describes as that state of mind which is equally removed from both extremes: σύμμετρον ὅπερ ἑκατέρου τῶν ἄκρων ἀπέχει.

The image of the balance provides a natural link to the second sense in which the word symmetry is used in modern times: *bilateral symmetry,* the symmetry of left and right, which is so conspicuous in the structure of the higher animals, especially the human body. Now this bilateral symmetry is a strictly geometric and, in contrast to the vague notion of symmetry discussed before, an absolutely precise concept. A body, a spatial configuration, is symmetric with respect to a given plane E if it is

---

[1] Dürer, *Vier Bücher von menschlicher Proportion*, 1528. To be exact, Dürer himself does not use the word symmetry, but the "authorized" Latin translation by his friend Joachim Camerarius (1532) bears the title *De symmetria partium*. To Polykleitos the statement is ascribed (περὶ βελοποιϊκῶν, IV, 2) that "the employment of a great many numbers would almost engender correctness in sculpture." See also Herbert Senk, Au sujet de l'expression συμμετρία dans Diodore I, 98, 5-9, in *Chronique d'Egypte 26* (1951), pp. 63-66. Vitruvius defines: "Symmetry results from proportion . . . Proportion is the commensuration of the various constituent parts with the whole." For a more elaborate modern attempt in the same direction see George David Birkhoff, *Aesthetic measure*, Cambridge, Mass., Harvard University Press, 1933, and the lectures by the same author on "A mathematical theory of aesthetics and its applications to poetry and music," *Rice Institute Pamphlet*, Vol. 19 (July, 1932), pp. 189-342.

carried into itself by reflection in E. Take any line l perpendicular to E and any point p on l: there exists one and only one point p' on l which has the same distance from E but lies on the other side. The point p' coincides with p only if p is on E. Reflection in E is that mapping of space

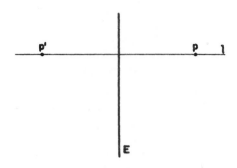

FIGURE 1—Reflection in E.

upon itself, $S: p \rightarrow p'$, that carries the arbitrary point p into this its mirror image p' with respect to E. A mapping is defined whenever a rule is established by which every point p is associated with an image p'. Another example: a rotation around a perpendicular axis, say by 30°, carries each point p of space into a point p' and thus defines a mapping. A figure has rotational symmetry around an axis l if it is carried into itself by all rotations around l. Bilateral symmetry appears thus as the first case of a geometric concept of symmetry that refers to such operations as reflections or rotations. Because of their complete rotational symmetry, the circle in the plane, the sphere in space were considered by the Pythagoreans the most perfect geometric figures, and Aristotle ascribed spherical shape to the celestial bodies because any other would detract from their heavenly perfection. It is in this tradition that a modern poet [2] addresses the Divine Being as "Thou great symmetry":

> *God, Thou great symmetry,*
> *Who put a biting lust in me*
> *From whence my sorrows spring,*
> *For all the frittered days*
> *That I have spent in shapeless ways*
> *Give me one perfect thing.*

Symmetry, as wide or as narrow as you may define its meaning, is one idea by which man through the ages has tried to comprehend and create order, beauty, and perfection.

The course these lectures will take is as follows.[3] First I will discuss

[2] Anna Wickham, "Envoi," from *The Contemplative Quarry*, Harcourt, Brace and Co., 1921.

[3] [The first two lectures are given here. Lecture 3 deals with ornamental symmetry, Lecture 4 with crystals and the general mathematical idea of symmetry. ED.]

bilateral symmetry in some detail and its role in art as well as organic and inorganic nature. Then we shall generalize this concept gradually, in the direction indicated by our example of rotational symmetry, first staying within the confines of geometry, but then going beyond these limits through the process of mathematical abstraction along a road that will finally lead us to a mathematical idea of great generality, the Platonic idea as it were behind all the special appearances and applications of symmetry. To a certain degree this scheme is typical for all theoretic knowledge: We begin with some general but vague principle (symmetry in the first sense), then find an important case where we can give that notion a concrete precise meaning (bilateral symmetry), and from that case we gradually rise again to generality, guided more by mathematical construction and abstraction than by the mirages of philosophy; and if we are lucky we end up with an idea no less universal than the one from which we started. Gone may be much of its emotional appeal, but it has the same or even greater unifying power in the realm of thought and is exact instead of vague.

I open the discussion on bilateral symmetry by using this

FIGURE 2

noble Greek sculpture from the fourth century B.C., the statue of a praying boy (Figure 2), to let you feel as in a symbol the great significance of this type of symmetry both for life and art. One may ask whether the aesthetic value of symmetry depends on its vital value: Did the artist discover the symmetry with which nature according to some inherent law has endowed its creatures, and then copied and perfected what nature presented but in imperfect realizations; or has the aesthetic value of symmetry an independent source? I am inclined to think with Plato that the mathematical idea is the common origin of both: the mathematical laws governing nature are the origin of symmetry in nature, the intuitive realization of the idea in the creative artist's mind its origin in art; although I am ready to admit that in the arts the fact of the bilateral symmetry of the human body in its outward appearance has acted as an additional stimulus.

Of all ancient peoples the Sumerians seem to have been particularly fond of strict bilateral or heraldic symmetry. A typical design on the famous silver vase of King Entemena, who ruled in the city of Lagash around 2700 B.C., shows a lion-headed eagle with spread wings *en face*, each of whose claws grips a stag in side view, which in its turn is frontally

FIGURE 3

FIGURE 4

attacked by a lion (the stags in the upper design are replaced by goats in the lower) (Figure 3). Extension of the exact symmetry of the eagle to the other beasts obviously enforces their duplication. Not much later the eagle is given two heads facing in either direction, the formal principle of symmetry thus completely overwhelming the imitative principle of truth to nature. This heraldic design can then be followed to Persia, Syria, later to Byzantium, and anyone who lived before the First World War will remember the double-headed eagle in the coats-of-arms of Czarist Russia and the Austro-Hungarian monarchy.

Look now at this Sumerian picture (Figure 4). The two eagle-headed men are nearly but not quite symmetric; why not? In plane geometry reflection in a vertical line *l* can also be brought about by rotating the plane in space around the axis *l* by 180°. If you look at their arms you would say these two monsters arise from each other by such rotation; the overlappings depicting their position in space prevent the plane picture from having bilaterial symmetry. Yet the artist aimed at that symmetry by giving both figures a half turn toward the observer and also by the arrangement of feet and wings: the drooping wing is the right one in the left figure, the left one in the right figure.

The designs on the cylindrical Babylonian seal stones are frequently ruled by heraldic symmetry. I remember seeing in the collection of my former colleague, the late Ernst Herzfeld, samples where for symmetry's sake not the head, but the lower bull-shaped part of a god's body, rendered in profile, was doubled and given four instead of two hind legs. In Christian times one may see an analogy in certain representations of the Eucharist as on this Byzantine platen (Figure 5), where two symmetric

Christs are facing the disciples. But here symmetry is not complete and has clearly more than formal significance, for Christ on one side breaks the bread, on the other pours the wine.

Between Sumeria and Byzantium let me insert Persia: These enameled sphinxes (Figure 6) are from Darius' palace in Susa built in the days of Marathon. Crossing the Aegean we find these floor patterns (Figure 7) at the Megaron in Tiryns, late helladic about 1200 B.C. Who believes strongly in historic continuity and dependence will trace the graceful designs of marine life, dolphin and octopus, back to the Minoan culture of Crete, the heraldic symmetry to oriental, in the last instance Sumerian, influence. Skipping thousands of years we still see the same influences at work in this plaque (Figure 8) from the altar enclosure in the dome of Torcello, Italy, eleventh century A.D. The peacocks drinking from a pine well among vine leaves are an ancient Christian symbol of immortality, the structural heraldic symmetry is oriental.

For in contrast to the orient, occidental art, like life itself, is inclined to mitigate, to loosen, to modify, even to break strict symmetry. But seldom is asymmetry merely the absence of symmetry. Even in asymmetric designs one feels symmetry as the norm from which one deviates under the influence of forces of non-formal character. I think the riders

FIGURE 5

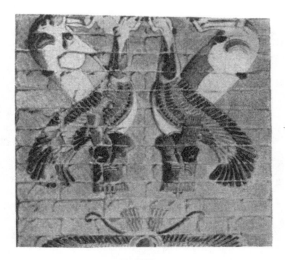

FIGURE 6

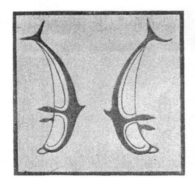

FIGURE 7

from the famous Etruscan Tomb of the Triclinium at Corneto (Figure 9) provide a good example. I have already mentioned representations of the Eucharist with Christ duplicated handing out bread and wine. The central group, Mary flanked by two angels, in this mosaic of the Lord's Ascension (Figure 12) in the cathedral at Monreale, Sicily (twelfth century), has almost perfect symmetry. [The band ornaments above and below the mosaic will demand our attention in the second lecture.] The principle of symmetry is somewhat less strictly observed in an earlier mosaic from San Apollinare in Ravenna (Figure 10), showing Christ surrounded by an angelic guard of honor. For instance Mary in the Monreale mosaic raises both hands symmetrically, in the *orans* gesture; here only the right hands

FIGURE 8

are raised. Asymmetry has made further inroads in the next picture
(Figure 11), a Byzantine relief ikon from San Marco, Venice. It is a
Deësis, and, of course, the two figures praying for mercy as the Lord
is about to pronounce the last judgment cannot be mirror images of each
other; for to the right stands his Virgin Mother, to the left John the
Baptist. You may also think of Mary and John the Evangelist on both
sides of the cross in crucifixions as examples of broken symmetry.

Clearly we touch ground here where the precise geometric notion of
bilateral symmetry begins to dissolve into the vague notion of *Ausge-
wogenheit*, balanced design with which we started. "Symmetry," says
Dagobert Frey in an article *On the Problem of Symmetry in Art*,[4]
"signifies rest and binding, asymmetry motion and loosening, the one
order and law, the other arbitrariness and accident, the one formal rigidity
and constraint, the other life, play and freedom." Wherever God or Christ
are represented as symbols for everlasting truth or justice they are given
in the symmetric frontal view, not in profile. Probably for similar reasons
public buildings and houses of worship, whether they are Greek temples
or Christian basilicas and cathedrals, are bilaterally symmetric. It is, how-

4 Studium Generale, p. 276.

FIGURE 9

ever, true that not infrequently the two towers of Gothic cathedrals are different, as for instance in Chartres. But in practically every case this seems to be due to the history of the cathedral, namely to the fact that the towers were built in different periods. It is understandable that a later time was no longer satisfied with the design of an earlier period; hence one may speak here of historic asymmetry. Mirror images occur where there is a mirror, be it a lake reflecting a landscape or a glass mirror into which a woman looks. Nature as well as painters make use of this motif. I trust, examples will easily come to your mind. The one most familiar to me, because I look at it in my study every day, is Hodler's *Lake of Silvaplana.*

While we are about to turn from art to nature, let us tarry a few minutes and first consider what one may call the *mathematical philosophy of left and right.* To the scientific mind there is no inner difference, no polarity between left and right, as there is for instance in the contrast of male and female, or of the anterior and posterior ends of an animal. It requires an arbitrary act of choice to determine what is left and what is right. But after it is made for one body it is determined for every body. I must try to make this a little clearer. In space the distinction of left and right concerns the orientation of a screw. If you speak of turning left you mean that the sense in which you turn combined with the upward direction from foot to head of your body forms a left screw. The daily rotation of the earth together with the direction of its axis from South to North Pole is

FIGURE 10

FIGURE 11

a left screw, it is a right screw if you give the axis the opposite direction. There are certain crystalline substances called optically active which betray the inner asymmetry of their constitution by turning the polarization plane of polarized light sent through them either to the left or to the right; by this, of course, we mean that the sense in which the plane rotates while the light travels in a definite direction, combined with that direction, forms a left screw (or a right one, as the case may be). Hence when we said above and now repeat in a terminology due to Leibniz, that left and right are *indiscernible*, we want to express that the inner structure of space does not permit us, except by arbitrary choice, to distinguish a left from a right screw.

I wish to make this fundamental notion still more precise, for on it depends the entire theory of relativity, which is but another aspect of symmetry. According to Euclid one can describe the structure of space by a number of basic relations between points, such as ABC lie on a straight line, $ABCD$ lie in a plane, AB is congruent CD. Perhaps the best way of describing the structure of space is the one Helmholtz adopted: by the

FIGURE 12

single notion of *congruence* of figures. A mapping S of space associates with every point p a point p' : $p \rightarrow p'$. A pair of mappings S, S' : $p \rightarrow p'$, $p' \rightarrow p$, of which the one is the inverse of the other, so that if S carries p into p' then S' carries p' back into p and vice versa, is spoken of as a pair

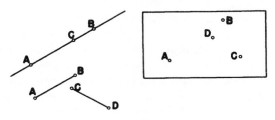

FIGURE 13

of one-to-one mappings or *transformations*. A transformation which preserves the structure of space—and if we define this structure in the Helmholtz way, that would mean that it carries any two congruent figures into two congruent ones—is called an *automorphism* by the mathematicians. Leibniz recognized that this is the idea underlying the geometric concept of similarity. An automorphism carries a figure into one that in Leibniz' words is "indiscernible from it if each of the two figures is considered by itself." What we mean then by stating that left and right are of the same essence is the fact that *reflection in a plane is an automorphism*.

Space as such is studied by geometry. But space is also the medium of all physical occurrences. The structure of the physical world is revealed by the general laws of nature. They are formulated in terms of certain basic quantities which are functions in space and time. We would conclude that the physical structure of space "contains a screw," to use a

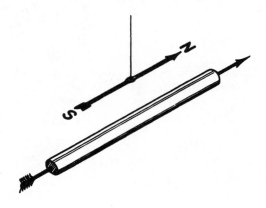

FIGURE 14

suggestive figure of speech, if these laws were not invariant throughout with respect to reflection. Ernst Mach tells of the intellectual shock he received when he learned as a boy that a magnetic needle is deflected in a certain sense, to the left or to the right, if suspended parallel to a wire through which an electric current is sent in a definite direction (Figure 14). Since the whole geometric and physical configuration, including the electric current and the south and north poles of the magnetic needle, to all appearances, are symmetric with respect to the plane E laid through the wire and the needle, the needle should react like Buridan's ass between equal bundles of hay and refuse to decide between left and right, just as scales of equal arms with equal weights neither go down on their left nor on their right side but stay horizontal. But appearances are sometimes deceptive. Young Mach's dilemma was the result of a too hasty assumption concerning the effect of reflection in E on the electric current and the positive and negative magnetic poles of the needle: while we know a priori how geometric entities fare under reflection, we have to learn from nature how the physical quantities behave. And this is what we find: under reflection in the plane E the electric current preserves its direction, but the magnetic south and north poles are interchanged. Of course this way out, which re-establishes the equivalence of left and right, is possible only because of the essential equality of positive and negative magnetism. All doubts were dispelled when one found that the magnetism of the needle has its origin in molecular electric currents circulating around the needle's direction; it is clear that under reflection in the plane E such currents change the sense in which they flow.

The net result is that in all physics nothing has shown up indicating an intrinsic difference of left and right. Just as all points and all directions in space are equivalent, so are left and right. Position, direction, left and right are *relative* concepts. In language tinged with theology this issue of relativity was discussed at great length in a famous controversy between Leibniz and Clarke, the latter a clergyman acting as the spokesman for Newton.[5] Newton with his belief in absolute space and time considers motion a proof of the creation of the world out of God's arbitrary will, for otherwise it would be inexplicable why matter moves in this rather than in any other direction. Leibniz is loath to burden God with such decisions as lack "sufficient reason." Says he, "Under the assumption that space be something in itself it is impossible to give a reason why God should have put the bodies (without tampering with their mutual distances and relative positions) just at this particular place and not somewhere else; for instance, why He should not have arranged everything in the opposite order by turning East and West about. If, on the other hand,

---

[5] See G. W. Leibniz, *Philosophische Schriften*, ed. Gerhardt (Berlin 1875 seq.), VII, pp. 352–440, in particular Leibniz' third letter, §5.

space is nothing more than the spatial order and relation of things then the two states supposed above, the actual one and its transposition, are in no way different from each other . . . and therefore it is a quite inadmissible question to ask why one state was preferred to the other." By pondering the problem of left and right Kant was first led to his conception of space and time as forms of intuition.[6] Kant's opinion seems to have been this: If the first creative act of God had been the forming of a left hand then this hand, even at the time when it could be compared to nothing else, had the distinctive character of left, which can only intuitively but never conceptually be apprehended. Leibniz contradicts: According to him it would have made no difference if God had created a "right" hand first rather than a "left" one. One must follow the world's creation a step further before a difference can appear. Had God, rather than making first a left and then a right hand, started with a right hand and then formed another right hand, He would have changed the plan of the universe not in the first but in the second act, by bringing forth a hand which was equally rather than oppositely oriented to the first-created specimen.

Scientific thinking sides with Leibniz. Mythical thinking has always taken the contrary view as is evinced by its usage of right and left as symbols for such polar opposites as good and evil. You need only think of the double meaning of the word *right* itself. In this detail from Michelangelo's famous *Creation of Adam* from the Sistine Ceiling (Figure 15) God's right hand, on the right, touches life into Adam's left.

People shake right hands. *Sinister* is the Latin word for left, and heraldry still speaks of the left side of the shield as its sinister side. But *sinistrum* is at the same time that which is evil, and in common English only this figurative meaning of the Latin word survives.[7] Of the two malefactors who were crucified with Christ, the one who goes with Him to paradise is on His right. St. Matthew, Chapter 25, describes the last judgment as follows: "And he shall set the sheep on his right hand but the goats on the left. Then shall the King say unto them on his right hand, Come ye, blessed of my Father, inherit the Kingdom prepared for you from the foundation of the world. . . . Then he shall say also unto them on the left hand, Depart from me, ye cursed, into everlasting fire, prepared for the devil and his angels."

I remember a lecture Heinrich Wölfflin once delivered in Zurich on "Right and left in paintings"; together with an article on "The problem of inversion (Umkehrung) in Raphael's tapestry cartoons," you now find it printed in abbreviated form in his *Gedanken zur Kunstgeschichte*, 1941.

[6] Besides his "Kritik der reinen Vernunft" see especially §13 of the *Prolegomena zu einer jeden künftigen Metaphysik.* . . .

[7] I am not unaware of the strange fact that as a *terminus technicus* in the language of the Roman augurs *sinistrum* had just the opposite meaning of propitious.

FIGURE 15

By a number of examples, as Raphael's *Sistine Madonna* and Rembrandt's etching *Landscape with the three trees*, Wölfflin tries to show that right in painting has another *Stimmungswert* than left. Practically all methods of reproduction interchange left and right, and it seems that former times were much less sensitive than we are toward such inversion. (Even Rembrandt did not hesitate to bring his Descent from the Cross as a converse etching upon the market.) Considering that we do a lot more reading than the people, say, of the sixteenth century, this suggests the hypothesis that the difference pointed out by Wölfflin is connected with our habit of reading from left to right. As far as I remember, he himself rejected this as well as a number of other psychological explanations put forward in the discussion after his lecture. The printed text concludes with the remark that the problem "obviously has deep roots, roots which reach down to the very foundations of our sensuous nature." I for my part am disinclined to take the matter that seriously.[8]

In science the belief in the equivalence of left and right has been upheld even in the face of certain biological facts presently to be mentioned which seem to suggest their inequivalence even more strongly than does the deviation of the magnetic needle which shocked young Mach. The same

[8] Cf. also A. Faistauer, "Links und rechts im Bilde," Amicis, *Jahrbuch der österreichischen Galerie*, 1926, p. 77; Julius v. Schlosser, "Intorno alla lettura dei quadri," *Critica 28*, 1930, p. 72; Paul Oppé, "Right and left in Raphael's cartoons," *Journal of the Warburg and Courtauld Institutes 7*, 1944, p. 82.

problem of equivalence arises with respect to *past and future*, which are
interchanged by inverting the direction of time, and with respect to *positive and negative electricity*. In these cases, especially in the second, it is
perhaps clearer than for the pair left-right that a priori evidence is not
sufficient to settle the question; the empirical facts have to be consulted.
To be sure, the role which past and future play in our consciousness
would indicate their intrinsic difference—the past knowable and unchange-
able, the future unknown and still alterable by decisions taken now—and
one would expect that this difference has its basis in the physical laws of
nature. But those laws of which we can boast a reasonably certain knowledge are invariant with respect to the inversion of time as they are with
respect to the interchange of left and right. Leibniz made it clear that the
temporal modi past and future refer to the *causal structure* of the world.
Even if it is true that the exact "wave laws" formulated by quantum
physics are not altered by letting time flow backward, the metaphysical
idea of causation, and with it the one way character of time, may enter
physics through the statistical interpretation of those laws in terms of
probability and particles. Our present physical knowledge leaves us even
more uncertain about the equivalence or non-equivalence of positive and
negative electricity. It seems difficult to devise physical laws in which they
are not intrinsically alike; but the negative counterpart of the positively
charged proton still remains to be discovered.

This half-philosophical excursion was needed as a background for the
discussion of the left-right symmetry in nature; we had to understand that
the general organization of nature possesses that symmetry. But one will
not expect that any special object of nature shows it to perfection. Even
so, it is surprising to what extent it prevails. There must be a reason for
this, and it is not far to seek: a state of equilibrium is likely to be symmetric. More precisely, under conditions which determine a unique state
of equilibrium the symmetry of the conditions must carry over to the state
of equilibrium. Therefore tennis balls and stars are spheres; the earth
would be a sphere too if it did not rotate around an axis. The rotation
flattens it at the poles but the rotational or cylindrical symmetry around
its axis is preserved. The feature that needs explanation is, therefore, not
the rotational symmetry of its shape but the deviations from this symmetry as exhibited by the irregular distribution of land and water and
by the minute crinkles of mountains on its surface. It is for such reasons
that in his monograph on the left-right problem in zoology Wilhelm Ludwig says hardly a word about the origin of the bilateral symmetry prevailing in the animal kingdom from the echinoderms upward, but in great
detail discusses all sorts of secondary asymmetries superimposed upon the
symmetrical ground plan.[9] I quote: "The human body like that of the

[9] W. Ludwig, *Rechts-links-Problem im Tierreich und beim Menschen*, Berlin 1932.

other vertebrates is basically built bilateral-symmetrically. All asymmetries occurring are of secondary character, and the more important ones affecting the inner organs are chiefly conditioned by the necessity for the intestinal tube to increase its surface out of proportion to the growth of the body, which lengthening led to an asymmetric folding and rolling-up. And in the course of phylogenetic evolution these first asymmetries concerning the intestinal system with its appendant organs brought about asymmetries in other organ systems." It is well known that the heart of mammals is an asymmetric screw, as shown by the schematic drawing of Figure 16.

FIGURE 16

If nature were all lawfulness then every phenomenon would share the full symmetry of the universal laws of nature as formulated by the theory of relativity. The mere fact that this is not so proves that *contingency* is an essential feature of the world. Clarke in his controversy with Leibniz admitted the latter's principle of sufficient reason but added that the sufficient reason often lies in the mere will of God. I think, here Leibniz the rationalist is definitely wrong and Clarke on the right track. But it would have been more sincere to deny the principle of sufficient reason altogether instead of making God responsible for all that is unreason in the world. On the other hand Leibniz was right against Newton and Clarke with his insight into the principle of relativity. The truth as we see it today is this: The laws of nature do not determine uniquely the one world that actually exists, not even if one concedes that two worlds arising from each other by an automorphic transformation, i.e., by a transformation which preserves the universal laws of nature, are to be considered the same world.

If for a lump of matter the overall symmetry inherent in the laws of nature is limited by nothing but the accident of its position P then it will assume the form of a sphere around the center P. Thus the lowest forms of animals, small creatures suspended in water, are more or less spherical. For forms fixed to the bottom of the ocean the direction of gravity is an important factor, narrowing the set of symmetry operations from all rotations around the center P to all rotations about an axis. But for animals capable of self-motion in water, air, or on land both the postero-anterior direction in which their body moves and the direction of gravity are of decisive influence. After determination of the antero-posterior, the dorso-ventral, and thereby of the left-right axes, only the distinction between left and right remains arbitrary, and at this stage no higher symmetry than the bilateral type can be expected. Factors in the phylogenetic evolution that tend to introduce inheritable differences between left and right are likely to be held in check by the advantage an animal derives from the bilateral formation of its organs of motion, cilia or muscles and limbs: in case of their asymmetric development a screw-wise instead of a straight-forward motion would naturally result. This may help to explain why our limbs obey the law of symmetry more strictly than our inner organs. Aristophanes in Plato's *Symposium* tells a different story of how the transition from spherical to bilateral symmetry came about. Originally, he says, man was round, his back and sides forming a circle. To humble their pride and might Zeus cut them into two and had Apollo turn their faces and genitals around; and Zeus had threatened, "If they continue insolent I will split them again and they shall hop around on a single leg."

The most striking examples of symmetry in the inorganic world are the crystals. The gaseous and the crystalline are two clear-cut states of matter which physics finds relatively easy to explain; the states in between these two extremes, like the fluid and the plastic states, are somewhat less amenable to theory. In the gaseous state molecules move freely around in space with mutually independent random positions and velocities. In the crystalline state atoms oscillate about positions of equilibrium as if they were tied to them by elastic strings. These positions of equilibrium form a fixed regular configuration in space.[10]. . . While most of the thirty-two geometrically possible systems of crystal symmetry involve bilateral symmetry, not all of them do. Where it is not involved we have the possibility of so-called enantiomorph crystals which exist in a laevo- and dextro-form, each form being a mirror image of the other, like left and right hands. A substance which is optically active, i.e., turns the plane of polarized light either left or right, can be expected to crystallize in such asymmetric forms. If the laevo-form exists in nature one would assume that the dextro-form exists likewise, and that in the average both occur with equal frequencies. In 1848 Pasteur made the discovery that when the sodium ammonium salt of optically inactive racemic acid was recrystallized from an aqueous solution at a lower temperature the deposit consisted of two kinds of tiny crystals which were mirror images of each other. They were carefully separated, and the acids set free from the one and the other proved to

---

[10] [In a later lecture Weyl explains how the visible symmetry of crystals derives from their regular atomic arrangement. ED.]

have the same chemical composition as the racemic acid, but one was optically laevo-active, the other dextro-active. The latter was found to be identical with the tartaric acid present in fermenting grapes, the other had never before been observed in nature. "Seldom," says F. M. Jaeger in his lectures *On the principle of symmetry and its applications in natural science*, "has a scientific discovery had such far-reaching consequences as this one had."

Quite obviously some accidents hard to control decide whether at a spot of the solution a laevo- or dextro-crystal comes into being; and thus in agreement with the symmetric and optically inactive character of the solution as a whole and with the law of chance the amounts of substance deposited in the one and the other form at any moment of the process of crystallization are equal or very nearly equal. On the other hand nature, in giving us the wonderful gift of grapes so much enjoyed by Noah, produced only one of the forms, and it remained for Pasteur to produce the other! This is strange indeed. It is a fact that most of the numerous carbonic compounds occur in nature in one, either the laevo- or the dextro-form only. The sense in which a snail's shell winds is an inheritable character founded in its genetic constitution, as is the "left heart" and the winding of the intestinal duct in the species *Homo sapiens*. This does not exclude that inversions occur, e.g. *situs inversus* of the intestines of man occurs with a frequency of about 0.02 per cent; we shall come back to that later! Also the deeper chemical constitution of our human body shows that we have a screw, a screw that is turning the same way in every one of us. Thus our body contains the dextro-rotatory form of glucose and laevo-rotatory form of fructose. A horrid manifestation of this genotypical asymmetry is a metabolic disease called phenylketonuria, leading to insanity, that man contracts when a small quantity of laevo-phenylalanine is added to his food, while the dextro-form has no such disastrous effects. To the asymmetric chemical constitution of living organisms one must attribute the success of Pasteur's method of isolating the laevo- and dextro-forms of substances by means of the enzymatic action of bacteria, moulds, yeasts, and the like. Thus he found that an originally inactive solution of some racemate became gradually laevo-rotatory if *Penicillium glaucum* was grown in it. Clearly the organism selected for its nutriment that form of the tartaric acid molecule which best suited its own asymmetric chemical constitution. The image of lock and key has been used to illustrate this specificity of the action of organisms.

In view of the facts mentioned and in view of the failure of all attempts to "activate" by mere chemical means optically inactive material,[11] it is understandable that Pasteur clung to the opinion that the production

[11] There is known today one clear instance, the reaction of nitrocinnaminacid with bromine where circular-polarized light generates an optically active substance.

of single optically active compounds was the very prerogative of life. In 1860 he wrote, "This is perhaps the only well-marked line of demarkation that can at present be drawn between the chemistry of dead and living matter." Pasteur tried to explain his very first experiment where racemic acid was transformed by recrystallization into a mixture of laevo- and dextro-tartaric acid by the action of bacteria in the atmosphere on his neutral solution. It is quite certain today that he was wrong; the sober physical explanation lies in the fact that at lower temperature a mixture of the two oppositely active tartaric forms is more stable than the in-active racemic form. If there is a difference in principle between life and death it does not lie in the chemistry of the material substratum; this has been fairly certain ever since Wöhler in 1828 synthesized urea from purely mineral material. But even as late as 1898 F. R. Japp in a famous lecture on "Stereochemistry and Vitalism" before the British Association upheld Pasteur's view in the modified form: "Only the living organisms, or the living intelligence with its conception of symmetry can produce this result (i.e. asymmetric compounds)." Does he really mean that it is Pasteur's intelligence that, by devising the experiment but to its own great surprise, *creates* the dual tartaric crystals? Japp continues, "Only asymmetry can beget asymmetry." The truth of that statement I am willing to admit; but it is of little help since there is no symmetry in the accidental past and present set-up of the actual world which begets the future.

There is however a real difficulty: Why should nature produce only one of the doublets of so many enantiomorphic forms the origin of which most certainly lies in living organisms? Pascual Jordan points to this fact as a support for his opinion that the beginnings of life are not due to chance events which, once a certain stage of evolution is reached, are apt to occur continuously now here now there, but rather to an event of quite singular and improbable character, occurring once by accident and then starting an avalanche by autocatalytic multiplication. Indeed had the asym-metric protein molecules found in plants and animals an independent origin in many places at many times, then their laevo- and dextro-varieties should show nearly the same abundance. Thus it looks as if there is some truth in the story of Adam and Eve, if not for the origin of mankind then for that of the primordial forms of life. It was in reference to these biological facts when I said before that if taken at their face value they suggest an intrinsic difference between left and right, at least as far as the constitution of the organic world is concerned. But we may be sure the answer to our riddle does not lie in any universal biological laws but in the accidents of the genesis of the organismic world. Pascual Jordan shows one way out; one would like to find a less radical one, for instance by reducing the asymmetry of the inhabitants on earth to some inherent, though accidental, asymmetry of the earth itself, or of the light received on earth from the sun. But neither the earth's rotation nor the combined magnetic fields of earth and sun are of immediate help in this regard. Another possibility would be to assume that development actually started from an equal distribution of the enantiomorph forms, but that this is an unstable equilibrium which under a slight chance disturbance tumbled over.

From the phylogenetic problems of left and right let us finally turn to their ontogenesis. Two questions arise: Does the first division of the fertilized egg of an animal into two cells fix the median plane, so that one of the cells contains the potencies for its left, the other for its right half? Secondly what determines the plane of the first division? I begin with the second question. The egg of any animal above the protozoa possesses from the beginning a polar axis connecting what develops into the animal and the vegetative poles of the blastula. This axis together with the point where the fertilizing spermatozoon enters the egg determines a plane, and it would be quite natural to assume that this is the plane of the first division. And indeed there is evidence that it is so in many cases. Present opinion seems to incline toward the assumption that the primary polarity as well as the subsequent bilateral symmetry come about by external factors actualizing potentialities inherent in the genetic constitution. In many instances the direction of the polar axis is obviously determined by the attachment of the oozyte to the wall of the ovary, and the point of entrance of the fertilizing sperm is, as we said, at least one, and often the most decisive, of the determining factors for the median plane. But other agencies may also be responsible for the fixation of the one and the other. In the sea-weed *Fucus* light or electric fields or chemical gradients determine the polar axis, and in some insects and cephalopods the median plane appears to be fixed by ovarian influences before fertilization.[12] The underlying constitution on which these agencies work is sought by some biologists in an intimate preformed structure, of which we do not yet have a clear picture. Thus Conklin has spoken of a spongioplasmic framework, others of a cytoskeleton, and as there is now a strong tendency among biochemists to reduce structural properties to fibers, so much so that Joseph Needham in his Terry Lectures on *Order and life* (1936) dares the aphorism that biology is largely the study of fibers, one may expect them to find that that intimate structure of the egg consists of a framework of elongated protein molecules or fluid crystals.

[12] Julian S. Huxley and G. R. de Beer in their classical *Elements of embryology* (Cambridge University Press, 1934) give this formulation (Chapter XIV, Summary, p. 438): "In the earliest stages, the egg acquires a unitary organization of the gradient-field type in which quantitative differentials of one or more kinds extend across the substance of the egg in one or more directions. The constitution of the egg predetermines it to be able to produce a gradient-field of a particular type; however, the localization of the gradients is not predetermined, but is brought about by agencies external to the egg."

We know a little more about our first question whether the first mitosis of the cell divides it into left and right. Because of the fundamental character of bilateral symmetry the hypothesis that this is so seems plausible enough. However, the answer cannot be an unqualified affirmation. Even if the hypothesis should be true for the normal development we know from experiments first performed by Hans Driesch on the sea urchin that a single blastomere isolated from its partner in the two-cell stage develops into a whole gastrula differing from the normal one only by its smaller size. Here are Driesch's famous pictures. It must be admitted that this is not so for all species. Driesch's discovery led to the distinction between the actual and the potential destiny of the several parts of an egg. Driesch himself speaks of prospective significance (prospektive Bedeutung), as against prospective potency (prospektive Potenz); the latter is wider than the former, but shrinks in the course of development. Let me illustrate this basic point by another example taken from the determination of limb-buds of amphibia. According to experiments performed by R. G. Harrison, who transplanted discs of the outer wall of the body representing the buds of

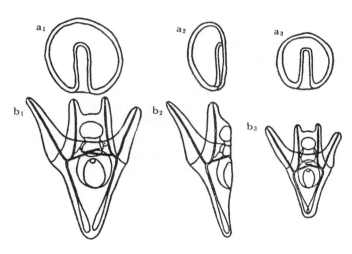

FIGURE 17

Experiments on pluripotence in "Echinus."
  a_1 *and* b_1. *Normal gastrula and normal pluteus.*
  a_2 *and* b_2. *Half-gastrula and half-pluteus, expected by Driesch.*
  a_3 *and* b_3. *The small but whole gastrula and pluteus, which he actually*
    obtained.

future limbs, the antero-posterior axis is determined at a time when transplantation may still invert the dorso-ventral and the medio-lateral axes; thus at this stage the opposites of left and right still belong to the prospec-

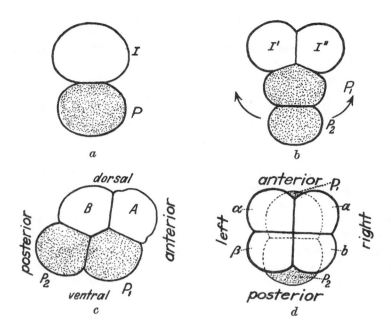

FIGURE 18

tive potencies of the discs, and it depends on the influence of the surrounding tissues in which way this potency will be actualized.

Driesch's violent encroachment on the normal development proves that the first cell division may not fix left and right of the growing organism for good. But even in normal development the plane of the first division may not be the median. The first stages of cell division have been closely studied for the worm *Ascaris megalocephala*, parts of whose nervous system are asymmetric. First the fertilized egg splits into a cell I and a smaller P of obviously different nature (Figure 18). In the next stage they divide along two perpendicular planes into $I' + I''$ and $P_1 + P_2$ respectively. Thereafter the handle $P_1 + P_2$ turns about so that P_2 comes into contact with either I' or I''; call the one it contacts B, the other A. We now have a sort of rhomboid and roughly AP_2 is the antero-posterior axis and BP_1 the dorsal-ventral one. Only the next division which along a plane perpendicular to the one separating A and B splits A as well as B into symmetric halves $A = a + a$, $B = b + \beta$, is that which determines left and right. A further slight shift of the configuration destroys this bilateral symmetry. The question arises whether the direction of the two consecutive shifts is a chance event which decides first between anterior and posterior and then between left and right, or whether the constitution of the egg in its one-cell stage contains specific agents which determine the direction of these shifts. The hypothesis of the mosaic egg favoring the second hypothesis seems more likely for the species *Ascaris*.

There are known a number of cases of genotypical inversion where the genetic constitutions of two species are in the same relation as the atomic constitutions of two enantiomorph crystals. More frequent, however, is phenotypical inversion. Left-handedness in man is an example. I give another more interesting one. Several crustacea of the lobster type have two morphologically and functionally different claws, a bigger A and a smaller a. Assume that in normally developed individuals of our species, A is the right claw. If in a young animal you cut off the right claw, inversive regeneration takes place: the left claw develops into the bigger form A while at the place of the right claw a small one of type a is regenerated. One has to infer from such and similar experiences the bipotentiality of plasma, namely that all generative tissues which contain the potency of an asymmetric character have the potency of bringing forth both forms, so however that in normal development always one form develops, the left or the right. Which one is genetically determined, but abnormal external circumstances may cause inversion. On the basis of the strange phenomenon of inversive regeneration Wilhelm Ludwig developed the hypothesis that the decisive factors in asymmetry may not be such specific potencies as, say, the development of a "right claw of type A," but two R and L (right and left) agents which are distributed in the organism with a certain gradient, the concentration of one falling off from right to left, the other in the opposite direction. The essential point is that there is not one but that there are two opposite gradient fields R and L. Which is produced in greater strength is determined by the genetic constitution. If, however, by some damage to the prevalent agent the other previously suppressed one becomes prevalent, then inversion takes place. Being a mathematician and not a biologist I report with the utmost caution on these matters, which seem to me of highly hypothetical nature. But it is clear that the contrast of left and right is connected with the deepest problems concerning the phylogenesis as well as the ontogenesis of organisms.

TRANSLATORY, ROTATIONAL, AND RELATED SYMMETRIES

From bilateral, we shall now turn to other kinds of geometric symmetry. Even in discussing the bilateral type I could not help drawing in now and then such other symmetries as the cylindrical or the spherical ones. It seems best to fix the underlying general concept with some precision beforehand, and to that end a little mathematics is needed, for which I ask your patience. I have spoken of transformations. A mapping S of space associates with every space point p a point p' as its image. A special such mapping is the identity I carrying every point p into itself. Given two mappings S, T, one can perform one after the other: if S carries p into p' and T carries p' into p'' then the resulting mapping, which we denote by ST, carries p into p''. A mapping may have an inverse S' such that $SS' = I$ and $S'S = I$; in other words, if S carries the arbitrary point p into p' then S' carries p' back into p, and a similar condition prevails with S' performed in the first and S in the second place. For such a one-to-one mapping S the word transformation was used in the first lecture; let the inverse be denoted by S^{-1}. Of course, the identity I is a transformation, and I itself is its inverse. Reflection in a plane, the basic operation of bilateral symmetry, is such that its iteration SS results in the identity; in other words, it is its own inverse. In general composition of mappings is not commutative; ST need not be the same as TS. Take for instance a point o in a plane and let S be a horizontal translation carrying o into o_1 and T a rotation around o by 90°. Then ST carries o into the point o_2 (Figure 19), but TS carries o into o_1. If S is a transformation with the inverse S^{-1}, then S^{-1} is also a transformation and its inverse is S. The composite of two transformations ST is a transformation again, and $(ST)^{-1}$ equals $T^{-1}S^{-1}$ (in this order!). With this rule, although perhaps not with its mathematical expression, you are all familiar. When you dress, it is not immaterial in which order you perform the operations; and when in dressing you start with the shirt and end up with the coat, then in undressing you observe the opposite order; first take off the coat and the shirt comes last.

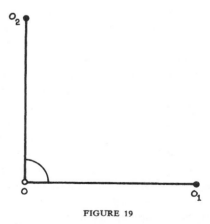

FIGURE 19

I have further spoken of a special kind of transformations of space called similarity by the geometers. But I preferred the name of automorphisms for them, defining them with Leibniz as those transformations which leave the structure of space unchanged. For the moment it is immaterial wherein that structure consists. From the very definition it is clear that the identity I is an automorphism, and if S is, so is the inverse S^{-1}. Moreover the composite ST of two automorphisms S, T is again an automorphism. This is only another way of saying that (1) every figure is similar to itself, (2) if figure F' is similar to F then F is similar to F', and (3) if F is similar to F' and F' to F'' then F is similar to F''. The mathematicians have adopted the word *group* to describe this situation and therefore say that the *automorphisms form a group*. Any totality, any set Γ of transformations form a group provided the following conditions are satisfied: (1) the identity I belongs to Γ; (2) if S belongs to Γ then its inverse S^{-1} does; (3) if S and T belong to Γ then the composite ST does.

One way of describing the structure of space, preferred by both Newton and Helmholtz, is through the notion of congruence. Congruent parts of space V, V' are such as can be occupied by the same rigid body in two of its positions. If you move the body from the one into the other position the particle of the body covering a point p of V will afterwards cover a certain point p' of V', and thus the result of the motion is a mapping $p \rightarrow p'$ of V upon V'. We can extend the rigid body either actually or in imagination so as to cover an arbitrarily given point p of space, and hence the congruent mapping $p \rightarrow p'$ can be extended to the entire space. Any such congruent transformation—I call it by that name because it evidently has an inverse $p' \rightarrow p$—is a similarity or an automorphism; you can easily convince yourselves that this follows from the very concepts. It is evident moreover that the congruent transformations form a group, a subgroup of the group of automorphisms. In more detail the situation is this. Among the similarities there are those which do not change the dimensions of a body; we shall now call them congruences. A congruence is either proper, carrying a left screw into a left and a right one into a right, or it is improper or reflexive, changing a left screw into a right one and vice versa. The proper congruences are those transformations which a moment ago we called congruent transformations, connecting the positions of points of a rigid body before and after a motion. We shall now call them simply motions (in a nonkinematic geometric sense) and call the improper congruences reflections, after the most important example: reflection in a plane, by which a body goes over into its mirror image. Thus we have this step-wise arrangement: similarities → congruences = similarities without change of scale → motions = proper congruences. The congruences form a subgroup of the similarities, the motions form a subgroup of the group of congruences, of index 2. The latter addition means that if B is any given improper congruence, we obtain all improper congruences in the form BS by composing B with all possible proper congruences S. Hence the proper congruences form one half, and the improper ones another half, of the group of all congruences. But only the first half is a group; for the composite AB of two improper congruences A, B is a proper congruence.

A congruence leaving the point O fixed may be called *rotation* around O; thus there are proper and improper rotations. The rotations around a

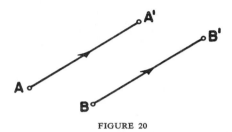

FIGURE 20

given center O form a group. The simplest type of congruences are the *translations*. A translation may be represented by a vector $\overrightarrow{AA'}$; for if a translation carries a point A into A' and the point B into B' then BB' has the same direction and length as AA', in other words the vector $BB' = AA'$.[13] The translations form a group; indeed the succession of the two translations $\overrightarrow{AB}$, $\overrightarrow{BC}$ results in the translation $\overrightarrow{AC}$.

What has all this to do with symmetry? It provides the adequate mathematical language to define it. Given a spatial configuration $\mathfrak{F}$, those automorphisms of space which leave $\mathfrak{F}$ unchanged form a group Γ, and *this group describes exactly the symmetry possessed by* $\mathfrak{F}$. Space itself has the full symmetry corresponding to the group of all automorphisms, of all similarities. The symmetry of any figure in space is described by a subgroup of that group. Take for instance the famous pentagram (Figure 21) by which Dr. Faust banned Mephistopheles the devil. It is carried into itself by the five proper rotations around its center O, the angles of which are multiples of $360°/5$ (including the identity), and then by the five reflections in the lines joining O with the five vertices. These ten operations form a group, and that group tells us what sort of symmetry the pentagram possesses. Hence the natural generalization which leads from bilateral symmetry to symmetry in this wider geometric sense consists in replacing reflection in a plane by any group of automorphisms. The circle in a plane with center

---

[13] While a segment has only length, a vector has length and direction. A vector is really the same thing as a translation, although one uses different phraseologies for vectors and translations. Instead of speaking of the translation $\mathfrak{a}$ which carries the point A into A' one speaks of the vector $\mathfrak{a} = \overrightarrow{AA'}$; and instead of the phrase: the translation $\mathfrak{a}$ carries A into A' one says that A' is the end point of the vector $\mathfrak{a}$ laid off from A. The same vector laid off from B ends in B' if the translation carrying A into A' carries B into B'.

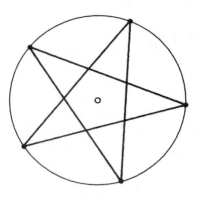

FIGURE 21

O and the sphere in space around O have the symmetry described by the group of all plane or spatial rotations respectively.

If a figure $\mathfrak{F}$ does not extend to infinity then an automorphism leaving the figure invariant must be scale-preserving and hence a congruence, unless the figure consists of one point only. Here is the simple proof. Had we an automorphism leaving $\mathfrak{F}$ unchanged, but changing the scale, then either this automorphism or its inverse would increase (and not decrease) all linear dimensions in a certain proportion $a : 1$ where a is a number greater than 1. Call that automorphism S, and let α, β be two different points of our figure $\mathfrak{F}$. They have a positive distance d. Iterate the transformation S,

$$S = S^1, \; SS = S^2, \; SSS = S^3, \; \ldots \ldots$$

The n-times iterated transformation S^n carries α and β into two points α_n, β_n of our figure whose distance is $d \cdot a^n$. With increasing exponent n this distance tends to infinity. But if our figure $\mathfrak{F}$ is bounded, there is a number c such that no two points of $\mathfrak{F}$ have a distance greater than c. Hence a contradiction arises as soon as n becomes so large that $d \cdot a^n > c$. The argument shows another thing: Any finite group of automorphisms consists exclusively of congruences. For if it contains an S that enlarges linear dimensions at the ratio $a : 1$, $a > 1$, then all the infinitely many iterations S^1, S^2, S^3, $\cdots$ contained in the group would be different because they enlarge at different scales a^1, a^2, a^3, $\cdots$. For such reasons as these we shall almost exclusively consider groups of congruences— even if we have to do with actually or potentially infinite configurations such as band ornaments and the like.

After these general mathematical considerations let us now take up some special groups of symmetry which are important in art or nature. The operation which defines bilateral symmetry, mirror reflection, is essentially a one-dimensional operation. A straight line can be reflected in any of its points O; this reflection carries a point P into that point P' that has the same distance from O but lies on the other side. Such reflections are the only improper congruences of the one-dimensional line, whereas its only proper congruences are the translations. Reflection in O followed by the translation OA yields reflection in that point A_1 which halves the distance OA. A figure which is invariant under a translation t shows what in the art of ornament is called "infinite rapport," i.e. repetition in a regular spatial rhythm. A pattern invariant under the translation t is also invariant under its iterations t^1, t^2, t^3, $\cdots$, moreover under the identity $t^0 = I$, and under the inverse t^{-1} of t and its iterations t^{-1}, t^{-2}, t^{-3}, $\cdots$ If t shifts the line by the amount a then t^n shifts it by the amount

$$na \qquad (n = 0, \pm 1, \pm 2, \cdots).$$

Hence if we characterize a translation t by the shift a it effects then the iteration or power t^n is characterized by the multiple na. All translations carrying into itself a given pattern of infinite rapport on a straight line are in this sense multiples na of one basic translation a. This rhythmic may be combined with reflexive symmetry. If so the centers of reflections follow each other at half the distance $\tfrac{1}{2}a$. Only these two types of symmetry, as illustrated by Figure 22, are possible for a one-dimensional pattern or "ornament." (The crosses $\times$ mark the centers of reflection.)

FIGURE 22

Of course the real band ornaments are not strictly one-dimensional, but their symmetry as far as we have described it now makes use of their longitudinal dimension only. Here are some simple examples from Greek art. The first (Figure 23) which shows a very frequent motif, the palmette, is of type I (translation + reflection). The next (Figure 24) are without reflections (type ii). This frieze of Persian bowmen from Darius' palace in Susa (Figure 25) is pure translation; but you should notice that the basic translation covers twice the distance from man to man because the costumes of the bowmen alternate. Once more I shall point out the Monreale mosaic of the Lord's Ascension (Figure 10), but this time drawing your attention to the band ornaments framing it. The widest, carried out in a peculiar technique, later taken up by the Cosmati, displays the translatory symmetry only by repetition of the outer contour of the basic tree-like motif, while each copy is filled by a different highly

FIGURE 23

FIGURE 24

symmetric two-dimensional mosaic. The palace of the doges in Venice (Figure 26) may stand for translatory symmetry in architecture. Innumerable examples could be added.

As I said before, band ornaments really consist of a two-dimensional strip around a central line and thus have a second transversal dimension. As such they can have further symmetries. The pattern may be carried into itself by reflection in the central line *l*; let us distinguish this as longitudinal reflection from the transversal reflection in a line perpendicular to *l*. Or the pattern may be carried into itself by longitudinal reflection combined with the translation by ½*a* (longitudinal slip reflection). A fre-

FIGURE 25

FIGURE 26

quent motif in band ornaments are cords, strings, or plaits of some sort, the design of which suggests that one strand crosses the other in space (and thus makes part of it invisible). If this interpretation is accepted, further operations become possible; for example, reflection in the plane of the ornament would change a strand slightly above the plane into one below. All this can be thoroughly analyzed in terms of group theory as is for instance done in a section of Andreas Speiser's book, *Theorie der Gruppen von endlicher Ordnung.*

In the organic world the translatory symmetry, which the zoologists call metamerism, is seldom as regular as bilateral symmetry frequently is. A maple shoot and a shoot of *Angraecum distichum* (Figure 27) may serve as examples.[14] In the latter case translation is accompanied by longitudinal slip reflection. Of course the pattern does not go on into infinity (nor does a band ornament), but one may say that it is potentially infinite at least in one direction, as in the course of time ever new segments separated from each other by a bud come into being. Goethe said of the tails of vertebrates that they allude as it were to the potential infinity of organic existence. The central part of the animal shown in this picture, a scolopendrid (Figure 28), possesses fairly regular translational, combined with bilateral, symmetry, the basic operations of which are translation by one segment and longitudinal reflection.

In one-dimensional *time* repetition at equal intervals is the musical principle of *rhythm.* As a shoot grows it translates, one might say, a slow temporal into a spatial rhythm. Reflection, inversion in time, plays a far less important part in music than rhythm does. A melody changes its character to a considerable degree if played backward, and I, who am a poor musician, find it hard to recognize reflection when it is used in the construction of a fugue; it certainly has no such spon-

FIGURE 27

FIGURE 28

[14] This and the next picture are taken from *Studium Generale*, p. 249 and p. 241 (article by W. Troll, "Symmetriebetrachtung in der Biologie").

taneous effect as rhythm. All musicians agree that underlying the emotional element of music is a strong formal element. It may be that it is capable of some such mathematical treatment as has proved successful for the art of ornaments. If so, we have probably not yet discovered the appropriate mathematical tools. This would not be so surprising. For after all, the Egyptians excelled in the ornamental art four thousand years before the mathematicians discovered in the group concept the proper mathematical instrument for the treatment of ornaments and for the derivation of their possible symmetry classes. Andreas Speiser, who has taken a special interest in the group-theoretic aspect of ornaments, tried to apply combinatorial principles of a mathematical nature also to the formal problems of music. There is a chapter with this title in his book, "Die mathematische Denkweise," (Zurich, 1932). As an example, he analyzes Beethoven's pastoral sonata for piano, opus 28, and he also points to Alfred Lorenz's investigations on the formal structure of Richard Wagner's chief works. Metrics in poetry is closely related, and here, so Speiser maintains, science has penetrated much deeper. A common principle in music and prosody seems to be the configuration $a\ a\ b$ which is often called a bar: a theme a that is repeated and then followed by the "envoy" b; strophe, antistrophe, and epode in Greek choric lyrics. But such schemes fall hardly under the heading of symmetry.[15]

We return to symmetry in space. Take a band ornament where the individual section repeated again and again is of length a and sling it around a circular cylinder, the circumference of which is an integral multiple of a, for instance $25a$. You then obtain a pattern which is carried over into itself through the rotation around the cylinder axis by $\alpha = 360°/25$ and its repetitions. The twenty-fifth iteration is the rotation by $360°$, or the identity. We thus get a finite group of rotations of order 25, i.e. one consisting of 25 operations. The cylinder may be replaced by any surface of cylindrical symmetry, namely by one that is carried into itself by all rotations around a certain axis, for instance by a vase. Figure 29 shows an attic vase of the geometric period which displays quite a number of simple ornaments of this type. The principle of symmetry is the same, although the style is no longer "geometric," in this Rhodian pitcher (Figure 30), Ionian school of the seventh century B.C. Other illustrations are such capitals as these from early Egypt (Figure 31). Any finite group of proper rotations around a point O in a plane, or around a given axis in space, contains a primitive rotation t whose angle is an aliquot part $360°/n$ of the full rotation by $360°$, and consists of its iterations t^1, t^2, . . . ,t^{n-1}, $t^n =$ identity. The order n completely characterizes this group. The result follows from the analogous fact that any group of translations

---

[15] The reader should compare what G. D. Birkhoff has to say on the mathematics of poetry and music in the two publications cited in note 1.

FIGURE 29   FIGURE 30

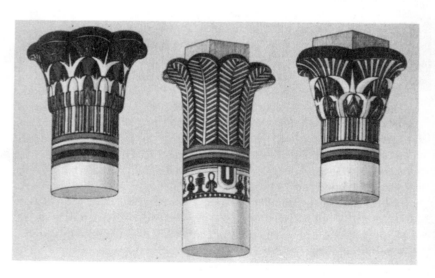

FIGURE 31

of a line, provided it contains no operations arbitrarily near to the identity except the identity itself, consists of the iterations va of a single translation $a(v = 0, \pm1, \pm2, \cdot\cdot\cdot)$.

The wooden dome in the Bardo of Tunis, once the palace of the Beys of Tunis (Figure 32), may serve as an example from interior architecture.

FIGURE 32

The next picture (Figure 33) takes you to Pisa; the Baptisterium with the tiny-looking statue of John the Baptist on top is a central building in whose exterior you can distinguish six horizontal layers each of rotary symmetry of a different order n. One could make the picture still more impressive by adding the leaning tower with its six galleries of arcades all having rotary symmetry of the same high order and the dome itself, the exterior of whose nave displays in columns and friezes patterns of the lineal translatory type of symmetry while the cupola is surrounded by a colonnade of high order rotary symmetry.

An entirely different spirit speaks to us from the view, seen from the rear of the choir, of the Romanesque cathedral in Mainz, Germany

FIGURE 33

(Figure 34). Yet again repetition in the round arcs of the friezes, octagonal central symmetry ($n = 8$, a low value compared to those embodied in the several layers of the Pisa Baptisterium) in the small rosette and the three towers, while bilateral symmetry rules the structure as a whole as well as almost every detail.

Cyclic symmetry appears in its simplest form if the surface of fully cylindrical symmetry is a plane perpendicular to the axis. We then can limit ourselves to the two-dimensional plane with a center O. Magnificent examples of such central plane symmetry are provided by the rose windows of Gothic cathedrals with their brilliant-colored glasswork. The richest I remember is the rosette of St. Pierre in Troyes, France, which is based on the number 3 throughout.

FIGURE 34

Flowers, nature's gentlest children, are also conspicuous for their colors and their cyclic symmetry. Here (Figure 35) is a picture of an iris with its triple pole. The symmetry of 5 is most frequent among flowers. A page like the following (Figure 36) from Ernst Haeckel's *Kunstformen der Natur* seems to indicate that it also occurs not infrequently among the lower animals. But the biologists warn me that the outward appearance of these echinoderms of the class of *Ophiodea* is to a certain degree deceptive; their larvae are organized according to the principle of bilateral symmetry. No such objection attaches to the next picture from the same source (Figure 37), a *Discomedusa* of octagonal symmetry. For the coelentera occupy a place in the phylogenetic evolution where cyclic has not yet given way to bilateral symmetry. Haeckel's extraordinary work, in which his interest in the concrete forms of organisms finds expression in

FIGURE 35

countless drawings executed in minutest detail, is a true nature's codex of symmetry. Equally revealing for Haeckel, the biologist, are the thousands and thousands of figures in his *Challenger Monograph*, in which he describes for the first time 3,508 new species of radiolarians discovered by him on the Challenger Expedition, 1887. One should not forget these accomplishments over the often all-too-speculative phylogenetic constructions in which this enthusiastic apostle of Darwinism indulged, and over his rather shallow materialistic philosophy of monism, which made quite a splash in Germany around the turn of the century.

Speaking of *Medusae* I cannot resist the temptation of quoting a few lines from D'Arcy Thompson's classic work on *Growth and Form*, a masterpiece of English literature, which combines profound knowledge in geometry, physics, and biology with humanistic erudition and scientific insight of unusual originality. Thompson reports on physical experiments

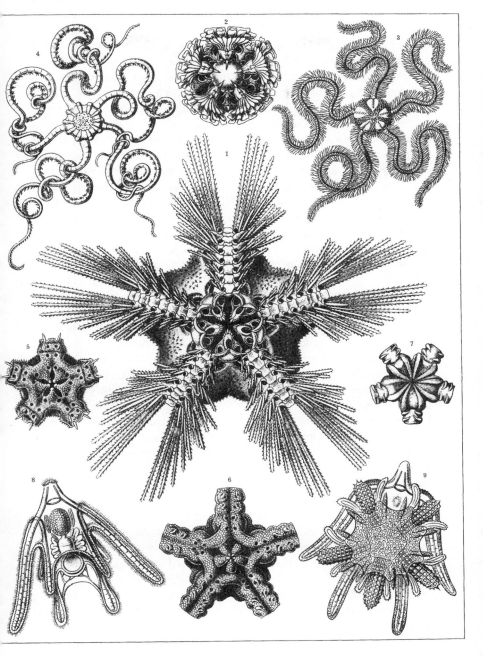

FIGURE 36

with hanging drops which serve to illustrate by analogy the formation of medusae. "The living medusa," he says, "has geometrical symmetry so marked and regular as to suggest a physical or mechanical element in the little creatures' growth and construction. It has, to begin with, its vortex-like bell or umbrella, with its symmetrical handle or manubrium. The bell is traversed by radial canals, four or in multiples of four; its edge is beset with tentacles, smooth or often beaded, at regular intervals or of graded sizes; and certain sensory structures, including solid concretions or 'otoliths,' are also symmetrically interspersed. No sooner made, then it begins to pulsate; the bell begins to 'ring.' Buds, miniature replicas of the parent-organism, are very apt to appear on the tentacles, or on the manubrium or sometimes on the edge of the bell; we seem to see one vortex producing others before our eyes. The development of a medusoid deserves to be studied without prejudice from this point of view. Certain it is that the tiny medusoids of *Obelia*, for instance, are budded off with a rapidity and a complete perfection which suggests an automatic and all but instantaneous act of conformation, rather than a gradual process of growth."

While pentagonal symmetry is frequent in the organic world, one does not find it among the most perfectly symmetrical creations of inorganic nature, among the crystals. There no other rotational symmetries are possible than those of order 2, 3, 4, and 6. Snow crystals provide the best known specimens of hexagonal symmetry. Figure 38 shows some of these little marvels of frozen water. In my youth, when they came down from heaven around Christmastime blanketing the landscape, they were the delight of old and young. Now only the skiers like them, while they have become the abomination of motorists. Those versed in English literature will remember Sir Thomas Browne's quaint account in his *Garden of Cyrus* (1658) of hexagonal and "quincuncial" symmetry which "doth neatly declare how nature Geometrizeth and observeth order in all things." One versed in German literature will remember how Thomas Mann in his *Magic Mountain* [16] describes the "hexagonale Unwesen" of the snow storm in which his hero, Hans Castorp, nearly perishes when he falls asleep with exhaustion and leaning against a barn dreams his deep dream of death and love. An hour before when Hans sets out on his unwarranted expedition on skis he enjoys the play of the flakes "and among these myriads of enchanting little stars," so he philosophizes, "in their hidden splendor, too small for man's naked eye to see, there was not one like unto another; an endless inventiveness governed the development and unthinkable differentiation of one and the same basic scheme, the equilateral, equiangled hexagon. Yet each in itself—this was the uncanny, the antiorganic, the

[16] I quote Helen Lowe-Porter's translation, Knopf, New York, 1927 and 1939.

FIGURE 37

life-denying character of them all—each of them was absolutely symmetrical, icily regular in form. They were too regular, as substance adapted to life never was to this degree—the living principle shuddered at this perfect precision, found it deathly, the very marrow of death—Hans Castorp felt he understood now the reason why the builders of antiquity purposely and secretly introduced minute variation from absolute symmetry in their columnar structures." [17]

Up to now we have paid attention to proper rotations only. If improper rotations are taken into consideration, we have the two following possibilities for finite groups of rotations around a center O in plane geometry, which correspond to the two possibilities we encountered for ornamental symmetry on a line: (1) the group consisting of the repetitions of a single proper rotation by an aliquot part $\alpha = 360°/n$ of $360°$; (2) the group of these rotations combined with the reflections in n axes forming angles of $\frac{1}{2}\alpha$. The first group is called the cyclic group C_n and the second the dihedral group D_n. Thus these are the only possible central symmetries in two-dimensions:

(1) $C_1, C_2, C_3, \cdot \cdot \cdot; D_1, D_2, D_3, \cdot \cdot \cdot.$

C_1 means no symmetry at all, D_1 bilateral symmetry and nothing else. In architecture the symmetry of 4 prevails. Towers often have hexagonal symmetry. Central buildings with the symmetry of 6 are much less frequent. The first pure central building after antiquity, S. Maria degli Angeli in Florence (begun 1434), is an octagon. Pentagons are very rare. When once before I lectured on symmetry in Vienna in 1937 I said I knew of only one example and that a very inconspicuous one, forming the passageway from San Michele di Murano in Venice to the hexagonal Capella Emiliana. Now, of course, we have the Pentagon building in Washington. By its size and distinctive shape, it provides an attractive landmark for bombers. Leonardo da Vinci engaged in systematically determining the possible symmetries of a central building and how to attach chapels and niches without destroying the symmetry of the nucleus. In abstract modern terminology, his result is essentially our above table of the possible finite groups of rotations (proper and improper) in two dimensions.

So far the rotational symmetry in a plane had always been accompanied by reflective symmetry; I have shown you quite a number of examples for the dihedral group D_n and none for the simpler cyclic group C_n. But this is more or less accidental. Here (Figure 39) are two flowers, a *gera-*

---

[17] Dürer considered his canon of the human figure more as a standard from which to deviate than as a standard toward which to strive. Vitruvius' *temperaturae* seem to have the same sense, and maybe the little word "almost" in the statement ascribed to Polykleitos and mentioned in note 1 points in the same direction.

FIGURE 38

nium (I) with the symmetry group D_5 while *Vinca herbacea* (II) has the more restricted group C_5 owing to the asymmetry of its petals. Figure 40 shows what is perhaps the simplest figure with rotational symmetry, the tripod ($n = 3$). When one wants to eliminate the attending reflective symmetry, one puts little flags unto the arms and obtains the triquetrum,

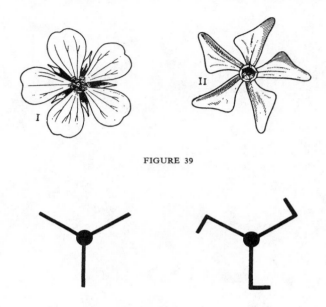

FIGURE 39

FIGURE 40

an old magic symbol. The Greeks, for instance, used it with the Medusa's head in the center as the symbol for the three-cornered Sicily. (Mathematicians are familiar with it as the seal on the cover of the *Rendiconti del Circolo Matematico di Palermo*.) The modification with four instead of three arms is the swastika, which need not be shown here—one of the most primeval symbols of mankind, common possession of a number of apparently independent civilizations. In my lecture on symmetry in Vienna in the fall of 1937, a short time before Hitler's hordes occupied Austria, I added concerning the swastika: "In our days it has become the symbol of a terror far more terrible than the snake-girdled Medusa's head"—and a pandemonium of applause and booing broke loose in the audience. It seems that the origin of the magic power ascribed to these patterns lies in their startling incomplete symmetry—rotations without reflections. Here (Figure 41) is the gracefully designed staircase of the pulpit of the Stephan's dome in Vienna; a triquetrum alternates with a swastika-like wheel.

So much about rotational symmetry in two dimensions. If dealing with

FIGURE 41

potentially infinite patterns like band ornaments or with infinite groups, the operation under which the pattern is invariant is not of necessity a congruence but could be a similarity. A similarity in one dimension that is not a mere translation has a fixed point O and is a dilatation s from O in a certain ratio $a : 1$ where $a \neq 1$. It is no essential restriction to assume $a > 0$. Indefinite iteration of this operation generates a group Σ consisting of the dilatations

$$(2) \qquad\qquad s^n \qquad (n = 0, \pm 1, \pm 2, \cdots).$$

A good example of this type of symmetry is shown by the shell of *Turritella duplicata* (Figure 42). It is really quite remarkable how exactly the widths of the consecutive whorls of this shell follow the law of geometric progression.

The hands of some clocks perform a continuous uniform rotation, others jump from minute to minute. The rotations by an integral number of

FIGURE 42

minutes form a discontinuous subgroup within the continuous group of all rotations, and it is natural to consider a rotation s and its iterations (2) as contained in the continuous group. We can apply this viewpoint to any similarity in 1, 2, or 3 dimensions, as a matter of fact to any transformation s. The continuous motion of a space-filling substance, a "fluid," can mathematically be described by giving the transformation $U(t,t')$ which carries the position P_t of any point of the fluid at the moment t over into its position $P_{t'}$ at the time t'. These transformations form a one-parameter group if $U(t,t')$ depends on the time difference $t' - t$ only, $U(t,t') = S(t' - t)$, i.e. if during equal time intervals always the same motion is repeated. Then the fluid is in "uniform motion." The simple group law

$$S(t_1)S(t_2) = S(t_1 + t_2)$$

expresses that the motions during two consecutive time intervals t_1, t_2 result in the motion during the time $t_1 + t_2$. The motion during 1 minute leads to a definite transformation $s = S(1)$, and for all integers n the motion $S(n)$ performed during n minutes is the iteration s^n: the discontinuous group Σ consisting of the iterations of s is embedded in the continuous group with the parameter t consisting of the motions $S(t)$. One could say that the continuous motion consists of the endless repetition of the same infinitesimal motion in consecutive infinitely small time intervals of equal length.

We could have applied this consideration to the rotations of a plane disc as well as to dilatations. We now envisage any proper similarity s, i.e. one which does not interchange left and right. If, as we assume, it is not a mere translation, it has a fixed point O and consists of a rotation about O combined with a dilatation from the center O. It can be obtained as the stage $S(1)$ reached after 1 minute by a continuous process $S(t)$ of combined uniform rotation and expansion. This process carries a point $\neq O$ along a so-called logarithmic or equiangular spiral. This curve, therefore, shares with straight line and circle the important property of going over into itself by a continuous group of similarities. The words by which James Bernoulli had the *spira mirabilis* adorned on his tombstone in the

Münster at Basle, "Eadem mutata resurgo," are a grandiloquent expression of this property. Straight line and circle are limiting cases of the logarithmic spiral, which arise when in the combination rotation-plus-dilatation one of the two components happens to be the identity. The stages reached by the process at the times

(3) $t = n = \cdots, -2, -1, 0, 1, 2, \cdots$

form the group consisting of the iterations (2). The well-known shell of *Nautilus* (Figure 43) shows this sort of symmetry to an astonishing per-

FIGURE 43

fection. You see here not only the continuous logarithmic spiral, but the potentially infinite sequence of chambers has a symmetry described by the discontinuous group Σ. For everybody looking at this picture (Figure 44) of a giant sunflower, *Helianthus maximus*, the florets will naturally arrange themselves into logarithmic spirals, two sets of spirals of opposite sense of coiling.

The most general rigid motion in three-dimensional space is a screw motion s, combination of a rotation around an axis with a translation along that axis. Under the influence of the corresponding continuous uniform motion any point not on the axis describes a screw-line or helix which, of course, could say of itself with the same right as the logarithmic spiral: *eadem resurgo*. The stages P_n which the moving point reaches at the equidistant moments (3) are equidistributed over the helix like stairs on a winding staircase. If the angle of rotation of the operation s is a fraction μ/ν of the full angle $360°$ expressible in terms of small integers

FIGURE 44

μ, ν then every νth point of the sequence P_n lies on the same vertical, and μ full turnings of the screw are necessary to get from P_n to the point $P_{n+\nu}$ above it. The leaves around the shoot of a plant often show such a regular spiral arrangement. Goethe spoke of a spiral tendency in nature, and under the name of *phyllotaxis* this phenomenon, since the days of Charles Bonnet (1754), has been the subject of much investigation and more speculation among botanists.[18] One has found that the fractions μ/ν representing the screw-like arrangement of leaves quite often are members of the "Fibonacci sequence"

---

[18] This phenomenon plays also a role in J. Hambidge's constructions. His *Dynamic symmetry* contains on pp. 146–157 detailed notes by the mathematician R. C. Archibald on the logarithmic spiral, golden section, and the Fibonacci series.

(4) $\frac{1}{1}$, $\frac{1}{2}$, $\frac{2}{3}$, $\frac{3}{5}$, $\frac{5}{8}$, $\frac{8}{13}$, $\frac{13}{21}$, $\frac{21}{34}$, $\cdots$,

which results from the expansion into a continued fraction of the irrational number $\frac{1}{2}(\sqrt{5} - 1)$. This number is no other but the ratio known as the *aurea sectio*, which has played such a role in attempts to reduce beauty of proportion to a mathematical formula. The cylinder on which the screw is wound could be replaced by a cone; this amounts to replacing the screw motion s by any proper similarity—rotation combined with dilatation. The arrangement of scales on a fir-cone falls under this slightly more general form of symmetry in phyllotaxis. The transition from cylinder over cone to disc is obvious, illustrated by the cylindrical stem of a plant with its leaves, a fir-cone with its scales, and the discoidal inflorescence of *Helianthus* with its florets. Where one can check the numbers (4) best, namely for the arrangement of scales on a fir-cone, the accuracy is not too good nor are considerable deviations too rare. P. G. Tait, in the *Proceedings of the Royal Society of Edinburgh* (1872), has tried to give a simple explanation, while A. H. Church in his voluminous treatise *Relations of phyllotaxis to mechanical laws* (Oxford, 1901–1903) sees in the arithmetics of phyllotaxis an organic mystery. I am afraid modern botanists take this whole doctrine of phyllotaxis less seriously than their forefathers.

Apart from reflection all symmetries so far considered are described by a group consisting of the iterations of one operation s. In one case, and that is undoubtedly the most important, the resulting group is finite, namely if one takes for s a rotation by an angle $\alpha = 360°/n$ which is an aliquot part of the full rotation $360°$. For the two-dimensional plane there are no other finite groups of proper rotations than these; witness the first line, C_1, C_2, C_3, $\cdots$ of Leonardo's table (1). The simplest figures which have the corresponding symmetry are the regular polygons: the regular triangle, the square, the regular pentagon, etc. The fact that there is for every number $n = 3, 4, 5, \cdots$ a regular polygon of n sides is closely related to the existence for every n of a rotational group of order n in plane geometry. Both facts are far from trivial. Indeed, the situation in three dimensions is altogether different: there do not exist infinitely many regular polyhedra in 3-space, but not more than five, often called the Platonic solids because they play an eminent role in Plato's natural philosophy. They are the regular tetrahedron, the cube, the octahedron, moreover the pentagondodecahedron, the sides of which are twelve regular pentagons, and the icosahedron bounded by twenty regular triangles. One might say that the existence of the first three is a fairly trivial geometric fact. But the discovery of the last two is certainly one of the most beautiful and singular discoveries made in the whole history of mathematics. With a fair amount of certainty, it can be traced to the colonial Greeks in southern Italy. The suggestion has been made that they abstracted the regular dodecahedron from the crystals of pyrite, a sulphurous mineral abundant in Sicily. But as mentioned before, the symmetry of 5 so characteristic for the regular dodecahedron contradicts the laws of crystallography, and indeed one finds that the pentagons bounding the dodecahedra in which pyrite crystallizes have 4 edges of equal, but one of different, length. The first exact construction of the regular pentagondodecahedron is probably due to Theaetetus. There is some evidence that dodecahedra were used as dice in Italy at a very early time and had some religious significance in Etruscan culture. Plato, in the dialogue *Timaeus*, associates the regular pyramid, octahedron, cube, icosahedron, with the four elements of fire, air, earth, and water (in this order), while in the pentagondodecahedron he sees in some sense the image of the universe as a whole. A. Speiser has advocated the view that the construction of the five regular solids is the chief goal of the deductive system of geometry as erected by the Greeks and canonized in Euclid's *Elements*. May I mention, however, that the Greeks never used the word "symmetric" in our modern sense. In common usage σύμμετρος means *proportionate*, while in Euclid it is equivalent to our *commensurable*: side and diagonal of a square are incommensurable quantities, ἀσύμμετρα μεγέθη.

Here (Figure 45) is a page from Haeckel's *Challenger Monograph* showing the skeletons of several Radiolarians. Numbers 2, 3, and 5 are octahedron, icosahedron, and dodecahedron in astonishingly regular form; 4 seems to have a lower symmetry.

Kepler, in his *Mysterium cosmographicum*, published in 1595, long before he discovered the three laws bearing his name today, made an attempt to reduce the distances in the planetary system to regular bodies which are alternatingly inscribed and circumscribed to spheres. Here (Figure 46) is his construction, by which he believed he had penetrated deeply into the secrets of the Creator. The six spheres correspond to the six planets, Saturn, Jupiter, Mars, Earth, Venus, Mercurius, separated in this order by cube, tetrahedron, dodecahedron, octahedron, icosahedron. (Of course, Kepler did not know about the three outer planets, Uranus, Neptune, and Pluto, which were discovered in 1781, 1846, and 1930 respectively.) He tries to find the reasons why the Creator had chosen this order of the Platonic solids and draws parallels between the properties of the planets (astrological rather than astrophysical properties) and those of the corresponding regular bodies. A mighty hymn in which he proclaims his credo, "Credo spatioso numen in orbe," concludes his book. We still share his belief in a mathematical harmony of the universe. It has withstood the test of ever widening experience. But we no longer seek this harmony in static forms like the regular solids, but in dynamic laws.

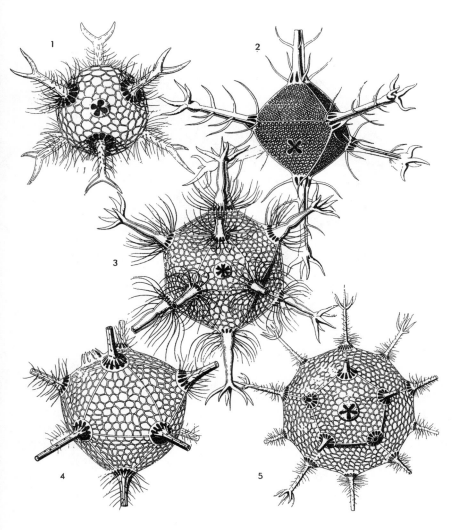

FIGURE 45

As the regular polygons are connected with the finite groups of plane rotations, so must the regular polyhedra be intimately related to the finite groups of proper rotations around a center O in space. From the study of plane rotations we at once obtain two types of proper rotation groups in space. Indeed, the group C_n of proper rotations in a horizontal plane around a center O can be interpreted as consisting of rotations in space around the vertical axis through O. Reflection of the horizontal plane in a line l of the plane can be brought about in space through a rotation

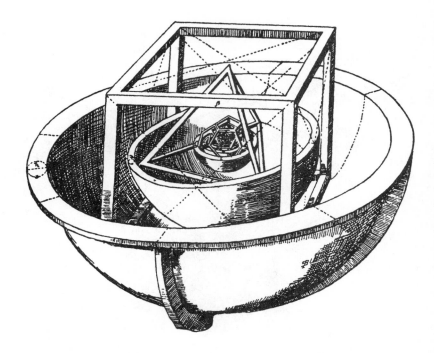

FIGURE 46

around l by 180° (*Umklappung*). You may remember that we mentioned this in connection with the analysis of a Sumerian picture (Figure 4). In this way the group D_n in the horizontal plane is changed into a group D'_n of proper rotations in space; it contains the rotations around a vertical axis through O by the multiples of $360°/n$ and the Umklappungen around n horizontal axes through O which form equal angles of $360°/2n$ with each other. But it should be observed that the group D'_1 as well as C_2 consists of the identity and the Umklappung around one line. These two groups are therefore identical, and in a complete list of the *different* groups of proper rotations in three dimensions, D'_1 should be omitted if C_2 is kept. Hence we start our list thus:

$$C_1, \ C_2, \ C_3, \ C_4, \ \cdots;$$
$$D'_2, \ D'_3, \ D'_4, \ \cdots.$$

D'_2 is the so-called four-group consisting of the identity of the Umklappungen around three mutually perpendicular axes.

For each one of the five regular bodies we can construct the group of those proper rotations which carry that body into itself. Does this give rise to five new groups? No, only to three, and that for the following reason. Inscribe a sphere into a cube and an octahedron into the sphere such that the corners of the octahedron lie where the sides of the cube touch the sphere, namely in the centers of the six square sides. (Figure 47 shows the two-dimensional analogue.) In this position cube and octa-

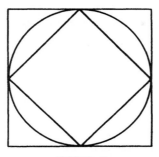

FIGURE 47

hedron are polar figures in the sense of projective geometry. It is clear that every rotation which carries the cube into itself also leaves the octahedron invariant, and vice versa. Hence the group for the octahedron is the same as for the cube. In the same manner pentagondodecahedron and icosahedron are polar figures. The figure polar to a regular tetrahedron is a regular tetrahedron the corners of which are the antipodes of those of the first. Thus we find three new groups of proper rotations, T, W, and P; they are those leaving invariant the regular tetrahedron, the cube (or octahedron), and the pentagondodecahedron (or icosahedron) respectively. Their orders, i.e. the number of operations in each of them, are 12, 24, 60 respectively.

It can be shown by a relatively simple analysis that with the addition of these three groups our table is complete:

$$\begin{aligned} &C_n & (n = 1, 2, 3, \cdots), \\ (5) \quad &D'_n & (n = 2, 3, \cdots); \\ &T, W, P. \end{aligned}$$

This is the modern equivalent to the tabulation of the regular polyhedra by the Greeks. These groups, in particular the last three, are an immensely attractive subject for geometric investigation.

What further possibilities arise if improper rotations are also admitted to our groups? This question is best answered by making use of one quite singular improper rotation, namely reflection in O; it carries any point P into its antipode P' with respect to O found by joining P with O and prolonging the straight line PO by its own length: $PO = OP'$. This operation Z commutes with every rotation S, $ZS = SZ$. Now let Γ be one of our finite groups of proper rotations. One way of including improper rotations is simply by adjoining Z, more precisely by adding to the proper rotations S of Γ all the improper rotations of the form ZS (with S in Γ). The order of the group $\overline{\Gamma} = \Gamma + Z\Gamma$ thus obtained is clearly twice that of Γ. Another way of including improper rotations arises from this situation: Suppose Γ is contained as a subgroup of index 2 in another group Γ' of proper rotations; so that one-half of the elements of Γ' lie in Γ, call them S, and one-half, S', do not. Now replace these latter by the improper rotations ZS'. In this manner you get a group $\Gamma'\Gamma$ which contains Γ while the other half of its operations are improper. For instance, $\Gamma = C_n$ is a subgroup of index 2 of $\Gamma' = D'_n$; the operations S' of D'_n not contained in C_n are the Umklappungen around the n horizontal axes. The corresponding ZS' are the reflections in the vertical planes perpendicular to these axes. Thus $D'_n C_n$ consists of the rotations around the vertical axis, the angles of which are multiples of $360°/n$, and of the reflections in vertical planes through this axis forming angles of $360°/2n$ with each other. You might say that this is the group formerly denoted by D_n. Another example, the simplest of all: $\Gamma = C_1$ is contained in $\Gamma' = C_2$. The one operation S' of C_2 not contained in C_1 is the rotation by $180°$ about the vertical axis; ZS' is reflection in the horizontal plane through O. Hence $C_2 C_1$ is the group consisting of the identity and of the reflection in a given plane; in other words, the group to which bilateral symmetry refers.

The two ways described are the only ones by which improper rotations may be included in our groups. Hence this is the complete table of all finite groups of (proper and improper) rotations:

$$C_n, \quad \overline{C}_n, \quad C_{2n}C_n \qquad\qquad\qquad (n = 1, 2, 3, \cdots)$$
$$D'_n, \quad \overline{D}_n, \quad D'_n C_n, \qquad D'_{2n}D'_n \quad (n = 2, 3, \cdots)$$
$$T, \quad W, \quad P; \qquad \overline{T}, \quad \overline{W}, \quad \overline{P}; \qquad WT.$$

The last group WT is made possible by the fact that the tetrahedral group T is a subgroup of index 2 of the octahedral group W.

Index

About James R. Newman

JAMES R. NEWMAN *was born in New York City in 1907 and died in Washington, D.C., in 1966. After taking his law degree at Columbia University he became a member of the New York Bar, practicing in New York from 1929 to 1941. During and after World War II he held several Government positions, including that of Chief Intelligence Officer at the United States Embassy in London, Special Assistant to the Under Secretary of War, and counsel to the United States Committee on Atomic Energy.*

He was a contributor to many publications and in 1948 became a member of the board of editors of Scientific American. *Mr. Newman was a visiting lecturer at Yale Law School and a Guggenheim Fellow. He was the author or co-author of several books, among them* The Tools of War, The Control of Atomic Energy, Mathematics and the Imagination *(a standard work in the popular scientific field, written with Edward Kasner),* Science and Sensibility *and* The Rule of Folly *(an exposition of our fallacious nuclear defense policies), and edited* The Harper Encyclopedia of Science.

A CATALOG OF SELECTED
DOVER BOOKS
IN ALL FIELDS OF INTEREST

A CATALOG OF SELECTED DOVER
BOOKS IN ALL FIELDS OF INTEREST

CONCERNING THE SPIRITUAL IN ART, Wassily Kandinsky. Pioneering work by father of abstract art. Thoughts on color theory, nature of art. Analysis of earlier masters. 12 illustrations. 80pp. of text. 5⅜ x 8½. 23411-8 Pa. $4.95

ANIMALS: 1,419 Copyright-Free Illustrations of Mammals, Birds, Fish, Insects, etc., Jim Harter (ed.). Clear wood engravings present, in extremely lifelike poses, over 1,000 species of animals. One of the most extensive pictorial sourcebooks of its kind. Captions. Index. 284pp. 9 x 12. 23766-4 Pa. $14.95

CELTIC ART: The Methods of Construction, George Bain. Simple geometric techniques for making Celtic interlacements, spirals, Kells-type initials, animals, humans, etc. Over 500 illustrations. 160pp. 9 x 12. (Available in U.S. only.) 22923-8 Pa. $9.95

AN ATLAS OF ANATOMY FOR ARTISTS, Fritz Schider. Most thorough reference work on art anatomy in the world. Hundreds of illustrations, including selections from works by Vesalius, Leonardo, Goya, Ingres, Michelangelo, others. 593 illustrations. 192pp. 7⅛ x 10¼. 20241-0 Pa. $9.95

CELTIC HAND STROKE-BY-STROKE (Irish Half-Uncial from "The Book of Kells"): An Arthur Baker Calligraphy Manual, Arthur Baker. Complete guide to creating each letter of the alphabet in distinctive Celtic manner. Covers hand position, strokes, pens, inks, paper, more. Illustrated. 48pp. 8¼ x 11. 24336-2 Pa. $3.95

EASY ORIGAMI, John Montroll. Charming collection of 32 projects (hat, cup, pelican, piano, swan, many more) specially designed for the novice origami hobbyist. Clearly illustrated easy-to-follow instructions insure that even beginning papercrafters will achieve successful results. 48pp. 8¼ x 11. 27298-2 Pa. $3.50

THE COMPLETE BOOK OF BIRDHOUSE CONSTRUCTION FOR WOOD-WORKERS, Scott D. Campbell. Detailed instructions, illustrations, tables. Also data on bird habitat and instinct patterns. Bibliography. 3 tables. 63 illustrations in 15 figures. 48pp. 5¼ x 8½. 24407-5 Pa. $2.50

BLOOMINGDALE'S ILLUSTRATED 1886 CATALOG: Fashions, Dry Goods and Housewares, Bloomingdale Brothers. Famed merchants' extremely rare catalog depicting about 1,700 products: clothing, housewares, firearms, dry goods, jewelry, more. Invaluable for dating, identifying vintage items. Also, copyright-free graphics for artists, designers. Co-published with Henry Ford Museum & Greenfield Village. 160pp. 8¼ x 11. 25780-0 Pa. $10.95

HISTORIC COSTUME IN PICTURES, Braun & Schneider. Over 1,450 costumed figures in clearly detailed engravings—from dawn of civilization to end of 19th century. Captions. Many folk costumes. 256pp. 8⅜ x 11¾. 23150-X Pa. $12.95

CATALOG OF DOVER BOOKS

STICKLEY CRAFTSMAN FURNITURE CATALOGS, Gustav Stickley and L. &
J. G. Stickley. Beautiful, functional furniture in two authentic catalogs from 1910. 594
illustrations, including 277 photos, show settles, rockers, armchairs, reclining chairs,
bookcases, desks, tables. 183pp. 6½ x 9¼.   23838-5 Pa. $11.95

AMERICAN LOCOMOTIVES IN HISTORIC PHOTOGRAPHS: 1858 to 1949,
Ron Ziel (ed.). A rare collection of 126 meticulously detailed official photographs,
called "builder portraits," of American locomotives that majestically chronicle the
rise of steam locomotive power in America. Introduction. Detailed captions. xi+
129pp. 9 x 12.   27393-8 Pa. $13.95

AMERICA'S LIGHTHOUSES: An Illustrated History, Francis Ross Holland, Jr.
Delightfully written, profusely illustrated fact-filled survey of over 200 American lighthouses since 1716. History, anecdotes, technological advances, more. 240pp. 8 x 10¾.
                                                              25576-X Pa. $12.95

TOWARDS A NEW ARCHITECTURE, Le Corbusier. Pioneering manifesto by
founder of "International School." Technical and aesthetic theories, views of industry, economics, relation of form to function, "mass-production split" and much more.
Profusely illustrated. 320pp. 6⅛ x 9¼. (Available in U.S. only.)   25023-7 Pa. $9.95

HOW THE OTHER HALF LIVES, Jacob Riis. Famous journalistic record, exposing poverty and degradation of New York slums around 1900, by major social
reformer. 100 striking and influential photographs. 233pp. 10 x 7⅞.
                                                              22012-5 Pa. $11.95

FRUIT KEY AND TWIG KEY TO TREES AND SHRUBS, William M. Harlow.
One of the handiest and most widely used identification aids. Fruit key covers 120
deciduous and evergreen species; twig key 160 deciduous species. Easily used. Over
300 photographs. 126pp. 5⅜ x 8½.   20511-8 Pa. $3.95

COMMON BIRD SONGS, Dr. Donald J. Borror. Songs of 60 most common U.S.
birds: robins, sparrows, cardinals, bluejays, finches, more—arranged in order of
increasing complexity. Up to 9 variations of songs of each species.
                                              Cassette and manual 99911-4 $8.95

ORCHIDS AS HOUSE PLANTS, Rebecca Tyson Northen. Grow cattleyas and
many other kinds of orchids—in a window, in a case, or under artificial light. 63 illustrations. 148pp. 5⅜ x 8½.   23261-1 Pa. $5.95

MONSTER MAZES, Dave Phillips. Masterful mazes at four levels of difficulty.
Avoid deadly perils and evil creatures to find magical treasures. Solutions for all 32
exciting illustrated puzzles. 48pp. 8¼ x 11.   26005-4 Pa. $2.95

MOZART'S DON GIOVANNI (DOVER OPERA LIBRETTO SERIES),
Wolfgang Amadeus Mozart. Introduced and translated by Ellen H. Bleiler. Standard
Italian libretto, with complete English translation. Convenient and thoroughly
portable—an ideal companion for reading along with a recording or the performance
itself. Introduction. List of characters. Plot summary. 121pp. 5¼ x 8½.
                                                              24944-1 Pa. $3.95

TECHNICAL MANUAL AND DICTIONARY OF CLASSICAL BALLET, Gail
Grant. Defines, explains, comments on steps, movements, poses and concepts. 15-
page pictorial section. Basic book for student, viewer. 127pp. 5⅜ x 8½.
                                                              21843-0 Pa. $4.95

THE CLARINET AND CLARINET PLAYING, David Pino. Lively, comprehensive work features suggestions about technique, musicianship, and musical interpretation, as well as guidelines for teaching, making your own reeds, and preparing for public performance. Includes an intriguing look at clarinet history. "A godsend," *The Clarinet,* Journal of the International Clarinet Society. Appendixes. 7 illus. 320pp. 5⅜ x 8½. 40270-3 Pa. $9.95

HOLLYWOOD GLAMOR PORTRAITS, John Kobal (ed.). 145 photos from 1926-49. Harlow, Gable, Bogart, Bacall; 94 stars in all. Full background on photographers, technical aspects. 160pp. 8⅜ x 11¼. 23352-9 Pa. $12.95

THE ANNOTATED CASEY AT THE BAT: A Collection of Ballads about the Mighty Casey/Third, Revised Edition, Martin Gardner (ed.). Amusing sequels and parodies of one of America's best-loved poems: Casey's Revenge, Why Casey Whiffed, Casey's Sister at the Bat, others. 256pp. 5⅜ x 8½. 28598-7 Pa. $8.95

THE RAVEN AND OTHER FAVORITE POEMS, Edgar Allan Poe. Over 40 of the author's most memorable poems: "The Bells," "Ulalume," "Israfel," "To Helen," "The Conqueror Worm," "Eldorado," "Annabel Lee," many more. Alphabetic lists of titles and first lines. 64pp. 5³⁄₁₆ x 8¼. 26685-0 Pa. $1.00

PERSONAL MEMOIRS OF U. S. GRANT, Ulysses Simpson Grant. Intelligent, deeply moving firsthand account of Civil War campaigns, considered by many the finest military memoirs ever written. Includes letters, historic photographs, maps and more. 528pp. 6⅛ x 9¼. 28587-1 Pa. $12.95

ANCIENT EGYPTIAN MATERIALS AND INDUSTRIES, A. Lucas and J. Harris. Fascinating, comprehensive, thoroughly documented text describes this ancient civilization's vast resources and the processes that incorporated them in daily life, including the use of animal products, building materials, cosmetics, perfumes and incense, fibers, glazed ware, glass and its manufacture, materials used in the mummification process, and much more. 544pp. 6⅛ x 9¼. (Available in U.S. only.) 40446-3 Pa. $16.95

RUSSIAN STORIES/PYCCKNE PACCKA3bl: A Dual-Language Book, edited by Gleb Struve. Twelve tales by such masters as Chekhov, Tolstoy, Dostoevsky, Pushkin, others. Excellent word-for-word English translations on facing pages, plus teaching and study aids, Russian/English vocabulary, biographical/critical introductions, more. 416pp. 5⅜ x 8½. 26244-8 Pa. $9.95

PHILADELPHIA THEN AND NOW: 60 Sites Photographed in the Past and Present, Kenneth Finkel and Susan Oyama. Rare photographs of City Hall, Logan Square, Independence Hall, Betsy Ross House, other landmarks juxtaposed with contemporary views. Captures changing face of historic city. Introduction. Captions. 128pp. 8¼ x 11. 25790-8 Pa. $9.95

AIA ARCHITECTURAL GUIDE TO NASSAU AND SUFFOLK COUNTIES, LONG ISLAND, The American Institute of Architects, Long Island Chapter, and the Society for the Preservation of Long Island Antiquities. Comprehensive, well-researched and generously illustrated volume brings to life over three centuries of Long Island's great architectural heritage. More than 240 photographs with authoritative, extensively detailed captions. 176pp. 8¼ x 11. 26946-9 Pa. $14.95

NORTH AMERICAN INDIAN LIFE: Customs and Traditions of 23 Tribes, Elsie Clews Parsons (ed.). 27 fictionalized essays by noted anthropologists examine religion, customs, government, additional facets of life among the Winnebago, Crow, Zuni, Eskimo, other tribes. 480pp. 6⅛ x 9¼. 27377-6 Pa. $10.95

FRANK LLOYD WRIGHT'S DANA HOUSE, Donald Hoffmann. Pictorial essay of residential masterpiece with over 160 interior and exterior photos, plans, elevations, sketches and studies. 128pp. 9¼ x 10¾.   29120-0 Pa. $12.95

THE MALE AND FEMALE FIGURE IN MOTION: 60 Classic Photographic Sequences, Eadweard Muybridge. 60 true-action photographs of men and women walking, running, climbing, bending, turning, etc., reproduced from rare 19th-century masterpiece. vi + 121pp. 9 x 12.   24745-7 Pa. $12.95

1001 QUESTIONS ANSWERED ABOUT THE SEASHORE, N. J. Berrill and Jacquelyn Berrill. Queries answered about dolphins, sea snails, sponges, starfish, fishes, shore birds, many others. Covers appearance, breeding, growth, feeding, much more. 305pp. 5¼ x 8¼.   23366-9 Pa. $9.95

ATTRACTING BIRDS TO YOUR YARD, William J. Weber. Easy-to-follow guide offers advice on how to attract the greatest diversity of birds: birdhouses, feeders, water and waterers, much more. 96pp. 5³/₁₆ x 8¼.   28927-3 Pa. $2.50

MEDICINAL AND OTHER USES OF NORTH AMERICAN PLANTS: A Historical Survey with Special Reference to the Eastern Indian Tribes, Charlotte Erichsen-Brown. Chronological historical citations document 500 years of usage of plants, trees, shrubs native to eastern Canada, northeastern U.S. Also complete identifying information. 343 illustrations. 544pp. 6½ x 9¼.   25951-X Pa. $12.95

STORYBOOK MAZES, Dave Phillips. 23 stories and mazes on two-page spreads: Wizard of Oz, Treasure Island, Robin Hood, etc. Solutions. 64pp. 8¼ x 11.
                23628-5 Pa. $2.95

AMERICAN NEGRO SONGS: 230 Folk Songs and Spirituals, Religious and Secular, John W. Work. This authoritative study traces the African influences of songs sung and played by black Americans at work, in church, and as entertainment. The author discusses the lyric significance of such songs as "Swing Low, Sweet Chariot," "John Henry," and others and offers the words and music for 230 songs. Bibliography. Index of Song Titles. 272pp. 6½ x 9¼.   40271-1 Pa. $9.95

MOVIE-STAR PORTRAITS OF THE FORTIES, John Kobal (ed.). 163 glamor, studio photos of 106 stars of the 1940s: Rita Hayworth, Ava Gardner, Marlon Brando, Clark Gable, many more. 176pp. 8⅜ x 11¼.   23546-7 Pa. $14.95

BENCHLEY LOST AND FOUND, Robert Benchley. Finest humor from early 30s, about pet peeves, child psychologists, post office and others. Mostly unavailable elsewhere. 73 illustrations by Peter Arno and others. 183pp. 5⅜ x 8½. 22410-4 Pa. $6.95

YEKL and THE IMPORTED BRIDEGROOM AND OTHER STORIES OF YIDDISH NEW YORK, Abraham Cahan. Film Hester Street based on *Yekl* (1896). Novel, other stories among first about Jewish immigrants on N.Y.'s East Side. 240pp. 5⅜ x 8½.   22427-9 Pa. $7.95

SELECTED POEMS, Walt Whitman. Generous sampling from *Leaves of Grass.* Twenty-four poems include "I Hear America Singing," "Song of the Open Road," "I Sing the Body Electric," "When Lilacs Last in the Dooryard Bloom'd," "O Captain! My Captain!"–all reprinted from an authoritative edition. Lists of titles and first lines. 128pp. 5³/₁₆ x 8¼.   26878-0 Pa. $1.00

THE BEST TALES OF HOFFMANN, E. T. A. Hoffmann. 10 of Hoffmann's most important stories: "Nutcracker and the King of Mice," "The Golden Flowerpot," etc. 458pp. 5⅜ x 8½. 21793-0 Pa. $9.95

FROM FETISH TO GOD IN ANCIENT EGYPT, E. A. Wallis Budge. Rich detailed survey of Egyptian conception of "God" and gods, magic, cult of animals, Osiris, more. Also, superb English translations of hymns and legends. 240 illustrations. 545pp. 5⅜ x 8½. 25803-3 Pa. $13.95

FRENCH STORIES/CONTES FRANÇAIS: A Dual-Language Book, Wallace Fowlie. Ten stories by French masters, Voltaire to Camus: "Micromegas" by Voltaire; "The Atheist's Mass" by Balzac; "Minuet" by de Maupassant; "The Guest" by Camus, six more. Excellent English translations on facing pages. Also French-English vocabulary list, exercises, more. 352pp. 5⅜ x 8½. 26443-2 Pa. $9.95

CHICAGO AT THE TURN OF THE CENTURY IN PHOTOGRAPHS: 122 Historic Views from the Collections of the Chicago Historical Society, Larry A. Viskochil. Rare large-format prints offer detailed views of City Hall, State Street, the Loop, Hull House, Union Station, many other landmarks, circa 1904-1913. Introduction. Captions. Maps. 144pp. 9⅜ x 12¼. 24656-6 Pa. $12.95

OLD BROOKLYN IN EARLY PHOTOGRAPHS, 1865-1929, William Lee Younger. Luna Park, Gravesend race track, construction of Grand Army Plaza, moving of Hotel Brighton, etc. 157 previously unpublished photographs. 165pp. 8⅞ x 11¾. 23587-4 Pa. $13.95

THE MYTHS OF THE NORTH AMERICAN INDIANS, Lewis Spence. Rich anthology of the myths and legends of the Algonquins, Iroquois, Pawnees and Sioux, prefaced by an extensive historical and ethnological commentary. 36 illustrations. 480pp. 5⅜ x 8½. 25967-6 Pa. $10.95

AN ENCYCLOPEDIA OF BATTLES: Accounts of Over 1,560 Battles from 1479 B.C. to the Present, David Eggenberger. Essential details of every major battle in recorded history from the first battle of Megiddo in 1479 B.C. to Grenada in 1984. List of Battle Maps. New Appendix covering the years 1967-1984. Index. 99 illustrations. 544pp. 6½ x 9¼. 24913-1 Pa. $16.95

SAILING ALONE AROUND THE WORLD, Captain Joshua Slocum. First man to sail around the world, alone, in small boat. One of great feats of seamanship told in delightful manner. 67 illustrations. 294pp. 5⅜ x 8½. 20326-3 Pa. $6.95

ANARCHISM AND OTHER ESSAYS, Emma Goldman. Powerful, penetrating, prophetic essays on direct action, role of minorities, prison reform, puritan hypocrisy, violence, etc. 271pp. 5⅜ x 8½. 22484-8 Pa. $7.95

MYTHS OF THE HINDUS AND BUDDHISTS, Ananda K. Coomaraswamy and Sister Nivedita. Great stories of the epics; deeds of Krishna, Shiva, taken from puranas, Vedas, folk tales; etc. 32 illustrations. 400pp. 5⅜ x 8½. 21759-0 Pa. $12.95

THE TRAUMA OF BIRTH, Otto Rank. Rank's controversial thesis that anxiety neurosis is caused by profound psychological trauma which occurs at birth. 256pp. 5⅜ x 8½. 27974-X Pa. $7.95

A THEOLOGICO-POLITICAL TREATISE, Benedict Spinoza. Also contains unfinished Political Treatise. Great classic on religious liberty, theory of government on common consent. R. Elwes translation. Total of 421pp. 5⅜ x 8½. 20249-6 Pa. $10.95

CATALOG OF DOVER BOOKS

MY BONDAGE AND MY FREEDOM, Frederick Douglass. Born a slave, Douglass became outspoken force in antislavery movement. The best of Douglass' autobiographies. Graphic description of slave life. 464pp. 5⅜ x 8½. 22457-0 Pa. $8.95

FOLLOWING THE EQUATOR: A Journey Around the World, Mark Twain. Fascinating humorous account of 1897 voyage to Hawaii, Australia, India, New Zealand, etc. Ironic, bemused reports on peoples, customs, climate, flora and fauna, politics, much more. 197 illustrations. 720pp. 5⅜ x 8½. 26113-1 Pa. $15.95

THE PEOPLE CALLED SHAKERS, Edward D. Andrews. Definitive study of Shakers: origins, beliefs, practices, dances, social organization, furniture and crafts, etc. 33 illustrations. 351pp. 5⅜ x 8½. 21081-2 Pa. $10.95

THE MYTHS OF GREECE AND ROME, H. A. Guerber. A classic of mythology, generously illustrated, long prized for its simple, graphic, accurate retelling of the principal myths of Greece and Rome, and for its commentary on their origins and significance. With 64 illustrations by Michelangelo, Raphael, Titian, Rubens, Canova, Bernini and others. 480pp. 5⅜ x 8½. 27584-1 Pa. $9.95

PSYCHOLOGY OF MUSIC, Carl E. Seashore. Classic work discusses music as a medium from psychological viewpoint. Clear treatment of physical acoustics, auditory apparatus, sound perception, development of musical skills, nature of musical feeling, host of other topics. 88 figures. 408pp. 5⅜ x 8½. 21851-1 Pa. $11.95

THE PHILOSOPHY OF HISTORY, Georg W. Hegel. Great classic of Western thought develops concept that history is not chance but rational process, the evolution of freedom. 457pp. 5⅜ x 8½. 20112-0 Pa. $9.95

THE BOOK OF TEA, Kakuzo Okakura. Minor classic of the Orient: entertaining, charming explanation, interpretation of traditional Japanese culture in terms of tea ceremony. 94pp. 5⅜ x 8½. 20070-1 Pa. $3.95

LIFE IN ANCIENT EGYPT, Adolf Erman. Fullest, most thorough, detailed older account with much not in more recent books, domestic life, religion, magic, medicine, commerce, much more. Many illustrations reproduce tomb paintings, carvings, hieroglyphs, etc. 597pp. 5⅜ x 8½. 22632-8 Pa. $12.95

SUNDIALS, Their Theory and Construction, Albert Waugh. Far and away the best, most thorough coverage of ideas, mathematics concerned, types, construction, adjusting anywhere. Simple, nontechnical treatment allows even children to build several of these dials. Over 100 illustrations. 230pp. 5⅜ x 8½. 22947-5 Pa. $8.95

THEORETICAL HYDRODYNAMICS, L. M. Milne-Thomson. Classic exposition of the mathematical theory of fluid motion, applicable to both hydrodynamics and aerodynamics. Over 600 exercises. 768pp. 6⅛ x 9¼. 68970-0 Pa. $20.95

SONGS OF EXPERIENCE: Facsimile Reproduction with 26 Plates in Full Color, William Blake. 26 full-color plates from a rare 1826 edition. Includes "TheTyger," "London," "Holy Thursday," and other poems. Printed text of poems. 48pp. 5¼ x 7. 24636-1 Pa. $4.95

OLD-TIME VIGNETTES IN FULL COLOR, Carol Belanger Grafton (ed.). Over 390 charming, often sentimental illustrations, selected from archives of Victorian graphics—pretty women posing, children playing, food, flowers, kittens and puppies, smiling cherubs, birds and butterflies, much more. All copyright-free. 48pp. 9¼ x 12¼. 27269-9 Pa. $7.95

PERSPECTIVE FOR ARTISTS, Rex Vicat Cole. Depth, perspective of sky and sea, shadows, much more, not usually covered. 391 diagrams, 81 reproductions of drawings and paintings. 279pp. 5⅜ x 8½. 22487-2 Pa. $9.95

DRAWING THE LIVING FIGURE, Joseph Sheppard. Innovative approach to artistic anatomy focuses on specifics of surface anatomy, rather than muscles and bones. Over 170 drawings of live models in front, back and side views, and in widely varying poses. Accompanying diagrams. 177 illustrations. Introduction. Index. 144pp. 8⅜ x11¼. 26723-7 Pa. $9.95

GOTHIC AND OLD ENGLISH ALPHABETS: 100 Complete Fonts, Dan X. Solo. Add power, elegance to posters, signs, other graphics with 100 stunning copyright-free alphabets: Blackstone, Dolbey, Germania, 97 more—including many lower-case, numerals, punctuation marks. 104pp. 8⅛ x 11. 24695-7 Pa. $8.95

HOW TO DO BEADWORK, Mary White. Fundamental book on craft from simple projects to five-bead chains and woven works. 106 illustrations. 142pp. 5⅜ x 8. 20697-1 Pa. $5.95

THE BOOK OF WOOD CARVING, Charles Marshall Sayers. Finest book for beginners discusses fundamentals and offers 34 designs. "Absolutely first rate . . . well thought out and well executed."–E. J. Tangerman. 118pp. 7¾ x 10⅝. 23654-4 Pa. $7.95

ILLUSTRATED CATALOG OF CIVIL WAR MILITARY GOODS: Union Army Weapons, Insignia, Uniform Accessories, and Other Equipment, Schuyler, Hartley, and Graham. Rare, profusely illustrated 1846 catalog includes Union Army uniform and dress regulations, arms and ammunition, coats, insignia, flags, swords, rifles, etc. 226 illustrations. 160pp. 9 x 12. 24939-5 Pa. $10.95

WOMEN'S FASHIONS OF THE EARLY 1900s: An Unabridged Republication of "New York Fashions, 1909," National Cloak & Suit Co. Rare catalog of mail-order fashions documents women's and children's clothing styles shortly after the turn of the century. Captions offer full descriptions, prices. Invaluable resource for fashion, costume historians. Approximately 725 illustrations. 128pp. 8⅜ x 11¼. 27276-1 Pa. $11.95

THE 1912 AND 1915 GUSTAV STICKLEY FURNITURE CATALOGS, Gustav Stickley. With over 200 detailed illustrations and descriptions, these two catalogs are essential reading and reference materials and identification guides for Stickley furniture. Captions cite materials, dimensions and prices. 112pp. 6½ x 9¼. 26676-1 Pa. $9.95

EARLY AMERICAN LOCOMOTIVES, John H. White, Jr. Finest locomotive engravings from early 19th century: historical (1804–74), main-line (after 1870), special, foreign, etc. 147 plates. 142pp. 11⅜ x 8¼. 22772-3 Pa. $12.95

THE TALL SHIPS OF TODAY IN PHOTOGRAPHS, Frank O. Braynard. Lavishly illustrated tribute to nearly 100 majestic contemporary sailing vessels: Amerigo Vespucci, Clearwater, Constitution, Eagle, Mayflower, Sea Cloud, Victory, many more. Authoritative captions provide statistics, background on each ship. 190 black-and-white photographs and illustrations. Introduction. 128pp. 8⅞ x 11¾. 27163-3 Pa. $14.95

LITTLE BOOK OF EARLY AMERICAN CRAFTS AND TRADES, Peter Stockham (ed.). 1807 children's book explains crafts and trades: baker, hatter, cooper, potter, and many others. 23 copperplate illustrations. 140pp. 4⅝ x 6.
23336-7 Pa. $4.95

VICTORIAN FASHIONS AND COSTUMES FROM HARPER'S BAZAR, 1867–1898, Stella Blum (ed.). Day costumes, evening wear, sports clothes, shoes, hats, other accessories in over 1,000 detailed engravings. 320pp. 9⅜ x 12¼.
22990-4 Pa. $16.95

GUSTAV STICKLEY, THE CRAFTSMAN, Mary Ann Smith. Superb study surveys broad scope of Stickley's achievement, especially in architecture. Design philosophy, rise and fall of the Craftsman empire, descriptions and floor plans for many Craftsman houses, more. 86 black-and-white halftones. 31 line illustrations. Introduction 208pp. 6½ x 9¼. 27210-9 Pa. $9.95

THE LONG ISLAND RAIL ROAD IN EARLY PHOTOGRAPHS, Ron Ziel. Over 220 rare photos, informative text document origin (1844) and development of rail service on Long Island. Vintage views of early trains, locomotives, stations, passengers, crews, much more. Captions. 8⅞ x 11¾. 26301-0 Pa. $14.95

VOYAGE OF THE LIBERDADE, Joshua Slocum. Great 19th-century mariner's thrilling, first-hand account of the wreck of his ship off South America, the 35-foot boat he built from the wreckage, and its remarkable voyage home. 128pp. 5⅜ x 8½.
40022-0 Pa. $5.95

TEN BOOKS ON ARCHITECTURE, Vitruvius. The most important book ever written on architecture. Early Roman aesthetics, technology, classical orders, site selection, all other aspects. Morgan translation. 331pp. 5⅜ x 8½. 20645-9 Pa. $8.95

THE HUMAN FIGURE IN MOTION, Eadweard Muybridge. More than 4,500 stopped-action photos, in action series, showing undraped men, women, children jumping, lying down, throwing, sitting, wrestling, carrying, etc. 390pp. 7⅞ x 10⅝.
20204-6 Clothbd. $27.95

TREES OF THE EASTERN AND CENTRAL UNITED STATES AND CANADA, William M. Harlow. Best one-volume guide to 140 trees. Full descriptions, woodlore, range, etc. Over 600 illustrations. Handy size. 288pp. 4½ x 6⅜.
20395-6 Pa. $6.95

SONGS OF WESTERN BIRDS, Dr. Donald J. Borror. Complete song and call repertoire of 60 western species, including flycatchers, juncoes, cactus wrens, many more–includes fully illustrated booklet. Cassette and manual 99913-0 $8.95

GROWING AND USING HERBS AND SPICES, Milo Miloradovich. Versatile handbook provides all the information needed for cultivation and use of all the herbs and spices available in North America. 4 illustrations. Index. Glossary. 236pp. 5⅜ x 8½.
25058-X Pa. $7.95

BIG BOOK OF MAZES AND LABYRINTHS, Walter Shepherd. 50 mazes and labyrinths in all–classical, solid, ripple, and more–in one great volume. Perfect inexpensive puzzler for clever youngsters. Full solutions. 112pp. 8⅛ x 11.
22951-3 Pa. $5.95

PIANO TUNING, J. Cree Fischer. Clearest, best book for beginner, amateur. Simple repairs, raising dropped notes, tuning by easy method of flattened fifths. No previous skills needed. 4 illustrations. 201pp. 5⅜ x 8½. 23267-0 Pa. $6.95

HINTS TO SINGERS, Lillian Nordica. Selecting the right teacher, developing confidence, overcoming stage fright, and many other important skills receive thoughtful discussion in this indispensible guide, written by a world-famous diva of four decades' experience. 96pp. 5⅜ x 8½. 40094-8 Pa. $4.95

THE COMPLETE NONSENSE OF EDWARD LEAR, Edward Lear. All nonsense limericks, zany alphabets, Owl and Pussycat, songs, nonsense botany, etc., illustrated by Lear. Total of 320pp. 5⅜ x 8½. (AVAILABLE IN U.S. ONLY.) 20167-8 Pa. $7.95

VICTORIAN PARLOUR POETRY: An Annotated Anthology, Michael R. Turner. 117 gems by Longfellow, Tennyson, Browning, many lesser-known poets. "The Village Blacksmith," "Curfew Must Not Ring Tonight," "Only a Baby Small," dozens more, often difficult to find elsewhere. Index of poets, titles, first lines. xxiii + 325pp. 5⅝ x 8¼. 27044-0 Pa. $8.95

DUBLINERS, James Joyce. Fifteen stories offer vivid, tightly focused observations of the lives of Dublin's poorer classes. At least one, "The Dead," is considered a masterpiece. Reprinted complete and unabridged from standard edition. 160pp. 5 3/16 x 8¼. 26870-5 Pa. $1.00

GREAT WEIRD TALES: 14 Stories by Lovecraft, Blackwood, Machen and Others, S. T. Joshi (ed.). 14 spellbinding tales, including "The Sin Eater," by Fiona McLeod, "The Eye Above the Mantel," by Frank Belknap Long, as well as renowned works by R. H. Barlow, Lord Dunsany, Arthur Machen, W. C. Morrow and eight other masters of the genre. 256pp. 5⅜ x 8½. (Available in U.S. only.) 40436-6 Pa. $8.95

THE BOOK OF THE SACRED MAGIC OF ABRAMELIN THE MAGE, translated by S. MacGregor Mathers. Medieval manuscript of ceremonial magic. Basic document in Aleister Crowley, Golden Dawn groups. 268pp. 5⅜ x 8½. 23211-5 Pa. $9.95

NEW RUSSIAN-ENGLISH AND ENGLISH-RUSSIAN DICTIONARY, M. A. O'Brien. This is a remarkably handy Russian dictionary, containing a surprising amount of information, including over 70,000 entries. 366pp. 4½ x 6⅛. 20208-9 Pa. $10.95

HISTORIC HOMES OF THE AMERICAN PRESIDENTS, Second, Revised Edition, Irvin Haas. A traveler's guide to American Presidential homes, most open to the public, depicting and describing homes occupied by every American President from George Washington to George Bush. With visiting hours, admission charges, travel routes. 175 photographs. Index. 160pp. 8¼ x 11. 26751-2 Pa. $11.95

NEW YORK IN THE FORTIES, Andreas Feininger. 162 brilliant photographs by the well-known photographer, formerly with *Life* magazine. Commuters, shoppers, Times Square at night, much else from city at its peak. Captions by John von Hartz. 181pp. 9¼ x 10¾. 23585-8 Pa. $13.95

INDIAN SIGN LANGUAGE, William Tomkins. Over 525 signs developed by Sioux and other tribes. Written instructions and diagrams. Also 290 pictographs. 111pp. 6⅛ x 9¼. 22029-X Pa. $3.95

ANATOMY: A Complete Guide for Artists, Joseph Sheppard. A master of figure drawing shows artists how to render human anatomy convincingly. Over 460 illustrations. 224pp. 8⅜ x 11¼. 27279-6 Pa. $11.95

MEDIEVAL CALLIGRAPHY: Its History and Technique, Marc Drogin. Spirited history, comprehensive instruction manual covers 13 styles (ca. 4th century through 15th). Excellent photographs; directions for duplicating medieval techniques with modern tools. 224pp. 8⅜ x 11¼. 26142-5 Pa. $12.95

DRIED FLOWERS: How to Prepare Them, Sarah Whitlock and Martha Rankin. Complete instructions on how to use silica gel, meal and borax, perlite aggregate, sand and borax, glycerine and water to create attractive permanent flower arrangements. 12 illustrations. 32pp. 5⅜ x 8½. 21802-3 Pa. $1.00

EASY-TO-MAKE BIRD FEEDERS FOR WOODWORKERS, Scott D. Campbell. Detailed, simple-to-use guide for designing, constructing, caring for and using feeders. Text, illustrations for 12 classic and contemporary designs. 96pp. 5⅜ x 8½. 25847-5 Pa. $3.95

SCOTTISH WONDER TALES FROM MYTH AND LEGEND, Donald A. Mackenzie. 16 lively tales tell of giants rumbling down mountainsides, of a magic wand that turns stone pillars into warriors, of gods and goddesses, evil hags, powerful forces and more. 240pp. 5⅜ x 8½. 29677-6 Pa. $6.95

THE HISTORY OF UNDERCLOTHES, C. Willett Cunnington and Phyllis Cunnington. Fascinating, well-documented survey covering six centuries of English undergarments, enhanced with over 100 illustrations: 12th-century laced-up bodice, footed long drawers (1795), 19th-century bustles, 19th-century corsets for men, Victorian "bust improvers," much more. 272pp. 5⅝ x 8¼. 27124-2 Pa. $9.95

ARTS AND CRAFTS FURNITURE: The Complete Brooks Catalog of 1912, Brooks Manufacturing Co. Photos and detailed descriptions of more than 150 now very collectible furniture designs from the Arts and Crafts movement depict davenports, settees, buffets, desks, tables, chairs, bedsteads, dressers and more, all built of solid, quarter-sawed oak. Invaluable for students and enthusiasts of antiques, Americana and the decorative arts. 80pp. 6½ x 9¼. 27471-3 Pa. $8.95

WILBUR AND ORVILLE: A Biography of the Wright Brothers, Fred Howard. Definitive, crisply written study tells the full story of the brothers' lives and work. A vividly written biography, unparalleled in scope and color, that also captures the spirit of an extraordinary era. 560pp. 6⅛ x 9¼. 40297-5 Pa. $17.95

THE ARTS OF THE SAILOR: Knotting, Splicing and Ropework, Hervey Garrett Smith. Indispensable shipboard reference covers tools, basic knots and useful hitches; handsewing and canvas work, more. Over 100 illustrations. Delightful reading for sea lovers. 256pp. 5⅜ x 8½. 26440-8 Pa. $8.95

FRANK LLOYD WRIGHT'S FALLINGWATER: The House and Its History, Second, Revised Edition, Donald Hoffmann. A total revision–both in text and illustrations–of the standard document on Fallingwater, the boldest, most personal architectural statement of Wright's mature years, updated with valuable new material from the recently opened Frank Lloyd Wright Archives. "Fascinating"–*The New York Times*. 116 illustrations. 128pp. 9¼ x 10¾. 27430-6 Pa. $12.95

PHOTOGRAPHIC SKETCHBOOK OF THE CIVIL WAR, Alexander Gardner. 100 photos taken on field during the Civil War. Famous shots of Manassas Harper's Ferry, Lincoln, Richmond, slave pens, etc. 244pp. 10⅝ x 8¼. 22731-6 Pa. $10.95

FIVE ACRES AND INDEPENDENCE, Maurice G. Kains. Great back-to-the-land classic explains basics of self-sufficient farming. The one book to get. 95 illustrations. 397pp. 5⅜ x 8½. 20974-1 Pa. $7.95

SONGS OF EASTERN BIRDS, Dr. Donald J. Borror. Songs and calls of 60 species most common to eastern U.S.: warblers, woodpeckers, flycatchers, thrushes, larks, many more in high-quality recording. Cassette and manual 99912-2 $9.95

A MODERN HERBAL, Margaret Grieve. Much the fullest, most exact, most useful compilation of herbal material. Gigantic alphabetical encyclopedia, from aconite to zedoary, gives botanical information, medical properties, folklore, economic uses, much else. Indispensable to serious reader. 161 illustrations. 888pp. 6½ x 9¼. 2-vol. set. (Available in U.S. only.) Vol. I: 22798-7 Pa. $9.95
Vol. II: 22799-5 Pa. $9.95

HIDDEN TREASURE MAZE BOOK, Dave Phillips. Solve 34 challenging mazes accompanied by heroic tales of adventure. Evil dragons, people-eating plants, blood-thirsty giants, many more dangerous adversaries lurk at every twist and turn. 34 mazes, stories, solutions. 48pp. 8¼ x 11. 24566-7 Pa. $2.95

LETTERS OF W. A. MOZART, Wolfgang A. Mozart. Remarkable letters show bawdy wit, humor, imagination, musical insights, contemporary musical world; includes some letters from Leopold Mozart. 276pp. 5⅜ x 8½. 22859-2 Pa. $7.95

BASIC PRINCIPLES OF CLASSICAL BALLET, Agrippina Vaganova. Great Russian theoretician, teacher explains methods for teaching classical ballet. 118 illus-trations. 175pp. 5⅜ x 8½. 22036-2 Pa. $6.95

THE JUMPING FROG, Mark Twain. Revenge edition. The original story of The Celebrated Jumping Frog of Calaveras County, a hapless French translation, and Twain's hilarious "retranslation" from the French. 12 illustrations. 66pp. 5⅜ x 8½. 22686-7 Pa. $3.95

BEST REMEMBERED POEMS, Martin Gardner (ed.). The 126 poems in this superb collection of 19th- and 20th-century British and American verse range from Shelley's "To a Skylark" to the impassioned "Renascence" of Edna St. Vincent Millay and to Edward Lear's whimsical "The Owl and the Pussycat." 224pp. 5⅜ x 8½. 27165-X Pa. $5.95

COMPLETE SONNETS, William Shakespeare. Over 150 exquisite poems deal with love, friendship, the tyranny of time, beauty's evanescence, death and other themes in language of remarkable power, precision and beauty. Glossary of archaic terms. 80pp. 5³⁄₁₆ x 8¼. 26686-9 Pa. $1.00

BODIES IN A BOOKSHOP, R. T. Campbell. Challenging mystery of blackmail and murder with ingenious plot and superbly drawn characters. In the best tradition of British suspense fiction. 192pp. 5⅜ x 8½. 24720-1 Pa. $6.95

CATALOG OF DOVER BOOKS

THE WIT AND HUMOR OF OSCAR WILDE, Alvin Redman (ed.). More than 1,000 ripostes, paradoxes, wisecracks: Work is the curse of the drinking classes; I can resist everything except temptation; etc. 258pp. 5⅜ x 8½. 20602-5 Pa. $6.95

SHAKESPEARE LEXICON AND QUOTATION DICTIONARY, Alexander Schmidt. Full definitions, locations, shades of meaning in every word in plays and poems. More than 50,000 exact quotations. 1,485pp. 6½ x 9¼. 2-vol. set.
Vol. 1: 22726-X Pa. $17.95
Vol. 2: 22727-8 Pa. $17.95

SELECTED POEMS, Emily Dickinson. Over 100 best-known, best-loved poems by one of America's foremost poets, reprinted from authoritative early editions. No comparable edition at this price. Index of first lines. 64pp. 5³⁄₁₆ x 8¼.
26466-1 Pa. $1.00

THE INSIDIOUS DR. FU-MANCHU, Sax Rohmer. The first of the popular mystery series introduces a pair of English detectives to their archnemesis, the diabolical Dr. Fu-Manchu. Flavorful atmosphere, fast-paced action, and colorful characters enliven this classic of the genre. 208pp. 5³⁄₁₆ x 8¼. 29898-1 Pa. $2.00

THE MALLEUS MALEFICARUM OF KRAMER AND SPRENGER, translated by Montague Summers. Full text of most important witchhunter's "bible," used by both Catholics and Protestants. 278pp. 6⅝ x 10. 22802-9 Pa. $12.95

SPANISH STORIES/CUENTOS ESPAÑOLES: A Dual-Language Book, Angel Flores (ed.). Unique format offers 13 great stories in Spanish by Cervantes, Borges, others. Faithful English translations on facing pages. 352pp. 5⅜ x 8½.
25399-6 Pa. $8.95

GARDEN CITY, LONG ISLAND, IN EARLY PHOTOGRAPHS, 1869–1919, Mildred H. Smith. Handsome treasury of 118 vintage pictures, accompanied by carefully researched captions, document the Garden City Hotel fire (1899), the Vanderbilt Cup Race (1908), the first airmail flight departing from the Nassau Boulevard Aerodrome (1911), and much more. 96pp. 8⅞ x 11¾. 40669-5 Pa. $12.95

OLD QUEENS, N.Y., IN EARLY PHOTOGRAPHS, Vincent F. Seyfried and William Asadorian. Over 160 rare photographs of Maspeth, Jamaica, Jackson Heights, and other areas. Vintage views of DeWitt Clinton mansion, 1939 World's Fair and more. Captions. 192pp. 8⅞ x 11. 26358-4 Pa. $12.95

CAPTURED BY THE INDIANS: 15 Firsthand Accounts, 1750-1870, Frederick Drimmer. Astounding true historical accounts of grisly torture, bloody conflicts, relentless pursuits, miraculous escapes and more, by people who lived to tell the tale. 384pp. 5⅜ x 8½. 24901-8 Pa. $8.95

THE WORLD'S GREAT SPEECHES (Fourth Enlarged Edition), Lewis Copeland, Lawrence W. Lamm, and Stephen J. McKenna. Nearly 300 speeches provide public speakers with a wealth of updated quotes and inspiration–from Pericles' funeral oration and William Jennings Bryan's "Cross of Gold Speech" to Malcolm X's powerful words on the Black Revolution and Earl of Spenser's tribute to his sister, Diana, Princess of Wales. 944pp. 5⅜ x 8⅜. 40903-1 Pa. $15.95

THE BOOK OF THE SWORD, Sir Richard F. Burton. Great Victorian scholar/adventurer's eloquent, erudite history of the "queen of weapons"–from prehistory to early Roman Empire. Evolution and development of early swords, variations (sabre, broadsword, cutlass, scimitar, etc.), much more. 336pp. 6⅛ x 9¼.
25434-8 Pa. $9.95

AUTOBIOGRAPHY: The Story of My Experiments with Truth, Mohandas K. Gandhi. Boyhood, legal studies, purification, the growth of the Satyagraha (nonviolent protest) movement. Critical, inspiring work of the man responsible for the freedom of India. 480pp. 5⅜ x 8½. (Available in U.S. only.) 24593-4 Pa. $8.95

CELTIC MYTHS AND LEGENDS, T. W. Rolleston. Masterful retelling of Irish and Welsh stories and tales. Cuchulain, King Arthur, Deirdre, the Grail, many more. First paperback edition. 58 full-page illustrations. 512pp. 5⅜ x 8½. 26507-2 Pa. $9.95

THE PRINCIPLES OF PSYCHOLOGY, William James. Famous long course complete, unabridged. Stream of thought, time perception, memory, experimental methods; great work decades ahead of its time. 94 figures. 1,391pp. 5⅜ x 8½. 2-vol. set.
Vol. I: 20381-6 Pa. $14.95
Vol. II: 20382-4 Pa. $14.95

THE WORLD AS WILL AND REPRESENTATION, Arthur Schopenhauer. Definitive English translation of Schopenhauer's life work, correcting more than 1,000 errors, omissions in earlier translations. Translated by E. F. J. Payne. Total of 1,269pp. 5⅜ x 8½. 2-vol. set.
Vol. 1: 21761-2 Pa. $12.95
Vol. 2: 21762-0 Pa. $12.95

MAGIC AND MYSTERY IN TIBET, Madame Alexandra David-Neel. Experiences among lamas, magicians, sages, sorcerers, Bonpa wizards. A true psychic discovery. 32 illustrations. 321pp. 5⅜ x 8½. (Available in U.S. only.) 22682-4 Pa. $9.95

THE EGYPTIAN BOOK OF THE DEAD, E. A. Wallis Budge. Complete reproduction of Ani's papyrus, finest ever found. Full hieroglyphic text, interlinear transliteration, word-for-word translation, smooth translation. 533pp. 6½ x 9¼.
21866-X Pa. $12.95

MATHEMATICS FOR THE NONMATHEMATICIAN, Morris Kline. Detailed, college-level treatment of mathematics in cultural and historical context, with numerous exercises. Recommended Reading Lists. Tables. Numerous figures. 641pp. 5⅜ x 8½.
24823-2 Pa. $11.95

PROBABILISTIC METHODS IN THE THEORY OF STRUCTURES, Isaac Elishakoff. Well-written introduction covers the elements of the theory of probability from two or more random variables, the reliability of such multivariable structures, the theory of random function, Monte Carlo methods of treating problems incapable of exact solution, and more. Examples. 502pp. 5³/₈ x 8¹/₂. 40691-1 Pa. $16.95

THE RIME OF THE ANCIENT MARINER, Gustave Doré, S. T. Coleridge. Doré's finest work; 34 plates capture moods, subtleties of poem. Flawless full-size reproductions printed on facing pages with authoritative text of poem. "Beautiful. Simply beautiful."–*Publisher's Weekly*. 77pp. 9¼ x 12. 22305-1 Pa. $7.95

NORTH AMERICAN INDIAN DESIGNS FOR ARTISTS AND CRAFTSPEOPLE, Eva Wilson. Over 360 authentic copyright-free designs adapted from Navajo blankets, Hopi pottery, Sioux buffalo hides, more. Geometrics, symbolic figures, plant and animal motifs, etc. 128pp. 8⅜ x 11. (Not for sale in the United Kingdom.) 25341-4 Pa. $9.95

SCULPTURE: Principles and Practice, Louis Slobodkin. Step-by-step approach to clay, plaster, metals, stone; classical and modern. 253 drawings, photos. 255pp. 8⅛ x 11.
22960-2 Pa. $11.95

CATALOG OF DOVER BOOKS

THE INFLUENCE OF SEA POWER UPON HISTORY, 1660–1783, A. T. Mahan. Influential classic of naval history and tactics still used as text in war colleges. First paperback edition. 4 maps. 24 battle plans. 640pp. 5⅜ x 8½. 25509-3 Pa. $14.95

THE STORY OF THE TITANIC AS TOLD BY ITS SURVIVORS, Jack Winocour (ed.). What it was really like. Panic, despair, shocking inefficiency, and a little heroism. More thrilling than any fictional account. 26 illustrations. 320pp. 5⅜ x 8½. 20610-6 Pa. $8.95

FAIRY AND FOLK TALES OF THE IRISH PEASANTRY, William Butler Yeats (ed.). Treasury of 64 tales from the twilight world of Celtic myth and legend: "The Soul Cages," "The Kildare Pooka," "King O'Toole and his Goose," many more. Introduction and Notes by W. B. Yeats. 352pp. 5⅜ x 8½. 26941-8 Pa. $8.95

BUDDHIST MAHAYANA TEXTS, E. B. Cowell and others (eds.). Superb, accurate translations of basic documents in Mahayana Buddhism, highly important in history of religions. The Buddha-karita of Asvaghosha, Larger Sukhavativyuha, more. 448pp. 5⅜ x 8½. 25552-2 Pa. $12.95

ONE TWO THREE . . . INFINITY: Facts and Speculations of Science, George Gamow. Great physicist's fascinating, readable overview of contemporary science: number theory, relativity, fourth dimension, entropy, genes, atomic structure, much more. 128 illustrations. Index. 352pp. 5⅜ x 8½. 25664-2 Pa. $9.95

EXPERIMENTATION AND MEASUREMENT, W. J. Youden. Introductory manual explains laws of measurement in simple terms and offers tips for achieving accuracy and minimizing errors. Mathematics of measurement, use of instruments, experimenting with machines. 1994 edition. Foreword. Preface. Introduction. Epilogue. Selected Readings. Glossary. Index. Tables and figures. 128pp. 5³⁄₈ x 8¹⁄₂. 40451-X Pa. $6.95

DALÍ ON MODERN ART: The Cuckolds of Antiquated Modern Art, Salvador Dalí. Influential painter skewers modern art and its practitioners. Outrageous evaluations of Picasso, Cézanne, Turner, more. 15 renderings of paintings discussed. 44 calligraphic decorations by Dalí. 96pp. 5⅜ x 8½. (Available in U.S. only.) 29220-7 Pa. $5.95

ANTIQUE PLAYING CARDS: A Pictorial History, Henry René D'Allemagne. Over 900 elaborate, decorative images from rare playing cards (14th–20th centuries): Bacchus, death, dancing dogs, hunting scenes, royal coats of arms, players cheating, much more. 96pp. 9¼ x 12¼. 29265-7 Pa. $12.95

MAKING FURNITURE MASTERPIECES: 30 Projects with Measured Drawings, Franklin H. Gottshall. Step-by-step instructions, illustrations for constructing handsome, useful pieces, among them a Sheraton desk, Chippendale chair, Spanish desk, Queen Anne table and a William and Mary dressing mirror. 224pp. 8⅛ x 11¼. 29338-6 Pa. $13.95

THE FOSSIL BOOK: A Record of Prehistoric Life, Patricia V. Rich et al. Profusely illustrated definitive guide covers everything from single-celled organisms and dinosaurs to birds and mammals and the interplay between climate and man. Over 1,500 illustrations. 760pp. 7½ x 10¼. 29371-8 Pa. $29.95

Prices subject to change without notice.

Available at your book dealer or write for free catalog to Dept. GI, Dover Publications, Inc., 31 East 2nd St., Mineola, N.Y. 11501. Dover publishes more than 500 books each year on science, elementary and advanced mathematics, biology, music, art, literary history, social sciences and other areas.